THE GREEK ALPHABET

Α	α	Alpha
Β	β	Beta
Γ	γ	Gamma
Δ	δ	Delta
Ε	ε	Epsilon
Ζ	ζ	Zeta
Η	η	Eta
Θ	θ	Theta
Ι	ι	Iota
Κ	κ	Kappa
Λ	λ	Lambda
Μ	μ	Mu
Ν	ν	Nu
Ξ	ξ	Xi
Ο	ο	Omicron
Π	π	Pi
Ρ	ρ	Rho
Σ	σ	Sigma
Τ	τ	Tau
Υ	υ	Upsilon
Φ	φ	Phi
Χ	χ	Chi
Ψ	ψ	Psi
Ω	ω	Omega

CELL BIOLOGY: STRUCTURE, BIOCHEMISTRY, AND FUNCTION

CELL BIOLOGY: STRUCTURE, BIOCHEMISTRY, AND FUNCTION

Phillip Sheeler
Professor of Biology
California State University Northridge

Donald E. Bianchi
Dean, School of Science and Mathematics
Professor of Biology
California State University Northridge

John Wiley & Sons, New York Chichester Brisbane Toronto

Library of Congress Cataloging in Publication Data:
Sheeler, Phillip.
Cell biology.
Includes index.
1. Cytology. I. Bianchi, Donald E., 1933–joint author. II. Title.
[DNLM: 1. Cells. 2. Molecular biology. QH581.2 S541c]
QH581.2.S53 574.8'7 79-2552
ISBN 0-471-78220-3

Printed in the United States of America

10 9 8 7 6 5 4 3 2 1

To my parents Barnett and Deborah,
to my wife Annette,
and to my children Wendy, Donna,
Lindsey, Paul and Carly.

P. S.

To my parents Ernest and Florence,
to my wife Georgia,
and to my children Diana, Dave and Bill.

D. E. B.

PREFACE

This book was written for junior and senior level courses in cell biology, molecular biology, and cellular physiology. We consider in some detail the fine structure of eucaryotic and procaryotic cells (and viruses), the chemical composition and organization of cells, cell metabolism and bioenergetics, and for each major cell organelle or structural compoenent, its particular molecular and supermolecular organization and its functions. Special attention has also been paid to a description of the major research tools used by cell biologists to further our knowledge of cell structure, biochemistry, and function.

In preparing this book for use as a text by students, we have drawn on more than thirty years of collective university teaching experience in courses in cell biology, cell physiology, and molecular biology. We are hopeful that our early training in different areas of cell biology and the continuous maintenance of active research programs in these areas has helped us to present a balanced approach.

Certain assumptions have necessarily been made with regard to the background of students reading this book. It is assumed, for example, that the student has had courses in introductory biology and introductory chemistry. Most portions of the book dealing with physical or biochemical principles or concepts are preceded with some discussion of the prerequisite fundamentals.

We feel strongly about the value of illustrations for clear depiction of concepts, mechanisms, structure, and interactions and, for this reason, the book is generously illustrated with drawings and photomicrographs. A number of the illustrations are presented in stereoscopic form, and although individual members of each "stereo pair" may be viewed without an optical aid, the perspective effect requires the use of a stereo viewer. Plastic-frame viewers with adjustable lenses are available in most campus bookstores or where graphic supplies are sold. The inclusion of stereoscopy in books dealing with chemistry and biology is becoming increasingly popular, and the student's purchase of a viewer is worthwhile.

Many individuals provided invaluable help and guidance during the preparation, writing, and production of this book. We are indebted to them. Selected chapters were read and critiqued, suggestions were added, and revised drafts reread by George Lefevre and Marvin Cantor (California State University, Northridge). Most of the chapters were reviewed by Thomas James (University of California, Los Angeles), Carl Nordahl (University of Nebraska at Omaha), John Roberts (North Carolina State University), and E. D. Salmon (University of North Carolina). Professor James' humor, support, and critical assistance were inspirational as well as exceedingly helpful to us, and to him we owe a special debt. A great number of fine microscopists provided us with original photographs. We are particularly grateful to Daisy Kuhn, Edward Pollock, and Richard Chao for letting us select prints from their collections. The photomicrograph on the cover, provided by Dr. Pollock, is a unique view of the *Fucus* egg cell. Our students also acted as our critics; we thank Mark Doolittle and Norbert Herzog in particular, as well as the students in our recent classes in cell biology.

Diane Grondin, our typist extraordinaire, kept us consistent, the manuscript in order, and our assignments on schedule. The production staff at Wiley was most helpful, encouraging, and talented, especially Frederick C. Corey, our principal editor, John Balbalis, our illustrator, Lilly Kaufman, our production supervisor, and Deborah Herbert and Elaine Miller, editors of the final manuscript. We most gratefully thank each of them.

We apologize for any errors that we allowed into print and should like to thank our readers in advance for bringing them to our attention.

Northridge
1979

Phillip Sheeler
Don Bianchi

TABLE OF CONTENTS

PART 1
CELLS AND CELL GROWTH

Chapter 1
The Cell: an Introduction

Development of the Cell Doctrine 3
Microscopy 5
Fundaments of Light Microscopy and Transmission Electron Microscopy 5
Preparation of Materials for Microscopy 8
Specialized Applications of Transmission Electron Microscopy 9
The Scanning Electron Microscope 18
Stereo Microscopy (Stereoscopy) 20
Cell Structure: A Preview 23
Eucaryotic Cells: The Generalized Animal Cell 24
Eucaryotic Cells: The Generalized Plant Cell 34
Procaryotic Cells: Bacteria 37
Procaryotic Cells: Blue-Green algae 39
Procaryotic Cells: PPLO 41
Viruses 41
Structure of Viruses 41
Proliferative Cycle of a Virus 43
Classification of Viruses and the Nature of Viral Nucleic Acids 43
Summary 46
References and Suggested Reading 47

Chapter 2
Cell Growth and Proliferation

The Population Growth Cycle 50
Exponential Growth 50
Generation Time 51
The Lag Phase of Growth 52
The Stationary Phase 52
The Death (Declining) Phase 52
Contact Inhibition 53
The Quantitation of Cells 53
Optical Enumeration of Cells 53
Electronic Enumeration of Cells 53
The Continuous Culture of Cells 54
Synchronous Cell Cultures 55
Synchrony by Induction 55
Synchrony by Selection 56
Culture Fractionation 57
The Growth of Individual Cells: The Cell Cycle 57
Phases of the Cell Cycle 58
Summary 60
References and Suggested Reading 60

PART 2
MOLECULAR CONSTITUENTS OF CELLS

Chapter 3
Cellular Chemistry, Molecules, and Ions

Water 63
Salts, Ions, and Gases 67
Acids, Bases, and Buffers 67
Chemical Bonds 70
Ligands and Chelates 71
Special Compounds 71
Summary 75
References and Suggested Reading 75

Chapter 4
The Cellular Macromolecules: Proteins

The Amino Acids 78
The Peptide Bond 80
Helical Polypeptides 83
Fundamental Properties of Helices 83
The Hydrogen Bond 83
The Alpha Helix 84
Other Polypeptide Helices 84
Helices of Fibrous Proteins 86
Collagen 86
Silk 86
Levels of Protein Structure 86
Primary Protein Structure 86
Secondary Protein Structure 88
Tertiary Protein Structure 89
Quaternary Protein Structure 90
Establishment of Secondary, Tertiary, and Quaternary Structure 93
Anatomy of the Vertebrate Hemoglobins 95
Association of Globin Chain with Heme 96
Function and Action of Hemoglobin: Cooperativity in Proteins 99
Evolution in Proteins 101
Ontogeny and Phylogeny of Hemoglobin 101
103
Chromoproteins 103
Glycoproteins 103
Lipoproteins 104
Nucleoproteins 105
Summary 105
References and Suggested Reading 106

Chapter 5
The Cellular Macromolecules: Polysaccharides

Monosaccharides 109
Pyranoses and Furanoses 109
Disaccharides 112
Polysaccharides 113
Cellulose 113
Chitin 114
Hyaluronic Acid 114
Inulin 115
Glycogen 115
Starch 117
Other Polysaccharides 118
Summary 118
References and Suggested Reading 119

Chapter 6
The Cellular Macromolecues: Lipids

Fatty Acids 121
Neutral Fats 121
Glycerophosphatides 121
Plasmalogens 122
Sphingolipids 122
Glycolipids 123
Steroids 124
Summary 124
References and Suggested Reading 125

Chapter 7
The Cellular Macromolecules: Nucleic Acids

Cellular Roles of the Nucleic Acids 127
The Discovery of DNA 127

Transformation of Bacterial Types 128
Virus Reproduction 128
Composition and Structure of the Nucleic Acids 131
Structure of DNA 133
Replication of DNA 135
''Single-stranded'' DNA 136
Structure of RNA 136
Synthesis of RNA 136
Replication of RNA Viruses 137
Types of Cellular RNA 138
Summary 139
References and Suggested Reading 140

PART 3
CELL METABOLISM

Chapter 8
Enzymes

Molecularity of Chemical Reactions 143
Reaction Kinetics 144
Effect of Enzyme on Reaction Rate 145
The Kinetics of Enzyme Action 146
Effects of Inhibitors on Enzyme Activity 150
Mechanics of Enzyme Catalysis 151
The Active Site 153
Lysozyme 156
Ribonuclease 158
Chymotrypsin 159
Cofactors 161
Isoenzymes 162
Zymogens 162
The Regulation of Enzyme Activity 163
Number of Polypeptide Chains and Number of Binding Sites of an Enzyme 165
Model for Allosteric Enzyme Function 165
Cooperativity in Enzymes 166
Allosterism, Cooperativity, and Michaelis-Menten Enzyme Kinetics 167
Control of Metabolic Processes by Allosteric Enzymes: The Alternate Pathways for Glycogen Synthesis and Degradation 169
Summary 169
References and Suggested Reading 171

Chapter 9
Bioenergetics

Energy 173
The Laws About Energy and Energy Changes 174
Coupled Reactions 177
Intracellular Phosphate Turnover 179
Redox Couples 179
Light and Chemical Transductions 180
Other Transductions 184
Summary 184
References and Suggested Reading 185

Chapter 10
Metabolism

Analysis of Metabolic Pathways 188
Marker and Tracer Techniques 188
Enzyme Techniques 190
Enzyme Production and Inhibition 190
Carbohydrate Metabolism 190
Glycolysis 193
Anaerobic Respiration and Fermentation 193
Oxidation of Pyruvate 193
Other Pathways of Carbohydrate Catabolism 194
Phosphogluconate Pathway 194
The Glyoxylate Pathway 194
Gluconeogenesis 196
Synthesis of Glycogen and Starch 199
Lipid Metabolism 199
Triglycerides 199
Nitrogen Metabolism 202
Cancer Cell Metabolism 203
Functions of Metabolic Pathways 203
Calculations of Energy Change 204
Summary 205
References and Suggested Reading 206

Chapter 11
Metabolic Regulation

Regulation by Mass Action 207
Regulation by Enzyme Activity 208

Substrate Concentration Effectors 208
Allosteric Effectors 209
Covalent Bond Modification of Enzyme Activity 211
Regulation by Number of Enzyme Molecules 212
Isozymes (Isoenzymes) 212
Regulation of Enzyme Synthesis 213
Constitutive and Induced Enzymes 213
Enzyme Repression 214
Catabolic Repression 214
Repressors and Transcription 214
The Operon 215
Subsequent Modifications of the Jacob-Monod Operon Model 216
Translational Control 216
Regulation of Enzyme Production in Eucaryotic Cells 217
Enzyme Induction by Hormones 218
Repression in Eucaryotes 218
Model of Gene Regulation in Eucaryotes 218
Compartmentalization 219
Summary 219
References and Suggested Readings 221

PART 4
TOOLS AND METHODS OF CELL BIOLOGY

Chapter 12
Fractionation of Tissues and Cells

Methods for Disrupting Tissues and Cells 225
Centrifugation 226
Theory of Centrifugation 226
Sedimentation Rate and Coefficient 227
The Analytical Ultracentrifuge 228
Differential Centrifugation 231
Density Gradient Centrifugation 233
Zonal Centrifugation 235
Dynamically Unloaded Zonal Rotors 238
Reograd Zonal Rotors 238
Examination and Analysis of Separated Cell Fractions 241
Methods for Separating Whole Cells 241
Tissue Disaggregation 241
Adherence and Filtration 242
Conventional and Zonal Centrifugation 242
Centrifugal Elutriation (Counter-Streaming Centrifugation) 242
Unit Gravity Separation 243
Countercurrent Distribution 244
Electrophoresis 244
Fluorescence-Activated Cell Sorting 244
Harvesting Cells and Subcellular Components: Continuous-Flow Centrifugation 245
Summary 246
References and Suggested Reading 247

Chapter 13
The Isolation and Characterization of Cellular Macromolecules

Salting In and Salting Out 250
Isoelectric Precipitation 251
Dialysis and Ultrafiltration 251
Ultracentrifugation 252
Electrophoresis 253
Moving-Boundary Electrophoresis 255
Zone Electrophoresis 255
Disc or Discontinuous Electrophoresis 256
Immunoelectrophoresis 259
Isoelectric Focusing 260
Countercurrent Distribution 261
Paper Chromatography 263
Thin-Layer Chromatography 265
Ion-Exchange Column Chromatography 266
Affinity Chromatography 269
Gel Filtration (Molecular Sieving) 270
Gas Chromatography 271
Summary 272
References and Suggested Reading 273

Chapter 14
The Utilization of Radioactive Isotopes as Tracers in Cell Biology

Advantages of the Radioisotope Technique 275
Determination of Molecular Fluxes Under Conditions of Zero Net Exchange 275
Simplification of Chemical Analyses 276
"Isotope Dilution" Methods 277
Precursor-Product Relationships 277
Properties of Radioactive Isotopes 278
Types of Radiation Emitted by Radioisotopes 279
Energy of Radiation and Its Interaction with Matter 279
Half-life 280
Detection and Measurement of Radiation 282
Geiger-Muller Counters 282
Solid Scintillation Counters 283
Liquid Scintillation Counters 284
Autoradiography 285
Summary 287
References and Suggested Reading 288

PART 5
STRUCTURE AND FUNCTION OF THE MAJOR CELL ORGANELLES

Chapter 15
The Plasma Membrane

Early Studies on the Chemical Organization of the Plasma Membrane 291
Existence of Lipid in the Membrane 292
The Langmuir Trough 292
Gorter and Grendel's Bimolecular Lipid Leaflet Model 293
The Danielli-Davson Membrane Model 294
Robertson's Unit Membrane 294
The Fluid-Mosaic Model of Membrane Structure 298
Freeze-Fractured Membranes 298
Membrane Proteins 300
Peripheral (Extrinsic) Proteins 300
Integral (Intrinsic) Proteins 300
Asymmetric Distribution of Membrane Proteins 301
Mobility of Membrane Proteins 302
Enzymatic Properties of Membrane Proteins 302
Isolation and Characterization of Membrane Proteins 302
Membrane Lipids 303
Mobility of Membrane Lipids 303
Lipid Asymmetry 305
Membrane Carbohydrate 305
Possible Function of Membrane Carbohydrate 305
Lectins, Antibodies, and Antigens and the Plasma Membrane 306
Lectins 306
Antigens and Antibodies 307
Origins of the Plasma Membrane Protein and Lipid Asymmetry 308
Special Cell Surface Properties Revealed by Erythrocytes 309
Intercellular Junctions and Other Specializations of the Plasma Membrane 310
Passive Movements of Materials Through Cell Membranes 313
Osmosis and Diffusion Across Membranes 314
The Gibbs-Donnan Effect 320
Facilitated (Mediated) Diffusion through the Cell Membrane 321
Active Transport 322
The Na^+/K^+ Exchange Pump 322
Cotransport: The Electrogenic Pump 324
"Simple" Active Transport 324
Bulk Transport Into and Out of Cells 324
Endocytosis 325
Exocytosis 327
Summary 327
References and Suggested Reading 330

Chapter 16
The Mitochondrion

Discovery of Mitochondria 331
Structure of Mitochondrion 333
Tricarboxylic Acid Reactions 343
Summary of the TCA Cycle 347
Electron Transport System 347

Oxidation-Reduction Reactions 348
Classes of Electron Transfer System Compounds 349
Electron Transport Pathway 350
Balance of Electrons from Glycolysis and TCA Cycle Metabolism 351
The Energetics of Electron Transport 352
Transport of Protons 352
Electron Transport Inhibitors 352
NAD[P]$^+$ Transhydrogenases 353
Oxidative Phosphorylation 353
Molecular Events in Oxidative Phosphorylation 353
The Chemical-Coupling Hypothesis 355
The Conformational-Coupling Hypothesis 355
The Chemiosmotic-Coupling Hypothesis 356
Other Functions of Mitochondria 357
The Glyoxylate Cycle 357
Fatty Acid Oxidation 357
Fatty Acid Chain Elongation 357
Superoxide Dismutase and Catalase 357
Amphibolic and Anaplerotic Reactions 358
Permeability of the Inner Membrane 358
Cytosol-Matrix Exchange of NADH and NADPH 360
Total Energy Production from Catabolism of Glucose 360
Summary 360
References and Suggested Reading 362

Chapter 17 The Chloroplast

Fine Structure of the Chloroplast 365
Structure of the Thylakoid 367
Stroma Structures 367
Chemical Composition of Chloroplasts 368
The Chlorophylls 369
The Carotenoids 371
Location and Arrangement of the Pigment 372
Development of Chloroplasts 372
Photosynthesis—Historical Background 374
Photosynthesis—Photochemical (Light) Reactions 375
The Absorption of Light by Chlorophyll 375
Primary Photochemical Events in Photosynthesis 376
Two Photosystems 376
Sequence of Energy (Electron) Flow 377
Redox Reactions 378
Cyclic and Noncyclic Photophosphorylation 378
Summary of the Light Reactions 380
Photosynthesis—Synthetic (Dark) Reactions 380
Other CO_2-Fixation Pathways 383
Bacterial Photosynthesis 385
Other Plastids 386
Summary 386
References and Suggested Reading 387

Chapter 18 The Golgi Apparatus

Structure of the Golgi Apparatus 390
Origin of Golgi Structures 390
Development of the Golgi Apparatus 396
Functions of the Golgi Apparatus 396
Cell-Specific Functions of the Golgi Apparatus 399
Formation of the Plant Cell Plate and Cell Wall 399
Neurosecretions 400
Interrelationship Between Golgi, Lysosomes, and Vacuoles 401
Acrosome Development in Sperm 402
Summary 402
References and Suggested Reading 402

Chapter 19 Lysosomes and Microbodies

Lysosomes 403
Structure and Forms of Lysosomes 405
Formation and Function of Lysosomes 408
Distribution of Lysosomes 410
Lysosome Precursors in Bacteria 411
Regulation of Lysosome Production 411
Disposition and Action of the Lysosomal Hydrolases 412
Microbodies 412
Peroxisomes 412
Glyoxysomes 413
Summary 414
References and Suggested Reading 415

Chapter 20 The Cell Nucleus

Chromatin 417
Structure and Function of Chromatin 418

Sites of DNA Replication 419
Sites of Transcription 420
The Nucleolus 421
The Nuclear Envelope 421
Other Nuclear Structures 425
Chromosomes 425
Mitosis 425
Ultrastructure of the Chromosome 431
Metaphase Chromosome-Chromosome Associations 431
Polytene Chromosomes 431
Bacterial and Viral Chromosomes 431
Replication 434
Replication as a Semiconservative Process 434
Replication by the Addition of Nucleotides in the 5' →3' Direction 434
Replication in One or Two Directions 436
''Swivel'' Mechanism 437
Enzymatic Replication 437
RNA Primers 438
Replication in Procaryotes 438
Replication in Eucaryotes 439
Replication of Viral DNA 439
The Virus Life Cycle 440
(λ) *Phage* 440
Plasmids and Episomes 441
Recombination 441
Recombinant DNA 441
Meiosis 443
Cytokinesis 449
Summary 451
References and Suggested Reading 451

Chapter 21
Ribosomes and the Synthesis of Proteins

Protein Turnover in Cells 453
A Preliminary Overview of Protein Biosynthesis 454
Structure, Composition, and Assembly of Ribosomes 455
Procaryotic Ribosomes 459
Assembly of Procaryotic Ribosomes 460
Model of Procaryotic Ribosomes 462
Genes for Ribosomal RNA and Protein 463
Eucaryotic Ribosomes 463
Model of Eucaryotic Ribosomes 469
Free and Attached Ribosomes 469
Ribosomes of Organelles 474
Chloroplast Ribosomes 475
Mitochondrial Ribosomes 475
Protein Synthesis in Chloroplasts and Mitochondria 475
Mechanism of Protein Synthesis 476
Linearity and Direction of Polypeptide Chain Assembly 477
Processing and Structure of Transfer RNA 481
Activation of Amino Acids 484
Formation of the Initiation Complex 484
Chain Elongation 490
Chain Termination 491
Polyribosomes (Polysomes) 494
Visualization of Translation 498
Co-Translational and Post-Translational Protein Modification 498
Transfer RNA Specialization 500
Inhibitors of Protein Synthesis 502
Inhibitors of Both *Procaryotic and Eucaryotic Protein Synthesis* 502
Inhibitors Specific for Procaryotes 503
Inhibitors Specific for Eucaryotes 503
Inhibitors of Organellar Protein Synthesis 503
Summary 504
References and Suggested Reading 505

Chapter 22
Cilia, Flagella, Microfilaments, and Microtubules

Distribution and Functions of Microfilaments 510
Muscle Cells 510
Cytokinesis 510
Plasma Membrane Movement 511
Distribution and Functions of Microtubules 514
Centrioles 514
Cilia and Flagella 517
The Mitotic Spindle 520
Other Cell Movements 522
Summary 522
References and Suggested Reading 523

Chapter 23
Cell Differentiation and Specialization

Nucleocytoplasmic Relationships 528
Serial Transplantation of Frog Embryo Nuclei 528
Experiments with Acetabularia 529
Environmental Effects on Differentiation 531
Red Blood Cells 532
Erythropoiesis 532
Genetic and Molecular Basis of Erythrocyte Differentiation 533
Morphological and Physiological Specialization of Red Blood Cells 534
Muscle Cells 535
Muscle Contraction 539
Nerve Cells or Neurons 541
Action Potentials 542
Ion Gradients Across the Membrane 542
Initiation of the Action Potential 543
Conduction of the Action Potential 543
Synaptic Transmission 544
Summary 544
References and Suggested Reading 545

Glossary 547

Index 559

CELL BIOLOGY: STRUCTURE, BIOCHEMISTRY, AND FUNCTION

Part 1
CELLS AND CELL GROWTH

Chapter 1
THE CELL: AN INTRODUCTION

Development of the Cell Doctrine

Briefly summarized, the **cell doctrine** states that the cell is the fundamental unit of both *structure* and *function* in all living things; that all forms of life (animal, plant, and microbial) are composed of cells and their secretions; and that cells arise only from *preexisting* cells, each cell having a life of its own in addition to its integrated role in multicellular organisms.

This statement seems both elementary and obvious to any student with some background in the biological sciences. Nevertheless, it took several centuries for this concept to be developed and accepted. The very existence of cells was not even suspected until the seventeenth century, because most cells are too small to be discerned with the naked eye, and because instruments for significantly magnifying small objects did not exist. However, with the introduction of the first crude light *microscopes,* investigators began to examine small organisms, tissues cut from animals and plants, and the "animalcules" in pond water. The invention of the microscope and its gradual improvement went hand-in-hand with the development of the cell doctrine. It finally became apparent that a fundamental similarity existed in the structural organization of all the living things studied.

What follows is a brief description of but a few of the historical highlights that culminated in the cell doctrine. Although a great many individuals made contributions of varying significance to the development of this concept, the works of a certain small number of men stand out as milestones.

In 1558, the works of *Conrad Gesner* (Swiss, 1516–1565) on the structure of foraminifera were published. His sketches included so much detail that they could only have been made if Gesner had the assistance of some form of magnifying lens. This appears to be the earliest recorded use of a magnifying instrument in biology.

Francis and *Zacharias Janssens,* who manufactured spectacles in Holland, are generally credited with the construction of the first compound microscopes in 1590. Their microscopes had magnifying powers between 10X and 30X and were used primarily to examine small whole organisms such as fleas and other insects. The first microscopes were in fact referred to as "flea-glasses."

Although noted principally for his contributions in the fields of astronomy and physics, *Galileo Galilei* (Italian, 1564–1642) produced several important biological works. His own microscopes were constructed at about the same time as that of the Janssens (around 1610) and were employed for several extensive studies on the arrangements of the *facets* of the compound eyes of insects.

Among the earliest descriptions of the microanatomy of tissues were those of *Marcello Malpighi* (Italian, 1628–

1694), one of the first great animal and plant anatomists. He was the first to describe the existence of the capillaries, thereby completing the work on the circulation of the blood started by the great English physiologist William Harvey. Malpighi was among the first to use a microscope to examine and describe thin slices of animal tissues from such organs as the brain, liver, kidney, spleen, lungs, and tongue. His published works also include descriptions of the development of the chick embryo. In his later years, Malpighi turned to investigators of plant tissues and suggested that these were composed of structural units which he called "utricles."

Antony van Leeuwenhoek (Dutch, 1632–1723) was one of the most distinguished of all the early microscopists. Although it was only an avocation, Leeuwenhoek became an expert lens grinder and built numerous microscopes, some with magnifications approaching 300X. Leeuwenhoek was the first to describe microscopic organisms in rainwater collected from tubes inserted into the soil during rainfall. His sketches included numerous bacteria (bacilli, cocci, spirilla, etc.), protozoa, rotifers, and hydra. Leeuwenhoek was the first to describe sperm cells (of humans, dogs, rabbits, frogs, fish, and insects) and observe the movements of blood in the web capillaries of the frog's foot and the rabbit's ear. He described the blood cells of mammals, birds, amphibians, and fish, noting that those of fish and amphibians were oval in shape and contained a central body (i.e., the nucleus), while those of humans and other mammals were round. Leeuwenhoek's observations were recorded in a series of reports that he sent to the Royal Society of London.

Many of Leeuwenhoek's observations were confirmed in experiments conducted by *Robert Hooke* (English 1635–1703), an architect and scientist employed by the Royal Society. Hooke popularized the use of microscopes among contemporary biologists in England and built several compound microscopes. On one occasion, Hooke examined a thin slice cut from a piece of dried cork. In his descriptions, Hooke wrote that he found the sections to be "all perforated and porous, much like a honeycombe" and referred to the boxlike structures as "cells." Thus, it is Hooke who introduced the term **cell** to biology. What he observed, of course, were not cork cells but rather the empty spaces left behind after the living portion of the cells had disintegrated.

Nehemiah Grew (English, 1641–1712), together with Marcello Malpighi, is recognized as one of the founders of plant anatomy. His publications included accounts of the microscopic examination of sections through the flowers, roots, and stems of plants and clearly indicate that he recognized the cellular nature of plant tissue. Grew was also the first to recognize that flowers are the sexual organs of plants.

In 1824, *Rene Dutrochet* (French, 1776–1847) wrote that all animal and plant tissues were "aggregates of globular cells" and in 1831, *Robert Brown* (English, 1773–1858) noted that the cells of plant epidermis, pollen grains, and stigmas contained certain "constant structures" which he called **nuclei,** thereby introducing this term to biology. Brown is also credited with the first description of the physical phenomenon now referred to as "Brownian motion." *Johannes E. Purkinje* (Czech, 1787–1869) coined the term *protoplasm* to describe the contents of cells.

Mathias J. Schleiden (German, 1804–1881) and *Theodor Schwann* (German, 1810–1882) are often credited, albeit incorrectly, with the first formal statement of a general cell theory. Their contributions to the development of the cell doctrine reside in the generalizations that they made based principally on the works of their predecessors. Schleiden and Schwann were particularly influential with their contemporaries and did, therefore, gain popular acceptance for the developing cell doctrine. Schleiden, a botanist, extended the studies begun by Robert Brown on the structure and function of the cell nucleus (which Schleiden called a "cytoblast") and was the first to describe nucleoli. Schleiden's writings clearly indicate his appreciation of the individual nature of cells. In 1838, he wrote that each cell leads a double life—one independent, pertaining to its own development, and another as an integral part of a plant. Schwann studied both plant and animal tissues. His work with connective tissues such as bone and cartilage led him to modify the evolving cell theory to include the notion that living things are composed of both cells *and the products of cells.* Schwann is also credited with the introduction of the term **metabolism** to describe the activities of cells.

Rudolf Virchow (German, 1821–1902) was a pathologist and recognized the cellular basis of disease. His writings, often in Latin, also reveal his appreciation of the cellular basis of life's continuity, as summarized in his famous expression *omnis cellula e cellula,* "all cells arise from preexisting cells."

In the last part of the 1800s and certainly by the turn of the century, the light microscope approached its limit in terms of resolving power, and nearly all major cellular structures had at least been described.

Microscopy

Fundamentals of Light Microscopy and Transmission Electron Microscopy

Until the 1940s, most of our knowledge concerning the structure and organization of cells was obtained by light microscopy, and major structures and *organelles,* including the cell wall, nuclei, chromosomes, chloroplasts, mitochondria, vacuoles, centrioles, flagella, and cilia, had been described.

The smallest distance, *d,* between two points resolvable as separate points when viewed through lenses is given by the relationship

$$d = \frac{0.6\lambda}{n \sin \alpha} \tag{1–1}$$

In this equation, λ is the *wavelength* of the light (radiation) employed to illuminate the specimen; n is the *refractive index* of the air or liquid between the specimen and the lens; and α is the *aperture angle*. The product, $n \sin \alpha$, is called the lens *numerical aperture,* and for a good microscope lens, it would be about 1.4.

Equation 1–1 also shows that the *resolving power* of a microscope varies with the wavelength of the source of illumination. The human eye cannot directly detect light having a wavelength of less than about 400 nm (see Table 1–1 for metric measurements). Therefore, in the case of the light microscope, the maximum resolving power is about 0.6(400/1.4), or about 0.17 μ. That is, points less than about 0.2 μ apart cannot be distinguished as separate points by light microscopy (in practice, the limit is closer to .05 μ). Using glass optics of the finest quality, it is possible to observe cells at a magnification of about 2000X. Resolution is improved when sources emitting rays that have shorter wavelengths are employed. For example, the resolving power of the ultraviolet light microscope (which requires quartz optics because glass does not transmit ultraviolet light) is approximately double that of the light microscope.

Much greater resolution has been obtained with the *electron microscope,* developed in the 1930s, with which magnifications of several hundred thousands are possible. The wavelength of radiations used with the electron microscope is typically about 0.005 nm (0.05 Å), although resolution of the order of an angstrom or less is theoretically possible. This is many thousand times greater than that attainable using microscopes with glass optics. The basic features of the *transmission* electron microscope (often simply abbreviated TEM) are shown in Figure 1–1, and a comparison between the component parts of the TEM and the light microscope is depicted diagrammatically in Figure 1–2. In recent years, the *scanning* electron microscope (SEM) has become an increasingly important tool of the cell biologist. The SEM employs quite different principles than the TEM and will be considered separately later.

In both the light and electron microscopes, the source of radiation is an electrically heated tungsten filament. In the light microscope, the light emitted from the glowing filament is focused by a *condenser* onto the specimen to be observed. In the transmission electron microscope, the condenser focuses *electrons* emitted by the excited tungsten atoms into a beam and directs this onto the specimen. While the condenser of a light microscope consists of one or a few glass lenses, the condenser of the electron microscope consists of several large, circular electromagnets. Indeed, all "lenses" of the electron microscope are electromagnets. In both microscopes, the radiation passes through the specimen and is then refocused by the *objective* lenses. The last lens of the light microscope is the *ocular,* through which the image may be viewed with the eye. The image of the electron microscope is viewed after its magnetic projection onto a zinc sulfide screen. The molecules of the screen are excited by the impinging electrons and emit visible light during their return to the ground state. Alternatively, the image may be captured on photographic film housed in a special camera mounted below the movable zinc sulfide screen.

The lenses of the light microscope have a fixed *focal length* and are focused by moving them nearer to or further from the specimen. In the electron microscope, focusing is accomplished by manipulating the amount of current flowing through the windings of the series of electromagnet lenses. This alters the electromagnetic force through which the electron beam must pass. The column through which the electron beam passes must be evacuated of air. If the vacuum is inadequate, the electrons would be scattered by collisions with gas molecules. Consequently, the specimen, the filament, the electromagnets, and the zinc sulfide screen

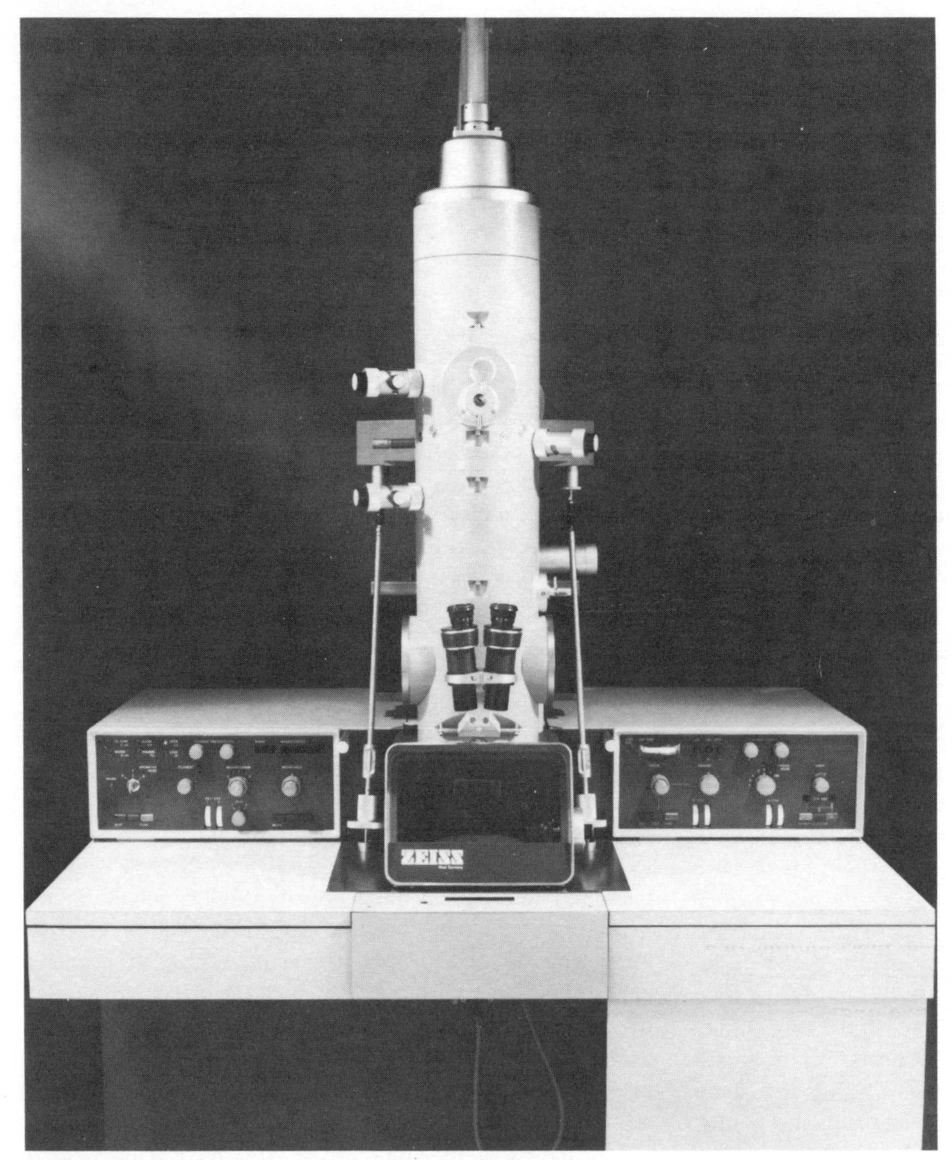

Figure 1–1
A transmission electron microscope. (Courtesy Carl Zeiss, Inc.)

are all mounted within a sealed compartment connected to a vacuum pump. In order to avoid excessive scattering or absorption of electrons by the specimen itself, the material to be examined must be cut into extremely thin sections.

Two special forms of light microscopy warrant further description because of their widespread use and special applications: phase-contrast microscopy and fluorescence microscopy.

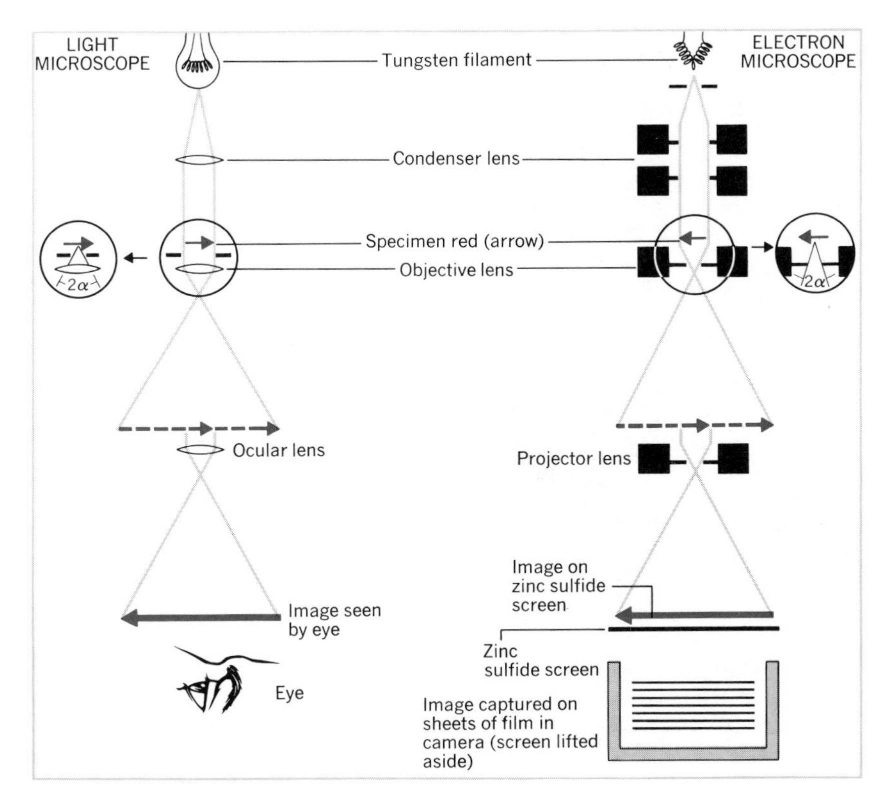

Figure 1–2
A comparison of the basic features of the light microscope and TEM

Phase-Contrast Microscopy. Although most regions of an unstained cell are transparent, they may have different densities and therefore different refractive indexes. Consequently, light rays travel through these regions at different velocities and may be refracted or bent to different extents. The phases of light rays that pass directly through an object and those that pass across its edges (i.e., at the interface where the refractive index changes) will necessarily be altered. The phase change increases the contrast between the object in focus and its surroundings. In the phase-contrast microscope, the phases of light rays entering the object are shifted by an annular diaphragm below the condenser. The phases of rays passing through and around the object are shifted again by a phase plate in the objective lens. The result is a striking increase in the contrast of the object as certain regions appear much brighter (owing to additive effects of rays brought into phase), while other regions appears much darker (owing to the cancelling effects of rays shifted further out of phase). The effect can be seen in Figure 1–39.

Fluorescence Microscopy. Certain chemical substances emit visible light when they are illuminated with ultraviolet light. The effect is termed **fluorescence** and is put to use in the fluorescence microscope in which ultraviolet light rays are focused on the specimen. Some cellular components (e.g., cytoplasm, mitochondria, and certain granules) possess a natural fluorescence and appear in various colors. Other, nonfluorescing structures can be made to fluoresce by staining them with fluorescent dyes *(fluorochromes)*. One of the most popular contemporary uses of fluorescence microscopy involves the preparation of *antibodies* that will bind to specific cellular proteins (see Chapter 4). The antibodies are first complexed with *fluorescein* (a fluorescent dye), and the *fluorescein-labeled* antibody is then applied to the cells. Cell structures containing the specific proteins

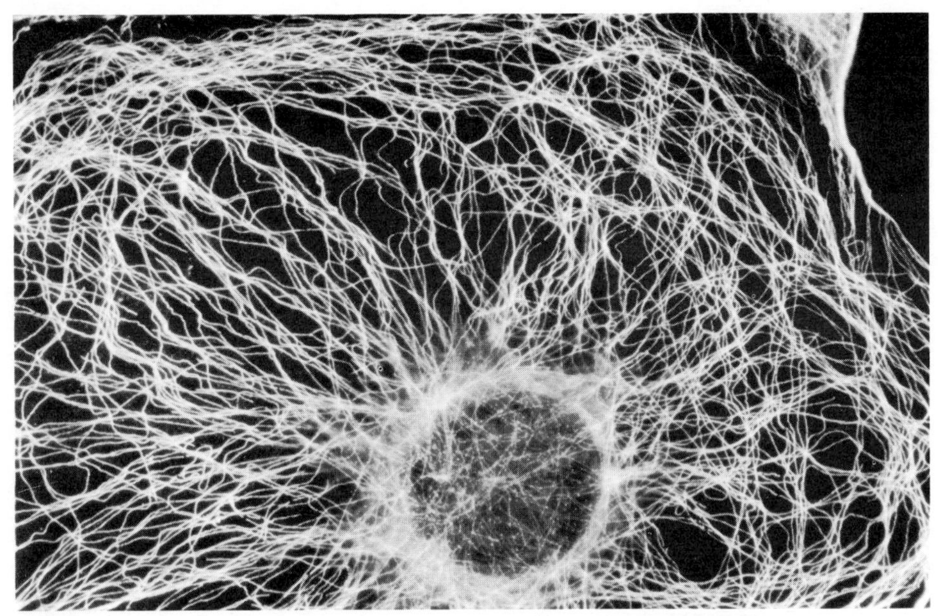

Figure 1–3
The network of microtubules present in this cell is clearly revealed using the combined techniques of fluorescent antibody labeling and fluorescence microscopy. (Photomicrograph courtesy of Drs. M. Osborn and K. Weber.)

capable of binding the fluorescein-labeled antibody are caused to fluoresce when examined with the fluorescence microscope, dramatically revealing their detail (Fig. 1–3).

Preparation of Materials for Microscopy

The preparation of biological material for examination with either the light microscope or the transmission electron microscope involves a series of physical and chemical manipulations that include (1) fixation, (2) embedding, (3) sectioning, and (4) mounting.

Fixation. One notable advantage of the light microscope is the ability to observe whole, *living* cells. It is also possible to employ ''vital stains,'' which improve contrast but do not interfere with normal cell activity. More frequently, however, the cells are first killed and fixed. The fixation step is intended to preserve the structure of the material by preventing the growth of bacteria in the sample and by precluding postmortem changes. Formaldehyde and osmium tetroxide (OsO_4) are examples of fixatives most often employed for light microscopy. OsO_4 has a very high electron density, and since this gives contrast to the resulting image, OsO_4 has also found widespread use as a fixative in electron microscopy. Other popular fixatives include potassium permanganate and glutaraldehyde. After fixation for the required length of time, the samples are dehydrated by successive exposures to increasing concentrations of alcohol or acetone.

Embedding. Cells or tissues to be examined by light microscopy are usually embedded in warm, liquid paraffin wax. The wax, which both surrounds the tissue and infiltrates it, hardens upon cooling, thereby supporting the tissue externally and internally. The resulting solid paraffin block is then trimmed to the appropriate shape before being sectioned. The ultrathin sections required for electron microscopy necessitate the use of harder embedding and infiltrating materials, such as methacrylate, Epon, or Vestopal. These initially are in liquid form and are poured into small molds containing pieces of the fixed tissue; upon heating, they polymerize into hard plastics (Fig. 1–4).

Sectioning. The trimmed blocks containing the embedded samples are sectioned using a *microtome* (Fig. 1–5). In this instrument, the block is sequentially swept over the blade of a knife that cuts the block into a series of thin sections, forming a ribbon. Between each stroke, the block is advanced a short distance toward the knife. For light microscopy, the microtome knives are usually constructed of polished steel and can provide sections several microns thick. The sections for electron microscopy must be much thinner (typically 100 to 500 Å) and require more elaborate microtomes (called ultramicrotomes), such as the one shown in Figure 1–5, which either mechanically or thermally advance the plastic block much shorter distances with each stroke. Morever, either diamond knives or knives prepared by fracturing plate glass are used in place of polished steel. Figure 1–6 illustrates the preparation of a ribbon of sections during ultramicrotomy.

Mounting. Sections prepared for light microscopy are mounted on glass slides and may be stained with dyes of various colors that specifically attach to different molecular constituents of the cells. Sections to be examined with the electron microscope are generally not stained (no colors are seen with the electron microscope), although contrast may be improved by ''poststaining'' with electron-dense materials such as uranyl acetate, uranyl nitrate, and lead citrate. The sections are mounted on copper ''grids'' (small disks perforated with numerous openings) that have been coated with a thin (sometimes monomolecular) film of Formvar or carbon (Fig. 1–7). The grid supports the film, which in turn supports the thin section. Thus the beam of electrons must pass through the spaces of the grid, the supporting film, and the section before striking the fluorescent screen. A comparison of photomicrographs obtained with the light microscope and the TEM is given in Figure 1–8.

Specialized Applications of Transmission Electron Microscopy

Shadow Casting. In shadow casting, the sample (usually containing small particles such as viruses or macromolecules) is spread on a coated grid, which is then placed in an evacuated chamber. A chromium or platinum wire is heated until the metal is vaporized, and the vapor is deposited onto the sample *at a precise angle*. The metal piles up in front of the sample particles but leaves clear areas behind them. If the resulting electron photomicrographs are printed in reverse, the areas containing the electron-dense metal that had piled up against the particles appear light, while the electron-transparent areas behind the particles appear as dark shadows. Because the vaporized metal atoms

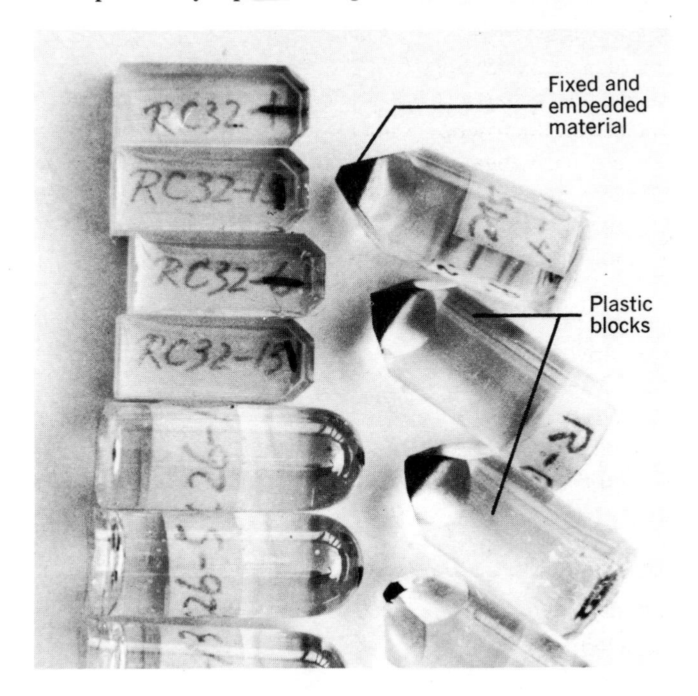

Figure 1–4
Plastic blocks of various shapes containing fixed and embedded tissue. (Photo courtesy of R. Chao.)

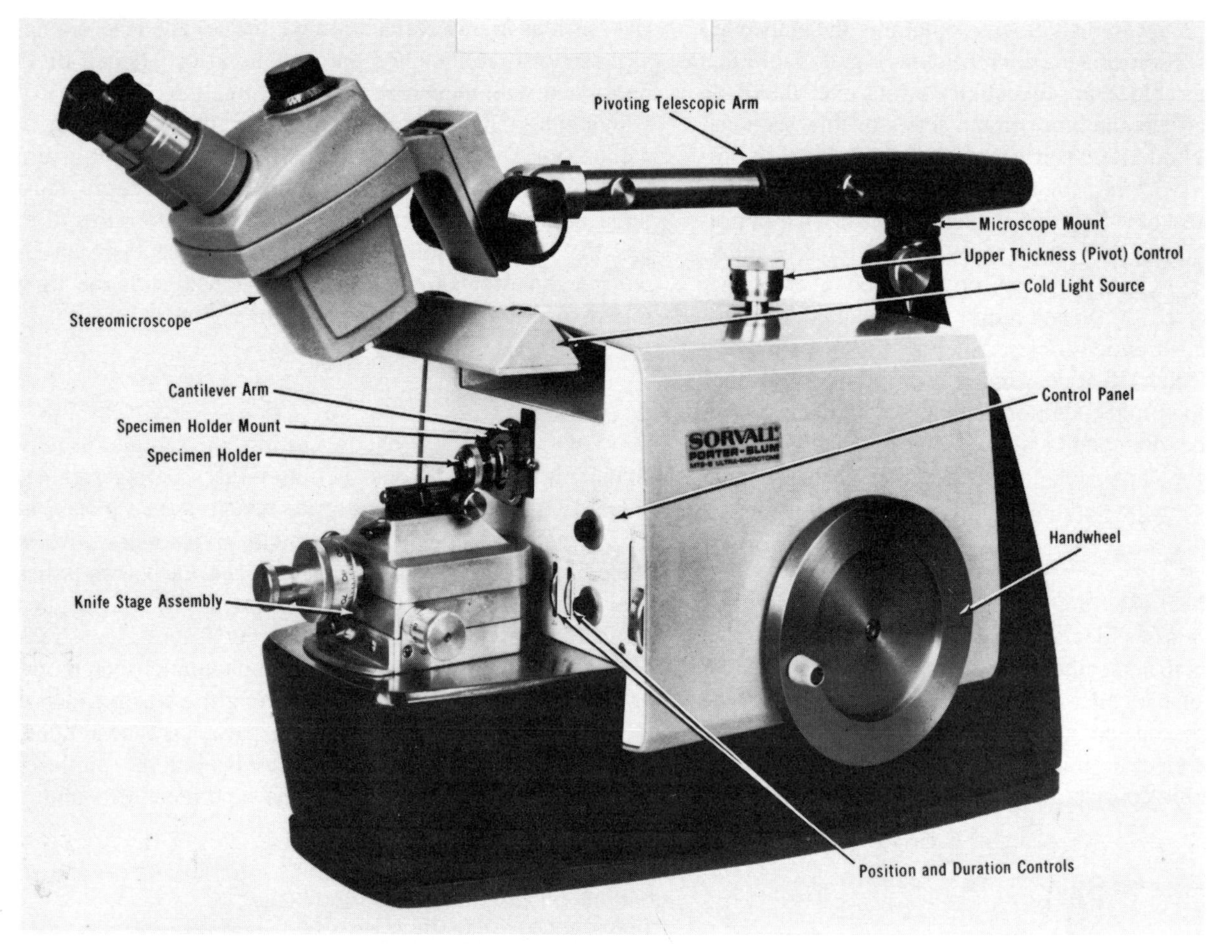

Figure 1–5
An ultramicrotome. (Photo courtesy of DuPont Instruments.)

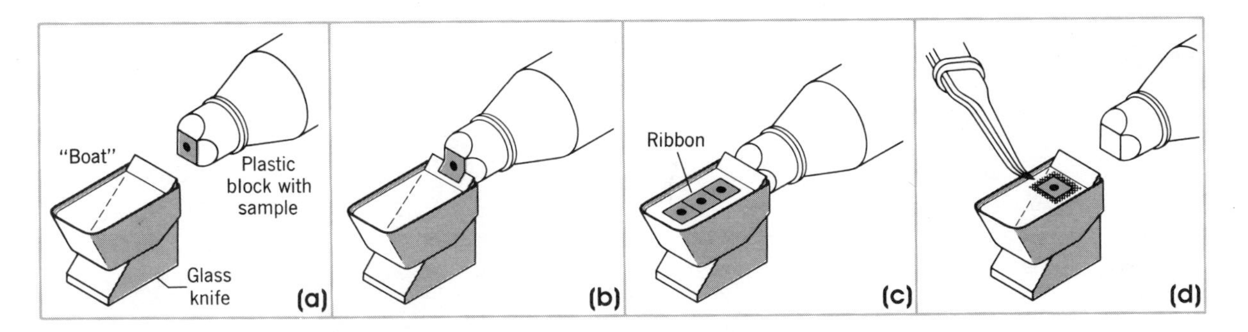

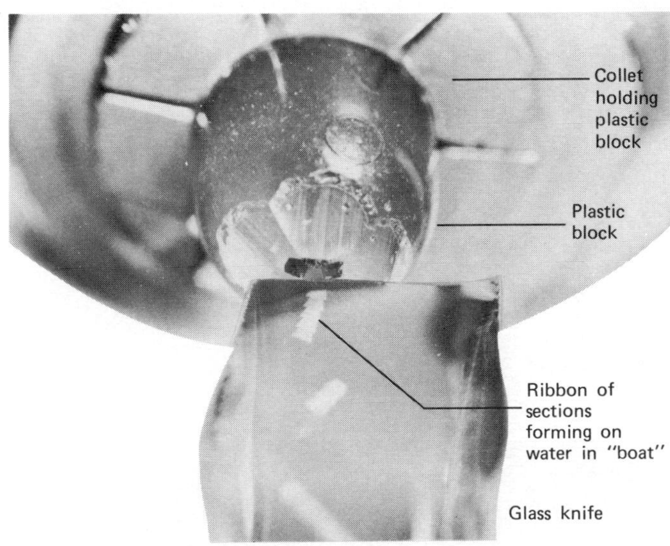

Figure 1–6
Sectioning of embedded tissue to form ribbon of thin sections floating on water in "boat" attached to glass microtome knife.

Figure 1–7
Various grids used for mounting sections for electron microscopy. (Photo courtesy of R. Chao.)

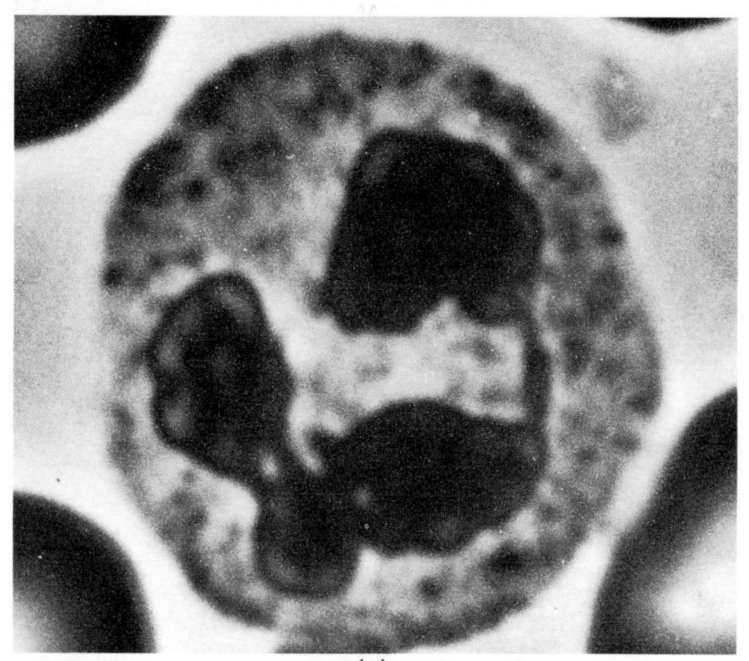

(a)

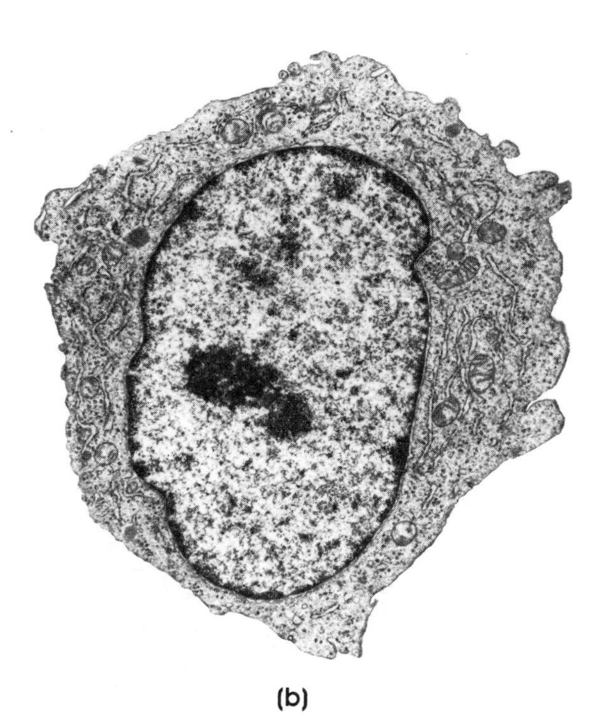

(b)

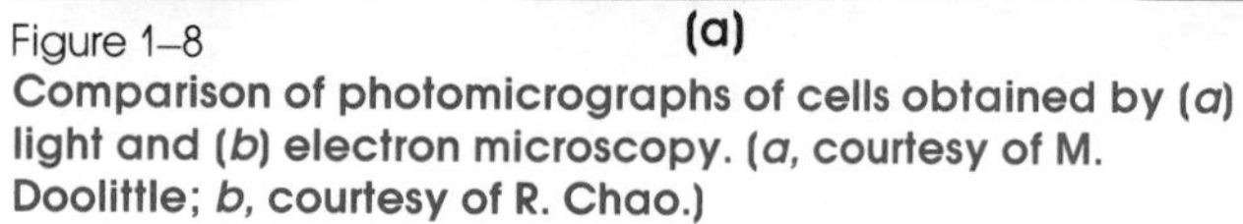

Figure 1–8
Comparison of photomicrographs of cells obtained by (*a*) light and (*b*) electron microscopy. (*a*, courtesy of M. Doolittle; *b*, courtesy of R. Chao.)

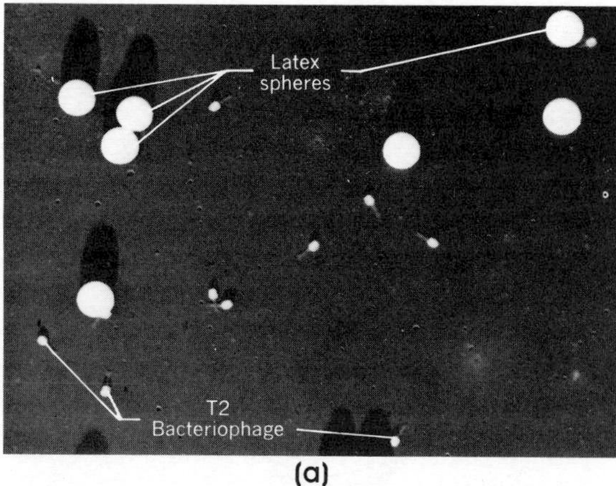

(a)

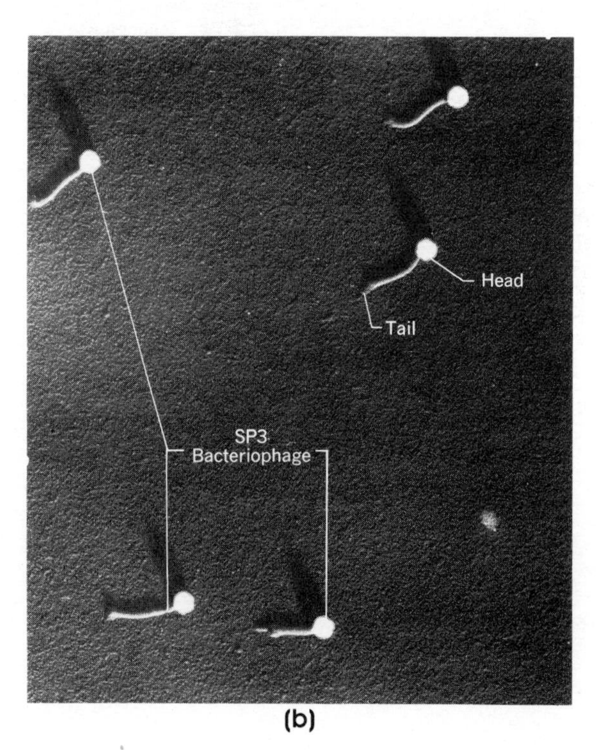

(b)

Figure 1–9
Shadow casting. (*a*) Metal shadowed mixture of latex spheres and viruses (T2 bacteriophage). The latex spheres are of known size (0.3 μ) and concentration, thereby permitting ready determination of virus size and concentration. (*b*) Shadowed SP3 viruses. From the shadowing angle and shadow contours, the sizes and shapes of the viral parts may be determined. (Courtesy of Dr. F. A. Eiserling.)

tend to be projected in a straight line, the shadows are cast at precise angles. in this manner, the general shape and profile of a particle may be discerned (Fig. 1–9).

Negative Staining. In the negative-staining procedure, the sample (again small particles such as viruses or macromolecules) is surrounded by an electron-dense material, such as phosphotungstic acid, that permeates the open superficial interstices of the sample. When the excess material is removed, the sample particles appear as light (i.e., electron-transparent) areas surrounded by a dark background (Fig. 1–10).

Figure 1–10
Glycogen particles from liver tissue visualized by negative staining (see also Chapter 5). Magnification 51,000.

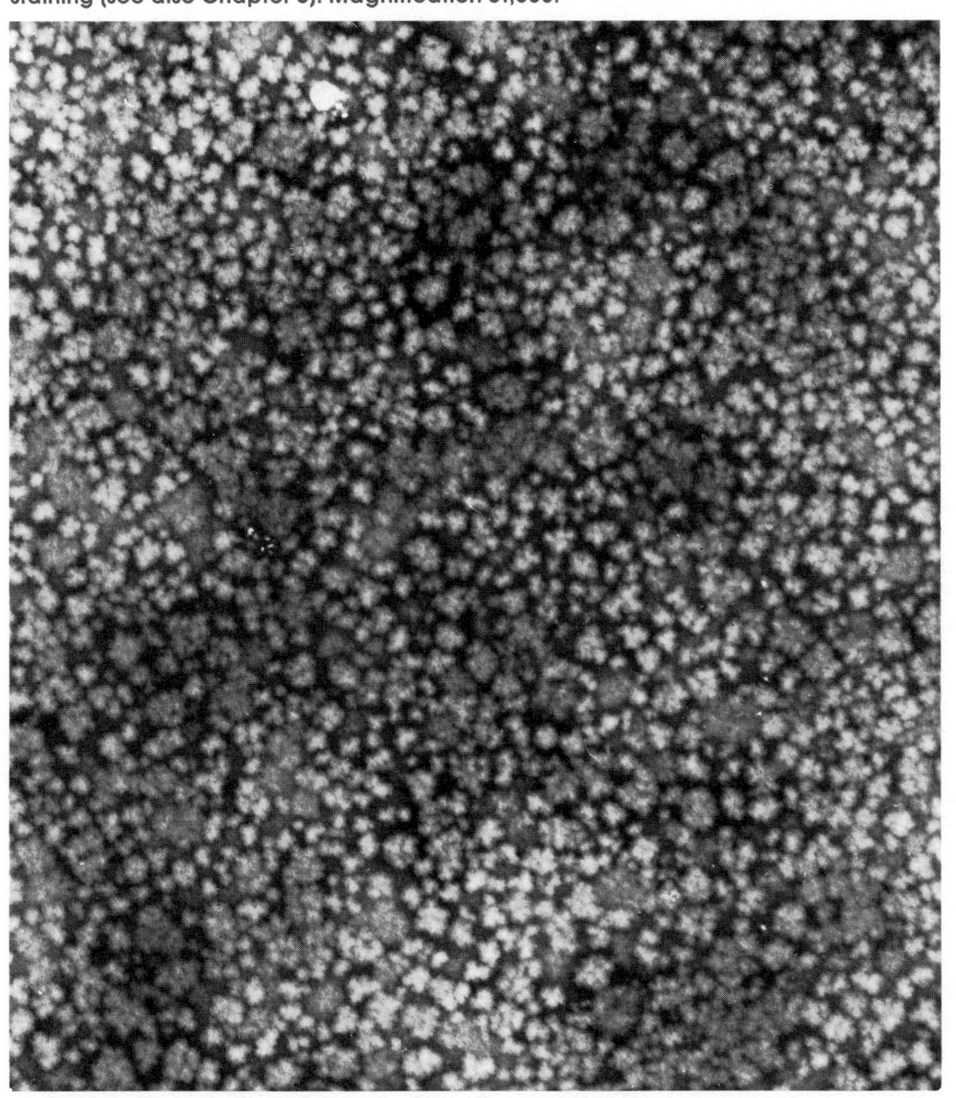

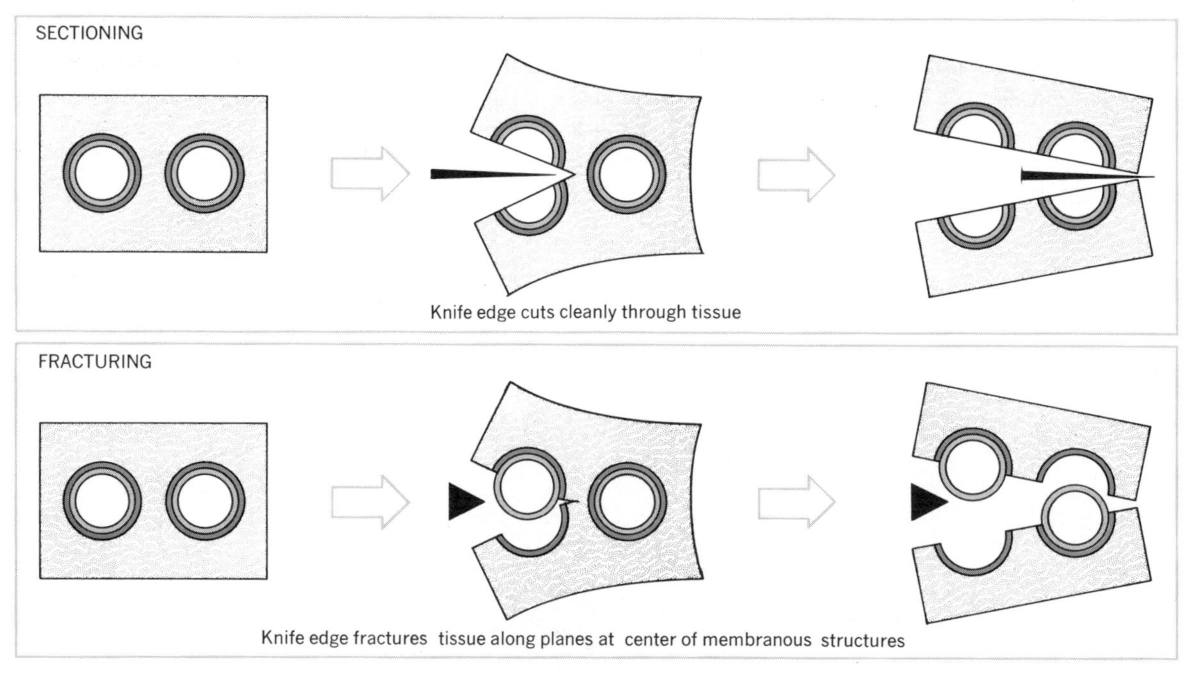

Figure 1–11
A comparison of sectioning and fracturing tissue.

Freeze Fracturing. Freeze fracturing is a technique in which the tissue is first *fractured* (i.e., cracked) along planes of natural weakness that run through each cell. These planes generally occur between the two layers of molecules that comprise the limiting membrane around the cells' various vesicular organelles (see Chapter 15). Figure 1–11 depicts the basic differences between sectioning and fracturing.

The tissue to be freeze-fractured is first impregnated with glycerol and then frozen at about −130°C in liquid Freon. The frozen tissue is transferred to an evacuated chamber containing a microtome and steel knife (also maintained at about −100°C using liquid nitrogen). The microtome knife is used to produce a fracture plane across the tissue (Fig. 1–12*a* and *b*). When the plane of the fracture intersects the

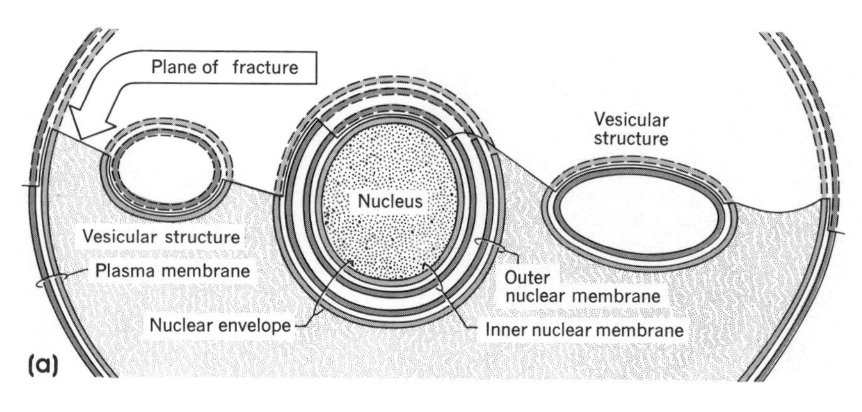

Figure 1–12
Stages in the freeze-fracturing procedure (see text for explanation).

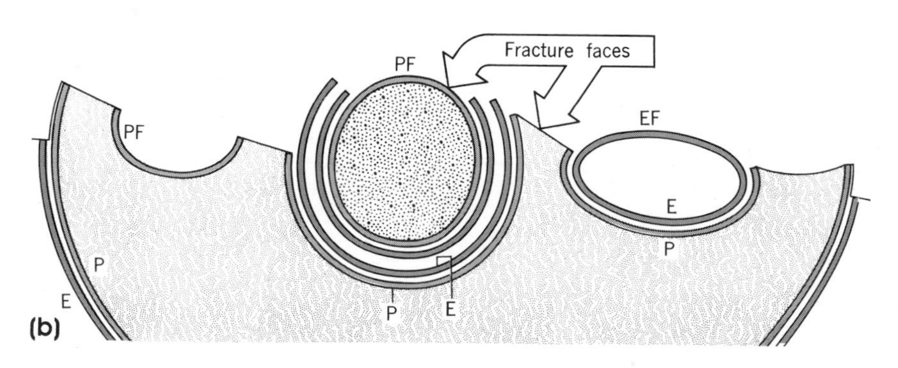
Fracture faces
PF
PF
EF
E
P
P
E
P
E
(b)

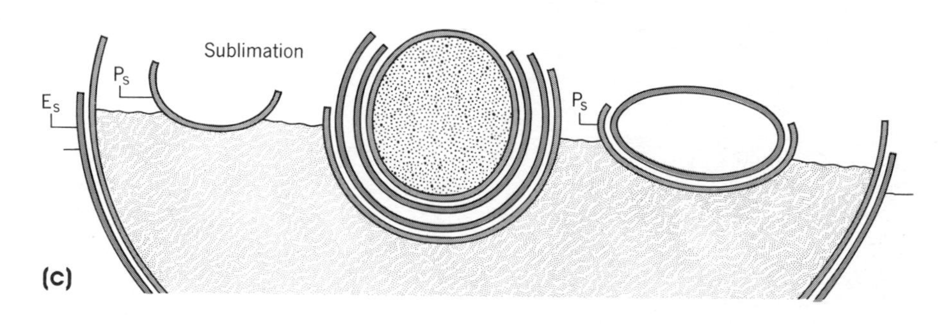
Sublimation
P_s
E_s
P_s
(c)

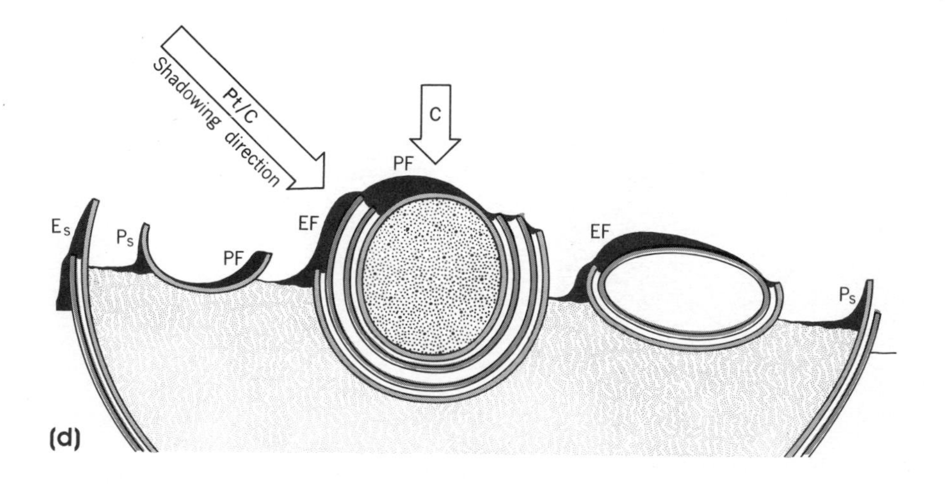
Pt/C
Shadowing direction
C
PF
E_s
P_s
PF
EF
EF
P_s
(d)

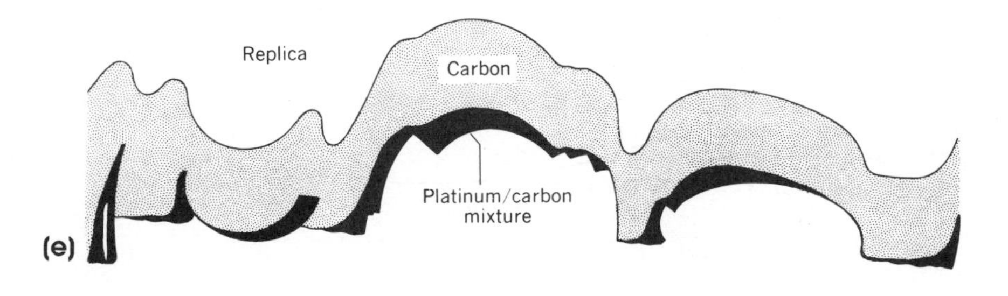
Replica
Carbon
Platinum/carbon mixture
(e)

Figure 1–13
A comparison of electron photomicrographs of similar regions of cells obtained by sectioning and by freeze-fracturing. (*a* and *b*) Micrographs of liver tissue (N, nucleus; P, pores; rer, rough endoplasmic reticulum; ser, smooth endoplasmic reticulum; Go, Golgi body; m, mitochondria; Mb, microbody; BC, bile canaliculus; Gl, glycogen). (*c* and *d*) Cortical regions of egg cells (pm, plasma membrane; cg, cortical granule). The direction of shadowing is indicated by the circled arrow. (*a* and *b*, courtesy of Dr. L. Orci; *c* and *d*, courtesy of Dr. E. G. Pollock.)

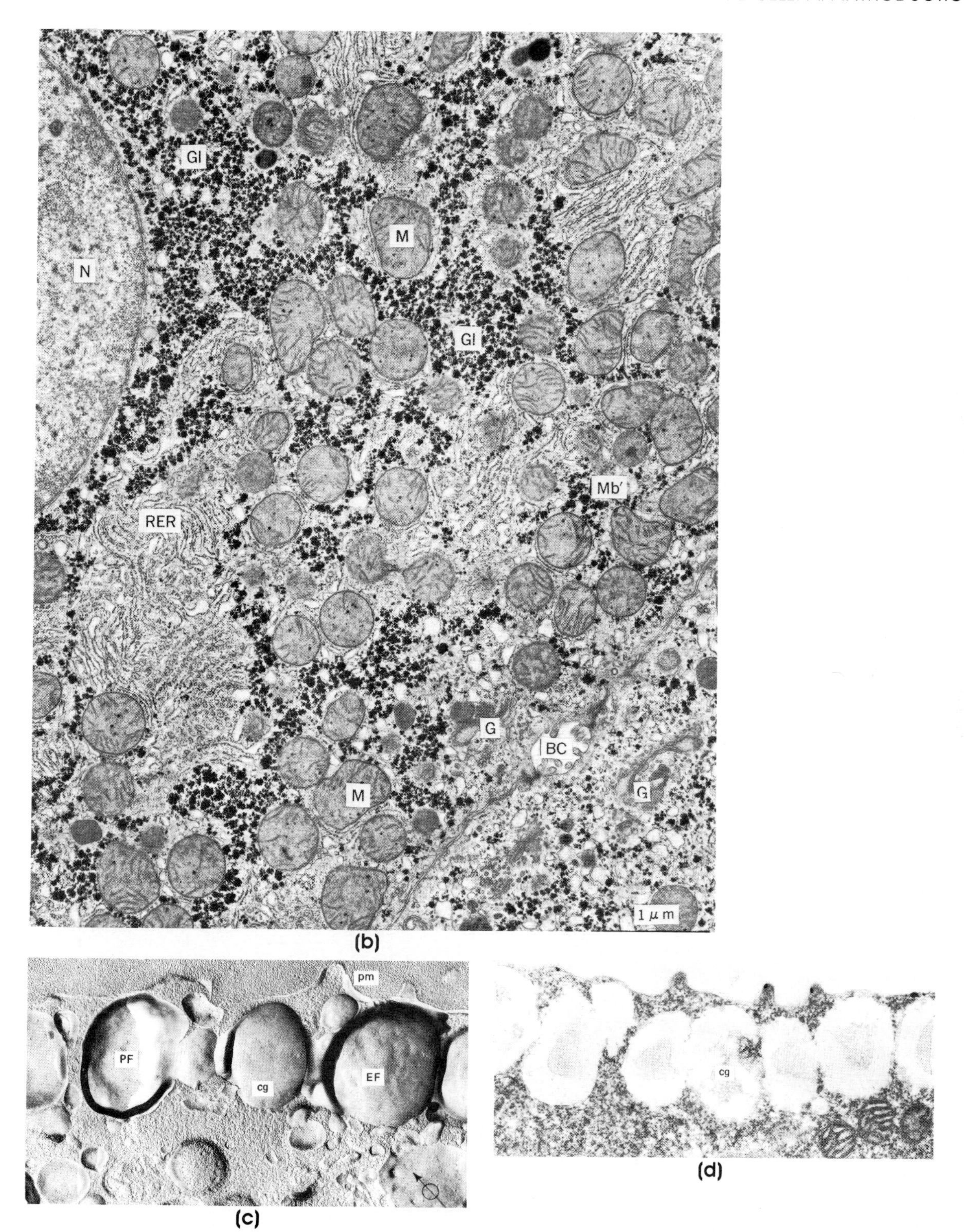
Gl
M
N
Gl
Mb'
RER
G
BC
M
G
1 μm
(b)
pm
PF
cg
EF
(c)
cg
(d)

membrane of a vesicular structure (e.g., nucleus, mitochondrion, vacuole, etc.), the membrane is split along its center, producing two "half-membranes." These are called the E half and the P half. The E half formerly faced the cell's external phase (see later), and the P half faced the internal phase (cytosol). One surface of each half-membrane is the original membrane surface (the **E** and **P** surfaces of Fig. 1–12*c* and *d*), while the other surface is the newly exposed fracture face (the **EF** and **PF** surfaces of Fig. 1–12*c* and *d*). The vacuum of the chamber is then used to sublimate water on the cut surface to a depth of several hundred angstroms. New membrane faces exposed by sublimation are termed $\mathbf{E_s}$ and $\mathbf{P_s}$ (Fig. 1–12*c*). An electron-dense combination of metal (usually platinum) and carbon is then deposited on the cut surface at an angle and piles up in front of and behind projections from the surface, as well as in pits and depressions (Fig. 1–12*d*). Additional carbon is added to form an electron-transparent backing.

The shadowed and coated tissue is removed from the chamber, and the tissue itself is either floated off or dissolved away, thereby leaving only the carbon-platinum "replica" (Fig. 1–12*e*). The replica is trimmed to the proper size, placed on a grid, and examined with the transmission electron microscope. Note that the replica is actually a templatelike impression of the distribution of particles in the original specimen. The electron beam readily passes through portions of the replica containing the carbon but is absorbed by the areas containing the platinum. The resulting images, which have a three-dimensional impact, are considerably different from those obtained with sectioned material (Fig. 1–13).

The Scanning Electron Microscope

Scanning electron microscopy has become an increasingly popular technique since its introduction as a biological tool in the 1960s. With this technique, the *surface topography* of a specimen may be examined in considerable detail. At the present time, resolution is of the order of 50 Å, so that most specimens examined are the sizes of whole cells or clusters of cells. The organization of the scanning electron microscope (SEM) is shown in Figure 1–14 and is basically similar to the TEM. However, instead of the electron beam passing through (i.e., being "transmitted" by) the specimen, the interaction of the electrons of the beam (called "primary" electrons) with the surface of the specimen causes the emission of "secondary" electrons from the surface. The beam rapidly scans back and forth over the surface of the specimen, thereby producing bursts of secondary electrons. Greater numbers of secondary electrons are produced when the beam strikes projections from the specimen surface than when the beam enters a pit or depression in the surface. Hence, the number of secondary electrons produced at each point on the specimen surface, as well as the direction in which scattering occurs, depends upon the surface topography. Therefore, there are quantitative and qualitative differences in the secondary electron bursts produced by the scanning electron beam. These ultimately give rise to an image in the following way.

Secondary electrons ejected at each point on the specimen surface are accelerated toward a positively charged scintillator located at the side of the specimen. Light scintillations created upon impact of these electrons with the scintillator

Figure 1–14
Essential components of scanning electron microscope (SEM).

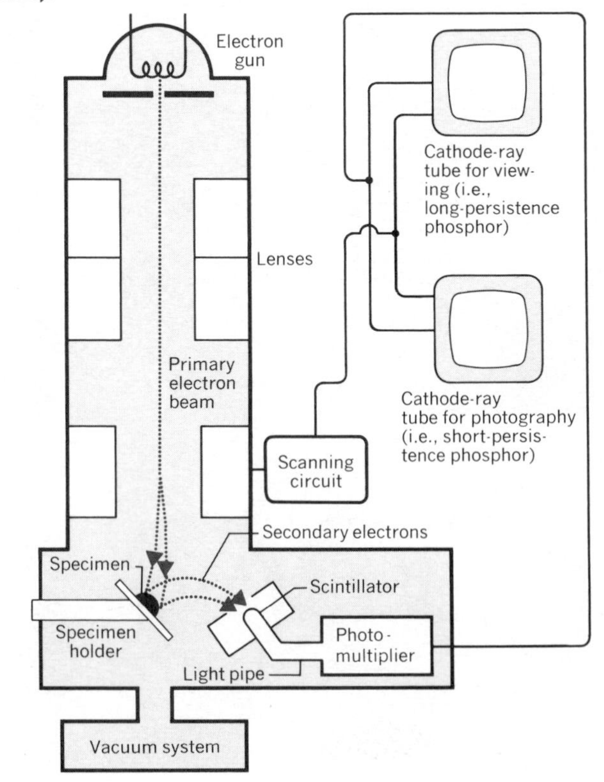

are conducted by a light guide to the photocathode of a photomultiplier tube. Electrical pulses produced in the photomultiplier tube are then amplified, and the resulting signal is relayed to a cathode-ray tube. The result is an image much like that of a television, consisting of light and dark spots. The scanning of the specimen surface by the primary electron beam is synchronized with the projection of a beam on the television screen in such a way that each portion of the specimen is reproduced in a corresponding region of the television image.

Samples to be examined are usually coated first with a metal (typically a gold-palladium alloy), forming a layer about 500 Å thick, and then are affixed to a supporting disk that is placed in the beam path. The metal coating efficiently reflects the primary electrons of the beam and also produces large numbers of secondary electrons. Figure 1–15 contains

Figure 1–15
Examples of scanning electron photomicrographs. (*a*) Field of teratoma tumor cells. (*b*) Red blood cells. (*c*) Higher magnification of view of the teratoma cell. (*d*) Lymphocyte (white blood cell) with platelets and red blood cells. Note the numerous surface projections (*microvilli*) from the teratoma cells and lymphocyte. (*a* and *c*, courtesy of S. B. Oppenheimer, E. G. Pollock, and R. Brennerman; *b*, courtesy of Dr. M. A. Lichtman; *d*, courtesy of R. Chao.)

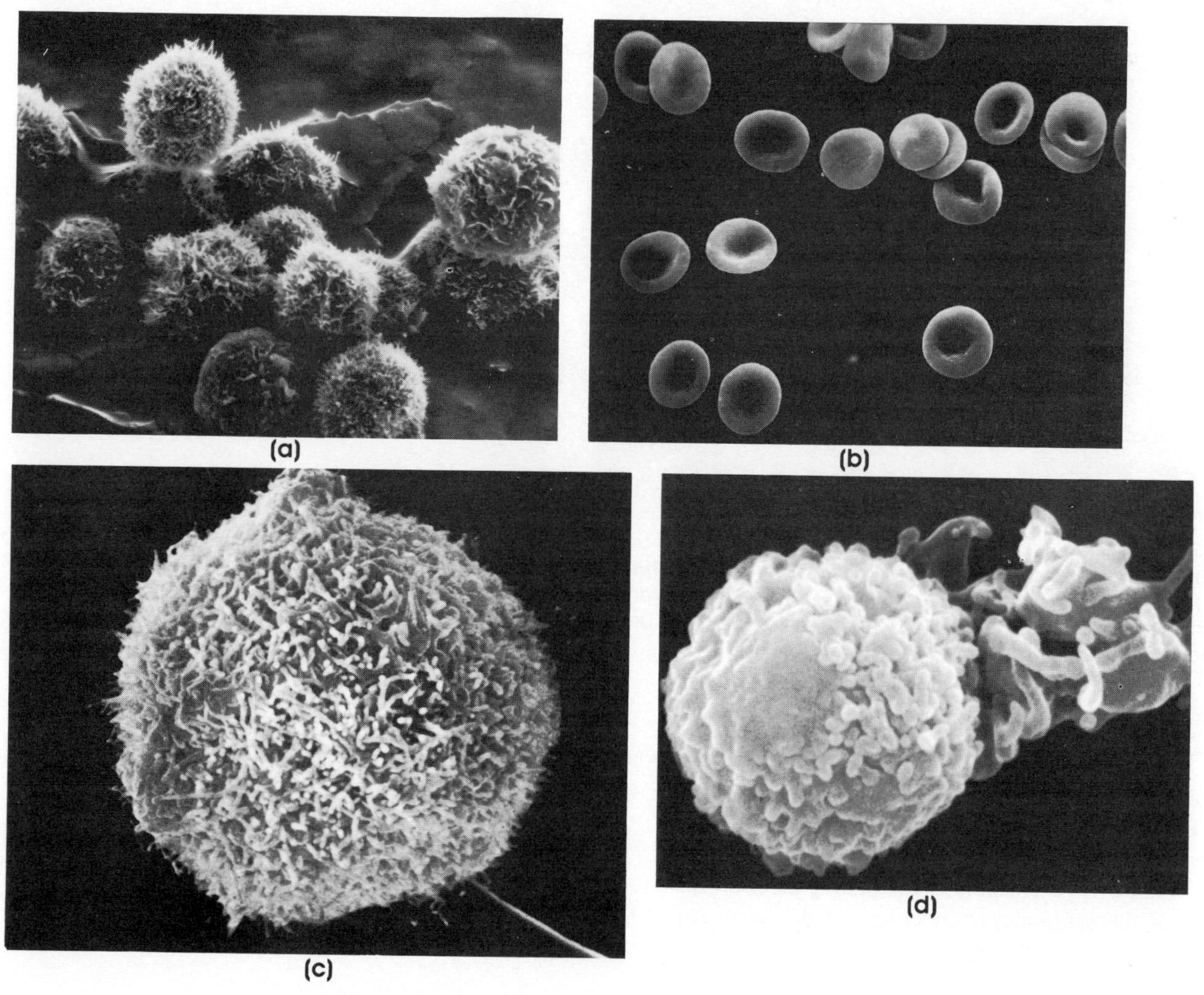

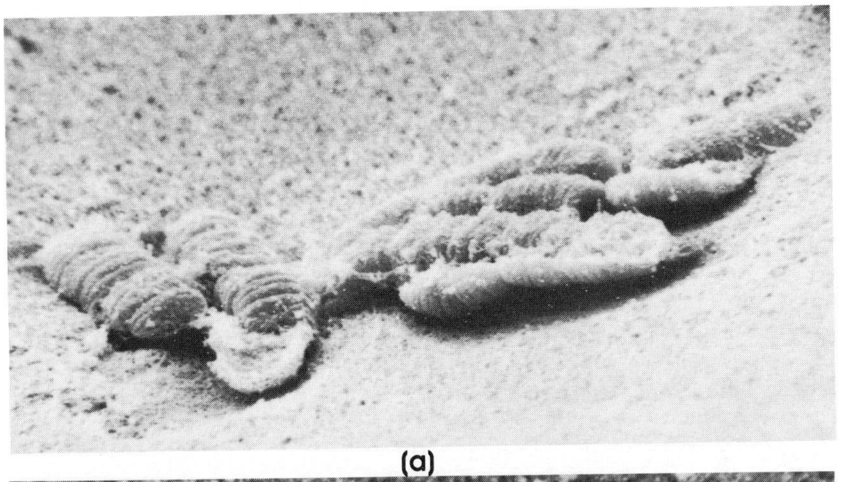

(a)

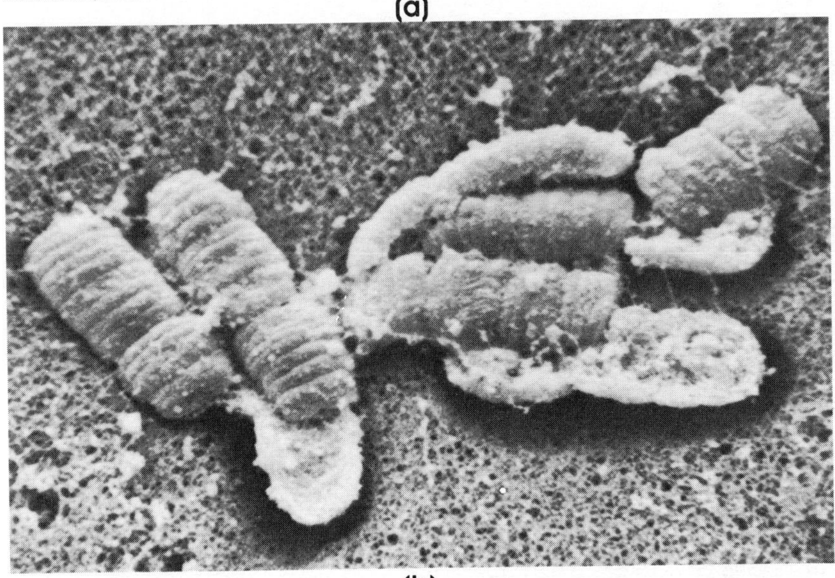

(b)

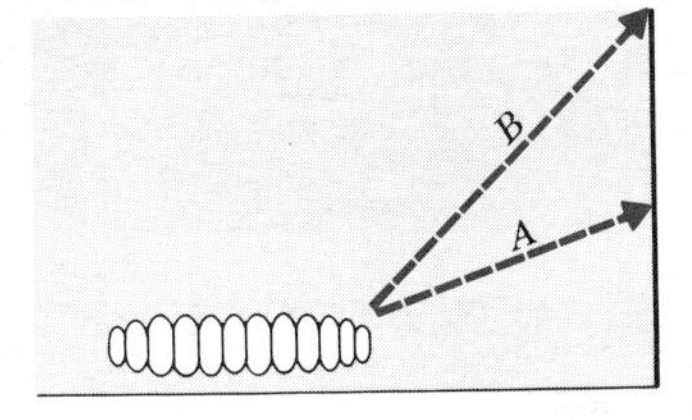

Figure 1–16
Scanning electron micrographs of chains of bacterial cells (*Simonsiella*) viewed from two (i.e., *a* and *b*) different angles showing the varying perspectives attainable. Magnification 4000 ×. (Courtesy of Drs. D. A. Kuhn, J. Pangborn, and J. R. Woods.)

examples of photomicrographs obtained with the SEM.

Since the specimen being examined with the SEM can be rotated, it is possible to obtain views from different angles. This provides additional information about the size, shape, and organization of the material being studied. For example, Figure 1–16 contains two scanning electron micrographs of the same cluster of chains of the bacterium *Simonsiella* taken from different angles.

Stereo Microscopy (Stereoscopy)

True three-dimensional (i.e., stereoscopic) images of the specimen being studied can be obtained if one photomicrograph is taken as though the specimen were being viewed with the left eye *only*, and a second is obtained representing the right eye view. (The two views are obtained by a minor tilting of the sample in the horizontal plane.) When the two

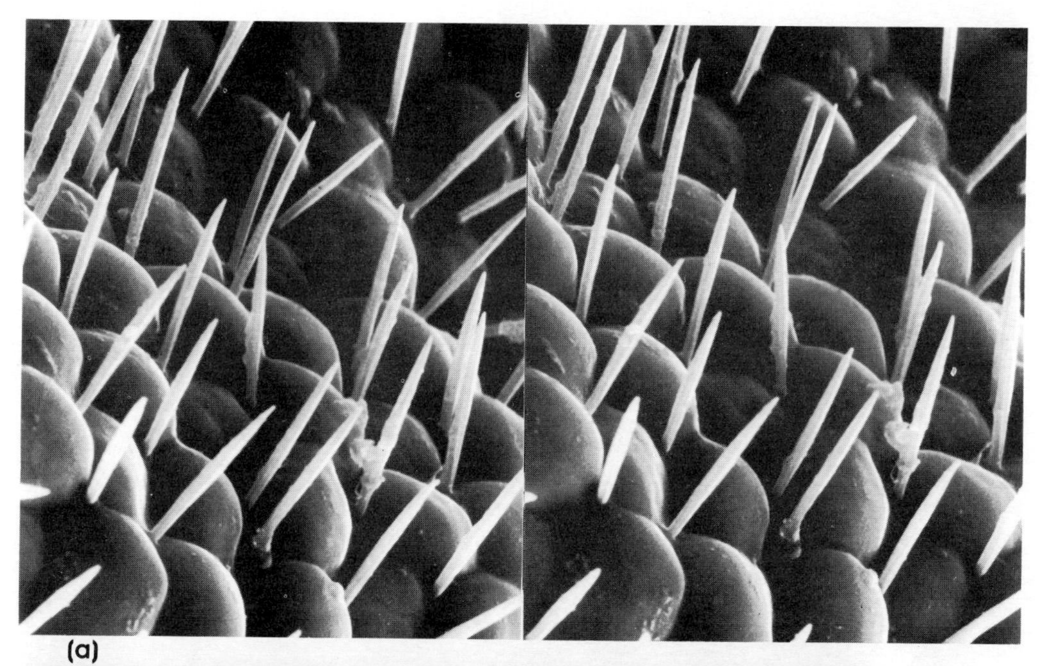

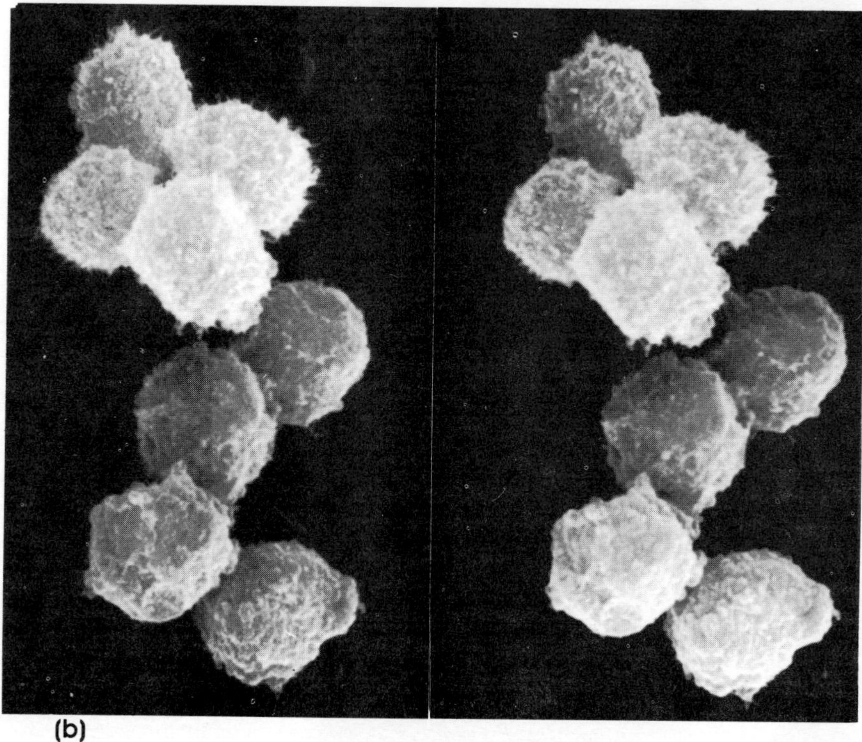

Figure 1–17
Stereo electron micrographs obtained using SEM. (*a*) Multiple lenses (*ommatidia*) of an insect eye (the spinelike structures are fine hairs); (*b*) a cluster of teratoma tumor cells; (*c*) a single tumor cell; (*d*) erythrocytes and a single lymphocyte (white blood cell); (*e*) lymphocyte. These stereo pairs must be viewed with a stereo viewer in order to obtain the three-dimensional perspective. (Photos courtesy of R. Chao.)

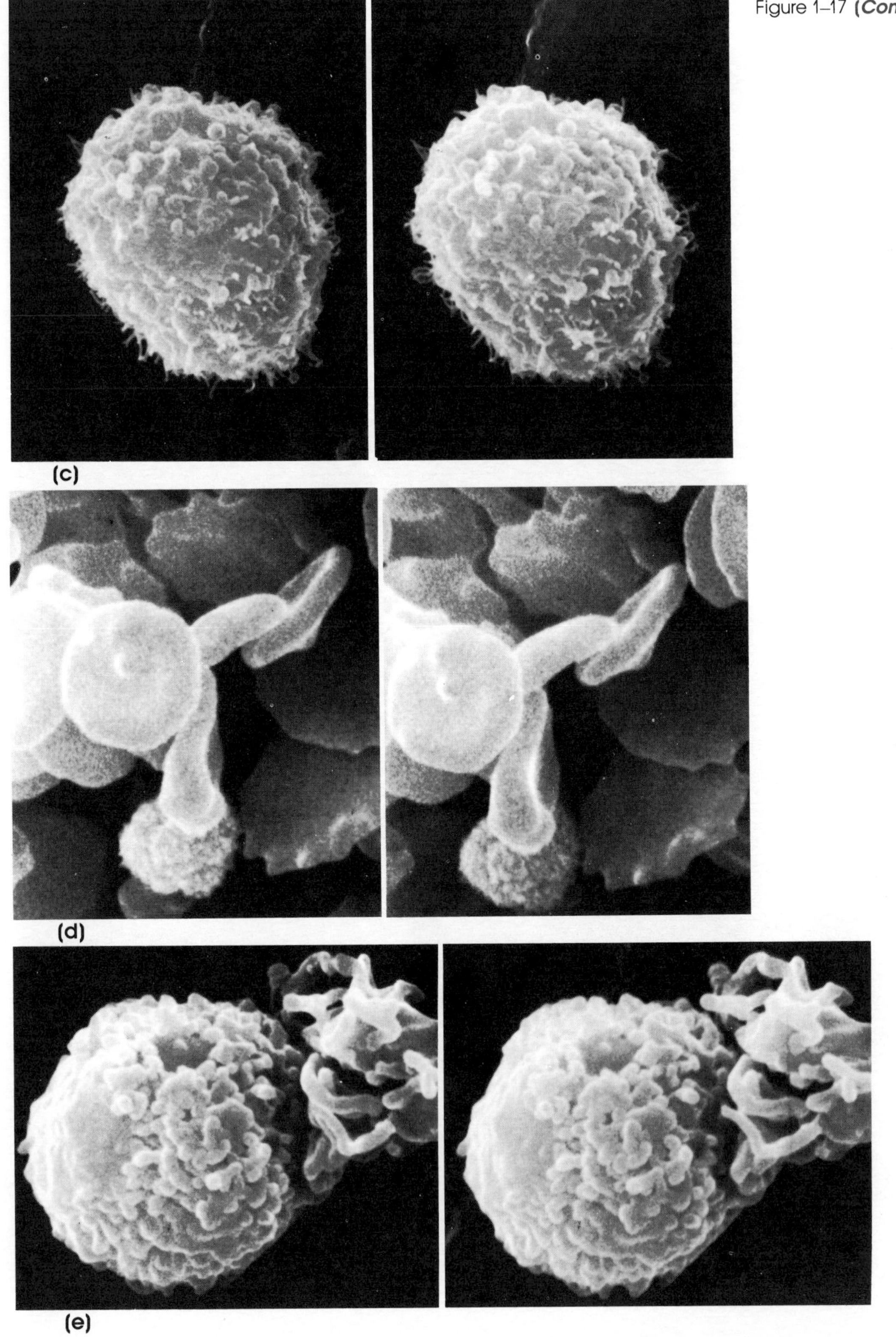

Figure 1–17 (***Continued***)

micrographs are placed side by side and the *stereo pair* is viewed through the appropriate pair of lenses (called "stereo viewers"), a striking three-dimensional impression is seen, revealing details and geometric relationships that cannot be discerned from a single photomicrograph. Stereo views of the surface topography of tissues and cells are readily obtained with specimens prepared for SEM study, and illustrations are presented in Figure 1–17.

The internal organization of cells is revealed in three dimensions by *high-voltage transmission electron stereoscopy*. In this procedure, cells are placed or cultured on a conventional grid, are then fixed and dehydrated, and the grid and cells are sandwiched between layers of carbon. The samples are examined in a TEM in which the accelerating voltage is great enough to penetrate the entire thickness of the cell (about 1 million volts). The cells are photographed at various tilt angles to produce the stereo pairs needed for the three-dimensional image (see Fig. 1–18).

The viewing of stereomicrographs may present some difficulties, especially for the novice. Generally, fewer problems are encountered with SEM stereoscopic views (i.e., Fig. 1–17), since the objects in the photomicrographs are opaque (i.e., certain objects are clearly in front of others). Transmission stereomicrographs are more difficult to assimilate and interpret because most of the objects are translucent. However, no other procedure provides direct images of the three-dimensional morphology of the cell's interior. A single stereo pair can reveal the entire population of mitochondria, lysosomes, or other organelles (see below) distributed through the cell.

Cell Structure: A Preview

Free-living cells and the cells of multicellular organisms are subdivided into two major classes—**eucaryotes** (i.e., "true nucleus") and **procaryotes** (i.e., "before nucleus"). In eucaryotes, the constituents of the cell nucleus (chromosomes, DNA, etc.) are separated from the rest of the cell by a boundary membrane, whereas in procaryotes these materials are not separated. Although the presence or ab-

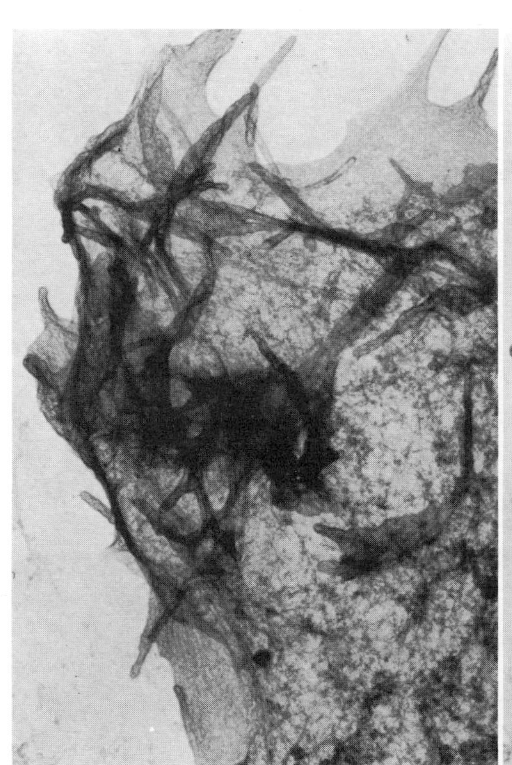

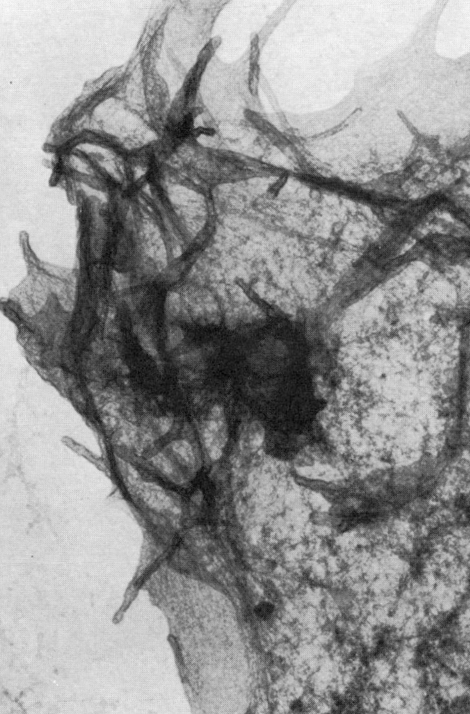

Figure 1–18
High-voltage stereo transmission electron photomicrograph. In this view, the ruffles and microvilli that cover the surface of the cell are clearly seen. A fine, irregular lattice of *microtrabeculae* (filamentous structures) can be seen to occupy the space between the upper and lower cell surfaces as well as the surface projections. Magnification, 11,000 ×. (Photomicrograph courtesy of Drs. J. J. Wolosewick and K. R. Porter.)

sence of a true nucleus is the most obvious distinction between eucaryotic and procaryotic cells, it will soon become clear that these two groups of cells also differ in many other important respects. Essentially all animal and plant cells are eucaryotic, whereas procaryotic cells include bacteria, blue-green algae, and the so-called pleuropneumonia-like organisms (PPLO) or mycoplasmas.

Eucaryotic Cells: The Generalized Animal Cell

Animal cells vary considerably in size, shape, organelle composition, and physiological roles. Consequently, there is no "typical" cell that can serve as an example of *all* animal cells. There are, however, a number of cell structures common to the majority of animal cells that are similar or

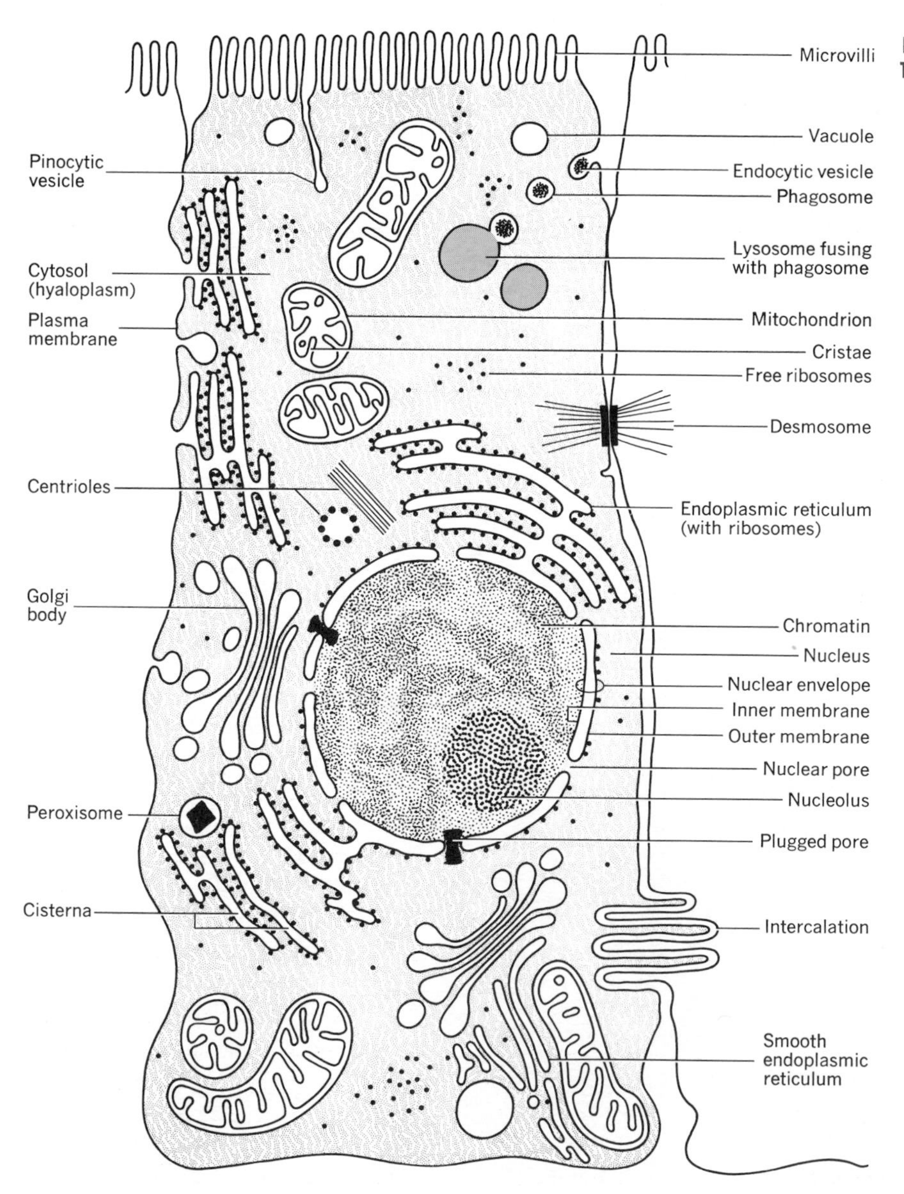

Figure 1–19
The *generalized* animal cell.

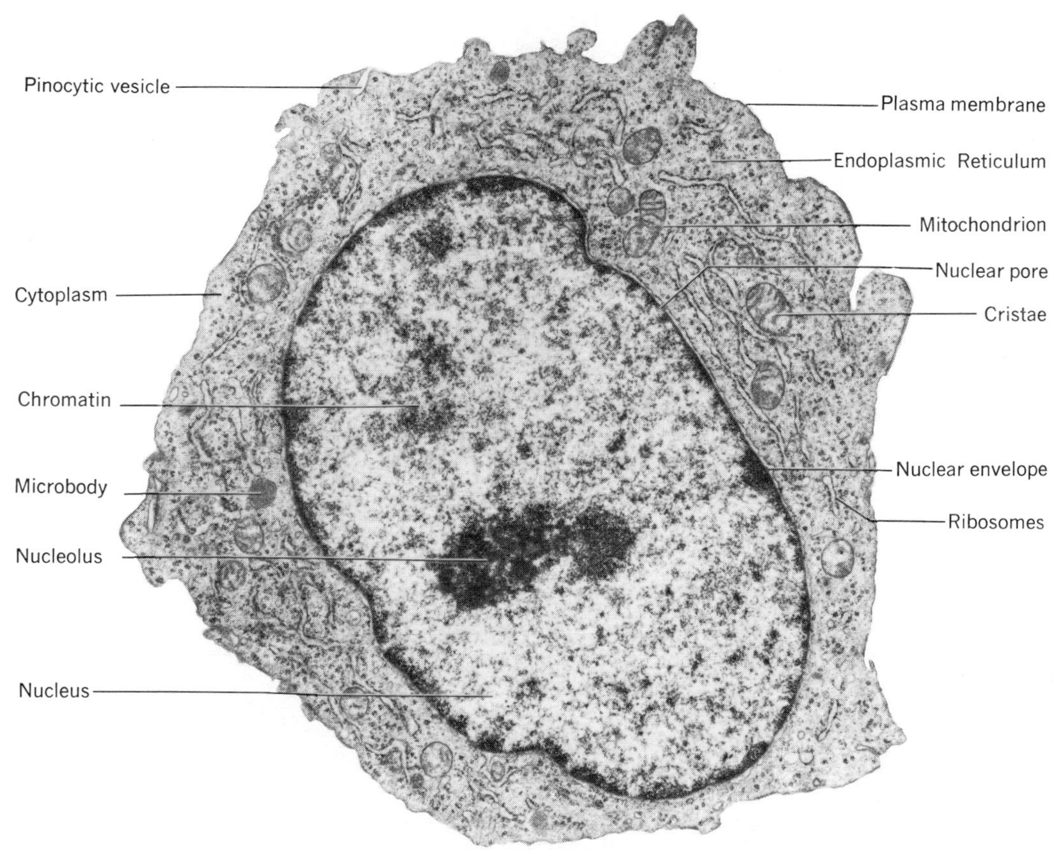

Figure 1–20
Electron micrograph of whole animal cell. Magnification, 13,000×. (Courtesy of R. Chao.)

identical in organization. These structures are depicted in the *generalized* animal cell diagrammed in Figure 1–19 and described briefly in the following sections. They are dealt with in greater detail in later chapters that are individually devoted to the structure and functions of cell organelles. Figure 1–20 is an electron photomicrograph of an animal cell containing many of the structures to be discussed.

The Plasma Membrane. The contents of the cell (*cytoplasm* and cytoplasmic organelles) are separated from the external surroundings by a limiting membrane, the *plasma membrane* (also called *cell membrane* or *plasmalemma*), which is composed of protein, lipid, and carbohydrate. This structure regulates the passage of materials between the cell and its surroundings and in some tissues is involved in intercellular communication (e.g., nerve tissue). These subjects are treated in Chapters 15 and 23. In some tissue cells, a portion of the plasma membrane is modified to form a large number of fingerlike projections called *microvilli* because of their resemblance to the much larger villi of the small intestine. The microvilli greatly increase the surface area of the cell and provide for the more quantitative passage of materials across the plasma membrane. When a large number of cells are in close contact with one another (as, for example, in a tissue), it is not unusual to observe special

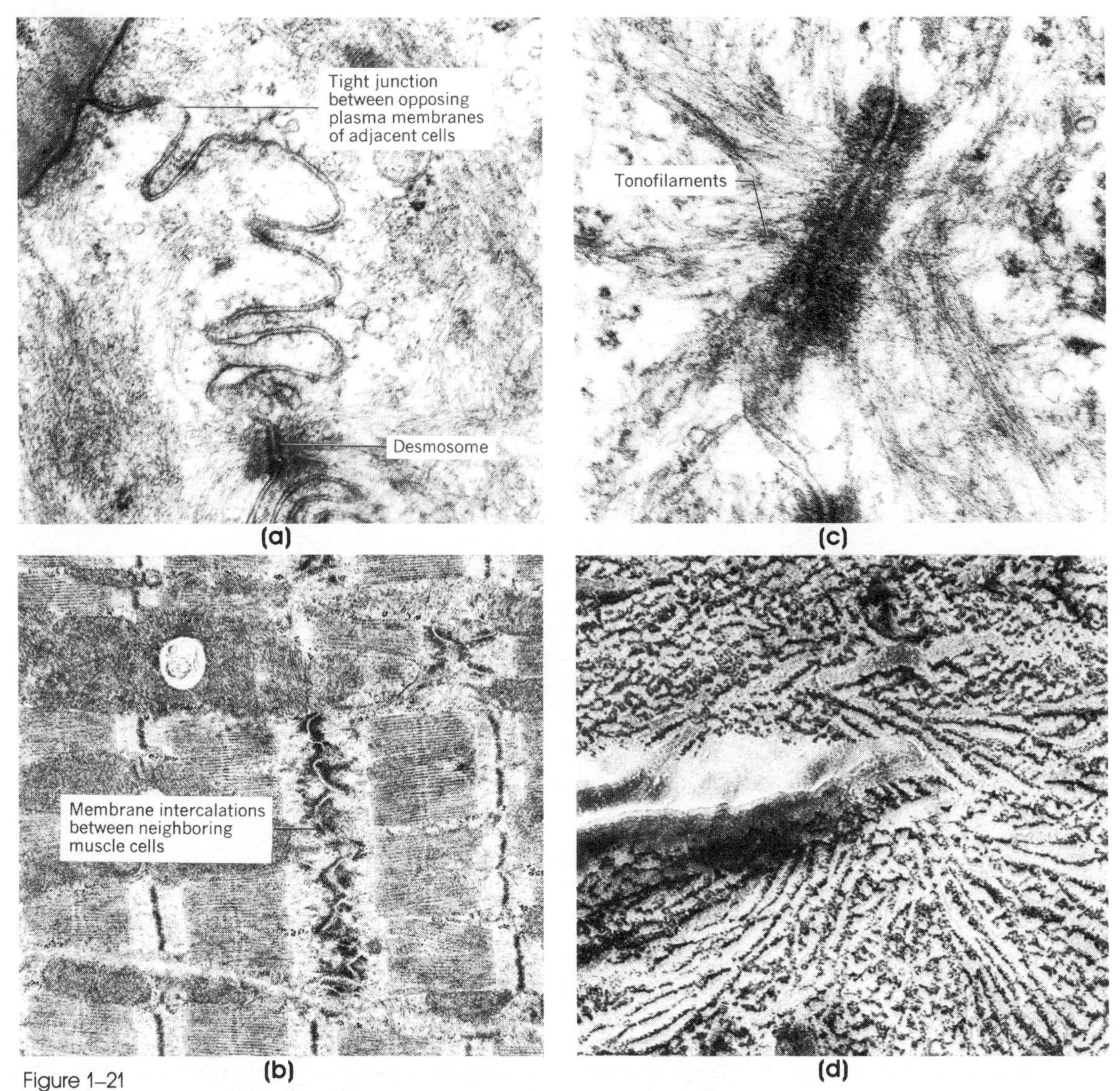

Figure 1–21
Junctions between opposing plasma membranes of neighboring cells. (*a*) Tight junction and spot desmosome. (*b*) Intercalations in heart tissue—a special form of desmosome. (*c*) Detail of spot desmosome showing *tonofilaments* (see Chapter 15). (*d*) freeze-fracture view of spot desmosome. Magnifications, (*a*) 26,000 ×; (*b*) 7,200 ×; (*c*) 29,000 ×; (*d*) 31,000 ×. (Photos *a*, *b*, and *c* courtesy of R. Chao. *d*, courtesy of M. Doolittle.)

forms of junctions between opposing plasma membranes. These take the form of *tight junctions, desmosomes,* and *gap junctions*. Some of these are illustrated in Figure 1–21 and are considered in greater depth in Chapter 15.

The plasma membrane should not be thought of as uniform or as having the same composition over its entire surface. Instead, the composition and organization vary in different regions of the membrane. Some areas of the plasma membrane of a liver cell, for example, face the plasma membranes of adjoining cells in the tissue; other areas face the bile channels *(bile canaliculi)* into which substances are secreted by the liver cell. Still other portions of the plasma membrane face the epithelial lining of *capillaries* from which substances are absorbed. Each of these regions of the plasma membrane is differently composed and differently organized and, in fact, is continually undergoing change and reorganization.

The Endoplasmic Reticulum and Ribosomes. Within the cytoplasm of most animal cells is an extensive network of branching and anastomosing membrane-limited channels or cisternae collectively called the *endoplasmic reticulum* (Fig. 1–22). The membranes of the endoplasmic reticulum (usually abbreviated ER) divide the cytoplasm into two phases: the *lumenal* phase and the *hyaloplasmic* phase or *cytosol*. The lumenal phase consists of the material enclosed within the cisternae of the endoplasmic reticulum, while the cytosol surrounds the ER membranes.

In the cytosol are large numbers of small particles called *ribosomes*. These particles are distributed either along the hyaloplasmic surface of the endoplasmic reticulum (''attached'' ribosomes) or free in the hyaloplasm (''free'' ribosomes). There is some evidence that the free ribosomes are interconnected by fine filaments. Endoplasmic reticulum with associated ribosomes is called *rough ER* (RER),

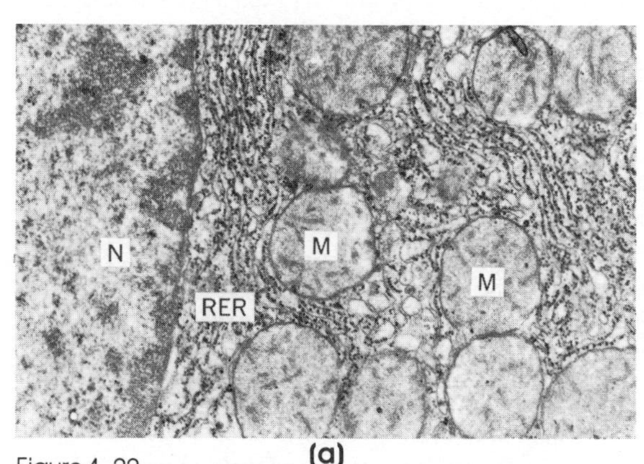

(a)

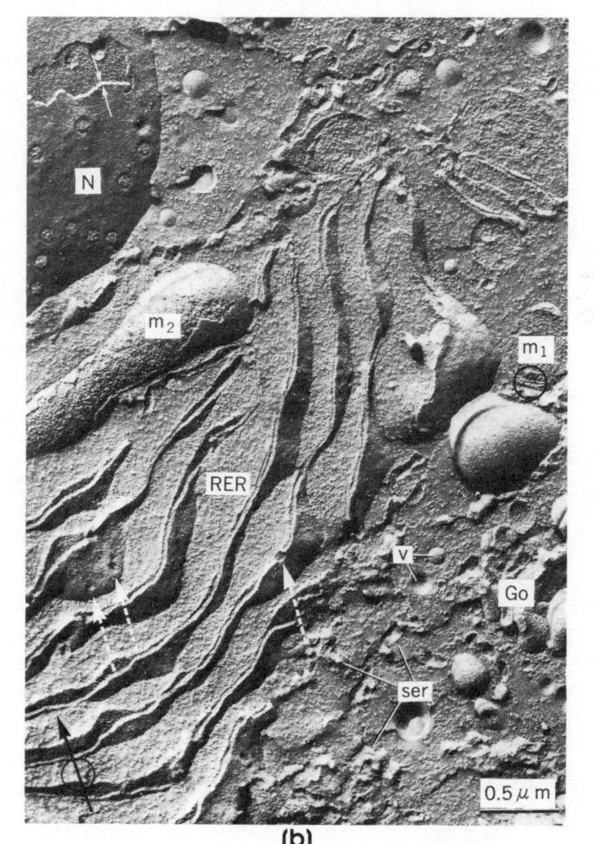

(b)

Figure 1–22
The endoplasmic reticulum. (*a*) and (*b*) Thin section and freeze-fracture through comparable regions of the cell. (*c*) High magnification view. N, nucleus; M, mitochondria; R, ribosomes; C, cisternae; H, hyaloplasm (cytosol); RER, rough endoplasmic reticulum; SER, smooth endoplasmic reticulum; Go, Golgi body; v, vesicle. In (*b*) the dashed arrows note perforations in the ER membranes. Also, one of the mitochondria (m_2) has been fractured in such a way that faces of both the outer and inner mitochondrial membranes may be seen. The space separating the two nuclear membranes is indicated by the white arrows. The direction of shadowing is indicated by the circled arrow. Magnifications: (*a*) 6,000 × ; (*b*) 20,000 × ; (*c*) 21,000 × . (*b*, courtesy of Dr. L. Orci; c, courtesy of R. Chao.)

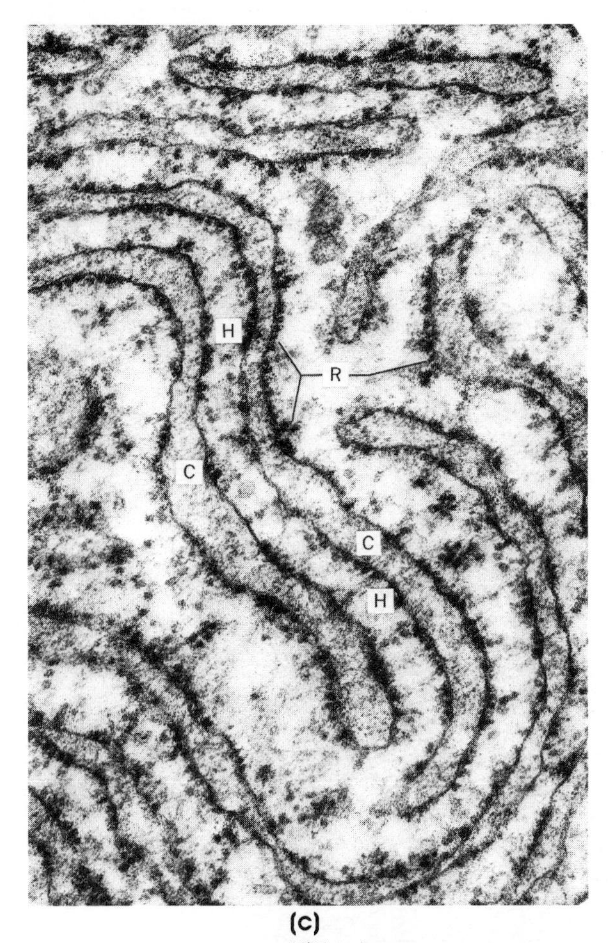

(c)

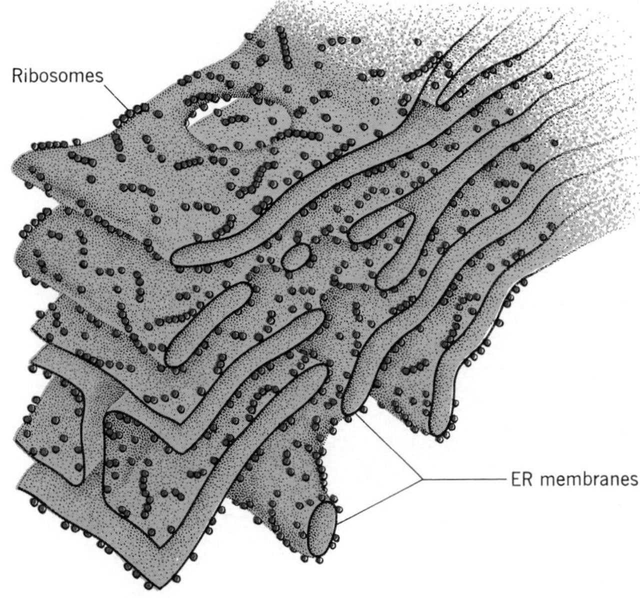

Figure 1–22 **(Continued)**

whereas *smooth ER* (SER) is devoid of attached ribosomes. Ribosomes carry out the synthesis of the cell's proteins, a subject covered at some length in Chapter 21 (which also deals with the molecular architecture of ribosomes). Certain portions of the endoplasmic reticulum may form a continuum with the plasma membrane and the nuclear envelope.

Mitochondria. Within the cytoplasm are numerous vesicular organelles called *mitochondria* (Figs. 1–13*a* and *b*, 1–22, and 1–23). Each mitochondrion is bordered by a double membrane. The outer membrane is smooth and continuous, and the inner membrane displays numerous infoldings called *cristae* (or *cristal membranes*). These greatly increase the surface area of the inner membrane. The space between neighboring cristae is called the mitochondrial *matrix* and often contains crystal-like inclusions. Mitochondria are engaged in numerous metabolic functions in the cell, including energy-producing phases of carbohydrate and fat metabolism (called respiration) and porphyrin biosynthesis. The mitochondrion and its functions are considered in greater detail in Chapter 16.

The Golgi Apparatus. The *Golgi apparatus* (also called Golgi *body* or Golgi *complex*) consists of a unique network of cisternae similar to and possibly continuous with the lumenal phase of the endoplasmic reticulum. The cisternae of a Golgi body are often stacked together in parallel rows, and in this state the body is referred to as a *dictyosome*. The Golgi apparatus is frequently surrounded by vesicles of various sizes that apparently are discharged from the margins of the main body of the organelle (Fig. 1–24). As related in Chapter 18, a variety of functions are ascribed to the Golgi apparatus, including secretory activity (especially the secretion of enzymes) and the proliferation of additional membranes for the cell.

Lysosomes. Many cells contain vesicular structures that are generally smaller than mitochondria and are called *lysosomes* (Fig. 1–25). The lysosomes are bounded by a single membrane and contain quantities of various hydrolytic enzymes capable of digesting protein, nucleic acid, polysaccharide, and other materials. Under normal conditions, these enzymes are inactive and isolated from the cytoplasm. However, if the lysosomal membranes are ruptured, the released enzymes can degrade the cell. Lysosomes are be-

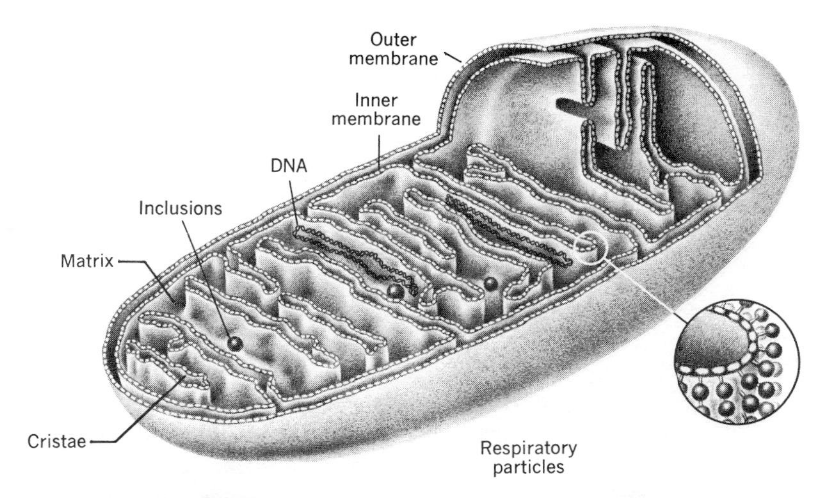

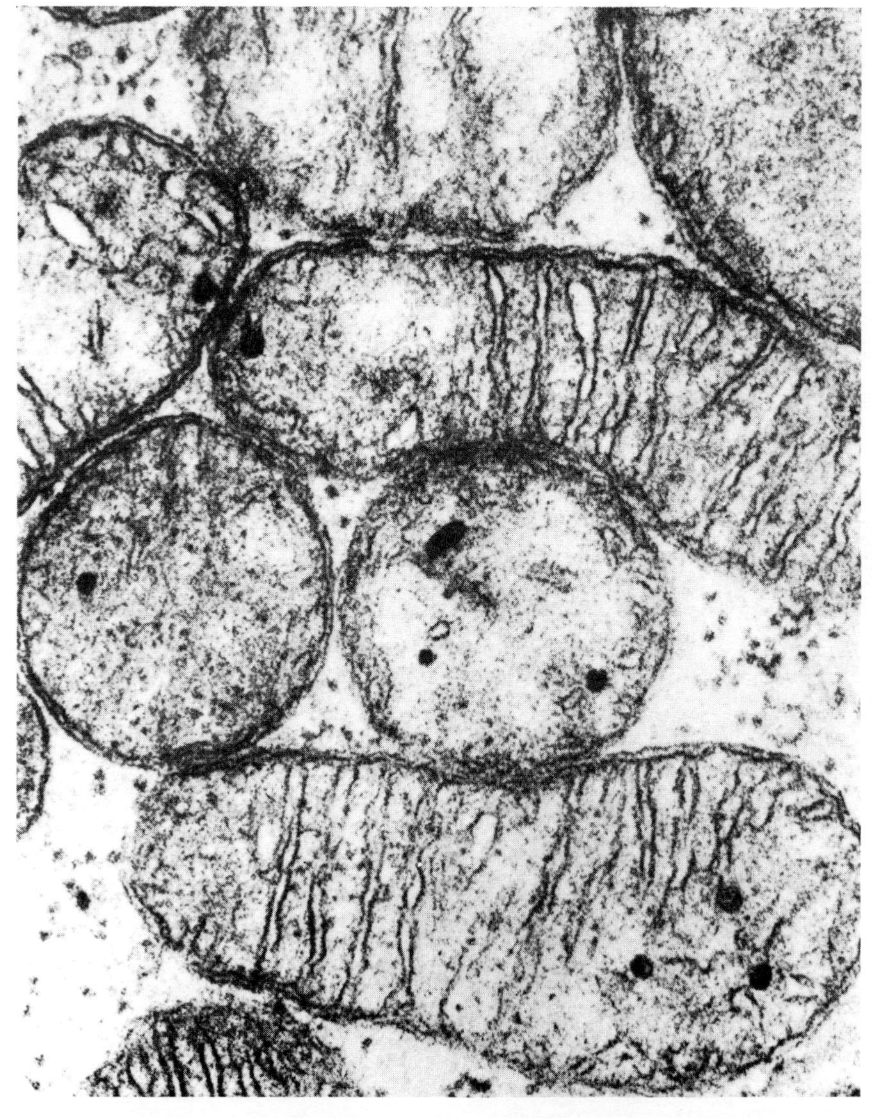

Figure 1–23
Mitochondria. (Electron photomicrograph courtesy of R. Chao.)

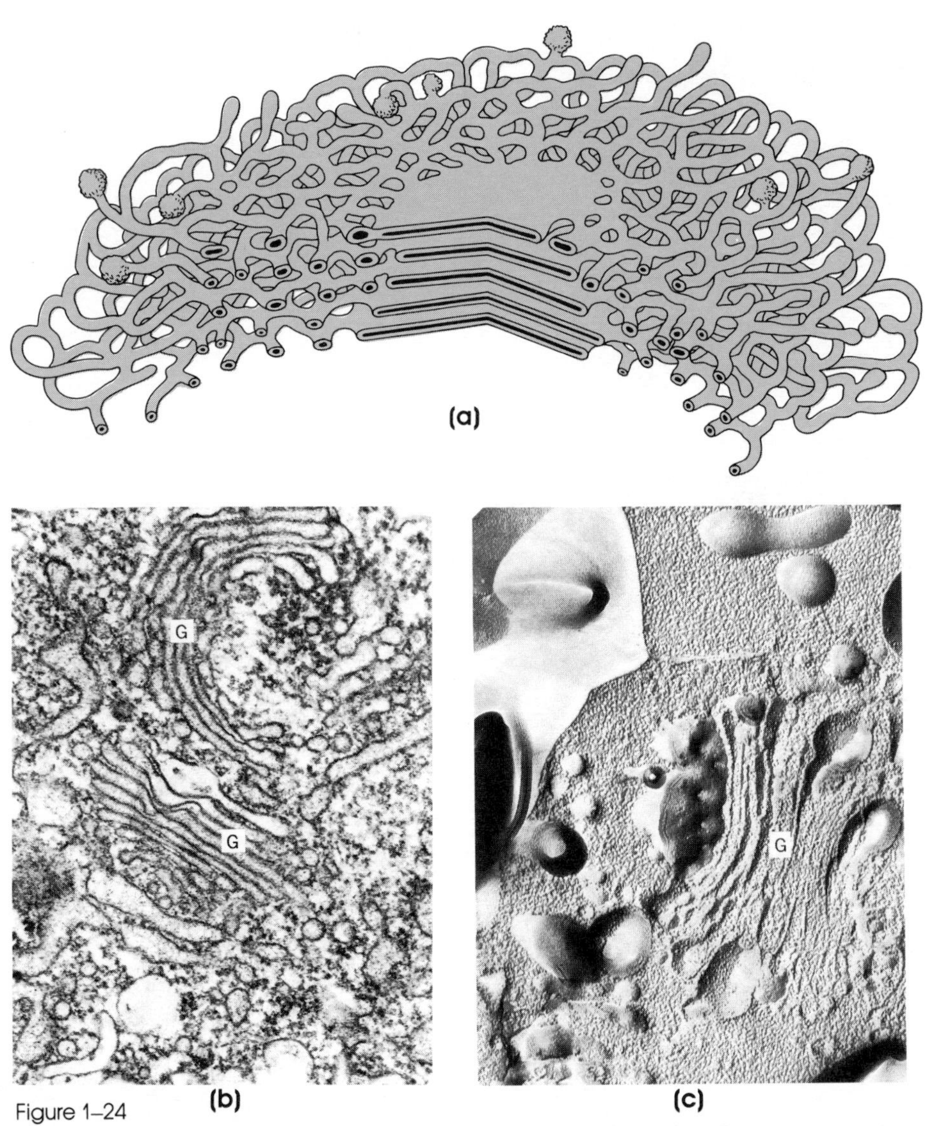

Figure 1–24

The Golgi apparatus. (*a*) Diagram depicting the three-dimensional relationship among the cisternae, channels, and peripheral vesicles. (*b*) Typical appearance of Golgi bodies (G) in thin sections. (*c*) Appearance of Golgi bodies in freeze-fractured cells. (Photomicrographs courtesy of Dr. E. G. Pollock.)

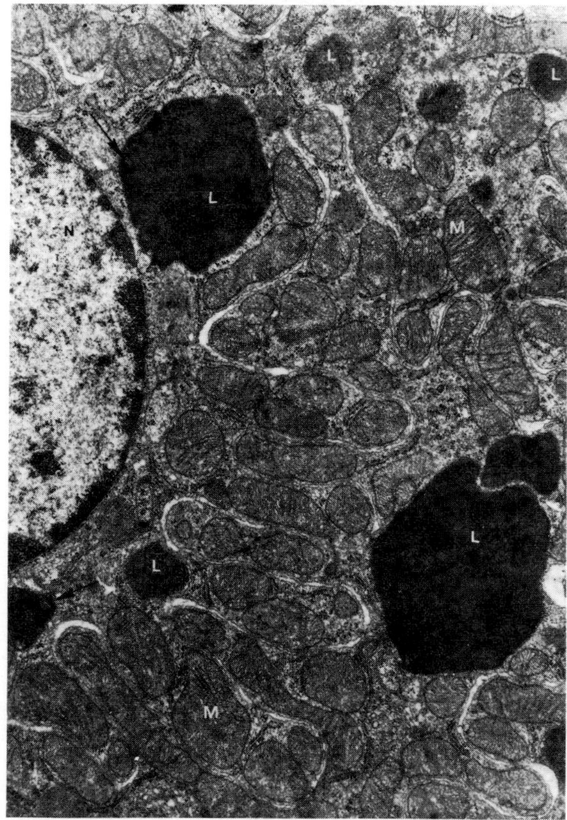

Figure 1–25
Lysosomes (L) appear as electron-dense structures in tissues especially treated to reveal these organelles (see also Chapter 19). Magnification, 8,000 ×. (Courtesy of Dr. E. G. Pollock.)

lieved to be responsible for the intracellular digestion of food particles ingested by the cell, the scavenging of worn and therefore poorly functioning organelles, and a number of other cell functions. The lysosome and other *microbodies* are considered in depth in Chapter 19.

Peroxisomes and Glyoxysomes. Many cells contain small numbers of *peroxisomes* and/or *glyoxysomes*. These small organelles, which are bounded by a single membrane, contain a number of enzymes whose functions are related to the metabolism of hydrogen peroxide and glyoxylic acid (see Chapter 19).

The Nucleus. The nucleus is a relatively large structure frequently but not always located near the center of the cell. The contents of the nucleus are separated from the cytoplasm by *two* membranes that together form the *nuclear envelope*. At various positions, the outer membrane of the envelope (membrane 1) fuses with the inner membrane (membrane 2) to form *pores* (Fig. 1–26). Nuclear pores provide a measure of continuity between the *cytosol* and the contents of the nucleus. Occasionally, the nuclear pores are plugged by a dense material (Fig. 1–27). The outer nuclear membrane often bears ribosomes on its hyaloplasmic side (Fig. 1–28) and may form continuities with the membranes of the endoplasmic reticulum. Since the latter may be continuous with the plasma membrane, the *perinuclear space* (i.e., the space between the inner and outer membranes of the nuclear envelope) corresponds to the lumenal phase and may be considered external to the cell (see Fig. 1–29).

Figure Figure 1–26
Thin section through cell nucleus. Magnification, 10,000 ×. (Photomicrograph courtesy of R. Chao.)

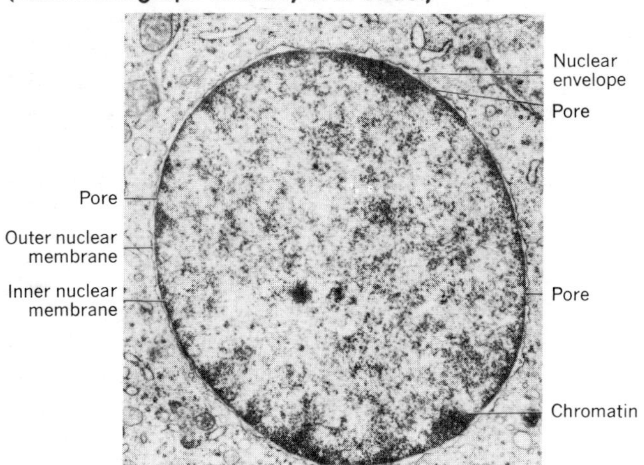

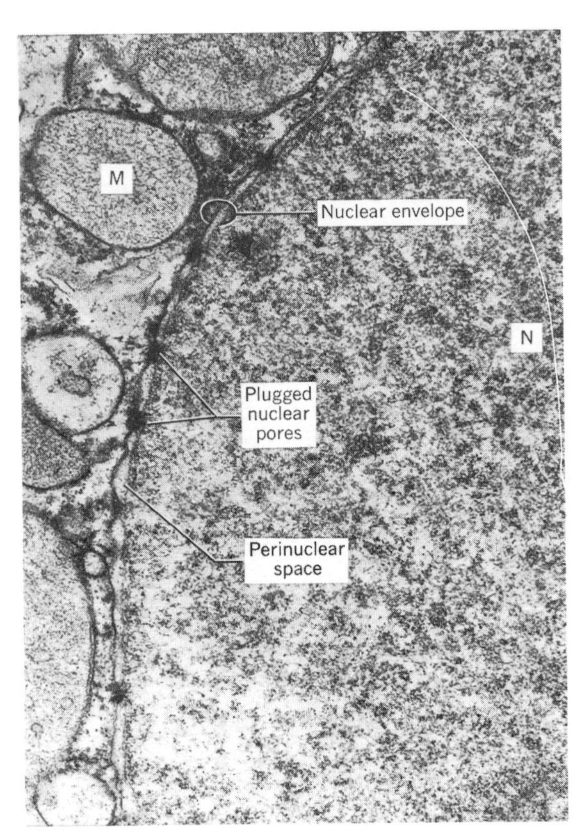

Figure 1–27
Plugged pores of the nuclear envelope. N, nucleus; M, mitochondrion. Magnification, 30,000×. (Photomicrograph courtesy of Dr. E. G. Pollock.)

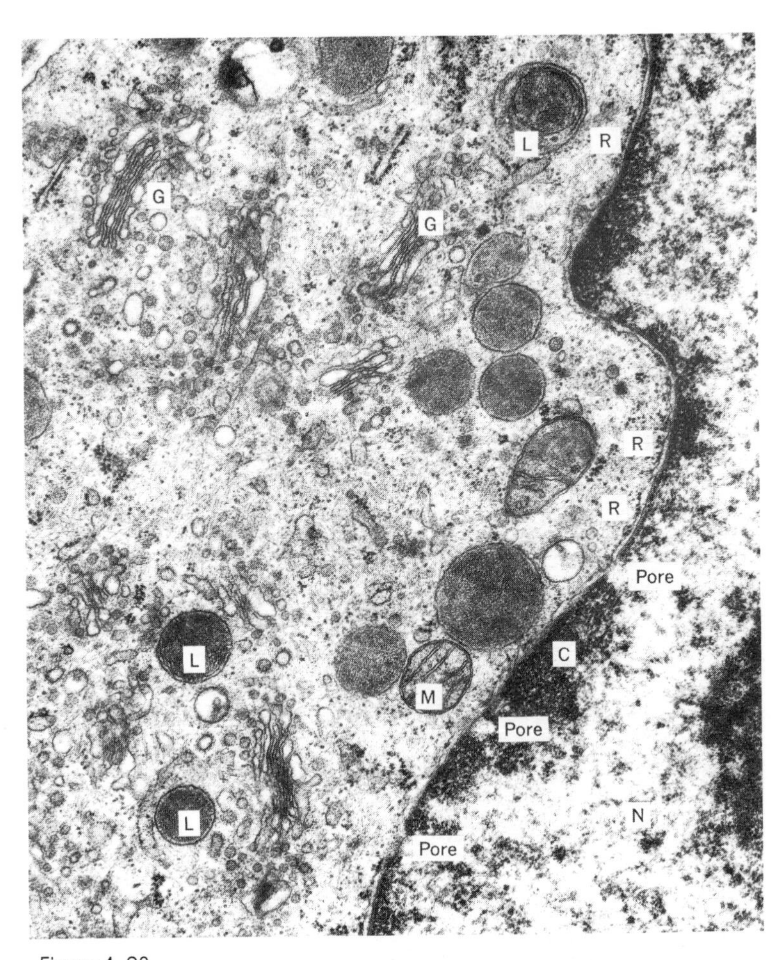

Figure 1–28
In this electron micrograph, one can clearly discern ribosomes (R) attached to the outer membrane of the nuclear envelope. Also seen are nuclear pores, mitochondria (M), lysosomes (L), Golgi bodies (G), and nuclear chromatin (C). Magnification, 20,000× (Courtesy of R. Chao.)

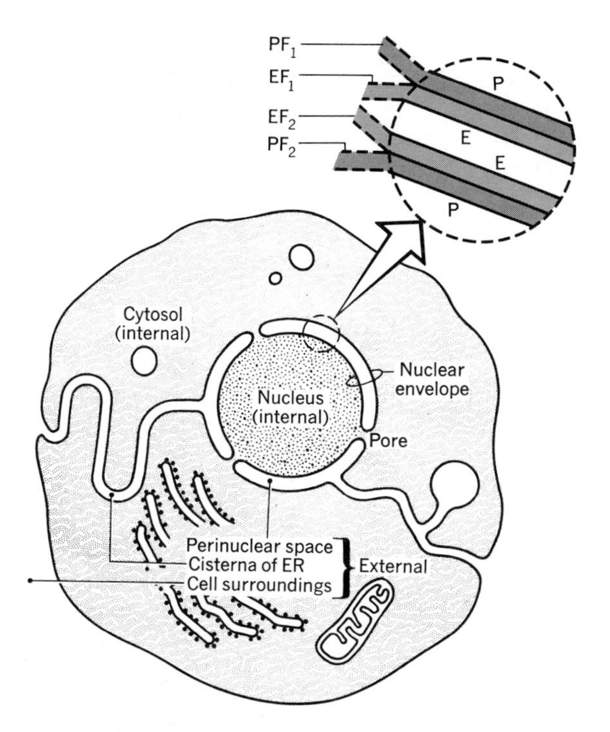

Figure 1–29
"Internal" and "external" phases of the cell. Region of the nuclear envelope is enlarged to illustrate the relationship among various faces of the inner and outer nuclear membranes. This illustration should be compared with that of Figure 1–12.

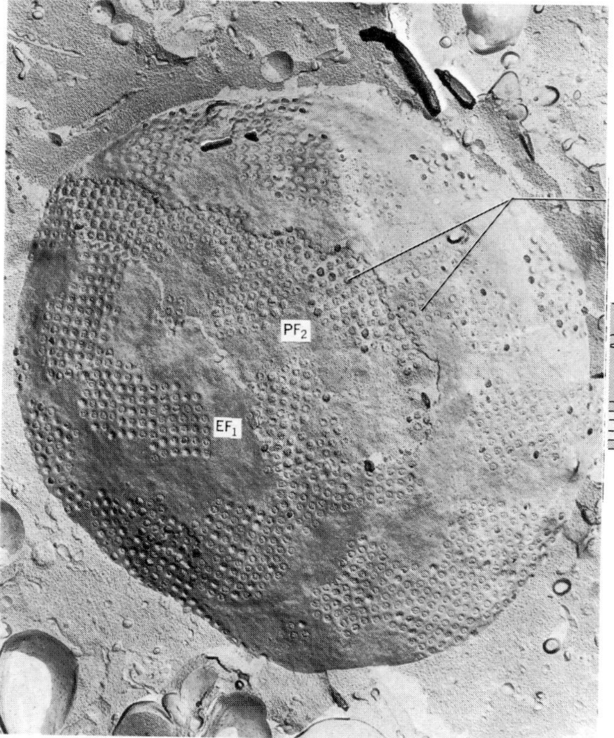

Figure 1–30
Freeze-fracture through the nuclear envelope exposing the EF_1 and PF_2 fracture faces. This freeze fracture view is especially unusual because the fracture plane has passed through so large a region of the cell nucleus. Magnification, 13,000 ×. (Courtesy of Dr. E. G. Pollock.)

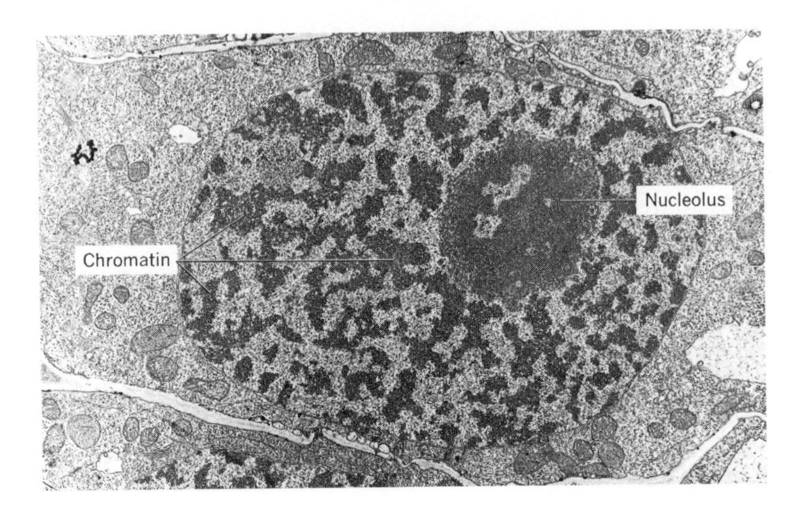

Figure 1–31
Dense, granular *nucleolus* in the nucleus of a plant cell. Magnification, 6,000 ×. (Photomicrograph courtesy of Dr. E. G. Pollock.)

The nuclear envelope and the pores that penetrate it are dramatically revealed in freeze-fracture preparations (Fig. 1–30). The cytosol-contacting half of the outer nuclear membrane is fractured away, exposing the inner half of that membrane (i.e., face $\mathbf{EF}_1$ of Fig. 1–29). Also fractured away are pieces of the inner nuclear membrane (the half-membrane that faced the perinuclear space), leaving only the half-membrane that faced the cytosol (i.e., $\mathbf{PF}_2$ of Fig. 1–29). The nuclear pores penetrate both membranes and in Fig. 1–30 appear to be nonrandomly distributed in the nuclear envelope.

The nucleus contains the genetic machinery of the cell (chromosomal DNA, histones, etc.), which is discussed in Chapter 20. Using either the light microscope or the electron microscope, the nucleus often reveals one or more dense, granular structures called *nucleoli* (Fig. 1–31). Nucleoli are not bounded by a membrane and appear to be formed in part from localized concentrations of ribosomal materials (see Chapter 21).

Flagella and Cilia. Many free-living cells (such as protozoa and other microorganisms) possess *locomotor* organelles that project from the cell surface. In animal cells these are either *flagella* or *cilia* (Fig. 1–32). The tissue cells of multicellular animals may also contain cilia, but they are employed here to advance a substrate across the cell surface (such as mucus in the respiratory tract or the egg cell during its passage through the oviduct) and not for self-locomotion. The organelles are called cilia when they are short but present in large numbers and are called flagella when long but few in number. Each cilium or flagellum is covered by an extension of the plasma membrane. Internally, these organelles contain a specific array of microtubules which run from the basal plate toward the tip of the structure. This array consists of two central microtubules and nine pairs of peripheral (outer) microtubules (Fig. 1–32*b*).

Other structural elements commonly found in animal cells include *microfilaments,* which may participate in intracellular movement and communication. These may be scattered through the cytoplasm, but many are located just under the plasma membrane and may be anchored to it. *Basal bodies* are found at the base of locomotor organelles and may give rise to the microtubules of these structures. In animal cells, pairs of *centrioles* are observed near the cell nucleus and may be involved in the mechanics of cell division. Cilia, flagella, and cytoplasmic microfilamentous and microtubular structures are treated in detail in Chapter 22.

Eucaryotic Cells: The Generalized Plant Cell

All the organelles described in the preceding section as regular constituents of animal cells are also found in similar form in many plant cells. Several other organelles are unique to plant tissues and include the *cell wall, plasmodesmata, chloroplasts* and large *vacuoles*. A generalized plant cell is depicted in Figure 1–33.

The Cell Wall. The *cell wall* is a thick polysaccharide-containing structure immediately surrounding the plasma membrane (Fig. 1–33). In multicellular plants, the plasma membranes of neighboring cells are separated by these walls, and adjacent plant cells have their walls fused together by a layer of material called the *middle lamella*. The cell wall serves both a protective and a supportive function for the plant. The degree to which the cell wall may be involved in the regulation of the exchange of materials between the plant cell and its surroundings is difficult to assess but is most likely restricted to macromolecules of considerable size. As in animal cells, most of the regulation of exchanges between the cytoplasm and the extracellular surroundings of plant cells is a function of the plasma membrane.

Plasmodesmata. At intervals the plant cell wall may be interrupted by cytoplasmic bridges between one cell and its neighbor (Fig. 1–34). These bridges are called *plasmodesmata* and represent regions in which channel-like extensions of the plasma membranes of neighboring cells merge. The channels serve in intercellular circulation of materials.

Chloroplasts. The ability to use light as a source of energy for sugar synthesis from water and carbon dioxide distinguishes animal cells and certain plant cells. This process, termed *photosynthesis,* is carried on in the *chloroplasts* (Fig. 1–35). These organelles are commonly ovoid structures bounded by an outer membrane but also containing a number of internal membranes. Internally, the chloroplast consists of a series of membranes arranged in *lamellae* (parallel sheets) and supported in a homogeneous matrix called the *stroma*. The membranes are arranged as thin sacs (called *thylakoids*) that contain chlorophyll and may be stacked on top of one another, forming structures called

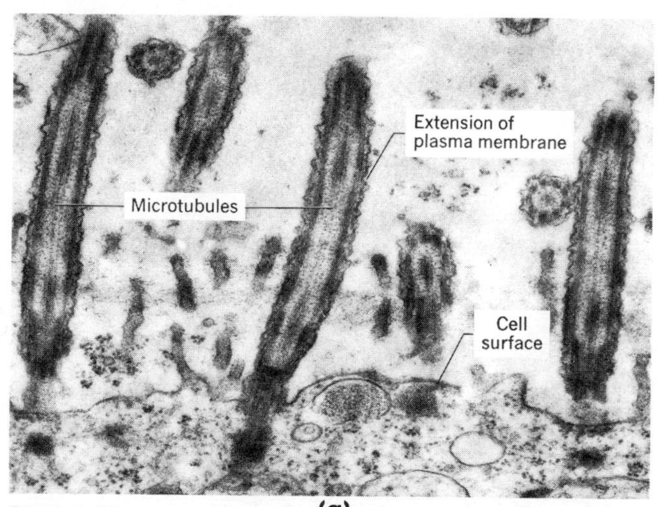

(a)

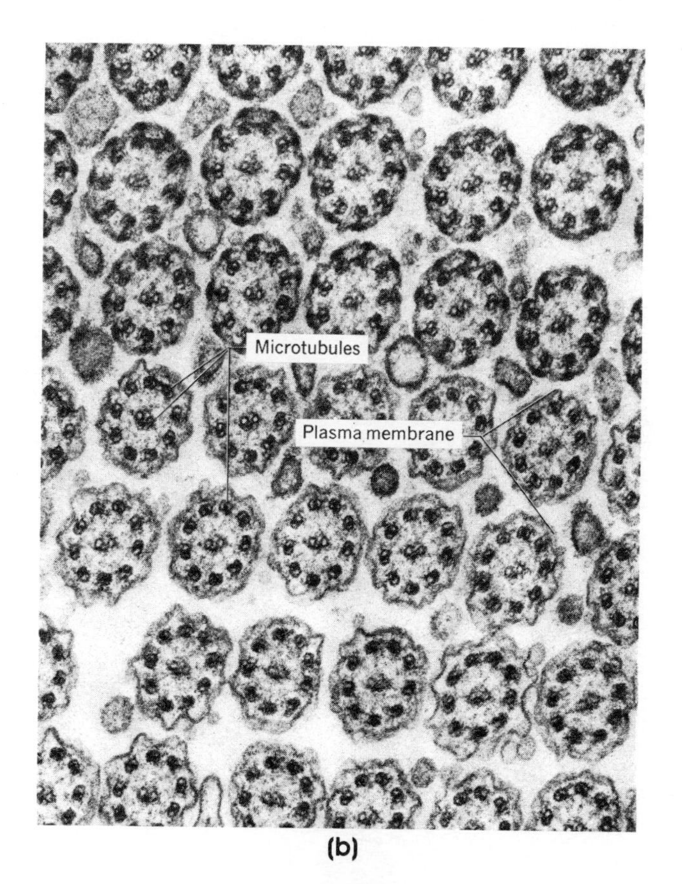

(b)

Figure 1–32
(*a*) Electron photomicrograph showing tangentially sectioned cilia projecting from the cell surface. (*b*) Cross section through numerous cilia, with long axis perpendicular to plane of the page. Note the arrangements of microtubules. Magnifications: (*a*) 25,000×; (*b*) 40,000×. (Photomicrographs courtesy of R. Chao.)

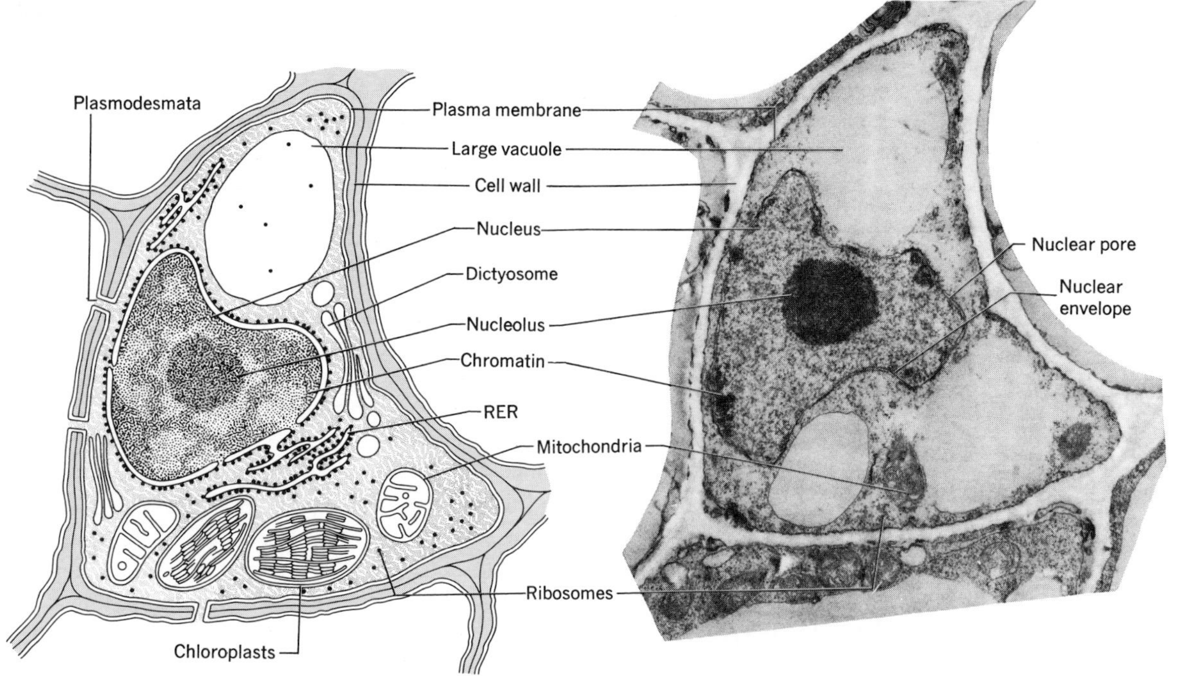

Figure 1–33
The generalized plant cell. (Photomicrograph courtesy of R. Chao.)

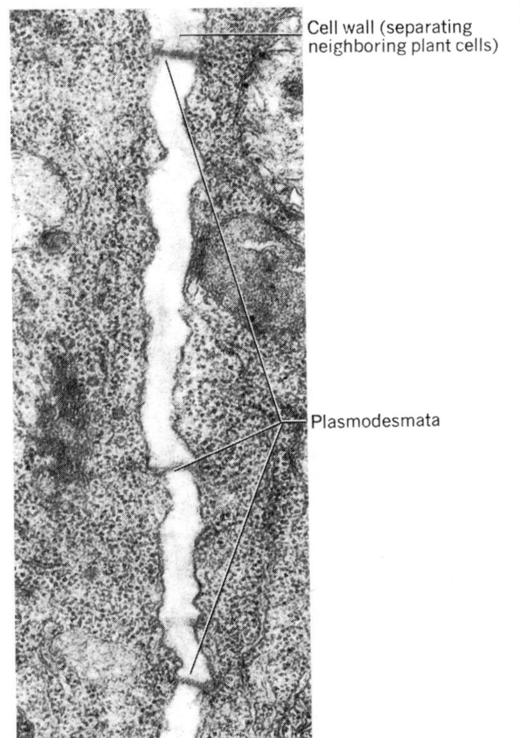

Figure 1–34
Plasmodesmata. (Photomicrograph courtesy of N. Herzog.)

Figure 1–35
Chloroplast of sugar beet. (Electron photomicrograph courtesy of Dr. W. Laetsch.)

grana. Lamellar membranes connecting the grana are called *stroma lamellae* (Fig. 1–35). A more detailed description of the structure of chloroplasts and the metabolic pathways of photosynthesis is found in Chapter 17.

Vacuoles. Although *vacuoles* are present in both animal and plant cells, they are particularly large and abundant in plant cells (Fig. 1–33), often occupying a major portion of the cell volume and forcing the remaining cell structures into a thin peripheral layer. These vacuoles are bounded by a single membrane and are formed by the coalescence of smaller vacuoles during the plant's growth and development. Vacuoles serve as sites for the storage of water and cell products or metabolic intermediates.

Procaryotic Cells: Bacteria

The bacteria are structurally distinct from eucaryotic microorganisms such as protozoa and contain a number of unique cellular organelles. The typical bacterial cell is about the size of a mitochondrion of an animal or plant cell, and in view of this small size, it is to be expected that the organelles of bacteria would be correspondingly smaller. The generalized structure of a bacterium is shown in Fig. 1–36.

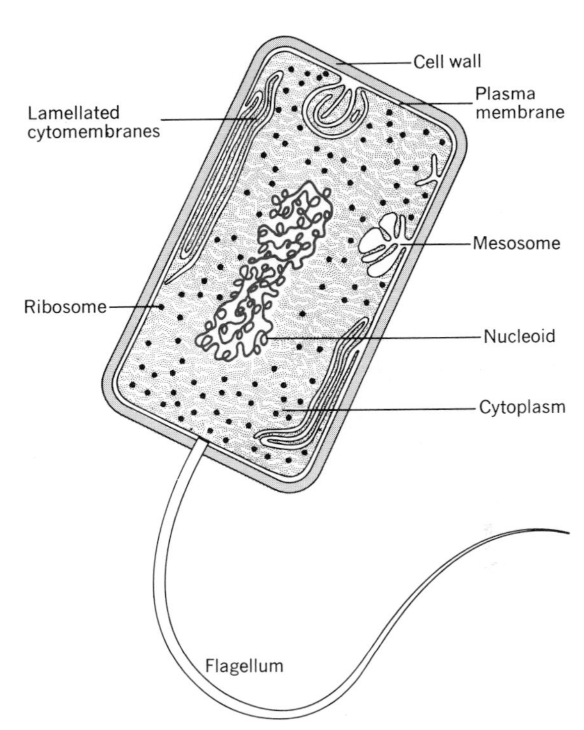

Figure 1–36
The generalized bacterium.

The Bacterial Cell Wall. The bacterial cell is enclosed within a wall that differs chemically from the cell wall of plants in that it contains protein and lipid as well as polysaccharide. Its content of a particular "mucopeptide" (a protein-carbohydrate complex) has been the basis of the histo-chemical classification of bacteria, being high in the so-called "gram-positive" bacteria (such as *Bacillus subtilis*) and low in the gram-negative bacteria (such as *Escherichia coli* and the *Simonsiella*). In some bacteria, the cell wall is surrounded by an additional structure called a *capsule*. The cell wall and capsule confer shape and form to the bacterium and also act as an osmotic barrier.

Mesosomes. Infoldings of the plasma membrane of gram-positive bacteria give rise to structures called *mesosomes* (or *chondrioids*) (Fig. 1–37). These structures appear to have many properties similar to those of the mitochondria of animal and plant cells yet they are not thought to have the same physiological role. Mesosomes may be absent in gram-negative bacteria in which the plasma membrane itself possesses respiratory activity. Intrusions of the plasma membrane form the photosynthetic organelles *(chromatophores)* of the photosynthetic bacteria.

Lamellate Cytomembranes. In some bacteria, there is a lamellar arrangement of membranes within the cytoplasm that may arise from the plasma membrane. However, there are no structures comparable to the endoplasmic reticulum of animal and plant cells. Whereas bacteria are often packed with ribosomes, these are free in the cytoplasm and are not attached to membranes. Although bacterial ribosomes, like the ribosomes of eucaryotic cells, are the sites of protein synthesis, considerable differences exist between the organelles of these two groups (see Chapter 21). Lamellate membranes are particularly abundant in the *autotrophic* bacteria, which support their growth through photosynthesis or similar processes.

Nucleoids. In bacteria the contents of the nucleus are not separated from the cytoplasm by membranes, although the

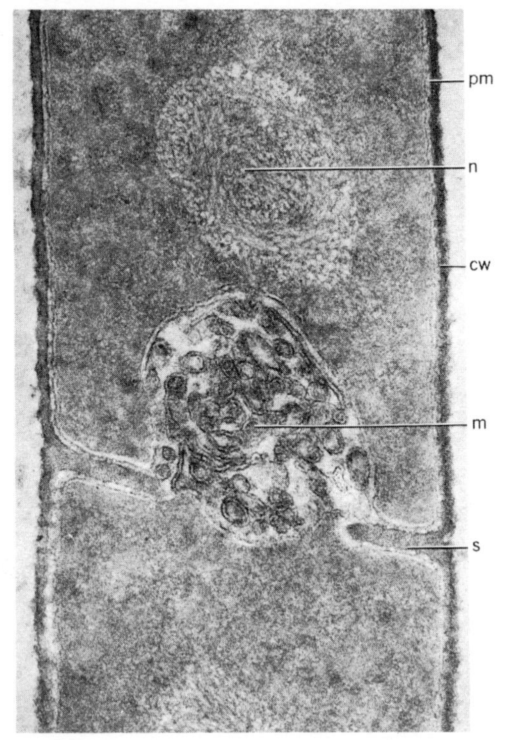

Figure 1–37
Electron photomicrograph of the bacterium *Bacillus subtilis.* cw, cell wall; pm, plasma membrane; n, nucleoid; m, mesosome (intracytoplasmic membrane inclusion). Note the formation of the transverse septum, s, as the cell divides. Magnification, 27,000×. (Courtesy of Dr. F. A. Eiserling.)

nuclear material may be confined to a specific region of the cell. The nuclear "area" of the bacterium is sometimes referred to as a *nucleoid*. During bacterial cell division, the nuclear materials are distributed to the daughter cells without formation of observable chromosomes. Nucleoli are not present in the nucleus.

Bacterial Flagella. Many bacteria contain one or more *flagella* employed for cellular locomotion. These organelles arise from a small basal granule in the cytoplasm and penetrate the plasma membrane and cell wall. Bacterial flagella are smaller than those of animal and plant cells and are simpler in organization, containing a single filament of globular proteins (called flagellin) surrounded by a sheath. The multiflagellated *(peritrichous)* bacterium *Proteus mirabilus* is shown in Figure 1–38.

Some bacteria, such as *E. coli, P. mirabilus,* and *B. subtilis,* occur as separate, individual cells. However, in a number of groups, the daughter cells remain attached following division, so that chains (e.g., *streptococci*) or filaments are formed. An example of a filamentous genus is *Simonsiella,* which colonizes the mucosal epithelial surface of the mouth (see Fig. 1–16). The individual cells of some filamentous bacteria reveal a dorsal-ventral differentiation; that is, the ventral surface (which in *Simonsiella* attaches to and glides along the epithelium) is structured differently than the dorsal surface (which faces away from the epithe-

Figure 1–38
Some bacteria such as *Proteus mirabilus* shown here are multiflagellated (*peritrichous*). (Courtesy of Dr. D. A. Kuhn and G. Patane.)

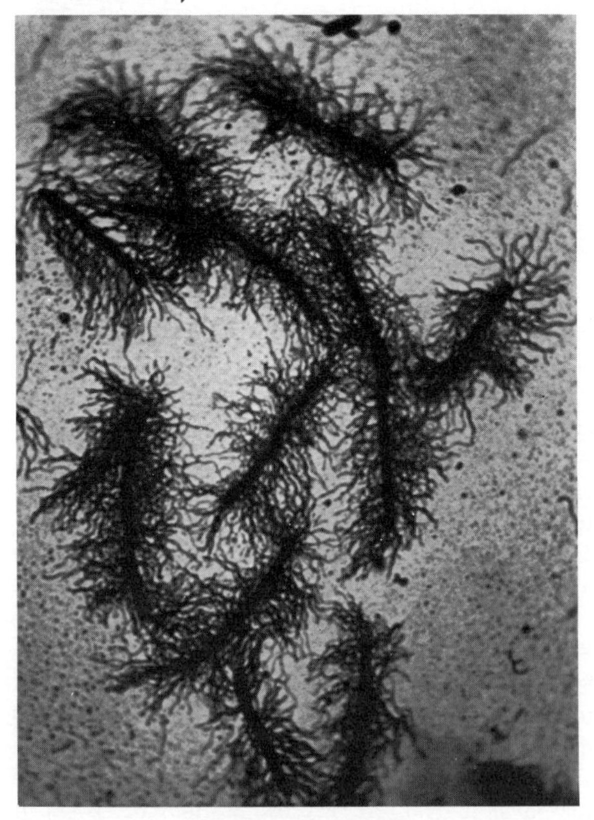

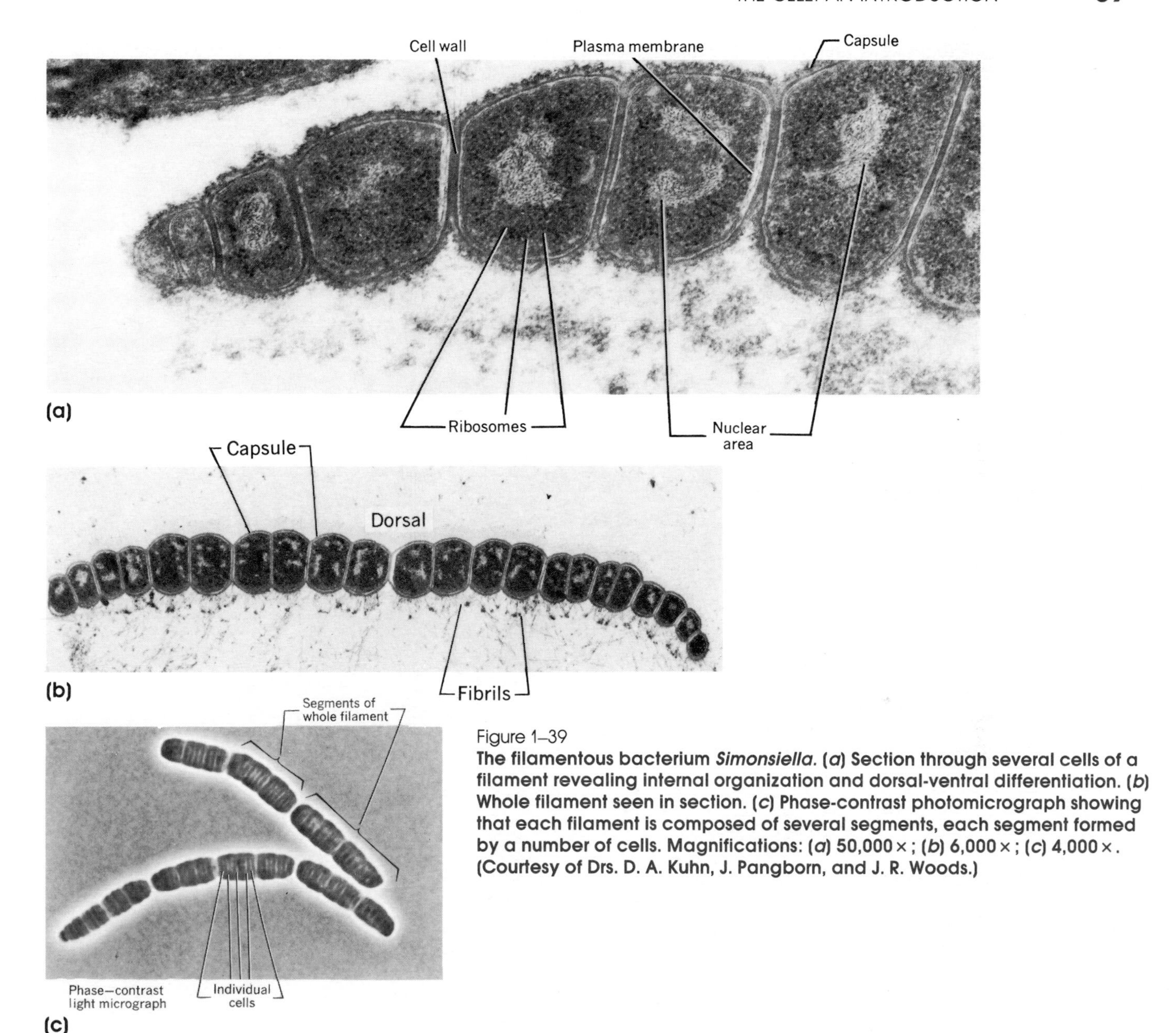

Figure 1–39
The filamentous bacterium *Simonsiella*. (*a*) Section through several cells of a filament revealing internal organization and dorsal-ventral differentiation. (*b*) Whole filament seen in section. (*c*) Phase-contrast photomicrograph showing that each filament is composed of several segments, each segment formed by a number of cells. Magnifications: (*a*) 50,000 ×; (*b*) 6,000 ×; (*c*) 4,000 ×. (Courtesy of Drs. D. A. Kuhn, J. Pangborn, and J. R. Woods.)

lium). This is apparent not only in the scanning electron micrographs of whole filaments shown in Figure 1–16 but also in transmission electron micrographs of thin sections through the filaments (Fig. 1–39), which reveal an internal differentiation. Individual cells of a filament exhibit features common to single-cell (i.e., nonfilamentous) forms like *E. coli* and *B. subtilis*.

Procaryotic Cells: Blue-Green Algae

The *blue-green algae* more closely resemble bacteria than they do algae and often are classified with the bacteria. Although most blue-green algae are blue-green in color, they do occur in a wide range of colors. Their name may be attributed to the fact that the first species recognized as members of the group were blue-green. The blue-green algae are photosynthetic procaryotes and occur as individual cells, as small clusters or colonies of cells, or as long, filamentous chains (Figs. 1–40 and 1–41). Blue-green algae lack locomotor organelles, and a gelatinous sheath replaces the capsule typical of bacteria. The photosynthetic apparatus consists of lamellae lined with pigment granules sometimes referred to as *phycobilosomes*.

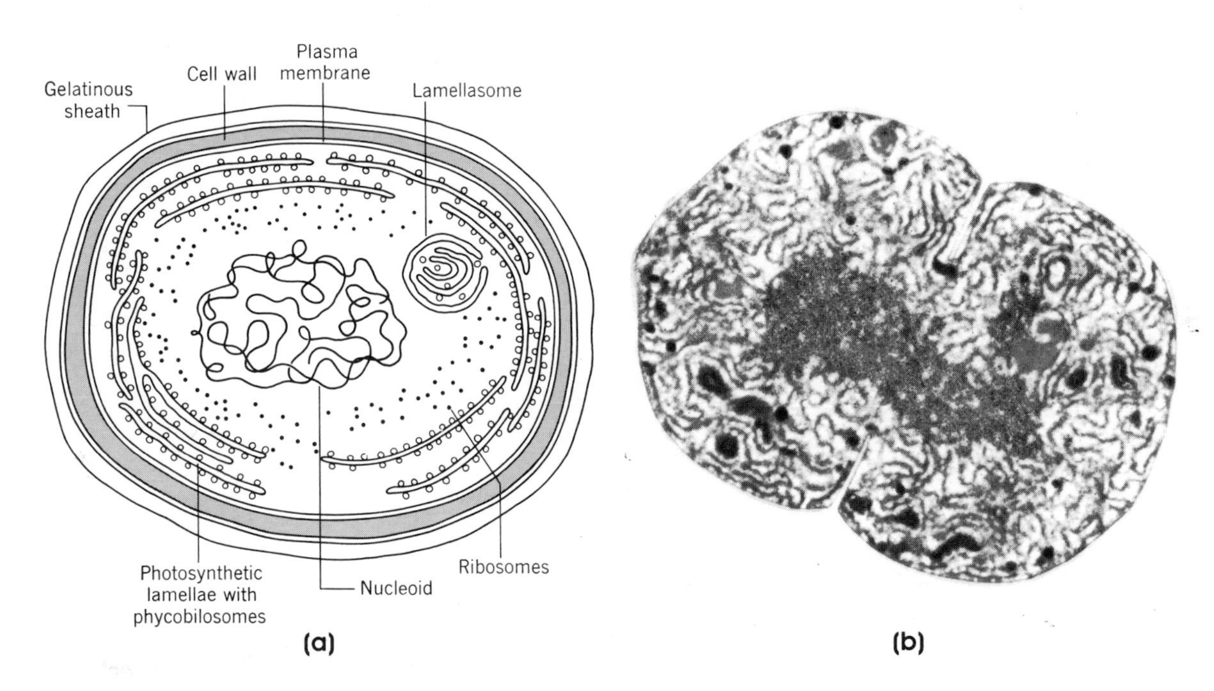

Figure 1–40
(*a*) Generalized blue-green alga cell. (*b*) Electron photomicrograph of dividing *Anabaena.* Note the formation of the septum. Magnification, 15,000 ×. (Courtesy of Drs. H. W. Beams and R. G. Kessel.)

Figure 1–41
Scanning electron micrograph of a filament of the blue-green alga *Anabaena* (compare with the thin section of Figure 1–40*b*). Magnification, 6,000 ×. (Courtesy of Drs. F. A. Eiserling and S. Eipert.)

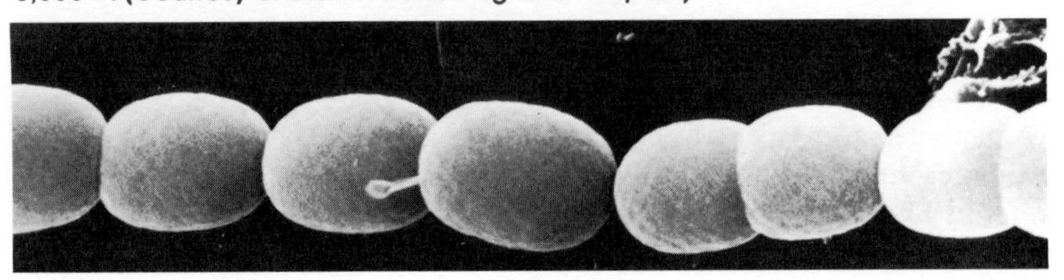

Procaryotic Cells: PPLO

The PPLO (i.e., pleuropneumonia-like organism) or mycoplasmas, which cause a number of diseases in humans and other animals, are the smallest (i.e., about 0.1 μ in diameter) of all free-living cells. They are smaller even than some of the larger viruses. The PPLO is bounded at its surface by a membrane composed of protein and lipid, but internally the cell's composition is more or less diffuse. The only microscopically discernible features within the cell are its genetic complement, which consists of a double-helical strand of circular DNA, and a number of ribosomes (Fig. 1–42). The PPLO appears to contain the bare minimum of structural organization required for a viable, free-living cell and may represent a form intermediate between viruses and bacteria. The relative sizes of typical eucaryotic cells, bacteria, PPLOs and viruses are compared in Figure 1–43, which dramatizes the differences that exist.

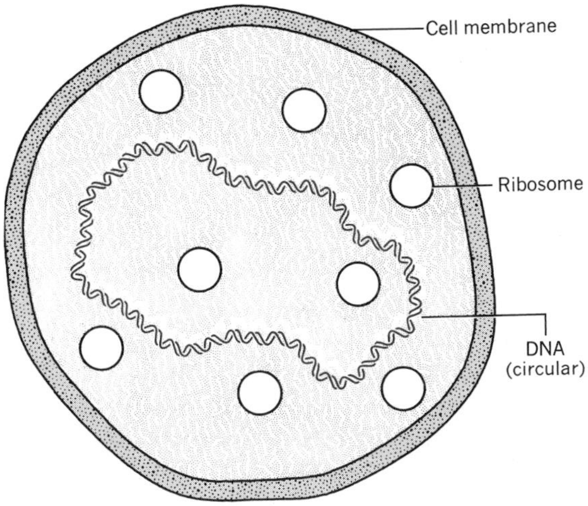

Cell membrane
Ribosome
DNA (circular)

Figure 1–42
Structure of a PPLO.

Viruses

So far, the descriptions in this chapter have been restricted to various kinds of cells. In this section, we are concerned with the organization and activities of viruses, but it should be emphasized at the outset that *viruses are not cells* and that it is debatable whether viruses constitute living systems. Viruses are described here because of their intimate association with cells and because of their contributions to our understanding of certain cellular phenomena. It will become apparent as we deal in later chapters with the structure and interactions of nucleic acids and proteins and with gene expression that much of our present-day understanding is based on studies initiated with viruses.

Structure of Viruses

Although all viruses or **virions** are extremely small, they are diverse in size and in organization. Generally, viruses range in diameter (or length) from about 20 nm to about 200 nm. Thus the largest viruses are actually larger than the smallest cells. However, even the smallest of cells (bacteria, PPLOs, etc.) are subject to infection by viruses. Among those viruses that attack animal cells, the most notorious are the viruses that cause diseases in humans. Smallpox, chicken pox, rabies, poliomyelitis, mumps, measles, influenza, hepatitis, and the "common cold" are all

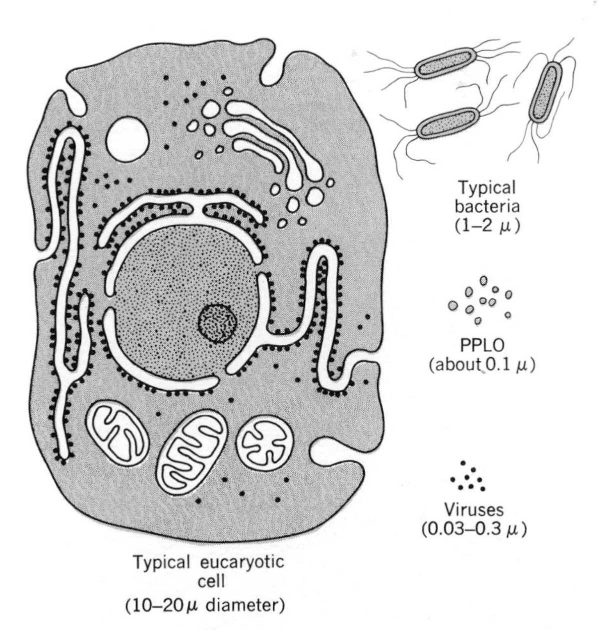

Figure 1–43
Relative sizes of eucaryotic cells, bacteria, PPLO, and viruses.

produced by viruses. Even certain leukemias and cancers are of viral origin.

Most virions are either rod-shaped or quasi-spherical and contain a nucleic acid *core* surrounded by a specific geometric array of protein molecules that form a coat or *capsid* (Fig. 1–44). The proteins of the capsid are arranged to form either a *helical* pattern (when the virus is rodlike) or an *isometric* pattern (when the virus is globular). In the latter state, the virus appears much like a *polyhedron*. The viruses causing chicken pox, mononucleosis, fever blisters, and colds are examples of virions having polyhedral capsids. Helical capsids are more common among viruses that infect plant cells and bacteria. The *tobacco mosaic virus* (TMV), which infects the leaves of the tobacco plant, is among the most extensively studied viruses and exhibits the helical capsid pattern.

Figure 1–44
Viruses. *(a)* Negatively stained preparation of a mixture of tobacco mosaic virus (TMV) and the T4 and φ X174 bacteriophages. *(b)* The SPα bacteriophage of *Bacillus subtilis*. *(c)* Type *D* bacteriophage of *B. thuringiensis*; note the hexagonal array of the base plate (compare with Figure 1–45). Magnifications: *(a)* 98,000×; *(b)* 190,000×; *(c)* 297,000×. (*a* and *b*, courtesy of Dr. F. A. Eiserling; *c*, courtesy of Dr. H. W. Ackermann.)

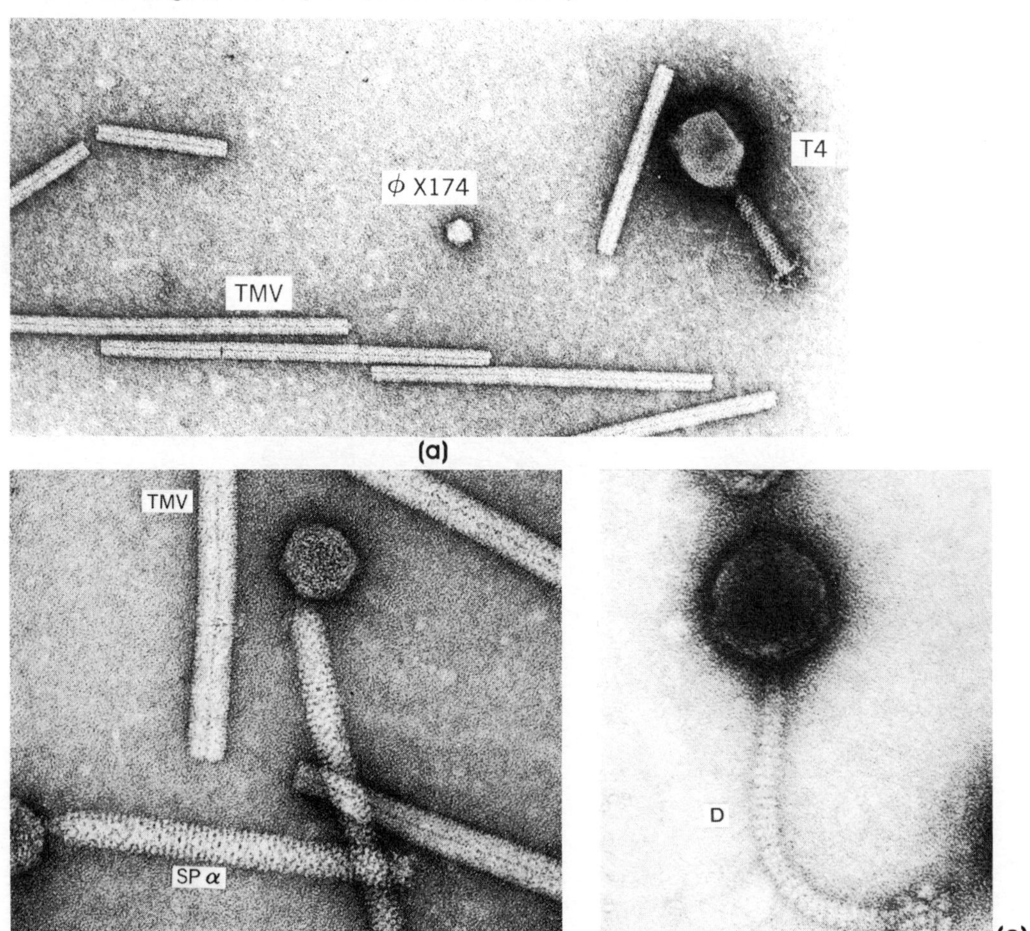

In many animal viruses and in some plant viruses, a lipoprotein *envelope* surrounds the capsid (e.g., influenza virus, herpesvirus, and smallpox virus). Among the largest and most complex virions are those that attack bacteria (i.e., the *bacteriophages*). Most extensively studied among these are the T2, T4, and T6 (i.e., the "T-even") bacteriophages (Fig. 1–44). These bacteriophages have a tail-like structure emerging from the capsid (Fig. 1–45). The tail is enclosed in a sheath of proteins arranged in a helical pattern, while the head of the virus is polyhedral. The end of the tail frequently reveals specialized structures (Fig. 1–45) involved in attachment to the surface of the host cell (see below).

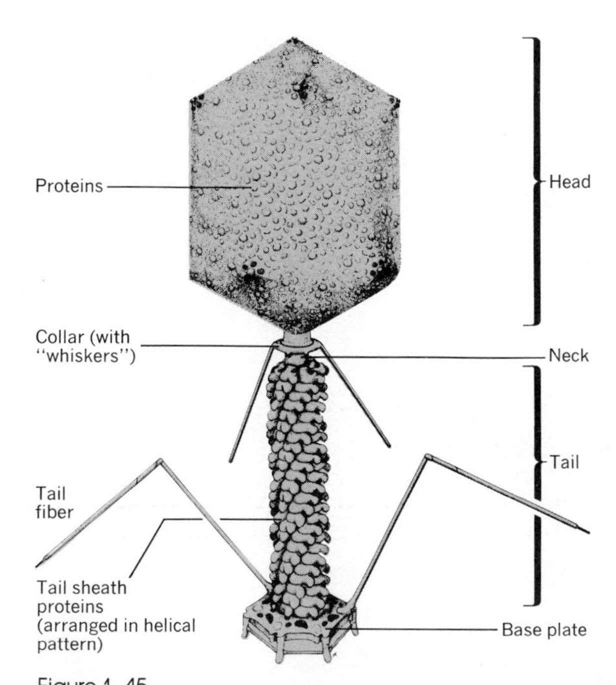

Figure 1–45
Diagram of the T4 bacteriophage (compare with Figure 1–44). Only two of the six tail fibers and collar whiskers are depicted. (Courtesy of Dr. F. A. Eiserling.)

Proliferative Cycle of a Virus

In the free or isolated state, viruses exhibit no metabolism and are incapable of proliferation. Proliferation of viruses requires a *host* cell and in its simplest and most direct form takes the following pattern. One or more viruses attach to specific sites on the surface of the host cell and insert their core nucleic acid into the host. Release of the core nucleic acid from a virus can be achieved experimentally; Figure 1–46 dramatically reveals the uncoiled DNA molecule released by a T4 bacteriophage. Once inside the host cell, the viral nucleic acid redirects the metabolism of the host so that new viral proteins and new viral nucleic acids are formed. These viral components combine in the host to form large numbers of new virions that egress from the cell by disruption of its plasma membrane (i.e., cell lysis). The cycle of infection then repeats itself. The proliferative cycle of a virus is best understood for the bacteriophages and is depicted diagrammatically in Figure 1–47. The electron photomicrographs of Figure 1–48 show stages of the process. The virion envelope that characterizes many animal viruses is acquired from a portion of the plasma membrane of the host cell as the virus emerges.

On some occasions and only for certain viruses, the injected nucleic acid does not cause proliferation and release of new virions. Instead, the injected nucleic acid is incorporated into the host's genetic material, and the host cell continues to function in its normal manner. However, duplication of the host's genetic material prior to cell division is accompanied by duplication of the incorporated viral nucleic acid. Several generations of cells may be produced, each containing a copy of the viral nucleic acid. Viruses exhibiting this phenomenon are called *temperate* viruses, because they do not cause the death of the *immediate* host. Viruses that engage only in the cycle described earlier and that kill the host cell are called *virulent* viruses. The dormant viral nucleic acid within the host is referred to as a *provirus,* and the infected cell is said to be *lysogenic,* because sooner or later, in one of the generations of host cells, the provirus nucleic acid *will* begin to direct the replication of new virions, and this in turn will lead to cell lysis and release of new infective virus particles.

Classification of Viruses and the Nature of Viral Nucleic Acids

The classification of viruses poses certain problems, and several different approaches have been used. One method is to classify the virus according to the type of host cell. Hence, there are *animal viruses, plant viruses,* and *bacterial viruses* (i.e., bacteriophages). This method is not always satisfactory; for example, a few viruses infect both animals

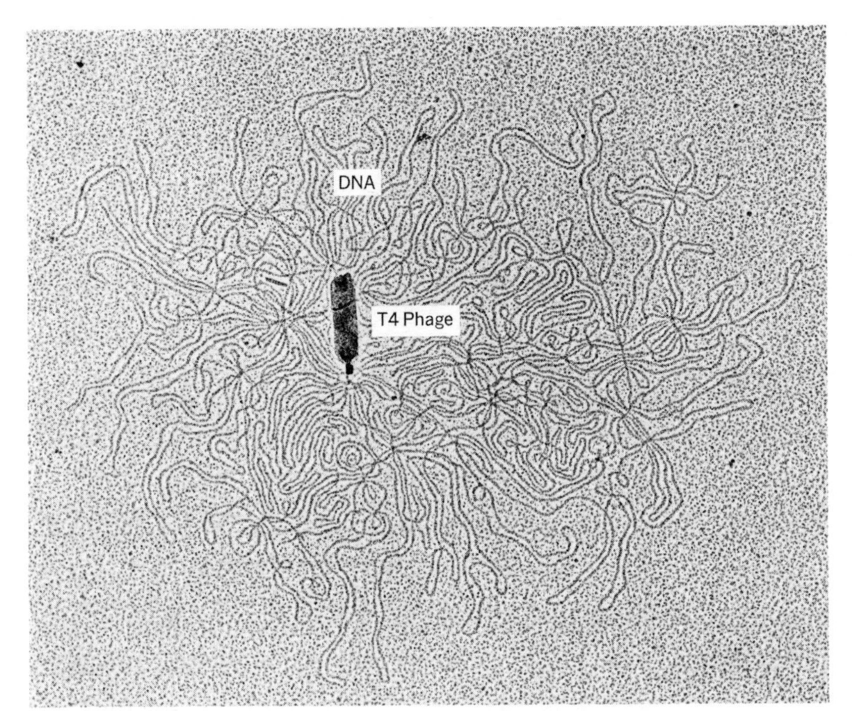

Figure 1–46
DNA molecule released from a "giant" T4 bacteriophage particle. The single DNA molecule is more than 150 μ long. Magnification, 20,000×. (Courtesy of Dr. F. A. Eiserling.)

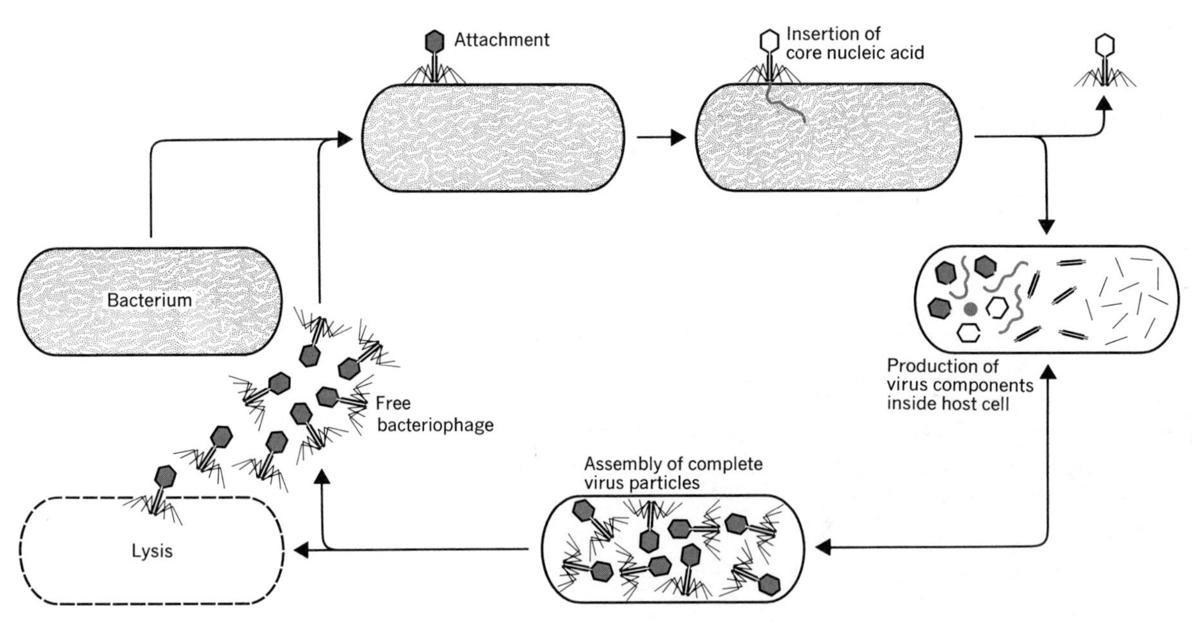

Figure 1–47
Proliferative cycle of the bacteriophage.

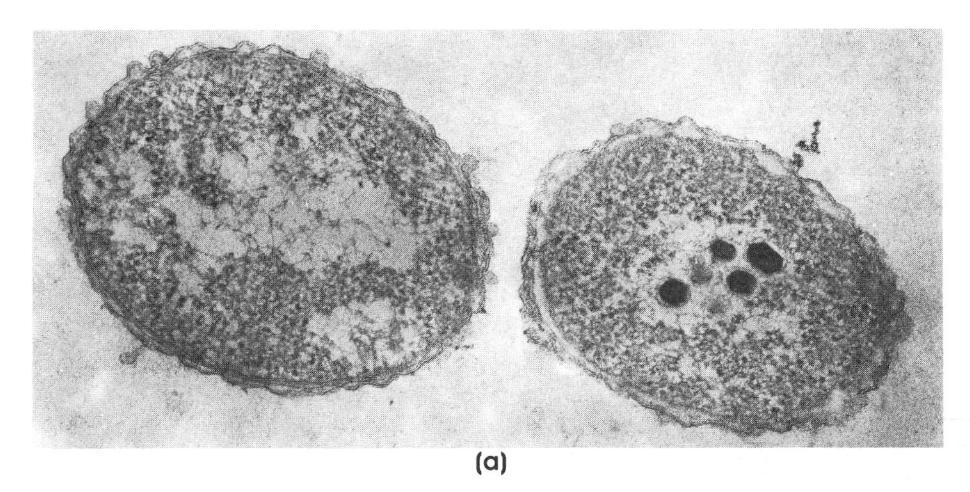

(a)

Figure 1–48
Stages in the proliferative cycle of a virus revealed by electron microscopy. (*a*) Two *E. coli* cells infected by T4 phage. In one cell, the thin section reveals viral DNA condensing within the head membranes of several T4 particles. (*b*) In this negatively stained preparation, more than 100 T4 particles are being released from the lysing *E. coli* cell. Along the upper right edge of the cell, several empty (contracted) phages which began the infection by injecting their DNA into the host cell can be seen. Magnifications: (*a*) 68,000×; (*b*) 26,000×. (*a,* courtesy of Dr. F. A. Eiserling; *b,* courtesy of Drs. L. Cañedo and F. A. Eiserling.)

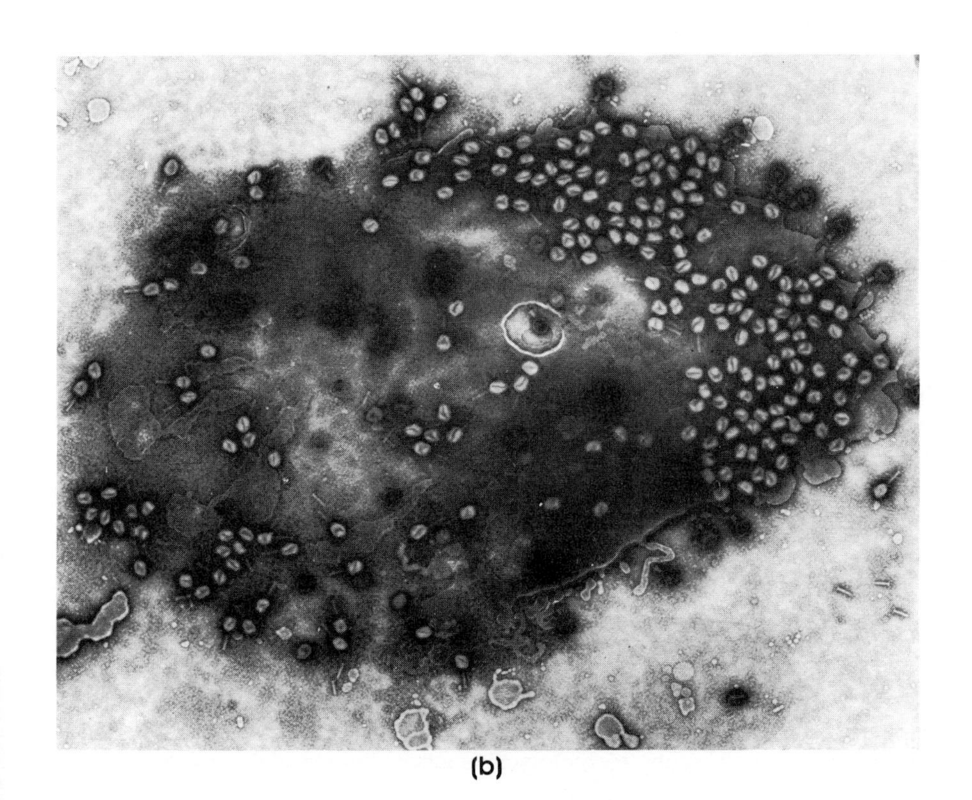

(b)

and plants. Another method employs comparisons of virus morphology (e.g., capsid shape and geometric symmetry). An interesting approach is to classify viruses according to the nature of their genetic material. Even the beginning student of biology becomes quickly aware of the functional relationship between DNA, RNA, and proteins in cells. Genetic information stored in molecules of DNA is transcribed or copied into a corresponding RNA molecule (called a messenger), and the messenger then directs the synthesis of a specific protein. The protein then acts as an enzyme or structural component of the cell. *In some viruses, however, the genetic information comprising the virion core is RNA and not DNA*. Among the RNA viruses are TMV and the virions causing polio, mumps, measles, influenza, and colds. While RNAs of cells are single-stranded molecules (Chapter 7), viral RNAs can be single stranded (e.g., TMV, RNA) or double stranded (e.g., the reoviruses). Like eucaryotic and procaryotic cells, the genetic information of many viruses is encoded in a DNA core. The DNA viruses include the T-even bacteriophages and those that cause chicken pox, herpes blisters, infectious mononucleosis, and shingles. However, the DNA may be *double-stranded linear* molecules (e.g., T5 and T7 bacteriophages), *double-stranded circular* molecules (polyoma and SV 40 viruses), *single-stranded linear* molecules (the parvoviruses), or *single-stranded circular* molecules (ϕX174 bacteriophage). The chemistry of these types of nucleic acids is considered in Chapter 7. Whatever form the nucleic acid takes, it includes the genetic information for the synthesis of the variety of proteins that either become components of new viruses or are involved in the redirection of the host cell's metabolism.

It was noted at the beginning of this discussion that while viruses themselves are not cells, the study of viruses has yielded a wealth of information about cells. Research with certain viruses has provided crucial and at times astounding information about the chemistry of and interactions among nucleic acids and proteins—subjects that are dealth with at length in Chapters 7, 20, and 21—and these viruses should be specifically noted.

TMV. Studies involving the tobacco mosaic virus (Fig. 1–44) began nearly a century ago and represent the starting point in the field of *virology*. In the 1950s, TMV was at the focal point of research that verified that nucleic acids and not proteins compose the genetic apparatus and that genetic information can be encoded in RNA as well as in DNA.

ϕX174. Studies with the ϕX174 bacteriophage which infects *E. coli* revealed that the information for viral proliferation can be encoded in a *single* strand of DNA and does not require a double strand (i.e., double helix). Moreover, the structure of ϕX174 DNA has now been completely analyzed, and the entire base sequence is known. A most astounding finding yielded by these studies is that the coding sequences for several of the virus' proteins are included *within* sequences for other proteins. That is, certain genes *overlap*. Similar findings are being reported for other viruses, including simian virus 40 (SV 40), that possess double-stranded DNA and have both temperate and virulent phases.

Reoviruses. Reoviruses (e.g., Rous sarcoma virus, avian leukemia virus, and other cancer-causing viruses) have RNA as their core nucleic acid, but unlike ϕX174, TMV, and other RNA viruses, replication within the host cells requires that the inserted RNA be used for the preliminary synthesis of DNA, following which transcription and translation take the conventional pattern. The reoviruses have demonstrated that transcription can take place in the *reverse direction*, that is, from RNA to DNA. Our current understanding of the molecular mechanisms for storing or expressing genetic information is based to a large degree on studies using viruses. Some of the concepts discussed above only in introductory terms are treated more fully in later chapters.

Table 1–1
Metric Measurements of Size

1 meter (m)	=	39.4 inches (in.)
1 meter (m)	=	100 centimeters (cm)
1 centimeter (cm)	=	10 millimeters (mm)
1 millimeter (mm)	=	1000 micrometers (μm) or microns (μ)
1 micrometer (μm)	=	1000 nanometers (nm) or millimicrons (mμ)
1 nanometer (nm)	=	10 angstroms (Å)

Summary

In the 300 years that followed the introduction of microscopy to biological science, the concept evolved that the cell is the fundamental unit of structure and function in all living things. This notion is referred to as the **cell doctrine.** Microscopy remains one of the cell biologist's most important and powerful research tools as new variations of light and electron microscopy have appeared, notably freeze-fracture transmission electron microscopy, scanning electron microscopy, and stereoscopic electron microscopy. With these tools the detailed structure and organization of nearly all subcellular organelles have been revealed.

Free-living cells and the cells of multicellular organisms are divided into two classes: **eucaryotes** (nearly all animal and plant cells) and **procaryotes** (bacteria, blue-green algae, mycoplasmas, etc.). Eucaryotic cells are characterized by a number of discrete organelles—especially the nucleus, mitochondria, Golgi bodies, lysosomes, peroxisomes, rough and smooth endoplasmic reticulum, and in plant cells, chloroplasts and cell wall. Like eucaryotic cells, many procaryotes possess a limiting (plasma) membrane and ribosomes, but lack the true nucleus and other discrete organelles characteristic of eucaryotes.

Although **viruses** are not cells, they are intimately associated with cells, and their study has made invaluable contributions to our understanding of cell function. In the isolated state, viruses are incapable of metabolizing or proliferating. Proliferation requires preliminary infection of a host cell, which takes the form of insertion of the viral nucleic acid (DNA or RNA). Following this, the metabolism of the host is redirected to make the components of the virus. Assembly of new virus particles within the host is ultimately followed by their egress to begin a new cycle of infection.

Most of the remaining chapters in the book are devoted to a detailed description of the biochemistry, structure, and physiological functions of cells and their organelles.

References and Suggested Reading

Articles and Reviews

Albersheim, P., The wall of growing plant cells. *Sci. Am. 232*(4), 80 (April 1975).

Allison, A., Lysosomes and disease. *Sci. Am. 217*(5), 62 (Nov. 1967).

Brachet, J., The living cell. *Sci. Am. 205*(3), 50 (Sept. 1961).

Butler, P. J. G., and Klug, A., The assembly of a virus. *Sci. Am. 239*(5), 62 (Nov. 1978).

Campbell, A. M., How viruses insert their DNA into the DNA of the host cell. *Sci. Am. 235*(6), 102 (Dec. 1976).

deDuve, C., The lysosome. *Sci. Am. 208*(5), 64 (May 1973).

Fiddes, J. C., The nucleotide sequence of viral DNA. *Sci. Am. 237*(6), 54 (Dec. 1977).

Fox, C. F., The structure of cell membranes. *Sci. Am. 226*(2), 30 (Feb. 1972).

Morowitz, H. J., and Tourtellotte, M. E., The smallest living cells. *Sci. Am. 206*(3), 30 (Mar. 1962).

Neutra, M., and Leblond, C. P., The Golgi apparatus. *Sci. Am. 220*(2), 100 (Feb. 1969).

Nomura, M., Ribosomes. *Sci. Am. 221*(4), 28 (Oct. 1969).

Orci, L., Matter, A., and Rouiller, C., A comparative study of freeze-etch replicas and thin sections of rat liver. *J. Ultr. Research 35,* 1 (1971).

Osborn, M., and Weber, K., The display of microtubules in transformed cells. *Cell 12,* 561 (1977).

Pangborn, J., Kuhn, D. A., and Woods, J. R., Dorsal-ventral differentiation in *Simonsiella* and other aspects of its morphology and ultrastructure. *Arch. Microbiol. 113,* 197 (1977).

Racker, E., The membrane of the mitochondrion. *Sci. Am. 218*(2), 32 (Feb. 1968).

Ravazzola, M., and Orci, L., Intercellular junctions in the rat parathyroid gland: A freeze-fracture study. *Rev. Biol. Cellulaire 28,* 137 (1977).

Satir, P., Cilia. *Sci. Am. 204*(2), 108 (Feb. 1961).

Satir, B., The final steps in secretion. *Sci. Am. 233*(4), 28 (Oct. 1975).

Schopf, J. W., The evolution of the earliest cells. *Sci. Am. 239*(3), 110 (Sept. 1978).

Sharon, N., The bacterial cell wall. *Sci. Am. 220*(5), 92 (May 1969).

Staehelin, L. A., and Hull, B. E., Junctions between living cells. *Sci. Am. 238*(5), 140 (May 1978).

Stolinski, C., Freeze-fracture replication in biological research: development, current practice and future prospects. *Micron 8,* 87 (1977).

Books, Monographs and Symposia

Bodenheimer, F. S., *The History of Biology,* Wm. Dawson and Sons, Ltd., London, 1958.

Echlin, P., The blue-green algae, in *Cellular and Organismal Biology* (D. Kennedy, ed.), W. H. Freeman and Co., San Francisco, 1974.

Fawcett, D. W., *Anatomy of Fine Structure,* W. B. Saunders, Philadelphia, 1966.

Florkin, M., and Stotz, E. H. (editors), *Comprehensive Biochemistry,* Vol. 32, *A History of Biochemistry.* Part IV. *Early Studies on Biosynthesis.* Elsevier Scientific Publishing Co., Amsterdam, 1977.

Gardner, E. L., *History of Biology,* Burgess Publishing Co., Minneapolis, 1965.

Haggis, G. H., *The Electron Microscope in Molecular Biology,* John Wiley & Sons, Inc., New York, 1968.

Haggis, G. H., Michie, D., Muir, A. R., Roberts, K. B., and Walker, P. M. B., *Introduction to Molecular Biology,* John Wiley & Sons, Inc., New York, 1964.

Ham, A. W., and Leeson, T. S., *Histology* (4th ed.), J. B. Lippincott Co., Philadelphia, 1961.

Hayat, M. A., *Principles and Techniques of Electron Microscopy,* Vol. I., Van Nostrand Reinhold Co., New York, 1970.

Jensen, W. A. and Park, R. B., *Cell Ultrastructure,* Wadsworth Publishing Co., Belmont, Calif., 1967.

Luria, S. E., Darnell, J. E., Baltimore, D., and Campbell, A., *General Virology* (3rd ed.). John Wiley & Sons, Inc., New York, 1978.

Murray, R. G. E., The organelles of bacteria, in *The General Physiology of Cell Specialization* (D. Mazia and A. Tyler, eds.), McGraw-Hill Book Col., New York, 1963.

Nordenskiold, E., *The History of Biology,* Tudor Publishing Co., New York, 1928.

Orci, L., and Perrelet, A., *Freeze-Etch Histology.* Springer-Verlag, New York, 1977.

Singer, C. J., *A History of Biology,* Abelard-Schuman, London, 1959.

Swift, J. A., *Electron Microscopes,* Kogan-Page, London, 1970.

Chapter 2

CELL GROWTH AND PROLIFERATION

All the cells that make up the tissues and organs of a multicellular animal or plant are derived initially from a single cell through growth and division. The cells of many tissues (e.g., epithelium, blood-cell-forming tissues, liver tissue, etc.) continue to grow and divide for most of the life of the organism; however, in some tissues, such as muscle and nerve, cell division ceases some time after birth, and subsequent tissue growth results from individual cell growth without division. If microorganisms such as bacteria or protozoa are placed in an appropriate nutrient medium, they may grow and divide until the medium is teeming with these cells. Similarly, single cells teased from tissues of higher plants or animals can be grown on artificial nutrient media to form what is called a **tissue culture.** Much of our current knowledge concerning the kinetics and mechanics of cell growth and division has been obtained with such cultures of microorganisms and tissue cells, and cell-culturing techniques have become increasingly important tools of the cell biologist.

Mathematical models may be derived that describe the rate at which a population of cells grows—that is, increases in numbers. Consider as an example the following hypothetical situation: Suppose that we begin with a culture medium containing a single cell that grows for some time period and then divides to yield two *daughter cells;* these cells in turn grow and divide after an identical period of time to yield four cells, and so on. In such a situation, the number of cells present in the population would change *exponentially* in the following manner: 1, 2, 4, 8, 16, 32 . . . etc. (Fig. 2–1*a*). That is, the population would *double* with each generation. Consequently, after any specific length of time (i.e., from time t_1 to time t_x), the number of cells present in the population would be given by the equation

$$N_x = N_1 \times 2^g \qquad (2\text{–}1)$$

where N_1 is the original number of cells present (at t_1), N_x is the number of cells present at time t_x, and g is the number of generations that have occurred during the time interval $t_x - t_1$. The exponential character of the increase in population size is seen by comparing Figures 2–1*a* and 2–1*b*.

The *ideal* situation described above does not exist in nature, except perhaps for some limited time following the fertilization of an egg cell; however, it is approximated in a number of instances. If a typical expanding population of cells is examined at any instant in time, some cells would be observed to be dividing, others would have just completed division, still others would be preparing to divide, and so on. Divisions of all cells present would not occur at exactly the same time. The hypothetical situation described above is approached artificially under conditions of

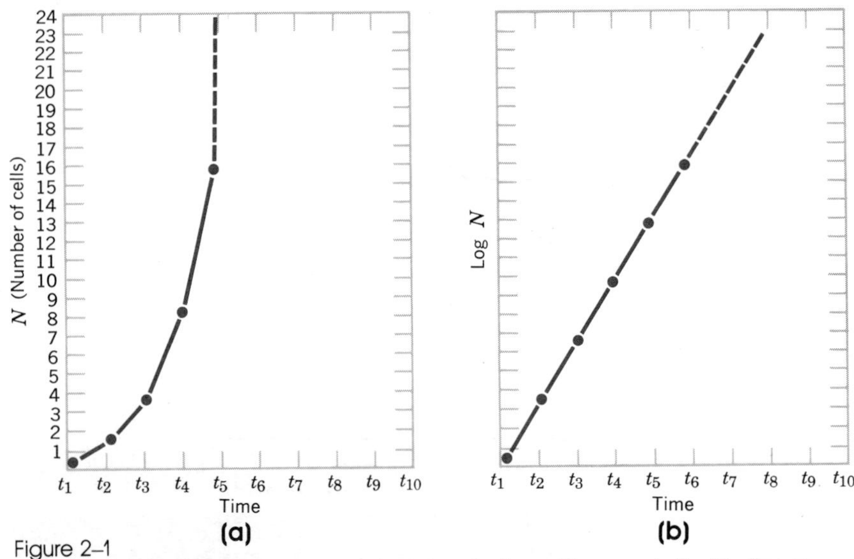

Figure 2–1
(*a*) A geometric expansion of numbers beginning with one plotted in linear terms as a function of time. (*b*) The same geometric expansion plotted in semilog terms. Note that the latter plot yields a straight line.

synchronous growth, which will be considered later in the chapter.

The Population Growth Cycle

Exponential Growth

If a large number of cells are cultured together in what is called a "batch culture," the individual cells will be found in a variety of stages in their *growth-division cycle* or *cell cycle*. The increase with time in the number of cells in the batch culture is proportional to the number of cells present. This, of course, presumes a steady state in which the necessary cell nutrients are always available in adequate supply, and in which cellular waste products excreted into the cells' environment do not interfere with the maintenance of normal growth and division. The growth of such a random culture is described by the following differential equation:

$$dN/dt = k\,N \qquad (2\text{–}2)$$

where N is the number of cells present, t is time, dN/dt is the change in cell numbers with time, and k is a constant. This equation may be solved by integration to yield the algebraic expression

$$2.3 \log_{10} \left(\frac{N_2}{N_1}\right) = k\,(t_2 - t_1) \qquad (2\text{–}3)$$

If we let N_1 = the number of cells present in the population at time t_1, and N_2 = the number of cells present at time t_2, then equation 2–2 is solved as follows.

First, transpose to collect like terms so that

$$dN/N = k\,dt$$

and integrate between the limits N_1 and N_2 and t_1 and t_2.

$$\int_{N = N_1}^{N = N_2} dN/N = k \int_{t = t_1}^{t = t_2} dt$$

Thus,

$$\ln N_2 - \ln N_1 = k\,(t_2 - t_1)$$

or

$$\ln (N_2/N_1) = k(t_2 - t_1).$$

$$N_2/N_1 = e^{k(t_2 - t_1)}$$

where e (the natural base) equals 2.72. By converting to the more familiar logarithmic base 10, this last equation takes the form of equation 2–3.

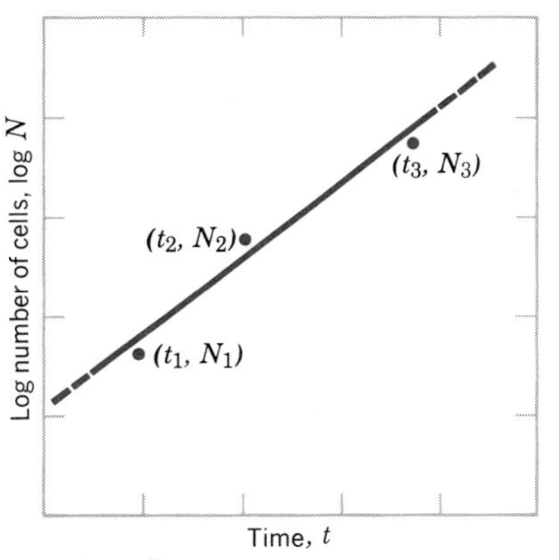

Figure 2–2
A hypothetical exponential increase of cell numbers. The generation time may be determined by the substitution of values for t and N into equation 2–3.

Equation 2–3 indicates that the growth of a cell population (i.e., the rate at which the number of cells in the population increases) is *logarithmic* or *exponential*.

Generation Time

Although the number of cells in a population increases exponentially with time, different types of cell populations (i.e., different species of microorganisms or cells from different tissues) grow at different rates. Even populations of the same types of cells may grow at different exponential rates if the temperature, nutrients, or other growth conditions vary. Differences in growth rates are reflected in the value of the constant k of equation 2–3. A convenient value that describes the specific rate of growth of a population of cells under a specified set of conditions is the **generation time.** The generation time is defined as the time required for the numbers of cells in the population to exactly double during exponential growth. An equation for the generation time may be derived as follows.

After a time interval equal to the generation time has elapsed, then the ratio (N_2/N_1) equals 2; therefore, from equation 2–3,

$$2.3 \log 2 = kT \quad (2\text{–}4)$$

where T is the generation time, $t_2 - t_1$. Hence,

$$0.693 = kT \quad (2\text{–}5)$$

and

generation time →

$$T = 0.693/k \quad (2\text{–}6)$$

The actual value for k or T may easily be determined when experimental data are used to make a semilogarithmic plot of cell number versus time (Fig. 2–2).

Sample Problem. Suppose that at time t_1, the number of cells in a population (i.e., N_1) is 62,400 and at time t_2, 18.5 hours later, there are 473,000 cells. What is the generation time for this population of cells?

From equation 2–3,

$$2.3 \log (473{,}000/62{,}400) = k(18.5)$$
$$2.3 \log 7.58 = 18.5k$$
$$k = 0.110$$

The dimensions of k in this instance are hr^{-1}; that is, the population density increases by 11% per hour. Now, from equation 2–6,

$$T = 0.693/0.110\text{hr}^{-1}$$

and

$$T = 6.34 \text{ hours}$$

Therefore, during exponential growth, the number of cells in the population doubles every 6.34 hours

The Lag Phase of Growth

Typically, when cells are placed in a nutrient medium that favors their growth and multiplication, exponential growth does *not* begin immediately. Instead, there is a short interval in which there is little or no increase in the number of cells present in the population. This time interval that precedes exponential growth is called the **lag phase.** The length of the lag phase is quite variable, even for cultures of the same type of cell. A number of factors are believed to influence the length of the lag phase of the growth cycle.

Experiments with bacteria and other microorganisms have shown that variations in the concentrations of certain constituents of the growth medium, such as carbon dioxide, and certain cations, such as H^+ (i.e., pH) markedly influence the length of the lag phase. Therefore, the chemical composition of the nutrient medium influences the time interval that precedes exponential population growth.

The cells used to initiate the growth of a culture are called the *inoculum*. Usually, the cells of an inoculum are obtained from a previous culture at some particular stage of the growth cycle. The stage of the parent culture used to provide the inoculum also influences the length of the ensuing lag phase. For example, the lag phase of cultures of the bacterium *Aerobacter aerogenes* is longer when the inoculum is drawn from a parental culture in early exponential growth and shorter when drawn from a culture in late exponential growth. For the protozoan *Paramecium caudatum,* little or no lag period is observed when the inoculum consists of cells from a culture that had been growing exponentially; however, when the inoculum consists of cells from the *stationary phase* (see below), a lag period is observed.

Generally, the greater the number of cells in the inoculum, the shorter will be the lag period. It has been suggested that this is due to the more rapid accumulation of a diffusible metabolic intermediate in the nutrient medium that is required during exponential growth and that reaches a critical concentration earlier when there is a large number of initial cells. The potential influence of such an intermediate might also account for the shorter lag period observed when the inoculum contains late logarithmic phase cells.

The Stationary Phase

If cells are growing in a medium the size and contents of which are initially fixed, then it is apparent that growth cannot continue indefinitely. The nutrients of the medium would eventually become depleted, and potentially harmful metabolic waste products excreted by the cells would accumulate in high concentrations in the medium. Either or both of these factors, or perhaps others (such as changes in the pH of the medium), cause the cell population to reach some limiting size. Following this, the numbers of cells in the population no longer increase and may even decrease. (It should be noted that the attainment of a constant population density does not mean that cells are no longer growing and dividing, but rather that any additional cells produced by division are more or less equally compensated for by the death and disruption of other cells.) The period during which the number of cells in the population remains fairly constant is called the **stationary phase.** Cells in healthy cultures that reach a stationary phase have long been considered to exist in a ''maintenance-only'' state.

The Death (Declining) Phase

Stationary cultures eventually enter a **death** or **declining phase** in which the number of cells lost by death and degradation is greater than that produced by cell divisions. The most obvious contributor to the onset of the death phase is the exhaustion of nutrients in the medium. The various phases of the growth cycle are summarized in Figure 2–3.

Figure 2–3
Various phases of a cell population growth cycle.

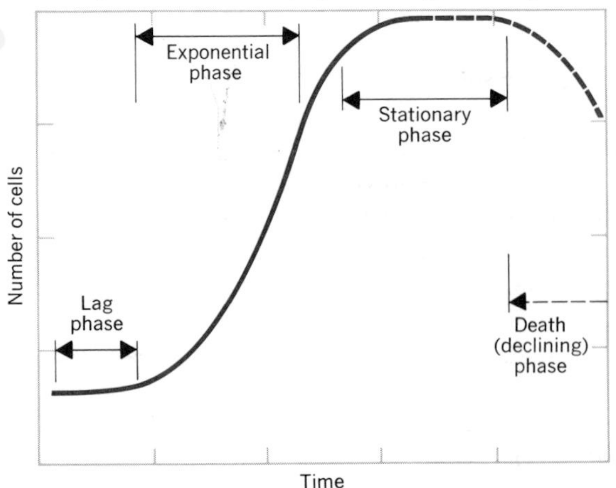

Contact Inhibition

When cells are cultured on a substratum (i.e., rather than in a suspension), the increase in numbers often halts when the population density is high enough for the cells to physically come into contact with one another. This phenomenon is called **contact inhibition** or **density-dependent inhibition.**

The Quantitation of Cells

Basic to quantitative studies of the kinetics of cell population growth and to the accurate measurement of changes in the cell's physical or chemical parameters are methods for determining either the mass or the number of cells present in the population. A variety of techniques are available, and among those that are easily applied are estimations of cell mass based on the total fresh (i.e., wet) weight or dry weight of the sample and the estimation of cell numbers by determining the relative turbidity of the cell suspension. However, these methods lack the level of accuracy generally considered necessary for quantitative studies. More frequently, the precise number of cells present in a sample is determined by the direct microscopic examination of an aliquot of cell suspension drawn from the culture or by electronic enumeration using an instrument especially designed to count cells. Although the latter two procedures are far more accurate than weight determinations or turbidity methods, they require that cells occur separately and, therefore, cannot be applied to tissues.

Optical Enumeration of Cells

The direct microscopic enumeration of cells requires a glass counting chamber (often referred to as *hemacytometer* because of its widespread application to the enumeration of the various types of *blood* cells). This apparatus consists of a glass slide, the surface of which is divided into a number of squares of known dimensions. A drop of cell suspension is placed between the counting area and an overlying cover glass; since the distance between the undersurface of the cover glass and the surface of the counting area is known, the total number of cells present in a known volume of sample can be readily determined. By extrapolation, the number of cells in the entire culture (or in any portion of it) can easily be calculated. The enumeration of cells by direct optical examination is tedious but extremely useful, especially if the suspensions contain different types of cells (as in the case of blood) that may be distinguished visually. Hemacytometric methods are still employed in many laboratories.

Electronic Enumeration of Cells

A more rapid and widely employed procedure for the enumeration of cells is by "electronic gating." The most popular instruments are the Coulter Counter™ (after the inventor, Wallace Coulter) and the Celloscope,™ both of which apply the same basic principles. The main features of an electronic cell gating device are shown in Figure 2–4. A glass probe containing a small aperture at its tip is submerged in a sample of cell suspension containing electrolyte; the aperture tube is also filled with electrolyte. Current is caused to flow from a platinum electrode inside the probe through the small aperture to a second electrode immersed

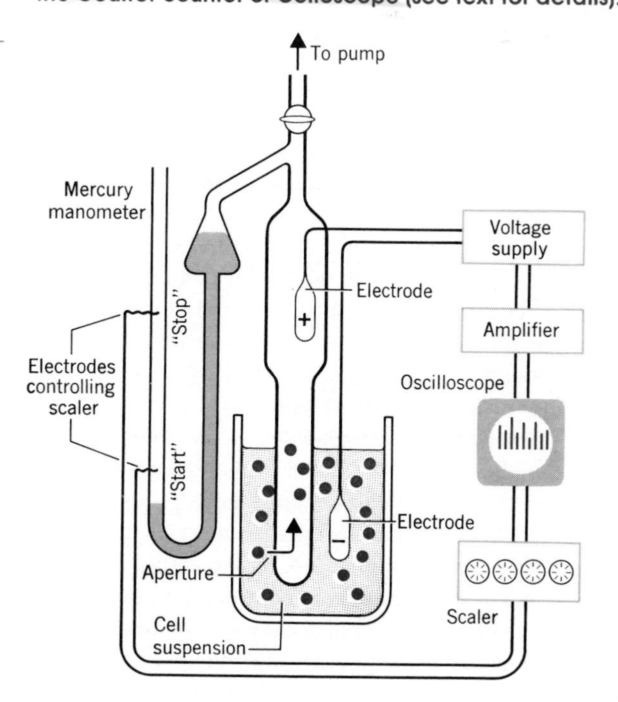

Figure 2–4
Basic components of an electronic cell counter such as the Coulter counter or Celloscope (see text for details).

in the cell suspension. At the same time, the suspension is drawn through the aperture and into the probe by a small pump. The passage of a cell through the aperture temporarily interferes with the flow of current through the aperture, since the cell displaces a volume of electrolyte and generally is not as conductive as the electrolyte. The resulting sudden drop in voltage is recorded by the instrument as a pulse (which may be displayed on an associated oscilloscope) and is counted by a scaling unit. A mercury manometer connected to the probe (and therefore also to the pump) operates two switches through two electrodes inserted into the manometer along its length. With the pump on and a vacuum applied to the aperture tube, not only is cell suspension drawn through the aperture but the mercury in the left arm of the manometer is also drawn downward below the first (i.e., "start") electrode. When the vacuum is then halted (by closing the stopcock), the mercury automatically rises in the left arm of the manometer, turning on the scaler as it passes the "start" electrode and turning off the scaler as it passes the "stop" electrode. With the stopcock closed, the movement of mercury upward in the left arm of the manometer draws cell suspension through the aperture, so that the cells present in a volume of suspension precisely equal to the volume of mercury between the "start" and "stop" electrodes are enumerated. Since this volume is known (i.e., the manometer is calibrated), the number of cells present per unit volume of suspension is readily determined. With an electronic cell counter, it is possible to enumerate thousands of cells in just a few seconds. However, if the suspension contains different types of cells, those of similar size cannot be distinguished by this procedure. In such a case, chemical methods may be used to eliminate certain types of cells. For example, during a clinical measurement of the white cell count in a blood sample, it is customary to eliminate the red blood cells by adding a specific lysing agent to the cell suspension.

The magnitude of each pulse recorded by the electronic counter is directly proportional to the size (i.e., volume) of the cell passing through the aperture. Two pulse-height threshold controls (upper and lower) may be separately adjusted on the instrument so that only those pulses whose peak height falls above the lower threshold and below the upper threshold will be counted. Consequently, in a cell suspension in which cell sizes vary considerably, only those cells within a selected size range may be counted. By simultaneously increasing both the lower and upper threshold values by the same increment, it is possible to obtain a pulse-height distribution for the sample of cells. This pulse-height distribution is analogous to a cell size distribution and is particularly useful to cell biologists studying the growth of individual cells, for it reveals for any selected phase of a population growth curve the relative numbers of cells of all sizes (and therefore ages) present in the population. Thus, with an electronic counter it is possible not only to rapidly enumerate the cells in the population but also to obtain measurements of the sizes of the cells. It would be difficult to overstate the value of this instrument to the cell biologist.

The Continuous Culture of Cells

When cells are cultured in a container of fixed volume, both the population density (i.e., the number of cells per unit volume of culture) and the population size (i.e., the total number of cells in the culture) increase at the same rate during the exponential phase of growth. If we consider the *biological unit* in such a culture as a cell plus some volume of surrounding medium, then this unit is undergoing continuous change. The medium surrounding each cell is being successively depleted in nutrients, while metabolic waste products diffusing from the cell accumulate in the medium, and the cell itself may be increasing (or decreasing) in size. In multicellular organisms, the cells of the various tissues and organs are bathed in a body fluid that continuously provides fresh nutrients while removing the cellular waste products. This condition differs markedly from that of a batch culture but may be approximated experimentally by methods of *continuous culture* in which an effort is made to keep the biological unit constant.

A variety of techniques may be employed for the continuous culture of cells. One popular method involves the use of an instrument known as a **chemostat** (Fig. 2–5), introduced by A. Novick and L. Szilard. In the chemostat, fresh culture medium is continuously pumped into the culture chamber, while old medium and cells are eliminated from the chamber at such a rate that the population size and population density remain constant. With the chemostat, it is possible to maintain the cell population in the exponential phase of growth for a greatly extended period of time.

A similar effect can be achieved by the addition to the medium of an essential substrate at such a slow rate that the growth of the population becomes a function of the rate of supply of the substrate (which, of course, can be main-

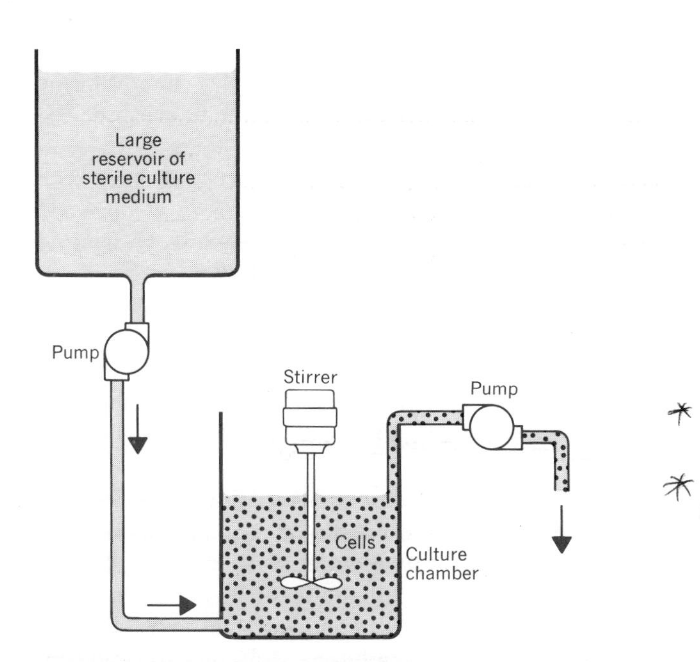

Figure 2–5
Simplified version of a chemostat. Cell-free culture medium enters the culture chamber at the same rate as medium containing cells is withdrawn. The rates are adjusted so that the population density of the culture is kept constant. Although the basic principles are represented in this diagram, most laboratory chemostats are considerably more complex.

tained constant). Another method simply involves the inoculation of a small number of cells into a very large volume of culture medium, so that for many generations the cells will have little influence on the contents of the medium.

The continuous culture of cells offers numerous advantages over growth in fixed volumes, since a balanced, steady-state growth rate can be maintained and physiological studies can be carried out with cells little influenced by fluctuations in the environment.

Synchronous Cell Cultures

The cells growing in a batch culture represent a heterogeneous collection at various stages of their growth-division cycle or *cell cycle* (see below). Some cells are dividing, some have just completed division, some are about to begin division, and so on. Any study of the progressive changes in the chemical composition of cells during growth, or changes in cell physiology or morphology, is difficult, if not impossible, when the cells are randomly distributed with respect to age. This problem can be avoided by studying the growth of individual cells, but cells are usually too small for individual examination or analysis. In order to resolve this problem, methods have been devised in which the cells in a culture can be brought to a similar stage of their growth-division cycle. Consequently, the entire population may be studied as though it were a single cell. This condition is known as *synchronous cell growth*.

The degree of synchrony achieved can be seen by determining the *mitotic index*, which is a measure of the percentage of cells undergoing division at any instant in time and is given by the formula

$$I_M = \frac{N_M}{N} \tag{2–7}$$

where I_M is the mitotic index, N_M is the number of cells visibly undergoing division, and N is the total number of cells present.

During the exponential expansion of a batch culture, I_M remains constant; however, during synchronous growth, I_M changes from some minimal value (ideally, 0) to some maximum value (ideally, 1.0) during a short time interval. The efficiency of the synchronization procedure can be evaluated from a comparison of the mitotic index, the time required for the population to double during synchronous division, and the generation time observed in a batch culture. A comparison of the growth curves for a batch culture and a synchronous culture is shown in Figure 2–6, and the corresponding mitotic indices are shown in Figure 2–7.

A variety of methods have been devised for inducing the synchronous growth of a cell culture; these fall into two major catagories: *synchrony by induction* and *synchrony by selection*.

Synchrony by Induction

The most frequently employed methods for inducing synchrony involve temperature cycles (temperature shocks), light cycles, and chemical manipulations. When temperature is used to induce synchrony, the cells are subjected to alternating cold and warm periods. Little cell division oc-

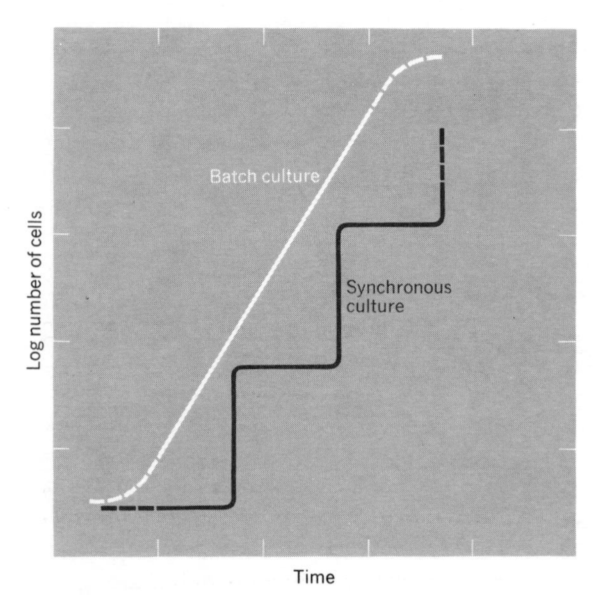

Figure 2–6
A comparison of the growth curves for idealized batch and synchronous cell cultures.

curs during the cold periods, but upon entry into the warm periods, cell division occurs. For example, cultures of the flagellate protozoan *Polytomella agilis* can be synchronized by a repetitive temperature cycle of 22 hours at 9°C, followed by 2 hours at 25°C. Similar procedures have been successful with other protozoa, including *Astasia longa* and *Tetrahymena pyriformis*. Synchrony may also be induced by a rapid succession of short cold and warm periods. Following this sequence, the cell population enters synchronous division.

Certain cultures of photosynthetic cells can be induced to divide synchronously by exposing them to alternating periods of light and dark, the population doubling with each light cycle.

If a specific substrate required for cell division is withheld from the culture medium for some time, then the division of the cell population follows addition of the substrate. Also, agents that chemically inhibit division may be added to the culture. Cell growth continues, but division is arrested. Removal of the inhibitor is followed by a burst of synchronous divisions and growth. These are but a few examples of a variety of chemical procedures in which cyclic manipulation of the culture medium's chemistry results in cell synchrony.

Figure 2–7
A comparison of mitotic indices during the growth of batch and synchronous cell cultures.

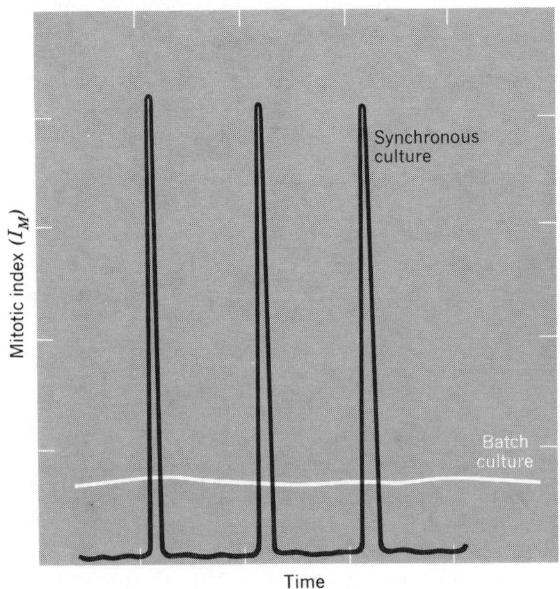

Synchrony by Selection

Selection techniques involve the mechanical isolation of cells of similar age from a random culture; these cells are then inoculated into fresh medium, where they grow and divide synchronously for some time. Among the methods frequently employed for mechanically isolating cells of similar age is *filtration*. In this procedure, a cell culture is filtered to separate the large and small cells; the small cells are then used to inoculate a fresh culture medium.

Cells of similar age may also be isolated from a random culture by *sedimentation*. Since young cells are generally smaller than older cells, they sediment less rapidly and may be isolated for subculture.

Another technique used in tissue culture is the ''grow-off'' method in which the cells are adsorbed onto some surface (such as filter paper); during cell division, one of the two daughter cells produced detaches from the surface and can be collected for subculturing.

Regardless of the procedure employed, the effect is to alter the random age distribution of the cells so that for several subsequent generations, all divisions occur over a short interval of time and all cells will be at a similar stage of their growth-division cycle. Of the various procedures employed, synchrony by selection is to be preferred, since chemical or temperature variations are unnatural and may have undesirable effects upon the cell population. The synchrony achieved by any of these methods is frequently observed to decay after several generations of synchronous growth.

Culture Fractionation

Another approach to selecting cells at the same stage of development in the growth-division cycle is through the technique of *culture fractionation*. This procedure avoids the potential problems of synchronization techniques where the inducing factor may itself alter or distort the normal events of the cell cycle. In culture fractionation, a random culture of cells (i.e., cells at various stages of their individual growth cycles) is sorted into subpopulations of varying cell size by centrifugation through liquid *density gradients* (see Chapter 12) and is followed by collection of the gradient with the entrained cells as a series of fractions. During centrifugation, the larger cells sediment further through the gradient than do the smaller cells. The rationale behind this approach is that a relationship exists between cell age and cell size (i.e., age and size are *directly* related) so that the population is sorted into a linear sequence of fractions of increasing cell age. By studying and comparing chemical, physiological, or morphological properties of the cells in each of the separated fractions, specific events occurring during the cell cycle can be assigned an age. Thus, changes taking place during cell growth may be followed even though a synchronous population is not being employed. Culture fractionation can also be extended to exponentially expanding populations that are given a preliminary treatment with a *radioactive tracer* (see Chapter 14) or some chemical inhibitor. Following treatment, the cells are rapidly fixed, separated into age classes by centrifugation, and analyzed.

Culture fractionation requires the processing of very large quantities of material so that all the collected fractions contain enough cells of the same age (i.e., size) for valid physical and/or chemical measurements. For this reason, the cell cultures are centrifugally fractionated in high-capacity *zonal rotors* (described in Chapter 12).

The Growth of Individual Cells: The Cell Cycle

It has already been noted that it is possible to determine the distribution of cell sizes (i.e., volumes) in a random population of cells. Since cell size is related to cell age, an age distribution is also obtained. If cells were produced in a manner similar to the commercial production of small spherical objects (such as ball bearings or marbles), then their sizes would be distributed normally about some mean size. However, in a random exponentially expanding population of cells, the distribution of size or age is not normal. Instead, a distribution exists in which there are more young (small) cells than old (large) cells. This is due to the fact that two young (small) cells are produced by each division of one old (large) cell. In an *ideal* population (i.e., one in which a cell divides equally into two progeny, each of which grows to the size of the parent before dividing), the age or size distribution takes the form shown in Figures 2–8. This curve indicates that there is an exponential *decrease* in the relative numbers of cells with age or size. This occurs because the absolute number of cells entering division is constantly increasing in an exponential manner, so that each new generation of cells is exponentially greater in number than the one that preceded it. It should be emphasized that

Figure 2–8
Theoretical, mathematically derived age distribution of an *ideal* population of growing cells.

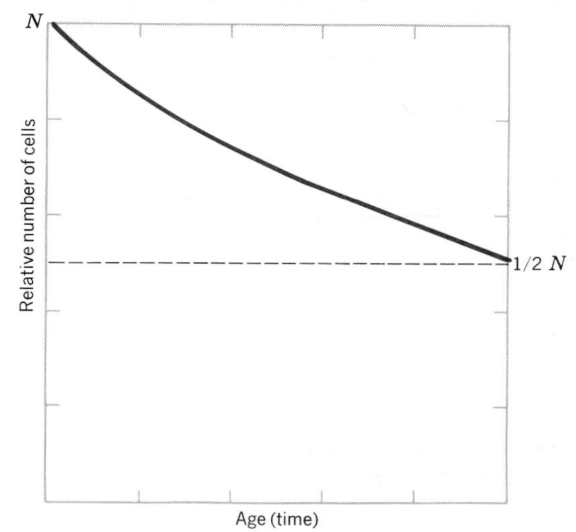

the curve of Figure 2–8 is derived mathematically and applies the assumption that the cells have identical generation times.

The distribution described above probably does not exist in nature, since cells display variations in individual generation times and do not divide into progeny of exactly equal volume. Thus the distribution of cell size ranges beyond V and $2V$ (where $2V$ is the size of the ideal parent cell, and V is the size of the equal progeny). Furthermore, it has been shown that the small cells produced by unequal or early division do not, in turn, give rise to smaller progeny; instead, there is a random fluctuation about some inherited average size. A *size* distribution typical for a random population of cells is shown in Figure 2–9.

Although several possibilities exist for the rate of cell growth between divisions, most studies of cell volume distributions in either batch or synchronized cultures indicate that cells grow either linearly or exponentially. For example, *Tetrahymena pyriformis* is believed to grow exponentially between divisions, while *Chlorella ellipsoidea* appears to grow linearly. The question of the rate of individual cell growth remains a controversial area in spite of the large number of reported studies.

Figure 2–9
A typical distribution of cell sizes in a random culture of cells. The distribution is not "normal" but is skewed toward larger cell size. Note that this size distribution differs from the age distribution of the ideal population of Figure 2–8.

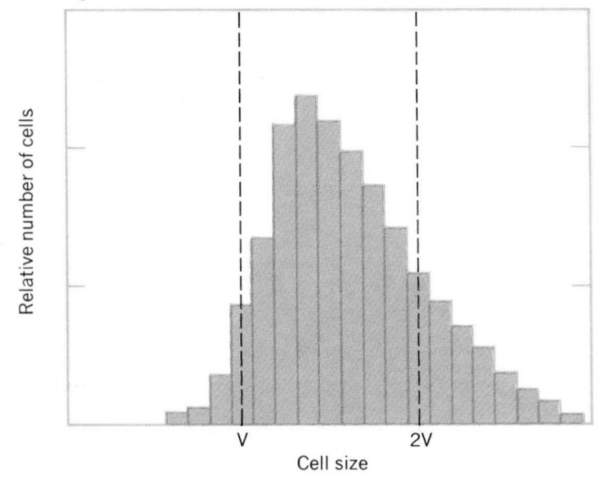

In some cells (such as bacteria and *Tetrahymena*), the cell volume distribution remains constant throughout the exponential phase of population expansion. In other cells (such as the flagellate *Polytomella agilis* and the yeast *Schizosaccharomyces pombe*), the cell volume distribution shifts toward smaller size during the exponential phase. The latter observation indicates that the growth of individual cells occurs less rapidly than the growth (in numbers of cells) of the entire population.

The sizes attained by cells in culture also depend on other factors, such as the temperature at which the cells are maintained. In *T. pyriformis,* average cell size increases with increasing culture temperature between 28° and 34°C. In *P. agilis,* cells cultured at lower temperatures have a greater average size than cells cultured at higher temperatures in the range 9° to 25°C. It is apparent that, depending upon the conditions of culture, the events that result in cell division may occur more or less rapidly than the events that result in individual cell growth.

Phases of the Cell Cycle

The various phases of the growth and reproduction of cells constitute what is called the **cell cycle** (Fig. 2–10), a complete cycle taking place in one generation time. The principal signposts of the cell cycle are the replication of the nuclear DNA and its distribution among the progeny cells. The replication of the nuclear DNA occurs in that portion of the cell cycle known as the **interphase,** whereas the distribution of the replicated DNA and the physical division of the parent cell into two daughter cells occur in the "period of division" referred to as the **M phase.** The symbol M stands for "mitosis," the process of nuclear division familiar to even the beginning biology students and taken up at some length in Chapter 20. The entire cell cycle of most higher animal and plant cells is about 10 to 25 hours, of which only about 1 hour is spent in the M phase. (Procaryotic cells that lack a nucleus and whose DNA is generally considerably smaller in content and is differently organized may have a cell cycle of only 20 to 30 minutes.)

The replication of the nuclear DNA occurs in a rather specific portion of the interphase called the **S phase** (S for "synthesis"). This was definitely demonstrated for the first time in the 1950s using radioisotopes and the technique of *autoradiography* (see Chapter 14). Cells provided with a

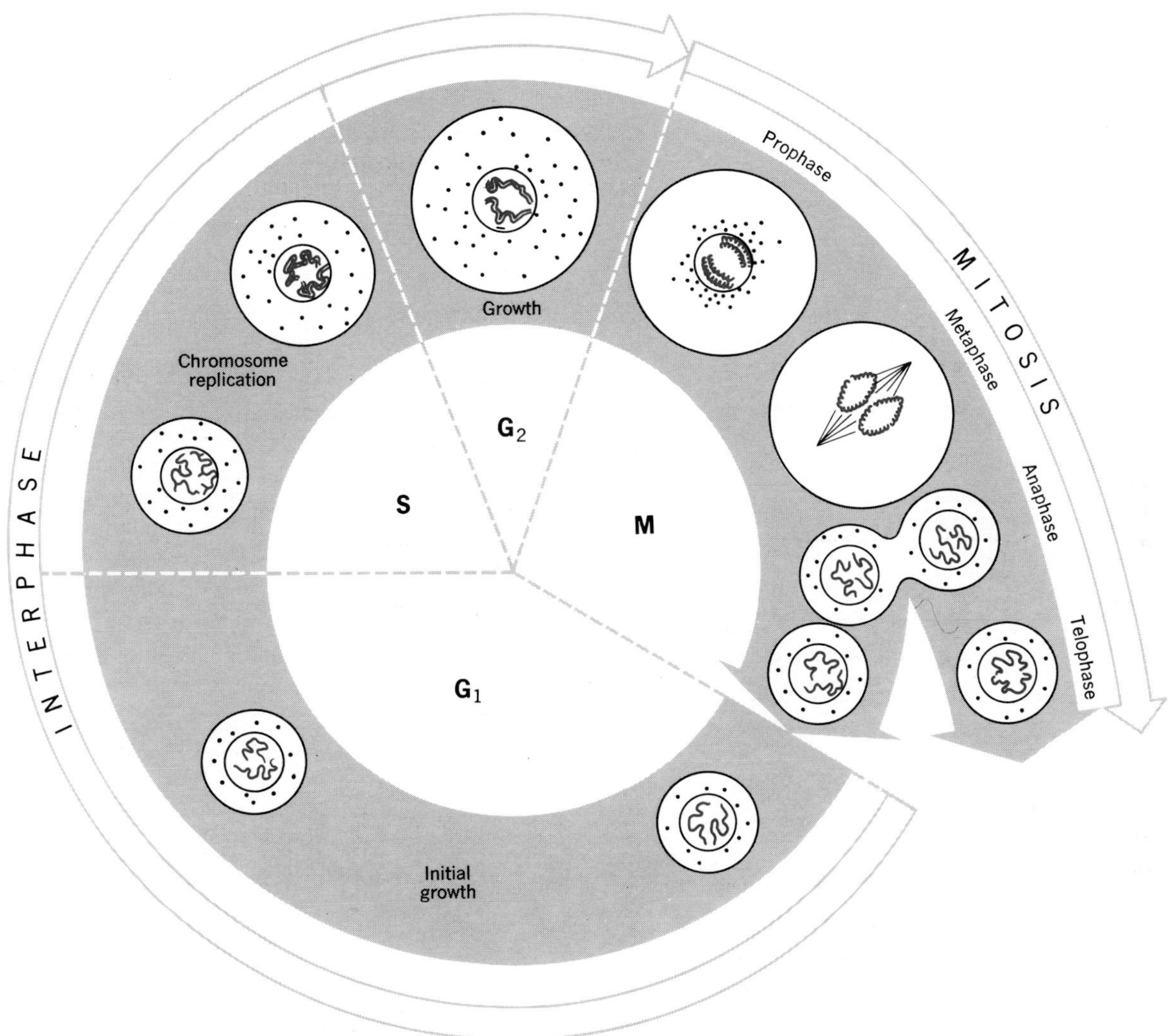

Figure 2–10
Various phases of the cell cycle, which generally lasts about 10 to 25 hours in higher animals and plants. In the diagram, the length of the M phase is deliberately exaggerated to show mitosis and cytokinesis.

supply of radioactive substrates from which new DNA could be synthesized were found to accumulate radioactivity in the cell nucleus during a small and specific portion of the interphase.

Some time elapses between the initial formation of a daughter cell at the conclusion of the cycle of the parent cell and the onset of the S phase. This interval is known as the **G_1 phase** (*G* for "gap"). Completion of the S phase is usually not immediately followed by the M phase. The gap between the end of the S phase and the onset of the M phase is called the **G_2 phase.** In both G phases, there is a considerable amount of cell growth.

The lengths of the G_1, S, G_2, and M phases vary among different types of cells, but variations between cells of the same type are usually very small. The greatest variation is seen in the G_1 phase. When the cell cycle is very long, most of the prolongation can be accounted for by a lengthened G_1 phase. In contrast, in egg cells where the cell cycle is very short, the G_1 phase is very brief or even nonexistent. The cells of animal embryos undergo a rapid sequence of divisions with little intervening cell growth. The cell cycle may last only an hour with no G_1 phase and with DNA synthesis occurring during mitosis or soon after mitosis is completed.

During the M phase, the nuclear contents undergo a series of changes and rearrangements that begin with the condensation of the nuclear chromatin into the clearly visible *chromosomes* and end with the distribution of the chromosomes to the daughter cells and cell division or *cytokinesis* itself. For convenience only (since the M phase is continuous and does not occur in discrete steps), the period of mitosis is subdivided into the classical *prophase, metaphase, anaphase,* and *telophase*. A detailed discussion of the nuclear changes that occur during these phases is deferred to Chapter 20, which deals at length with the structure, organization, and functions of the cell nucleus.

Summary

All cells of multicellular animals and plants are derived initially from a single cell through growth and division. Mathematical models may be derived that describe the growth of cell populations, and proliferation may be studied experimentally using cell cultures. A number of alternative methods are used to culture cells and to measure population growth. The growth of a random or **batch** culture of cells characteristically includes **lag, exponential, stationary,** and **declining** phases. The extrapolation of observations made using an entire population of cells to the cycle of events occurring within an individual cell requires that the cell culture be *synchronous*. The cycle of individual cells (i.e., the **cell cycle**) takes a characteristic form and is divided into the G_1, S, G_2, and M phases. The lengths of these phases vary from one type of cell to another but are characterized by the same physiological events and culminate in the division of one cell into two.

References and Suggested Reading

Articles and Reviews

Johnson, B. F., Morphometric analysis of yeast cells. *Exp. Cell Res. 49,* 59 (1968).

Mattern, C. F. T., Bracket, F. S., and Olson, B. J., Determination of number and sizes of particles by electrical gating. *J. Appl. Physiol. 10,* 56 (1957).

Mazia, D., The cell cycle. *Sci. Am. 230* (1), 54 (Jan. 1974).

Pardee, A. B., Dubrow, R., Hamlin, J. L., and Kletzien, R. E., Animal cell cycle, in *Annual Reviews of Biochemistry,* Vol. 47 (E. E. Snell et al., eds.), Annual Reviews Inc., Palo Alto, Calif., 1978.

Scherbaum, O., and Rausch, G., Cell size distribution and single cell growth in *Tetrahymena pyriformis* GL. *Acta Pathol. Microbiol. Scand. 41,* 161 (1957).

Schmid, P., Temperature adaptation of the growth and division of *Tetrahymena pyriformis. Exp. Cell Res. 45,* 471 (1967).

Wells, J. R., and James, T. W., Cell cycle analysis by culture fractionation. *Exp. Cell Res. 75,* 465 (1972).

Books, Monographs, and Symposia

Dean, A. C. R., and Hinshelwood, C., *Growth, Function and Regulation in Bacterial Cells,* Oxford University Press, Oxford, 1966.

Mitchison, J. M., *The Biology of the Cell Cycle,* Cambridge University Press, New York, 1972.

Zeuthen, E., *Synchrony in Cell Division and Growth,* Interscience Publishers, New York, 1964.

Part 2
MOLECULAR CONSTITUENTS OF CELLS

Chapter 3
CELLULAR CHEMISTRY, MOLECULES, AND IONS

Chemically, the cell is composed of water, proteins, lipids, salts, nucleic acids, carbohydrates, and minute quantities of a variety of organic compounds such as vitamins and growth factors. The proteins, lipids, nucleic acids, and carbohydrates are relatively large molecules called **macromolecules.** The macromolecules and their components are described in the following chapters. Water, the salts that are present in the form of ions in the cell, vitamins, growth factors, and other organic compounds may complex with the macromolecules in the cell or remain as free, relatively smaller **micromolecules.**

The quantities of each of these classes of compounds vary widely from one type of cell to another and from one organism to another (Table 3–1); however, water is the most common molecule, and proteins are the most prevalent organic constituents.

Water

Water is the solvent necessary for life. In part, this is because of a unique set of physical and chemical characteristics that protect living systems and are necessary to the structure and function of cells. Water has a high melting point, heat of fusion, boiling point, heat of vaporization, specific heat, and surface tension. Each of these properties prevents extreme temperature fluctuations and serves to retain water in the liquid form that is necessary for life. Table 3–2 compares water with several other common solvents. One gram of water will absorb more energy for each degree rise in temperature than comparable solvents. Water thus acts as a moderator of temperature change. Likewise, in order to convert water from a liquid state to a vapor state at 100°C, an additional 540 calories (cal) of energy per gram of water must be absorbed—a factor that tends to keep water in the liquid state. The high surface tension is a related factor that stabilizes this state. At the other extreme, large amounts of energy, 80 cal per gram of water, must be lost to reduce water from the liquid to the solid state.

Water is also know as the "universal" solvent. In fact, it does not dissolve all substances, but it does dissolve most salts and other ionic compounds, as well as nonionic polar compounds such as sugars, alcohols, and other molecules containing hydroxyl, aldehyde, and ketone groups. In addition, water will form **micelles** of many compounds that contain both a highly polar and a large nonpolar side group. A micellar arrangement is not a true solution but a *dispersion* or *suspension*. Soaps in water are common examples of micelles. The soap molecules consist of a long-chain nonpolar hydrocarbon (Fig. 3–1) containing a terminal polar carboxyl group ionically bonded to a metal. Suspended in water, the polar carboxyl group forms weak but significant bonds (hydrogen bonds) with the water molecules. The

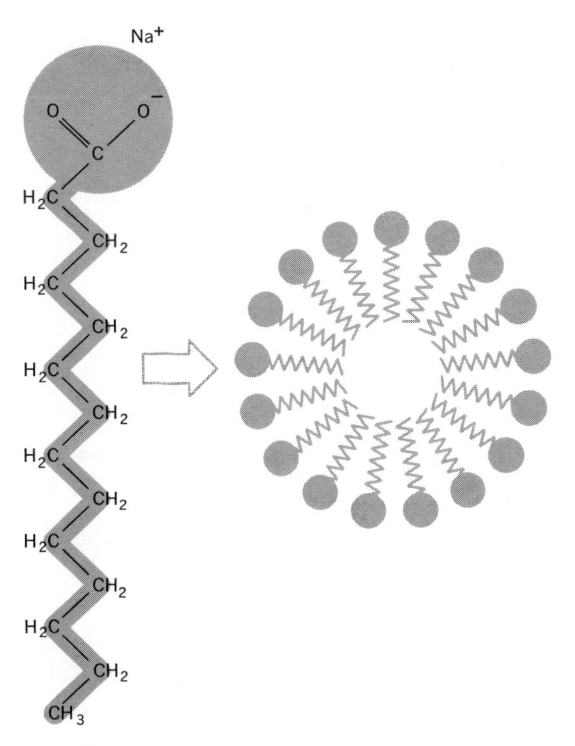

Figure 3–1
Soaps (fatty acids bonded to metals) coalesce into micelles in aqueous suspensions with nonpolar ends of molecules oriented together.

nonpolar portions of the molecules repel the water molecules and tend to aggregate together. Small spheres of soap molecules thus form with their polar groups on the surface of the micelle and the nonpolar chains oriented to the center (Fig. 3–1).

The kinetic properties and the unique solubility properties both result from the strong cohesive properties of water and the structure of the molecule. The two hydrogen atoms are bonded to the oxygen atom at an angle of 104.5°. The strongly electronegative oxygen atom produces electrical asymmetry in the molecule by withdrawing the electrons from the hydrogens. As a result, the oxygen atom has a local negative charge, while each of the two hydrogen atoms is positive. As illustrated in Figure 3–2 the water molecule acts as a dipole even though it has no *net* charge.

Because of their dipole nature, water molecules tend to bond with each other and form a latticelike structure, as shown in Figure 3–2. The connection between the molecules is a *hydrogen bond,* which forms by the attraction of the electronegative oxygen of one atom for the positively charged hydrogen of another. These bonds are also discussed in connection with protein structure in Chapter 4. As the bond forms, there is a redistribution of electronic charges enhancing the bond. Each oxygen may form two hydrogen bonds, thus becoming bonded to four hydrogens that form a tetrahedron about the oxygen (Fig. 3–3). The angle of the bonds and the distance between the atoms account for the latticelike structure. As kinetic energy is removed and the temperature drops, the decrease in movement of the molecules allows for more extensive formation of hydrogen bonds and lattice development. Because of the positioning of the molecules, more space develops between the molecules as the freezing point is reached. The freezing water and ice become less dense and rise to the surface.

When salts or polar compounds are dissolved or suspended in water, the orientation of the water molecules to each other changes. Sodium chloride is a salt common to most biological fluids. In solution the salt ionizes and the Na^+ and Cl^- ions attract and orient the water molecules about them, forming hydration spheres (Fig. 3–4). These hydration spheres may help suspend many larger molecules such as proteins in water. As a general rule, ionic compounds are stronger binders of water than nonionic polar compounds. This property is frequently used to separate macromolecules in a suspension. Soluble proteins are held in suspension by hydrogen bonds to water and encompassing spheres of hydration. When salt is added to the mixture, the salt ions attract and form more extensive hydration spheres than the proteins. The reorganization of water prevents the continued suspension of the large protein molecules, and they precipitate out of solution. This technique is called *salting out*.

Another important property of water is its **ionization.** Hydrogen atoms in water, as described, are covalently bonded to the oxygen of one water molecule and form a hydrogen bond with the oxygen of another water molecule. With a measurable frequency, the internal kinetics of one molecule may favor the breakage of the covalent bond and a closer association of the hydrogen with the oxygen to which it was hydrogen bonded.

$$H_2O \cdots H{-}O{-}H \longrightarrow H_2O^+{-}H + OH^-$$

Table 3–1
Components of Various Tissues and Organisms (Percentage of Total Fresh Weight)

Constituent	Rat Liver	Rat Skeletal Muscle	Sea Urchin Eggs	Bacteria (*E. coli*)	Corn Seed
Water	69–72	76.0	77.3	73.0	13.0
Protein	16–22	21.0	15.8	19.4	8.8
Carbohydrate	2.3	0.53–0.58	1.4	1.1	73.0
Lipid	5.0	3.7–10.6	4.8	1.1	4.0
Nucleic acids	0.75–1.57	0.75–1.63	0.4	3.5	
Salts	1.4–1.6	1.3	0.3	2.3	1.2

Figure 3–2
Structure of water.

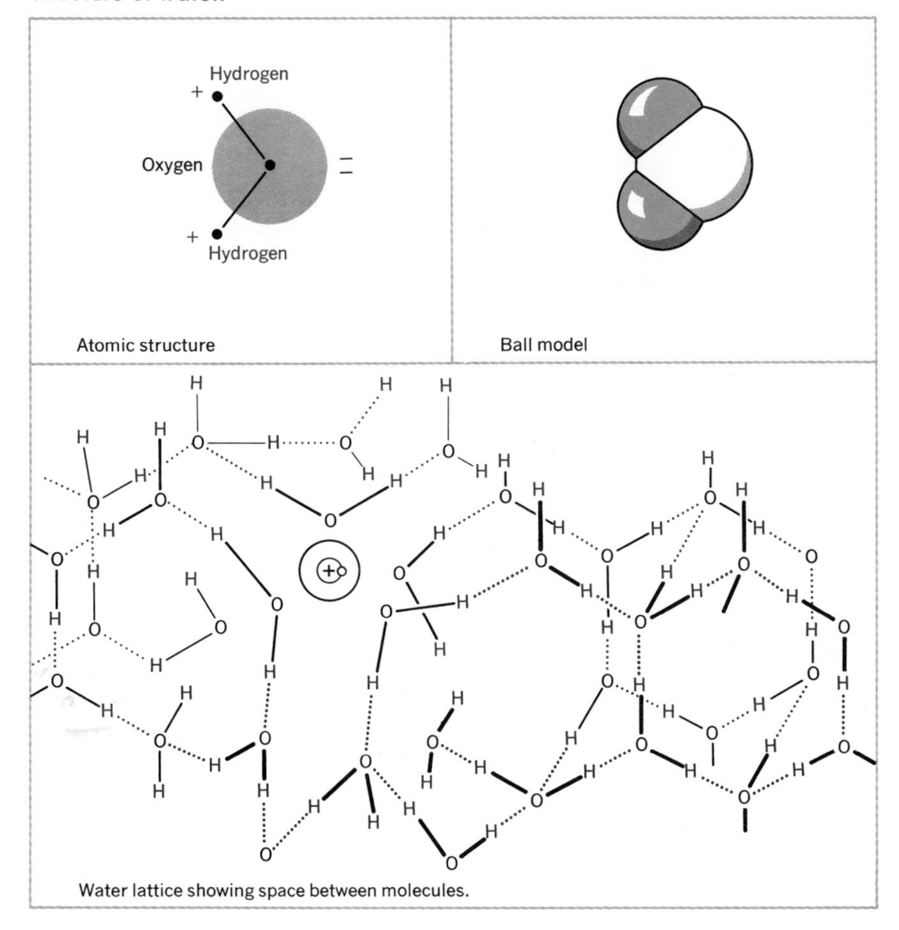

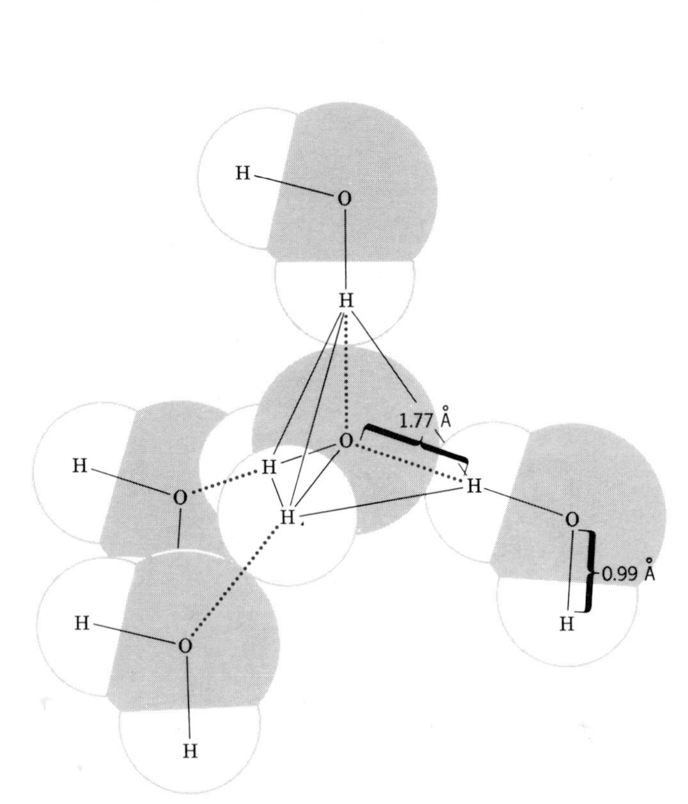

Figure 3–3
Tetrahedral arrangement of hydrogen around oxygen as a result of the formation of two hydrogen bonds between oxygen and hydrogens of adjacent water molecules.

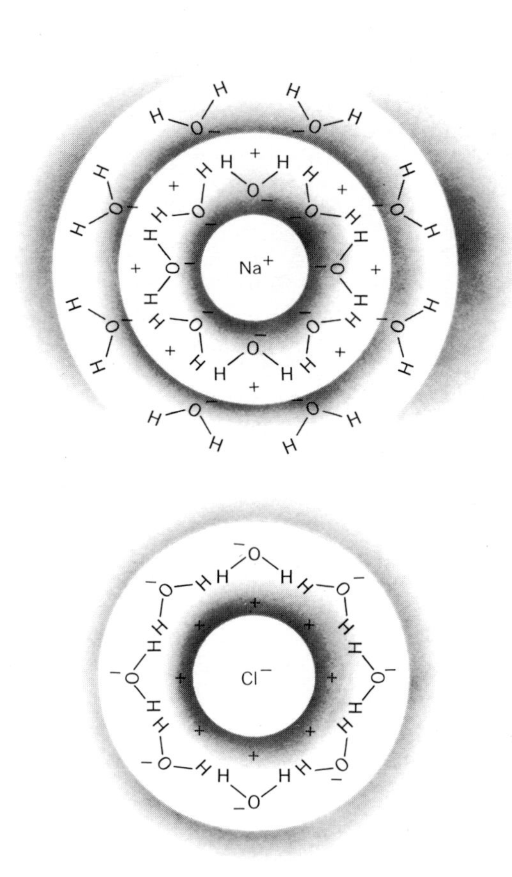

Figure 3–4
Hydration spheres about ions in solution.

Table 3–2
Physical Constants of Several Liquids

Property	Water	Methanol	Acetone	Ammonium Hydroxide
Melting point (°C)	0	−97.8	−95	−77.6
Heat of fusion (cal/g)	80	22.0	19.6	83.9
Boiling point (°C)	100	64.65	56.5	−20
Heat of vaporization (cal/g)	540	263	125	317.6
Specific heat (cal/g/°C)	1.0	0.566	0.506	0.999 (50%)
Surface tension (dynes/cm)	72.8	22.6	23.7	73.5

The water has thus ionized, forming a hydroxide anion (OH^-) and a hydronium cation (H_30^+). At standard temperature and pressure, pure water dissociates to the extent of 1.0×10^{-7} moles per liter. Both the hydronium ions and the hydroxide ions will form hydrogen bonds with other water molecules. The notation H_30^+ is rarely used; instead H^+, a hydrogen ion or proton designation, is more frequently used, even though free protons do not generally exist in aqueous solutions but are bonded to water as hydronium ions.

Salts, Ions, and Gases

Salts are found in all cells and exist in an ionic form. The ions may be free in solution or specifically bound to other molecules such as proteins or lipids. The salts have two broad roles in the cell. One is osmotic, in that the total concentration of salts affects the flux of water across membranes. The second is the more specific role of individual ions in determining the structure of macromolecules. The specific functions of many ions are known; several examples are outlined in Table 3–3.

Gases from the environment or produced by the cell dissolve in the fluid of the cytoplasm. The most common gases in air are nitrogen (78.03% at sea level), oxygen (20.99%), and carbon dioxide (0.03%). Both nitrogen and oxygen dissolve in water in their molecular form. The solubility of these two gases is low; at 25°C and at 1 atmosphere pressure 2.83 ml of oxygen and 1.43 ml of nitrogen will dissolve in 100 ml of distilled water.

Carbon dioxide behaves differently. Some of the dissolved carbon dioxide reacts with water, forming carbonic acid, which in turn ionizes:

$$CO_2 + H_2O \rightleftharpoons \underset{\text{carbonic acid}}{H_2CO_3} \rightleftharpoons H^+ + \underset{\text{bicarbonate}}{HCO_3^-}$$

Carbon dioxide in the cell is usually in the bicarbonate or carbonate form. The gas is highly soluble in water; 75.9 ml of the gas will dissolve in 100 ml of distilled water at 25°C and at 1 atmosphere pressure.

Acids, Bases, and Buffers

Acids can be defined as *proton donors* or *potential electron-pair acceptors*. A **base** is a *proton acceptor* or a *potential*

Table 3–3
Functions and Molecular Associations of Selected Ions in Cells

Ion	Function/Association
Phosphates ($PO_4^{=}$, $HPO_4^{=}H_2PO_4^-$)	Natural buffer, component of nucleic acids, structural proteins, and lipids
Carbonates ($CO_3^{=}$, HCO_3^-)	Natural buffer
$Fe^{++(+++)}$	Hemoglobin (binding of oxygen), chlorophyll, cytochromes, peroxidases, histidine decarboxylase
Cu^{++}	Tyrosinase, ascorbic acid oxidase
Zn^{++}	Carbonic anhydrase, peptidase
Mg^{++}	Phosphatases, ATP enzymes
Mn^{++}	Peptidases
Co^{++}	Peptidase
K^+	ATP-pyruvate transphosphorylase
Ca^{++}	Actomyosin, malt amylase
B	Arabinose isomerase
NO_3^-	Peroxidase
Mo	Nitrate reductase

electron-pair donor. For example, when dissolved in water, acetic acid ionizes:

$$\underset{\text{(acetic acid)}}{CH_3COOH} + H_2O \rightleftharpoons \underset{\text{(acetate anion)}}{CH_3COO^-} + H_3O^+$$

The proton of acetic acid has been donated to water. In the reverse reaction, the acetate anion is a potential proton acceptor and, as such, acts as a base. Acetic acid and acetate may be referred to as a *conjugate acid-base pair*. Protons are attracted to both the conjugate base and water. An acid that readily gives up protons to water is called a *strong acid*; an acid that does not readily give up protons to water is termed a *weak acid*. Each acid may be characterized by its tendency to dissociate or by its *dissociation constant*, K, which can be defined by

$$K = \frac{[H^+][A^-]}{[HA]} \qquad (3\text{–}1)$$

where [HA] is the concentration of undissociated acid, $[A^-]$ is the concentration of conjugate base, and $[H^+]$ is the hydrogen ion concentration. The dissociation constant is affected by temperature as well as by concentration and ionic strength. Customarily in physiology and biochemistry, an *apparent dissociation constant, K′*, is used that is based on measured concentrations of reactants and products. Table 3–4 lists some common apparent dissociation constants.

The concentration of H_3O^+ (H^+ or protons) in solution in cells varies widely. The range may exceed 0.1 to 10^{-10} M. Although the concentrations are not great, even small changes in the concentration over this range may produce noticeable effects in cell function. Rather than use awkward notations such as 1.0×10^{-7} M for concentration, the pH scale proposed by Sørensen simplifies the system and is used universally.

$$\text{pH} = -\log_{10}[H^+] \qquad (3\text{–}2)$$

At a hydrogen ion concentration of 1.0×10^{-7} M at 25°C, there are equal amounts of OH^- and H^+ present, and the solution is at neutrality; that is,

$$\text{pH} = -\log_{10}[1.0 \times 10^{-7}] = 7.0 \qquad (3\text{–}3)$$

Table 3–5 shows the usual range of the pH scale. pH measurements are generally made with a special electrode that is immersed in the solution. An electrical potential develops between this so-called *glass electrode* and a *reference electrode*. When properly calibrated, the potential can be used to directly indicate the pH in a solution.

A buffer solution is a mixture of a weak acid and its conjugate base. These mixtures are effective in slowing the rate of change in pH over a limited range when the buffer is *titrated* with a strong acid or base. As is shown in Figure 3–5, titration of H_3PO_4 with NaOH causes an initial sharp rise in pH with the addition of over 1.0 ml of base; in contrast, the addition of the next 8 ml produces a gradual increase to pH 2.5. This "plateau" was caused by the conjugate acid-base buffer system of H_3PO_4 and NaH_2PO_4. (The NaOH reacted with the H_3PO_4, forming NaH_2PO_4 and H_2O.)

$$\underline{Na^+ + OH^-} + \underline{H^+ + H_2PO_4^-} \longrightarrow \underline{Na^+ + H_2PO_4^-} + H_2O$$

Following another sharp rise in the titration curve, a second buffering "plateau" is produced by the conjugate acid-base buffer system of NaH_2PO_4 and Na_2HPO_4. The third "buffer plateau" is regulated by Na_2HPO_4 and Na_3PO_4.

Just as hydrogen ion concentration can be defined in terms

Table 3–4
Dissociation Constants for Some Common Acids

Acid	K'	pK'
Formic acid	1.78×10^{-4}	3.75
Acetic acid	1.74×10^{-5}	4.76
Propionic acid	1.35×10^{-5}	4.87
Lactic acid	1.38×10^{-4}	3.86
Succinic acid	6.16×10^{-5}	4.21
Phosphoric acid ($PO_4^{=}$)	7.25×10^{-3}	2.14
Monobasic phosphate ($H_2PO_4^-$)	6.31×10^{-8}	7.20
Dibasic phosphate ($HPO_4^{=}$)	3.98×10^{-13}	12.4
Carbonic acid	1.70×10^{-4}	3.77

Table 3–5
The pH Range

H^+ concentration (M)	pH	Examples	Condition
1.0	0		
0.1	1	Stomach(gastric) fluid(humans)	
0.01	2	Orange juice	
0.001	3	Grapefruit juice	Acid
0.0001	4	Pineapple juice	
10^{-5}	5	Tomato juice	
10^{-6}	6	Blood (human = 6.7)	
10^{-7}	7	Most body fluids	Neutral
10^{-8}	8	Sea water	
10^{-9}	9		
10^{-10}	10	Alkaline desert ponds	
10^{-11}	11		Basic
10^{-12}	12		
10^{-13}	13		
10^{-14}	14		

Figure 3–5
Titration curve of H_3PO_4 with NaOH showing buffering plateaus.

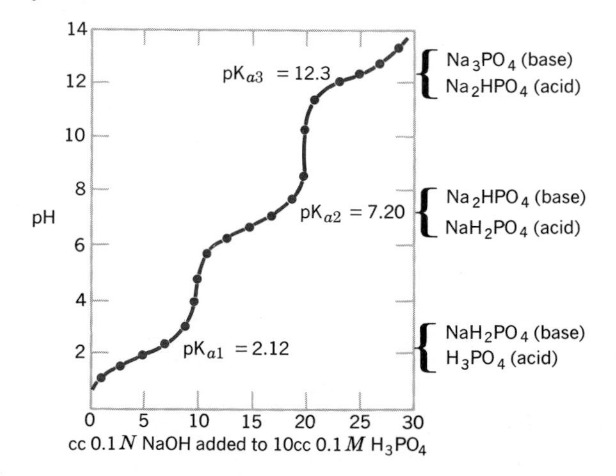

of pH or negative logarithms, the dissociation constant for the weak acid, K_a, can be similarly defined:

$$pK_a = -\log K_a \quad (3\text{–}4)$$

The pK_a for an acid-base pair when plotted on a titration curve occurs at the midpoint of a buffer plateau. The most effective buffering range usually extends 0.5 pH units to each side of the pK_a. The relationship between pH and pK_a is expressed in the Henderson-Hasselbach equation:

$$\text{pH} = pK_a + \log_{10}\frac{[A^-]}{[HA]} \quad (3\text{–}5)$$

where the weak acid is expressed as [HA] and its conjugate base by [A^-]. When the concentrations of these two compounds are equal, as at the midpoint of the titration plateau, the last term of equation 3–5 becomes zero (i.e., $\log_{10} 1 = 0$), and therefore, pH = pK_a. The Henderson-Hasselbach equation can be used to calculate the pH if the concentrations of the acid and base are known. It is also useful for calculating the necessary combinations of acid and conjugate base to mix in order to prepare a buffer with a desired pH. However, it should be pointed out that the equation is an approximation and works best when almost equal proportions of the two compounds are mixed. Several commonly used buffers are shown in Table 3–6.

Chemical Bonds

Atoms bond together to form stable molecules; the association of atoms requires less energy than that required to keep them apart. Energy is required to break the bonds. When a large amount of energy is required to break or form a bond, the bond is called a *strong bond*. If low levels of energy are required, the bond is a *weak* bond. *Covalent bonds* are generally strong bonds. These bonds are formed by the sharing of one or more electron pairs among the atoms comprising the molecule. Hydrogen with its one proton and one electron tends to release energy and bond with other hydrogen atoms so that the two electrons are paired between them. This single covalent bond has an energy of 104 kcal per mole; the double covalent bond between oxygen and carbon, C=0, has 170 kcal per mole. (**Calories** are basic units of energy defined as the amount of heat required to raise the temperature of one gram of water at 15.0°C to 16.0°C. **Kilocalories** are large calories equal to 1000 cal.)

Weak bonds have only a few kilocalories of energy per mole. The more common weak bonds in biological systems are hydrogen bonds (~5 kcal per mole), ionic bonds (~5 kcal per mole), ionic-dipole interactions (~1 kcal per mole), van der Waals forces (~2 kcal per mole), and hydrophobic bonds (not true bonds).

Hydrogen bonds occur by sharing protons (H^+) between electronegative atoms. Oxygen and nitrogen attract electrons very strongly. Even when the electrons are covalently shared, they are attracted more closely to these atoms and therefore form a weak dipole in the molecule. The electronegative atoms of adjacent molecules tend to repel one another; however, protons (H^+) are attracted and held between the electronegative atoms forming this weak bond. Hydrogen bonds most commonly occur between water molecules but also occur between two nitrogens, two oxygens, or between an oxygen and a nitrogen (see also Chapter 4).

Ionic bonds form by the attractive force of opposite charges and approach the strength of covalent bonds. In

Table 3–6
Common Buffer Systems

Compound	pK_{a1}	pK_{a2}	pK_{a3}	pK_{a4}
Acetic acid	4.7			
Ammonium chloride	9.3			
Carbonic acid	6.4	10.3		
Citric acid	3.1	4.7	5.4	
Diethanolamine	8.9			
Ethanolamine	9.5			
Fumaric acid	3.0	4.5		
Glycine	2.3	9.6		
Glycylglycine	3.1	8.1		
Histidine	1.8	6.0	9.2	
Maleic acid	2.0	6.3		
Phosphoric acid	2.1	7.2	12.3	
Pyrophosphoric acid	0.9	2.0	6.7	9.4
Triethanolamine	7.8			
Tris-(hydroxymethyl) amino methane	8.0			
Veronal (sodium diethylbarbiturate)	8.0			
Versene (ethylenediaminotetraacetic acid)	2.0	2.7	6.2	10.3

solution, however, ions form hydration spheres that keep the ions apart. The bond energies are thus weakened by the distance between the ions.

Ionic-dipole interactions were briefly described in the discussion of hydration spheres. Ions tend to associate with that portion of the dipole that is oppositely charged; such bonds are easily broken. Van der Waals forces (interactions) are extremely weak attractions between all uncharged atoms or molecules.

Hydrophobic bonds are associations of molecules or portions of molecules that have nonpolar side groups. They are not chemical bonds in the usual sense. Because such groups are not attracted to water but instead repel water, the attraction of water to itself predominates and causes the nonpolar groups to aggregate and thus minimize the surface contact (i.e., interface) with the surrounding water.

Ligands and Chelates

A special kind of bond may form between certain metals or metal ions and oxygen or nitrogen. The electrons are shared much as in a covalent bond, but in this case, both electrons of the electron pair are donated by the nitrogen (or oxygen) atom (i.e., not one electron from each atom forming the bond). Molecules containing nitrogen or oxygen atoms capable of donating such electron pairs are called **ligands.** A molecule that has two or more ligand atoms is called a **chelate.**

Small divalent and trivalent metal ions are most commonly involved, such as Fe, Cu, Ni, Co, Mn, Mg, and Ca. The metal ion in a chelate may be held between two or more ligand atoms, as is cobalt in the molecule:

$$\begin{array}{lll} & H_2N & \text{---} CH_2 \\ \overset{+3}{Co} & & \quad\;\; | \\ & \underset{H_2}{N} & \text{---} CH_2 \end{array}$$

The metal is thus said to be *sequestered*. The energies of such bonds are of the order of 60 kcal per mole. Water is also capable of forming ligands with metals or metal ions, but a sequestered metal in solution may be more easily removed from its ligand than one might expect from this high bond energy level. Hemoglobin is a good example of an iron chelate (Fig. 4–23). Enzymes may be activated through chelating with a metal (e.g., see cofactors, Table 9–3).

Special Compounds

There are a great number of metabolically important molecules found in small amounts in cells that are not normally classified as macromolecules but whose size cannot be considered small. These are the mono-, di-, and trinucleotides, such as adenosine monophosphate (AMP), adenosine diphosphate (ADP), and adenosine triphosphate (ATP), the pyridine nucleotides such as NAD and NADP, the vitamins, and other growth factors.

The basic structure of the nucleotides is discussed in Chapter 7, and the reader is referred to that chapter for detailed chemical formulas. The adenosine nucleotides (ATP, ADP), as well as the guanosine nucleotides (GTP, GDP) and to a lesser extent the uridine nucleotides (UTP, UDP), are important "energy-rich" compounds formed by the cell during metabolism and used as a source of energy for cell processes such as protein synthesis, carbohydrate synthesis, muscle contraction, and active transport. These compounds in effect transfer energy from energy-yielding (i.e., *exergonic*) reactions to energy-requiring (i.e., *endergonic*) reactions. The structure of ATP reveals why these compounds are good carriers of energy:

$$\text{adenine—ribose—O—}\overset{O^-}{\underset{O}{\overset{|}{\underset{\|}{P}}}}\text{—O—}\overset{O^-}{\underset{O}{\overset{|}{\underset{\|}{P}}}}\text{—O—}\overset{O^-}{\underset{O}{\overset{|}{\underset{\|}{P}}}}\text{—O}^-$$

The molecule has three terminal phosphates bonded in sequence (ADP has two phosphates, and AMP has one phosphate). ATP can have four negatively charged oxygen atoms that strongly repel each other. When ATP hydrolyzes,

$$ATP^{-4} + H_2O \longrightarrow ADP^{-3} + HPO_4^{-}$$

the two resulting anions strongly repel each other, thereby making it exceedingly difficult for the reaction to be reversed and indicating the greater than expected amount of energy associated with bonding. Biologists and biochemists often refer to these phosphate ester linkages as *high-energy phosphate bonds* and denote them by a wiggly line:

adenine—ribose—P ~ P ~ P

Actually, the energy in the bond is not great, but more importantly, it is readily available. A more appropriate term would be *energy-rich* phosphate bond. AMP, while occasionally a product in energy transfer reactions, has a regulatory function in its cyclic form in which the phosphate is covalently bonded to both the third and fifth carbon atoms of ribose. Cyclic AMP (cAMP) has been implicated as a secondary messenger stimulated by an epinephrine in the blood during "flight or fight" stimulation and is believed to induce the breakdown of glycogen to glucose in liver and muscle. There is also evidence linking cAMP with gene expression.

The pyridine nucleotides nicotinamide adenine dinucleotide (NAD; formerly DPN) and nicotinamide adenine dinucleotide phosphate (NADP; formerly TPN) usually function as *coenzymes* in dehydrogenation reactions. These nucleotides accept the hydrogen removed from a compound by the dehydrogenase enzymes. The hydrogens are usually removed in pairs, with the first hydrogen binding to the pyridine nucleotide enzymes as a hydride (Fig. 3–6), while the second is carried as a hydrogen ion.

Vitamins are organic molecules present in small amounts in cells and are utilized by all organisms for essential phases of their metabolism. Lack of any one vitamin may produce characteristic deficiency symptoms. Not all organisms or cells are capable of synthesizing all of their essential vitamins and therefore must have an external supply in their diets. Most of the known vitamins have a role associated with enzyme function. They make up part of the nonprotein portion of an enzyme called the *coenzyme*. Table 3–7 lists many of the more common vitamins, the coenzymes of which they are a part, and the function played by the coenzyme and/or enzyme. The vitamin *niacin* is required in the synthesis of the coenzymes NAD and NADP. The position of this moiety in the coenzyme is shown in Figure 3–7. Two other examples of vitamins important to metabolic enzymes are pantothenic acid and riboflavin (vitamin B_2); their positions in the associated coenzymes are shown in Figure 3–8.

Plants usually produce all their required vitamins, although there has not been clear evidence that the fat-soluble vitamins are required by plants. Vitamin K has been re-

Table 3–7
Vitamins and Coenzymes

Vitamin	Associated Coenzyme	Function of Coenzyme and/or Enzyme
Niacin	Nicotinamide nucleotide coenzymes (NAD^+, $NADP^+$)	Coenzyme is part of dehydrogenases that act in biological oxidations
Riboflavin	Flavin coenzymes (FMN, FAD)	Coenzyme is part of flavoproteins that catalyze oxidation-reductions
Biotin	Biocytin	Coenzyme is part of enzymes associated with carboxylations
Thiamin	Thiamin pyrophosphate	Coenzyme associated with transketolases, some oxidases, and decarboxylases
Vitamin B_6	Pyridoxal phosphate Pyridoxamine phosphate	Coenzymes are part of enzymes involved with amino acid metabolism
Vitamin B_{12}	Coenzyme B_{12}	C-C, C-O, C-N bond cleavage
Pantothenic acid	Coenzyme A	Enzymatic acetylation

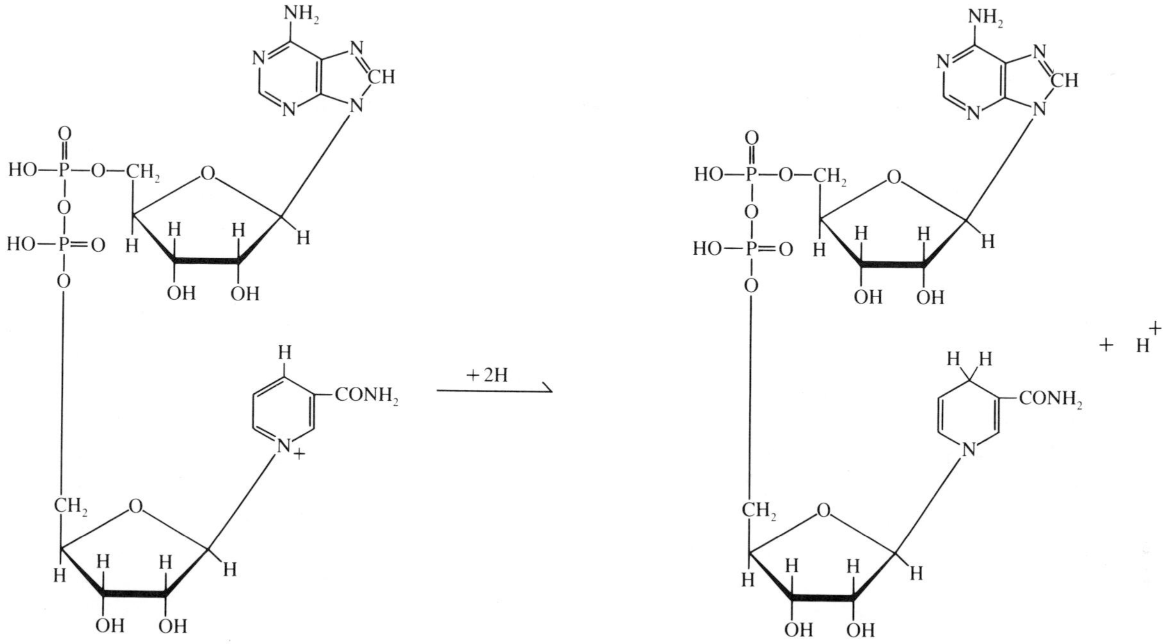

Nicotinamide adenine dinuclecotide (NAD^+)
Diphosphopyridine nucleotide (DPN^+)
or Coenzyme I

Figure 3–6
Reduction of NAD^+ and $NADP^+$

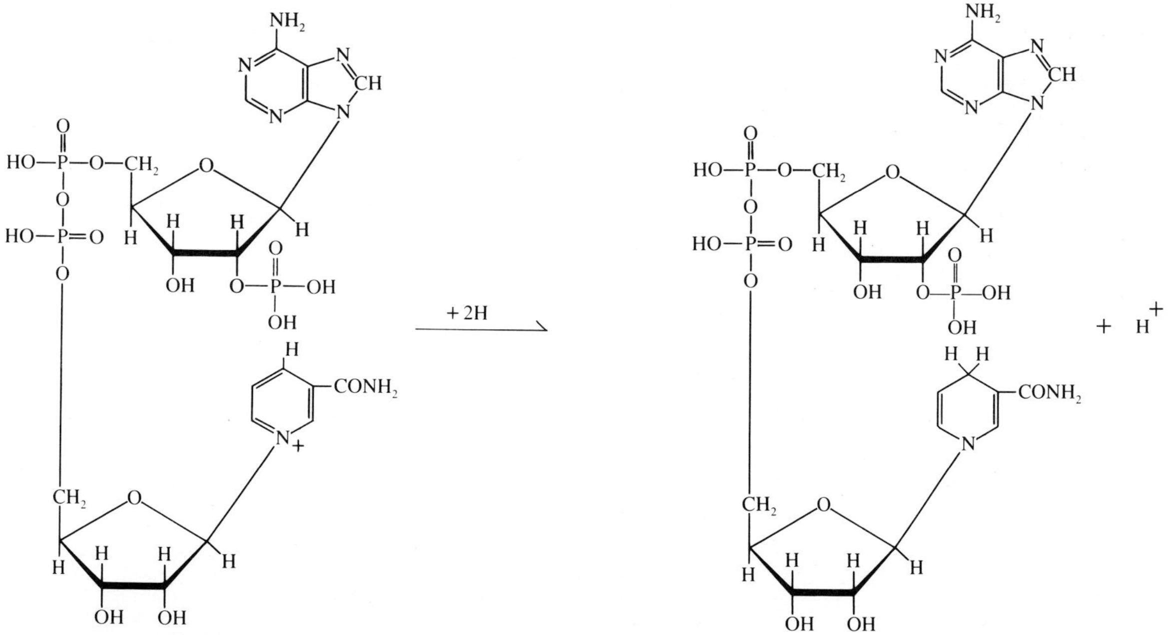

Nicotinamide adenine dinucleotide phosphate ($NADP^+$)
Triphosphopyridine nucleotide (TPN^+)
or Coenzyme II

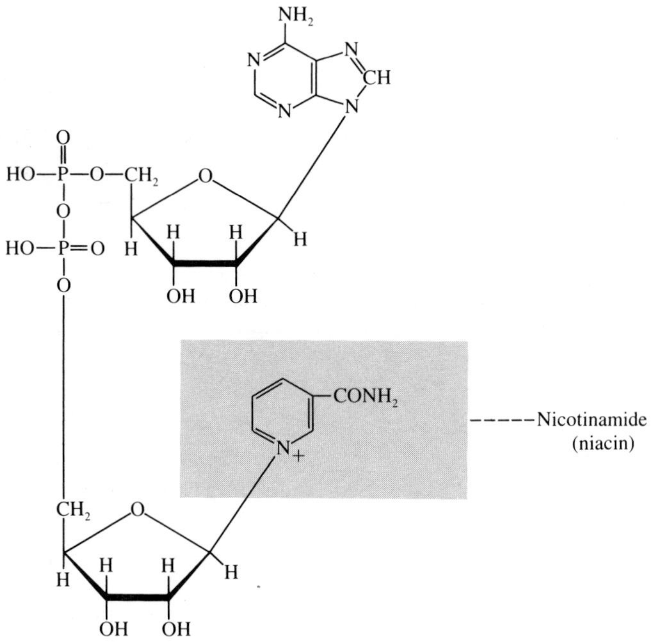

Figure 3–7
Niacin showing position in NADP[v]

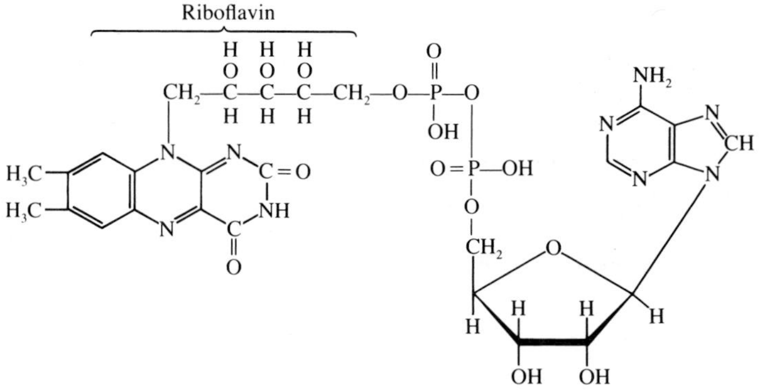

Nicotinamide (niacin)

Riboflavin

Flavin adenine dinucleotide (FAD)

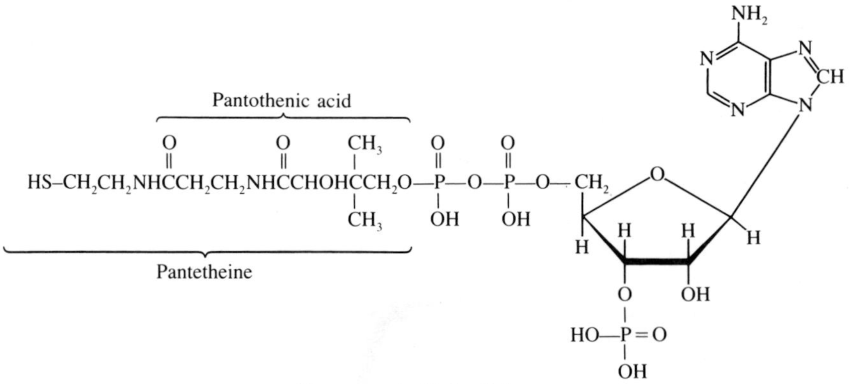

Pantothenic acid

HS–CH_2CH_2NHCCH_2CH_2NHCCHOHCCH_2O—P—O—P—O—CH_2

Pantetheine

Coenzyme A (CoA—SH)

Figure 3–8
Coenzyme A and FAD showing the positions of the associated vitamins.

ported in the chloroplasts of plants and may function in electron transport (see Chapter 17). A plant synthesizes most of the required vitamins in its green tissues. Vitamin requirements in plants have been demonstrated in tissue cultures where portions of root or other tissue are removed and cultivated separately under controlled nutritional conditions.

Summary

Although chemical components may differ from cell to cell, all cells are characterized by the presence of macromolecules of **carbohydrates, lipids, proteins,** and **nucleic acids.** In addition, there are a wide variety of smaller molecules, especially **salts, ions,** and **water,** as well as special molecules present in very small quantities, such as **vitamins** and **hormones.**

The physical and chemical properties of water make it the most suitable solvent for cell systems. The high **melting point, heat of fusion, boiling point, heat of vaporization, specific heat,** and **surface tension** of water tend to maintain the liquid state necessary for life. Because of its polar character, water dissolves ionic and polar compounds and often forms micelles of many large compounds that may be partially polar and nonpolar. The association of water with other molecules (including other water molecules) usually takes the form of **hydrogen bonds.** These bonds involve relatively little energy but are vital in providing a changeable structure to cells.

Salts are normally present in ionic form and frequently associate with macromolecules, where they may stabilize structure and impart **specificity.** Ions are also important in **osmotic** properties of cells and in **buffering** actions.

Some of the more important special molecules that are present in relatively small amounts are the vitamins, hormones, and nucleotides. The vitamins usually act as **coenzymes**, attaching to the protein portion of an enzyme and providing specificity for the complex. Hormones also activate reactions by attachment to cell membranes or other structures. Nucleotides (ATP, UTP, GTP, etc.) are important energy-rich compounds formed and consumed during metabolism. Some nucleotides such as cyclic AMP act as **regulatory compounds,** changing reaction rates or altering gene expression. Other nucleotides like NAD and NADP may function as coenzymes.

The associations of molecules in cells are stabilized by **chemical bonds.** These bonds may be strong (e.g., **covalent**) or weak (e.g., **ionic bonds** and **hydrogen bonds**). In hydrogen bonds, protons are shared between electronegative atoms such as oxygen and nitrogen. Metals associated with oxygen or nitrogen through electron donation are called **ligands.** Two or more ligands in a molecule form a **chelate.**

References and Suggested Reading

Articles and Reviews

Morton, B. A., The vitamin concept. *Vitam. Horm. 32,* 155 (1974).

Bitensky, M. W., and Gorman, R. E., Cellular responses to cyclic AMP. *Prog. Biophys. 26,* 409 (1973).

Books, Monographs and Symposia

Barker, R., *Organic Chemistry of Biological Compounds,* Prentice-Hall, Englewood Cliffs, N.J., 1971.

Cohen, G. N., *Biosynthesis of Small Molecules,* Harper & Row, New York, 1967.

National Academy of Sciences, *Body Composition in Animals and Man,* NAS, Washington D.C., 1968.

Robinson, G. A., Butcher, R.W., and Sutherland, E. W., *Cyclic AMP,* Academic Press, New York, 1971.

Schutte, K. H., *The Biology of Trace Elements,* J. B. Lippincott Co., Philadelphia, 1964.

Segel, I. H., *Biochemical Calculations,* John Wiley & Sons, Inc., New York, 1976.

Chapter 4
THE CELLULAR MACROMOLECULES: PROTEINS

More than 90% of the total cell mass (excluding water) is represented by large molecules called **macromolecules.** The macromolecules vary in size from several hundred to several hundred million molecular weight units. There are four major classes of macromolecules: the *proteins, polysaccharides, lipids* and *nucleic acids*. The relative amounts of these materials in a "typical" cell are given in Table 4–1.

Of all macromolecules found in the cell, the proteins are probably the most chemically and physically diverse. The term **protein** was introduced to the biological and chemical literature in 1838 by the Dutch chemist Gerard Johannes Mulder (1802–1880), who recognized the primary importance of the substance in living matter (the word "protein" is derived from the Greek *proteios,* which means "of the first order or first rank"). Mulder's writings clearly indicate that he recognized the universality of protein in living things; however, he and his contemporaries believed protein to be a single substance; that is, all protein, regardless of its source, was essentially the same. Of course, this is not so, for today we recognize that there are a myriad of chemically different proteins within each cell.

Table 4–1
The Cellular Macromolecules

Substance	Percentage of Total Cell Weight
Water	80–85
Protein	10–20
Polysaccharide	1–5
Lipid	1–2
Nucleic acids	1

The proteins serve a great variety of biological roles but may be divided functionally into two major classes: *structural proteins* and *dynamic proteins*. The *intracellular* structural proteins form the mechanical framework of the cell, such as certain membrane proteins. *Extracellular* structural proteins are found in multicellular organisms and also play a supportive role; included here are proteins such as *collagen* of skin, tendons, and bone, and *keratin* found in epidermis, nails, and hair. The *dynamic proteins* include the *enzymes,* which serve as catalysts in intracellular and extracellular metabolism; also included in this class are certain *hormonal* proteins (insulin, thyroxin, erythropoietin, etc.), certain *blood pigments* (hemoglobin, hemocyanin, etc.), *contractile proteins* (actin, myosin, etc.), and other proteins the role of which is not fundamentally structural.

Proteins are sometimes also classified according to their molecular organization. Accordingly, there are the *fibrous* or threadlike proteins (such as collagen, fibrin, actin, myosin, etc.) and the more compact *globular* proteins (such as hemoglobin, myoglobin, the plasma proteins, and most enzymes). Although it is convenient to classify proteins according to their function or other properties, it should be recognized that all such systems of classification are at best artificial.

The Amino Acids

About 80 years ago, the German biochemist Emil Fischer (1852–1919) showed that proteins consist of chains of smaller units called **amino acids.** There are over 20 different amino acids that occur as regular constituents of proteins and the size (i.e., molecular weight), shape, and function of proteins is determined by the number, type, and distribution of the amino acids present in the molecule. Proteins occur in a wide spectrum of molecular sizes from small molecules such as the hormone *adrenocorticotrophic hormone* (ACTH), which consists of 39 amino acids and has a molecular weight of 4500, to extremely large proteins such as the invertebrate blood pigment *hemocyanin,* which consists of 8200 amino acids and has a molecular weight greater than 900,000 (see Table 4–2 for additional examples).

Amino acids conform to the following general chemical formula:

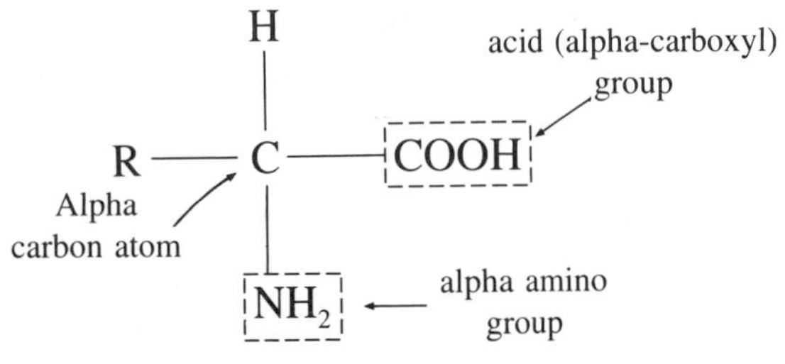

H (top); R—C—COOH; NH_2 (bottom)

acid (alpha-carboxyl) group

Alpha carbon atom

alpha amino group

The alpha carbon atom of each amino acid is covalently bonded to four groups: (1) a hydrogen atom, (2) an amino group, (3) an acid group, and (4) a side chain called an R group. It is the specific chemical nature of the R group that distinguishes one amino acid from another. The amino acids that most frequently occur in proteins are listed in Table 4–3 where they are classified according to the chemical structure of the R group. In contrast, Table 4–4 lists the amino acids according to the functional properties of the R groups—properties that determine their specific contributions to protein structure, as will be discussed later. As seen in the tables, the amidic forms of the amino acids aspartic acid (called asparagine) and glutamic acid (called glutamine) are also found in proteins. In addition to the common amino acids, there are a number of rare amino acids. Beta-alanine (found in the vitamin pantothenic acid), gamma-aminobutyric acid (found in brain tissue), and ornithine, citrulline, and homoserine are amino acids that occur regularly as metabolic intermediates, but these are rarely, if ever, included in proteins.

The four groups attached to the alpha carbon atom of an amino acid are all different except in the case of glycine, where R is a hydrogen atom, and all amino acids except glycine are therefore optically active. The four groups may be considered to lie at the four corners of a regular tetrahedron, with the alpha carbon atom at the center. Accordingly, these groups may be arranged in either of two ways, yielding the *L* and *D* forms (Fig. 4–1). Only the *L* forms of the amino acids have been identified in proteins. (The *D* forms may occur in some antibiotics.)

Table 4–2
Molecular Weight and Amino Acid Content of Some Representative Proteins

Protein	Number of Amino Acids	Molecular Weight
Adrenocorticotrophic hormone	39	4,500
Insulin	51	5,700
Ribonuclease	124	12,000
Cytochrome-C	140	15,600
Horse myoglobin	150	16,000
Trypsin	180	20,000
Hemoglobin	574	64,500
Urease	4,500	473,000
Snail hemocyanin	8,200	910,000

Table 4.3
The Common Amino Acids

Class	Name	R group
Neutral amino acids	Glycine	H—
	Alanine	CH_3—
	Valine	CH_3 CH— CH_3
	Leucine	CH_3 CH—CH_2— CH_3
	Isoleucine	CH_3—CH_2—CH— CH_3
	Serine	HO—CH_2—
	Threonine	OH CH_3—CH—
Acidic amino acids	Aspartic acid	HOOC—CH_2—
	Glutamic acid	HOOC—CH_2—CH_2—
Amidic amino acids	Asparagine	NH_2 O=C—CH_2—
	Glutamine	NH_2 O=C—CH_2—CH_2—
Basic amino acids	Histidine	HC=C—CH_2— HN N CH
	Arginine	NH_2 C—NH—$(CH_2)_3$— ‖ NH
Basic amino acids (cont'd)	Lysine	NH_2—$(CH_2)_4$—
Aromatic amino acids	Phenylalanine	CH—CH HC C—CH_2— CH=CH
	Tyrosine	CH—CH HO-C C—CH_2— HC=CH
	Tryptophan	C— CH—C CH HC C—NH CH=CH
Sulfur-containing amino acids	Cysteine	HS—CH_2—
	Methionine	CH_3—S—$(CH_2)_2$—
Secondary amino acids	Proline	H_2C—CH_2 H_2C CH—COOH NH
	Hydroxyproline	HO—CH—CH_2 H_2C CH—COOH NH

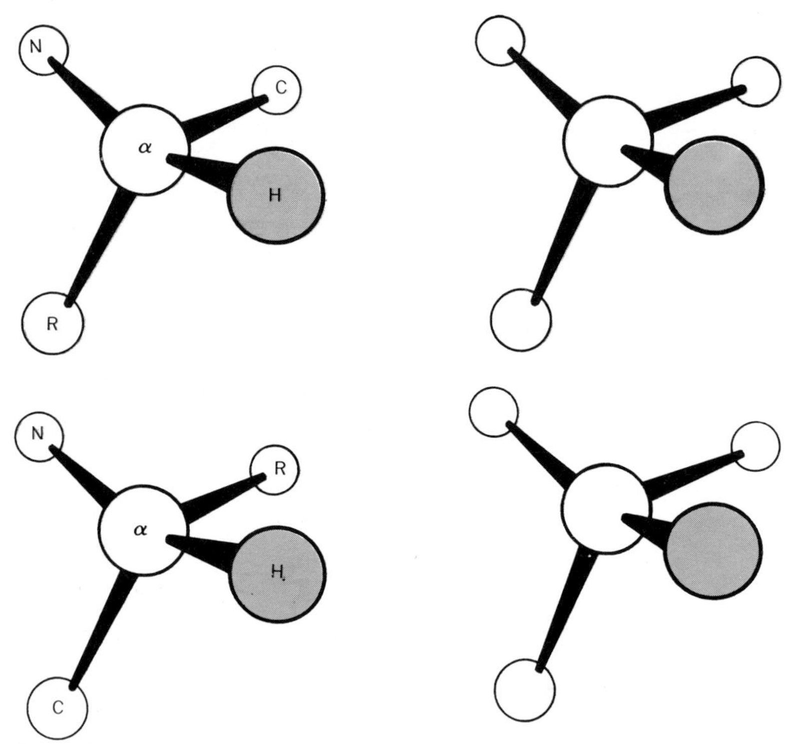

Figure 4–1
Stereoscopic diagrams of the arrangments of groups about the alpha carbon atom of L (*upper stereo pair*) and D (*lower stereo pair*) amino acids. *R*, *side chain; N, amino group; C, carboxyl group; H, hydrogen;* α, alpha carbon.

Table 4-4
Functional Classification of the Amino Acids

Hydrophilic amino acids
Acidic
Aspartic acid
Glutamic acid,
Tyrosine
Neutral
Serine
Threonine
Cysteine
Asparagine
Glutamine
Basic
Arginine
Lysine
Histidine
Hydrophobic amino acids
Glycine
Alanine
Valine
Leucine
Isoleucine
Phenylalanine
Methionine
Tryptophan
Proline

The Peptide Bond

Proteins are composed of one or more chains of amino acids called *polypeptides*. The neighboring amino acids of a polypeptide are covalently linked together by **peptide bonds;** these bonds are formed by the elimination of one molecule of water from the two amino acids sharing the bond, such that the alpha-carboxyl carbon atom of one amino acid is linked to the alpha-amino nitrogen atom of the neighboring amino acid. This is known as a *dehydration synthesis*. As will be seen later, the polymerization of amino acids during protein biosynthesis involves a number of intermediate steps.

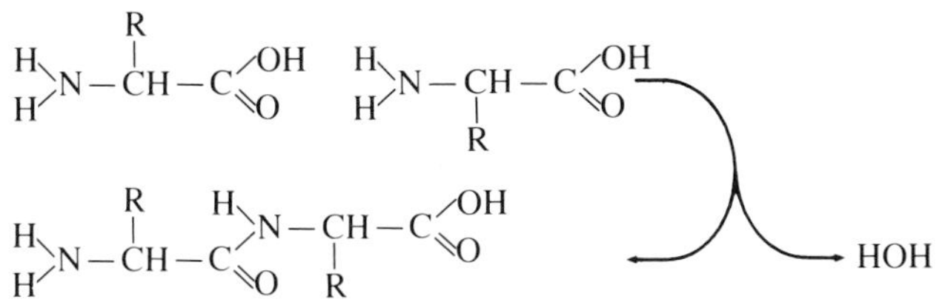

$$H_2N-CH(R)-COOH \quad H_2N-CH(R)-COOH$$

$$H_2N-CH(R)-CO-NH-CH(R)-COOH \quad + \quad HOH$$

The amino and carboxyl groups not directly bonded to the alpha carbon atoms (as in lysine, glutamic acid, aspartic acid, etc.) are not believed to form covalent bonds with neighboring amino acids.

The two atoms involved in the peptide linkage and the four adjacent atoms lie in the same plane in space and are referred to as a **planar group** (Fig. 4–2). Primarily through the pioneering work of L. Pauling and R. B. Corey, the specific interatomic distances and bond angles of the planar group are known; these are shown in Figure 4–3. Although a single bond, the peptide bond has double-bond character; hence, no rotation about this bond normally takes place. In contrast, the double bond between the alpha-carboxyl carbon atom and its oxygen has a single-bond character.

It should be noted that a planar group includes parts of two neighboring amino acids, and the alpha carbon atoms at each end of a planar group are also included as the ends of neighboring planar groups. The bond angle between planar units is 111°; this is the angle formed between the C—C bond of one planar group and the C—N bond of the next group. Note that this is an *intra*-amino acid bond angle! Figure 4–4 shows two successive planar groups arranged in such a manner that they are *coplanar* (i.e., both planar groups lie in the same plane). The alpha carbon atoms are identified simply by α. The important 111° angle formed

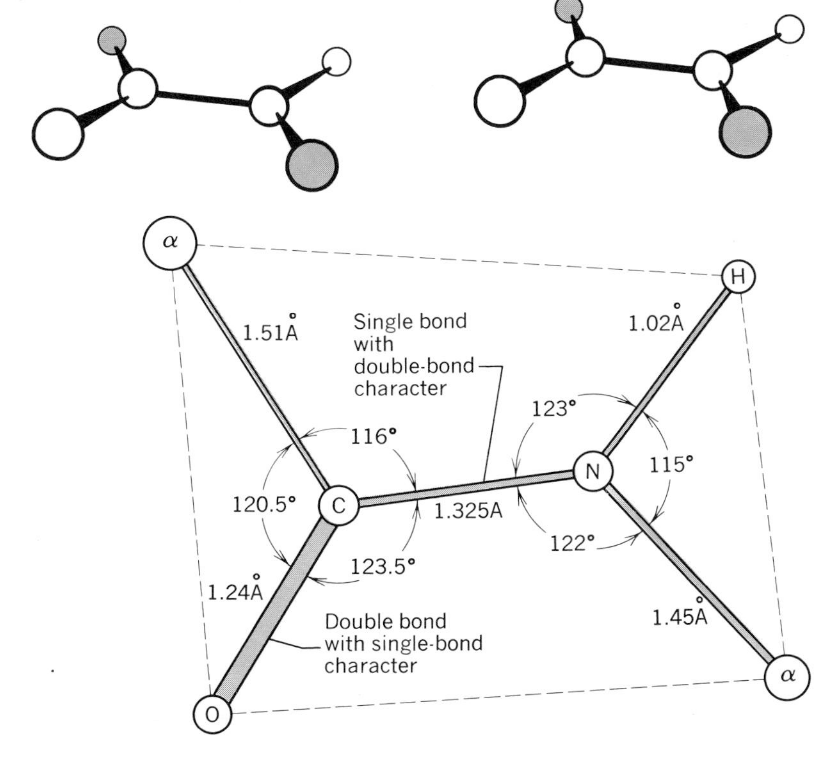

Figure 4–2
Stereoscopic diagrams of the arrangement of atoms in a *planar group.* The alpha carbons are shown in color.

Figure 4–3
Properties of the planar group.

between the C—C bond of one group (the upper group) and the C—N bond of the other (the lower group) is specifically identified. Clockwise rotation (when viewed from the position of the alpha carbon atom) of the upper planar group about the C—C bond sweeps out the angle called ψ (psi). Clockwise rotation (again when viewed from the position of the alpha carbon atom) of the lower planar group about the C—N bond sweeps out the angle termed ϕ (phi). It should be noted that the 111° angle between planar groups remains fixed regardless of the values of ψ and ϕ. The flexibility of a chain of amino acids (i.e., a polypeptide) results from the rotations of successive planar groups about these bonds yielding various values for ψ and ϕ and does *not* result from the linkages *between* amino acids. This point cannot be overemphasized if the structures assumable by polypeptides are to be properly understood. Figure 4–5 is a stereoscopic view showing two successive planar groups that are *not* coplanar.

The total length of two (or more) successive planar groups is maximized when ψ and ϕ are zero. Chain length is reduced as the ψ and ϕ angles are altered between 0° and 360°. If these angles are held constant within this region, the series of planar groups naturally assumes a helical structure. In theory, all kinds of helices can be formed using various values for ψ and ϕ. However, in actuality, most combinations of ψ and ϕ are not possible, because they would result in the overlap of atoms. Much of the data accumulated to date indicate that most polypeptides contain many regions of helical coiling and few regions of complete extension. The relative lengths of extended or helical segments of a polypeptide are determined by the specific sequence of amino acids present in the polypeptide (see below).

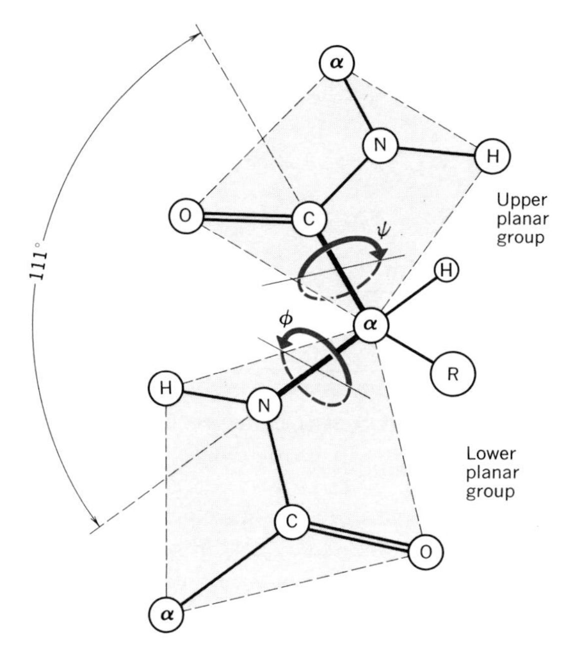

Figure 4–4
Two successive planar groups.

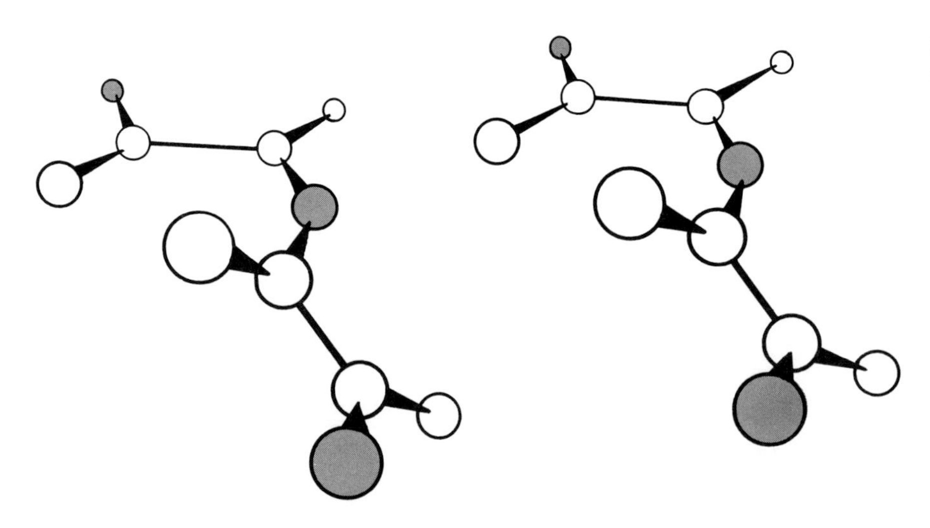
Figure 4–5
Stereoscopic diagram of two successive planar groups that are not coplanar. The alpha carbons are shown in color.

Helical Polypeptides

Fundamental Properties of Helices

Helices may differ from one another in *direction, pitch,* and *diameter* (Fig. 4–6). The **direction** of a helix can be *right-handed* (i.e., turning clockwise as it rotates about its linear axis) or *left-handed* (i.e., turning counterclockwise). The **pitch** is the angle formed between a tangent to the helix and a line drawn normal to the helix's linear axis.

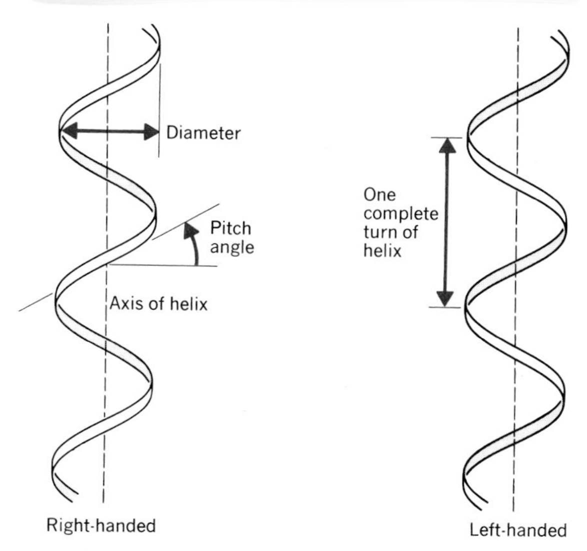

Figure 4–6
Fundamental properties of geometric helices.

The helices of Figure 4–6 are geometric and therefore ideal, but helices formed by polypeptide chains necessarily deviate from the ideal by virtue of the complex arrangements of the constituent atoms. Although rules determining direction remain the same, the diameter and pitch of a helical polypeptide require redefinition. **Diameter** is determined by the *number of amino acids (n) per turn of the helix* (where n is positive for right-handed helices and negative for left-handed helices), and **pitch** is the product of n and the *amount of linear translation achieved per amino acid (d)*. That is, $p = n \times d$. Some simple examples are presented in Figure 4–7.

The Hydrogen Bond

The stability of polypeptide helices results in part from the formation of weak electrostatic bonds called **hydrogen bonds** between amino acids. The attraction of atomic nuclei for orbiting electrons (referred to as *electrophilia*) is not equal for all atoms; it depends on the number of protons within the atomic nucleus. For example, in a carbonyl group, the oxygen atom is more electrophilic than the carbon atom; as a result, the shared electrons forming the double bond spend more time in the vicinity of the oxygen nucleus than the carbon nucleus. Consequently, there is a small negative charge associated with the oxygen atom and a small positive charge associated with the carbon atom.

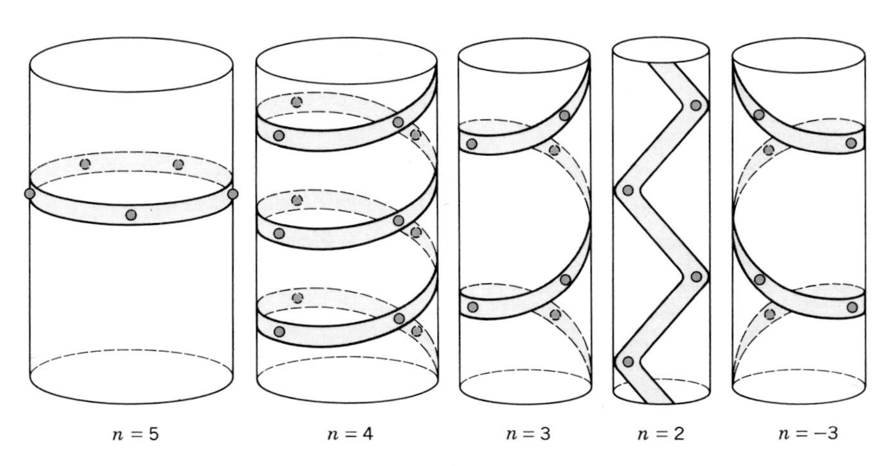

Figure 4–7
Examples of helical polypeptides having various values of *n*.

Similar *partial* charges (represented as δ+ and δ−) occur in amino and hydroxyl groups in which the hydrogen atoms are less electrophilic than nitrogen and oxygen. Weak electrostatic bonds may be formed between carbonyl oxygen atoms and amino hydrogen atoms or hydroxyl hydrogen atoms:

$$>\overset{\delta+}{C}=\overset{\delta-}{O}\ \text{-------------}\ \overset{\delta+}{H}-\overset{\delta-}{N}<$$

$$>\overset{\delta+}{C}=\overset{\delta-}{O}\ \text{-------------}\ \overset{\delta+}{H}-\overset{\delta-}{O}-$$

In certain helices, a number of such bonds are formed between the amino hydrogen atoms and carbonyl oxygen atoms of residues separated by various distances along the polypeptide. Although individual hydrogen bonds are weak, their large numbers in a helix provide a significant stabilizing influence.

The Alpha Helix

Probably the most commonly occurring helical structure in proteins, and the first whose properties were worked out, is the **alpha helix.** Principally through the work of Linus Pauling, it has been shown that this helix has 3.6 amino acids per turn, a pitch of 5.4 Å, and a resulting linear translation of 1.5 Å per amino acid. In the alpha helix, as in all polypeptide helices so far studied, the R groups of the amino acids are directed radially away from the axis of the helix. The helix is stabilized in part by the formation of hydrogen bonds between the carbonyl oxygen of one residue and the amino hydrogen of another—*four residues further along the polypeptide*. The hydrogen bonds bridge the carbonyl oxygens and amino hydrogens at the ends of sequences of 13 covalently bonded atoms in the polypeptide (see Figs. 4–8 and 4–9). The alpha helix is also referred to as the "3.6_{13}" helix, which reveals the value of *n* and the nature of the hydrogen bonding. Alpha helices may be right-handed or left-handed, although right-handed forms are the most common. A stereoscopic view of a portion of a right-handed alpha helix is presented in Figure 4–10.

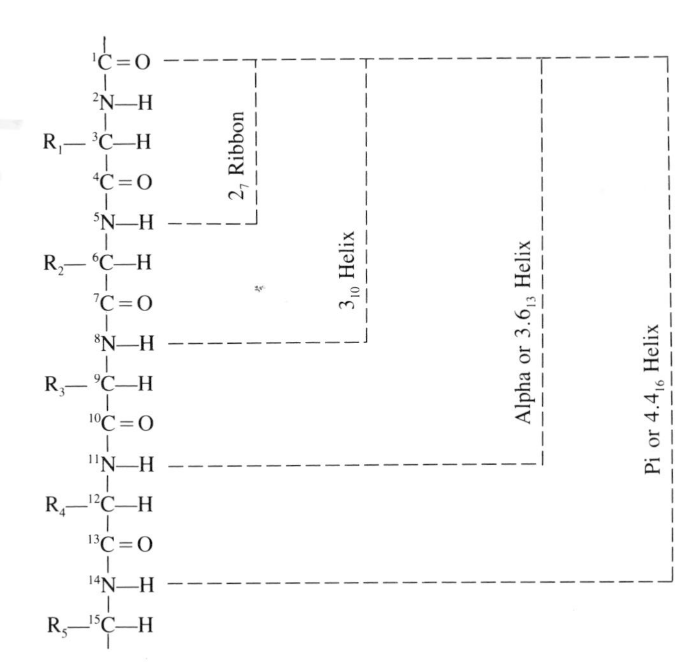

Figure 4–8
Hydrogen bonding patterns of the common polypeptide helices.

Other Polypeptide Helices

Among the other common polypeptide helices found in proteins are the right-handed forms of the *pi* (or 4.4_{16}) helix, the 3_{10} helix, and a helixlike structure called the 2_7 ribbon. Their hydrogen bonding patterns are compared with that of the alpha helix in Figure 4–8 (see also Table 4–5).

Table 4–5
Structural Parameters of the Four Common Right-Handed Helices

Name of Helix	*n*	*p*	*d*	H-bond Atoms
Alpha helix	3.6	5.4 Å	1.5 Å	13
Pi helix	4.4	3.5 Å	0.8 Å	16
3_{10} helix	3.0	6.0 Å	2.0 Å	10
2_7 ribbon	2.0	5.6 Å	2.8 Å	7

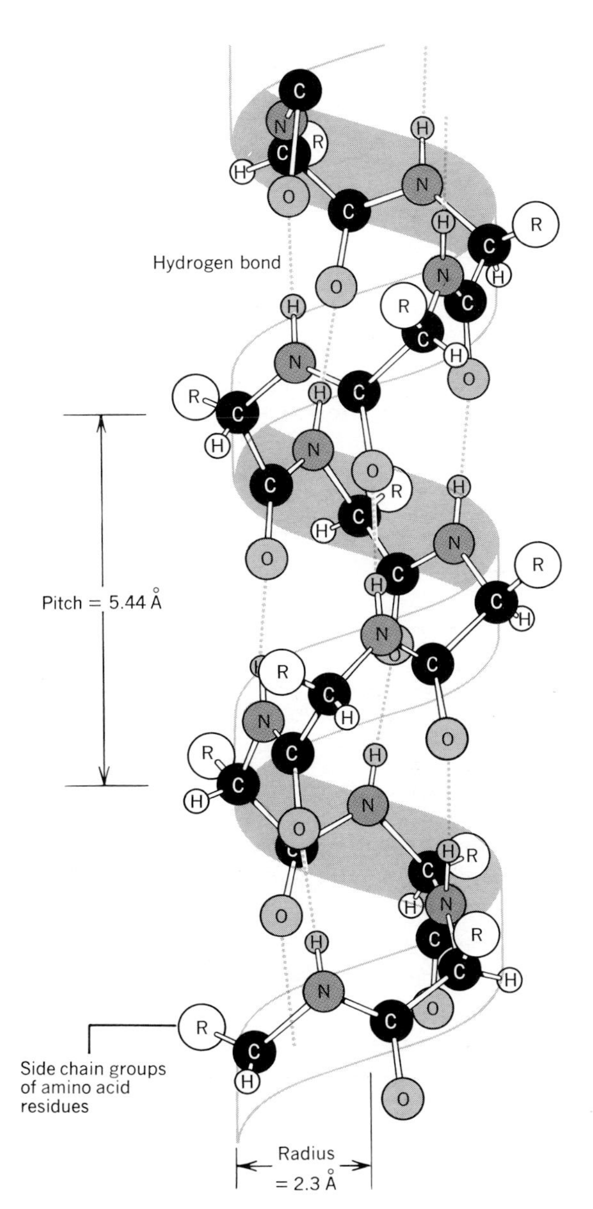

Figure 4–9
Dimensions of the alpha helix.

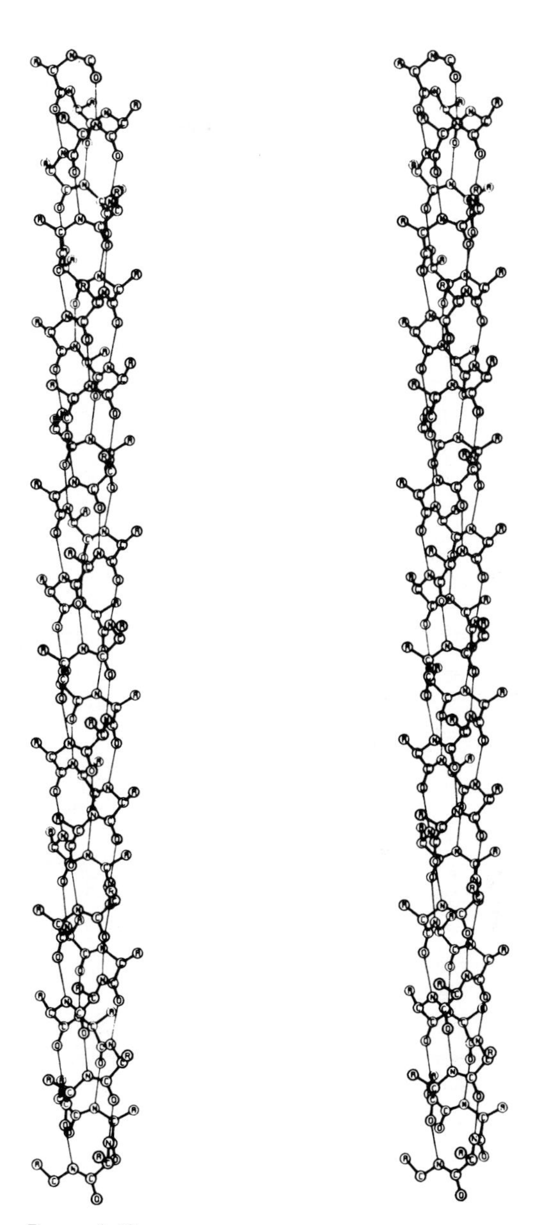

Figure 4–10
Stereoscopic diagrams of an alpha helix. (Courtesy of C. K. Johnson, Oak Ridge National Laboratory.)

Helices of Fibrous Proteins

All the helices described above are characteristic of globular proteins. While fibrous proteins such as keratin, wool, and myosin also contain alpha helices, some additional helices are peculiar to other fibrous proteins. Among these, we will consider those found in collagen and in silk.

Collagen

Collagen is a fibrous protein composed principally of the three amino acids—*glycine, proline,* and *hydroxyproline.* A. Rich, F. H. C. Crick, and G. Ramachandran have shown that a collagen molecule is formed by the intertwining of three parallel polypeptide chains. Each of the three chains is individually twisted to form a left-handed helix, but these are then twisted about one another to form a "super helix" that is *right-handed.* This arrangement is shown in Figure 4–11. The three chains are held together in part by hydrogen bonding *between* separate polypeptide chains (i.e., "interchain hydrogen bonding"). The arrangement of the helices accounts for the unique properties of collagen, namely rigidity (resistance to stretching) and strength.

Silk

Pauling and Corey have shown that silk fibers are composed of a number of parallel, extended polypeptide chains. Neighboring chains are opposite in *polarity* and are held together by interchain hydrogen bonding, as shown in Figure 4–12. Successive planar groups are not quite coplanar, so that the overall structure takes on a "pleated" appearance and is referred to as a *beta pleated sheet.* Since each polypeptide chain is nearly fully extended, silk fibers tend to resist stretching, although lateral flexibility between parallel chains exists.

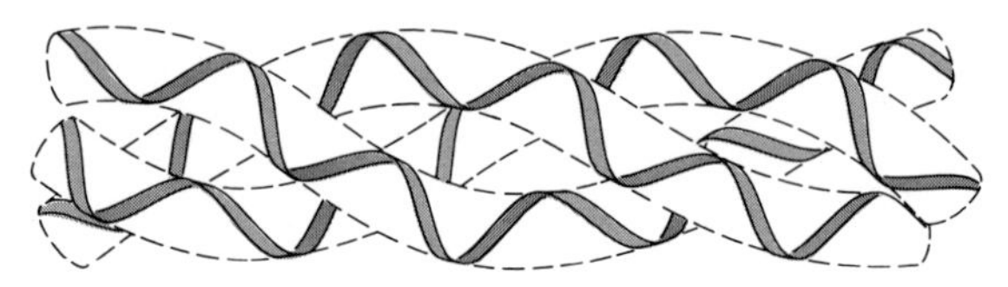

Figure 4–11
Collagen. Three left-handed polypeptide helices (solid) are intertwined to form a right-handed triple superhelix.

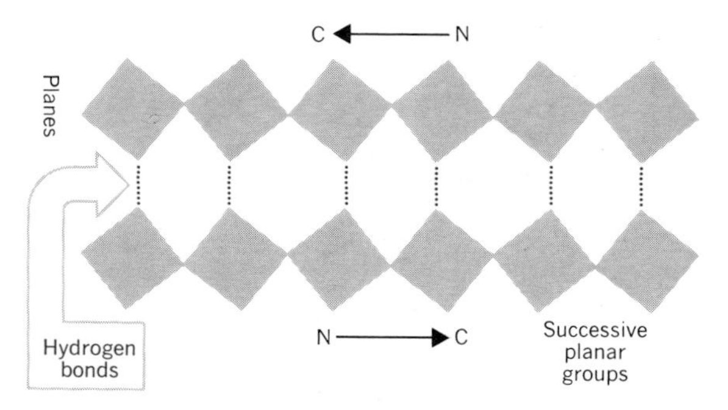

Figure 4–12
Silk. Chains of nearly coplanar planar groups forming a pleated-sheet pattern are linked together through hydrogen bonds.

Levels of Protein Structure

Primary Protein Structure

Four levels of protein structure may be delineated; these are called *primary, secondary, tertiary,* and *quaternary* protein structure. The primary structure of a protein is a specific and sequential delineation of the *covalent association of all of the amino acids in the protein.* So far, we have considered only the covalent peptide bond that links consecutive amino acids of the polypeptide chain. Another covalent bond called the disulfide bridge also occurs in certain proteins. The bond is formed between the sulfur atoms of two cysteine residues by the elimination of the sulfhydryl hydrogen atom of each; that is,

$$\underbrace{\vdash CH_2—SH \quad HS—CH_2 \dashv}_{\text{2 cysteines}} \Rightarrow \underbrace{\vdash CH_2—S—S—CH_2 \dashv}_{\text{cystine}} + 2H$$

Such bonds can occur between cysteine residues of the *same* polypeptide chain or *different* polypeptide chains. All the alpha-amino and alpha-carboxyl groups of the amino acids in a polypeptide chain participate in the formation of peptide bonds except two, which are located at either end of the polypeptide chain. The end of the polypeptide that bears the amino acid with the free carboxyl group is the *C-terminus,* while the end with the free amino group is the *N-terminus*. Primary structure is usually delineated beginning with the N-terminal amino acid of each of the polypeptide chains of the protein (if indeed there is more than one). The number, nature, and positions of any intrachain or interchain covalent bonds are also specified. Some of these relationships are depicted for a hypothetical protein in Figure 4–13.

An internationally recognized set of abbreviations is used for each amino acid to assist in the description of all or part of the primary structure. These abbreviations are shown in Table 4–6 and are used to delineate the primary structure of the hormone adrenocorticotrophin in Table 4–7.

Inherent Variety of Protein Primary Structures. The diversity of amino acids that may be included in proteins provides for an enormous variety of primary structures. Consider, for example, the mathematical variety possible in a polypeptide chain consisting of only 61 amino acids (and this would be considered a relatively small protein). Each of the 61 residue positions can be occupied by any one of 20 different amino acids. Therefore, altogether there would be 20^{61} possible polypeptide molecules (i.e., 20^{61} different primary structures are possible). Now, $20^{61} = 2.3 \times 10^{79}$, and since it has been estimated that the entire universe contains 0.9×10^{79} atoms, there is greater potential variety in a polypeptide chain 61 residues long than there are atoms in the universe!

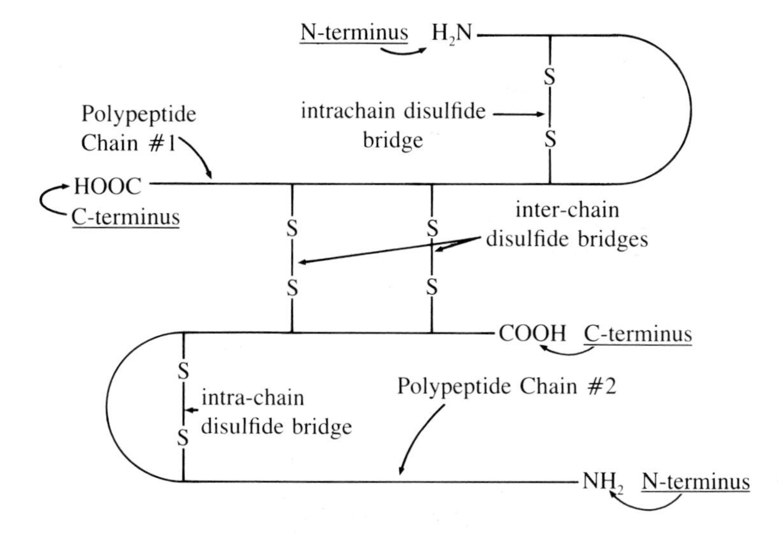

Figure 4–13
A hypothetical protein consisting of two polypeptide chains interlinked by disulfide bridges.

Table 4–6
Standard Abbreviations for the Common Amino Acids

Amino Acid	Abbreviation	Amino Acid	Abbreviation
Alanine	ala	Leucine	leu
Arginine	arg	Lysine	lys
Aspartic acid	asp	Methionine	met
Asparagine	asn or asp-NH_2	Phenylalanine	phen or phe
Cysteine	cys	Proline	pro
Glutamic acid	glu	Serine	ser
Glutamine	gln or glu-NH_2	Threonine	thr
Glycine	gly	Tryptophan	try
Histidine	his	Tyrosine	tyr
Isoleucine	ile	Valine	val

Table 4–7
Primary Structure of Adrenocorticotrophin (BOVINE)

One Polypeptide Chain No Intrachain Covalent Bonds	
Position	Residue
1	ser
2	tyr
3	ser
4	met
5	glu
6	his
7	phe
8	arg
9	try
10	gly
11	lys
12	pro
13	val
14	gly
15	lys
16	lys
17	arg
18	arg
19	pro
20	val
21	lys
22	val
23	tyr
24	pro
25	asp
26	gly
27	glu
28	ala
29	glu
30	asp
31	ser
32	ala
33	gln
34	ala
35	phe
36	pro
37	leu
38	glu
39	phe

Secondary Protein Structure

In the description of a protein's primary structure, the three-dimensional shape or arrangement of the amino acid sequence making up the polypeptide chain (or chains) is not considered. In contrast, the description of a protein's secondary structure identifies (1) the position and extent of those regions of the polypeptide chain (or chains) that are twisted to form helices, (2) the nature or type of helix present, and (3) the position, extent, and nature of any nonhelical regions. For convenience, the various segments of each polypeptide chain are assigned a specific nomenclature. Beginning at the N-terminus, the helical regions are denoted by the letters A, B, C, D, and so on, and the amino acids within each helix are assigned numbers (i.e., C1, C2, C3, etc.). The interhelical regions of each chain are denoted by the letters of the adjoining helices (i.e., nonhelical regions AB, BC, CD, etc.) and the amino acids within these are also assigned numbers (i.e., BC1, BC2, BC3, etc.). The nonhelical region at the amino terminus (if present) is referred to as NA, and the amino acids here are numbered consecutively (i.e., NA1, NA2, etc.). If there is a nonhelical segment at the C-terminus, it is identified on the basis of the last helix. For example, if the chain contains eight helices, the nonhelical region at the C-terminus

Figure 4–14
A hypothetical polypeptide illustrating the nomenclature used to describe secondary structure.

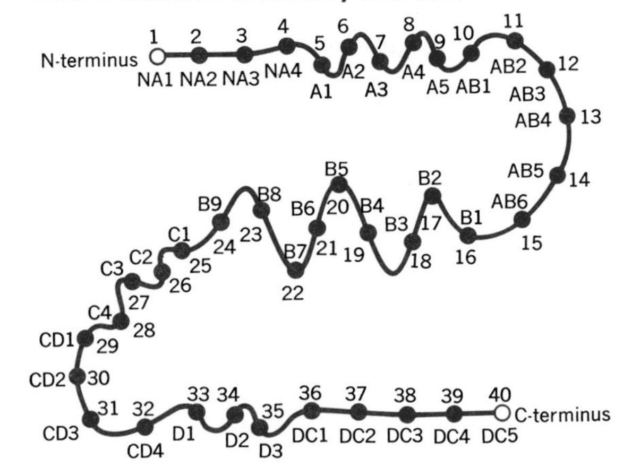

would be identified as HC (and the amino acids numbered HC1, HC2, etc.). Thus, a specific position in any polypeptide chain can readily be referred to. These relationships are shown in Figure 4–14.

From the above discussion, it can be seen that there are two ways to identify a particular amino acid in a polypeptide chain: (1) by its position in the primary structure, and (2) by its position in the secondary structure. For example, in the hypothetical protein illustrated in Figure 4–14, position 3 is also NA3, 8 is also A4, 14 is also AB5, 30 is also CD2, and so on. Two successive helical segments of a polypeptide may lack an intervening nonhelical region. For example, the protein in Figure 4–14 lacks a region BC.

Tertiary Protein Structure

Tertiary protein structure refers to the manner in which the helical and nonhelical regions of a globular polypeptide are *folded back on themselves* to add yet another order of shape to the protein molecule. In globular proteins, it is the nonhelical regions that permit folding. The folding of a polypeptide chain is not random but occurs in a specific fashion, thereby imparting certain specific steric properties to the protein. Specific folding is achieved and maintained by a variety of interactions between one part of the amino acid chain and another, and these interactions include (1) *electrostatic attractions,* (2) *hydrogen bonds,* (3) *hydrophobic bonds,* and (4) *disulfide bridges.*

Electrostatic Attractions.

Most amino acids exist in an ionized (dissociated) form in aqueous solutions. For example, most glycine molecules occur in the following form when dissolved in water:

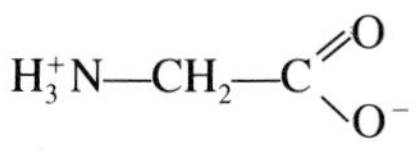

$$H_3^+N{-}CH_2{-}C{\overset{\displaystyle /\!\!/ O}{\underset{\displaystyle \backslash O^-}{}}}$$

In this form, a hydrogen ion has been dissociated from the alpha-carboxyl group, while another has been removed from the water by the alpha-amino group. The resulting ion is called a *zwitterion* because it bears different charges. While glycine carries both a positive and negative charge in this state, it has no *net* charge.

The acidic amino acid aspartic acid exists in the following zwitterionic form:

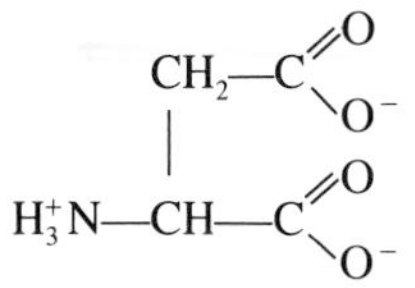

$$\begin{array}{c} CH_2{-}COO^- \\ | \\ H_3^+N{-}CH{-}COO^- \end{array}$$

In this case, aspartic acid bears one positive charge and two negative charges and therefore has a net negative charge (i.e., -1).

Finally, the basic amino acid lysine yields the following zwitterion in solution:

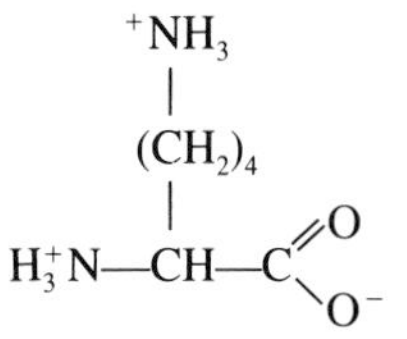

$$\begin{array}{c} ^+NH_3 \\ | \\ (CH_2)_4 \\ | \\ H_3^+N{-}CH{-}COO^- \end{array}$$

In this form, lysine carries two positive charges and one negative charge and has a net positive charge (i.e., $+1$).

In polypeptides, the alpha-amino and alpha-carboxyl groups of all the amino acids except those at the N- and C-terminals are involved in peptide linkages. Therefore, except at the ends of the polypeptide, these groups contribute no charge to the polypeptide. However, the side chains of acidic and basic amino acids (as well as certain others) may contribute positive and negative charges along the length of the polypeptide if either conditions of pH or the nature of other side chains in that region of the tertiary structure allow dissociation or protonation. **Electrostatic attractions** occur between oppositely charged side chains of the amino acids of a polypeptide and may bring these regions of the polypeptide closer together and stabilize their positions relative to one another. The bonds so formed are called *electrostatic bonds* (also *salt bonds* or *salt bridges*). This may be represented as follows:

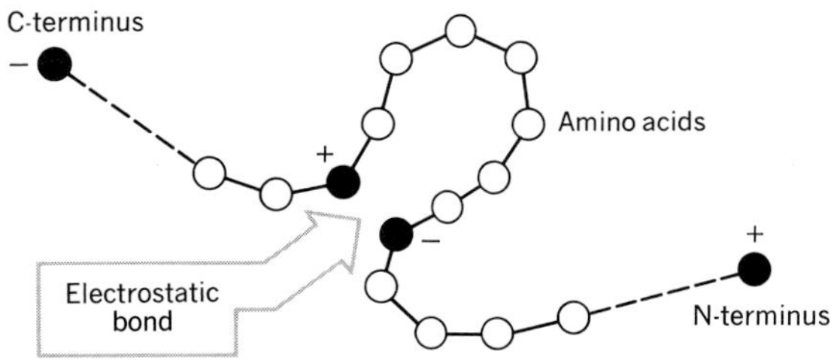

Hydrogen Bonds. Hydrogen bonds formed between alpha-amino hydrogen atoms and alpha-carboxyl oxygen atoms have already been discussed in connection with the stabilization of helices and the parallel chains of the beta pleated sheet. These bonds can also occur between *undissociated* side chains of the acidic amino acids and the *protonated* side chains of basic amino acids or the hydroxyl groups of tyrosine. Although individually weak, these bonds collectively contribute to the stability of a specific tertiary structure.

Hydrophobic Bonds. A third class of interactions that stabilize tertiary structure are hydrophobic bonds. These are attractions between the nonpolar side chains of the neutral, aromatic, and other hydrophobic amino acids. The side chains are drawn together by their mutual hydrophobic properties as they organize themselves in such a manner as to have minimal contact with the surrounding water. Placed in such close proximity to one another, the neighboring atoms of each R group undergo *van der Waals interaction* with each other, resulting in weak bonds. This may be represented as follows:

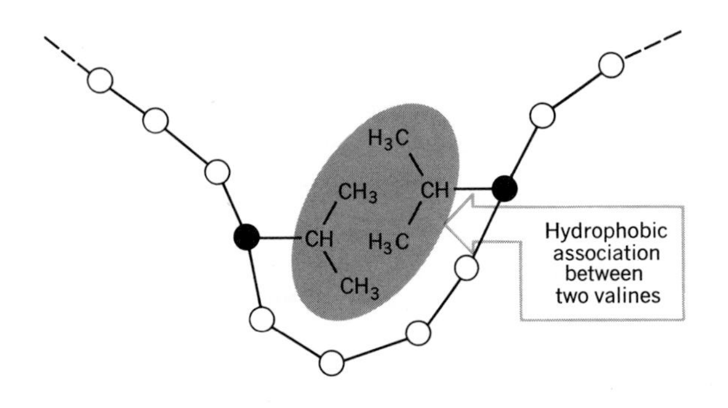

Disulfide Bridges. The strongest bond formed between one part of a polypeptide and another is the *covalent* disulfide bridge. The nature and formation of these bonds have already been discussed in connection with primary protein structure (see above). Such bonds can be formed between cysteine residues in different regions of the same polypeptide or between cysteine residues in different polypeptides. Disulfide bridges are quite common, and many proteins contain several such bonds. The four bonds discussed above are represented together in Figure 4–15.

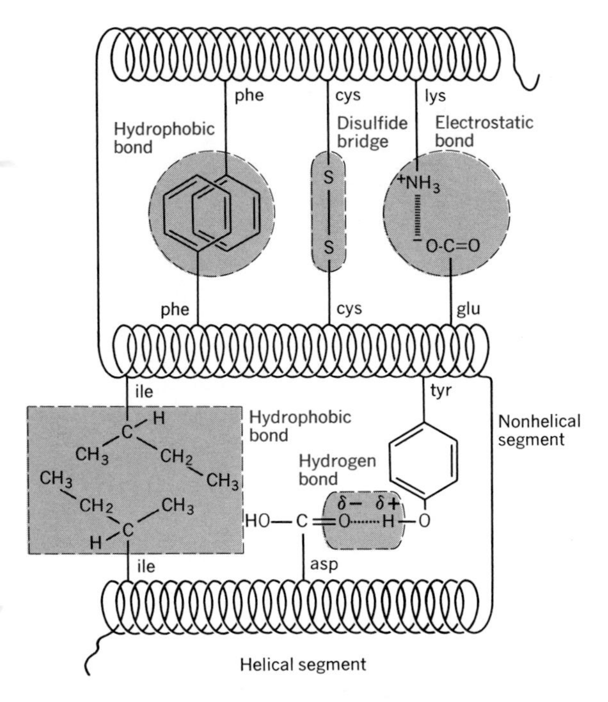

Figure 4–15
A hypothetical polypeptide illustrating bonds involved in the maintenance of tertiary structure.

Quaternary Protein Structure

Many proteins consist of more than one polypeptide chain. In proteins that have two or more polypeptides, the *quaternary structure* refers to the specific orientation of these chains with respect to one another. The individual polypeptides of the protein are usually referred to as **subunits.** Table 4–8 lists several proteins that consist of more than one subunit. As can be seen from this table, proteins may contain either a small number of large subunits (as in thyroglobulin), a large number of small subunits (as in apoferritin), or any intermediate combination. Moreover, some proteins consist of combinations of both large and small subunits.

The same kinds of interactions that contribute to the stability of a specific tertiary structure can also maintain a specific quaternary structure. This is an important point to be appreciated and understood and will be amplified by briefly considering the quaternary organization of three

Table 4–8
Representative Proteins Containing Two or More Subunits

Protein	Molecular Weight	Subunits	
		Number	Mol. Wt.
Hemoglobin (blood pigment)	64,000	4	16,000
Hemerythrin (invertebrate blood pigment)	107,000	8	13,500
Lactic dehydrogenase (enzyme)	150,000	4	37,500
Fumarase (enzyme)	194,000	4	48,500
Apoferritin (iron-storage protein)	480,000	20	24,000
Thyroglobulin (hormone precursor)	669,000	2	335,000
Chlorocruorin (invertebrate blood pigment)	2,750,000	12	250,000

well-studied proteins: *insulin*, *immunoglobulin*, and *hemoglobin*.

Insulin. Insulin is a hormonal protein produced and secreted by the islet cells of the pancreas. Its principal physiological role is in the regulation of sugar metabolism, especially the regulation of sugar transport between the bloodstream and the liver. Insulin is a small protein consisting of two polypeptide chains, *A* and *B*. It was the first protein to have its primary structure sequenced—a scientific feat painstakingly achieved in the 1950s by F. Sanger. A considerable proportion of each of the chains of insulin is arranged in the form of an alpha helix. The two chains are held together by a number of interactions including two disulfide bridges (Fig. 4–16). A third disulfide bridge occurs within the *A* chain. It is to be noted that interactions that maintain a specific tertiary and quaternary structure indirectly contribute to the stability of alpha helical regions by imposing restrictions on the tendency of the helix to unwind.

Figure 4–16
Diagrammatic representation of the insulin molecule.

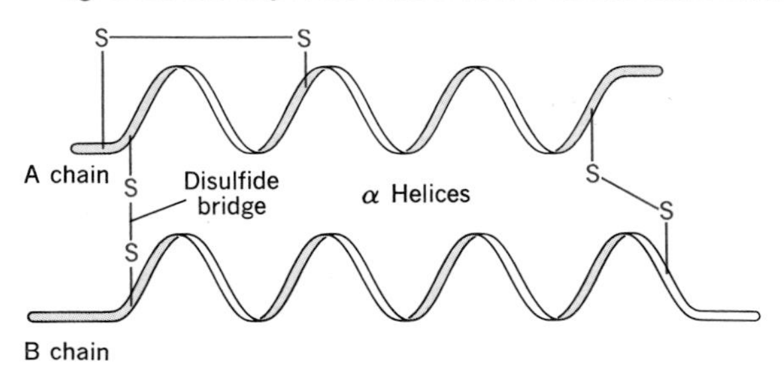

Immunoglobulin. One of the body's most important defense mechanisms against invasion by foreign elements is the rapid production of a class of proteins called **antibodies** by the reticuloendothelial tissues of the spleen and lymph glands. These antibody molecules *complex* with the surface antigens (usually also proteins or polysaccharides) present in certain parasites, bacteria, viruses, and the like, and in so doing, render these foreign invaders harmless. The antigen-antibody reaction is generally referred to as *agglutination* (since the antigen-bearing particles are often clumped together into large masses by the action of the antibodies) and the general phenomenon is called the *immune response*. This body defense mechanism appears to be of rather recent evolutionary origin, since it is characteristic only of vertebrates.

Antibodies are also known as **immunoglobulins** and circulate in the bloodstream as part of the gamma globulin fraction of plasma (see also Chapter 13). Many different immunoglobulins can be produced by the body, each designed to meet the agglutination requirements posed by the invading antigen source.

Although diverse, the immunoglobulin molecules are similarly constructed in that each consists of four polypeptide chains—two identical small *(light)* chains and two identical large *(heavy)* chains. The quaternary structure of the molecule (and the tertiary and secondary structure of the subunits) is maintained by the interchain and intrachain disulfide bridges. Weak noncovalent bonds between chains are also believed to occur, as segments of each chain exist in the beta pleated sheet conformation. The generalized form of an immunoglobulin molecule is represented diagrammatically in Figure 4–17, which also notes the presumed positions of antigen-binding.

Hemoglobin. Hemoglobin is a *conjugated* globular protein (i.e., it contains some nonprotein constituents) consisting of 574 amino acids (human hemoglobin) with a total molecular weight of about 64,500. The protein portion, called **globin,** is made up of four polypeptide chains, each of which is also globular in shape. In normal human hemoglobin, the four globin chains consist of two identical pairs: two *alpha* chains (141 amino acids each) and two *beta* chains (146 amino acids each). The nonprotein portion of hemoglobin consists of four **heme** groups—one associated with each of the four globin chains. The molecule is depicted diagrammatically in Figure 4–18.

Each alpha chain consists of two cysteine residues, and each beta chain has one cysteine. Yet, although disulfide bridges between cysteines play such an important role in the maintenance of secondary, tertiary, and quaternary structure in many proteins, there are no such linkages in hemoglobin. Instead, the tertiary structure of each chain and the quaternary structure of the whole tetramer are maintained by noncovalent linkages—primarily electrostatic and hydrophobic bonds. The analysis of the structure of hemoglobin has preoccupied vast numbers of biochemists in the past 30 years, and as a result, probably more is known about this single protein than any other. It is becoming apparent that the same factors that determine

Figure 4–17
Generalized structure of an immunoglobulin.

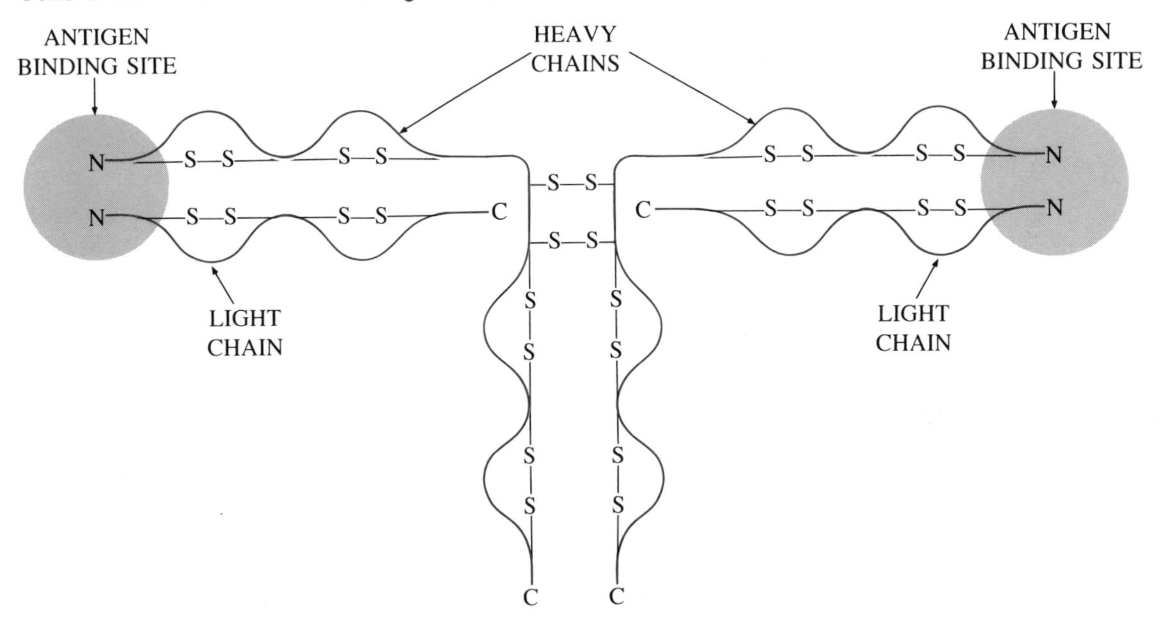

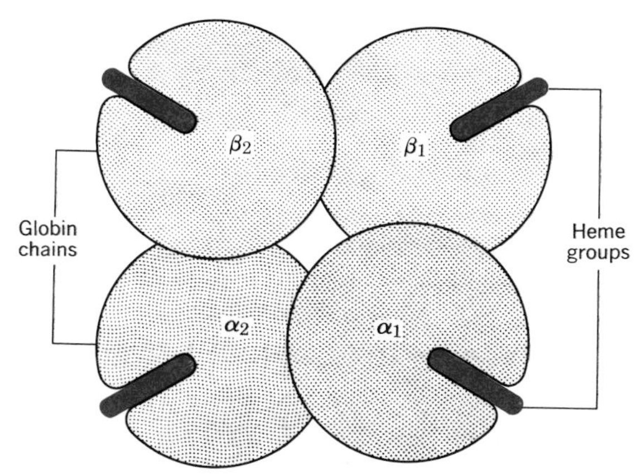

Figure 4–18
Arrangement of globin chains and hemes in hemoglobin.

regulate, and maintain the specific tertiary and quaternary structures of hemoglobin play similar if not identical roles in other proteins. For this reason, as well as for its physiological preeminence, we will consider this protein in considerably greater detail later in this chapter.

Many cellular enzymes are composed of several polypeptide chains, the resulting quaternary structure being of fundamental importance in the regulation of enzyme activity (see Chapter 8). For some of the higher molecular weight enzymes, electron microscopy has provided additional information about quaternary structure, for it is often possible to discern the numbers and orientations of the enzyme subunits in the molecule in negatively stained preparations (Fig. 4–19).

Although so far we have been using the terms primary, secondary, tertiary, and quaternary structure exclusively in connection with proteins, similar levels of organization also exist among other macromolecules, notably DNA and RNA.

Establishment of Secondary, Tertiary, and Quaternary Structure

Although it is reasonably clear just how a specific secondary, tertiary, and quaternary structure may be maintained, how such a specific arrangement is initially achieved is not completely understood. There is no evidence for the existence in cells of templates or special enzymes that function in molding a particular three-dimensional shape. It is possible that polypeptides spontaneously assume a specific structure from among the thousands of possibilites that might exist, especially if the native structure is also that which is most favored thermodynamically. Covalent bonds between atoms of the amino acids and those linking amino acids resist most environmental fluctuations to which they are exposed, and yet polypeptide chains are quite flexible. Thus, while the primary structure is held constant, the polypeptide is free to twist and fold, assuming a variety of shapes until one that is energetically most stable and most favored is established.

It is also possible that coded into the primary structure is the critical amino acid sequence that results in the assumption of the appropriate tertiary structure by the polypeptide as the primary structure is laid down. This notion is examined more fully in Chapter 21, which deals with protein synthesis. For proteins having quaternary structure, it has been suggested that the subunits may seek one another out by virtue of their respective unique tertiary structures and may orient themselves and combine in such a manner as to establish the proper (i.e., biologically active) quarternary structure. Some of the evidence supporting these ideas follows.

The side chains of amino acids such as lysine, glutamic acid, aspartic acid, and so on are often dissociated (or protonated), exposing positive or negative charges; these charges can be coordinated and stabilized by water molecules, which are also partially polar. Consequently, segments of a polypeptide containing such residues tend to become arranged at the surface of the protein, thereby facing the surrounding aqueous environment. The side chains of tyrosine, glutamine, and so forth, which are electrically neutral on the whole, contain atoms of nitrogen, and oxygen, which are partially polarized as a result of the different electrophilic properties of these atoms. Therefore, water is also attracted to these groups, although not to the same degree as fully polarized side chains. These attractions result from the polar nature of the water molecules which act like dipoles. By associating with these electrically charged groups, the water molecules minimize the strength of the electric fields surrounding these groups and stabilize the entire structure.

Amino acids such as leucine and phenylalanine are nonpolar and repel water (i.e., they are hydrophobic). This repulsion results from the fact that nonpolar groups disturb

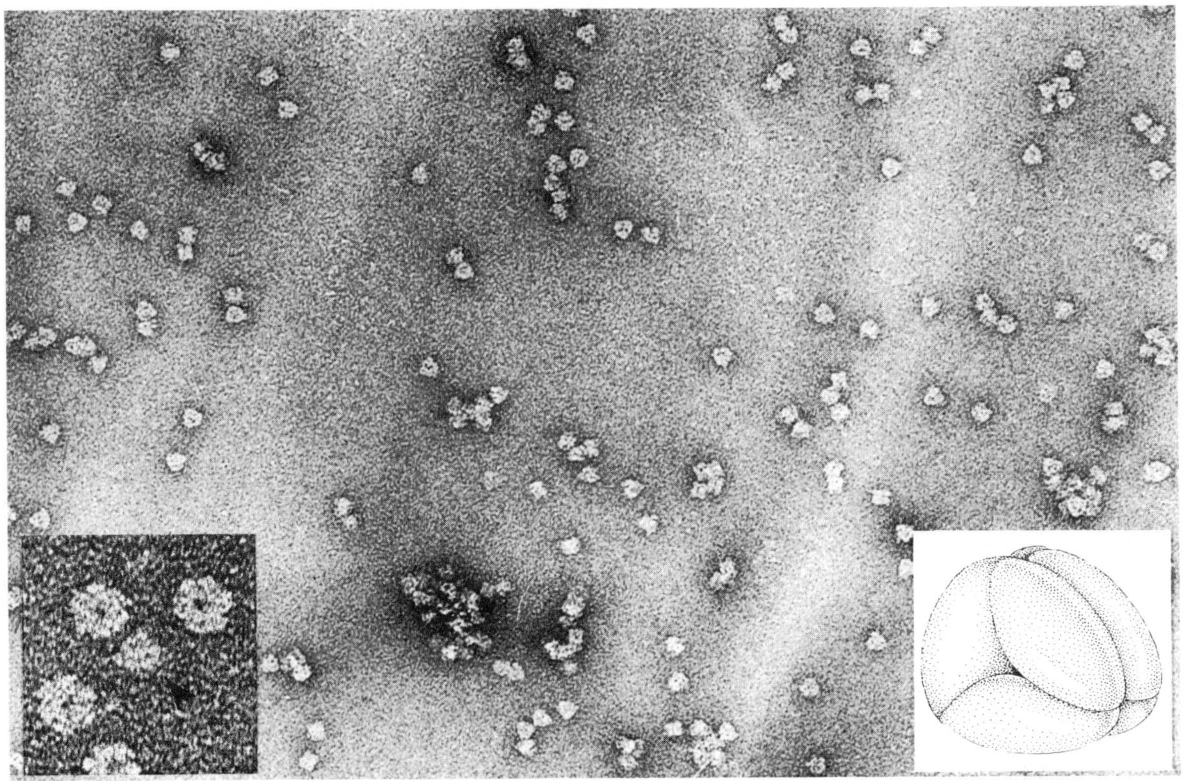

Figure 4–19
Molecules of the enzyme *L-arabinose isomerase* revealed by negative staining. The quaternary structure is formed from two stacks of three subunits each and is seen more clearly in the enlarged insert (the arrow identifies another enzyme, *glutamine synthetase,* present in the preparation) and in the diagram. Magnification 200,000 ×. (Courtesy of Drs. F. A. Eiserling and L. J. Wallace.)

the haphazard arrangements of the water molecules and impart order to the water (as in ice crystals). An increase in order would make the system less stable by reducing its entropy (which is a measure of the disorder of a system; see Chapter 9). Since entropy tends to be maximized, the side chains are turned away from the surrounding water.

Accordingly, in many of the proteins that have been extensively studied so far, amino acid side chains that are polar reside near the surface of the molecule, while the nonpolar side chains are confined to the interior of the molecule or form crevices in its surfaces so as to have minimal contact with the surrounding water.

The concept of spontaneous assumption of tertiary structure by a polypeptide is supported by observations with the protein ribonuclease. Ribonuclease is an enzyme produced in large amounts by the cells of the pancreas; it is then liberated into the small intestine, where it acts to degrade ingested ribonucleic acids. Ribonuclease contains 124 amino acids, has a molecular weight of about 12,000, and consists of a single polypeptide chain. The primary structure of the protein was worked out by C. B. Anfinsen and found to contain four cystine groups (i.e., four sulfide bridges), indicating that it is extensively folded (Fig. 4–20). As is the case with nearly all enzymes, the ability of ribonuclease to carry out its enzymatic activity depends upon the maintenance of the appropriate three-dimensional structure. In the presence of certain solvents, such as concentrated urea or guanidine, the disulfide bridges of the protein are broken, yielding cysteine residues where cystine existed before. The consequent unfolding of the protein can be detected by a loss in its enzymatic activity. The enzyme is said to be "denatured." If the urea or guanidine is removed and the cystine disulfide bridges permitted to re-form by reaction with atmospheric oxygen, it is found that almost all the material is reconverted into normal ribonuclease (see Fig. 4–20). Similar findings have been made with other proteins found capable of spontaneously reestablishing their original tertiary structure after undergoing extensive molecular disorganization. In these proteins, the three-dimensional structure crucial to biological function is directed by the primary structure, and one configuration of the polypeptide is overwhelmingly favored energetically over other possibilites. Polypeptides possess sufficient mobility to explore the various configurations and to select the intrinsically most stable of these. However, it should also be emphasized that in many cases, especially involving large proteins, denaturation is not reversible.

Figure 4–20
Denaturation and renaturation of ribonuclease.

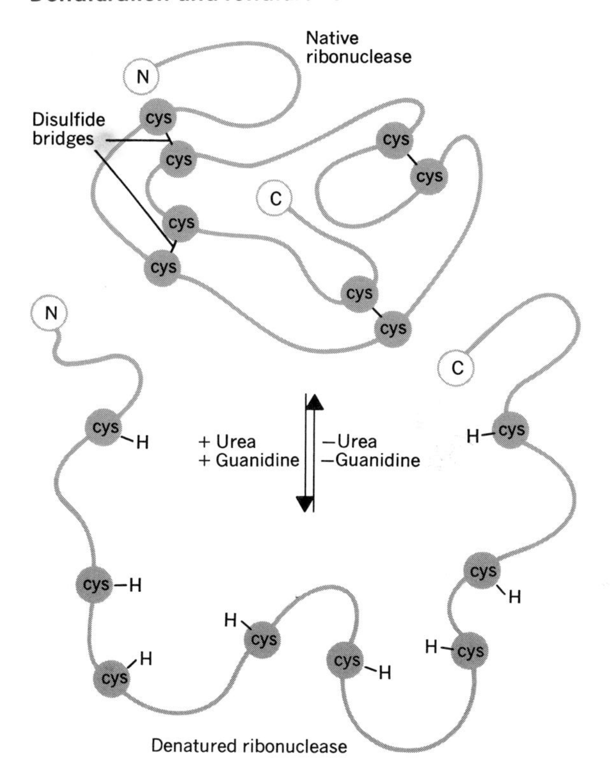

Observations with hemoglobin support the notion of spontaneous assumption of a specific and functional quaternary protein structure. Normal human hemoglobin, called hemoglobin A, may be represented as follows:

$$\alpha_2\,\beta_2$$

This notation indicates that the molecule contains two alpha and two beta chains. When a hemoglobin solution is made weakly acidic or weakly alkaline, the hemoglobin tetramers, undergo a stepwise dissociation into their constituent subunits; that is

$$\underset{\text{(tetramer)}}{\alpha_2\,\beta_2} \rightleftharpoons \underset{\text{(hybrid or asymmetric dimer)}}{2\,\alpha\beta} \rightleftharpoons \underset{\text{(monomers or subunits)}}{2\alpha + 2\beta}$$

As indicated in the above equation, the dissociation of the tetramer into monomers or subunits is reversible; that is, upon restoration of the normal pH, fully functional hemoglobin molecules are sequentially re-formed. This observation supports the idea that intrinsic properties of the individual subunits of a protein may be sufficient to promote the assumption of a prescribed quaternary structure. In the case just cited, it seems that the subunits of hemoglobin can seek one another out in solution and complex to form the biologically functional quaternary structure of the protein.

Anatomy of the Vertebrate Hemoglobins

Our understanding of the structure of hemoglobin has expanded rapidly during the past decade, primarily because information about this macromolecule has been obtained by a number of independent methods of investigation. These include (1) x-ray crystallographic studies providing atomic resolution on the order of only a few angstroms; (2) biochemical studies on amino acid sequences in globin chains and radioactive tracer studies of globin biosynthesis; and (3) electron spin resonance (ESR) studies of the heme groups and their association with the protein. Most notable among those who have contributed to our present understanding of the structure and function of this molecule is

M. F. Perutz, whose work with hemoglobin began more than 40 years ago.

The following discussion of the anatomy of hemoglobin molecules will illustrate a number of general features characterizing protein structure and function. There is, however, a more far-reaching lesson to be learned from the study of the hemoglobins. Hemoglobin, like many other proteins (especially enzymes), is widespread in nature, and the primary structure of the molecule is now known for many different species. Although differences in composition do exist, there are overriding similarities among all hemoglobins. These similarities are not merely chemical curiosities; instead, *they provide us with an insight into the nature of evolution at the molecular level.*

The hemoglobin molecule is highly symmetric. As noted earlier, the molecule can be divided into two identical halves, each consisting of an αβ dimer (called a ''hybrid'' or ''asymmetric'' dimer). The complete tetramer is similar to a flattened sphere, having a maximum diameter of about 60Å. The four polypeptides are arranged in such a manner that unlike chains have numerous stabilizing interactions, whereas like chains have few. The like chains face one another across an axis of symmetry such that rotation of any chain 180° about this axis would make it congruent with its identical partner. A cavity about 25 Å long and varying in width from 5 to 10 Å passes through the molecule along this axis. The molecule is depicted in Figure 4–21.

The interior of the hemoglobin molecule contains mostly nonpolar amino acids such as glycine, alanine, valine, leucine, and phenylalanine, and these form van der Waals contacts with one another. Amino acids whose side chains are ionized at neutral pH are located at the surface of the molecule. Amino acids with nonpolar side chains may be found along a polar segment at the surface; in such cases, the side chains of the amino acids are buried in a crevice in the surface of the molecule and thus produce minimum contact with the surrounding water. Contacts between like subunits are few in number but include electrostatic interactions between the ionized amino and carboxyl ends of the respective polypeptide chains.

Contacts between unlike chains are much more numerous; these contacts result primarily from hydrophobic bonds between nonpolar amino acid side chains, but there are also a few electrostatic bonds and one or two hydrogen bonds. No covalent bonds occur between the four polypeptide chains of hemoglobin.

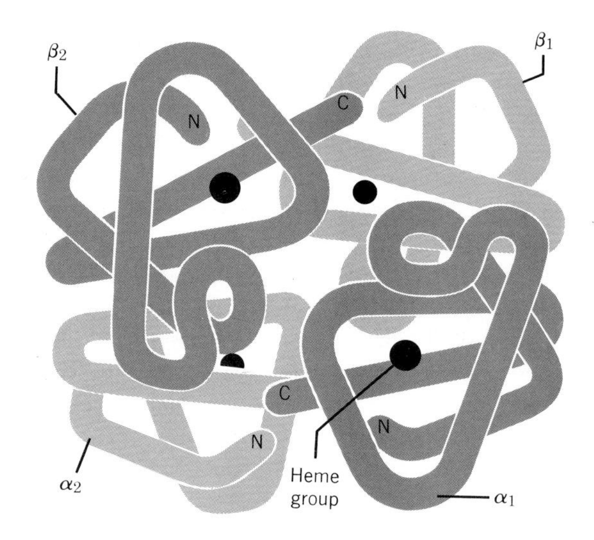

Figure 4–21
The hemoglobin molecule.

Association of Globin Chain with Heme

Each globin chain is made up of eight helical regions (primarily alpha helix with small excursions into pi and 3_{10} configurations) with nonhelical regions between these and at the carboxyl and amino terminals. The secondary structure of a globin chain is depicted in Figure 4–22.

The heme groups (all atoms of which lie in a single plane) are inserted into crevices in the surface of each of the globin chains. Heme (Fig. 4–23) is essentially nonpolar, and its association with globin is partially maintained by its hydrophobic interactions with nonpolar amino acids lining these crevices. Each iron atom can form six bonds, and two of these are coordinated by nitrogen atoms of two histidine residues located on each side of the plane of the heme group (Fig. 4–24). The bond between the ''distal'' histidine (amino acid number 58 (or E7) in alpha chains and number 63 (also E7) in beta chains) is broken when molecular oxygen is bound to hemoglobin (see below). The ''proximal'' histidine (amino acid number 87 (or F8) in alpha chains and number 92 (also F8) in beta chains) is directly bonded to the iron atom. The tertiary structure of the alpha globin chain with its heme group is depicted stereoscopically in Figure 4–25.

In addition to human hemoglobin, hemoglobins from about 30 other species have been completely analyzed, including those of the chimpanzee, gorilla, rabbit, mouse, horse, pig, cow, dog, sheep, camel, goat, llama, chicken, turtle, frog, carp, and lamprey. Several vertebrate *myoglobins* (conjugated oxygen-binding proteins closely related to the hemoglobins and consisting of a single polypeptide chain (also called globin) and heme group), including those of the sperm whale, porpoise, seal, dolphin, and horse, have also been actively studied. In all of these molecules, there is a remarkable similarity in the secondary and tertiary structures of the globin chains. Since the precise primary structure is known for many of these globins, it is possible to examine and analyze the relationships between certain critical amino acid positions and the maintenance of a common shape.

Invariant and "Semi-Invariant" Positions. **Invariant residues are specific amino acids that occur at structurally identical sites in all the hemoglobin (exclusive of *abnormal* human hemoglobins) and myoglobin molecules so far studied. *Semi-invariant* positions are those that are occupied by the same amino acid in *almost* every globin chain. Seven such positions have been identified (Table 4–9 and Fig.

Figure 4–22
Secondary structure of the beta chain of human hemoglobin. The seven invariant and semi-invariant residues discussed in the text are specifically noted.

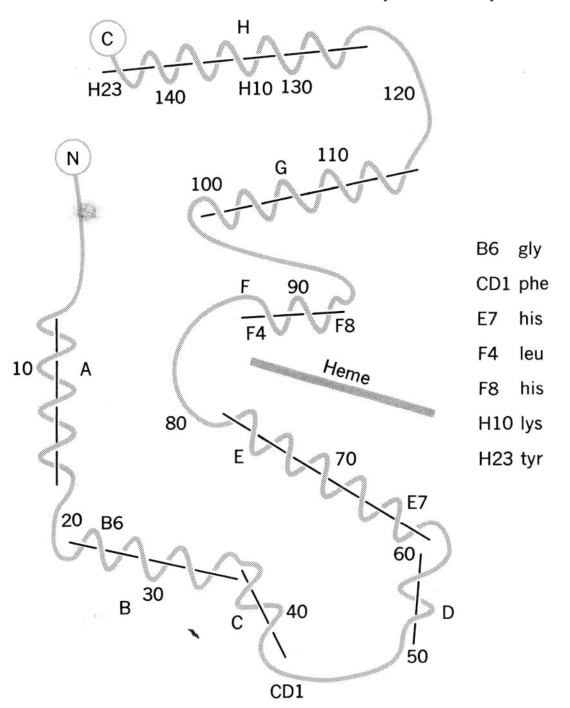

Figure 4–23
Heme (formula above, abbreviated form below).

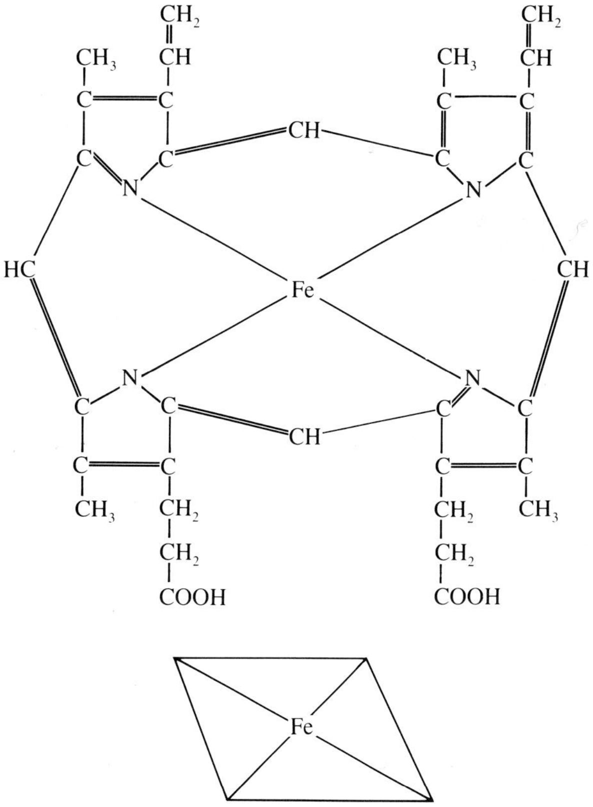

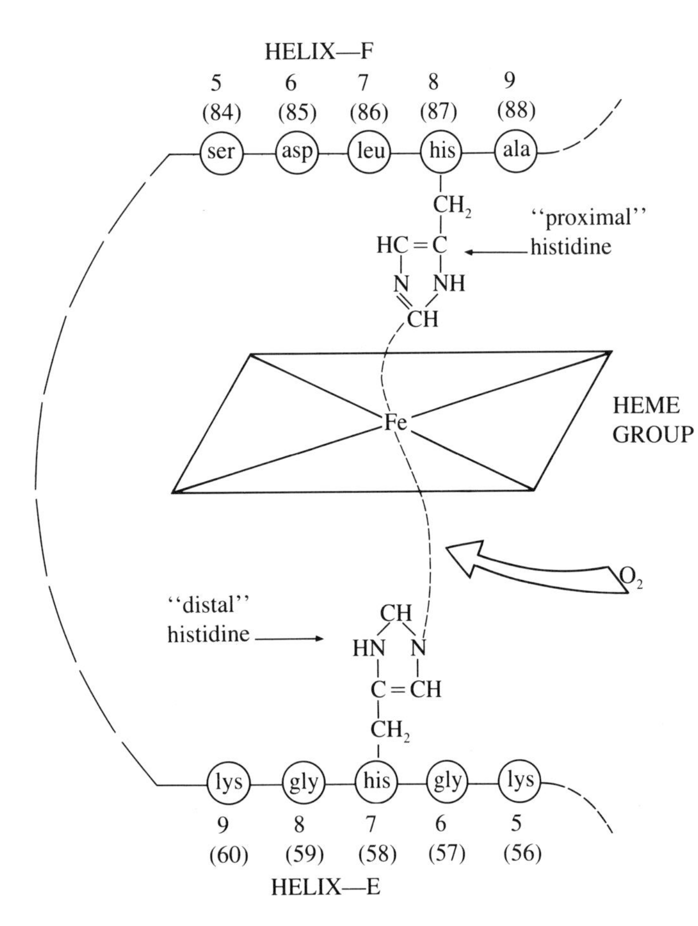

Figure 4–24
Coordination of heme iron by histidine residues in human hemoglobin. The central iron atom of each heme group forms bonds with nitrogen atoms of two histidine residues above and below the plane of the heme group. The bond between the distal histidine (E7) (amino acid 58 in alpha chains and 63 in beta chains) is replaced by molecular oxygen during the reversible oxygenation of hemoglobin. The proximal histidine (F8) (amino acid 87 in alpha chains and 92 in beta chains) is permanently bonded to the iron atom. Only the alpha chain is shown.

Figure 4–25
Stereoscopic pair showing the tertiary structure of the alpha globin chain of hemoglobin. (Courtesy of R. A. Dickerson and I. Geis.)

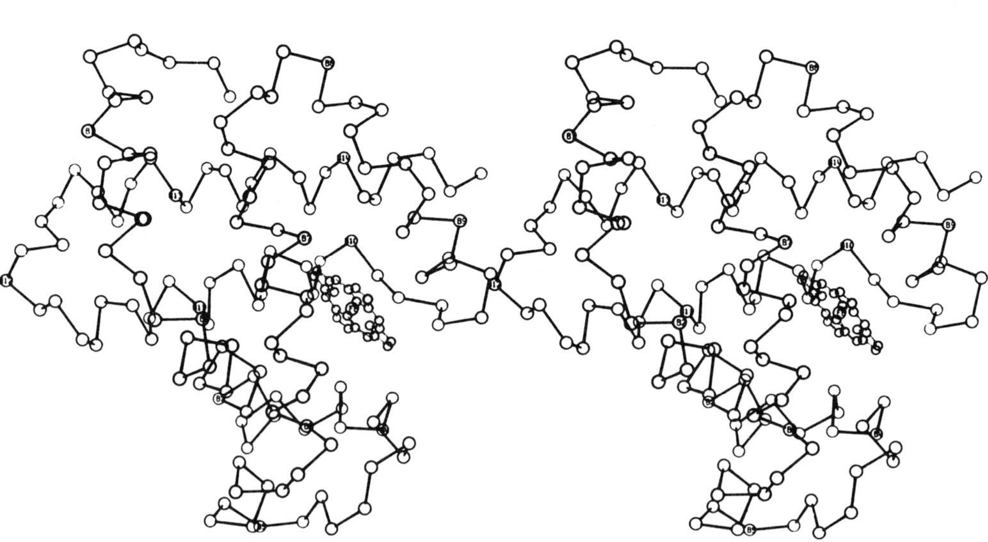

4–22). Seven is believed to be too small a number to alone *impose* a specific secondary and tertiary globin chain structure. However, it is conceded that these residues must play an important part in the *maintenance* of the specific secondary and tertiary structure of the globin chains. For some of these residues the role may readily be explained. For example, amino acids E7 and F8 are the distal and proximal histidine residues that coordinate the iron atom of the heme.

Invariably Ionized Positions. There are five positions in globin that invariably contain ionized amino acids. Three of these are invariably basic amino acids and two are invariably acidic amino acids. The positions containing basic amino acids (i.e., lys or arg) are B12, E5, and H10; the positions containing acidic amino acids (i.e., asp or glu) are A4 and B8. In view of the small number of such sites, invariably ionized positions probably do not play a major role in the maintenance of secondary and tertiary globin chain structure.

Invariably NonPolar Positions. There are 33 positions in the interior of the globin molecule that do not make contact with the surrounding environment; of these, 30 are invariably nonpolar. There are also 10 positions at the surface of the molecule or in crevices in the surface that are invariably nonpolar. Thus, more than one-fourth of the amino acids in globin are invariably nonpolar. The invariably nonpolar positions may be occupied by any of a variety of amino acids, or in some instances the amino acid may be more specific. For example AB1 may be ala, gly, or ser; C5 may be leu, gln, lys, or arg; F3 may be ala, ser, thr, gln, lys, or pro; in contrast, F4 is nearly always leu. It is generally accepted that the invariably nonpolar residues, together with the nonpolar portions of the heme group, may be decisive in determining the configuration of the globin chain. The nonpolar amino acids serve as an *internal molecular skeleton,* helping to maintain the particular secondary and tertiary structure of the polypeptide. The substitution through genetic mutation of a single polar amino acid in a position that is invariably nonpolar usually disrupts the normal organization of the polypeptide.

Function and Action of Hemoglobin: Cooperativity in Proteins

The function of hemoglobin is to reversibly bind molecular oxygen; that is,

$$\underset{\text{unoxygenated hemoglobin}}{\mathrm{Hb}} + \underset{\text{oxygen}}{4\,\mathrm{O_2}} \underset{\text{in tissues}}{\overset{\text{in lungs}}{\rightleftharpoons}} \underset{\text{Oxyhemoglobin}}{\mathrm{Hb(O_2)_4}}$$

The manner in which oxygen is bound by and released from hemoglobin reveals yet another widespread characteristic of protein (especially enzyme) action, namely that of **cooperativity.** Hemoglobin is contained within the red blood cells (erythrocytes) and is oxygenated as the blood circulates through the capillary networks of the lungs. Upon leaving the lungs, virtually every hemoglobin molecule is combined with four molecules of oxygen. In this state, the hemoglobin molecule is said to be 100% *saturated.* Later, when the blood is circulated to other body tissues, hemoglobin releases its bound oxygen. The amount of oxygen released is determined by the concentration of dissolved oxygen gas (i.e., the *partial pressure*) in the surrounding plasma and body fluid. In muscle, for example, it would not be unusual for the percent saturation of hemoglobin to fall to 40% or lower as the released oxygen diffuses from the erythrocytes into the plasma and then into the muscle tissue. As noted previously, the closely related and structurally similar oxygen binding protein myoglobin (Mb), which acts to temporarily store oxygen in certain muscle tissues, consists of a single polypeptide chain and heme group. Consequently, although hemoglobin can reversibly bind four molecules of oxygen, myoglobin binds only one. Figure 4–26 compares the oxygen association/dissociation curves of hemoglobin and myoglobin.

Referring to Figure 4–26, as the oxygen partial pressure rises between 0 and 20 mm Hg, myoglobin rapidly combines with oxygen and quickly approaches complete saturation. The association/dissociation curve exhibits *hyperbolic* kinetics. The behavior of hemoglobin is considerably different and is much more complex. At low oxygen partial

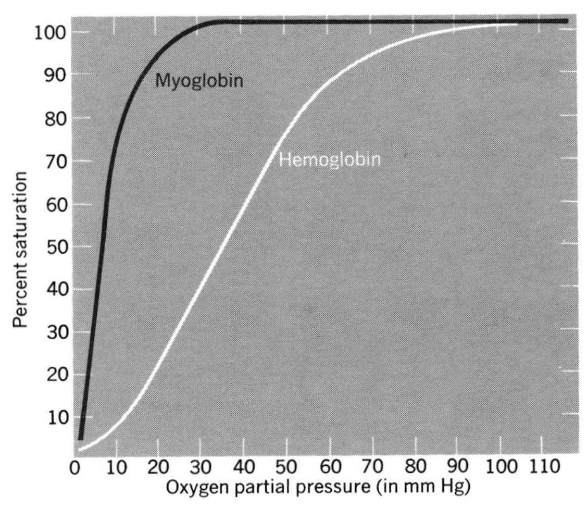

Figure 4–26
Oxygen association/dissociation curves of hemoglobin and myoglobin.

pressure, the affinity of hemoglobin for oxygen is considerably less than that of myoglobin. For example, at 20 mm Hg partial pressure, hemoglobin is only about 21% saturated. However, between 20 and 60 mm Hg oxygen partial pressure, the affinity of hemoglobin for oxygen is greatly increased and approaches saturation. Above 60 mm Hg, hemoglobin binds only small quantities of additional oxygen. The oxygen association/dissociation curve of hemoglobin exhibits *sigmoid* kinetics. From a physiological standpoint, the unique oxygen-binding characteristics of hemoglobin are crucial. The partial pressure of oxygen in the blood leaving the lung capillaries is generally in excess of 100 mm Hg, but by the time the blood reaches the capillaries of the various body tissues, the partial pressure has fallen below 80 mm Hg. An examination of the curve of Figure 4–26 shows that in this interval hemoglobin will have released only a small percentage of its oxygen. However, in the tissue capillaries, where the oxygen partial pressure often falls below 40 mm Hg, hemoglobin releases much of its bound oxygen. In actively exercising muscle, where the partial pressure may drop to 20 mm Hg, still greater quantities of oxygen would be released. Thus, within the range from 60 to 20 mm Hg (a range within which most of the tissues of the body operate), a relatively small decrease in oxygen partial pressure is accompanied by a quantitative release of hemoglobin-bound oxygen.

Cooperativity in Hemoglobin. The complex behavior of hemoglobin, which is precisely what is required of an efficient oxygen-transporting system, may be attributed to its quaternary structure, for when hemoglobin is dissociated into its four subunits, the separate subunits exhibit the oxygen-binding kinetics of myoglobin (i.e., hyperbolic kinetics). In the intact tetramer, the various subunits exhibit *cooperativity*. The binding of the first oxygen molecule to unoxygenated hemoglobin involves one of the alpha subunits and is achieved only slowly. It is followed by a minute configurational change in this alpha subunit (i.e., a change in its tertiary structure), which in turn induces a small change in the neighboring beta subunit (i.e., the other member of the asymmetric dimer). This change facilitates more rapid binding of the second oxygen molecule by this beta subunit. The configurational changes taking place in one asymmetric dimer then induce changes in the other half of the hemoglobin molecule (i.e., the other asymmetric dimer), causing its alpha and beta subunits to bind the third and fourth oxygen molecules even more rapidly. Effects in which the activity of one or more subunits of a protein alter the structure of other subunits in such a way as to modify their physiological behavior are called *cooperative effects*. In the case of hemoglobin, it may now be understood why the oxygen association/dissociation curve is sigmoid and not hyperbolic, for the individual alpha and beta subunits of the tetramer do not function independently in the binding of oxygen but instead influence one another. The importance of cooperative effects among proteins cannot be overstated, for we shall see in Chapter 8 that cooperativity is also manifested by a number of enzymes whose affinities for

Table 4–9
Invariant and Semi-Invariant Globin Chain Positions

Position	Amino Acid
1. Invariant	
CD1	phe
F8	his
2. Semi-invariant	
B6	gly
E7	his
F4	leu
H10	lys
H23	tyr

their substrates can be modulated through subtle changes in individual subunits of the enzyme's (i.e., the protein's) quaternary structure.

Evolution of Proteins

Ontogeny and Phylogeny of Hemoglobin

Functionally similar proteins exist in diverse animal and plant species. Many of these are believed to have a common evolutionary origin. The picture of "protein evolution" is probably clearest in the case of the hemoglobins, where it is possible to make accurate comparisons of hemoglobin structure in a large number of diverse animal species. Moreover, in humans (and in other vertebrates), various hemoglobins occur at different stages of development, and this **ontogeny** also sheds light on the phylogenetic picture. In addition to hemoglobin A, several other normally occurring hemoglobins have been identified in humans; these include hemoglobins A_2, F, Portland,-1, Gower-1, and Gower-2. Hemoglobin A_2 is found in normal adult blood but accounts for only about 2% of the total hemoglobin content (the remainder being hemoglobin A). Hemoglobin A_2 differs from A in that the beta chains are replaced by delta (δ) chains, which differ from the beta chains in their primary structure. Hemoglobin A_2 is denoted $\alpha_2\delta_2$. The blood cells of the maturing human fetus contain a form of hemoglobin that differs from both of those found in adults. It is designated hemoglobin F and is denoted $\alpha_2\gamma_2$, which indicates that there are two alpha chains (identical to those in adult hemoglobin) and two gamma (i.e., γ) chains (which differ in their primary structure from the beta and delta chains of adult hemoglobin). Hemoglobin Portland-1 is a very early embryonic hemoglobin consisting of two ζ (zeta) and two γ chains ($\zeta_2\gamma_2$). Hemoglobin Gower-1 also is an *early* embryonic hemoglobin and consists of two ζ and two ε (epsilon) chains ($\zeta_2\varepsilon_2$). Hemoglobin Gower-2, which appears somewhat later in embryonic development, contains two alpha chains and two epsilon chains ($\alpha_2\varepsilon_2$). The human hemoglobins and their subunit composition are summarized in Table 4–10.

The various human hemoglobin chains make a differential temporal appearance in the blood during embryonic and fetal development and account for varying percentages of the total hemoglobin (Fig. 4–27). The zeta, epsilon, and gamma chains appear first. Alpha chain production, once begun, continues throughout embryonic, fetal, and adult life. The zeta and epsilon chains are replaced by gamma chains during the early stages of fetal development. During the latter part of fetal development, gamma chain synthesis declines as beta chain synthesis increases, and as a result, hemoglobin A slowly replaces hemoglobin F. Delta chain synthesis occurs at a slow rate throughout late fetal life and adult life, accounting for the small quantity of hemoglobin A_2.

A consistent feature of hemoglobins G_2, F, A, and A_2 is that two members of each tetramer are always alpha chains. A similar situation exists in many other animals in which multiple hemoglobin forms have been identified. For example, the metamorphosis of the tadpole into a frog is

Table 4–10
The Human Hemoglobins

	Symbol	Chains
Embryonic hemoglobins		
Portland-1	P_1	$\zeta_2\gamma_2$
Gower-1	G_1	$\zeta_2\varepsilon_2$
Gower–2	G_2	$\alpha_2\varepsilon_2$
Fetal hemoglobin	F	$\alpha_2\gamma_2$
Adult hemoglobins		
Hemoglobin A	A	$\alpha_2\beta_2$
Hemoglobin A_2	A_2	$\alpha_2\delta_2$

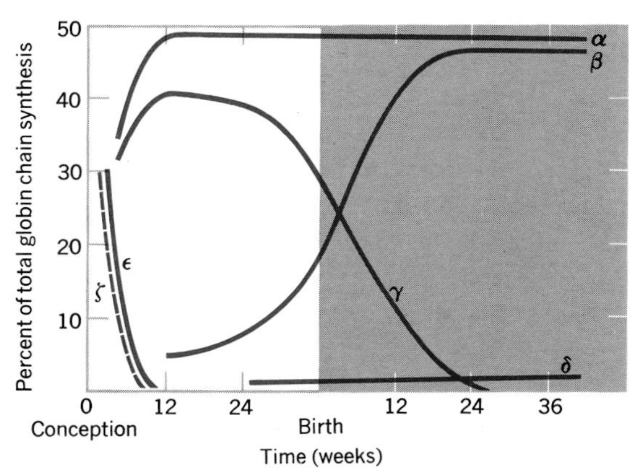

Figure 4–27
Ontogeny of the globin chains of human hemoglobins.

accompanied by a change in hemoglobin type. However, two members of each tetramer are the same in both tadpole and adult frog hemoglobins.

The carp has two different hemoglobins in the adult form, and these differ in only two of the globin chains of each tetramer; the other two chains are the same for both hemoglobin types. Identical situations exist in domestic fowl (which have three or more hemoglobin types), and so on. All of these observations play a critical role in the development of the current concept concerning the evolution of the hemoglobin molecule.

The primary structures of the alpha and beta chains of human hemoglobin are very similar. There are 62 positions in each chain occupied by the same amino acids (these positions are said to be *homologous*), and there are 84 differences. The frequency with which various amino acids are found in each chain is also very similar (Table 4–11). A common evolutionary origin for the two chains is suggested on the basis of their high degree of similarity.

An even greater homology exists between beta, gamma, and delta chains of human hemoglobin (Table 4–12). Beta and gamma chains have the same amino acids in 107 positions, with only 39 differences; beta and delta chains have the same amino acids in 136 positions and only 10 differences. Thus the beta, gamma, and delta chains are more closely related to each other than any of them are to the alpha chain. (Zeta and epsilon chain primary structures have not yet been determined.)

Just as striking are the similarities between human alpha and beta chains and those of other mammals (Table 4–13). No differences exist between the primary structures of chimpanzee and human alpha and beta globin chains. Gorilla and human globin chains differ in only two positions. Although the more distantly related mammals reveal greater numbers of differences, it has been consistently found that *alpha or beta chains of the various species are more closely related than the alpha or beta chains within a species*. The latter observation is basic to contemporary theory on the evolution of hemoglobin. Another consistent and important feature of hemoglobin studies is that species that are taxonomically more closely related have a greater similarity in primary globin chain structure.

The secondary and tertiary structure of myoglobin is almost identical to that of the globin chains of hemoglobin,

Table 4–11
Frequency of Occurrence of Amino Acids in Alpha and Beta Chains of Human Hemoglobin

Amino Acid	Alpha Chain	Beta Chain
ala	21	15
leu	18	18
val	13	18
ser	11	5
lys	11	11
his	10	9
asp	9	9
thr	9	7
gly	7	13
pro	7	7
phe	7	8
glu	4	7
tyr	3	3
arg	3	3
asn	3	4
met	2	1
gln	1	4
cys	1	2
try	1	2
ile	0	0

Table 4–12
Differences Between Human Globin Chains

	Alpha	Beta	Gamma	Delta
Alpha	—	84	89	85
Beta	84	—	39	10
Gamma	89	39	—	41
Delta	85	10	41	—

Table 4–13
Differences Between Human Globin Chains and Those of Other Mammals

Mammal	Alpha Chain	Beta Chain
Chimpanzee	0	0
Gorilla	2	2
Pig	18	17
Rabbit	18	18
Horse	18	25

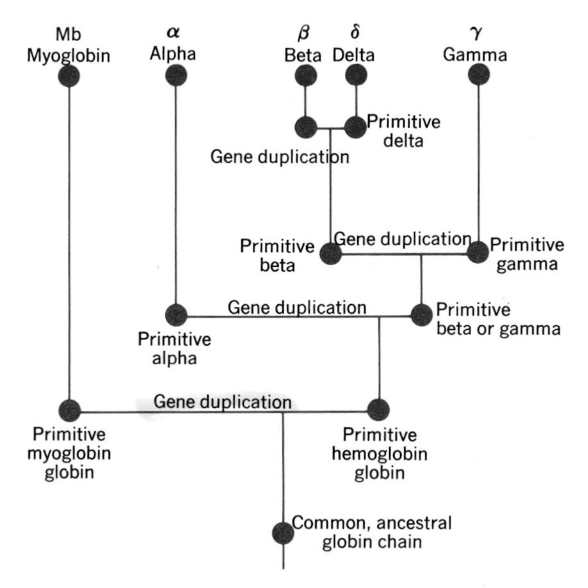

Figure 4–28
Evolution of the globin chains of human hemoglobin and myoglobin. Beginning with a common ancestral structural gene and corresponding globin chain, a series of gene duplications occurred. Following each duplication, the product genes underwent individual evolution by mutation, ultimately resulting in the variety of normal globin chains existing today. The zeta and epsilon genes are not shown, since the primary structures of their product globin chains are not yet known.

and its primary structure reveals an especially high degree of homology with alpha globin chains. In view of their similarity, myoglobin and hemoglobin are believed to have a common evolutionary origin. This notion is also supported by the observations that the hemoglobin of the lowest vertebrates may consist of a single polypeptide chain rather than a tetramer.

On the basis of the numerous studies of vertebrate myoglobins and hemoglobins, it is generally agreed that all the vertebrate globin chains have a common evolutionary origin. The proposed scheme for the evolution of human hemoglobin and myoglobin is shown in Figure 4–28. According to this scheme, the *structural genes* (see Chapters 20 and 21) for the various globin chains arose through gene duplication and subsequently underwent separate and independent evolution. The degree of similarity between the primary structures of the globin chains (Table 4–12) is used to establish the relative phylogenetic positions of the structural genes. Thus the greater degree of homology between the beta and delta globin chains than between beta and gamma chains is presumably the consequence of the more recent duplication of the primitive beta gene, and so on.

Based on the extent of homology between the primary structures of polypeptide chains from different species and the presumed frequencies with which gene mutations occur, it is possible to estimate the age (and the primary structure) of the ancestral protein. This has given rise to the field of biology called chemical (or molecular) paleogenetics. Phylogenetic relationships between species suggested on the basis of protein similarities are remarkably consistent with those predicted earlier using more conventional taxonomic criteria.

Conjugated Proteins

Hemoglobin and myoglobin belong to a major class of proteins called **conjugated proteins.** Conjugated proteins contain nonprotein constituents or *prosthetic groups* and can be subdivided into three major classes: (1) chromoproteins, (2) glycoproteins, and (3) lipoproteins.

Chromoproteins

The **chromoproteins** are a heterogeneous group of conjugated proteins related to each other only in that they possess color. The hemoglobins, myoglobins, and other heme-containing proteins such as the *cytochromes* and hemerythrins belong to this group.

Glycoproteins

Glycoproteins (from the Greek *glykys* meaning ''sweet'') are proteins containing varying amounts of carbohydrate (the chemistry of carbohydrates is dealt with in the next chapter). A number of important proteins fall in this category, including many of the blood plasma proteins and a large number of enzymes and hormones. The surfaces (i.e., plasma membranes) of most cells also contain quantities of glycoproteins, and these molecules serve there as antigenic determinants and as receptor sites. Virtually all the carbohydrate present in red blood cells occurs as membrane glycoproteins. Although over 100 different sugars (or *monosaccharides*) are known, only about 9 occur as regular constituents of glycoproteins (Table 4–14).

The amount of carbohydrate present in glycoproteins varies from less than 1% to more than 85% (Table 4–15). For example, in egg white *ovalbumin* (molecular weight 45,000), there is only one monosaccharide per molecule, while in *mucin* (a secretion of the submaxillary gland having a molecular weight of about 1 million), about 800 monosaccharides are present. The carbohydrate moieties of glycoproteins are usually bound to the protein through covalent bonds with either asparagine, threonine, hydroxylysine, serine, or hydroxyproline (see Figure 4–29). The carbohydrate bonded at each site of the protein may consist of a single monosaccharide unit (as in Figure 4–29) or a linear or branched chain of several monosaccharides (called an *oligosaccharide*), as depicted in Figure 4–30.

Lipoproteins

Lipid-containing proteins are called **lipoproteins.** This class includes some of the blood plasma proteins and also a large number of membrane proteins. The lipid content of lipoproteins is often very high, accounting for as much as 40 to 90% of the total molecular weight of the complex. In lipoproteins, the amount of lipid present markedly affects the density of the molecule, and this property is often used as the basis for lipoprotein classification. Whereas uncomplexed proteins have a density of about 1.35, lipoproteins vary in density down to 0.9 (i.e., a lipoprotein may be less dense than water).

Table 4–14
Carbohydrates Occurring as Regular Constituents of Glycoproteins

Glucose
Galactose
Mannose
Fucose
Acetylglucosamine
Acetylgalactosamine
Acetylneuraminic acid
Arabinose
Xylose

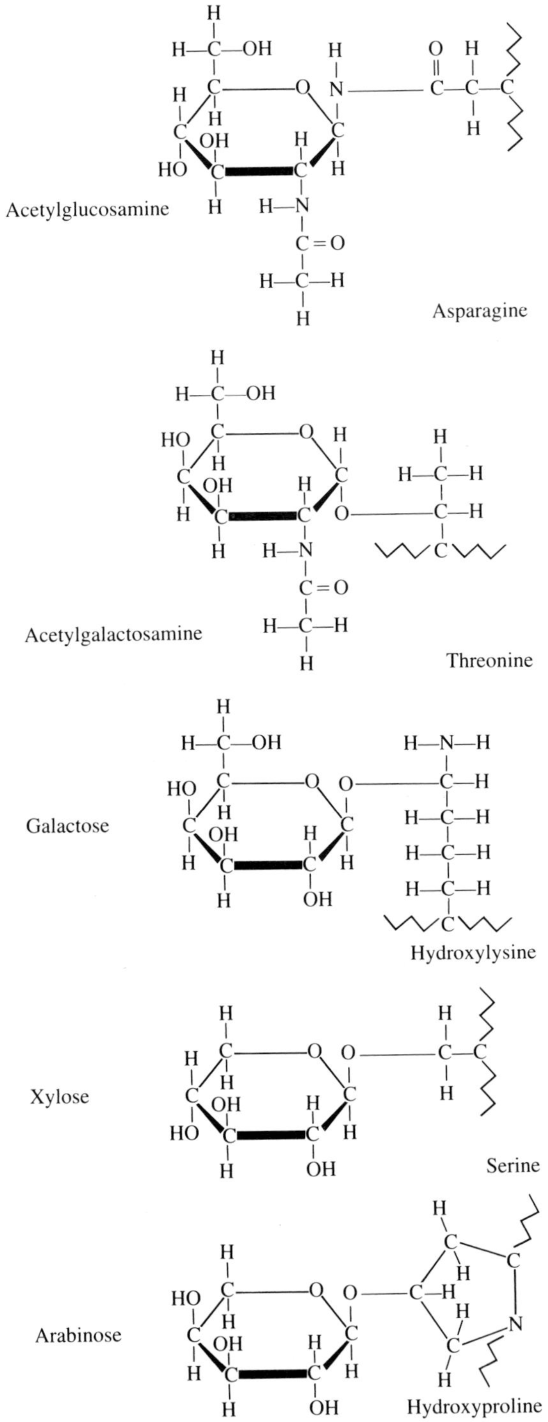

Acetylglucosamine — Asparagine

Acetylgalactosamine — Threonine

Galactose — Hydroxylysine

Xylose — Serine

Arabinose — Hydroxyproline

Figure 4–29
Covalent bonding between carbohydrate and amino acids in glycoproteins.

Table 4–15
Carbohydrate Content of Glycoproteins

Glycoprotein	Percentage of Carbohydrate	Function
Ovalbumin	1	Hen-egg food reserve
Follicle-stimulating hormone (FSH)	4	Hormone
Fibrinogen	5	Blood coagulation protein
Transferrin	6	Iron transport protein of blood plasma
Ceruloplasmin	7	Copper transport protein of blood plasma
Glucose oxidase	15	Enzyme
Peroxidase	18	Enzyme
Luteinizing hormone	20	Hormone
Haptoglobin	23	Hemoglobin-binding protein of blood plasma
Erythropoietin	33	Hormone
Mucin	50–60	Mucus secretion
Blood-group glycoproteins	85	Unknown

The association between the lipid and protein portions of a lipoprotein usually involves similar functional groups. For example, the hydrophobic portions of fatty acids, sterols, glycerides, and the like (see Chapter 6) form van der Waals interactions with the hydrophobic sidechains of the nonpolar amino acids. Covalent bonds are believed to occur between the phosphate moieties of certain phospholipids and the hydroxyl-containing side chains of amino acids like serine. Lipoproteins are discussed further in connection with membrane structure in Chapter 15.

Nucleoproteins

In eucaryotic cells, specific proteins are found intimately associated with the nuclear DNA to form **nucleoproteins,** and in procaryotes as well as eucaryotes, ribonucleoprotein complexes (i.e., protein plus RNA) occur. These are not usually classified with the conjugated proteins, since the nucleic acids involved cannot be regarded as prosthetic groups. Two types of proteins have been identified in nucleoproteins, the *histones* and *nonhistones*. Histones have a rather restricted amino acid composition (containing about 25% arginine and lysine) and are quite similar in all plant and animal cells. Their highly basic nature accounts for the close associations they form with the nucleic acids and lends credence to the notion that they are involved in the tight packing of DNA molecules during the condensation of chromatin to form chromosomes. The nonhistones are considerably more heterogeneous in amino acid composition and have acidic properties. There is much evidence to implicate the nonhistones in the regulation of gene expression. The histones and nonhistones are considered further in conjunction with the discussion of nuclear organization and function in Chapter 20.

Summary

Of the four major classes of macromolecules, the proteins are the most diverse and complex, being composed of one or more chains of **amino acids.** About 20 different amino

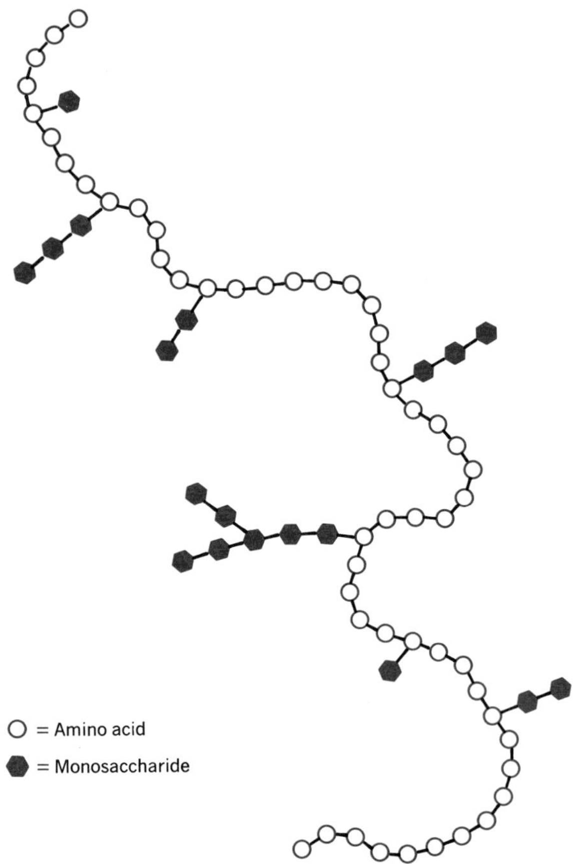

Figure 4–30
Organization of a glycoprotein molecule.

acids occur in proteins and are covalently linked together within each chain by **peptide bonds.** The specific sequence of covalent associations of all the amino acids in the protein is called **primary** structure. Each polypeptide chain may contain one or more regions twisted to form a variety of **helical** structures stablized by covalent, hydrophobic, electrostatic, and/or hydrogen bonds. This is called the **secondary** level of protein structure. The helical (and nonhelical) regions of a chain may be folded relative to one another to form the polypeptide's **tertiary** structure. In proteins that have two or more polypeptides, the specific orientation of the chains with respect to one another is called **quaternary** structure.

The secondary, tertiary, and quaternary structure of proteins may be established during assembly of the primary structure or spontaneously following completion of synthesis. Hydrophobic and hydrophilic amino acids are specifically distributed in relation to the final tertiary and quaternary structure of the molecule. Among all proteins studied, the structure, action, and evolution of the tetrameric hemoglobin molecule is best understood. All hemoglobin chains are descended from a common evolutionary ancestor through gene duplication and mutation.

Proteins containing **prosthetic groups** (such as the heme groups of hemoglobin) are called **conjugated** proteins. **Chromoproteins** are chemically heterogeneous but are related in that they possess color. In **glycoproteins** the prosthetic constituents consists of one or more branches of sugar chains covalently linked to the polypeptides. In the **lipoproteins,** the prosthetic groups consist of lipids linked to the protein through either covalent bonds or hydrophobic interactions.

References and Suggested Reading

Articles and Reviews

Dayhoff, M. O., Computer analysis of protein evolution. *Sci. Am. 221*(1), 86 (July 1969).

Dickerson, R. E., The structure and history of an ancient protein. *Sci. Am. 226*(4), 58 (April 1972).

Edelman, G. M., The structure and function of antibodies. *Sci. Am. 223*(2), 34 (Aug. 1970).

Fraser, R. D. B., Keratins. *Sci. Am. 221*(2), 86 (Aug. 1969).

Gross, J., Collagen. *Sci. Am. 204*(5), 120 (May 1961).

Kendrew, J. C., The three-dimensional structure of a protein molecule. *Sci. Am. 205*(6), 96 (Dec. 1961).

Kitchen H., and Boyer S. (editors) Hemoglobin: Comparative molecular biology models for the study of disease. *Ann. N.Y. Acad. Sci.*, Vol. 241, 1974.

Klotz, I. M., Protein subunits: A table. *Science* 155, 697 (1967).

Li, C. H., The ACTH molecule. *Sci. Am. 209*(1), 46 (July 1963).

Perutz, M. F., Electrostatic effects in proteins. *Science 201*, 1187 (1978).

Perutz, M. F., The hemoglobin molecule. *Sci. Am. 211*(5), 64 (Nov. 1964).

Perutz, M. F., Hemoglobin structure and respiratory transport. *Sci. Am. 239* (6), 92 (Dec. 1978).

Perutz, M. F., Structure and function of haemoglobin. I. A tentative atomic model of horse oxyhaemoglobin. *J. Mol. Biol. 13,* 646 (1965).

Perutz, M. F., X-ray analysis, structure and function of enzymes. *Eur. J. Biochem. 8,* 455 (1969).

Perutz, M. F., Structure and mechanism of haemoglobin. *Br. Med. Bull. 32,* 195 (1976).

Perutz, M. F. Kendrew, J. C., and Watson, H. C., Structure and function of haemoglobin. II. Some relations between polypeptide chain configuration and amino acid sequence. *J. Mol. Biol. 13,* 669 (1965).

Perutz, M. F., Muirhead, H., Cox, J. M., and Goaman, L. C. G., Three-dimensional Fourier synthesis of horse oxyhaemoglobin at 2.8 Å resolution: The atomic model. *Nature 219,* 131 (1968).

Schroeder, W. A., The hemoglobins. *Annu. Rev. Biochem. 32,* 301 (1963).

Sharon, N., Glycoproteins. *Sci. Am. 230*(5), 78 (May 1974).

Tanford, C., The hydrophobic effect and the organization of living matter. *Science* 200, 1012 (1978).

Wood, W. G., Haemoglobin synthesis during human fetal development. *Br. Med. Bull.32,* 282 (1976).

Zuckerkandl, E., The evolution of hemoglobin. *Sci. Am. 212*(5), 110 (May 1965).

Books, Monographs, and Symposia

Anfinsen, C. B., *The Molecular Basis of Evolution,* John Wiley & Sons, Inc., New York, 1963.

Bunn, H. F., Forget, B. G., and Ranney, H. M., *Human Hemoglobins,* W. B. Saunders Co., Philadelphia, 1977.

Dickerson, R. E., and Geis, I., *The Structure and Action of Proteins,* Harper & Row, New York, 1969.

Gottschalk, A., Glycoproteins and glycopeptides, in *Comprehensive Biochemistry,* Vol. 8 (M. Florkin and E. A. Stotz, eds.), Elsevier Publishing Co., Amsterdam, 1963.

Haggis, G. H., Michie, D., Muir, A. R., Roberts, K. B., and Walker, P. M. B., *Introduction to Molecular Biology,* John Wiley & Sons, Inc., New York, 1964.

Ingram, V. M., *The Hemoglobins in Genetics and Evolution,* Columbia University Press, New York, 1963.

Ingram, V. M., *Biosynthesis of Macromolecules* (2nd ed.), W. A. Benjamin, Inc., Menlo Park, Calif., 1972.

Masters, C. J., and Holmes, R. S., *Haemoglobin, Isoenzymes and Tissue Differentiation,* North-Holland Publishing Co., Amsterdam, 1975.

McConkey, E. H., *Protein Synthesis,* Vol. I, Marcel Dekker, Inc., New York, 1971.

McGilvery, R. W., *Biochemical Concepts,* W. B. Saunders, Co., Philadelphia, 1975.

Putnam, F. W., *The Plasma Proteins.* Vols. 1 and 2, Academic Press, New York, 1960.

Snell, F. M., Shulman, S., Spencer, R. P., and Moos, C., *Biophysical Principles of Structure and Function,* Addison-Wesley, Reading, Mass., 1965.

Stryer, L., *Biochemistry,* W. H. Freeman and Co., San Francisco, 1975.

Timasheff, S. N., and Fasman, G. D., *Structure and Stability of Biological Macromolecules,* Marcel Dekker, Inc., New York, 1969.

Wold, F., *Macromolecules: Structure and Function,* Prentice-Hall, Inc., Englewood Cliffs, N. J., 1971.

Chapter 5
THE CELLULAR MACROMOLECULES: POLYSACCHARIDES

Carbohydrates play important roles in the physiology of cells, since they serve as structural components of certain organelles and are also stored as reserve energy sources. Carbohydrates are composed of simple sugar molecules called **monosaccharides,** which have the general chemical formula $C_nH_{2n}O_n$. Occasionally, nitrogen and sulfur groups are also present. Carbohydrates consisting of two simple sugars are called **disaccharides,** those containing three simple sugars are called **trisaccharides,** and so on. **Polysaccharides** contain very large numbers (often up to several millions) of simple sugars.

Monosaccharides

The most commonly occurring monosaccharides are either *aldoses* or *ketoses* and contain three to six carbon atoms forming an unbranched chain; those sugars containing three carbon atoms are *trioses,* those with four carbon atoms, *tetroses;* those with five carbon atoms, *pentoses;* and those with six carbon atoms, *hexoses* (Figs. 5–1 and 5–2). Note that only one carbon atom in the chain forms a *double* bond with oxygen. Each of the carbon atoms of the monosaccharide is assigned a number beginning with the end closest to the double-bonded oxygen. In glyceraldehyde, the middle carbon atom (i.e., carbon number 2) is *asymmetric;* that is, it has four chemically different groups attached to it, and it therefore has two optical isomers. The two isomers of glyceraldehyde are shown in Figure 5–3 and are identified as the D and L forms. By chemical convention, the hydroxyl group of the asymmetric carbon is to the right in the D form and to the left in the L form. All the aldoses may be considered derivatives of glyceraldehyde, while the corresponding ketoses are derivatives of dihydroxyacetone. D or L notation is determined by similarity to glyceraldehyde. The most common aldoses and ketoses found in nature (e.g., those in Figs. 5–1 and 5–2) are the D forms.

The number of possible isomers of a monosaccharide is determined by the number, n, of asymmetric carbon atoms present in the chain. Glucose ($n = 4$) has 2^4 or 16 possible isomers. In spite of the variety of possible monosaccharide isomers, only a few different forms occur naturally in any abundance. For example, there are only three naturally occuring isomers of glucose; these are (1) D-glucose, (2) D-mannose, and (3) D-galactose (see Fig. 5–1).

Pyranoses and Furanoses

Some of the monosaccharides can occur in two basic configurations: *open-chain* structures and *cyclic (ringed)* structures. Both configurations are in reversible equilibrium with one another in aqueous solution, although the cyclic forms are much more prevalent. The formation of the cyclic struc-

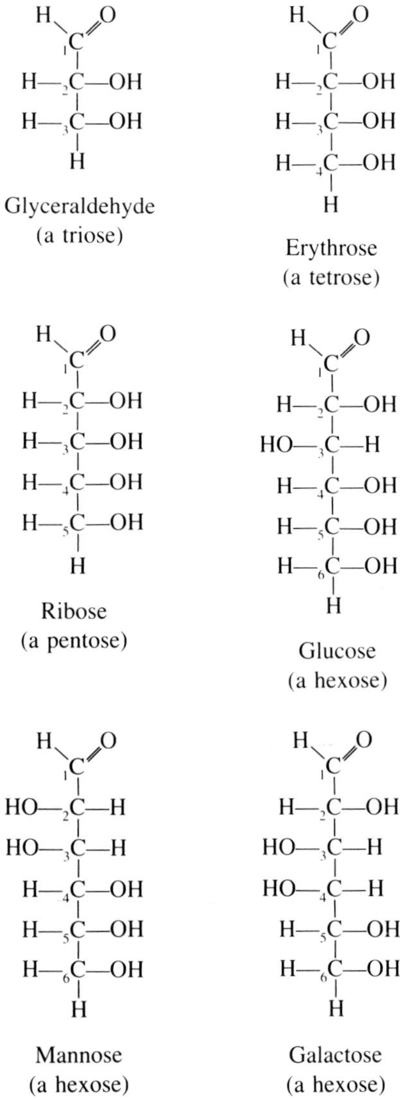

Glyceraldehyde
(a triose)

Erythrose
(a tetrose)

Ribose
(a pentose)

Glucose
(a hexose)

Mannose
(a hexose)

Galactose
(a hexose)

Figure 5–1
The common *aldoses.*

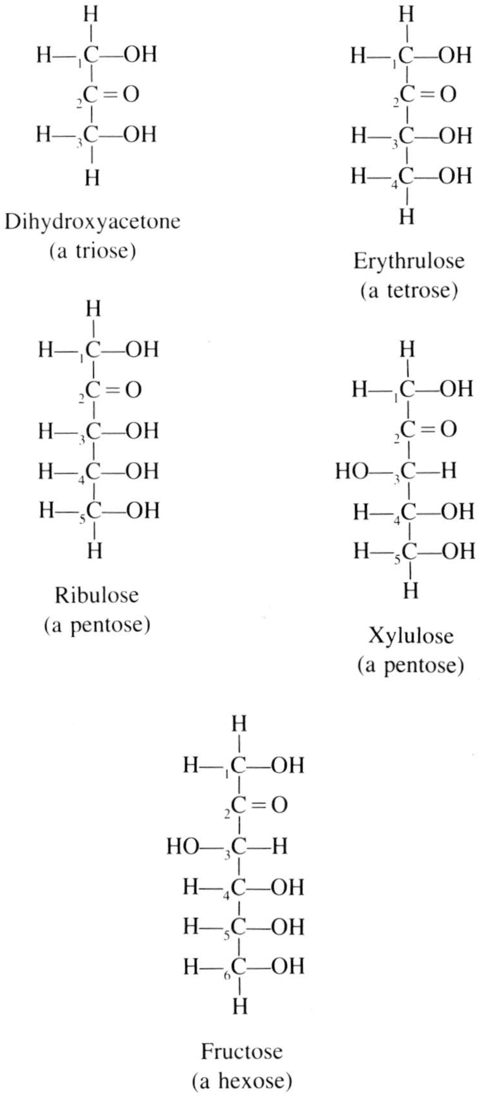

Dihydroxyacetone
(a triose)

Erythrulose
(a tetrose)

Ribulose
(a pentose)

Xylulose
(a pentose)

Fructose
(a hexose)

Figure 5–2
The common *ketoses.*

tures of D-glucose is shown in Figure 5–4. The aldehyde group of carbon number 1 reacts with the hydroxyl group of carbon number 5 to produce a six-membered ring containing oxygen. The formation of the ring form of glucose makes carbon number 1 asymmetric and thereby increases the number of isomers. The two ring forms of D-glucose are called the α and β *anomers*.

The three-dimensional arrangement of the constitutent atoms of the ring structure of monosaccharides is more readily understood when the structure is depicted using the

D-glyceraldehyde L-glyceraldehyde

Figure 5–3
D and L forms of glyceraldehyde.

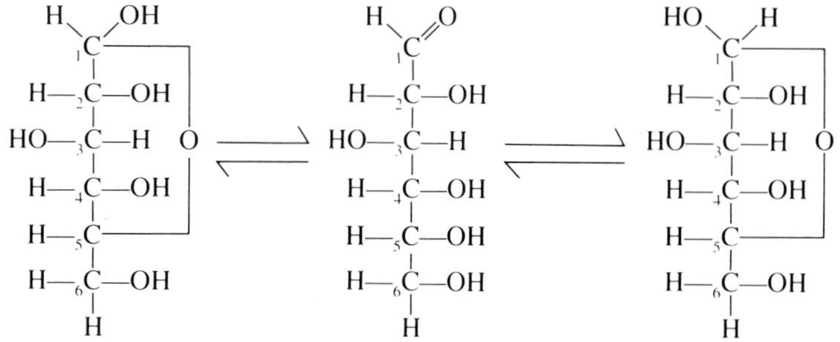

α-D-glucose (ring form) D-glucose (open chain) β-D-glucose (ring form)

Figure 5–4
Open-chain and cyclic forms of D-glucose

Haworth formulation (Fig. 5–5). In this formulation, the members of the ring are arranged in a plane that is perpendicular to the plane of the page. Groups to the right of the carbon skeleton in the open-chain configuration are depicted below the plane of the ring, and those to the left of the carbon skeleton extend above the plane of the ring. Certain bonds between members of the ring are deliberately shaded more heavily in order to assist the viewer's perception of the three-dimensional perspective. This is accepted chemical convention.

The Haworth formulation of glucose in which the members of the ring are depicted in the same plane is not entirely accurate. Carbon atoms 2, 3, and 5 and the oxygen atom do indeed lie in the same plane (to form the corners of a square), but carbon atom number 1 lies either slightly above or below the plane, while carbon atom number 4 always lies slightly above the plane. This gives rise to the "boat" and "chair" forms of glucose (Fig. 5–6) of which the chair form is the more stable. Other cyclic aldohexoses exhibit corresponding boat and chair forms.

For convenience and simplicity, the carbon atoms of the ring and their associated hydrogens are often omitted from the Haworth representation; that is,

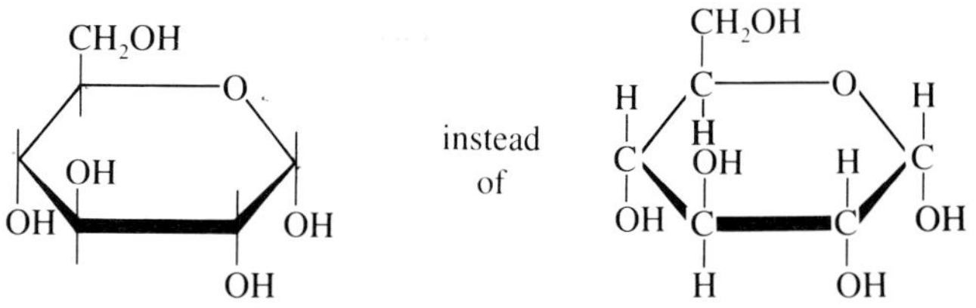

instead of

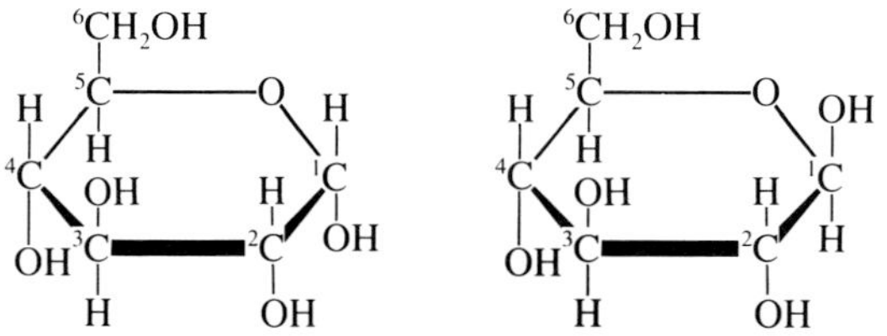

α-D-glucose β-D-glucose

Figure 5–5
Haworth formulations of glucose.

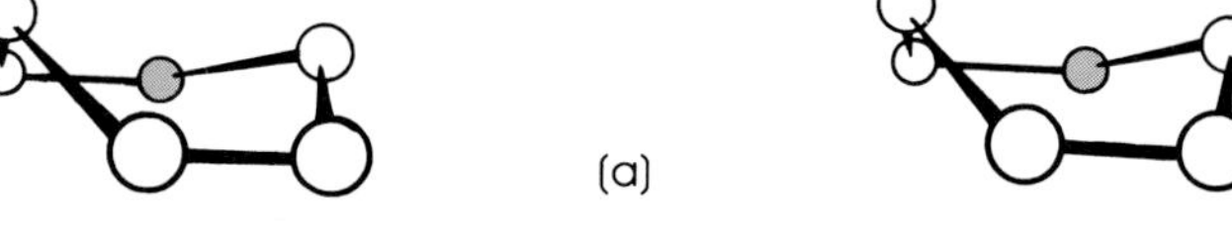

(a)

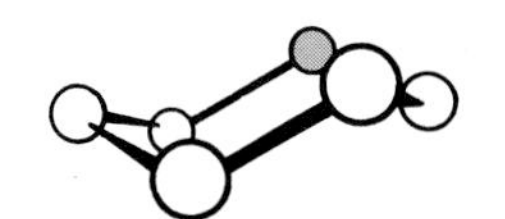

(b)

Figure 5–6
Stereoscopic views of the "boat" (*a*) and "chair" (*b*) forms of aldohexoses and the complete α anomer (*c*). The oxygen atom of the ring is shown in color.

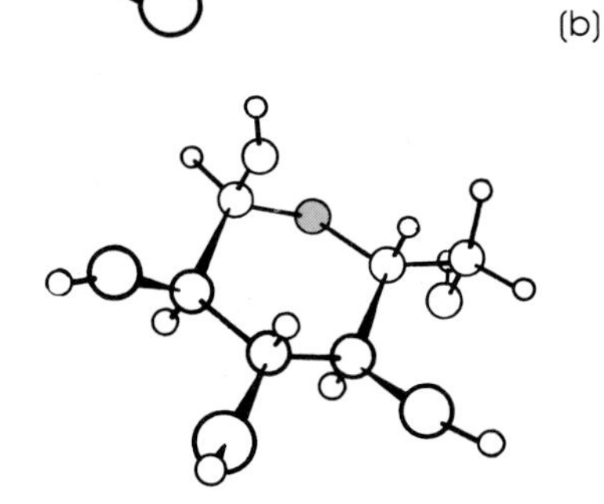
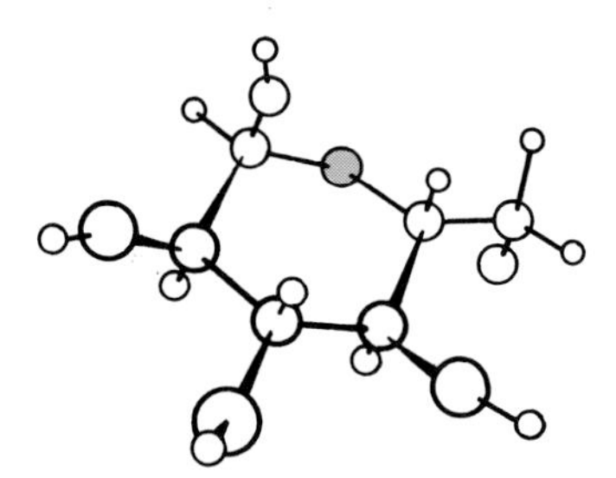

(c)

Other five- and six-carbon monosaccharides also occur in cyclic forms. Most important to our consideration are the cyclic forms of D-ribose and D-fructose:

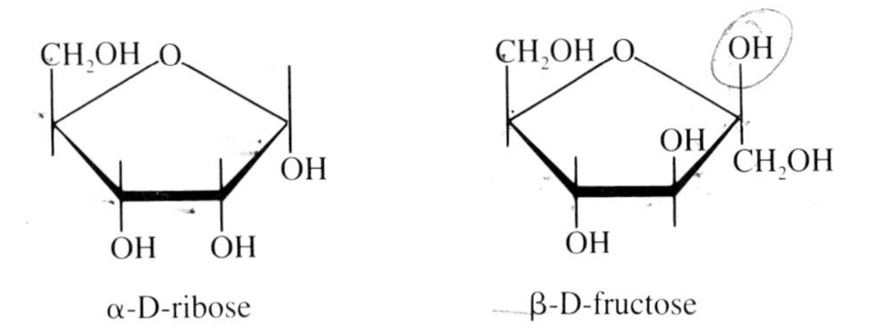

Cyclic forms that contain a six-member ring are called **pyranoses** and those containing a five-member ring are called **furanoses.** It should be noted that while both glucose and fructose are hexoses, glucose (an aldohexose) forms a pyranose, whereas fructose (a ketohexose) forms a furanose.

Disaccharides

Disaccharides consist of two ringed monosaccharides. The bonds uniting neighboring monosaccharides are called *glycosidic* bonds and are formed by the condensation of a hydroxyl group of carbon atom number 1 of one monosaccharide with the hydroxyl group of either the number 2, 4, or 6 carbon atom of another. The formation of the common disaccharide *maltose* from two glucoses is shown in Figure 5–7.

In maltose, the oxygen bridge is formed between the number 1 carbon atom of one α-D-glucose unit and the number 4 carbon atom of the other. The bond formed is referred to as an α 1 ⟶ 4 glycosidic bond. Another important disaccharide is *sucrose* (i.e., table sugar) and is formed by the condensation of a α-D-glucose and β-D-fruc-

CH2OH O HOCH2 O
OH HO
HO O
OH OH CH2OH

Sucrose

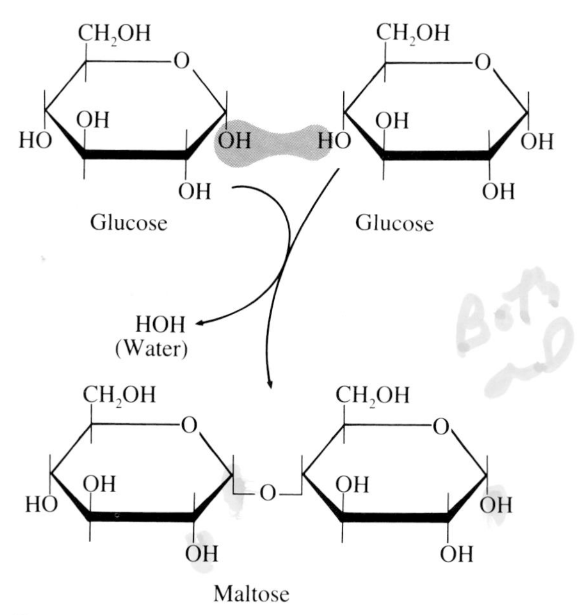

Figure 5–7
Condensation of two glucoses to form maltose.

tose. Milk contains the disaccharide *lactose*, which consists of the hexoses β-D-galactose and β-D-glucose:

Lactose

The glycosidic bond is of the *beta* variety; that is, β 1 ⟶ 4 (compare with maltose).

Polysaccharides

The polysaccharides are composed of long chains of sugars and can be divided into two major functional groups: the *structural* polysaccharides and the *nutrient* polysaccharides. The structural polysaccharides serve as structural components of certain cell organelles and also as interstitial (intercellular) supporting elements. Included in this group are *cellulose* (found in plant cell walls), *mannan* (yeast cell walls), *chitin* (arthropod shells, fungal cell walls), *chondroitin* (tendons, cartilage and bones), and *hyaluronic acid* (synovial fluid). The nutrient polysaccharides serve as reserves of monosaccharides and are in continuous metabolic turnover. Included in this group are *paramylum* (found in certain protozoa), *starch* (plant cells), and *glycogen* (animal cells).

On chemical bases, the polysaccharides can be divided into two broad classes: the **homopolysaccharides** and the **heteropolysaccharides.** In the homopolysaccharides, the constituent sugars are the same. Included in this class are cellulose, paramylum, glycogen, and starch. In the heteropolysaccharides, the constituent sugars may take several forms; included here are hyaluronic acid and chondroitin sulfate. A description of several of the more important polysaccharides follows.

Cellulose

Cellulose is the most abundant of all polysaccharides and is found primarily in plant cell walls, where it plays a structural role. Cellulose is an unbranched polymer of glucose in which the neighboring monosaccharides are joined by 1 ⟶ 4 glycosidic bonds. Chain lengths vary from several hundred to several thousand glucosyl units. Note that the "planes" of successive pyranose rings in Cellulose are rotated 180° relative to each other, so that the chain of sugars takes on a "flip-flop" appearance.

Cellulose

In plant cell walls, large numbers of cellulose molecules are organized into cross-linked, parallel fibers and bundles whose long axis is that of the individual chains. Cellulose has also been identified in algae and in fungi.

Chitin

Chitin is an extracellular structural polysaccharide found in large quantities in the body covering *(cuticle)* of arthropods and in smaller amounts in sponges, mollusks, and annelids. Chitin has also been identified in fungal cell walls. This polysaccharide is an unbranched polymer of *acetylglucosamine* and may contain several thousand units.

Hyaluronic Acid

Hyaluronic acid is an unbranched heteropolysaccharide containing repeating disaccharides of *N-acetylglucosamine* and *glucuronic acid*. Glucuronic acid is linked to *N*-acetylglucosamine in each disaccharide by a 1⟶ 3 glycosidic bond, but successive disaccharides are 1⟶ 4 linked. Hyaluronic acid is found in the synovial fluid of joints, the vitreous humor of the eye, and in the ground substance of the subcutaneous tissue.

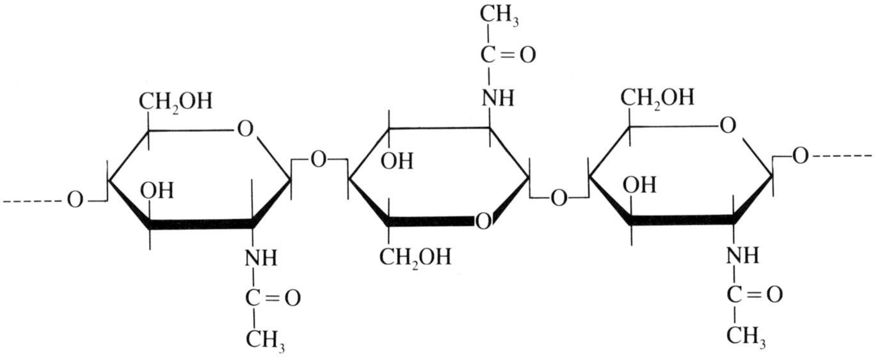

Chitin

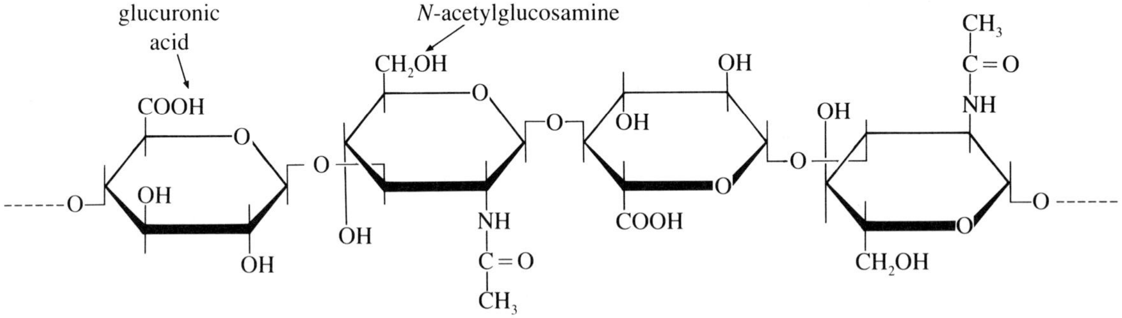

glucuronic acid

N-acetylglucosamine

Hyaluronic Acid

Inulin

Inulin is an unbranched nutrient polysaccharide found in the bulbs of such plants as artichokes and dandelions. It consists of repeating *fructose* units in 2 ⟶ 1 linkage.

HOCH$_2$ O O OH OH CH$_2$ HOCH$_2$ O O OH OH CH$_2$ HOCH$_2$ O O OH CH$_2$ OH

Inulin

Glycogen

Glycogen is a branched nutrient homopolysaccharide containing *glucose* in 1 ⟶ 4 and 1 ⟶ 6 linkages. It is found in nearly all animal cells and also in certain protozoa and bacteria. In view of its ubiquitous occurrence, it is an extremely important polysaccharide and has been the subject of numerous extensive studies. In humans and other vertebrates, glycogen is stored primarily in the liver and muscles and is the principal form of stored carbohydrate. In an unstarved animal, more than 10% of the liver weight may be glycogen. Glycogen undergoes an almost continuous biosynthesis and degradation, especially in liver tissue, and may be almost completely depleted during 24 hours of starvation. It is rapidly resynthesized from ingested carbohydrate upon refeeding.

Glycogen exists in a continuous spectrum of molecular sizes, with the largest molecules containing millions of glucosyl units. A small portion of a glycogen molecule is shown in Figure 5–8. The glucosyl units linked by 1 ⟶ 4 glycosidic bonds are organized into long chains; the chains are interconnected at branch points by 1 ⟶ 6 glycosidic bonds. This yields the "bush"- or "tree"-like structure depicted in Figure 5–9. It should be noted that in a single glycogen molecule, there is only one glucose unit whose number 1 carbon atom bears an hydroxyl group. All other 1-OH groups were involved in the formation of the 1 ⟶ 4 and 1 ⟶ 6 glycosidic bonds. The single free 1-OH group is called the "reducing end" of the molecule and is noted by the letter *R* in Figure 5–9. In contrast, numerous "nonreducing ends" are present (free 4-OH and 6-OH groups) at the terminals of the outermost chains.

In Figure 5–9, the individual glucose units are represented by circles, and the branch points (i.e., the 1 ⟶ 6 linkages), by heavier connections. In this model of glycogen, a number of different kinds of chains may be distinguished. *A chains* are attached to the molecule by a single 1 ⟶ 6 linkage (chains of open circles in Figure 5–9). *B chains* (shaded circles) bear one or more A chains. Each glycogen molecule contains only one *C chain* (colored circles); it is the chain that ends in the free reducing group. *Exterior chains* are those portions of individual chains between the nonreducing end groups and the outermost branch points. Finally, those parts of individual chains between branch points are called *interior chains*.

The exterior chains of glycogen are usually six to nine glucosyl units long, while interior chains contain only three to four glucosyl units. Approximately 8 to 10% of all glycosidic linkages are of the 1 ⟶ 6 type. It has also been found that the numbers of A and B chains are approximately equal and that the ratio of exterior chains to interior chains is constant. Most of the A chains are located near the periphery of the molecule; however, a small number of A chains may be buried within the interior.

Glycogen molecules are sufficiently large to be seen and studied by electron microscopy. This is usually carried out with material that has been negatively stained with phosphotungstic acid and osmium tetroxide (Fig. 5–10).

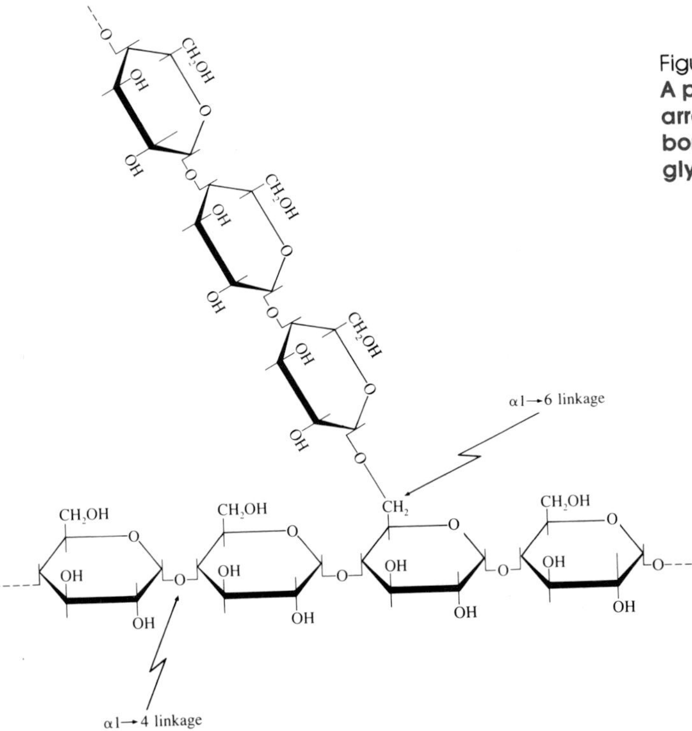

Figure 5–8
A portion of a glycogen molecule. Glucosyl units are arranged into long chains linked by 1 → 4 glycosidic bonds and these chains are interconnected by 1 → 6 glycosidic bonds.

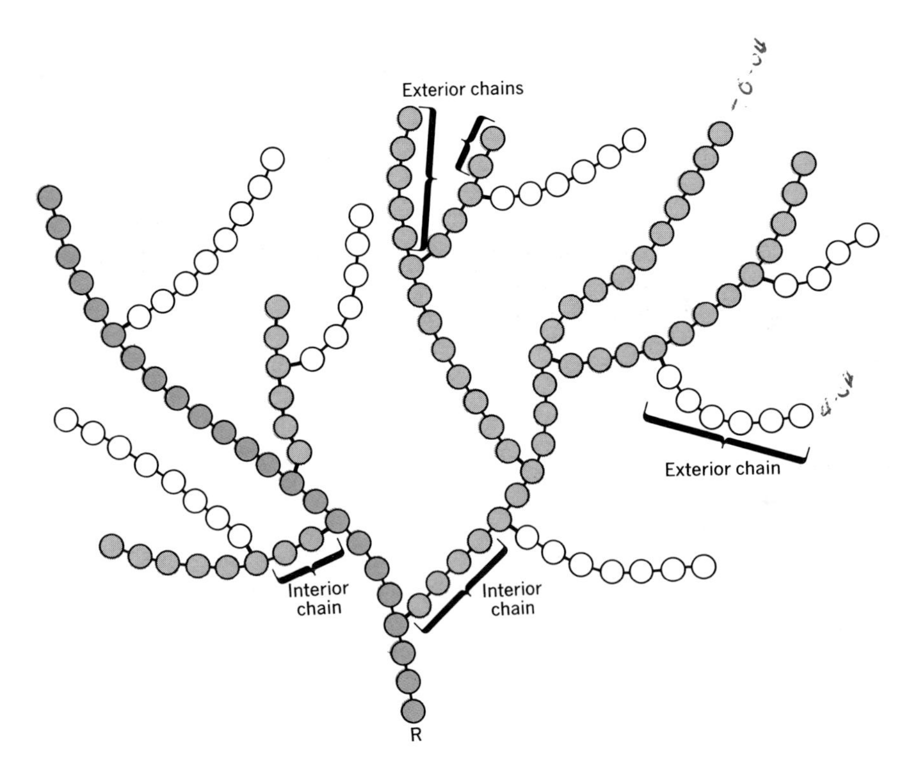

Figure 5–9
"Bush" or "tree"-like structure of the glycogen molecule. A chains are shown by open circles, B chains are shaded circles, and C chain is in color. The reducing end of the molecule is denoted by the letter R.

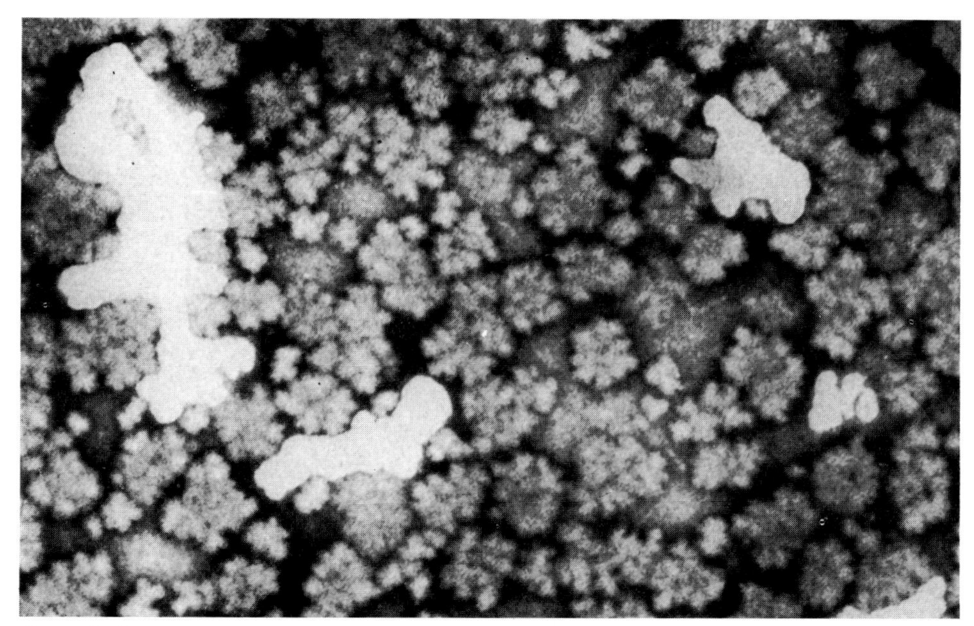

Figure 5–10
Electron photomicrograph of negatively stained glycogen particles (rosettelike structures) from liver tissue magnified 80,000 ×.

Starch

Starch is a nutrient polysaccharide found in plant cells and certain microorganisms and is similar in many respects to glycogen. (Glycogen is often referred to as "animal starch.") Starch usually occurs in cells in the form of visible granules. In plant cells (such as potato, corn, etc.) these granules may be several microns in diameter, whereas in microorganisms, their diameter may be only 0.5 to 2 μ. Starch granules contain a mixture of two different polysaccharides: *amylose* and *amylopectin*. The relative quantities of these two polysaccharides vary according to the source of the starch. In potato, corn, and many other plant starches, the amylopectin content is about 75 to 80% and the amylose 20 to 25%. Among the protozoa the percentage composition is much more variable, ranging from 0 to 45% amylose.

The amylose component of starch is an unbranched polymer of 1 ⟶ 4 linked glucose and may be several thousand glucosyl units long. The polysaccharide chain exists in the form of a helix. The familiar blue color produced when starch and iodine react is believed to result from the coordination of iodide ions within the helix, and the intensity of the blue color produced depends upon the relative amylose content of the starch sample.

Amylopectin is a branched polysaccharide containing 1⟶ 4 and 1⟶ 6 linked glucosyl units; in this respect, it is similar to glycogen. However, amylopectin has a more open structure with fewer 1 ⟶ 6 linkages and longer chain lengths. Some of the characteristics of amylopectin and glycogen are compared in Table 5–1.

Table 5–1
Some Chemical Properties of Amylopectin and Glycogen

	Amylopectin	Glycogen
Molecular weight	10^7	10^6–10^9
Average chain length	20–25	10–14
Percentages 1 ⟶ 6	4–5	8–10
linkages	12–17	6–9
Exterior chain length	5–8	3–4
Interior chain length		

Other Polysaccharides

In addition to the polysaccharides already described, several others should be briefly reviewed. *Chondroitin,* found in cartilage and bone, is similar in organization to hyaluronic acid, consisting of chains of glucosamine-acetylgalactosamine disaccharides. Occasionally, various hydroxyl groups of the galactosamine units are replaced by sulfate groups. *Mannan* is a branched structural polysaccharide found in the walls of yeast cells. It consists of repeating mannose units in $1 \longrightarrow 2$, $1 \longrightarrow 3$, and $1 \longrightarrow 6$ linkage. Finally, *paramylum* is a nutrient homopolysaccharide stored as large granules in certain protists (e.g., *Euglena*). It is unbranched and consists of chains of glucose units in $1 \longrightarrow 3$ linkage. Some of the properties of the various polysaccharides described in this chapter are summarized in Table 5–2.

Summary

The **polysaccharides** are macromolecules composed of chains of individual sugar units called **monosaccharides.** Among these, the most common is the hexose, **glucose.** Although monosaccharides may contain three or more carbon atoms and are derivatives of either glyceraldehyde or dihydroxyacetone, the six-carbon forms occur most often in polysaccharides. Individual sugars may take the open-chain configuration or a ringed structure, but only the ringed **pyranoses** and **furanoses** are incorporated into macromolecules. From a functional standpoint, polysaccharides may be divided into nutrient and structural types. Nutrient polysaccharides such as glycogen, starch, and paramylum are often stored in cells as discrete granules visible either by light or electron microscopy. The most abundant structural

Table 5–2
Polysaccharides

Class	Name	Source	Composition	Linkages
Structural polysaccharides	Cellulose	Plant cell walls	Glucose (beta linkage)	Unbranched $1 \longrightarrow 4$
	Mannan	Yeast cell walls	Mannose (beta linkage)	Branched $1 \longrightarrow 2$, $1 \longrightarrow 3$ and $1 \longrightarrow 6$
	Chitin	Arthropod shells, fungal cell walls	Acetylglucosamine (beta linkage)	Unbranched $1 \longrightarrow 4$
	Hyaluronic acid	Synovial fluid (joints), subcutaneous tissue	Acetylglucosamine and glucuronic acid (beta linkage)	Unbranched $1 \longrightarrow 3$ and $1 \longrightarrow 4$
	Chondroitin sulfate	Cartilage, bone, and tendons	Galactosamine and glucuronic acid (beta linkage)	Unbranched $1 \longrightarrow 3$
Nutrient polysaccharides	Inulin	Artichokes, dandelions	Fructose (beta linkage)	Unbranched $2 \longrightarrow 1$
	Paramylum	Certain protozoa (i.e., *Euglena*)	Glucose (beta linkage)	Unbranched $1 \longrightarrow 3$
	Glycogen	Certain protozoa (i.e., *Tetrahymena*) and most animals	Glucose (alpha linkage)	Branched $1 \longrightarrow 4$ and $1 \longrightarrow 6$
	Starch			
	Amylopectin	Plant cells and some protozoa (i. e., *Polytomella*)	Glucose (alpha linkage)	Branched $1 \longrightarrow 4$ and $1 \longrightarrow 6$
	Amylose		Glucose (alpha linkage)	Unbranched $1 \longrightarrow 4$

polysaccharide, **cellulose,** serves as the major supportive element in the cell walls of higher plants. Polysaccharides may be formed from unbranched chains of sugars (as in cellulose and the amylose component of starch) or from highly branched chains (as in glycogen and the amylopectin component of starch). All sugars are of the same type in the **homopolysaccharides** (e.g., starch, glycogen and cellulose) but vary in the **heteropolysaccharides** (e.g., hyaluronic acid).

References and Suggested Reading

Articles and Reviews

Manners, D. J., The molecular structure of glycogens. *Adv. Carbohyd. Chem. 13,* 261 (1957).

Manners, D. J., Enzymatic synthesis and degradation of starch and glycogen. *Adv. Carbohyd. Chem. 17,* 371 (1962).

Stetten, D., and Stetten, M. R., Glycogen metabolism. *Physiol. Rev. 40,* Suppl. 4, 505 (1960).

Books, Monographs, and Symposia

Conn, E. E., and Stumpf, P. K., *Outlines of Biochemistry* (3rd ed.), John Wiley & Sons, Inc., New York, 1972.

Lehninger, A. L., *Biochemistry* (2nd ed.), Worth Publishing, Inc. New York, 1975.

Mazur, A., and Harrow, B., *Textbook of Biochemistry,* W. B. Saunders Co., Philadelphia, 1971.

McGilvery, R. W., *Biochemical Concepts,* W. B. Saunders Co., Philadelphia, 1975.

Reithel, F. J., *Concepts in Biochemistry,* McGraw-Hill Book Co., New York, 1967.

Stryer, L., *Biochemistry,* W. H. Freeman and Co., San Francisco, 1975.

Chapter 6
THE CELLULAR MACROMOLECULES: LIPIDS

Lipids are a chemically heterogeneous collection of molecules insoluble in water but soluble in nonpolar (organic) solvents such as ether, chloroform, and benzene; it is because of their similar solubility properties that they are usually considered together. Like the carbohydrates, the lipids serve two major roles in cells: (1) they occur as constituents of certain structural components of cells, particularly membranous organelles; and (2) they may be stored in cells as reserve energy sources. The most common cell lipids include the **fatty acids, neutral fats, glycerophosphatides, sphingolipids, plasmalogens, glycolipids,** and **steroids.**

Fatty Acids

The *saturated* fatty acids consist of long hydrocarbon chains terminating in a carboxyl group and conform to the following general formula: $CH_3-(CH_2)_n-COOH$. In nearly all naturally occuring fatty acids, n is an even number, usually 14 (i.e., palmitic acid) or 16 (stearic acid). In *unsaturated* fatty acids, two or more of the carbon atoms of the hydrocarbon chain are linked by double bonds. The two most common unsaturated fatty acids are oleic acid and linoleic acid (Fig. 6–1).

Fatty acid molecules contain both hydrophilic and hydrophobic parts. In their dissociated states (not shown in the formulas of Fig. 6–1), the carboxyl ends of the molecules are soluble in water, while the long hydrocarbon chains repel water. As a result, fatty acids form monomolecular layers on water, with their polar ends facing into the water and their hydrophobic ends directed away from the water (Fig. 6–2). This phenomenon is shared by lipids other than fatty acids (see below) and contributes to the suitability of lipids in membrane formation, as discussed in Chapter 15.

Neutral Fats

The neutral fats or **triglycerides** are esters of glycerol and three fatty acids and have the general formula shown in Figure 6–3. In this formula, n, n', and n'' may be the same number or different numbers, and the three fatty acids may be saturated and/or unsaturated. Unlike the fatty acids, the neutral fats are entirely nonpolar. Neutral fats, which are used by the cell as sources of energy, represent the major type of stored lipid, and most of the lipid recovered in the soluble phase of disrupted cells takes this form.

Glycerophosphatides

The major members of this group of lipids are derivatives of *phosphatidic acid*. Phosphatidic acid is similar to a tri-

Saturated region (first); Saturated region (second)

$$CH_3\text{—}(CH_2)_7\text{—}CH{=}CH\text{—}(CH_2)_7\text{—}COOH$$

Oleic Acid

Saturated region (first); Saturated region (second)

$$CH_3\text{—}(CH_2)_4\text{—}CH{=}CH\text{—}CH_2\text{—}CH{=}CH\text{—}(CH_2)_7\text{—}COOH$$

Linoleic Acid

Figure 6–1
The unsaturated fatty acids—*oleic* and *linoleic* acid.

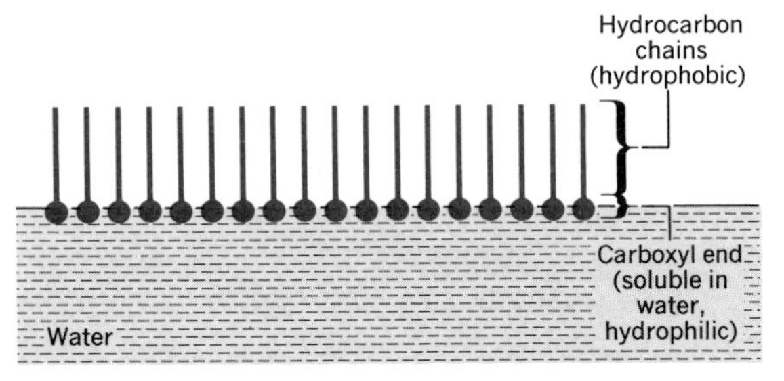

Figure 6–2
As the result of possessing hydrophilic and hydrophobic regions, fatty acids form monomolecular layers on water.

Figure 6–3
Generalized formula for a neutral fat.

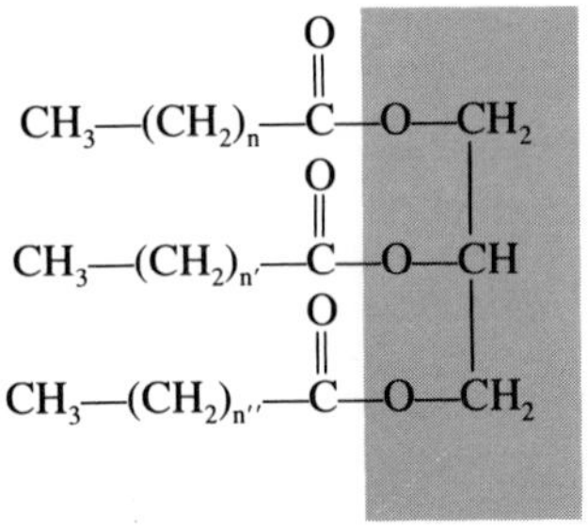

$$\begin{array}{l} CH_3\text{—}(CH_2)_n\text{—}\overset{O}{\overset{\|}{C}}\text{—}O\text{—}CH_2 \\ \qquad\qquad\qquad\qquad\quad | \\ CH_3\text{—}(CH_2)_{n'}\text{—}\overset{O}{\overset{\|}{C}}\text{—}O\text{—}CH \\ \qquad\qquad\qquad\qquad\quad | \\ CH_3\text{—}(CH_2)_{n''}\text{—}\overset{O}{\overset{\|}{C}}\text{—}O\text{—}CH_2 \end{array}$$

Glycerol backbone

glyceride except that one of the fatty acids is replaced by a phosphate group (Fig. 6–4). Phosphatidic acid and its derivatives are present in cell membranes, where they play an active role in membrane function (Chapter 15), in addition to serving as structural constituents. The most common derivatives of phosphatidic acid are *phosphatidyl choline* (also called *lecithin*), *phosphatidyl ethanolamine* (also called *cephalin*), *phosphatidyl serine*, and *phosphatidyl inositol* (Fig. 6–5). Free rotation about the single bonds of the glycerol backbone allow the hydrophilic phosphate and its derivatives to face away from the hydrophobic portion, as shown in the structural formulas of Figure 6–5. Consequently, glycerophosphatides (and other lipids discussed below) are *amphipathic*—that is, one end of the molecule is hydrophobic (i.e., the end containing the hydrocarbon chains), while the other end is extremely hydrophilic as a result of the charged nature of the dissociated phosphate group and other substituents. As will be seen in Chapter 15, these properties are important to the organization of glycerophosphatides in cell membranes.

Plasmalogens

Plasmalogens are a special class of lipids especially abundant in the membranes of nerve and muscle cells. Plasmalogens are also found in blood. A plasmalogen is similar to a glycerophosphatide except that an unsaturated ether is substituted at one of the glycerol positions (Fig. 6–6).

Sphingolipids

Sphingolipids are derivatives of *sphingosine*, an amino alcohol possessing a long, unsaturated hydrocarbon chain

Figure 6–4
Phosphatidic acid.

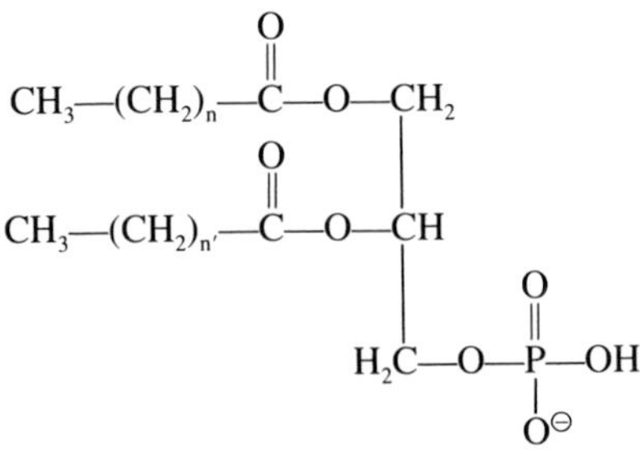

$$\begin{array}{l} CH_3\text{—}(CH_2)_n\text{—}\overset{O}{\overset{\|}{C}}\text{—}O\text{—}CH_2 \\ \qquad\qquad\qquad\qquad\quad | \\ CH_3\text{—}(CH_2)_{n'}\text{—}\overset{O}{\overset{\|}{C}}\text{—}O\text{—}CH \\ \qquad\qquad\qquad\qquad\quad | \\ \qquad\qquad\qquad\qquad H_2C\text{—}O\text{—}\underset{\underset{O^{\ominus}}{|}}{\overset{O}{\overset{\|}{P}}}\text{—}OH \end{array}$$

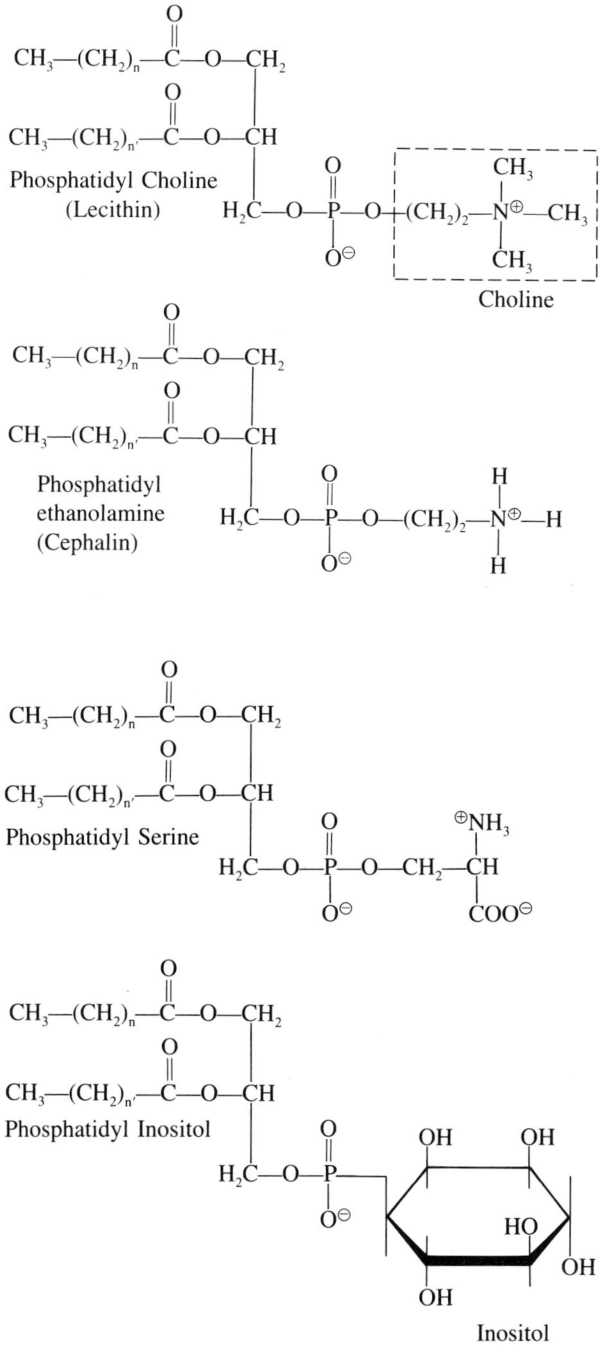

$CH_3{-}(CH_2)_n{-}\overset{O}{\overset{\|}{C}}{-}O{-}CH_2$

$CH_3{-}(CH_2)_{n'}{-}\overset{O}{\overset{\|}{C}}{-}O{-}CH$

Phosphatidyl Choline
(Lecithin)

$H_2C{-}O{-}\overset{O}{\overset{\|}{P}}(O^{\ominus}){-}O{-}(CH_2)_2{-}N^{\oplus}(CH_3)_3$

Choline

$CH_3{-}(CH_2)_n{-}\overset{O}{\overset{\|}{C}}{-}O{-}CH_2$

$CH_3{-}(CH_2)_{n'}{-}\overset{O}{\overset{\|}{C}}{-}O{-}CH$

Phosphatidyl
ethanolamine
(Cephalin)

$H_2C{-}O{-}\overset{O}{\overset{\|}{P}}(O^{\ominus}){-}O{-}(CH_2)_2{-}N^{\oplus}H_3$

$CH_3{-}(CH_2)_n{-}\overset{O}{\overset{\|}{C}}{-}O{-}CH_2$

$CH_3{-}(CH_2)_{n'}{-}\overset{O}{\overset{\|}{C}}{-}O{-}CH$

Phosphatidyl Serine

$H_2C{-}O{-}\overset{O}{\overset{\|}{P}}(O^{\ominus}){-}O{-}CH_2{-}CH(^{\oplus}NH_3)(COO^{\ominus})$

$CH_3{-}(CH_2)_n{-}\overset{O}{\overset{\|}{C}}{-}O{-}CH_2$

$CH_3{-}(CH_2)_{n'}{-}\overset{O}{\overset{\|}{C}}{-}O{-}CH$

Phosphatidyl Inositol

$H_2C{-}O{-}\overset{O}{\overset{\|}{P}}(O^{\ominus}){-}$ Inositol (OH, OH, HO, OH, OH)

Inositol

Figure 6–5
Phosphatidyl choline (lecithin), phosphatidyl ethanolamine (cephalin), phosphatidyl serine, and phosphatidyl inositol.

$CH_3{-}(CH_2)_n{-}CH{=}CH{-}O{-}CH_2$

$CH_3{-}(CH_2)_{n'}{-}\overset{O}{\overset{\|}{C}}{-}O{-}CH$

$H_2C{-}O{-}\overset{O}{\overset{\|}{P}}(O^{\ominus}){-}$ [Choline / Serine / Ethanolamine / Inositol]

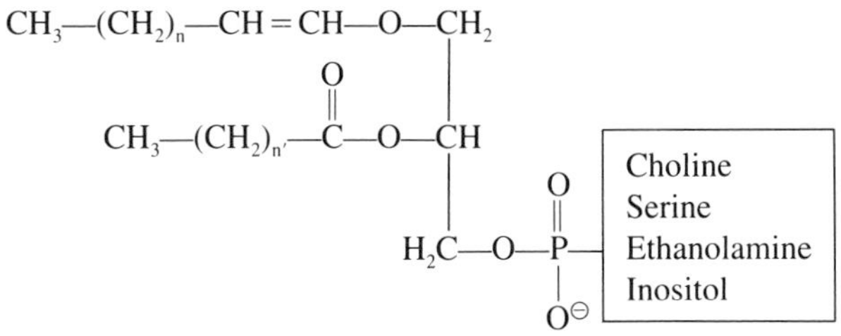

Figure 6–6
Generalized formula for a plasmalogen.

(Fig. 6–7*a*). Blood and the myelin sheath surrounding nerve cells are particularly rich in the sphingolipid sphingomyelin (Fig. 6–7*b*). In sphingomyelin, the amino group of the sphingosine skeleton is linked to a fatty acid, and the hydroxyl group is esterified to *phosphorylcholine*.

Glycolipids

The glycolipids are sugar-containing lipids similar to sphingosine. In glycolipids, the amino group of the sphingosine skeleton is acylated by a fatty acid (as in sphingomyelin),

Figure 6–7
The sphingolipids sphingosine (a) and sphingomyelin(b).

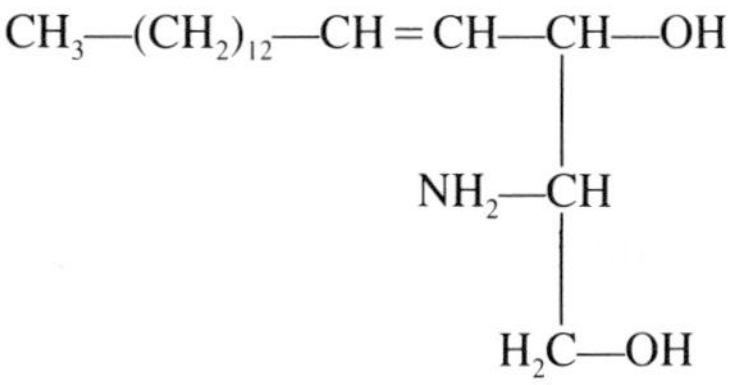

$CH_3{-}(CH_2)_{12}{-}CH{=}CH{-}CH{-}OH$

$NH_2{-}CH$

$H_2C{-}OH$

(a)

$CH_3{-}(CH_2)_{12}{-}CH{=}CH{-}CH{-}OH$

$CH_3{-}(CH_2)_n{-}\overset{O}{\overset{\|}{C}}{-}NH{-}CH$

$H_2C{-}O{-}\overset{O}{\overset{\|}{P}}(O^{\ominus}){-}O{-}(CH_2)_2{-}N^{\oplus}(CH_3)_3$

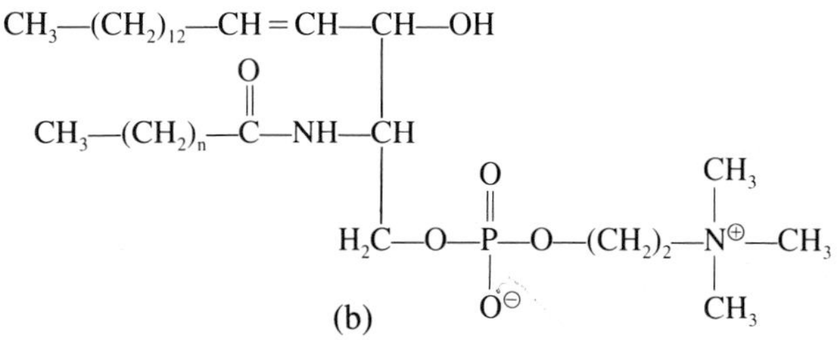

(b)

but the hydroxyl group is associated with one or more monosaccharides. The simplest glycolipids are the *cerebrosides* which, as their name suggests, are abundant in brain cells. However, cerebrosides are also present in kidney, liver, and spleen cells. The sugar of a cerebroside is either glucose or galactose (Fig. 6–8). *Gangliosides* are glycolipids found in cell membranes and are especially abundant in the membranes of gray matter cells of the brain. A chain of several sugar molecules (including galactose, glucose, and neuraminic acid) composes the carbohydrate portion of the molecule.

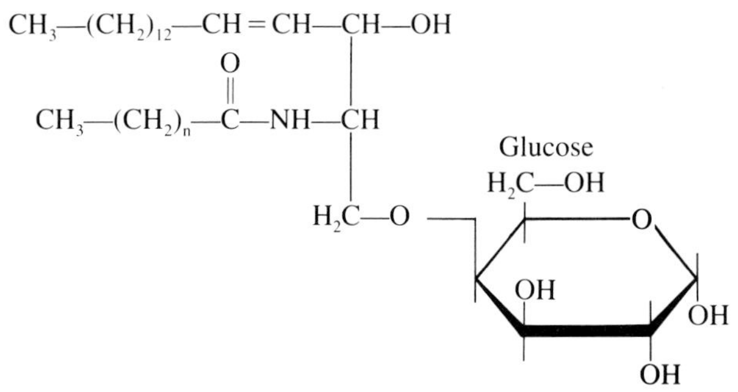

Steroids

The steroids are a physiologically important class of complex lipids consisting of a system of fused cyclohexane and cyclopentane rings. All are derivatives of *perhydrocyclopentanophenanthrene,* which consists of three fused cyclohexane rings (in a nonlinear arrangement) and a terminal cyclopentane ring (Fig. 6–9*a*).

Steroids have widely different physiological properties. The properties of the steroid derivatives are determined by the groups attached to the basic skeleton. Some are hormones (estrogen, progesterone, corticosterone, etc.), some are vitamins (i.e., vitamin D), and others are regular constituents of subcellular structures. Probably the best known steroid and an important constituent of the plasma membranes of certain animal cells is cholesterol (Fig. 6–9*b*).

Table 6–1 shows the distribution of lipid in the four major cell fractions obtained when liver tissue is dispersed and serially centrifuged (see Chapter 12). The distribution is believed to be fairly representative at least of animal cells. It should be noted that the mitochondrial and microsomal fractions of the cells are richest in lipid. Most of the lipid in these fractions is phospholipid (i.e., glycerophosphatides, etc.) serving as structural constituents of the mitochondrial and microsomal membranes, where it is combined with membrane protein (see Chapter 15). Phospholipids occur in similar quantities in the chloroplasts of plant cells. Most of the lipids found in the nonsedimenting soluble phase of the cell (or "cytosol") are neutral fats and are in rapid metabolic turnover. The neutral fat content of the cytosol fractions of tissues specialized for fat storage (e.g., adipose tissues) is much higher. We will return to a consideration of lipids in conjunction with several chapters that deal with the structure and function of the major cellular organelles.

Figure 6–8
A glycolipid. The molecule shown is a cerebroside and contains only one sugar, in this instance glucose. Cerebrosides may also contain galactose. In gangliosides, the hexose is replaced by a chain of sugars.

Summary

The **lipids** are a heterogeneous collection of macromolecules soluble in nonpolar solvents that play two major roles

Figure 6–9
Perhydrocyclopentanophenanthrene (*a*) and one of its derivatives, cholesterol (*b*).

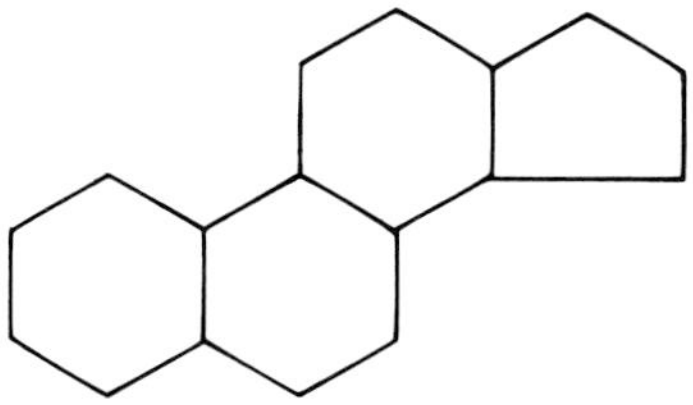

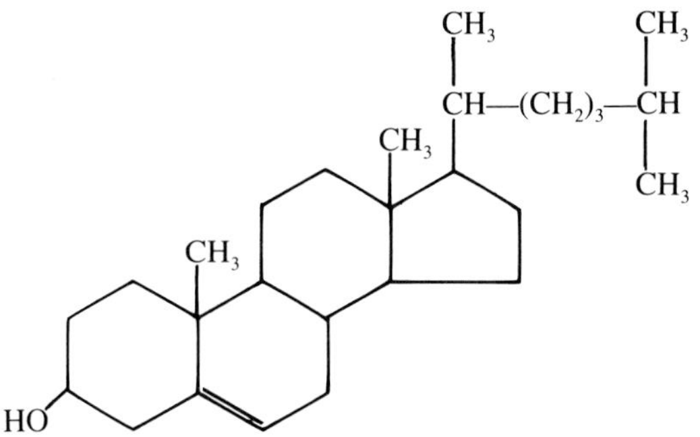

Table 6–1
Lipid Content of Rat Liver Cell Fractions

Fraction	Total Lipid (% dry weight)	Percentage of Total Lipid		
		Phospholipid	Steroid	Neutral Fat
Nuclei	16	90	5	3
Mitochondria	21	90	6	1
Microsomes	32	90	6	0
Soluble Phase	7	30	4	70

in cells and tissues: they are sources of reserve energy and structural constituents of cellular membranes. The simplest lipids, the saturated and unsaturated **fatty acids,** consist of unbranched hydrocarbon chains terminating in carboxyl groups. In the **neutral fats,** these carboxyl groups are esterified to the three carbon atoms of glycerol and represent the most abundant form of stored lipid. Neutral fats are in rapid metabolic turnover. **Glycerophosphatides,** in which one fatty acid chain of a neutral fat is replaced by the polar phosphate group or a phosphate derivative, are **amphipathic** and, as such, play an important part in the structure and organization of cell membranes. The most physiologically diverse lipids are the **steroids,** formed by a series of fused, ringed hydrocarbons. Steroids serve as hormones, as vitamins, and occasionally as constituents of cell membranes.

References and Suggested Reading

Articles and Reviews

Capaldi, R. A., A dynamic model of cell membranes. *Sci. Am. 230* (3), 26 (Mar. 1974).

Fieser, L. F., Steroids. *Sci. Am. 192* (1), 52 (Jan. 1955).

Hokin, L. E., and Hokin, M. R., The chemistry of cell membranes. *Sci. Am. 213* (4), 78 (Oct. 1965).

Rothman, J. E., and Lenard, J., Membrane asymmetry. *Science 195,* 743 (1977).

Books, Monographs, and Symposia

Conn. E. E., and Stumpf, P. K., *Outlines of Biochemistry* (3rd ed.), John Wiley & Sons, Inc., New York, 1972.

Lehninger, A. L., *Biochemistry* (2nd ed.), Worth Publishing, Inc., New York, 1975.

McGilvery, R. W., *Biochemical Concepts,* W. B. Saunders, Co., Philadelphia, 1975.

Stryer, L., *Biochemistry,* W. H. Freeman and Co., San Francisco, 1975.

Chapter 7
THE CELLULAR MACROMOLECULES: NUCLEIC ACIDS

The nucleic acids were discovered over 100 years ago, but their role in genetics and in the control of cellular activity has been elucidated only during the past 40 years. This is probably due in part to the great emphasis placed on the study of proteins during the first half of this century and to the mistaken belief by many scientists during this period that the proteins were endowed with the genetic information of the cell. This is understandable in view of the fact that the composition and organization of proteins is so diverse, whereas the chemical nature of the nucleic acids (and also their structure) is so much more regular and restricted. Consequently, it was difficult to reconcile the great diversity of life with the fundamental chemical similarities manifested by all the nucleic acids. We begin our consideration of the nucleic acids by describing some of the major observations and discoveries that ultimately led biologists to recognize that these molecules and not proteins were intimately involved in the transmission of genetic information and in the determination and control of the activities of the cell. In this respect, this chapter will differ from the previous chapters on cellular macromolecules, which were strictly chemically oriented.

Cellular Roles of the Nucleic Acids

The Discovery of DNA

Friedrich Miescher (Swiss, 1844–1895) is credited with the discovery of nucleic acid in 1869. He isolated nuclei from white blood cells present in pus, using dilute solutions of hydrochloric acid to dissolve away other cell structures and then added the protein-digesting enzyme *pepsin* to further degrade residual cell protein adhering to the nuclei. Nuclei isolated in this manner were then extracted with alkali, and the chemical composition of the extract was analyzed. In view of the unique chemical composition of the extract (which differed markedly from protein), Miescher called this material "nuclein." By comparing the chemical analyses reported by Miescher with those carried out more recently, it is clear that Miescher's nuclein was, in fact, the nucleic acid DNA (deoxyribonucleic acid). The term **nucleic acid** was introduced 20 years following Miescher's discovery by another biochemist, Richard Altmann.

Miescher also worked with the sperm cells of salmon, which contain particularly large nuclei (more than 90% of the cell mass is accounted for by the nucleus). In addition to isolating nuclein from the sperm nuclei, he used acid to extract an organic material having an unusually high nitrogen content. He called this substance "protamine" and suggested that this basic material formed an insoluble complex with the acidic nuclein in the nucleus. We now recognize that Miescher's protamine extract contained the histones with which the DNA of the nucleus is associated (see Chapter 20). The function of the cell nucleus was unknown in Miescher's time, and while Miescher was convinced of the fundamental importance of nuclein, especially during fertilization, it was not until 60 years after his discovery that a series of experiments was carried out that established the genetic role of the nucleic acids. Also during this period, it was shown that two types of nucleic acid occur in cells: deoxyribonucleic acid and ribonucleic acid (RNA). Most notable among the experiments that established the genetic role of the nucleic acids were those on (1) the *transformation of bacterial types* and (2) *virus reproduction*.

Transformation of Bacterial Types

Two types of pneumonia bacteria *(Diplococcus pneumoniae)* exist and are readily distinguished by the appearance of their colonies when cultured on agar plates; they are called "smooth" *(S)* and "rough" *(R)* types. The *S* type (which is the normal, *virulent* kind) is enclosed within a polysaccharide capsule and gives rise to smooth, shiny colonies. In contrast, the *R* type is noninfective (i.e., nonvirulent), is unable to synthesize the polysaccharide capsule, and gives rise to granular (rough-appearing) colonies. In 1928, F. Griffith showed that it was possible to transform the rough bacteria into the smooth type. He simultaneously injected small numbers of live *R* bacteria and large numbers of heat-killed *S* bacteria into mice, many of which subsequently died. When these mice were examined, they were found to contain live *S* bacteria. Since the *S* bacteria originally injected into the mice were incapable of reproducing, Griffith concluded that some of the *R* bacteria must have been transformed into the *S* type in the presence of the dead *S* cells. These observations were subsequently confirmed by a number of investigators who also ruled out the possibility of contamination of the inoculum by a few live *S* cells or the mutation of some of the *R* cells into the *S* type following injection.

In 1932, J. L. Alloway showed that similar transformations were possible in vitro, for when an extract of *S* cells was added to a culture of *R* cells, some of the latter were permanently transformed into the viable *S* type. Therefore, there appeared to be a substance in *S* cells that was capable of bringing about an inheritable change in the *R* cells; this unknown material was termed the *transforming principle*.

The transforming principle was finally identified in 1944 by O. T. Avery, C. M. MacLeod, and M. McCarty. Although crude extracts of heat-killed *S* cells were found to contain protein, polysaccharide, lipid, and nucleic acid, the removal of the protein, polysaccharide, and lipid by a combination of chemical procedures, including enzymatic hydrolysis, chloroform extraction, and alcohol fractionation, resulted in a product that retained the transforming activity. However, when the product was treated with the enzyme *deoxyribonuclease* (which degrades DNA), the capacity to transform *R* cells was lost. This evidence, together with chemical analyses, showed that the transforming principle was DNA. These experiments have been repeated several times since then, and similar transformations have been demonstrated in several other bacterial species.

Virus Reproduction

As noted in Chapter 1, viruses are composed of protein and nucleic acid. Depending on the type of virus, the nucleic acid is either DNA or RNA. The protein forms a coat around the head and tail structures of the particle and encloses the core of nucleic acid. In 1952, A. D. Hershey and M. W. Chase conducted a series of experiments to determine whether the viral protein or nucleic acid was required for virus reproduction. Hershey and Chase employed the bacterium *E. coli* and the T2 virus (a DNA-containing virus that infects *E. coli*) in two sets of experiments. In one set of experiments, *E. coli* was cultured in a medium containing the radioactive isotope of sulfur, ^{35}S. During the growth of the bacterial population, ^{35}S was incorporated into the cells. The culture was then infected with the T2 *bacteriophage*, and during the reproduction of the bacteriophage within the host cells, bacterial ^{35}S was used in the synthesis of phage protein (i.e., it was incorporated into cysteine and methionine residues). Nucleic acids do not contain sulfur. Following lysis of the bacterial cells, the ^{35}S-labeled viruses were collected and used to infect *E. coli* cultured on media devoid of ^{35}S. A Waring blender was then used to separate by physical agitation what was left of the attached viruses from the surfaces of the bacterial hosts. A comparison of the radioactive sulfur content of the bacteria and freed viruses (Chapter 14) revealed that nearly all the ^{35}S remained with the viruses and had not entered the bacterial cytoplasm.

In other experiments, *E. coli* was cultured in media containing the radioisotope ^{32}P prior to infection with the phage. The nucleic acids contain phosphorus, and therefore ^{32}P was incorporated into newly synthesized viral DNA. When labeled viruses were used to infect additional bacteria and the blender again was employed to shake off the attached viruses, it was found that the ^{32}P radioactivity was within the infected cells. The two sets of experiments are depicted in Figure 7–1. Hershey and Chase concluded that it was the DNA of the virus that entered the host cell during infection and that DNA was required for the reproduction of genetically identical virus particles by the metabolic machinery of the host cell. It should be noted that some viruses contain RNA instead of DNA; in these cases, it is the viral RNA that enters the host cell during infection.

The observations of Hershey and Chase also explained earlier findings by T. F. Anderson and R. M. Herriott that the T2 phage loses its ability to reproduce when distilled water is added to a suspension of the virus particles prior to their addition to a bacterial culture. Even though these viruses were still able to attack the bacterial host, the sudden osmotic shock caused them to empty their nucleic acid content into the suspending medium.

Tobacco mosaic virus (TMV), the virus that infects tobacco leaves, contains RNA instead of DNA. In the early 1950s, H. Fraenkel-Conrat separated the RNA and protein components of TMV and found that the RNA separately injected into the tobacco leaves could infect the plant and produce new virus particles containing both RNA and protein. Fraenkel-Conrat also found that when protein isolated

Figure 7–1
The experiments of Hershey and Chase (see text for explanation).

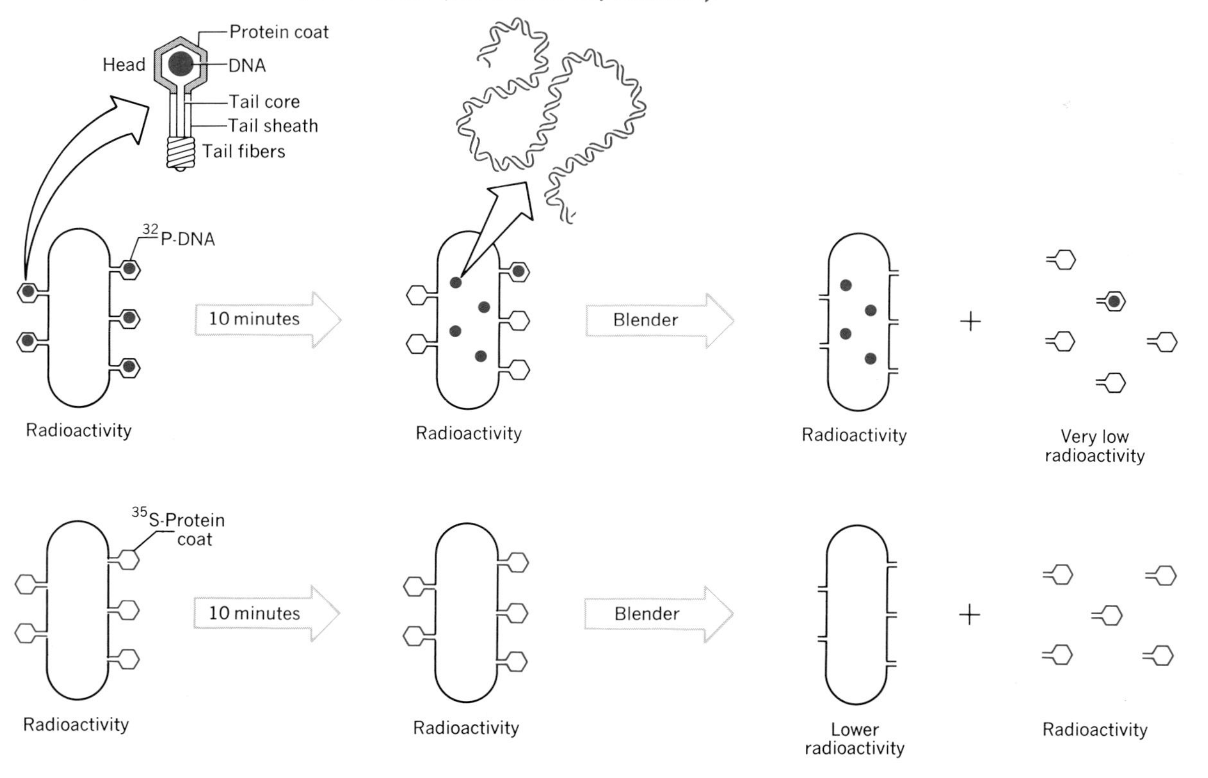

from one strain of TMV was mixed with RNA isolated from another strain, a reconstituted "hybrid" virus was produced that retained the ability to infect the tobacco leaves. The new viruses produced following infection by the hybrid were isolated and their RNA and protein components separated and analyzed. It was found that the type of protein in the virus coat was identical to that of the strain used as the source of RNA (see Fig. 7–2); that is, it was *not* the same as the protein of the hybrid viruses causing the infection. These observations support earlier conclusions concerning the genetic role of the nucleic acids and also prove that these molecules alone contain information that determines the specific nature of newly synthesized protein.

Although the most direct evidence supporting the notion that DNA (or in the case of some viruses, RNA) is the genetic material was obtained using microbial systems, observations supporting this idea were also made with higher organisms. Based on numerous studies of the mechanism of fertilization, including those by O. Hertwig (1865), H. Fol (1877), E. Strasburger (1884), A. Weismann (1892), and E. B. Wilson (1895), it was generally acknowledged by the turn of the century that the chromosomes contained within the cell nucleus were concerned with the transmission of heredity. Therefore, the development of staining reactions (such as the Feulgen reaction) that are specific for DNA and the subsequent microscopic localization of the DNA in the chromosomes implicated DNA as the genetic material. Further evidence was provided during the 1940s from quantitative chemical analyses of DNA present in measured quantities of cells (i.e., numbers, dry weight, etc.). These analyses revealed that the amount of DNA per cell was more or less constant within the various tissues of an organism. Moreover, it was also found that the total quantity of DNA present in the cell nucleus was related to its *ploidy* (i.e., the number of complete sets of chromosomes). The gametes (sperm and egg cells) of an organism were shown to have one-half as much DNA as the diploid somatic cells. This is precisely what would be expected if DNA served as the genetic material.

Although RNA appears to be the genetic material of some virus particles, there is no evidence that it plays a similar role in cells. Instead, RNA serves as an intermediary be-

Figure 7–2
Fraenkel-Conrat's reconstitution experiments demonstrating that RNA and not protein is the genetic material in tobacco mosaic virus (see text for details).

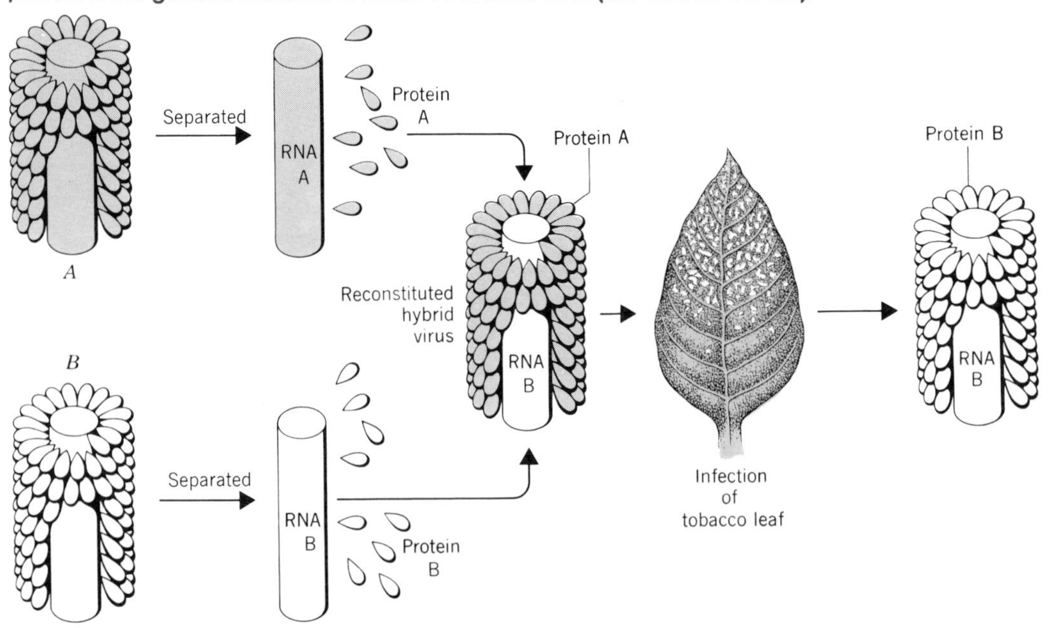

tween the genetic information of DNA, confined for the most part to the cell nucleus, and the expression of this information, primarily in the cytoplasm, as the synthesis of the cell's enzymes and other proteins (Chapter 21). Before we explore just how the relationship between the two nucleic acids can be effected and the mechanism by which DNA serves as the genetic material, it is first necessary to consider the chemistry of these two macromolecules.

Composition and Structure of the Nucleic Acids

As we have already noted, there are two major classes of nucleic acids: DNA and RNA. Both are composed of unbranched chains of subunits called **nucleotides,** each of which contains (1) a nitrogenous base (either a *purine* or *pyrimidine*), (2) a *pentose,* and (3) *phosphoric acid.* In RNA, the pentose is *ribose,* whereas in DNA it is *2-deoxyribose*. Both DNA and RNA contain the purine nitrogen bases *adenine* (abbreviated A) and *guanine* (G) and the pyrimidine *cytosine* (C), but in DNA a second pyrimidine is *thymine* (T), while in RNA it is *uracil* (U). A number of other nitrogenous bases have been identified in DNA and RNA, but these occur much less frequently. The phosphoric acid component of each nucleotide is, of course, chemically identical in both nucleic acids. These relationships are summarized in Table 7–1, and the chemical formulas of the five major nitrogen bases are shown in Figure 7–3.

The pentose of each nucleotide unit is simultaneously bonded through its number 1 carbon atom to the nitrogen base (forming a **nucleoside;** see Table 7–2) and through its number 5 carbon atom to phosphoric acid. The structures of two nucleotides together with the specific numbering system used to identify each constituent atom are shown in Figure 7–4. Successive nucleotides of DNA and RNA are joined together by ester linkages between the 5′-phosphate group of one unit and the 3′-hydroxyl group of the neighboring unit (Figure 7–5).

The "backbone" of a nucleic acid molecule is formed by the repeating sequence of pentose and phosphate groups, and this is the same in all molecules. What distinguishes one DNA (or RNA) molecule from another is the specific *sequence* of purine and pyrimidine bases present in the chain of nucleotides and the *total number* of nucleotides (i.e., the size of the molecule).

Table 7–1
Chemical Constituents of DNA and RNA

	DNA	RNA
Purines	Adenine Guanine	Adenine Guanine
Pyrimidines	Cytosine Thymine	Cytosine Uracil
Pentose	2-Deoxyribose Phosphoric acid	Ribose Phosphoric acid

Adenine
(6-aminopurine)

Guanine
(2-amino-6-oxypurine)

Thymine
(2, 6-dioxy-5-methyl pyrimidine)

Cytosine
(2-oxy-6-aminopyrimidine)

Uracil
(2, 6-dioxypyrimidine)

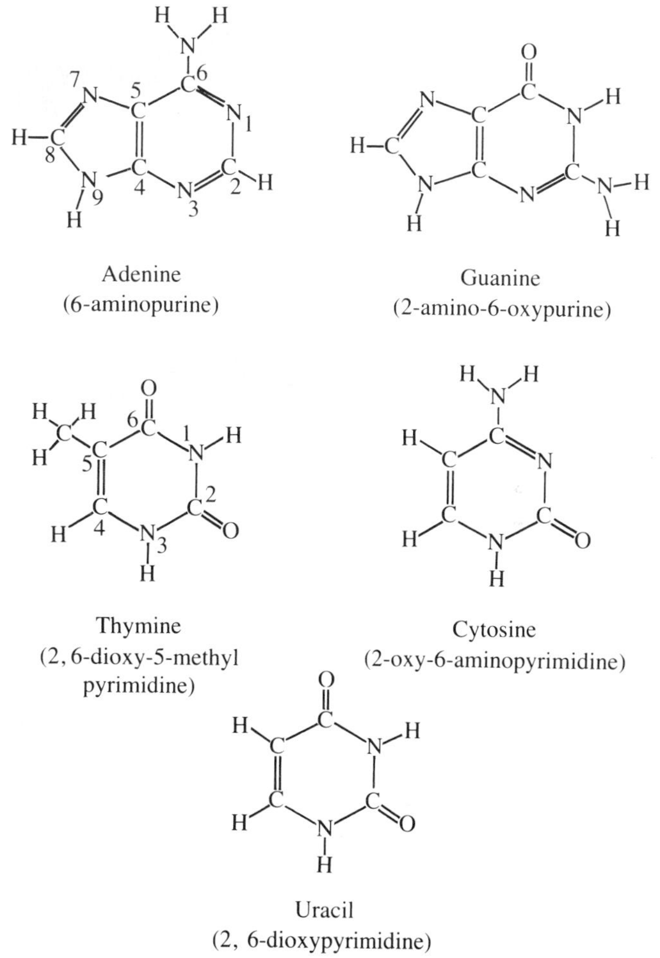

Figure 7–3
Purines and pyrimidines of DNA and RNA.

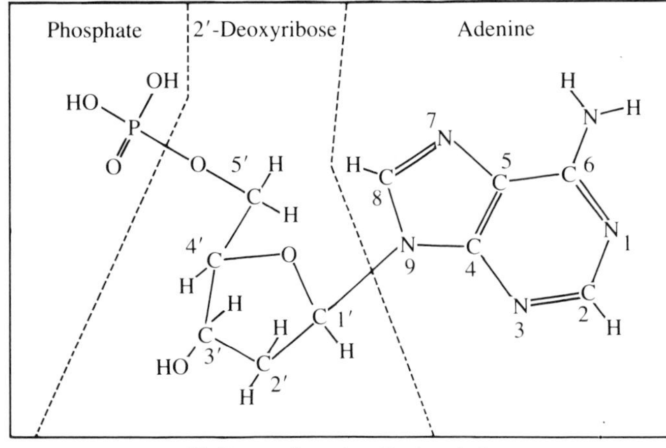

2′-Deoxyadenosine-5′-phosphate
(a nucleotide of DNA)

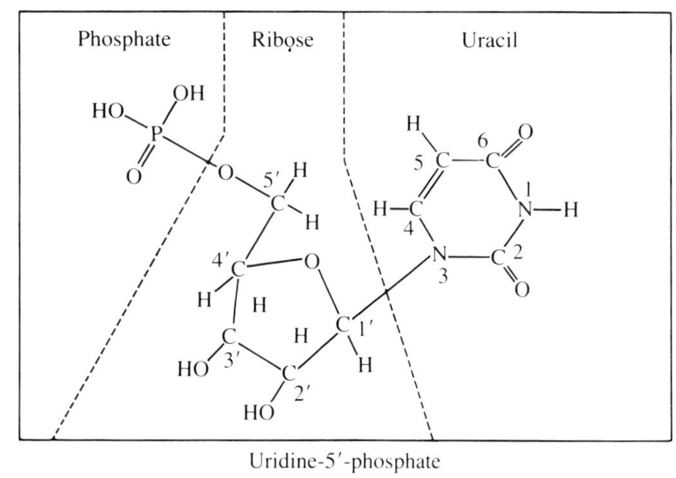

Uridine-5′-phosphate
(a nucleotide of RNA)

Figure 7–4
Chemical formulae of two representative nucleotides.

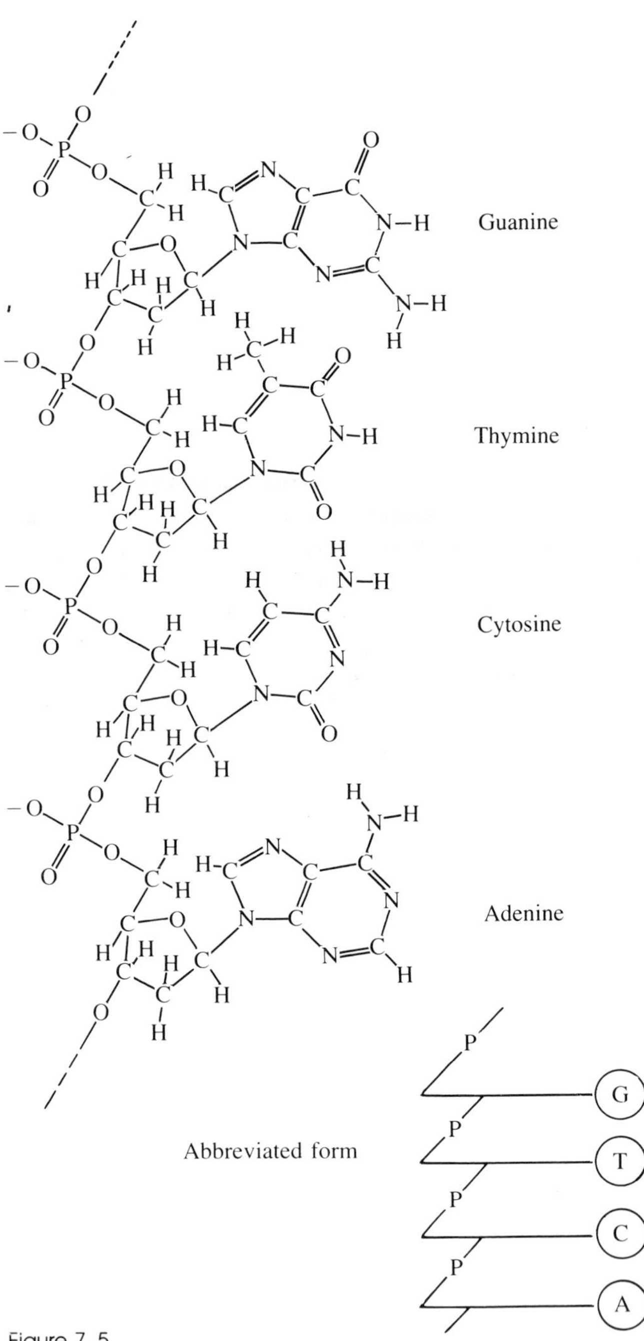

Figure 7–5
Segment of DNA molecule showing four successive nucleotide units. Except for the substitution of ribose for deoxyribose and uracil for thymine, the structure of RNA is the same. Insert at bottom right shows widely used abbreviated form.

Structure of DNA

At one time, it was believed that the four purines and pyrimidines of DNA occurred in approximately equal amounts in the molecule. However, the studies of E. Chargaff and others in the late 1940s showed that this was not the case. Instead, they found that the relative amounts of the nitrogen bases varied between species but were constant within a species. The constancy noted within a species was maintained regardless of the tissue or organ from which the DNA was isolated. Furthermore, the relative amounts of the nitrogen bases were similar in closely related species and quite different in unrelated species. Chargaff also made the following extremely important finding. *Regardless of the species used as the source of DNA, the molar ratios of adenine and thymine were always very close to unity, and the same was true for guanine and cytosine.* No such constant relationship could be demonstrated for any other combination of nitrogen base pairs. This implied that for some reason, every molecule of DNA contained equal amounts of adenine and thymine and also equal amounts of guanine and cytosine.

Using chemical information of this sort, together with the results of x-ray crystallographic studies of DNA, J. D. Watson, F. H. C. Crick, M. H. F. Wilkins, and R. Franklin proposed a model for the structure of DNA in the early

Figure 7–6
Formation of hydrogen bonds between guanine and cytosine and between thymine and adenine. Note that three bonds are formed between *G* and *C* and only two between *T* and *A*.

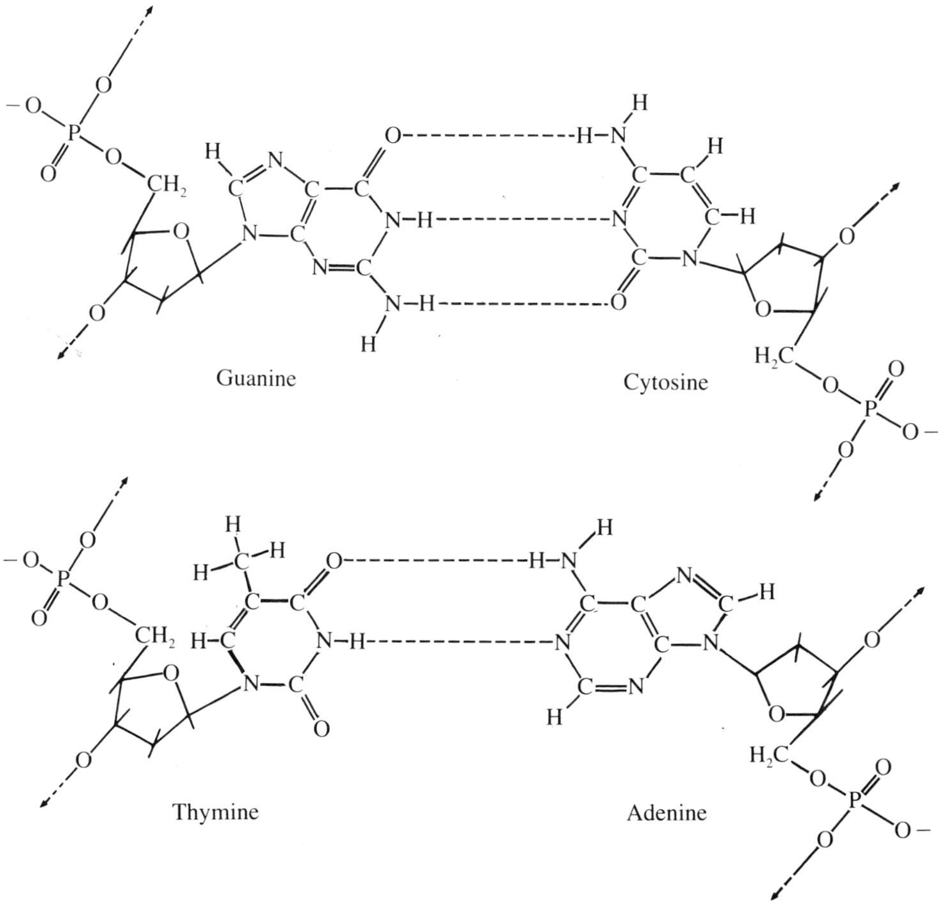

1950s. They suggested that a molecule of DNA consists of two helical polynucleotides wound around a common axis to form a right-handed "double helix." In contrast to the arrangement of amino acid side chains in helical polypeptides (where the side chains are directed radially away from the helix axis), the purine and pyrimidine bases of each polynucleotide chain were directed inward toward the center of the double helix so that they faced each other.

On the basis of stereochemical studies, Watson and Crick further suggested that the only possible arrangement of the nitrogen bases within the double helix that was consistent with its predicted dimensions was that in which a purine always faced a pyrimidine, for the diametric distance between the two polynucleotide chains is too small to accommodate two juxtaposed purines. Which purine was matched

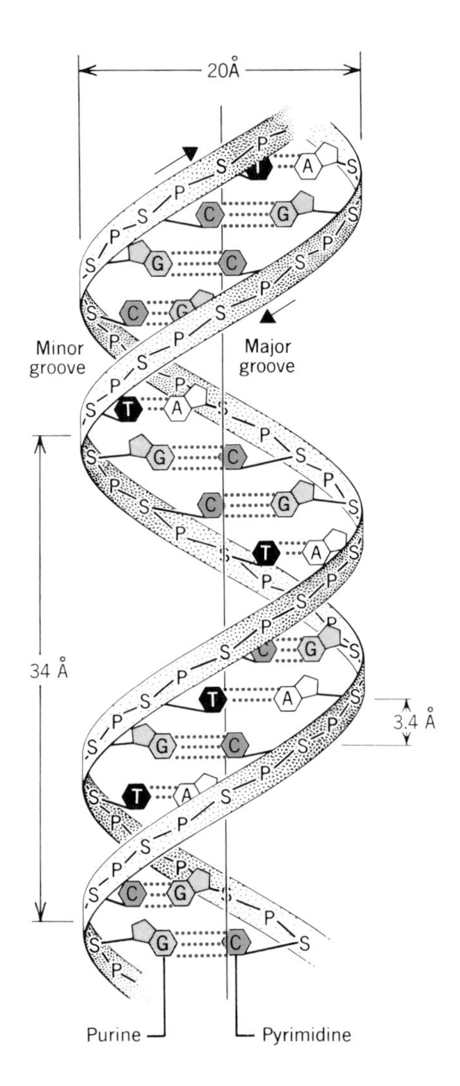

Figure 7–8
Diagram of DNA double helix showing complementary base-pairing, major and minor surface grooves, and certain molecular dimensions. The double helix has a diameter of 20 Å. Each complete turn of a helix accounts for 34 Å of linear translation and each nucleotide 3.4 Å. (Hence, there are 10 nucleotides per helical turn.) The usual symbols for the bases are used, with S for sugar and P for phosphate.

Figure 7–7
Use of abbreviated form of polynucleotide chain to demonstrate the antiparallel nature of the DNA double helix.

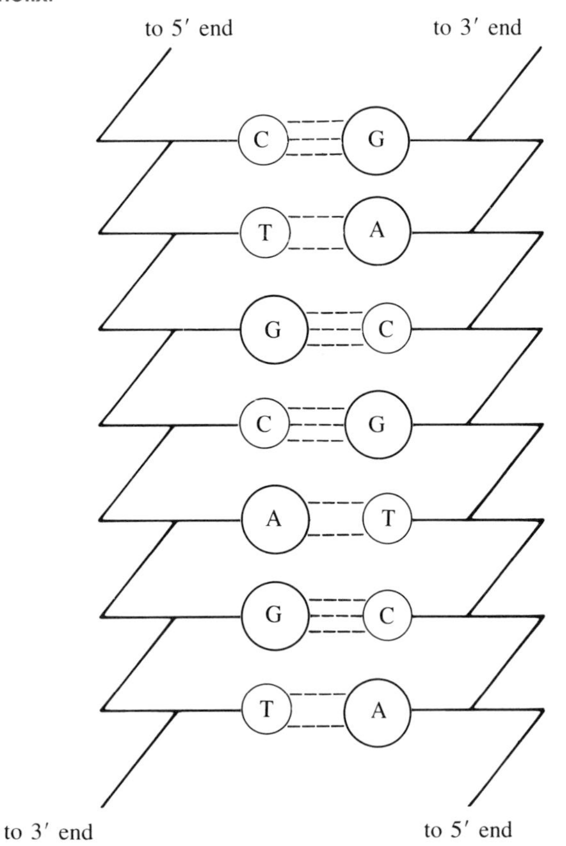

with which pyrimidine became clear from a consideration of which pairs would be able to form the hydrogen bonds necessary to stabilize the double-helical structure. Accordingly, Watson and Crick concluded that *adenine must be matched with thymine and guanine with cytosine*. This conclusion was, of course, in agreement with the chemical findings of Chargaff (see above)—in fact, Chargaff's data may have been critical to the development of Watson and Crick's proposals. The manner in which hydrogen bonds are formed between adenine and thymine and between guaning and cytosine is shown in Figure 7—6. Although individually weak, the great number of these bonds contributes appreciably to the stability of the double-helical structure. In addition, the double helix is stabilized by hydrophobic bonds between neighboring nitrogen bases of each polynucleotide chain.

Certain other features of the structure of DNA should be noted. The two polynucleotide chains that make up the molecule are *antiparallel*. That is, beginning at one end of the molecule and progressing toward the other, successive nucleotides of one chain are joined together by $3' \longrightarrow 5'$ phosphodiester linkages, whereas the complementary nucleotides of the other chain are joined by $5' \longrightarrow 3'$ phosphodiester linkages. This antiparallel arrangement is depicted diagrammatically in Figure 7–7. The two polynucleotides are twisted around one another in such a way as to produce two helical grooves in the surface of the molecule; these are called the *major* and *minor* grooves and are shown in Figure 7–8, together with some of the physical dimensions of the double helix.

Replication of DNA

One of the intrinsic properties of the genetic material is its capacity for replication. The manner in which DNA satisfies this requirement is apparent from the nitrogen base pairing required in the model. Since the sequence of bases in one polynucleotide chain automatically determines the sequence of bases in the other, it is clear that one-half of a molecule contains all the information necessary for constructing a whole molecule. For example, if we know that the sequence of bases along one polynucleotide chain of DNA is A T G C A C C G and so on, then the complementary sequence in the other chain must be T A C G T G G C, and so on. Therefore, if the double helix were unwound, each separate polynucleotide chain could act as a template for the production of a new, complementary chain. The result would be two identical double helices where there was only one before. Of course, one-half of each new double helix would be represented by one of the original polynucleotide chains. The basic features of this process are shown in Figures

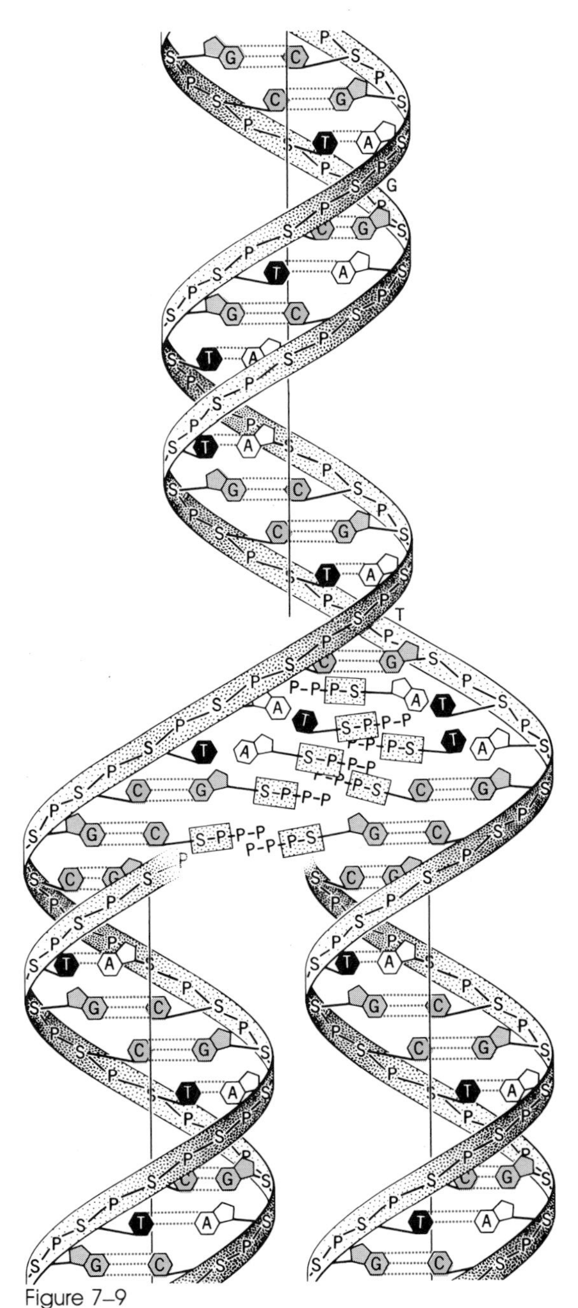

Figure 7–9

Model for the replication of DNA. The replication of the DNA molecule could be effected if the double helix unwound and the separated chains acted as templates for the production of two new complementary chains. In this figure, the original double helix has partially unwound, and two complementary chains are being formed in this region by the addition of nucleotide units.

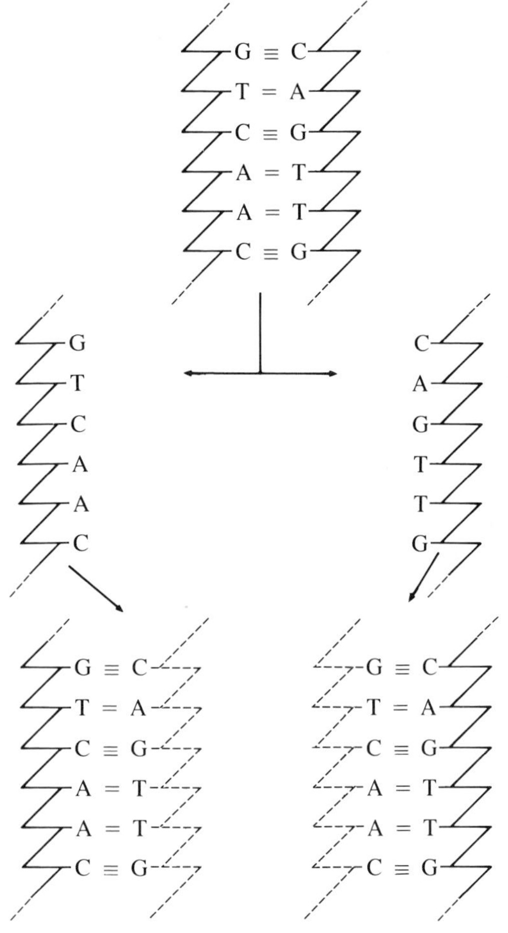

Figure 7–10
Replication of the base sequence of DNA. The complementary base sequence of a portion of a DNA molecule is shown at the top of the figure. The complementary bases in this region are separated as double helix unwinds and serve as templates for the production of two new polynucleotide chains (middle). Finally, two complete DNA molecules are produced with identical base sequences (bottom). One-half of each of these DNA molecules is represented by one newly produced chain (dashed) and by one original chain.

7–9 and 7–10. A detailed description of the mechanism by which the replication of DNA occurs, together with its experimental verification, is given in Chapter 20, which deals with the morphology, chemistry, and function of the cell nucleus.

"Single-Stranded" DNA

Although in nearly every case so far studied, DNA consists of two polynucleotide chains twisted about one another to form a double helix, it is now apparent that in a few bacterial viruses (i.e., the øX174 and S13 *E. coli* phages), DNA exists as a single polynucleotide chain. This was initially suspected when chemical analyses of the nitrogen base contents of these viral DNAs revealed that the amounts of adenine and thymine, as well as guanine and cytosine, were not equal. During reproduction of these viruses, the single-stranded DNA (referred to as the "+ strand") is injected into the host bacterial cell, where it acts as a template for the reproduction of a complementary polynucleotide chain (called the "− strand"); these two polynucleotides combine to form a conventional double helix, which then serves as the template for the production of additional + strands. The newly produced + strands are then enclosed in the viral protein coats to form new virus particles.

Structure of RNA

As noted earlier, RNA and DNA differ chemically in two notable ways: in RNA, *ribose* is the pentose (not deoxyribose as in DNA) and the pyrimidine *uracil* occurs in place of thymine (Fig. 7–4). Early analyses of the nitrogen base contents of RNAs from various sources revealed that the molar ratios of these bases were quite different from those of DNA and were also quite variable. On this basis, it was concluded that RNA occurs as a single polynucleotide chain. This contention has more recently been supported by other physicochemical studies, but it should be noted that there are some viral RNAs that are double stranded. Although only one polynucleotide chain is usually present, RNA does possess regions of double-helical coiling where the single chain loops back upon itself. These regions are stabilized by the formation of hydrophobic bonds between neighboring bases (as in DNA) and also by the formation of hydrogen bonds between opposing units of guanine and cytosine and between adenine and *uracil*. (In RNA, A and U may form two hydrogen bonds in a manner similar to the two bonds formed between A and T in DNA.)

Synthesis of RNA

Except perhaps in the case of the reproduction of certain RNA viruses (see below), the synthesis of RNA appears to be directed by DNA. The formation of the RNA polynucleotide takes place using the base sequence along only *one* of the two deoxyribonucleotide helices of DNA (producing a temporary RNA-DNA hybrid) and resulting in the release of a single, complementary polyribonucleotide chain in which the base uracil occurs in place of thymine (Figure 7–11).

Replication of RNA Viruses.

The viruses may be subdivided into two classes (1) viruses whose genetic complement consists of DNA, and (2) viruses whose genetic complement consists of RNA. In cells infected with DNA viruses, the infecting viral DNA is *replicated,* forming new viral DNA which is then transcribed into RNA; this RNA is then translated into viral protein.

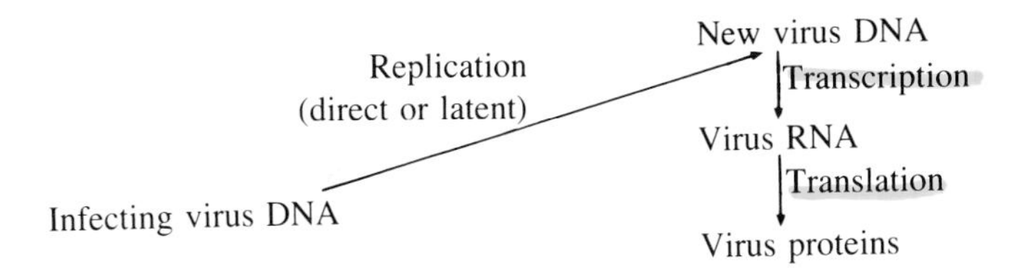

The newly produced viral DNA and viral proteins combine in the assembly of new, complete virus particles that are

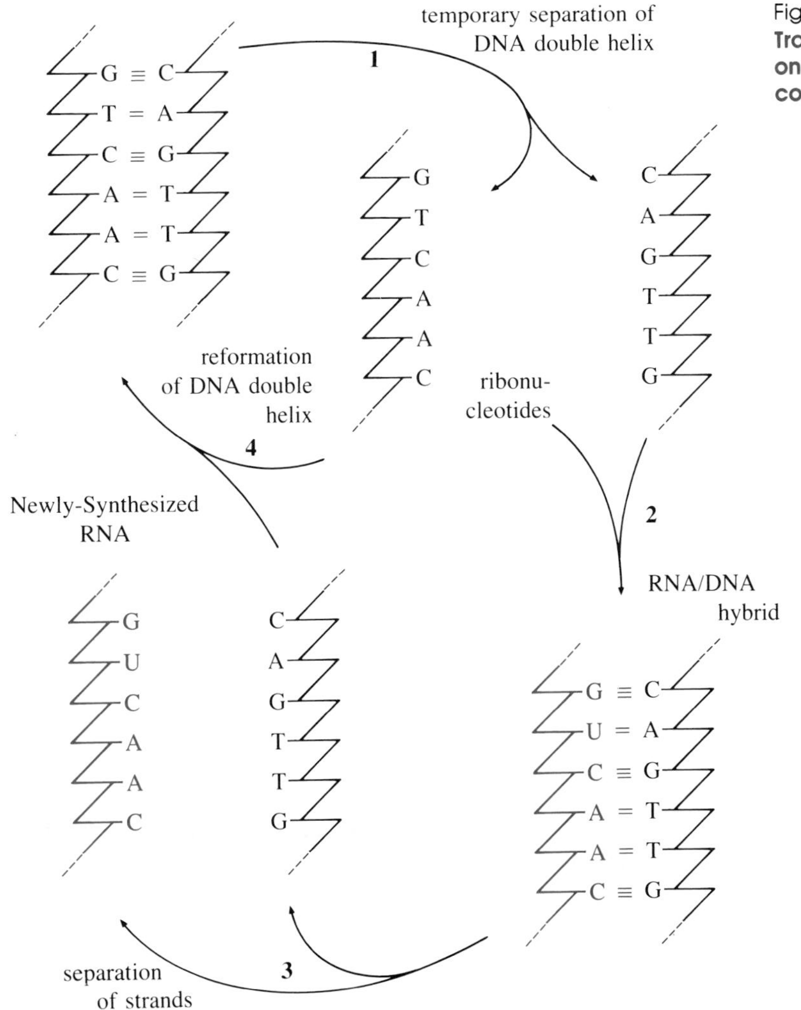

Figure 7–11
Transcription of the base sequence of one of the helices of DNA into the complementary base sequence of RNA.

released upon lysis of the host cell. A latent state (called a **provirus**) can also be established in which the viral DNA is incorporated into the host cell's genome, being replicated and distributed along with the host cell's native DNA, until it is transcribed once again into additional viral RNA and thence into viral proteins (see Chapter 1).

For most RNA viruses (e.g., poliomyelitis, influenza, common cold, etc.), DNA involvement is essentially bypassed. For example, during the infection of a cell with the polio virus, the single-stranded RNA (called a "+ strand") enters the host cell, where it acts as a template for the synthesis of complementary "− strands". The latter are then employed in the proliferation of new + strands, and these are translated into viral proteins.

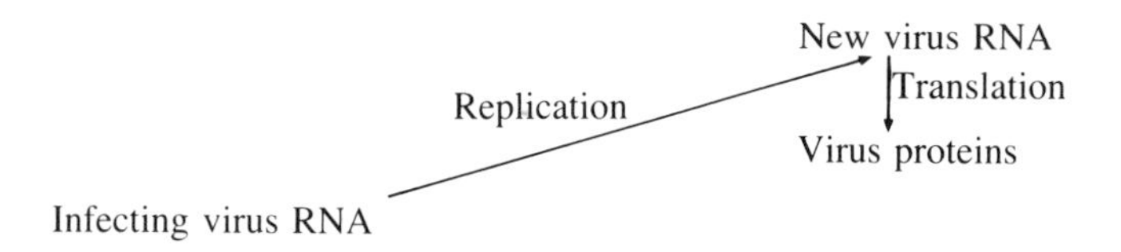

The mechanism described above is varied in several other viruses where the RNA is either double stranded (e.g., reovirus, in which only one of the two RNA strands produced during replication is transcribed) or where the infecting single RNA strand is complementary (rather than identical) to the newly produced viral RNAs that are to be translated into viral proteins (e.g., Sendai virus, Newcastle disease virus, etc.).

It has recently become clear that yet another mechanism exists in the case of the RNA tumor viruses (e.g., Rous sarcoma virus). These viruses do not transfer information from RNA to RNA, but rather from RNA to DNA and then to new RNA. The viral RNA is employed as a template for the synthesis of DNA by the infected cell (a phenomenon that has come to be called "reverse transcription"). Some of the resulting "viral DNA" may be incorporated into the genome of the host cell, thereby establishing the provirus state.

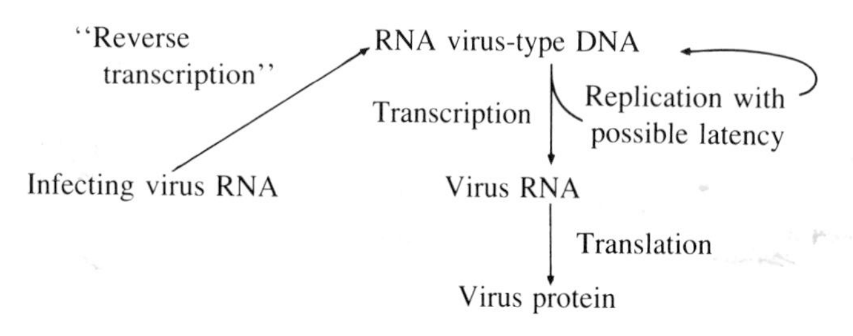

The latent or provirus state in which viral DNA is incorporated into the host cell genome has been suggested as the basis of a number of different RNA virus-induced and DNA virus-induced cancers. According to this view, one or more of the provirus genes—which are normally repressed by the host cell—may become derepressed and cause the production of an *oncogenic* (i.e., cancer-causing) substance that alters the cell's normal properties or behavior. Such a change may be delayed for a number of generations, depending upon the period of latency.

During the 1960s and early 1970s, the so-called "central dogma" of molecular biology was the orderly and *unidirectional* flow of information encoded in the base sequences of a cell's DNA to RNA and then to protein; that is

DNA ⟶ RNA ⟶ protein

The discovery of reverse transcription by certain RNA viruses in which the information of RNA is passed on to DNA has necessitated a reexamination of that dogma and raises the question of whether or not a similar interaction between RNA and DNA might normally occur in cells (i.e., cells not infected by viruses) under specific conditions (e.g., during cellular differentiation). The central dogma might more appropriately be represented as

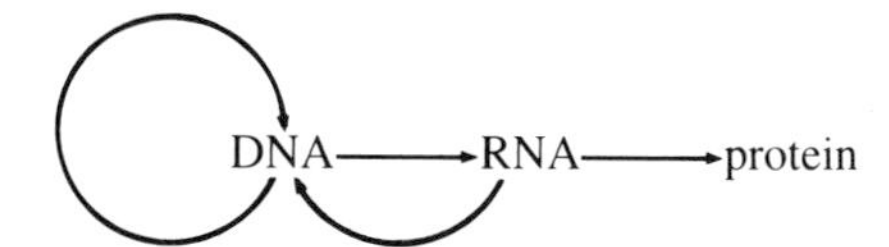

Types of Cellular RNA

Cells contain three major functional types of RNA: **ribosomal RNA** (abbreviated rRNA), **messenger RNA** (mRNA), and **transfer RNA** (tRNA). All of these are transcribed from nuclear DNA and are engaged in mediating the genetic message of DNA by participating in the synthesis of the cell's proteins, as discussed in Chapter 21. It has already been noted that RNA occasionally serves as the genetic material of viruses.

Of the cellular RNAs, rRNA is the most abundant, accounting for 50 to 80% of the total RNA of the cell. Only three or four different kinds of rRNA are present in cells, and these are confined for the most part to the cell's ribosomes. mRNA accounts for about 5 to 10% of the cell's RNA and is much more heterogeneous with respect to size and nitrogen base content than the rRNAs. This results from the relationship (see below) between the chain lengths and base sequences of mRNAs and the variable sizes and primary structures of polypeptides synthesized in a cell. Most mRNA occurs in the cytoplasm, where it transiently combines with ribosomes during protein synthesis.

About 10 to 20% of the cell's RNA is tRNA. All tRNA molecules are similar in size and typically contain 75 nucleotide units. In spite of these similarities, a single cell may contain about 60 species of tRNA differing in their base sequences. Because most of the tRNA is recovered in the cytoplasmic (i.e., soluble) phase of disrupted cells following centrifugation, tRNA is also called *soluble RNA* (i.e., sRNA). In view of its small size and relative ease of isolation, tRNA has been more extensively studied than the other two ribonucleic acids, and the specific primary, secondary, and tertiary structures of a number of tRNAs have already been determined (see Chapter 21). tRNAs contain moderate amounts of unusual nucleotides such as *ribothymidine, dihydrouridine, pseudouridine,* and *methylguanosine*. These are formed by modification of the four common RNA bases and play a crucial role in establishing the unique spatial organization of these molecules.

Although the mechanisms of DNA replication and protein synthesis are considered in depth in subsequent chapters, it is appropriate that a brief accounting of the functional relationships among the nucleic acids and between nucleic acids and proteins be made at this time. Inheritable information is encoded in the various nitrogen base sequences possessed by the cell's DNA, and by the processes of *transcription* and *translation,* these base sequences are employed to specify the primary structures of all proteins produced by the cell. Most important among these proteins are the enzymes that catalyze and regulate the myriad of chemical reactions characertizing the cell's metabolism. Therefore, the information of DNA confined essentially to the cell nucleus manifests itself primarily in the cytoplasm as the synthesis of a unique assemblage of proteins. The replication of DNA that precedes mitotic cell division and the equal distribution of the duplicated DNA among the progeny cells provides for the passage of complete sets of information from one generation of cells to another. In addition to serving as templates for their own replication, the nucleotide sequences of DNA are used during transcription to produce complementary base sequences of RNA. The resulting RNAs then serve as intermediaries in translating the original message into protein. Of paramount importance in this process are the mRNA molecules whose base sequences directly determine the primary structures of the polypeptides. These mRNA molecules leave the nucleus of the cell following their synthesis and attach in the cytoplasm to one (or several) ribosomes. The rRNA of each ribosome is believed to play a role in this attachment. tRNA molecules also produced in the nucleus enter the cytoplasm, where they combine with specially activated amino acids (distinct tRNAs exist for each species of amino acid). The resulting complexes, directed by the base sequences of mRNA attached to the ribosomes, sequentially deposit their amino acids in the growing polypeptide chains. The details of this process, described only superficially at this time, are examined in Chapter 21.

Summary

Although **nucleic acids** were discovered and chemically characterized more than 100 years ago, their genetic role was not appreciated until the phenomenon of bacterial transformation and the mechanism of virus reproduction were understood. The two major nucleic acids, DNA and RNA, are composed of unbranched chains of subunits called **nucleotides,** each nucleotide containing **phosphate,** a **pentose,** and a **nitrogenous base.** Specific differences exist in the nucleotides found in DNA and in RNA. In DNA, two **antiparallel** polynucleotide chains are twisted around one another to form a *double helix,* the helices being held together in part through hydrogen bonds between juxtaposed nitrogenous bases. Constraints imposed by the known shape and organization of the double helix and the nitrogenous bases capable of forming such stabilizing bonds suggest that the base sequence in one polynucleotide necessarily determines the sequence of bases in the other. Therein lies the capacity for replication, and DNA serves as the genetic material in all cells and in many viruses.

Most RNAs are formed by a single polynucleotide twisted about itself in certain regions, thereby forming a periodic double-helical structure. In certain viruses, RNA serves as the genetic material, but in procaryotic and eucaryotic cells, RNA serves three primary functions. The **ribosomal** RNAs are the most abundant and serve as functional components of the cell's ribosomes during protein synthesis. The **messenger** RNAs possess the sequence of nucleotides specifying the primary structures of the cell's proteins. **Transfer** RNAs function in the transport of amino acids to the messenger RNA–ribosome complex during protein synthesis or **translation.** The cell's RNAs are produced by transcription of its DNA.

References and Suggested Reading

Articles and Reviews

Baltimore, D., Viruses, polymerases and cancer. *Science 192,* 632 (1973).

Britten, R. J., and Kohne, D. E., Repeated segments of DNA. *Sci. Am.* 222 (4), 24 (April 1970).

Brown, D. D., The isolation of genes. *Sci. Am. 229*(2), 20 (Aug. 1973).

Campbell, A. M., How viruses insert their DNA into the DNA of the host cell. *Sci. Am. 235*(6), 102 (Dec. 1966).

Chan, H. W., Israel, M. A., Garon, C. F., *et al.*, Molecular cloning of polyoma virus DNA in *Escherichia coli:* plasmid vector system. *Science* 203, 883 (1979).

Crick, F. H. C., The structure of the hereditary material. *Sci. Am. 194* (4), 54 (Oct. 1954).

Fiddes, J. C., The nucleotide sequence of a viral DNA. *Sci. Am. 237*(6), 54 (Dec. 1977).

Grobstein, C., The recombinant-DNA debate. *Sci. Am. 237*(1), 22 (July 1977).

Holley, R. W., The nucleotide sequence of a nucleic acid. *Sci. Am. 214*(2), 30 (Feb. 1966).

Kornberg, A., The synthesis of DNA. *Sci. Am. 219*(4), 64 (Oct. 1968).

Mirsky, A. E., The discovery of DNA. *Sci. Am. 218*(6), 78 (June 1968).

Reddy, V. B., Thimmappaya, B., Dhar, R., et al., The genome of simian virus 40. *Science 200,* 494 (1978).

Sinsheimer, R. L., Single-stranded DNA. *Sci. Am. 207*(1), 109 (July 1962).

Stent, G. S., The multiplication of bacterial viruses. *Sci. Am. 188*(5), 36 (May 1953).

Temin, H. M., RNA-directed DNA synthesis. *Sci. Am. 226*(1), 24 (Jan. 1972).

Books, Monographs, and Symposia

Braun, W., *Bacterial Genetics,* W. B. Saunders Co., Philadelphia, 1953.

Goldstein, L., *The Control of Nuclear Activity,* Prentice-Hall, Inc., Englewood Cliffs, N.J., 1967.

Haggis, G. H., Michie, D., Muir, A. R., et al., *Introduction to Molecular Biology,* John Wiley & Sons, Inc., New York, 1965.

Lehninger, A. L., *Biochemistry,* Worth Publishers, Inc., New York, 1975.

Levine, R. P., *Genetics,* Holt, Rinehart and Winston, Inc., New York, 1962.

Luria, S. F., Darnell, J. E., Baltimore, D., and Campbell, A., *General Virology* (3rd ed.), John Wiley & Sons, Inc., New York, 1978.

McGilvery, R. W., *Biochemical Concepts,* W. B. Saunders Co., Philadelphia, 1975.

Stewart, P. R., and Letham, D. S. (editors), *The Ribonucleic Acids,* Springer-Verlag, New York, 1973.

Stryer, L., *Biochemistry,* W. H. Freeman and Co., San Francisco, 1975.

Taylor, J. H., *Selected Papers on Molecular Genetics,* Academic Press, New York, 1965.

Watson, J. D., *Molecular Biology of the Gene* (3rd ed.), W. A. Benjamin, Inc., Menlo Park, Calif., 1976.

Woodward, D. O., and Woodward, V. W., *Concepts of Molecular Genetics,* McGraw-Hill Book Co., New York, 1977.

Part 3
CELL METABOLISM

Chapter 8
ENZYMES

The metabolism of the cell is characterized by a myriad of simultaneously occurring chemical reactions. Nearly all of these reactions are catalyzed by a special class of proteins called **enzymes.** It has been estimated that an average cell contains about 3000 different enzymes. In the absence of these enzymes, the cellular reactions would proceed at a much slower, perhaps negligible rate. Enzymes integrate the cellular chemical reactions and provide the order without which the complex processes of life would not be possible. Enzyme molecules themselves are not consumed in the chemical reactions that they catalyze; instead, they are made available over and over again to catalyze additional reactions. Enzymes are characterized by a high degree of specificity; that is, they will catalyze one particular chemical reaction but not another.

More than a thousand different enzymes have been specifically identified, and of these, several hundred have already been isolated or crystallized. Most of the enzymes responsible for such diverse functions as alimentary digestion, blood coagulation, muscle contraction, carbohydrate and fat metabolism, and nucleic acid biosynthesis are known. Before considering the properties of enzymes and the mechanisms of enzymic catalysis, some general features of chemical reactions will be reviewed, including the various levels of **molecularity** and the associated **reaction kinetics**.

Molecularity of Chemical Reactions

Chemical reactions may be classified according to the **level of molecularity;** for example, *monomolecular* reactions, *bimolecular* reactions, *trimolecular* reactions, and so on. A monomolecular reaction may be written

$$A \rightleftharpoons P \tag{8–1}$$

and involves the conversion of one molecular species (i.e., A) to another (i.e., P) without the addition or removal of atoms. Instead, an intramolecular reorganization of the molecule occurs. The interconversion of the α and β isomers of glucose via the open-chain intermediate is an example of this type of reaction (see Chapter 5). In the case just cited, the molecular changes are spontaneous and do not require catalysis by an enzyme. In contrast, the conversion of glucose-6-phosphate to fructose-6-phosphate during glycolysis (Chapter 10) is an example of an enzyme-catalyzed monomolecular reaction. Enzymes catalyzing intramolecular reorganizations are termed **isomerases.**

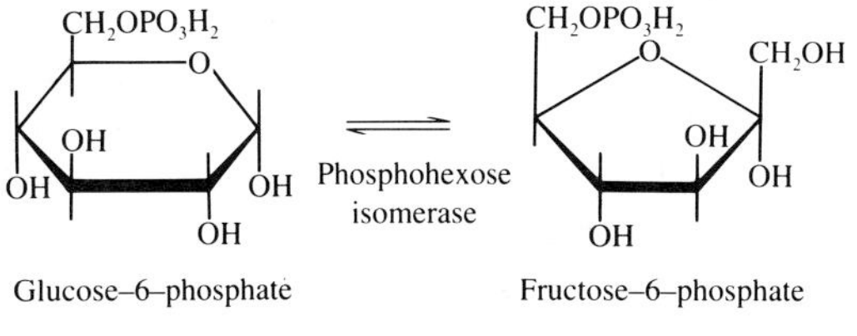

$CH_2OPO_3H_2$ / OH / OH / OH / OH — Glucose–6–phosphate $\rightleftharpoons$ (Phosphohexose isomerase) $CH_2OPO_3H_2$ / CH_2OH / OH / OH / OH — Fructose–6–phosphate

Monomolecular reactions are relatively uncommon in cells. Much more numerous are bimolecular reactions and reactions of higher order, involving two or more reactants and/or two or more products; for example,

$$A + B \rightleftharpoons P \qquad (8\text{–}2)$$

$$A \rightleftharpoons P_1 + P_2 \qquad (8\text{–}3)$$

$$A + B \rightleftharpoons P_1 + P_2 \qquad (8\text{–}4)$$

$$A + B + C \rightleftharpoons P \qquad (8\text{–}5)$$

Among the major classes of enzymes catalyzing bimolecular and higher order reactions are the **hydrolases, dehydrogenases, decarboxylases,** and **transferases.** Hydrolases catalyze reactions in which water is either added to or removed from the reactant(s). Dehydrogenases oxidize compounds by catalyzing the removal of hydrogen atoms. Decarboxylases remove carbon dioxide from carboxylic acids, and transferases remove reactive groups from one compound and transfer them to another. Numerous examples of these and other enzyme-catalyzed cellular reactions will be encountered in subsequent reading.

Reaction Kinetics

To begin our consideration of the kinetics of chemical reactions, consider the reaction described by the equation 8–2 in which *A* and *B* are the reactants and *P* is the single product of the reaction. One prerequisite for the reaction to occur is the collision of molecules *A* and *B*. However, a collision alone is not sufficient to guarantee the formation of molecule *P*. In addition, molecules *A* and *B* must collide with some minimal amount of kinetic energy and must be oriented with respect to one another at the time of collision such that the chemical bonds strained by the collision allow the shifts of orbital electrons necessary for the formation of the product.

The probability of a reaction occurring between *A* and *B* is determined in part by the probability of finding molecules *A* and *B* in the same region of space at the same time (i.e., this would constitute a collision). The probability of finding molecule *A* in a specific region of space is directly proportional to its concentration, and the same is true for molecule *B*. Therefore, the probability of finding both molecules *A* and *B* in the same region of space at the same instant in time is equal to the product of the probabilities of finding *either one* there. That is,

$$\text{Probability of collision} \sim (A)\,(B) \qquad (8\text{–}6)$$

where (A) is the concentration of molecule *A* and (B) is the concentration of molecule *B* in the solution. The same line of reasoning leads us to conclude that the rate of the chemical reaction, that is, the change in the concentrations of *A* and *B* with time as they are converted to *P* (i.e., $\Delta(A)/\Delta t$ or $\Delta(B)/\Delta t$) would be directly proportional to $(A)\,(B)$.

For a solution of any molecular species, not all the molecules will have the same kinetic energy. Instead, the energies are distributed as shown in Figure 8–1. This distribution, known as a Maxwell-Boltzmann distribution, shows that molecules in the solution have greatly varying kinetic energies. This variation results from random collisions of the molecules with one another during which some molecules gain kinetic energy while others lose kinetic energy. For each chemical reaction, there is a minimum kinetic energy that the participating molecules must have in order for the chemical reaction to occur. As Figure 8–1 shows, only a small percentage of the molecules in solution have this energy at any instant in time (those above E_1).

Even though a collision might involve two molecules with the requisite kinetic energies, it is still necessary that they be oriented with respect to one another in a specific manner at the instant of collision in order for the reaction to proceed. Thus, only a small percentage of the collisions involving molecules with the requisite kinetic energy will result in the appropriate chemical reaction. Hence, the actual rate at which the reaction takes place is equal to some constant, k_1, times the probability of a collision. This constant, called a **rate constant,** accounts for both molecular orientation and kinetic energy considerations. Thus,

$$\text{Rate of reaction} = \frac{\Delta(A)}{\Delta t} \quad \frac{\Delta(B)}{\Delta t} = k_1\,(A)\,(B) \qquad (8\text{–}7)$$

If the reaction is a reversible one, then

$$\frac{\Delta(P)}{\Delta t} = k_2\,(P) \qquad (8\text{–}8)$$

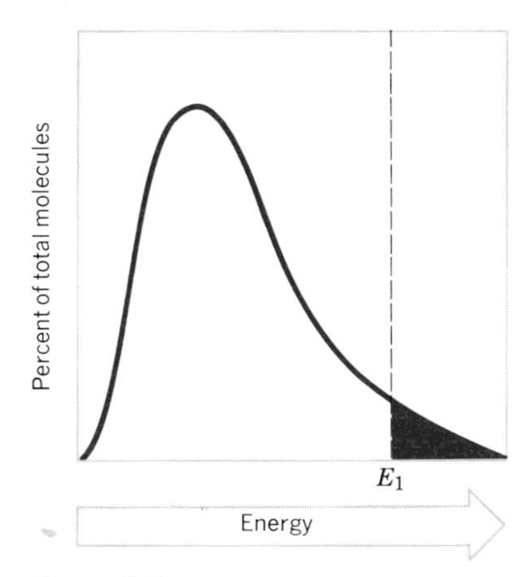

Figure 8–1
Maxwell-Boltzmann distribution of molecular kinetic energies. Not all molecules in solution have the same kinetic energy; instead, these energies are distributed as shown. Only those molecules having some minimal kinetic energy (shown at E_1) may participate in a particular chemical reaction.

Once a reversible reaction has reached equilibrium, no *net* change in the concentrations of the reactants or the product(s) occurs. Therefore, at this time

$$\frac{\Delta(A)}{\Delta t} = \frac{\Delta(B)}{\Delta t} = \frac{\Delta(P)}{\Delta t} \tag{8–9}$$

and

$$k_1\,(A)\,(B) = k_2\,(P) \tag{8–10}$$

Consequently,

$$\frac{k_2}{k_1} = \frac{(A)\,(B)}{(P)} = K \tag{8–11}$$

where K is the **equilibrium constant** for the reaction and describes the relative concentrations of all reacting molecular species at equilibrium.

Effect of Enzyme on Reaction Rate

Two general mechanisms are involved in enzyme catalysis. First, the presence of the enzyme increases the likelihood that the potentially reacting molecular species will encounter each other with the required orientations in space. This occurs because the enzyme has a high affinity for the reactants (more appropriately referred to as the **substrates**) and forms a temporary chemical union with them. The association of substrates with an enzyme is not arbitrary; instead the substrates are bound to the enzyme in such a manner that each substrate is oriented with respect to the other in precisely the manner required for the reaction to occur. Second, the formation of temporary bonds (mostly noncovalent) between the enzyme and substrate forces a redistribution of electrons within the substrate molecules, and this redistribution imposes a *strain* upon the maintenance of specific covalent bonds within the substrates—a strain that culminates in bond breakage. Biochemists refer to the introduction of bond strains in a substrate by association with an enzyme as ''substrate activation.'' The net effect of this is to greatly increase the percentage of molecules in the population that at any instant in time are sufficiently reactive to react with each other. This rather brief and general accounting of enzyme function is expanded later in the chapter as the specific structures and actions of certain representative enzymes are considered.

For a reversible enzyme-catalyzed reaction such as

$$A + B \overset{E}{\rightleftharpoons} P \tag{8–12}$$

where E is the enzyme (and can catalyze the reaction in both directions)

$$\frac{\Delta(A)}{\Delta t} = \frac{\Delta(B)}{\Delta t} = k_1(A)(B)e \tag{8–13}$$

In this equation, e is the ''enzyme factor''—a factor that accounts for the increase in reaction rate through catalysis—and is proportional to the concentration of the enzyme in solution. That is, if the reaction proceeds at some rate at a given enzyme concentration, then the reaction would proceed twice as rapidly at twice the enzyme concentration (Fig. 8–2).

For the reverse reaction,

$$\frac{\Delta(P)}{\Delta t} = k_2(P)e \tag{8–14}$$

It can be shown using equations 8–13 and 8–14 that the equilibrium constant for the reaction (i.e., k_2/k_1) is not altered by the presence of the enzyme. Thus, *the enzyme greatly affects (i.e., increases) the rate at which equilibrium is achieved, but it does not alter the respective equilibrium concentrations of the various molecular species.*

The marked effect of enzymes on reaction rates can be clearly appreciated by considering the *turnover number* of an enzyme. The turnover number is the number of moles of substrate converted to product per minute per mole of enzyme (when the enzyme is fully saturated with substrate). For example, the hydrolytic enzyme *urease,* which catalyzes the conversion of urea to ammonia, has a turnover number of about 10^6. In the absence of urease, the hydrolysis of urea is several orders of magnitude slower. *Amylase,* an enzyme involved in the breakdown of starch and other polysaccharides, can hydrolyze about 10^5 glycosidic bonds per minute, but *no* detectable hydrolysis occurs in the absence of amylase. The turnover numbers of most enzymes fall between 10^2 and 10^6 (Table 8–1).

Figure 8–2
Relationship between enzyme concentration and rate of enzyme-catalyzed reaction.

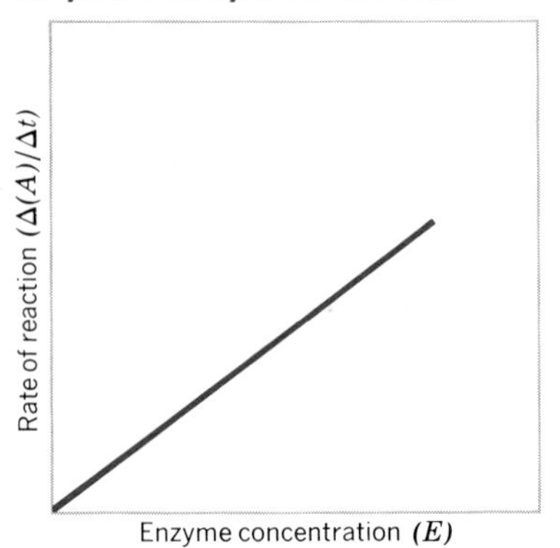

Table 8–1
Turnover Numbers of Some Enzymes

Enzyme	Turnover Number[a]
Carbonic anhydrase	3.6×10^6
Acetylcholinesterase	1.5×10^6
Urease	1.0×10^6
Amylase	1.0×10^5
Lactic dehydrogenase	6.0×10^4
Chymotrypsin	6.0×10^3
Lysozyme	3.0×10^1

[a]Moles of substrate converted to product per minute per mole of substrate-saturated enzyme.

The Kinetics of Enzyme Action

In most instances, the association of the enzyme with the substrate is so fleeting that the complex is extremely difficult to detect. In only a few dozen instances has the enzyme substrate complex been identified or isolated. Yet, as early as 1913, L. Michaelis and M. L. Menten postulated the existence of this transient complex. On the basis of their observations with the enzyme *invertase,* which catalyzes the hydrolysis of sucrose to glucose and fructose, they proposed that enzyme-catalyzed reactions were characterized by a sequence of phases that involves (1) the formation of a complex *(ES)* between the enzyme *(E)* and the substrate *(S)*; (2) the modification of the substrate to form the product *(P)* or products, which briefly remain associated with the enzyme *(EP)*; and (3) the release of the product or products from the enzyme; that is,

$$E + S \rightleftharpoons ES \rightleftharpoons EP \longrightarrow E + P \tag{8–15}$$

These events are more conventionally described by the equation

$$E + S \underset{k_2}{\overset{k_1}{\rightleftharpoons}} ES \xrightarrow{k_3} E + P \tag{8–16}$$

In this equation, it is assumed that the combination of enzyme and substrate is reversible; k_1 is the rate constant for the formation of ES (the dimensions of the rate constant are *seconds*$^{-1}$), and k_2 is the rate constant for the dissociation of ES. After the ES complex is formed, S is converted to P with the rate constant k_3. As long as the concentration of P remains negligible (as it would be at the outset of catalysis) or if P is in some manner removed from the system, then it is not necessary to consider the reverse flux from $E + P$ to EP to ES.

Figure 8–3 depicts the relationship that exists between substrate concentration and the rate at which reaction products appear for enzyme-catalyzed reactions of this type. The curve describes the *initial rate of product formation* at a fixed enzyme concentration when the substrate concentration is varied on successive trials. At low concentrations of substrate, the initial velocity of the reaction (i.e., v_o) is directly proportional to the substrate concentration (i.e., follows first-order kinetics). However, as the substrate concentration is increased, the reaction velocity levels off, approaching a maximum value. At this high substrate concentration, the reaction velocity is limited by the amount of available enzyme, almost all of which would be in the ES form. The curve now follows zero-order kinetics. The curve of Figure 8–3 is called a Michaelis-Menten curve.

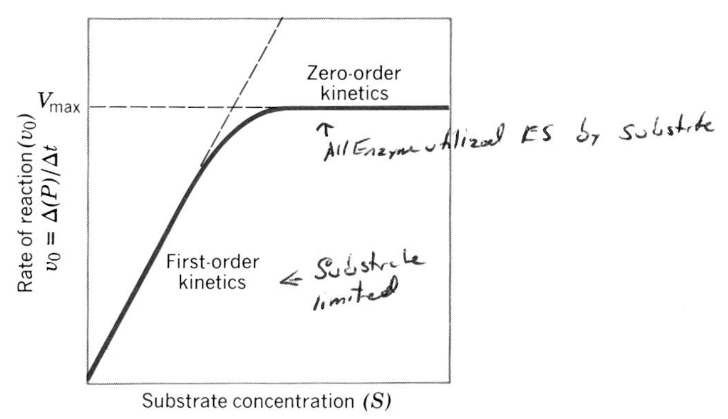

Figure 8–3
Relationship between substrate concentration and rate of enzyme catalyzed reaction. Curve describes *initial rate of product formation* when the enzyme concentration is held constant and the substrate concentration is increased on successive trials. At low substrate concentrations, the reaction flux follows first-order kinetics. As substrate concentration is increased, reaction rate approaches a maximum value (V_{max}) at which time zero-order kinetics is followed.

Michaelis and Menten are also credited with the first mathematical study of the relationship between substrate concentration and reaction rates. They introduced two particularly useful mathematical expressions which for any enzyme relate (S) to V (velocity) and permit quick comparisons of various enzyme-catalyzed reactions; the two expressions are now called the **Michaelis-Menten constant** and the **Michaelis-Menten equation.** They are derived as follows:

Let (S) = the concentration of free substrate, S (we may assume that the amount of available substrate is so great that the amount combined with the enzyme may be ignored in comparison. Hence, the total substrate concentration and the free substrate concentration are the same);

$(E)_T$ = the total concentration of available enzyme, E;

(E) = the concentration of free enzyme (that not complexed with substrate);

(ES) = the concentration of enzyme-substrate complex (since the total amount of enzyme present is assumed to be very small in comparison with the total amount of substrate, a significant proportion of the total enzyme may be involved in the ES complex. Hence, separate terms for the total and free enzyme concentrations are warranted);

(P) = the concentration of products, P.

Thus,

$$(E)_T = (E) + (ES) \tag{8–17}$$

For equation 8–16, the initial reaction rate, v_o, is given by

$$v_o = \frac{\Delta(P)}{\Delta t} = k_3(ES) \tag{8–18}$$

The change in the concentration of ES with time is equal to the rate at which ES is formed minus the rate at which it is being eliminated. That is,

$$\frac{\Delta(ES)}{\Delta t} = k_1(E)(S) - [k_2(ES) + k_3(ES)] \tag{8–19}$$

Therefore,

$$\frac{\Delta(ES)}{\Delta t} = k_1(E)(S) - (k_2 + k_3)(ES) \qquad (8\text{–}20)$$

If sufficient substrate is available, then $\Delta(ES)/\Delta t$ would be zero, since for every P leaving the complex, a new S would enter it. Under such conditions,

$$k_1(E)(S) = (k_2 + k_3)(ES) \qquad (8\text{–}21)$$

and

$$\frac{k_2 + k_3}{k_1} = \frac{(E)\,(S)}{(ES)} \qquad (8\text{–}22)$$

The ratio of constants given in equation 8–22 may be set equal to a new constant, K_M, which is the Michaelis-Menten constant. Thus, the Michaelis-Menten constant is a steady-state constant that relates the concentrations of enzyme, enzyme-substrate complex, and substrate.

Each enzyme-catalyzed reaction reveals a characteristic K_M value, and this value is a measure of the tendency of the enzyme and the substrate to combine with each other. In this sense, the K_M value is an index of the affinity of the enzyme for its particular substrate. Some representative K_M values are given in Table 8–2.

The relationships derived above are based on reactions in which a single substrate molecule is bound to the enzyme. However, they also apply in situations where more than one substrate is bound to the enzyme, as long as the concentrations of all but one substrate species are held constant or are not rate-limiting (i.e., present in large excess). For example, even the pioneering studies of Michaelis and Menten involved a bimolecular enzyme-catalyzed reaction. They employed the enzyme invertase, which forms a complex with one molecule of sucrose and one molecule of water and releases glucose and fructose as products. However, in this reaction (as in nearly all hydrolyses), the concentration of water remains virtually unaltered during the course of the reaction.

K_M values cannot usually be determined using the relationship given in equation 8–22, since the enzyme-substrate concentration (ES) cannot easily be measured. Some additional mathematical manipulations may be carried out in order to convert equation 8–22 into a more useful form.

Since $(E) = (E)_T - (ES)$, then

$$K_M = \frac{(E)\,(S)}{(ES)} = \frac{[(E)_T - (ES)]\,(S)}{(ES)} \qquad (8\text{–}23)$$

$$K_M\,(ES) + (S)(ES) = (E)_T(S) \qquad (8\text{–}24)$$

$$(ES)[K_M + (S)] = (E)_T\,(S) \qquad (8\text{–}25)$$

and

$$(ES) = \frac{(E)_T(S)}{K_M + (S)} \qquad (8\text{–}26)$$

Table 8–2
Some Representative K_M Values

Enzyme	Source	Substrate	K_M
Sucrase	Intestine	Sucrose	$2 \times 10^{-2}\ M$
Urease	Soybeans	Urea	$2.5 \times 10^{-2}\ M$
Catalase	Liver	Hydrogen peroxide	$2.5 \times 10^{-2}\ M$
Carbonic anhydrase	Blood	CO_2	$9 \times 10^{-3}\ M$
Chymotrypsin	Pancreas	Peptides	$5 \times 10^{-3}\ M$
Phosphatase	Bone	Glycerophosphate	$3 \times 10^{-3}\ M$
Hexokinase	Liver	Glucose	$1.5 \times 10^{-4}\ M$
Lysozyme	Egg white	Hexa-*N*-acetylglucosamine	$6 \times 10^{-6}\ M$

Since $v_o = \Delta(P)/\Delta t = k_3\,(ES)$, then

$$v_o = \frac{k_3(E)_T(S)}{K_M + (S)} \tag{8–27}$$

As seen in Figure 8–3, the initial reaction velocity increases with substrate concentration until the enzyme concentration becomes limiting, and at that time, the initial reaction velocity attains a maximum value (i.e., V_{max}). This occurs because all or nearly all the enzyme is maintained in the *ES* form. Thus, *(ES)* will equal $(E)_T$, and $k_3(E)_T$ corresponds to V_{max}. By substituting V_{max} for $k_3(E)_T$ in equation 8–27, we obtain

$$v_o = \frac{V\max(S)}{K_M + (S)} \tag{8–28}$$

Equation 8–28 is the Michaelis-Menten equation. From this equation, it may be seen that when the substrate concentration is numerically equal to the K_M value of the enzyme, then the reaction velocity is equal to one-half the maximum value. That is,

$$v_o = \frac{V\max(K_M)}{K_M + K_M} = \frac{V\max}{2} \tag{8–29}$$

Consequently, the Michaelis-Menten constant for an enzyme may be determined from the substrate concentration at which the reaction velocity proceeds at one-half its maximum value (Fig. 8–4).

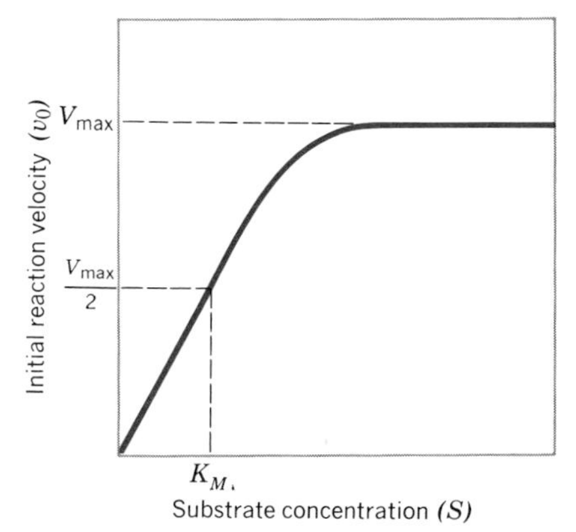

Figure 8–4
Relationship between substrate concentration and reaction rate. When the rate of reaction is one-half its maximum initial value, the substrate concentration is numerically equal to the Michaelis-Menten constant. All such curves are rectangular hyperbolas. Consequently, the lower the K_M value, the lower the substrate concentration at which the reaction proceeds at a maximum rate.

From the above, we may conclude that the limiting velocity of an enzyme catalyzed reaction depends on the affinity between the enzyme and its substrate (i.e., the K_M value). The higher the K_M value, the lower the affinity of the enzyme for the substrate; and the smaller the K_M value, the greater the affinity. In many instances, the same substrate may enter either of several different enzyme-catalyzed reactions occurring in cells. Which of the alternative reactions predominates in the cell depends in part on the K_M values of the respective enzymes and the concentration of available substrate. At very low substrate concentrations, the specific reaction catalyzed by the enzyme with the lowest K_M will dominate, whereas at higher substrate concentrations, the reaction catalyzed by the enzyme having the greatest K_M value will dominate. Therefore, which of several different metabolic pathways is actually followed by the initial substrate in the course of a series of enzyme-catalyzed reactions may be regulated by controlling the amount of available substrate. The relationship between reaction velocity and substrate concentration for two enzymes having the same substrate is depicted in Figure 8–5.

In order to obtain the K_M value of an enzyme experimentally, it is necessary to determine v_o for a series of substrate concentrations. In practice the evaluation of K_M from a plot similar to that in Figure 8–4 is difficult, since the precise value of V_{max} cannot be determined. This is because the curve is a rectangular hyperbola and approaches V_{max} asymptotically. In 1934, H. Lineweaver and D. Burk introduced a different form of the Michaelis-Menten equation (8–28) that simplifies the determination of K_M from experimentally obtained values for *(S)* and v_o. In this form, the acquired data are used to consruct a straight line rather

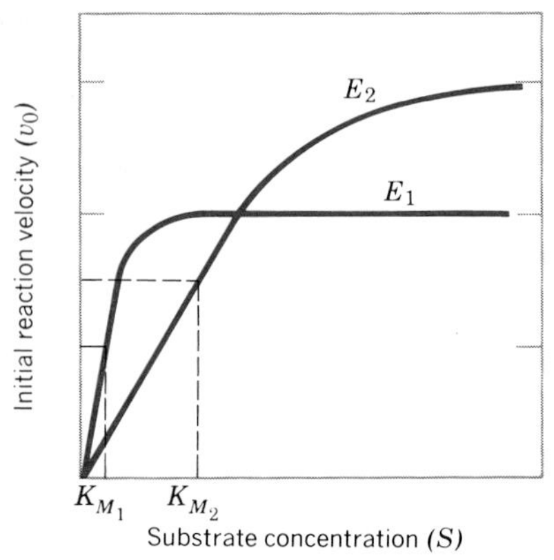

Figure 8–5
Relationship between substrate concentration and rate of reaction for two enzymes having the same substrate. enzyme 1 (E_1) has a lower K_M and attains maximum rate of reaction at lower substrate concentration. Enzyme 2 has a higher K_M value and produces maximum reaction rate at higher concentration of substrate. As a result, reaction catalyzed by enzyme 1 is favored at low substrate levels, while reaction catalyzed by enzyme 2 is favored at high substrate concentrations.

than a hyperbola. By taking the inverse of equation 8–28, we obtain,

$$\frac{1}{v_o} = \frac{K_M}{V\max(S)} + \frac{(S)}{V\max(S)} \quad (8\text{–}30)$$

or

$$\frac{1}{v_o} = \frac{K_M}{V\max}\frac{1}{(S)} + \frac{1}{V\max} \quad (8\text{–}31)$$

The terms are now arranged in the form for the general equation of a straight line, $y = ax + b$, in which a is the slope of the line and b is the intercept on the y-axis. Thus, for equation 8–31, $1/v_o$ serves as the y-axis and $1/(S)$ as the x-axis. Consequently, the y-intercept will be $1/V_{max}$, the slope will be K_M/V_{max}, and the x-intercept will be $-1/K_M$. The graph showing this relationship (known as a **double reciprocal plot**) is presented in Figure 8–6).

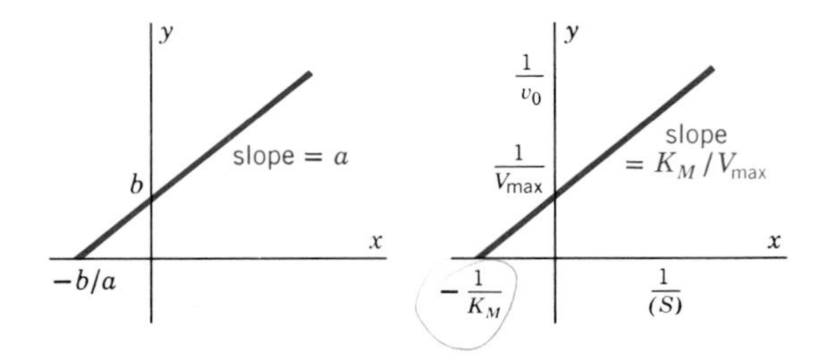

Figure 8–6
Lineweaver-Burk double reciprocal plot of $1/v_o$ versus $1/(S)$. Experimentally obtained values of v_o and (S) may be used to construct a line that may easily be extrapolated backward to find the K_M value for the enzyme.

Effects of Inhibitors on Enzyme Activity

The interaction between the substrate and the enzyme takes place in a particular region of the enzyme molecule called the **active site** (discussed later). In many instances, compounds other than the normal substrate for a particular enzyme-catalyzed reaction may become bound to the enzyme's active site, and this may have a significant effect on the kinetics of the normal reaction. One possible consequence of this phenomenon is the inhibition of normal

Figure 8–7
Effect of competitive inhibitor on normal enzyme kinetics. V_{max} is not altered, but K_M value is increased. At low substrate concentrations, the effect of inhibitor on reaction rate is marked but is completely reversed as substrate concentration is increased and $1/(S)$ approaches 0.

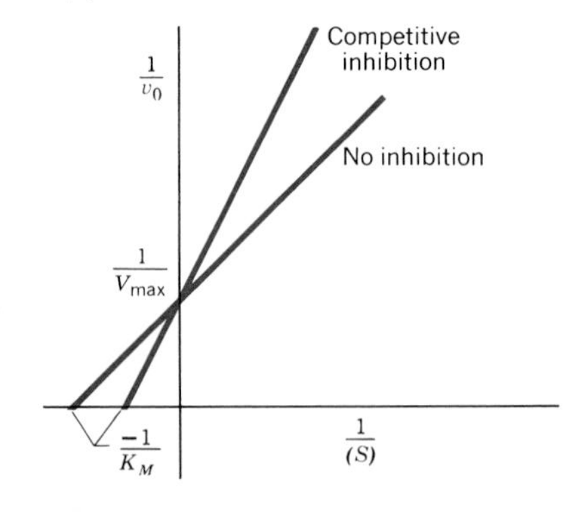

enzyme activity, and such compounds are therefore called **enzyme inhibitors.** (Usually, the inhibitor is unaltered by its interaction with the enzyme.) In some instances, the normal substrate *(S)* and the inhibitor *(I) compete* with each other for the active site of enzyme; the manner in which this affects the normal kinetics of the reaction is shown in Figure 8–7. V_{max} is not altered by the presence of a competitive inhibitor, but the K_M value is elevated. As can be seen in Figure 8–7, the effect of the inhibitor is maximal at low substrate concentration (i.e., when 1/ *(S)* is large) and minimal at high substrate concentration (i.e., when 1/ *(S)* approaches 0).

A classical example of this form of inhibition is the competition between succinic acid and malonic acid for the enzyme succinic acid dehydrogenase.

Succinic acid	Malonic acid
COOH–CH_2–CH_2–COOH	COOH–CH_2–COOH

In this instance, competition between these two compounds for the active site of the enzyme is understandable in view of their marked chemical similarity. Succinic acid is the normal substrate for the enzyme and, in the absence of the inhibitor, is converted to fumaric acid:

$$\underset{\text{Succinic acid}}{\text{COOH–CH}_2\text{–CH}_2\text{–COOH}} \xrightarrow[\text{dehydrogenase}]{\text{Succinic acid}} \underset{\text{Fumaric acid}}{\text{COOH–CH=CH–COOH}}$$

Enzyme inhibition can also be *noncompetitive* in that the binding of the inhibitor to the enzyme cannot be reversed by increasing the concentration of the normal substrate. A common example of negative inhibition is the action of heavy metals such as mercury on the active sites of enzymes containing a reactive sulfhydryl (i.e., –SH) group. In effect, the presence of the inhibitor prevents some percentage of the enzyme present from participating in normal catalysis. As a result, the maximum reaction velocity is depressed, even though the K_M value remains the same (Fig. 8–8).

The inhibition of enzyme activity by competitors also occurs naturally in cells and serves in the regulation of certain metabolic pathways. This is known as **feedback inhibition,** a special form of **allosterism,** and will be considered at length later in the chapter.

Mechanics of Enzyme Catalysis

Enzymes are proteins, and therefore their capacity for catalysis is intimately relative to a specific tertiary or quaternary molecular structure. If the tertiary or quaternary structure of an enzyme is altered, a loss of enzyme activity usually follows. Thus, environmental factors that modify protein structure also influence enzyme activity. Among the more common environmental factors that affect enzyme activity are pH and temperature. As discussed at some length in Chapter 4, the polar side chains of certain amino acids form electrostatic bonds with each other and with surrounding ions and water molecules; these interactions contribute in part to the specific tertiary and quaternary structure of the protein.

Whether or not a particular amino acid side chain bears

Figure 8–8
Effect of noncompetitive inhibitor on normal enzyme kinetics. K_M value is not altered, but V_{max} is reduced. Noncompetitive inhibition can be reduced but not reversed by increasing the substrate concentration.

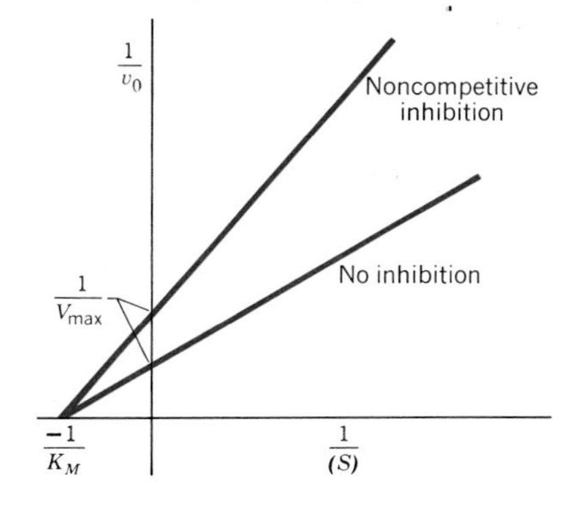

a charge is determined in part by the pH of the protein's environment. As the pH is lowered (i.e., the concentration of H^+ is increased), groups that may be negatively charged, such as the secondary COO^- of aspartic acid and glutamic acid and the O^- of tyrosine, become protonated, thereby neutralizing these negative charges. At the same time, some secondary amino groups, such as those of lysine and arginine, may accept additional protons, thereby imparting charge to these formerly neutral side chains. In contrast, as the pH is elevated (i.e., the concentration of OH^- is increased), positively charged side chains dissociate protons and are thereby neutralized, while the loss of protons from secondary COOH and OH groups renders these groups negative. Some of these relationships are depicted in Figure 8–9.

In addition to playing important parts in the maintenance of a specific tertiary or quaternary molecular structure, the polar side chains of some of the amino acids in the enzyme (i.e., those in the active site) may be involved in binding the substrate to the enzyme and thereby introducing bond strains into the substrate molecule. Consequently, most enzymes can operate only within a narrow pH range and thus display a pH optimum (Fig. 8–10). The pH optima of the various enzymes studied reveal a broad distribution; several examples are given in Table 8–3. On either side of the pH optimum, enzyme activity declines as the configuration of the protein is altered and its affinity for the substrate is correspondingly decreased.

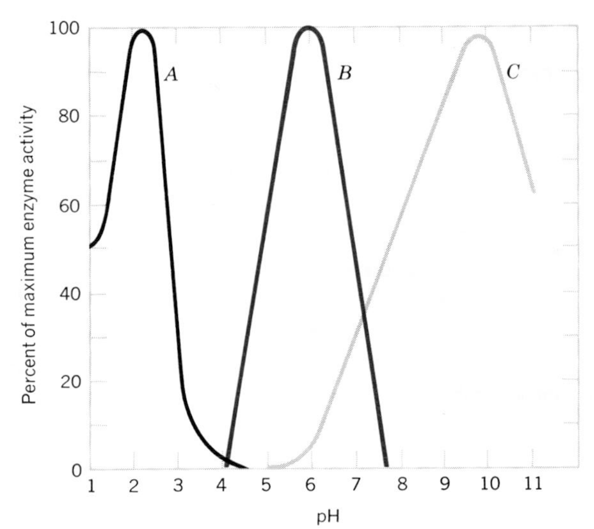

Figure 8–10
Effect of pH on enzyme activity. *(a)* Pepsin, *(b)* glutamic acid decarboxylase, and *(c)* arginase.

Figure 8–9
Influence of pH on neutral, negatively charged, and positively charged side chains of amino acids.

Aspartic acid residue

$$HC—CH_2—COO^{\ominus} \underset{OH^{\ominus}}{\overset{H^{\oplus}}{\rightleftharpoons}} HC—CH_2—COOH$$

Lysine residue

$$HC—(CH_2)_4—NH_2 \underset{OH^{\ominus}}{\overset{H^{\oplus}}{\rightleftharpoons}} HC—(CH_2)_4—NH_3^{\oplus}$$

Tyrosine residue

$$HC—CH_2—C_6H_4—O^{\ominus} \underset{OH^{\ominus}}{\overset{H^{\oplus}}{\rightleftharpoons}} HC—CH_2—C_6H_4—OH$$

Table 8–3
Optimum pH Values for Certain Enzymes

Enzyme	Optimum pH
Pepsin	2.0
Glutamic acid decarboxylase	5.9
Urease	6.7
Salivary amylase	6.8
Pancreatic lipase	7.0
Trypsin	9.5
Arginase	10.0

Temperature also influences enzyme activity, and most enzymes display a temperature optimum that corresponds to the normal temperature of the cell or organism possessing that enzyme. Accordingly, the temperature optima of plant cell enzymes and enzymes of poikilothermic animals inhabiting cold regions of the earth are usually lower than enzymes of homeothermic animals.

As the temperature is elevated above the optimum, there is a progressive increase in enzyme activity, and this is most likely due to the effect of temperature on the kinetic energy distribution of the substrate molecules. Above the temperature optimum, enzyme activity decreases, presumably the result of alteration of the enzyme's structure (called **denaturation**). Most enzymes are irreversibly denatured if maintained at temperatures above 55° to 65°C for an extended period of time.

The Active Site

The formation of the enzyme-substrate complex is not a random process. This was recognized as long ago as 1894 when Emil Fischer postulated that an enzyme allows only one or a few compounds to fit onto its surface. This is the "lock-and-key" hypothesis according to which the enzyme and its substrate have a complementary shape. The specific substrate molecules (and prosthetic groups, if any) are bound to a specific region of the enzyme molecule called its **active site.** The active site of an enzyme is formed by a number of amino acid residues whose side chains have two principal roles: (1) they serve to *attract* and *orient* the substrate in a specific manner within the site (such amino acids are called **contact residues** and contribute in large degree to substrate specificity); and (2) they participate in

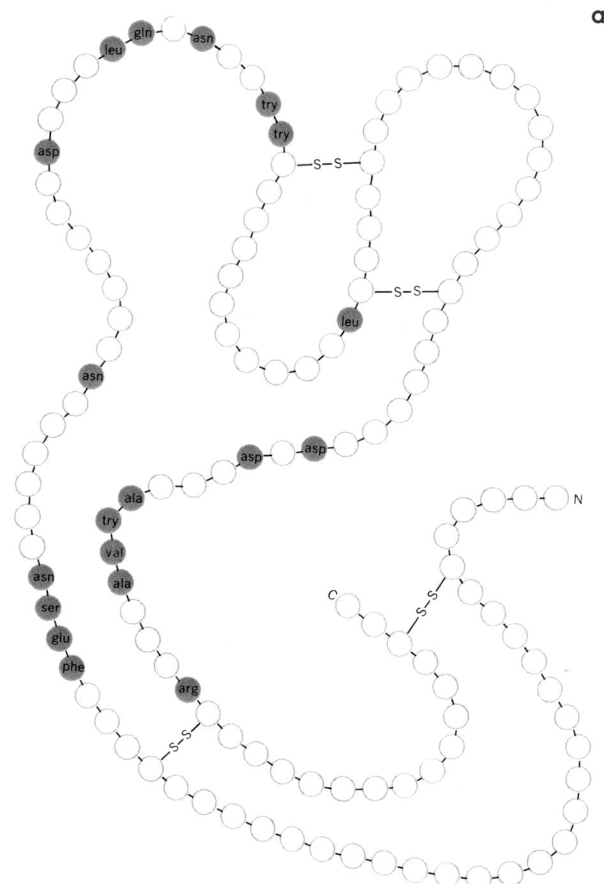

Figure 8–11
Structure of lysozyme showing distribution of amino acids forming the active site.

the *formation of temporary bonds* with the substrate molecule, bonds that polarize the substrate, introduce strain into certain of its bonds, and trigger the catalytic change (such amino acids are termed **catalytic residues**).

The contact and catalytic residues that make up the active site may be located in widely separated regions of the enzyme's primary structure, but as the consequence of stabilized polypeptide chain folding, they are brought into the appropriate juxtaposition. This is exemplified by the enzyme lysozyme (Fig. 8–11) in which the amino acids that form the active site in the folded structure (shaded residues) are widely separated in the primary structure. The bonds formed between a substrate and the amino acid side chains forming the active site may be either covalent or noncovalent.

Figure 8–12 depicts the interaction between enzyme and substrate according to the lock-and-key model. The substrate has polar (i.e., ⊕ and ⊖) and nonpolar (Ⓗ, hydrophobic) regions and is attracted to and associates with the active site which is complementary in both shape and charge distribution (Fig. 8–12*a* and *b*). Positive, negative, and hydrophobic regions of the active site are created by the side chains of the contact residues which align the substrate for interaction with the site's catalytic residues (△A and △B). Following catalysis (Fig. 8–12*c*), the products are released from the active site (Fig. 8–12*d*), thereby freeing the enzyme for another round of catalysis. The lock-and-key model of enzyme catalysis accounts for enzyme specificity, since compounds that lack the appropriate shape or are too large or too small (Fig. 8–12*e*) cannot be bound to the active site.

Although the lock-and-key model accounts for much of the substrate specificity data, certain observations about enzyme behavior do not fit or are difficult to explain by this model. For example, there are a number of instances in which compounds other than the true substrate bind to the enzyme even though they fail to form reaction products. Furthermore, for many enzyme-catalyzed reactions, substrates are bound to the active site in a specific temporal

Figure 8–12
Fischer's "lock-and-key" model of enzyme action (see text for explanation).

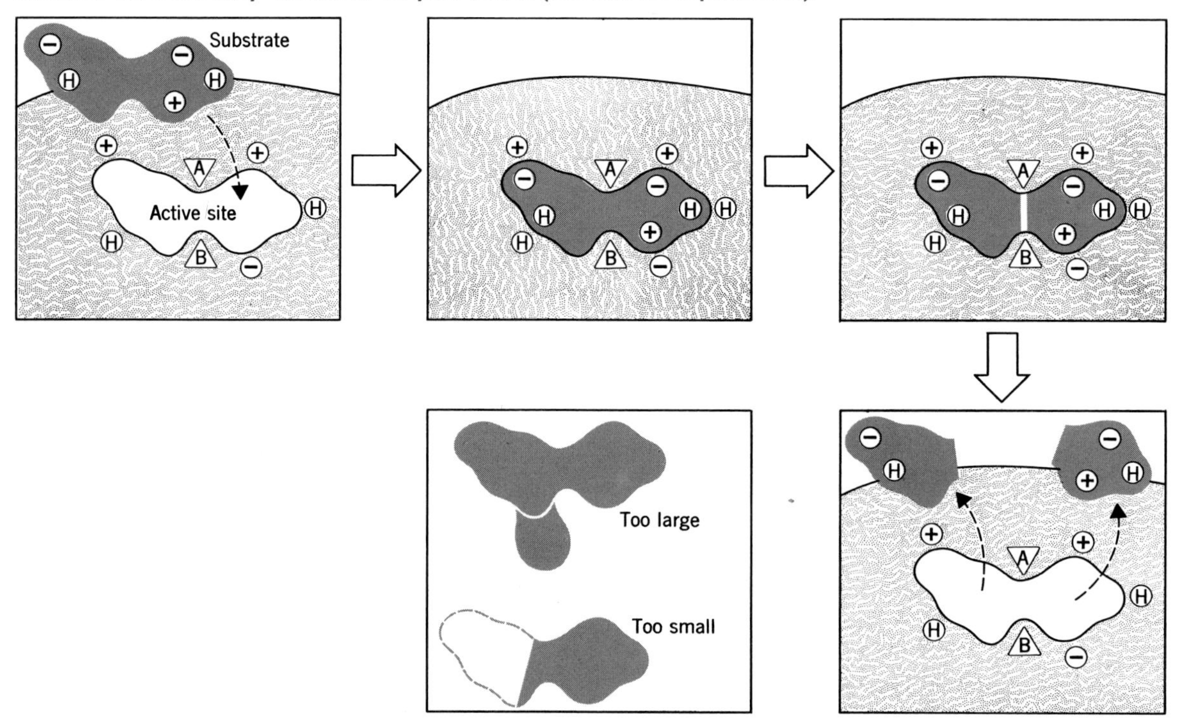

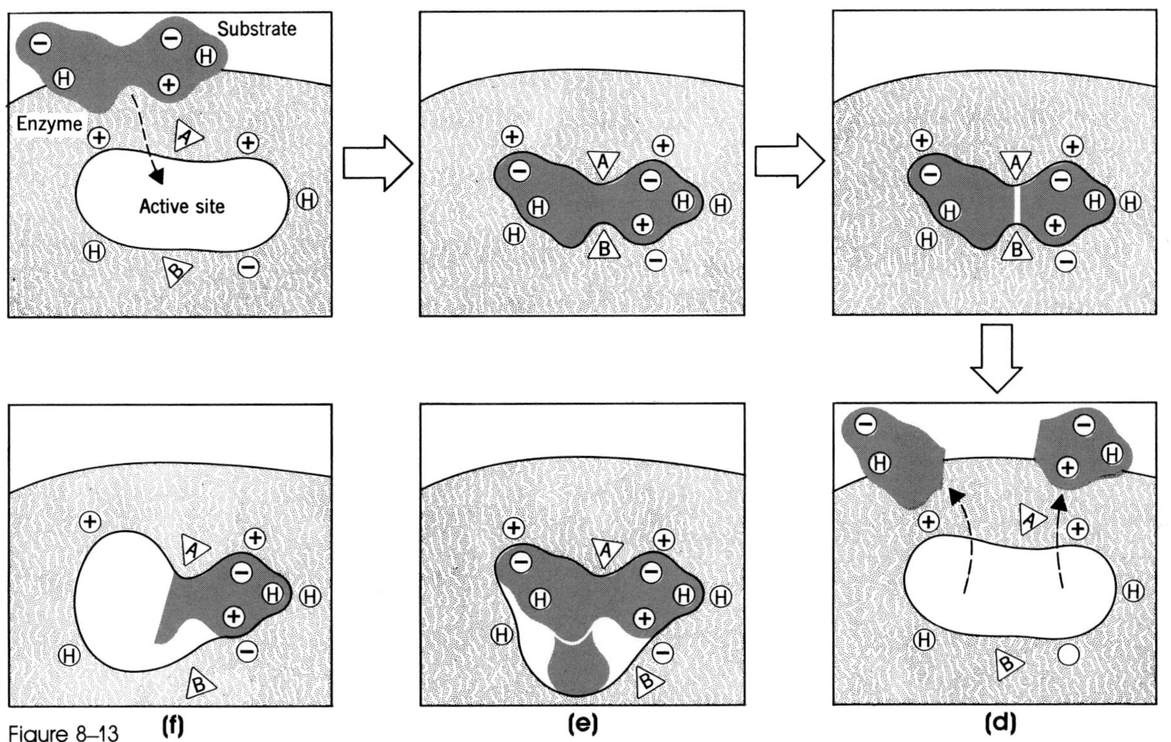

Figure 8–13
Koshland's "induced-fit" model of enzyme action (see text for explanation).

order. In the 1960s, Daniel Koshland proposed the ''induced-fit'' theory of enzyme action according to which the active site of the enzyme does not initially exist in a shape that is complementary to the substrate but is induced to assume the complementary shape as the substrate becomes bound. As Koshland puts it, the active site is induced to assume complementary shape ''in much the same way as a hand induces a change in the shape of a glove.'' Thus, according to this model, the enzyme (or its active site) is *flexible*.

The ''induced fit'' model is depicted diagrammatically in Figure 8–13. The active site and substrate initially have different shapes (Fig. 8–13*a*) but become complementary upon substrate binding (Fig. 8–13*b*). The shape change places the catalytic residues in position to alter the bonds in the substrate (Fig. 8–13*c*), following which the products are released (Fig. 8–13*d*) and the active site returns to its initial state. Although molecules that are larger or smaller than the true substrate or that have different chemical properties may nonetheless be bound to the active site, none succeed in inducing the proper alignment of catalytic groups, and no catalysis occurs (Fig. 8–13*e* and *f*). The induced-fit model explains the effects of certain competitive and noncompetitive inhibitors of enzyme action.

Before proceeding further, it should be acknowledged that some enzyme-catalyzed reactions are adequately explained by the lock-and-key model, so that a flexible active site is not a strict requirement for catalysis. By the same token, the possession of a flexible active site does not imply that just any molecule may become bound to the enzyme.

A change in the shape of the active site of an enzyme can also be induced by binding at sites on the enzyme's surface that are far removed from the active site. In such a case, the change is transmitted through the enzyme molecule from the site of binding to the active site. Such changes may either decrease or increase the enzyme's activity. The latter phenomenon will be discussed later in the chapter in connection with enzyme *cooperativity* and *allosterism*.

Extensive studies during the past decade have revealed

the precise atomic structure of a number of important enzymes including lysozyme, ribonuclease, chymotrypsin, trypsin, carboxypeptidase, and papain. A consideration of some of these reveals certain generalizations about the organization of enzyme molecules and provides insight into the mechanisms by which catalysis may be accomplished.

Lysozyme

Lysozyme is an enzyme produced by both animal and plant tissues and was the first to have its complete three-dimensional structure revealed (by C. C. F. Blake, D. C. Phillips, and A. C. T. North). The enzyme cleaves polysaccharide chains found in the cell walls of certain bacteria by hydrolyzing the glycosidic bonds between neighboring hexosyl residues. Egg white lysozyme has a molecular weight of about 14,600 and consists of a single polypeptide chain of 129 amino acids. The enzyme is oval in shape with a deep cleft across its midline that divides the molecule into two parts. The shape of lysozyme and the orientation of the substrate are depicted stereoscopically in Figure 8–14. The enzyme differs from the globin chains of hemoglobin and myoglobin in that there is much less helical structure (only three short helical regions) and a segment of the molecule is arranged in the form of a beta-pleated sheet. All the polar residues are located on the enzyme's surface, while nearly all uncharged residues are buried internally. Hydrogen bonds between one portion of the polypeptide chain and another are pronounced and appear to be vital in sustaining the active tertiary structure of the molecule.

The cleft at the center of the molecule contains the active site and is studded with hydrophobic groups that probably form van der Waals bonds with the substrate. The bacterial polysaccharide that acts as substrate for the enzyme consists of chains of *N*-acetylglucosamine (NAG) and *N*-ace-

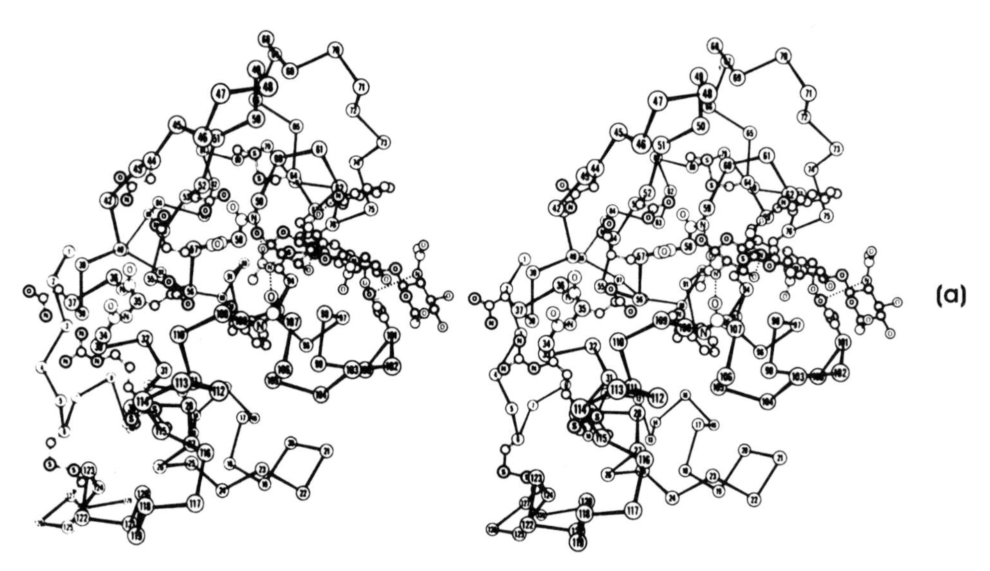

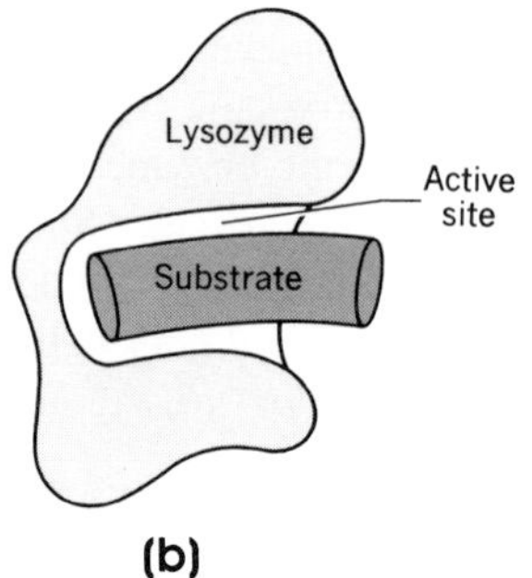

Figure 8–14
(a) **Stereoscopic diagrams of lysozyme (courtesy of R. E. Dickerson and I. Geis).** ***(b)*** **Diagram illustrating binding of substrate in deep cleft forming active site of lysozyme.**

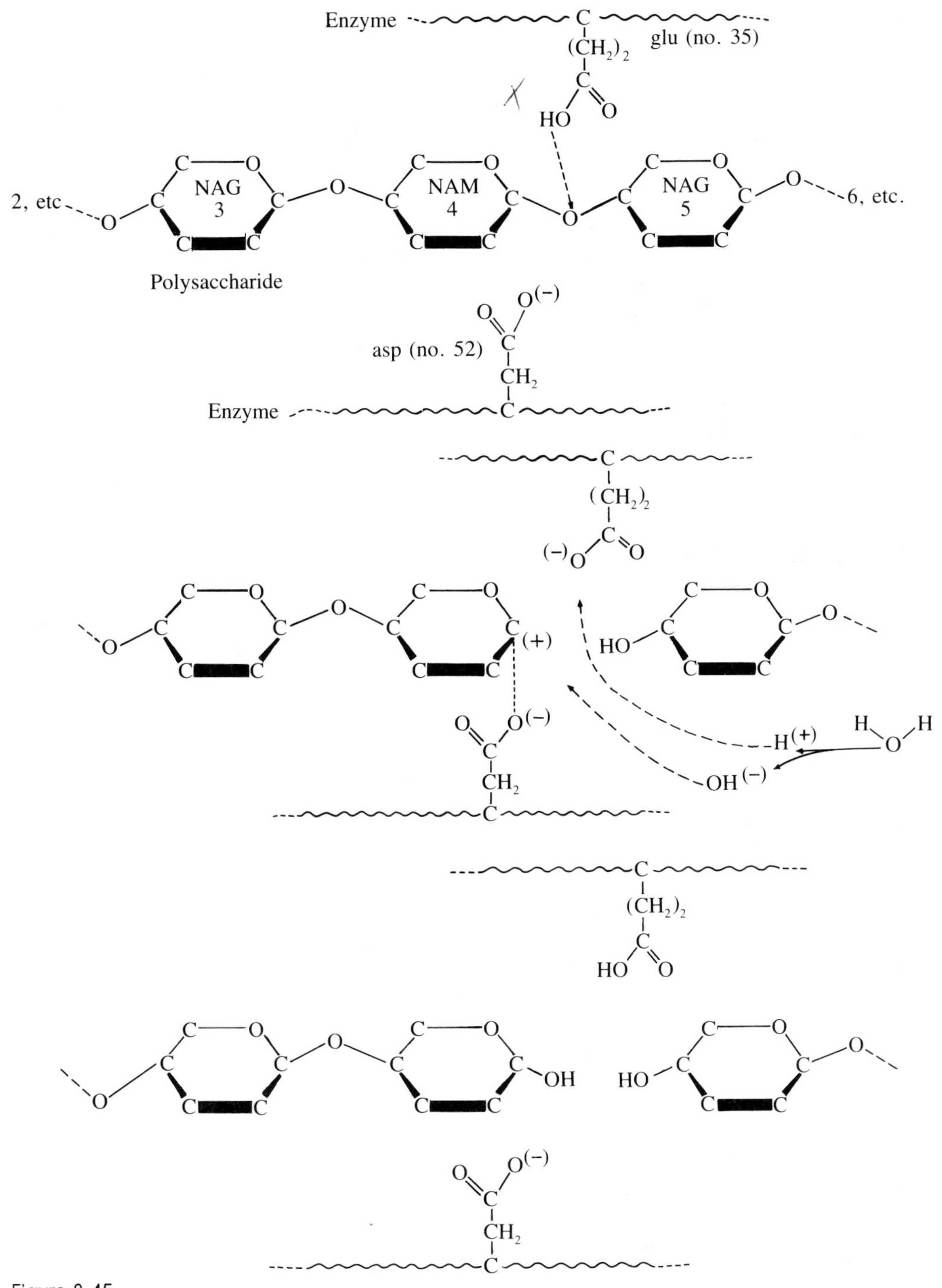

Figure 8–15

Proposed mechanism for the action of lysozyme. Only three of the five or six glucosyl units of the polysaccharide bound to the active site are shown (numbered 3, 4, and 5). Hydroxyl groups, other groups, and hydrogen atoms of the pyranose rings have been omitted for clarity. The side chains of the two catalytic residues are shown but not the remainder of the enzyme molecule. See text for details of the catalytic process.

tylmuramic acid (NAM) in which the two sugars occur at alternate positions. In the baterial cell wall, these chains are cross-linked by short polypeptides. Six hexosyl units of the substrate polysaccharide are simultaneously bound at the active site of lysozyme. A change in tertiary structure, which takes the form of a narrowing and deepening of the cleft in the enzyme, accompanies substrate binding and forces a modification of the conformation of one of the six hexosyl groups (i.e., NAM) bound at the active site. This conformational change involves a distortion of the "chair" form normally assumed by this sugar and primes this region of the substrate for catalytic alteration. It is the glycosidic bond between this group and its neighbor that is broken during catalysis. Of the 19 or more amino acids that compose the active site, only two have been identified as catalytic; these are asp (amino acid no. 52) and glu (amino acid no. 35). The remaining 17 residues align the substrate through hydrogen bonds and nonpolar interactions.

Asp 52 carries a dissociated carboxyl group (i.e., negative charge), but glu 35, being surrounded by the nonpolar side chains of other residues, is protonated. The sequence of events occuring during catalysis is depicted in Figure 8–15*a*, *b*, and *c* in which only 3 of the 6 hexosyl units bound to the active site (i.e., numbers 3, 4 and 5) are shown. The binding of the substrate to the enzyme and the conformational changes that occur in both as a result are followed by an attack by a hydrogen ion of the undissociated COOH group of glu 35 on the oxygen atom, forming the bridge between hexosyl units 4 and 5 (Fig. 8–15*a*). This, of course, is the specific region of the substrate placed under strain during binding. The glycosidic bond is broken, leaving carbon atom no. 1 of hexosyl unit 4 with a positive charge (i.e., a carbonium ion) (Fig. 8–15*b*). The carbonium ion is stabilized by the formation of a temporary ionic bond with the dissociated side chain of asp 52 (Fig. 8–15*b*). The dissociation of a neighboring molecule of water provides a hydroxyl group for attack upon the carbonium ion and a hydrogen for glu 35 (Fig. 8–15*c*). Once the reaction is completed, the two fragmentary polysaccharides leave the active site.

Ribonuclease

Ribonuclease is produced by the pancreas and secreted into the small intestine where it catalyzes the hydrolytic digestion of a polyribonucleotide chain. The enzyme consists of a single polypeptide chain of 124 amino acids and has a molecular weight of about 13,700. The reaction catalyzed by ribonuclease is shown in Figure 8–16 and involves the hydrolytic cleavage of the ester linkage between the

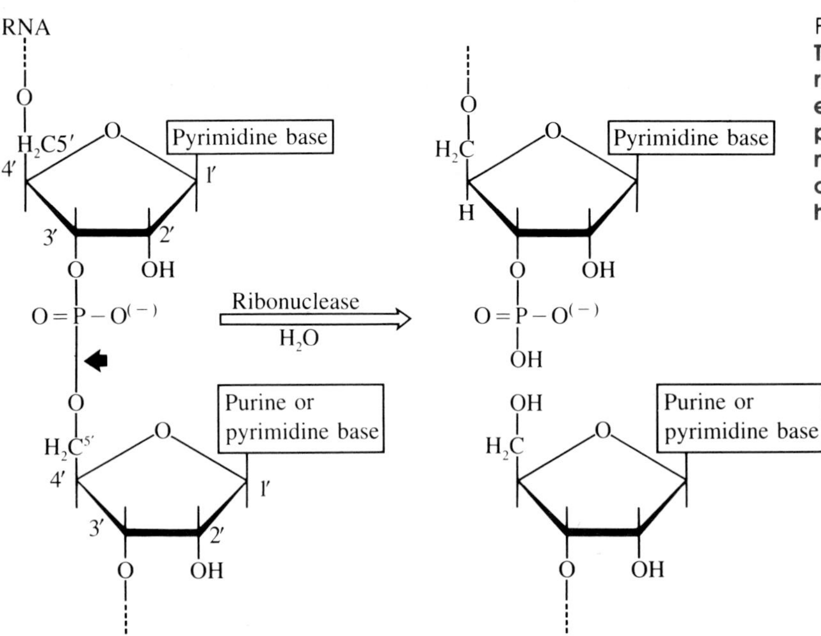

Figure 8–16
The chemical reaction catalyzed by ribonuclease. The enzyme cleaves the ester linkage between the 3′-phosphate group of a pyrimidine nucleotide and the 5′ position of the adjacent nucleotide. The specific bond hydrolyzed is shown by the heavy arrow.

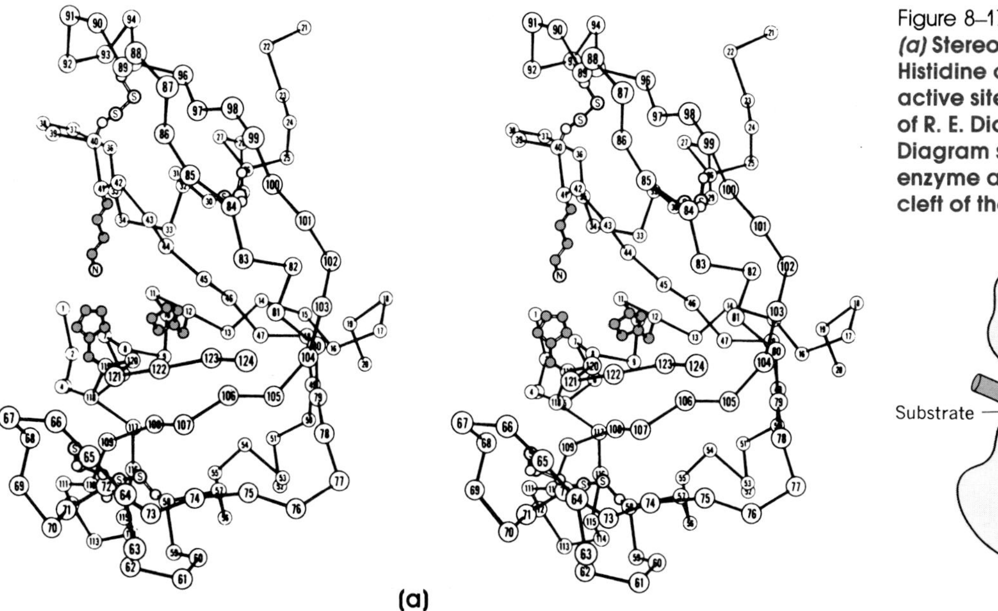

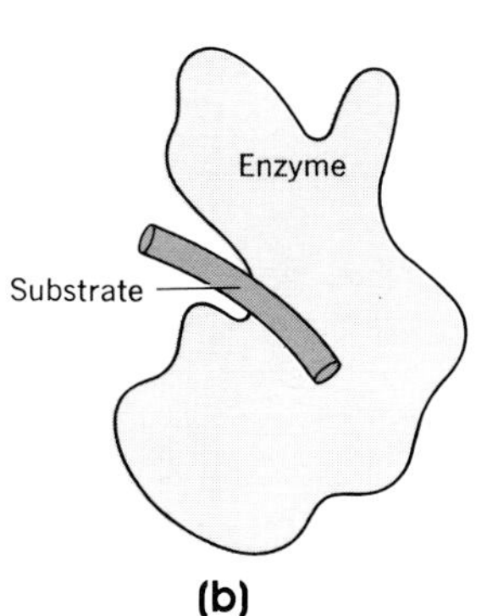

Figure 8–17
***(a)* Stereoscopic view of ribonuclease. Histidine and lysine residues of the active site are shown in color. (Courtesy of R. E. Dickerson and I. Geis.) *(b)* Diagram showing general outline of the enzyme and position of substrate in the cleft of the active site.**

3′-phosphate group of a pyrimidine nucleotide and the 5′ position of the adjacent nucleotide.

Like lysozyme, the active site of ribonuclease lies in a cleft in the enzyme's surface. The cleft is lined by a number of positively charged side chains of lysine and arginine residues that are believed to act as contact residues by forming ionic bonds with the negatively charged phosphate groups of the ribonucleic acid backbone (Fig. 8–16). Projecting into the cleft from either side are two histidine residues that are believed to hydrolyze the RNA substrate by general acid-base catalysis in which the P–O bond linking the ribose of one nucleotide with the ribose of a neighboring nucleotide is broken. Ribonuclease is depicted stereoscopically in Figure 8–17.

Chymotrypsin

The digestive enzyme chymotrypsin is derived from the inactive precursor polypeptide chymotrypsinogen produced in the pancreas and containing 245 amino acids. The activation of chymotrypsinogen involves the preliminary cleavage of four of its peptide bonds, resulting in three separate polypeptide chains cross-linked by two disulfide bridges; two dipeptides released in the activation process do not become a part of the functional enzyme. Chymotrypsin digests alimentary protein by hydrolyzing peptide bonds on the carboxyl side of amino acids with large hydrophobic side chains (i.e., phenylalanine, tyrosine, and tryptophan residues).

Unlike lysozyme and ribonuclease, the active site of chymotrypsin is not in a cleft but resides in a shallow dish-shaped cavity on the surface of this roughly spherical enzyme. In the active site is a hydrophobic pocket that is believed to act as the receptor of the hydrophobic side chain of the substrate amino acid residue and thereby imparts specificity to this enzyme. In addition to these and other contact residues, the active site contains three catalytic residues: asp 102, his 57, and ser 195. Figure 8–18 depicts the series of stages that are believed to be involved in the action of chymotrypsin. The imidazole group of histidine attracts the hydroxyl hydrogen atom of the serine, thereby rendering the serine oxygen strongly nucleophilic and particularly reactive toward amides and esters (Figure 8–18*a*). Binding of the substrate polypeptide to the enzyme (Fig. 8–18*b*) is followed by formation of a covalent bond between the serine oxygen and the alpha-carboxyl carbon atom of the substrate residue (Fig. 8–18*c*). This is followed by donation of the serine proton to the alpha-amino nitrogen atom of the neighboring substrate residue, the breakage of the

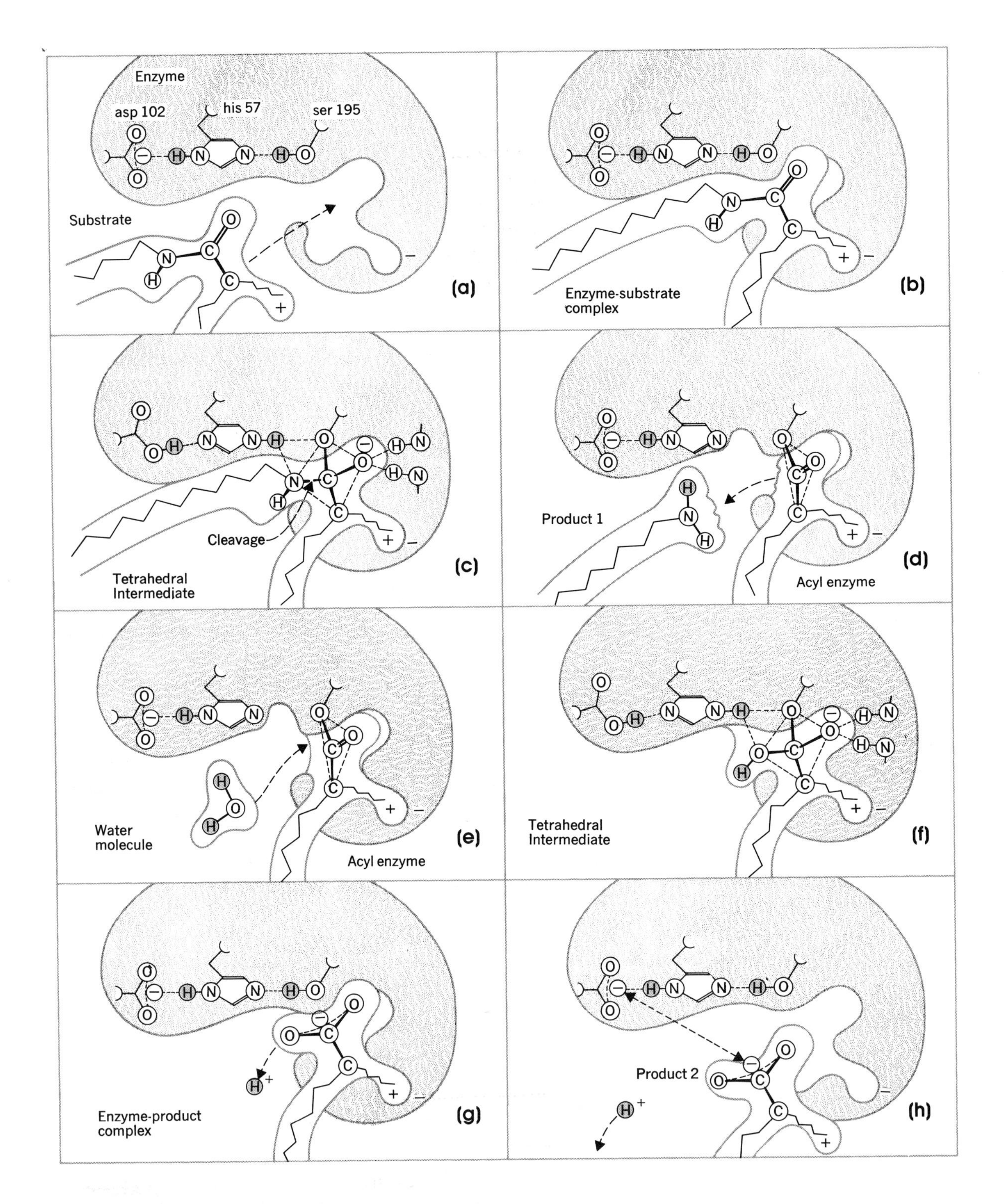

Figure 8–18
Stages involved in proteolysis by chymotrypsin.

peptide bond, and the release of the first product (Fig. 8–18*d*).

At this stage, a molecule of water enters the reaction (Fig. 8–18*e*), donating a hydrogen ion to serine and a hydroxyl group to the alpha-carboxyl carbon (Fig. 8–18*f*). Dissociation of this carboxyl group (Fig. 8–18*g*) leaves the carboxyl group negatively charged. Repulsive forces between this group and the negatively charged aspartic acid residue of the active site facilitate dissociation of the enzyme and final product (Fig. 8–18*h*).

The behavior of the catalytic serine residue is critical to the function of chymotrypsin. Other proteases, including trypsin, elastase, collagenase, thrombin, fibrinolysin, and some lysosomal proteases, share a similar overall tertiary structure and active site and cleave peptide bonds via the identical action of a catalytic serine residue. The different substrate specificities of these "serine proteases" reside in the nature and locations of the contact residues in their active sites. In view of their remarkable similarity, the serine proteases are believed to have a common evolutionary origin.

The accumulated information on lysozyme, ribonuclease, and the serine proteases, as well as several other enzymes, permits certain generalizations to be made concerning enzyme structure and action. The active site of the enzyme nearly always resides in a depression in its surface, and the binding of the substrate to the active site is followed by a conformational change in the enzyme, the substrate, or both. The binding and/or the conformational changes that follow place a strain on certain bonds in the substrate and render them particularly susceptible to chemical attack. The catalysis itself may be facilitated by as few as one or two amino acid side chains of the enzyme which, because of their spatial arrangement in the active site and the polar or hydrophobic nature of their surroundings, are especially reactive.

Cofactors

Enzymes are composed of one or several polypeptide chains. However, there are a number of cases in which nonprotein constituents called **cofactors** must be bound to the enzyme (in addition to the substrate) in order for it to be catalytically active. In these instances, the exclusively protein portion of the enzyme is called the **apoenzyme.** Three kinds of cofactors may be identified: *prosthetic groups, coenzymes,* and *metal ions.*

Prosthetic groups are organic compounds and are distinguished from other cofactors in that they are permanently bound to the apoenzyme. For example, in the enzymes peroxidase and catalase, which catalyze the breakdown of hydrogen peroxide to water and oxygen, *heme* is the prosthetic group and is a permanent part of the enzyme's active site.

Coenzymes are also organic compounds, but their association with the apoenzyme is only transient, usually occurring only during the course of catalysis. Furthermore, the same coenzyme may serve as the cofactor in a number of different enzyme-catalyzed reactions. In general, coenzymes not only assist enzymes in the cleavage of the substrate but also serve as temporary acceptors for one of the products of the reaction. The essential chemical components of many coenzymes are *vitamins.* For example, the coenzymes *nicotinamide adenine dinucleotide* (NAD) and *nicotinamide adenine dinucleotide phosphate* (NADP) contain the vitamin *niacin; coenzyme A* contains *pantothenic acid; flavin adenine dinucleotide* (FAD) contains *riboflavin* (i.e., vitamin B_2); *thiamine pyrophosphate* contains *thiamine* (i.e., vitamin B_1), and so on. The chemistry of some of these cofactors was discussed in Chapter 3.

A number of enzymes require **metal ions** for their activity. The metal ions form *coordination bonds* with specific side chains at the active site and at the same time form one or more coordination bonds with the substrate. The latter assist in the polarization of the substrate bonds to be cleaved by the enzyme. For example, zinc is a cofactor for the proteolytic enzyme *carboxypeptidase* and forms coordination bonds with the side chains of two histidines and one glutamic acid residue at the active site. A fourth bond is formed between zinc and the alpha-carboxyl group of the substrate amino acid, and it is here that the cleavage of the peptide occurs. Table 8–4 contains a list of some of the cofactor-requiring enzymes.

The observation that catalytic activity is lost when an enzyme is stripped of its cofactor testifies to the crucial role played by these atoms or molecules. The role, however, appears to be diverse:

1. In some cases, the cofactor completes the active site of the enzymes or modifies it in such a manner that substrate binding can ensue.

Table 8–4
Some Enzymes That Require Cofactors

Cofactor	Enzyme	Reaction
Prosthetic groups		
Heme	Catalase	$2\ H_2O_2 \longrightarrow 2\ H_2O + O_2$
Heme	Peroxidase	$2\ H_2O_2 \longrightarrow 2\ H_2O + O_2$
Flavin adenine dinucleotide (FAD)	Succinic dehydrogenase	Succinic acid $\longrightarrow$ fumaric acid
Coenzymes		
Flavin mononucleotide (FMN)	Some dehydrogenases	Removal of hydrogen atoms
Thiamine pyrophosphate (TPP)	Some decarboxylases	Removal of CO_2
Nicotinamide adenine dinucleotide (NAD)	Some dehydrogenases	Removal of hydrogen atoms
Lipoic acid	Some decarboxylases	Removal of CO_2
Metal ions		
Zn^{++}	Carboxypeptidase	Hydrolysis of proteins
Zn^{++}	Carbonic anhydrase	$CO_2 + H_2O \leftrightarrow H_2CO_3$
Cu^{++}	Ascorbic acid oxidase	Ascorbate $\leftrightarrow$dehydroascorbate
Cu^{++}	Uric acid oxidase	Uric acid $\longrightarrow$ allantoin + CO_2 + H_2O_2
Mg^{++}	Hexokinase	Glucose + ATP $\longrightarrow$ glucose phosphate

2. The cofactor acts as a donor of electrons or atoms to the substrate and following the reaction is returned to its former state.

3. Cofactors may also serve as temporary recipients of either one of the reaction products or simply an electron or proton, again being recycled to its former state some time after the main reaction is completed.

4. Finally, the cofactor together with the side chains of residues at the active site, may serve to polarize the substrate and prime it for catalytic alteration.

Isoenzymes

Occasionally, several different enzyme molecules, all of which appear to catalyze the same chemical reaction, have been isolated from a single tissue. Such families of enzymes are called **isoenzymes** or **isozymes** (see also Chapter 11). Among the various isoenzymes the *lactic dehydrogenases* have been most extensively studied, and five different forms have been identified. All are composed of four polypeptide chains of two types called *M* and *H* subunits. Thus the lactic dehydrogenase isoenzymes may take either of the following forms: M_4, M_3H, M_2H_2, MH_3, or H_4. Similar arrangements are believed to exist for other groups of isoenzymes. Isozymes should be distinguished from allelozymes (or alleloenzymes), which are multiple forms of single-polypeptide enzymes resulting from variations in a single allelic pair of genes. Although allelozymes act on the same substrate, and in this regard are similar to isozymes, isozymes result from combinations of polypeptide chain products of two or more separate pairs of alleles.

Zymogens

A number of enzymes possess an inactive form called a **zymogen** or **proenzyme.** Several of the alimentary digestive enzymes belong to this group including *pepsin, trypsin,* and *chymotrypsin*. The conversion of the zymogen to the active enzyme involves the preliminary cleavage of one or more of the zymogen's peptide bonds, followed occasionally by complete removal of a portion of the original protein molecule. This phenomenon may best be understood by considering a few examples.

Pepsin, the major protein-digesting enzyme of the stomach, is synthesized in the form of the precursor polypeptide *pepsinogen* (molecular weight 42,000). Upon entering the acidic gastric juice, the pepsinogen molecule is hydrolyzed at several positions to yield a number of small peptides and the active enzyme pepsin (molecular weight of 35,000). The activation of pepsinogen can also be carried out by pepsin itself.

Trypsin, another proteolytic digestive enzyme, is produced in the pancreas as the zymogen trypsinogen and secreted into the duodenum (the anterior portion of the small intestine). In the duodenum, another enzyme, *enterokinase*, catalyzes the removal of six amino acids from the *N*-terminus of trypsinogen, thereby yielding trypsin. Additional activation can also be effected by trypsin itself.

The enzyme chymotrypsin also participates in the alimentary digestion of protein and is produced in the pancreas as the inactive proenzyme *chymotrypsinogen*. The activation of this zymogen involves a series of peptide bond cleavages catalyzed by trypsin already present in the duodenum and also by chymotrypsin itself. These cleavages split the chymotrypsinogen molecule into three polypeptides that remain interconnected in the activated enzyme through disulfide bridges that were part of the molecule's primary structure.

The activation of the proenzyme *trypsinogen* is shown in Figure 8–19 and illustrates some of the general features of zymogen activation. The active site of the enzyme is devoid of binding and/or catalytic activity until peptide bonds of the zymogen are broken. Following this peptide cleavage, the remaining portion of the molecule is reorganized with a consequent unmasking of the active site which can now bind and act on the substrate.

Not only proenzymes but also other proteins may be activated by a preliminary proteolysis. For example, during the coagulation of blood, the formation of the matrix of the clot is brought about through a cascade of proteolytic activations that finally convert inactive, soluble protein monomers (fibrinogen) in blood plasma to the active, polymerizable form, which then produces the insoluble protein threads (fibrin). Some protein hormones are also synthesized as inactive precursors that are activated only upon peptide cleavage (e.g., the conversion of *proinsulin* to *insulin*).

The Regulation of Enzyme Activity

Literally thousands of different enzyme-catalyzed reactions characterize the metabolism of a cell, and these are organized into a number of interconnected (branching and converging) pathways. Some of the more important or universal pathways are considered in Chapter 10. It is clear that the orderly functioning of a cell or organism demands that controls be placed on these reactions so that specific metabolic pathways (and, therefore, specific cell functions) are active or operative at certain times and inactive at others. One way in which such regulation can be achieved is by *altering the activity of specific enzymes under specific circumstances*. (Other regulatory mechanisms are discussed in Chapter 11).

Consider as an example the following situation. Suppose that in the course of five enzyme-catalyzed reactions, sub-

Figure 8–19
The activation of trypsinogen. Proenzyme activation involves the cleavage of the peptide bond linking amino acids 15 and 16, thereby freeing a small peptide. Amino acid 16 of trypsinogen thereby becomes the N-terminus of trypsin and interacts ionically with asp 194. This alters the orientation of lys 145, which was previously neutralized by asp 194. Reorientation generates the specific binding site, allowing substrate binding and eventual cleavage (dashed line).

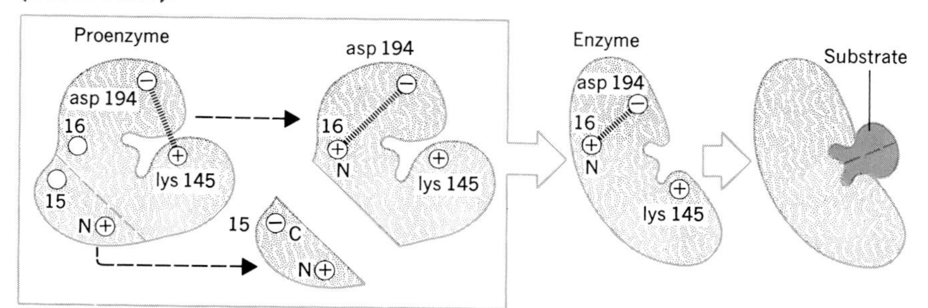

strate A is converted to end-product F; that is,

$$A \xrightarrow{E_1} B \xrightarrow{E_2} C \xrightarrow{E_3} D \xrightarrow{E_4} E \xrightarrow{E_5} F$$

where E_1, E_2, E_3, and so on are the enzymes involved in the pathway. If the end product F is able to bind to enzyme E_1 and in so doing render the enzyme inactive (i.e., unable to catalyze the conversion of A to B), then the synthesis of end product F can be regulated, for under conditions of low F concentration, the metabolic pathway leading to F would be functioning, while at high concentrations of F, E_1 inhibition by the end product would put a halt to the pathway leading to F. The phenomenon in which the end product of a metabolic pathway can regulate its own production by inhibition of this sort is called **feedback inhibition.**

Many metabolic pathways are characterized by a number of branch points. In the example below, substrate A may be converted either to end product F or end product I. In such a pathway, the enzyme affected by feedback inhibition would occur at the branch point; that is, enzyme E_3 would be inhibited by end product F. Consequently, although the conversion of A to F would be inhibited, the formation of I would continue. The continued synthesis of I would not be possible if feedback inhibition occurred before the branch point.

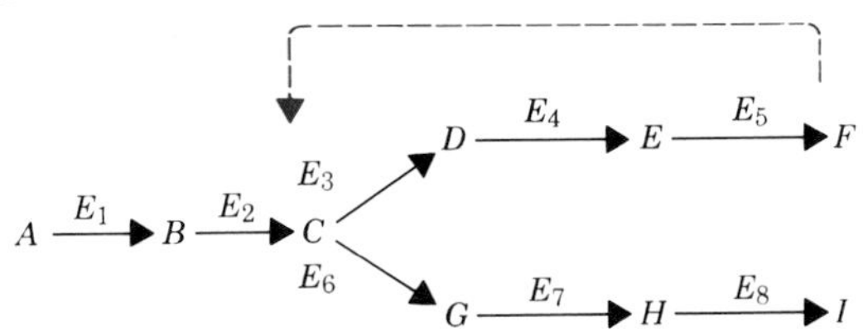

That the phenomenon of feedback inhibition actually occurs in cells was definitively established in the 1950s. One of the earliest discovered examples of feedback inhibition and a classic case is the effect of cytidine triphosphate (CTP, one of the final products formed during pyrimidine synthesis) on the enzyme *aspartate transcarbamylase* (Fig. 8–20). The biosynthesis of CTP begins with the interaction of aspartic acid and carbamyl phosphate to form carbamyl asparate. The reaction is catalyzed by aspartate transcarbamylase (ATCase) and is followed by four more reactions, culminating in the formation of CTP. When the free CTP concentration is low (i.e., when CTP is being consumed

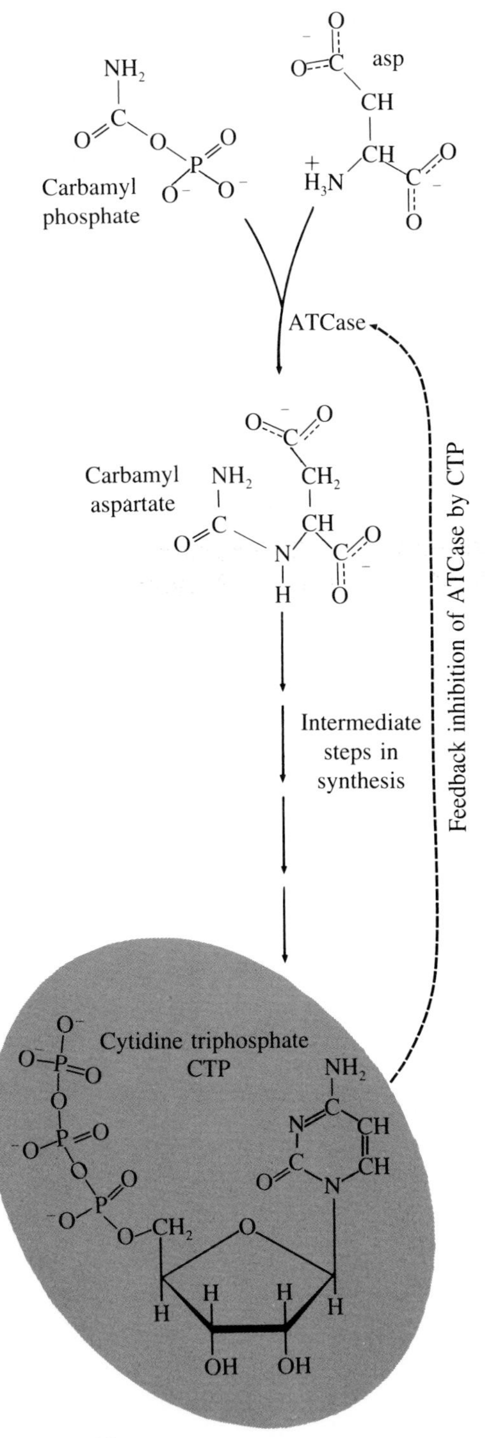

Figure 8–20
Feedback inhibition of aspartate transcarbamylase by cytidine triphosphate.

in the cell's metabolism), the formation of carbamyl asparate proceeds uninhibited. However, when the CTP level rises, carbamyl asparate synthesis is inhibited.

It is now clear that feedback inhibition is only one of several ways in which enzyme activity can be regulated. Not only end products but also other metabolites may be bound to an enzyme and in so doing alter its activity. The binding can take place at the active site or at other sites on the enzyme's surface. Indeed, in instances where the enzyme molecule has more than one active site (see below), the binding of substrate at one site can influence subsequent substrate binding at another. The various mechanisms of enzyme regulation now known to exist will be considered in the following section.

Number of Polypeptide Chains and Number of Binding Sites of an Enzyme

Many enzymes, including lysozyme and ribonuclease, are composed of a single polypeptide chain. Others consist of two or more chains. For example, the enzyme *phosphorylase* (involved in glycogen metabolism) consists of two polypeptide chains; *fumarase* (a Krebs cycle enzyme) contains four polypeptide chains; and *aspartyl transcarbamylase* (involved in the metabolic pathway leading to the synthesis of cytidine triphosphate) consists of 12 polypeptide chains. Each of the enzyme's constituent polypeptide chains is referred to as a *subunit,* the subunits being held together by electrostatic interactions, hydrogen bonds, van der Waals interactions or other noncovalent forces that provide and stabilize protein quaternary structure (Chapter 4).

Enzymes may have two *functionally different* (as well as topologically separate) binding sites. One type of site, the active site, binds the substrate of the enzyme and possesses catalytic activity, while the other type of site, the *allosteric* (*allo,* "other" + *steric,* "space") or *regulatory* site, lacks catalytic activity and binds an *effector* molecule. Such enzymes are usually referred to as **allosteric enzymes.** Depending upon the enzyme, active and allosteric sites may be on the same polypeptide or on separate subunits. Effector molecules that inhibit enzyme activity (as in feedback inhibition) are called *negative* effectors. In some cases, binding of an effector molecule enhances enzyme activity, and these molecules are known as *positive* effectors. A single enzyme may possess regulatory sites capable of binding either negative or positive effectors, and these may compete with each other for the same regulatory site or be bound at separate regulatory sites.

In some enzymes, two or more subunits may each possess an active site, and substrate binding to the active site of one subunit may influence substrate binding at another active site. Such enzymes exhibit *cooperativity*—a phenomenon already discussed in connection with the reversible oxygenation of hemoglobin (see Chapter 4). Enzymes that exhibit cooperativity and allosteric enzymes do not obey conventional Michaelis-Menten kinetics; the consequences of this are discussed later in the chapter.

Model for Allosteric Enzyme Function

A simple scheme depicting the influence of positive and negative effectors on allosteric enzyme activity is given in Figure 8–21, in which substrates and effectors and active and allosteric binding sites are represented as geometric figures or areas. The active and regulatory sites of the enzyme (in the absence of bound effectors) are depicted as circular areas. The binding of a positive effector (hexagon)

Figure 8–21
Simple model of allosteric enzyme function (see text for explanation).

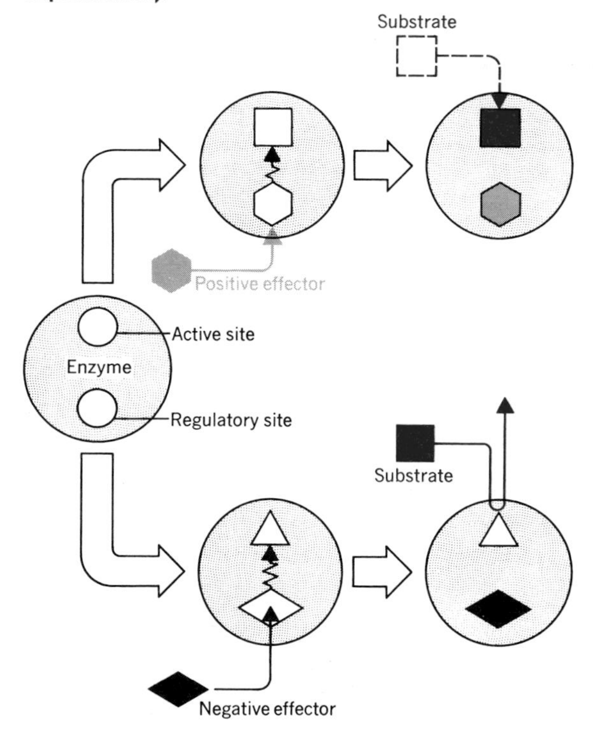

induces a conformational change in the enzyme molecule at the regulatory site (symbolized as a change from circular to hexagonal shape). The conformational shape change is transmitted through the molecule (zigzag arrow) until it reaches and alters the active site in such a way that the substrate is bound more readily (i.e., more readily than in the absence of such activation). In Figure 8–21, the alteration of the active site is represented as a change from circular to square shape. Binding of a negative effector (diamond) induces a different type of conformational change in the regulatory site (circular to diamond shape) which is transmitted through the enzyme's structure such that the resulting change at the active site (circular to triangular shape) prevents subsequent substrate binding. The negative effector may, of course, be the end product of the pathway involving this enzyme.

The behavior of aspartate transcarbamylase, discussed earlier in connection with feedback inhibition, serves as a good example of the mechanism depicted in Figure 8–21. Although aspartate transcarbamylase activity is inhibited by the end product CTP (i.e., CTP is a negative effector), ATCase activity is stimulated by ATP (i.e., ATP acts as a positive effector). It appears that CTP and ATP compete for the same allosteric site of the enzyme.

The model of Figure 8–21 deliberately fails to stipulate whether the allosteric enzyme is composed of one or more than one polypeptide chain, whether more than one active and regulatory site are present, or whether the sites are on the same or on different subunits. Although all of these possibilities exist, most allosteric enzymes consist of several polypeptide subunits with active and regulatory sites on *separate* subunits. The combination of two subunits having active and regulatory sites comprises a **protomer.** For example, ATCase contains four protomers, each composed of one catalytic and one regulatory polypeptide having respective molecular weights of 100,000 and 27,000 daltons. Table 8–5 lists some of the allosteric enzymes, their substrates, and negative and positive effectors.

Cooperativity in Enzymes

The protomers of allosteric enzymes may exhibit cooperativity. That is, binding of a substrate molecule to the active site of one protomer may influence binding of a second (third, etc.) substrate molecule to the active site of a second (third, etc.) protomer. The influence may be positive in that binding of the first substrate molecule facilitates binding of subsequent substrate molecules (called "positive cooperativity") or the influence may be negative in that binding of a second or subsequent substrate molecule occurs less readily than binding of the first (called "negative cooperativity"). In a sense, the substrate itself is acting as either a positive or negative effector for neighboring protomers of the enzyme. These relationships are depicted in Figure 8–22.

Cooperative effects are not restricted to enzymes but are observed with other proteins. Earlier (Chapter 4), we considered the positive cooperativity that exists among the globin chains of hemoglobin, a cooperativity that facilitates successive binding of oxygen molecules to the alpha and beta globin chains (i.e., positive cooperativity).

Table 8–5
Some Allosteric Enzymes

Enzyme	Substrate	Negative Effector	Positive Effector
Aspartate transcarbamylase	Aspartic acid and carbamyl phosphate	CTP	ATP
Threonine deaminase	Threonine	Isoleucine	Valine
Acetolactate synthetase	Pyruvic acid	Valine	—
Isocitric dehydrogenase	Isocitric acid	Alpha-ketoglutaric acid	Citric acid
Phosphofructokinase	Fructose-6-P and ATP	ATP	3′, 5′-AMP
Phosphorylase b	Glucose-1-P, glycogen and phosphate	ATP	5′-AMP
Glycogen synthetase	Uridine diphosphoglucose	—	Glucose-6-P
Homoserine dehydrogenase	Homoserine and aspartic semialdehyde	Threonine	Isoleucine and methionine

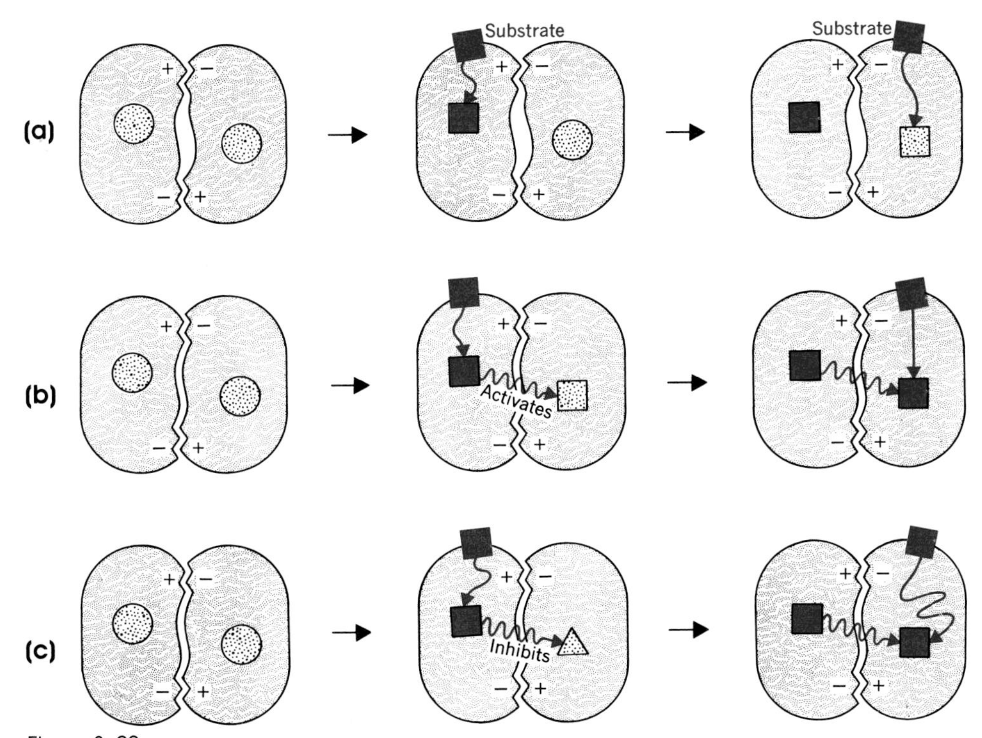

Figure 8–22
Cooperativity in enzymes. The enzyme depicted here consists of two subunits. In *(a)*, binding of the substrate to the active site of one subunit is without effect on the other subunit; i.e., *noncooperativity.* In *(b)*, substrate binding to one subunit activates the site of the other subunit (i.e., *positive cooperativity)*, thereby facilitating binding of the second substrate molecule (indicated by the straight path to the active site in contrast to the somewhat more devious path shown above in *a)*. In *(c)*, binding of the first substrate molecule to the active site of one subunit produces a conformational change in the enzyme that inhibits substrate binding at the other active site (i.e., *negative cooperativity)*.

Both cooperative effects involving active sites on neighboring subunits of an enzyme and true allosteric effects involving regulatory sites may occur in a single enzyme molecule. A case in point is that of cytidine triphosphate synthetase, an enzyme involved in nucleic acid metabolism and consisting of two protomers. One of the substrates of the enzyme is glutamine, but when glutamine is bound to one protomer a conformational change transmitted through the enzyme to the other protomer renders the latter's active site unable to bind glutamine (i.e., negative cooperativity). A regulatory (allosteric) site on each protomer binds the effector GTP. GTP bound to the regulatory site of one protomer has a positive effect on glutamine binding by that protomer but negatively effects GTP binding at the regulatory site of the other protomer. In other words, the effector GTP serves to activate catalysis but to inhibit further GTP binding.

Allosterism, Cooperativity, and Michaelis-Menten Enzyme Kinetics

As noted earlier, allosteric enzymes and enzymes that exhibit cooperative effects do not display conventional Michaelis-Menten kinetics. Figure 8–23 compares the Mi-

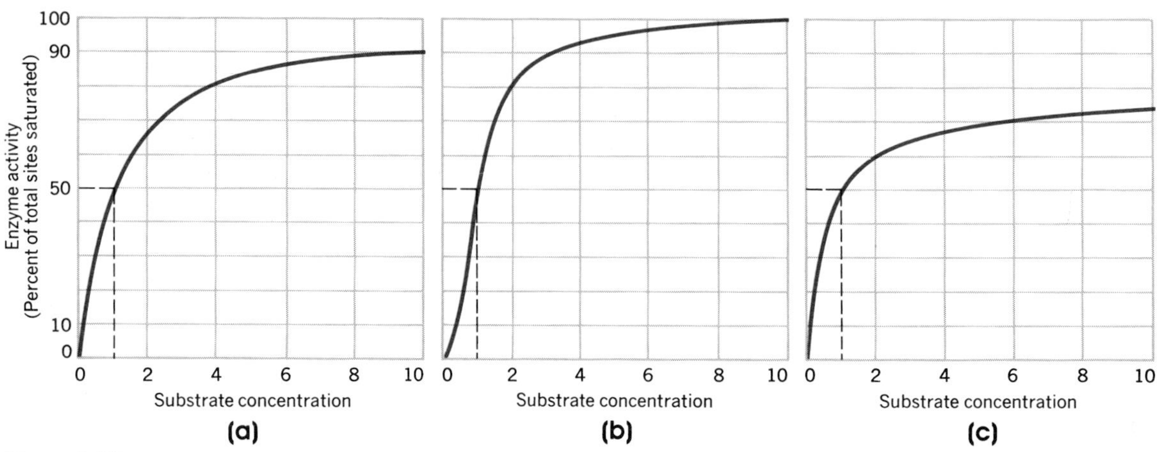

Figure 8–23
Michaelis-Menten curves for enzymes exhibiting non-cooperativity *(a)*, positive cooperativity *(b)*, and negative cooperativity *(c)*. See text for details.

chaelis-Menten curves for enzymes exhibiting noncooperativity, positive cooperativity, and negative cooperativity. Curve *A* shows the normal hyperbolic binding pattern exhibited by most enzymes. In this example, an 81-fold increase in substrate concentration is required in order to elevate enzyme activity from 10% to 90% of its maximum level. Curve *B* depicts the sigmoid pattern characteristic of positive cooperativity. Here, only a ninefold increase in substrate concentration elevates enzyme activity from 10% to 90% of maximum. Note the resemblance of this curve to that for hemoglobin oxygenation (Chapter 4). Negative cooperativity produces the curve shown in *C*. Although the curve appears hyperbolic, it actually is not. An increase in substrate concentration greater than 6000-fold would be required to elevate the enzyme activity from 10% to 90% of maximum. The curves have been drawn so that a substrate concentration of 1 unit corresponds to 50% of maximum enzyme activity.

A similar set of curves would be obtained for allosteric enzyme activity in the absence of effectors (i.e., curve *A*), in the presence of positive effectors (i.e., curve *B*) and in the presence of negative effectors (i.e., curve *C*).

As noted earlier, enzymes may exhibit cooperativity and also be affected by the binding of effectors to regulatory sites. ATCase, the enzyme discussed above, is a good example; is it negatively affected by CTP, positively affected by ATP, and shows positive cooperativity among its substrate binding sites. Figure 8–24 shows the effects of positive and negative effectors on ATCase.

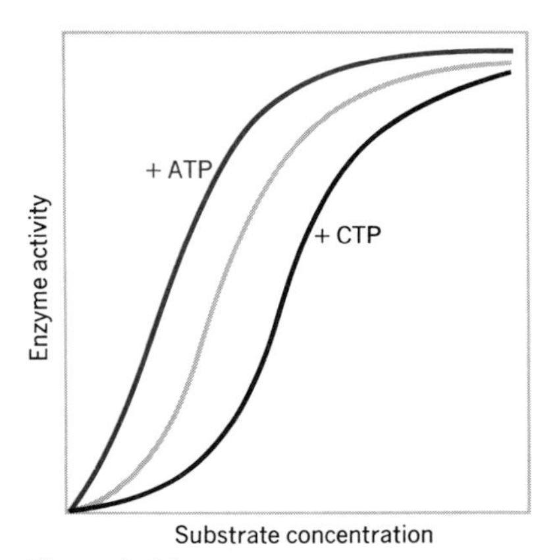

Figure 8–24
Effects of CTP (feedback inhibitor, negative effector) and ATP (positive effector) on the Michaelis-Menten curve of the positively cooperative enzyme aspartate transcarbamylase.

Control of Metabolic Processes by Allosteric Enzymes: The Alternate Pathways for Glycogen Synthesis and Degradation

The metabolism of a cell or an organism involves a large number of branching and converging metabolic pathways that might be likened to a complicated electrical circuit. The orderly functioning of a cell or organism requires control that permit certain pathways to be in operation during specific periods while others are temporarily halted. This control provides *direction* in space and in time for the system. In the absence of such control, all of the various pathways might be simultaneously in operation, devoid of direction and consequence, and much like a ''short circuit.'' Allosteric enzymes are essential elements in the control of metabolism.

An excellent example of allosteric enzyme regulation of metabolic processes is provided by a consideration of the interrelationship in animals between the metabolic pathways that result in the synthesis of glycogen from glucose and the oxidation of glucose to CO_2 and water. Nearly all the active body processes require energy in the form of ATP, and this, in turn, is derived principally from the oxidation of glucose. During periods of elevated activity (i.e., exercise), glycogen is broken down to yield glucose, which then enters the metabolic pathway converting it to CO_2 and water, with the consequent generation of ATP. In contrast, during periods of rest or low energy demand, absorbed glucose is stored in the form of glycogen. Thus, glucose has alternative fates in the body that might be depicted as follows:

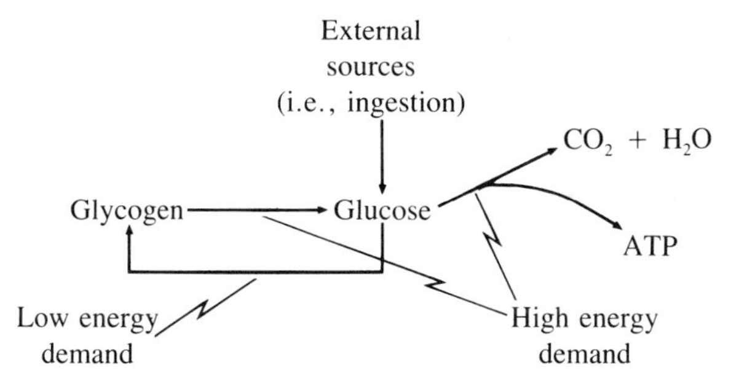

There are three allosteric enzymes involved in these interrelated pathways. One is *phosphofructokinase*, an enzyme required in the series of reactions that convert *glucose-6-phosphate* to CO_2 and water. A second is *glycogen synthetase*, which is involved in the incorporation of *glucose-1-phosphate* into glycogen. The third allosteric enzyme is *glycogen phosphorylase*, which removes glucose as glucose-1-phosphate from glycogen during glycogen catabolism.

When ATP levels are high (i.e., no great consumption of energy taking place in the body), glucose is diverted into glycogen (i.e., ''glycogenesis'' predominates). This is achieved because ATP acts as a negative effector of phosphofructokinase and glycogen phosphorylase and as a positive effector, along with glucose-6-phosphate, of glycogen synthetase (Fig. 8–25*a*).

When ATP levels are low (i.e., during exercise) and there is an increased demand for ATP, glycogen synthesis is halted as absorbed glucose is directly consumed in the production of ATP and additional glucose is made available through the catabolism of glycogen (i.e., ''glycogenolysis''). This pathway is activated by the positive effects on phosphofructokinase and glycogen phosphorylase of the ATP precursor *AMP*. The hormone *adrenaline (epinephrine)*, secreted into the bloodstream during periods of great activity, also has an effect on these metabolic pathways in muscle and in liver. Adrenaline activates phosphofructokinase and glycogen phosphorylase but inhibits glycogen synthetase (Fig. 8–25*b*).

The pathways just described are excellent illustrations of the mechanisms for turning allosteric enzymes on and off. In the absence of such mechanisms, both pathways might simultaneously be active so that their effects cancel one-another—a most unproductive state! Allosterism thus provides a basis for regulating the levels of activity of related metabolic pathways. As we shall see in Chapter 11, allosterism is only one of several mechanisms employed for the regulation of cellular chemical activity.

Summary

Regardless of the level of **molecularity,** nearly all reactions occuring in cells are catalyzed by the class of proteins called **enzymes.** Enzymes increase the likelihood that potentially reacting molecules (i.e., **substrates**) will encounter each

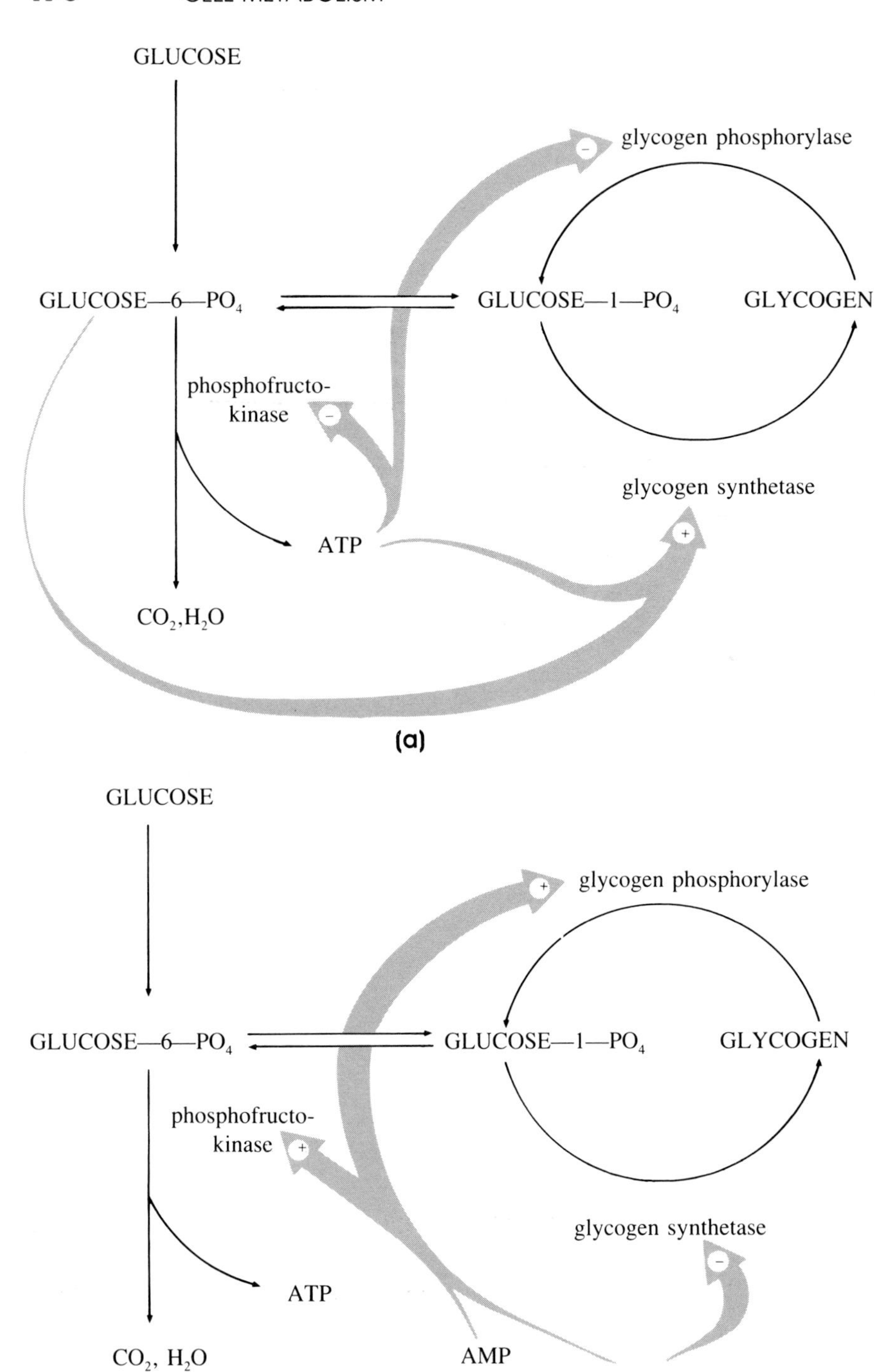

Figure 8–25
Allosteric regulation of the control of glycogen synthesis (glycogenesis) and glycogen degradation (glycogenolysis). *(a)* Glycogenesis. *(b)* Glycogenolysis. + and − symbols indicate whether effector acts positively or negatively.

other with the necessary orientation in space and "activate" their substrates by imposing strains upon certain bonds. Although greatly increasing reaction rates, enzymes do not alter equilibrium concentrations. In the presence of finite amounts of enzyme, reactions approach a limiting velocity, the kinetics of which may be used to measure the enzyme's **Michaelis-Menten constant;** this, in turn, is a measure of the tendency of the enzyme and substrate to combine with each other.

The catalytic properties of enzymes are intimately related to the enzyme's primary, secondary, tertiary, and quaternary structure, and changes in structure are often accompanied by loss of catalytic activity. Consequently, most enzymes operate within narrow ranges of pH and temperature. Association of the substrate takes place in the enzyme's **active site,** which contains **contact** and **catalytic** amino acids. The active site nearly always resides in a depression or cavity in the enzyme's surface, and the binding of substrate is followed by a conformational change in the enzyme and/or the substrate. Some enzymes require a nonprotein component called a **cofactor** in order to be catalytically active. Cofactors may be permanently bound to the enzyme (i.e., **prosthetic groups**) or may form transient associations (e.g., **coenzymes** and **metal ions**). In some organisms, several different enzyme molecules catalyze the same chemical reactions; such families of enzymes are called **isozymes.** Certain enzymes possess an inactive form called a **zymogen** or **proenzyme.**

In addition to the active site, some enzymes possess an **allosteric** or **regulatory** site. The binding of a positive **effector** to the allosteric site increases the activity of the enzyme, whereas the binding of a negative effector decreases enzyme activity. Changes in the level of activity of allosteric enzymes are fundamental to the control of many metabolic pathways through feedback mechanisms. Enzymes composed of more than one polypeptide chain may contain two or more active sites, and the binding of substrate to one site influences binding of additional substrate to other sites. Such influence, which can be positive or negative, is called **cooperativity.**

References and Suggested Reading

Articles and Reviews

Bell, R. M., and Koshland, D. E., Covalent enzyme-substrate intermediates. *Science 172,* 1253 (1971).

Hammes, G. G., and Wu, C. – W., Regulation of enzyme activity. *Science 172,* 1205 (1971).

Koshland, D. E., Correlation of structure and function in enzyme action. *Science 142,* 1533 (1963).

Koshland, D. E., Protein shape and biological control. *Sci. Am. 229*(4), 52 (Oct. 1973).

Koshland, D. E., and Neet, K. E., The catalytic and regulating properties of enzymes. *Annu. Rev. Biochem. 37,* 359 (1968).

Neurath, H., Protein-digesting enzymes. *Sci. Am. 211*(6), 68 (Dec. 1964).

Perutz, M. F., X-ray analysis, structure and function of enzymes. *Eur. J. Biochem. 8,* 455 (1969).

Phillips, D. C., The three-dimensional structure of an enzyme molecule. *Sci. Am. 215*(5), 78 (Nov. 1966).

Stroud, R. M., A family of protein-cutting proteins. *Sci. Am. 231*(1), 74 (July 1974).

Books, Monographs, and Symposia

Bernhard, S. A., *The Structure and Function of Enzymes,* W. A. Benjamin, Inc., New York, 1968.

Boyer, P. D. (editor), *The Enzymes* (3rd ed.), Academic Press Inc. New York, 1970.

Dickerson, R. E., and Geis, I., *The Structure and Action of Proteins,* Harper & Row, New York, 1969.

Fersht, A., *Enzyme Structure and Mechanism,* W. H. Freeman and Co., San Francisco, 1977.

Lehninger, A. L., *Biochemistry,* Worth Publishers, Inc., New York, 1975.

McGilvery, R. W., *Biochemical Concepts,* W. B. Saunders, Philadelphia, 1975.

Snell, F. M., Shulman, S., Spencer, R. P., and Moos, C., *Biophysical Principles of Structure and Function,* Addison-Wesley, Reading, Mass., 1965.

Stryer, L., *Biochemistry,* W. H. Freeman and Co., San Francisco, 1975.

Chapter 9
BIOENERGETICS

In the study of physiology and biochemistry, cells are frequently thought of as machines in which all events are explainable as one or more chemical reactions, as the result of fluid dynamics, as electrical fluxes across partitions, or as the absorption or emission of light. In other words, the functions of cells can be explained in terms of known principles of physics and chemistry, for cell actions conform to the laws of the physical sciences.

This chapter discusses those laws of physics and chemistry that are important for an understanding of cell metabolism, including the breakdown of some molecules (catabolism) and the synthesis of others (anabolism), the absorption of light as in vision and photosynthesis, propagation of electrical charges such as nerve impulses, and fluid dynamics as they relate to diffusion and osmotic phenomena.

Energy

All the chemical reactions and physical events in cells are related to energy changes. For example, the synthesis of new membranes in a growing cell requires or consumes energy. That energy must be ultimately absorbed from the environment in some form, such as light energy or chemical energy, and then transformed by the cell into forms that can provide the energy for the membrane-synthesizing reactions. Depending upon the source of the environment's energy, cells and organisms are divided into two basic groups. If the organism requires organic chemicals (foods) from the environment, it is classified as a **heterotroph.** If the organism can survive with only inorganic compounds and energy sources such as light, it is called an **autotroph.** Microbiologists make a further distinction according to the need for environmental energy or raw materials. By their classification system, an organism requiring a chemical source of energy supply is termed a **chemotroph,** but if only light is required for energy, the organism is a **phototroph.** A further requirement for organic raw materials for synthesis would classify the organism as a **organotroph,** while a requirement for only inorganic raw materials would designate a **lithotroph.** For example, a green plant requires only light for energy (phototroph) and inorganic compounds for growth (lithotroph) and could therefore be classified as a photolithotroph. Most animals by this system would be classified as chemo-organotrophs.

The primary sources of energy and raw materials for heterotrophs are proteins, lipids, and carbohydrates. Organisms remove these compounds from the environment and break them down in progressive stages called the catabolic reactions of metabolism. As these compounds are chemically degraded into smaller units, their energy is both released in the form of heat and used to form new chemical bonds, as in the attachment of a phosphate to ADP to form

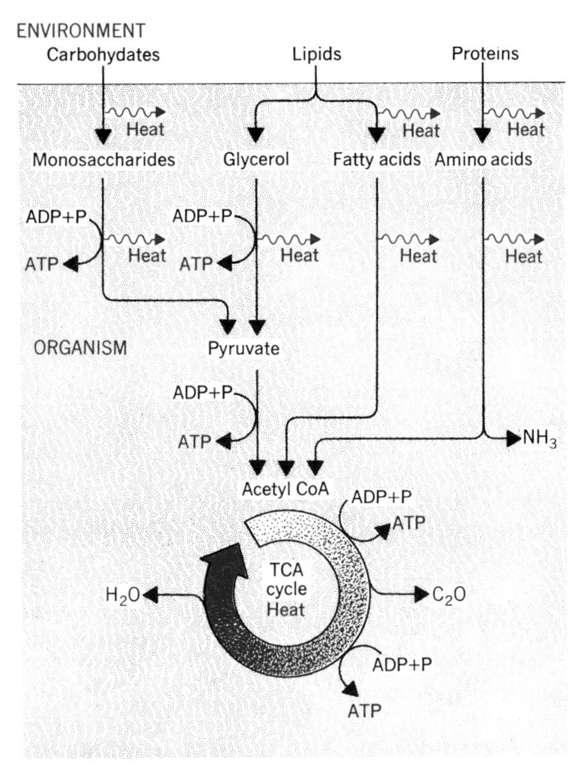

Figure 9–1
Energy flow in living systems during catabolic reactions.

ATP (Fig. 9–1). The ultimate primary products of catabolism are NH_3, CO_2, and H_2O.

Although autotrophic organisms absorb CO_2, H_2O, and small nitrogen-containing compounds from the environment, these small compounds do not contain sufficient energy to maintain the organisms. Consequently, autotrophs absorb energy in the form of light, and using the light energy, they synthesize simple organic acids from CO_2 and water (photosynthesis) as well as phosphorylate ADP and synthesize amino acids from the organic acids by incorporating NH_3 (amination). From these simpler molecules more complex molecules such as proteins, lipids, and polysaccharides are formed (Fig. 9–2); however, the synthesis of these molecules requires further energy consumption which the cell supplies from its pool of ATP, and with each change, energy in the form of heat is lost from the cell.

In general, autotrophic organisms not only contain the enzymes for the anabolic reactions just described but also possess catabolic enzyme systems similar to those of the heterotrophs and produce ATP upon the breakdown of carbohydrates, lipids, and proteins. Heterotrophs have anabolic enzyme systems that require ATP and are capable of synthesizing macromolecules much like the autotrophs, but they are unable to carry out photosynthesis. Cells tend to cycle compounds internally, and compounds are also cycled between organisms (Fig. 9–3); with each change, there is a corresponding change in energy.

The Laws About Energy and Energy Changes

The laws that relate to energy in general, and that therefore apply to cellular energy as well, are the two **laws of thermodynamics.** The **first law** is concerned with the conservation of energy and requires no modifying statements when applied to biological systems. The first law states the following:

Energy cannot be created or destroyed but can be converted from one form into another.

This definition applies to cells, to organelles, and even to single chemical reactions. In practice, the measurement of energy in a cell or similar units is difficult, since energy may escape from the cell into the surrounding environment during the measurement. Similarly, energy may be gained from the environment; for example, a photosynthesizing cell absorbs energy in the form of light. The escape or absorption of energy by a body should not be confused with the destruction or creation of energy, which by the first law of thermodynamics does not occur. Usually the term ''system'' is used to delimit the matter in which the observer is interested. In theoretical discussions, energy is assumed not to enter or leave the system. In practical measurements the system is isolated as much as possible, and the exchange of energy between the system and the surroundings is measured.

Applied to the cell, the first law of thermodynamics indicates that the cell has a finite amount of energy. That energy may be (1) **potential,** as in the bonds of molecules or in the pressure-volume relationships within the cell membranes; (2) **electrical,** as in the distribution of charges across membranes; or (3) **kinetic,** as in the thermal activity of

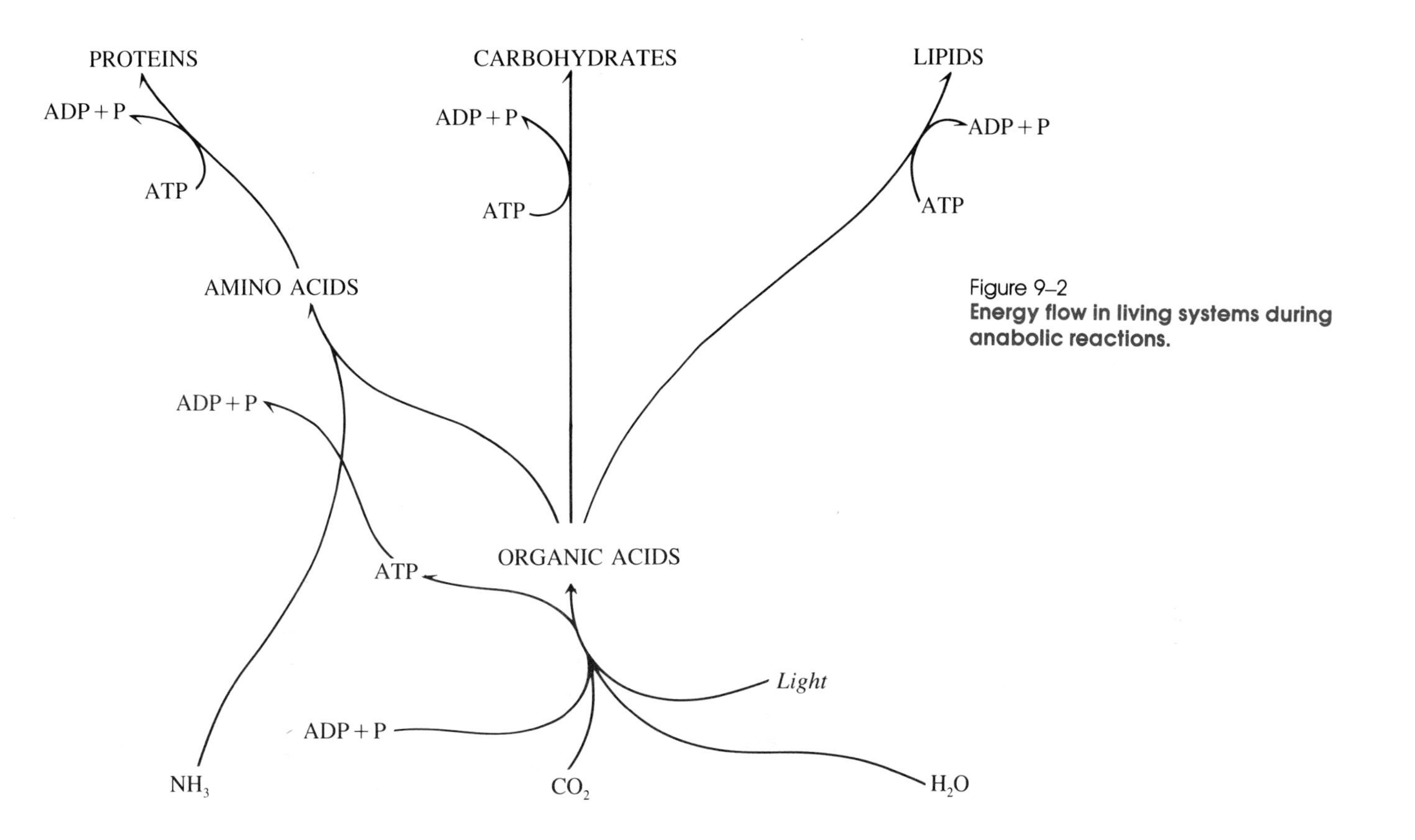

Figure 9–2
Energy flow in living systems during anabolic reactions.

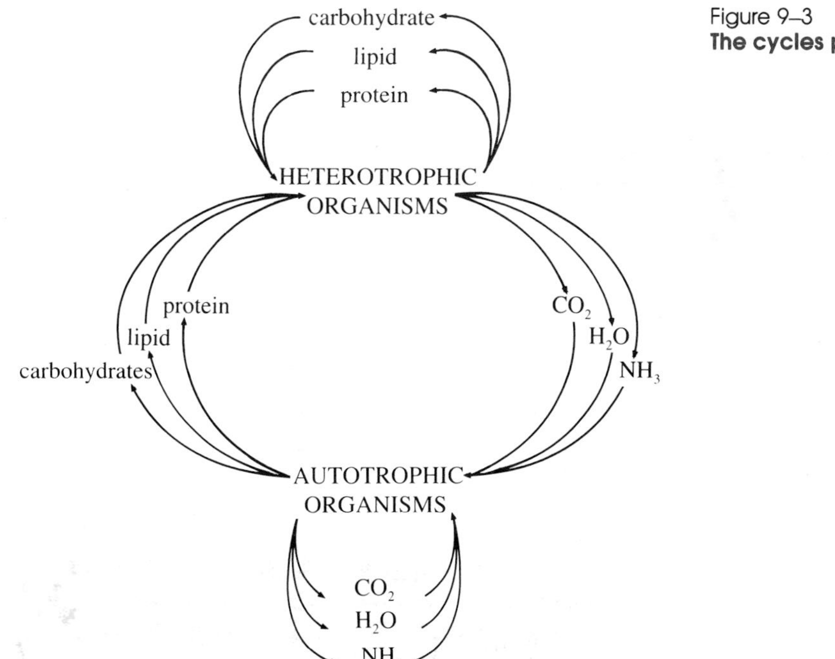

Figure 9–3
The cycles produced by the interchange of foodstuff.

molecules. The first law also states that these energies may be converted from one form into another; in other words, the potential chemical energy could decrease and, in so doing, be converted into electrical or kinetic energy. However, the *total* energy within the cell when isolated as a system remains constant. It is not possible for the cell to make energy out of matter or to convert energy into matter or to destroy any energy. For example, in the breakdown of a polysaccharide into CO_2 and H_2O, some of the potential energy in the carbohydrate is transferred to ADP and phosphate to make potential energy in the form of ATP, and some of the energy is converted to kinetic energy (which is of little practical use to the cell); however, none of the energy is lost or destroyed, and all the energy originally in the polysaccharide should be accountable for in other forms within the system.

The **second law of thermodynamics** explains the direction of the energy changes:

> **In all processes of energy change, the entropy of the system increases until equilibrium is achieved.**

Entropy is an expression of the energy in a system that is of no value for performing work and that cannot be converted into useful or reactive forms of energy. For example, a molecule of sucrose at 25°C requires a certain amount of internal energy just to be at that temperature and to maintain the thermal agitation of the molecule as well as the oscillations of the component atoms and their particles. This is unavailable energy, since it cannot be given up without a change in temperature. This unavailable energy changes with a change in temperature and is thus expressed by the paired symbols TS, or T for absolute temperature and S, which is called the **entropy.**

In catabolism of sucrose in a cell, many molecules of energy-rich ATP may be formed. Although, superficially it may appear that useful energy has increased in the form of ATP, within the whole cell (system) the total amount of useful energy has decreased and the unavailable energy has increased. Some of the potential energy that was present in the sucrose has been converted into useful potential energy in the ATP, but some has also been converted into kinetic energy, which raises the temperature of the cell and the entropy *(TS)*. In addition, when a large molecule such as sucrose is catabolized to many smaller molecules the energy for agitation of the many small molecules is greater than that for the large molecule, thus contributing to the increase in entropy.

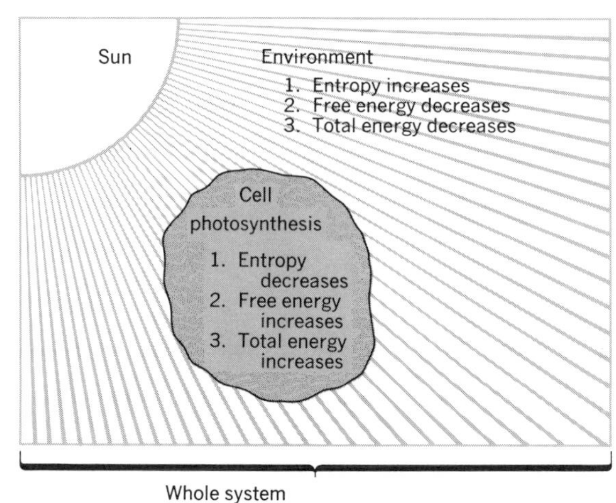

Figure 9–4
Relationship of entropy and free energy between various parts of a system and the net changes in energy.

Suggestions that cells can decrease entropy by photosynthesis are misleading. Although it is true that photosynthesis can convert small molecules with little potential energy (CO_2 and H_2O) into large molecules with considerable potential energy (sugars) and that the entropy of the cell does decrease, ''high-grade'' energy (i.e., light) was absorbed from the surrounding environment (outside the system). If the light energy consumed is included as part of the system, then the *net* energy changes will reveal an increase in entropy and a decrease in useful energy. These relationships are expressed in Figure 9–4.

Not *all* the energy is changed in a system each time there is a chemical reaction (or similar phenomenon). For example, when sucrose is hydrolyzed to glucose and fructose, much of the potential energy of the original source still remains in the resulting glucose and fructose molecules. The *change* in energy, rather than *total energy,* is a commonly used measure. The change in total energy, ΔH, is called the **enthalpy** and is composed of two factors: (1) the change in useful energy (or free energy) capable of performing work, ΔG; and (2) the entropy factor $T\ \Delta S$, which is the absolute temperature times the change in entropy. For example,

$$\Delta H = \Delta G + T\,\Delta S \qquad (9\text{–}1)$$

Since the change in entropy is difficult to determine while the important aspect of reactions in cells is the change in amount of useful or free energy (ΔG), equation 9–1 is more frequently written

$$\Delta G = \Delta H - T\,\Delta S \qquad (9\text{–}2)$$

The change in free energy (ΔG) can also be defined as the total amount of free energy in the products minus the total amount of free energy in the reactants; that is,

$$\Delta G = G_{\text{(products)}} - G_{\text{(reactants)}} \qquad (9\text{–}3)$$

A reaction that has a negative ΔG value (i.e., the products have less energy than the reactants) will occur spontaneously. A reaction with a positive ΔG value requires an input of energy from some outside source. The hydrolysis of sucrose,

$$\text{Sucrose} + H_2O \longrightarrow \text{glucose} + \text{fructose}$$

has a negative value. If 5 moles of sucrose are mixed with water the reaction proceeds spontaneously and the ΔG can be determined. The value is, of course, greater than if 4 or 2 moles of sucrose were used. The ΔG is dependent upon the concentration of reactants and products. Since the ΔG values vary with concentration, it is difficult to compare changes in free energy between various reactions. As a result, the concept of standard free energy, $\Delta G°$, was established by convention. This value represents the change in free energy when the reactants are present in 1.0 molal concentrations (1.0 molar is frequently substituted) at 25°C and a pressure of 1.0 atmosphere. $\Delta G°'$ is the change in standard free energy at pH 7.0, a value more useful in biological systems.

The $\Delta G°$ value for the hydrolysis of sucrose is -7.0 kcal; that is, the reaction will proceed spontaneously and release 7 kcal of free energy if 1 mole of sucrose is mixed with 1000 ml of H_2O. However, in cells concentrations of reactants and products vary and the actual ΔG should not be confused with the $\Delta G°$.

$\Delta G°$ is calculated from the equilibrium constant:

$$\Delta G° = -RT \ln K, \qquad (9\text{–}4)$$
$$= -2.303\,RT \log_{10} K \qquad (9\text{–}5)$$

where R is the gas constant (1.98 cal degree^{-1} mole^{-1}), T is the absolute temperature (degrees Kelvin), and K is the equilibrium constant. Table 9–1 lists a number of $\Delta G°$ values for common reactions. The equilibrium constant is determined analytically.

In the reaction $A + B \longrightarrow C + D$,

$$K = \frac{[C]\ [D]}{[A]\ [B]} \text{ at equilibrium} \qquad (9\text{–}6)$$

$[A]$ and $[B]$ are the concentrations of reactants, and $[C]$ and $[D]$ are concentrations of products. If the equilibrium constant is 1.0, then the $\Delta G° = 0$. If the equilibrium constant is less than 1.0, then the $\Delta G°$ is positive (e.g., $+$ 2.73 kcal mole^{-1} for $K = .01$), and the reaction is said to be endergonic because it requires an additional supply of energy to make it proceed. A K value greater than 1.0 would indicate a negative $\Delta G°$ (-1.36 kcal mole^{-1} for $K = 10$), and the reaction would be called exergonic because energy is spontaneously released in the reaction.

Coupled Reactions

In the cell, an endergonic reaction, one with a positive $\Delta G°$ value, would normally not be expected to occur spontaneously. However, two common events may cause the reaction to take place. One possibility is that the equilibrium is shifted by additional reactions in the cell that further convert or remove the products of the reaction, thereby altering the ΔG value (remember the difference between ΔG and $\Delta G°$). For example, the reaction

$$\text{Glucose-6-phosphate} \xrightarrow{\text{isomerase}} \text{fructose-6-phosphate}$$

has a $\Delta G° = +0.4$ kcal mole^{-1}. The reaction would not be expected to occur spontaneously as written. However, in the cell is a second enzyme that catalyzes the breakdown of the product,

Table 9–1
Standard Free Energy Changes of Common Biochemical Reactions at pH 7.0 and 25°C

Reaction	$\Delta G°$ kcal mol^{-1}
Hydrolysis:	
Acid anhydrides:	
Acetic anhydride + H_2O → 2 acetate	−21.8
Pyrophosphate + H_2O → 2 phosphate	− 8.0
Esters:	
Ethyl acetate + H_2O → ethanol + acetate	− 4.7
Glucose-6-phosphate + H_2O → glucose + phosphate	− 3.3
Amides:	
Glutamine + H_2O → glutamate + NH_4^+	− 3.4
Glycylglycine + H_2O → 2 glycine	− 2.2
Glycosides:	
Sucrose + H_2O → glucose + fructose	− 7.0
Maltose + H_2O → 2 glucose	− 4.0
Esterification:	
Glucose + phosphate → glucose-6-phosphate + H_2O	+ 3.3
Rearrangement:	
Glucose-1-phosphate → glucose-6-phosphate	− 1.7
Fructose-6-phosphate → glucose-6-phosphate	− 0.4
Elimination:	
Malate → fumarate + H_2O	+ 0.75
Oxidation:	
Glucose + $6O_2$ → $6CO_2$ + $6H_2O$	−686
Palmitic acid + $23O_2$ → $16CO_2$ + $16H_2O$	−2338

Source. From A. L. Lehninger, *Biochemistry* (2nd ed.), Worth Publishers, New York, 1975, p. 397.

$$\text{Fructose-6-phosphate} + \text{ATP} \xrightarrow{\text{fructokinase}} \text{fructose-1,6-diphosphate} + \text{ADP}$$

The equilibrium of the latter reaction lies far to the right; $\Delta G° = -3.40$ kcal mole^{-1}. The two reactions thus become coupled; the product of the first is removed, the equilibrium is thereby shifted, and both reactions proceed in the direction written.

A second type of coupling occurs when an endergonic reaction and an exergonic reaction are catalyzed by the same enzyme. In the two reactions

$$\text{ATP} + H_2O \longrightarrow \text{ADP} + \text{phosphate} \quad (\Delta G° = -7.3 \text{ kcal mole}^{-1})$$

$$\text{Glucose} + \text{phosphate} \longrightarrow \text{glucose-6-phosphate} \quad (\Delta G° = +3.3 \text{ kcal mole}^{-1})$$

the enzyme *glucokinase* catalyzes the transfer of the phosphate from ATP to glucose.

$$\text{Glucose} + \text{ATP} \longrightarrow \text{glucose-6-phosphate} + \text{ADP} \quad (\text{net } \Delta G° = -7.3 + 3.3 = -4.0 \text{ kcal mole}^{-1})$$

The exergonic reaction is thus coupled to the endergonic reaction, and the free energy of one drives the other.

Intracellular Phosphate Turnover

In all cells, most of the more significant energy changes and reaction couplings occur with ADP and ATP. This is true in the cytoplasm, mitochondria, ribosomes, nucleus, and other organelles. The cell has a number of nucleotide-phosphate pools. Although

$$\text{ADP} + \text{phosphate} \rightleftharpoons \text{ATP}$$

exchanges are most common,

$$\text{ATP} \longrightarrow \text{AMP} + \text{pyrophosphate}$$

reactions occur in lipid metabolism as does

$$\text{CTP} \longrightarrow \text{CDP} + \text{phosphate}$$

reactions. Uridine triphosphate (UTP) is utilized in polysaccharide synthesis; guanosine triphosphate is required in protein synthesis. The deoxy derivatives, dATP, dGTP, dUTP, and dCTP, are all used in DNA synthesis.

Regeneration of each of the triphosphates is catalyzed by the relatively nonspecific nucleotide diphosphate kinase:

$$\text{CDP} + \text{ATP} \longrightarrow \text{CTP} + \text{ADP}$$

$$\text{UDP} + \text{ATP} \longrightarrow \text{UTP} + \text{ADP}$$

$$\text{GDP} + \text{ATP} \longrightarrow \text{GTP} + \text{ADP}$$

$$\text{dCDP} + \text{ATP} \longrightarrow \text{dCTP} + \text{ADP}$$

$$\text{dADP} + \text{GTP} \longrightarrow \text{dATP} + \text{GDP}$$

The total concentration of nucleotides in cells in a steady state is relatively constant. In a relatively inactive steady-state cell, most of the adenylate system is in the form of ATP. The major catabolic pathways operate at a level that keeps the system filled with phosphates. When cellular activity increases, the level of ATP decreases and the ADP + ATP level increases. The change in levels of these compounds initiates an acceleration in reaction sequences, such as glycolysis (Chapter 10), which generates ATP. In a similar manner, when activity decreases, the level of ATP will quickly increase, causing in turn the slowing of ATP generating systems. The regulation of ATP synthesis is discussed in Chapter 11; the mechanism of regulation is through allosteric enzymes that are modified by ATP, ADP, and AMP.

Although ATP hydrolysis is a major source of energy in the cell and pools of ATP are relatively predictable, ATP does not form a substantial energy pool. The amount of ATP in the cell will last only a short time during periods of increased activity. Most cells contain additional reservoirs of energy-rich compounds that can quickly be converted into ATP. Skeletal muscle, smooth muscle, and nerve cells contain phosphocreatine reservoirs. When ATP levels fall, creatine kinase is activated, catalyzing the reaction,

$$\text{Phosphocreatine} + \text{ADP} \longrightarrow \text{creatine} + \text{ATP}$$
$$(\Delta G^{\circ} = -3.0 \text{ kcal mole}^{-1})$$

Phosphoarginine and polymetaphosphate act in a similar way in invertebrate and bacterial cells, respectively.

Redox Couples

In addition to the exchange of energy between substrates in a reaction, energy can also be changed during electron transfer between oxidants and reductants, or **redox reactions.** An oxidant or oxidizing agent is a substance that loses electrons to a reductant or reducing agent. Thus, a reductant is a substance that absorbs electrons from an oxidant. Substances have various potentials to retain or lose electrons. Hydrogen is known to dissociate,

$$H_2 \rightleftharpoons 2H^+ + 2e^-$$

and is used as a standard against which other substances may be compared for their ability to absorb electrons from hydrogen (positive potential) or lose electrons to hydrogen (negative potential). The measurements are made using an electrode standardized against hydrogen. By placing the electrode in a solution of the substance to be measured, the standard electrode potential, E_0, can be determined in volts. Oxidizing agents (substances capable of absorbing electrons from hydrogen) have positive potentials. Agents capable of reducing hydrogen (substances giving up electrons to hydrogen) have negative potentials (Table 9–2). E_0 values represent potentials at pH O and 25°C; more commonly, measurements are made at other pH values, especially pH

Table 9–2
Standard Redox Potentials at pH 7.0 and 25°–37° C

Reductant	Oxidant	E_o' (volts)
Pyruvate	Acetate + $2H^+$ + $2e^-$	− 0.70
Acetaldehyde	Acetate + $2H^+$ + $2e^-$	− 0.58
H_2	$2H^+$ + $2e^-$	− 0.42[a]
$NADH_2$	NAD^+ + $2H^+$ + $2e^-$	− 0.32
Ethanol	Acetaldehyde + $2H^+$ + $2e^-$	− 0.197
Lactate	Pyruvate + 2H + $2e^-$	− 0.185
Succinate	Fumarate + 2H + $2e^-$	− 0.031
Ubiquinol	Ubiquinone + $2H^+$ + $2e^-$	+ 0.10
2 Cytochrome $b_{(ox)}$	2 Cytochrome $b_{(red)}$ + $2e^-$	+ 0.030
2 Cytochrome $c_{(ox)}$	2 Cytochrome $c_{(red)}$ + $2e^-$	+ 0.254
2 Cytochrome $a_{3(ox)}$	2 Cytochrome $a_{3(red)}$ + $2e^-$	+ 0.385
H_2O	½ O_2 + $2H^+$ + $2e^-$	+ 0.816

[a] E_o = 0.0 volts.

7.0, and these measurements are noted by the symbol E_o'.

Any substance with a more positive E_o' value than another contains a potential for oxidizing that substance (removing electrons from the substance with the more negative E_o' value). The greater the difference in potentials the greater the energy changes involved. The change in standard free energy, $\Delta G°$, is related to E_o as follows

$$\Delta G° = -nF\Delta E_o \qquad (9\text{–}7)$$

where n is the number of electrons exchanged, F is the *Faraday* (23,040 cal/volt) and ΔE_o is the difference in redox potential between the more positive and more negative of the *redox couple*. For example, the oxidized form of cytochrome *c* can oxidize the reduced form of cytochrome *b*. The difference between the E_o of the two is

$$+0.254 - (+0.030) = +0.251 \text{ volts.}$$

Therefore,

$$\Delta G° = -2\ (23{,}040 \text{ cal/volt})\ (0.251 \text{ volt}) = 11.57 \text{ kcal}$$

The coupling of redox reactions such as these to synthesis of ATP in the cell is discussed in Chapter 16.

Light and Chemical Transductions

The conversion of energy from one form to another is called **transduction.** Light photons are converted into potential chemical bond energy in such seemingly diverse processes as photosynthesis and vision. Chemical bond energy is transformed into light energy in *bioluminescence*, such as the light emitted by fireflies and the light emitted by microorganisms in the sea when agitated by a passing boat.

Natural electromagnetic radiations comprise more than the spectrum of colors perceived by the human eye. The range of natural radiations is described by the electromagnetic spectrum (Table 9–3). Each of these radiations may be envisioned as a stream of moving packets of energy called photons. The stream of photons move in a wavelike manner as illustrated in Figure 9–5. The distance from the crest of one wave to the crest of the next (or from a point on one wave to an equivalent point on an adjacent wave) is defined as the **wavelength.** Among the differences between radiations in the electromagnetic spectrum are differences in wavelengths. At one extreme are the long radio waves or television waves with a wavelength of 1 to 10 m. At the opposite end of the spectrum are the gamma rays with wavelengths as short as 0.0001×10^{-9} m (10^{-4} nm). Since most of the important physiological wavelengths are

Table 9–3
Wavelengths of the Radiations of the Electromagnetic Spectrum

Type of Radiation	Photon Energy (ergs × 10^{-12})	Wavelength (nm[a])
Television waves	2.0×10^{-10}–2.0×10^{-8}	10^{9}–10^{11}
Radar waves	2.0×10^{-8}–2.0×10^{-5}	10^{6}–10^{9}
Radio waves	2.0×10^{-11}–2.0×10^{-5}	10^{6}–10^{12}
Infrared (far)	0.005–0.99	2×10^{3}–4×10^{5}
Infrared (near)	0.99–2.5	780–2×10^{3}
Red light	2.5–3.2	620–780
Orange light	3.2–3.4	590–620
Yellow light	3.4–3.6	545–590
Green light	3.6–4.1	490–545
Blue light	4.1–4.6	430–490
Violet light	4.6–5.1	390–430
Ultraviolet (long)	5.1–6.6	300–390
Ultraviolet (short)	6.6–9.9	200–300
Ultraviolet (very short)	9.9–132.4	15–200
X-rays (soft)	99.3–2.0×10^{4}	0.1–20
X-rays (hard)	2.0×10^{4}–3.9×10^{6}	0.0005–0.1
Gamma rays	1.4×10^{4}–19.9×10^{6}	0.0001–0.14

[a]1 nm (nanometer) = 10^{-9}m (meter); 1 nm = 1 mμ (millimicron); 1 nm = 10 Å (angstrom).

Figure 9–5
Particulate, wavelike characteristics of electromagnetic radiations.

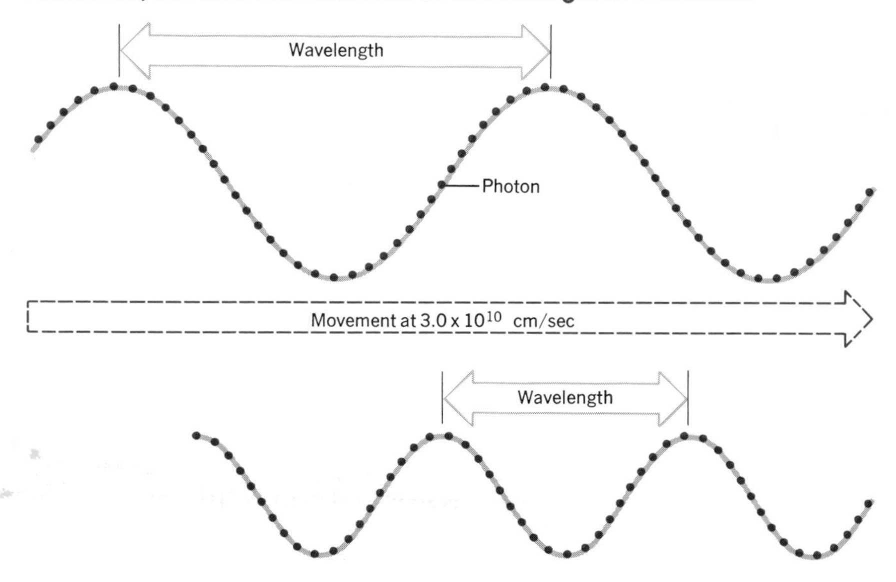

small, the nanometer (10^{-9} m) is the most frequently used unit (1 nm = 1 mμ = 10 Å). Each of the photons in these radiations travels at the same speed, namely 3×10^{10} cm sec^{-1}.

The radiations in the electromagnetic spectrum also differ in the amount of energy each contains. Although radiations behave as though each is composed of discrete packets of energy called **quanta,** the energy in a quantum is related to the wavelength, as defined by **Planck's law:**

$$q = h\nu = \frac{hc}{\lambda} \qquad (9\text{–}8)$$

where q is the quantum in *ergs,* h is Planck's constant (6.624×10^{-27} erg-seconds), ν is the frequency of the light wave, c is the speed of light (3.0×10^{10} cm/sec), and λ is the wavelength in centimeters. Radiations with shorter wavelengths contain more energy than those with longer wavelengths.

In addition to the electromagnetic radiations, particulate radiations carry energy and affect biological systems. Alpha rays and beta rays are considered particulate radiations because they consist of streams of energy containing particles that move at comparable but varying speeds. An alpha ray consists of a stream of helium nuclei, and a beta ray is composed of a stream of electrons (see Chapter 14).

Radiations are absorbed at random in a substance by individual molecules or atoms. Individual molecules (atoms) will absorb the energy only in units of 1 quantum. This one molecule (atom)/one quantum proportion is known as the **Einstein-Starck law** or the **primary reaction.** The energy of the electromagnetic radiation is absorbed by an electron in the molecule. The energy from the radiation causes the electron either to be transferred to an outer orbital at a higher energy level or to spin at a faster rate. Since the allowable orbitals and speeds of electrons fall within limited ranges and whole quanta of energy must be absorbed at a time, not all molecules absorb all radiations. Only those radiations containing the specific energy per quantum to raise an electron to a higher defined energy state or to increase the spin rate to allowable limits will be absorbed by specific molecules. Other radiations will be diffracted or completely pass through the substance. Very high energy radiations such as x-rays and gamma rays may cause an electron to be displaced from the molecule, thereby forming an ion (ionization).

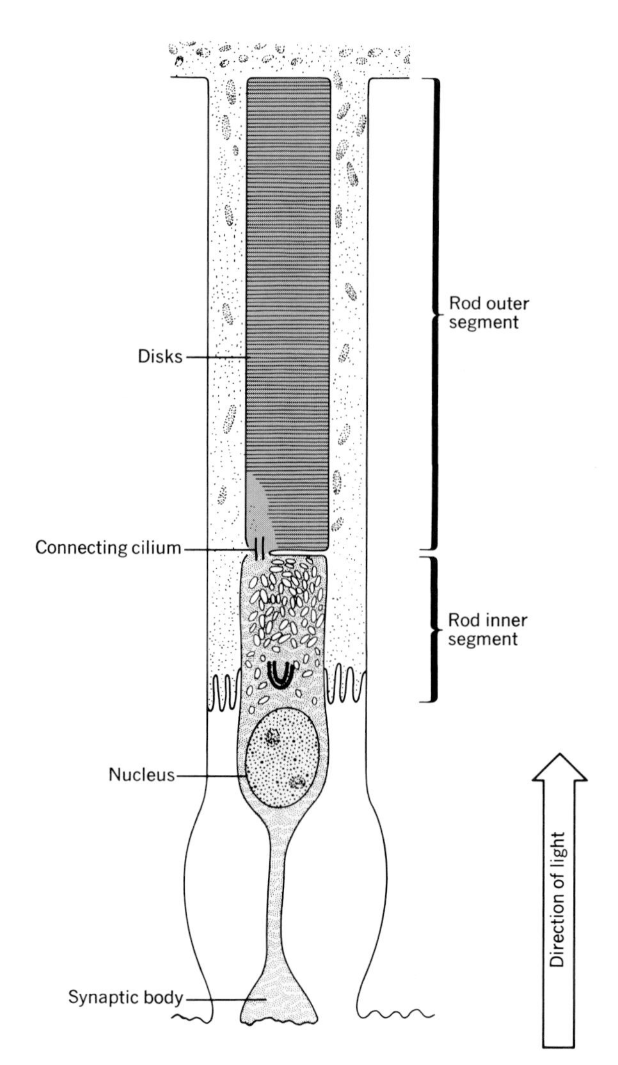

Figure 9–6
Rod cell of vertebrate eye.

Electrons raised to higher energy levels are very unstable and return to their original, stable orbitals or *ground state* in a fraction of a second (generally, less than 10^{-9} seconds). The energy released by the molecule in returning to the ground state is a *secondary reaction,* which may be (1) the emission of light or fluorescence, (2) the emission of heat, or (3) the activation of another molecule. It is this third possibility that induces biological reactions such as those

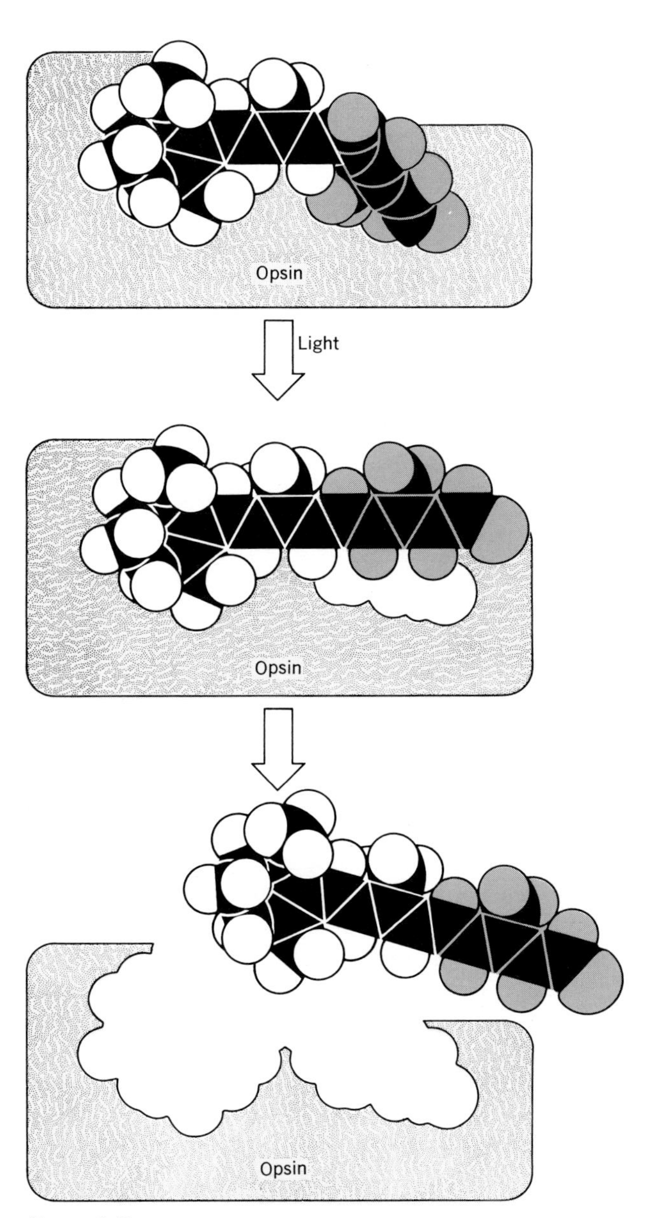

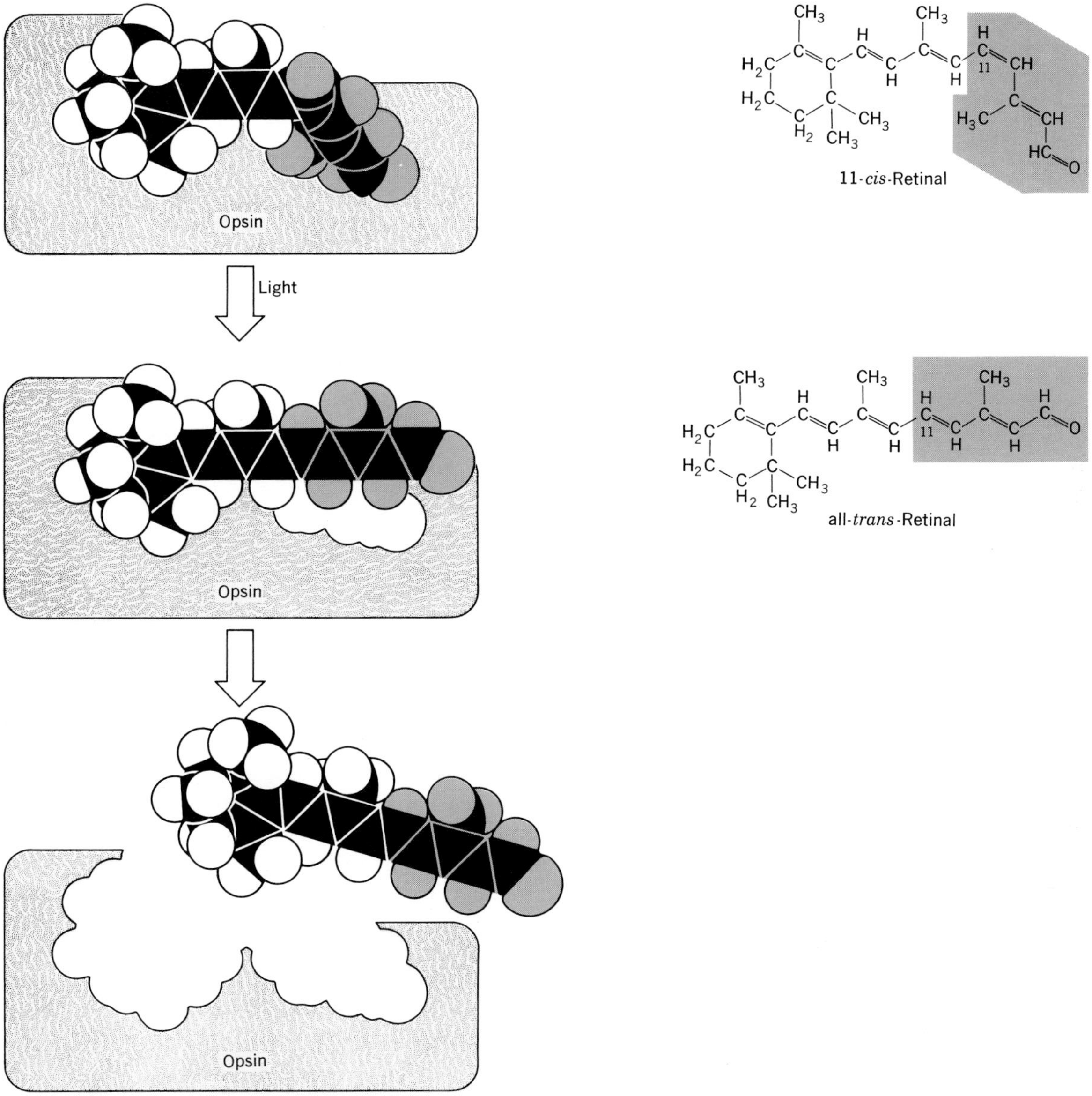

11-*cis*-Retinal

all-*trans*-Retinal

Figure 9–7
Light-induced change in rhodopsin.

in vision and photosynthesis. Light absorption in photosynthesis is discussed in Chapter 17.

A number of carotenoid pigments act as light receptors in the eyes of animals. *Rhodopsin,* the pigment found in the rod cells of the vertebrate eye, has been well studied. This pigment consists of a protein called *opsin* and a compound, *retinal,* that is related to vitamin A. The pigments are arranged in the outer segment of the rod cell and are associated with the *disks* of stacked membranes (Fig. 9–6). Light entering the eye is absorbed by the rhodopsin and induces a change in the retinal from the *cis* isomeric configuration to the *trans* form (Fig. 9–7). This change alters the bonding between the retinal and the opsin, causing the eventual release of the retinal from the protein. The separation of these compounds in turn alters the permeability of the outer segment disk membranes to ions such as Na^+,Ca^{++}, and K^+ and establishes a change in potential across the outer membrane. The final result is a change in the rate at which a transmitter is released from the inner rod segment to the neurons leading to the brain.

Other Transductions

The use of chemical energy to transport material across cell membranes against concentration gradients or to maintain pools of potential energy within a cell in the form of high concentrations of material are common transductions in cells. These changes, which are related to diffusion, osmosis, and active transport, are more fully described in Chapter 15.

Energy potentials in the form of concentration gradients in cells are also believed to be convertible to chemical or molecular energy. The development of an ionic potential across the membranes of mitochondria is a possible energy source for the production of ATP in mitochondria. This mechanism is described in Chapter 16.

The active transport of ions across membranes requires the consumption of molecular energy, usually in the form of ATP. The electrical potential that can develop across the membrane as a result of selective ion transport is an example of transduction from chemical energy to electrical energy. The action potentials developed in nerve cells and the potential that initiates muscle contraction (Chapter 23) are good examples of this type of conversion.

Finally, conversion of molecular energy into mechanical energy is displayed in the conversion of ATP energy into contraction and sliding of proteinaceous muscle filaments past one another during muscle contraction. This process is also fully described in Chapter 23.

Summary

Energy changes regulate the chemical and physical transitions in cells. **Autotrophic** cells derive their energy from light. **Heterotrophic** cells derive their energy from catabolism of proteins, lipids, and carbohydrates. The reactions that break down large molecules, thereby yielding energy, are called **exergonic** reactions. In most instances, energy released from exergonic reactions is used to attach phosphate to ADP by an energy-rich bond, thereby forming ATP. ATP and its energy-rich bonds are used by the cell to ''drive'' energy-requiring **(endergonic)** reactions.

These and other energy changes conform to the **laws of thermodynamics.** Specifically, these are:

First law: **Energy cannot be created or destroyed but can be converted from one form into another.**

Second law: **In all processes involving energy changes, the *entropy* of the system increases until equilibrium is achieved.**

The energy changes in cellular reactions obey these laws and can be expressed by the relationship:

$$\Delta G = \Delta H - T\Delta S$$

where the change in useful, free energy is equal to the change in total energy less the entropy factor or the product of the absolute temperature and the change in entropy.

Two reactions tied together by energy exchange in which the energy released from the exergonic reaction drives the endergonic reaction are said to be **coupled.** The coupling can be mediated by ATP or by an equilibrium shift in which the exergonic reaction consumes the product of the endergonic reaction. Energy can be exchanged during electron transfer between **oxidants** and **reductants (redox reactions)** as well as between substrates.

Conversion of energy from one form to another is **transduction.** Light contains energy in packets called **quanta.** The energy in a quantum is proportional to the speed of light *(c)* and inversely proportional to the wavelength (λ).

$$q = \frac{hc}{\lambda}$$

Light energy is converted to chemical energy in photosynthesis and in vision. Another transduction is the conversion of the electrical potential across a membrane (due to an unequal distribution of ions) into chemical energy in the form of ATP. Muscle contraction is an example of a transduction of chemical energy into mechanical energy.

References and Suggested Reading

Articles and Reviews

Ingraham, L. L., and Pardee, A. B., Free energy and entropy in metabolism, in *Metabolic Pathways* (D. M. Greenberg, ed.), Vol. 1, Academic Press, New York, 1967, p. 2.

Lipmann, F., Metabolic generation and utilization of phosphate bond energy. *Adv. Enzymol. 18,* 99 (1941).

Books, Monographs, and Symposia

Blum, H. F., *Time's Arrow and Evolution,* Harper, New York, 1962.

Florkin, M., and Stotz, E. H., Bioenergetics, in *Comprehensive Biochemistry,* Vol. 22, American Elsevier Co., New York, 1967.

Lehninger, A. L., *Bioenergetics,* W. A. Benjamin Inc., Menlo Park, Calif., 1972.

Lehninger, A. L., *Biochemistry,* Worth Publishers, Inc., New York, 1975.

Wall, F. T., *Chemical Thermodynamics,* W. H. Freeman and Co., San Francisco, 1965.

Chapter 10
METABOLISM

The metabolic processes in cells include all the individual chemical reactions and sequences of reactions that convert various substrates into products. The individual reactions may be spontaneous and energy yielding (the exergonic reactions), or they may be energy consuming (endergonic reactions). Commonly, the primary reactants are converted into products by means of a sequence of reactions, each reaction enzymatically catalyzed and each subsequent reaction in the sequence requiring the products of the prior reaction as a substrate.

The overall reaction sequences may be classified as **catabolic** (i.e., degradative) if the ultimate products of the reaction sequence are considered to be subunits or parts of the initial substrate. Alternatively, if the products are a result of the combining of two or more different substrates, the sequence is considered to be **anabolic** (i.e., synthetic). A sequence of reactions is usually referred to as a **metabolic pathway** or simply a **pathway.** Some pathways are common to all living organisms or cells. Some pathways function as very active reaction sequences from which less active pathways may branch or join. The more active pathways are usually referred to as **central pathways** of metabolism.

Figure 10–1 diagrammatically shows some of the relationships between the catabolic and anabolic pathways of the major groups of compounds. Intermediates in the breakdown of carbohydrates can be diverted to lipid synthesis or to the formation of nitrogen compounds such as nucleotides and amino acids. Lipids in microbial and plant (and to a limited extent animal) cells can be converted into carbohydrates and nitrogen compounds. Likewise, nitrogen compounds, once identified, can be converted into lipids or carbohydrates. All the compounds may be broken down, their catabolism acting as sources of energy for ATP synthesis or to provide reduced pyridine nucleotides ($NADH_2$ and $NADPH_2$) for other coupling reactions.

It is possible to identify specific sites within a cell or an organelle where particular metabolic pathways are operative. For example, the enzymes necessary for the tricarboxylic acid cycle or Krebs' cycle reactions are located in the matrix of the mitochondria; the primary reactions in cholesterol (sterol) synthesis are associated with the microsomes; fatty acids are oxidized (β-oxidation) by reaction sequences in the mitochondria and are synthesized in the fluid of the cytoplasm (cytosol); proteins are formed on the ribosomes.

The intermediates, as well as the products of a pathway, may be drawn off and used in other pathways. For example, in the breakdown of carbohydrates, a large number of intermediate compounds are formed before the ultimate products, CO_2 and H_2O, are formed. It is possible for some of the intermediates to be ''pulled away'' from the catabolic process and used in the formation of fatty acids. Other

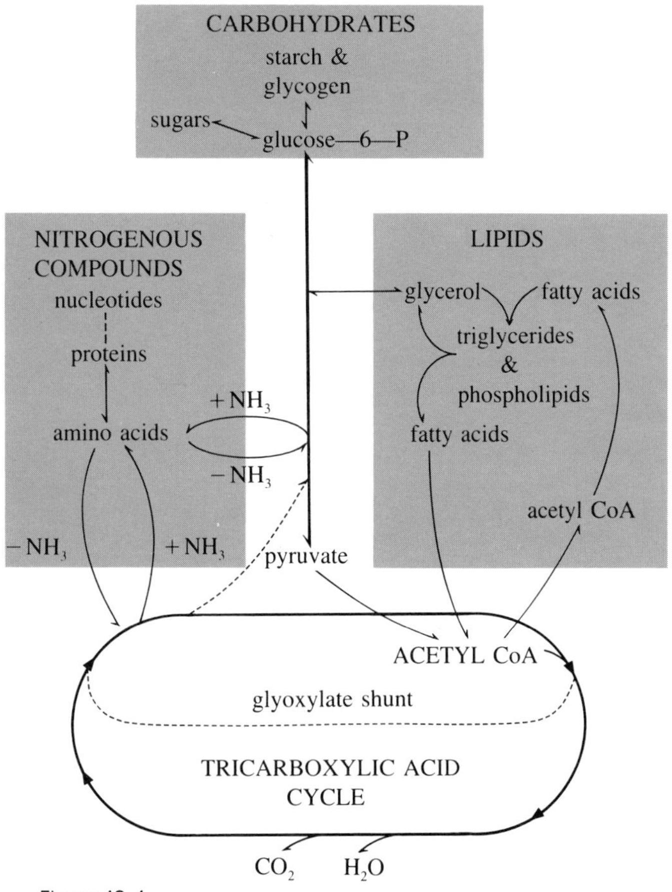

Figure 10–1
Pathways of conversion of major metabolic intermediates.

intermediates may be used in the formation of amino acids. A number of natural mechanisms have been found that enable the cell to regulate the activity of the pathways. These *metabolic control mechanisms* are discussed in the next chapter.

In this chapter, some of the major pathways that are common to most cells are discussed. For convenience the major pathways of carbohydrate metabolism will be discussed first, followed by those of lipid and nitrogen metabolism. Although described separately, it should be remembered that in the cell various pathways may be operative at the same time, and one pathway may influence the rate of metabolic reactions of another pathway.

Analysis of Metabolic Pathways

Our current knowledge of metabolic pathways was assembled from a variety of observations and experiments. Initially, pathways are identified by observing the consumption of reactants and accumulation of products. For example, the consumption of sugar and the production of carbon dioxide and alcohol during *alcoholic fermentation* is a pathway that has been known for centuries; i.e.

$$C_6H_{12}O_6 \longrightarrow 2CO_2 + 2\ C_2H_5OH \text{ (ethyl alcohol)} \quad (10\text{–}1)$$

By analysis, it is possible to measure the amount of sugar consumed and the amount of CO_2 and ethyl alcohol produced. From such an analysis, it is learned that all the carbon in the sugar is converted to these two products. Therefore, one may conclude that no other products are formed from the reactants and that the reaction can be balanced as written in equation 10–1.

Quantitative analysis of reactants and products does not reveal the *steps* or individual reactions that bring about the overall reaction; nor does such an experiment reveal *coupled reactions* such as the formation of ATP during fermentation.

Marker and Tracer Techniques

A frequently used technique for identifying the steps in a pathway is to incorporate a radioactive isotope (or in rare instances a "heavy" isotope) as one or more of the atoms in a substrate (Table 10–1). Such a substrate is said to be labeled with a **tracer.** The labeled substrate is then fed to the cell, and metabolism is allowed to proceed for a period of time. Afterward the reaction is stopped by heating, addition of enzymatic inhibitors, or some other means (Chapter 8), and the cell is extracted to remove potential intermediates. By chromatographic or related techniques (Chapter 12) the components are separated, their isotopic activity measured from their radioactivity (see Chapter 14), and they are chemically identified. Those components with labeled atoms are then studied, and possible reaction sequences are postulated to account for the formation of the marked intermediates.

Because in a reaction sequence such as

$$\text{Substrate}^* \longrightarrow B^* \longrightarrow C^* \longrightarrow D^* \longrightarrow E^* \longrightarrow \text{product}^* \quad (10\text{–}2)$$

Table 10–1
Isotopes Commonly Used as Tracers

Isotopes[a]	Type of Radiation	Half-life
$^{2}_{1}H$ ("deuterium")	None	Stable
$^{3}_{1}H$ ("tritium")	β-particles	12.3 years
$^{13}_{6}C$	None	Stable
$^{14}_{6}C$	β-particles	5.57×10^{3} years
$^{15}_{7}N$	None	Stable
$^{18}_{8}O$	None	Stable
$^{24}_{11}Na$	β-particles; γ-rays	15.0 hours
$^{32}_{15}P$	β-particles	14.3 days
$^{35}_{16}S$	β-particles	87.2 days
$^{36}_{17}Cl$	β-particles	3.0×10^{5} years
$^{42}_{19}K$	β-particles; γ-rays	12.5 hours
$^{45}_{20}Ca$	β-particles	164 days
$^{59}_{26}Fe$	β-particles; γ-rays	45.1 days
$^{131}_{53}I$	β-particles; γ-rays	8.1 days

[a] $^{\text{Atomic mass}}_{\text{Atomic number}}\text{ELEMENT}$

the reaction forming B must occur before the reaction forming C, which in turn must occur before reaction D and so forth, the time factor can be used in the elucidation of the sequence. The labeled substrate (*) is given to the cell at a designated time, and the reaction is allowed to proceed for only a short interval before all reactions are stopped. In such a short time span, one or a few, but not all, intermediates formed will contain the labeled compound. Separation, identification, and analysis of these intermediates would proceed as before. The experiment is then repeated with labeled substrate fed to fresh cells, and the experiment is allowed to proceed for a somewhat longer period of time before being halted for analysis. The experiment is repeated as many times and for as long a duration as is necessary to reveal each of the successive intermediates formed.

When it is suspected that a substrate may be split into two or more products

$$\underset{\ddagger}{{}^{*}\text{Substrate}} \longrightarrow \underset{\ddagger}{\overset{*}{B}} \begin{cases} \overset{*}{C} \\ \underset{\ddagger}{D} \end{cases} \qquad (10\text{–}3)$$

atoms at various positions in the substrate (* and ‡) may be specifically labeled so that each of the products formed may be followed and identified. Because some intermediates are not formed in sufficient amounts to easily allow extraction and identification, additional nonlabeled intermediates may be added. This material mixes with the labeled intermediate being formed and provides sufficient "carrier" for extraction and analysis; in effect it traps some of the labeled compound. The technique is known as **isotopic trapping.**

$$\text{Substrate}^* \longrightarrow B^* \longrightarrow C^* \longrightarrow D^* \longrightarrow E^* \longrightarrow \text{product}^* \quad (10\text{–}4)$$

$$\begin{array}{c} \uparrow \\ C \end{array} \quad \text{Carrier added}$$

Another method for clarifying the role of an intermediate is called **overloading** and involves two parallel experiments. In one experiment, the substrate is labeled, and the amount of label appearing in the product is followed (i.e., via reaction 10–2). In the second experiment, the substrate is labeled as before, but a significant amount of unlabeled intermediate (i.e., C in reaction 10–4) is also added to the system. If C is indeed an intermediate in the pathway, then the amount of label in the product formed during the second experiment should be less than that measured in the product during the first experiment.

Enzyme Techniques

There are relatively few metabolic reactions in cells that are not catalyzed by enzymes. Thus, one would normally expect to be able to identify an enzyme for each metabolic reaction postulated or identified in a cell. In studies of the cell enzymes, the cells are first disrupted and a cell-free extract prepared. After supplying the cell-free extract with selected substrates and cofactors, enzymatic activity can be demonstrated by measuring the rate of disappearance of the substrate or rate of appearance of a product. The cell-free extract can be fractioned by centrifugation into its components, thereby isolating the mitochondria, cell membranes, ribosomes, other organelles, or the cytosol. Each of these fractions can also be tested for enzymatic activity. In addition, using a variety of chromatographic techniques applicable to protein isolations (Chapter 13), the enzymes can be specifically purified. Ultimately, an enzyme in its pure crystalline state may be obtained.

Enzyme Production and Inhibition

A third approach to determining metabolic reaction sequences is by the use of specific enzyme inhibitors or genetic mutant cells that fail to produce a specific enzyme. Numerous enzyme inhibitors are known that block specific reaction steps. In the conversion of precursor A to product E, that is,

$$A \xrightarrow{1} B \xrightarrow{2} C \xrightarrow{3} D \xrightarrow{4} E \quad (10\text{–}5)$$

an inhibitor that blocks the enzyme of reaction 2 would prevent the formation of final product E. But one should still be able to make two additional observations: (1) intermediate B should still be formed from A, since step 1 is not inhibited; in fact, B may even be found to accumulate in excess; and (2) addition of an exogenous source of C or D should allow for the continued production of final product E. The identification of an inhibitor for a metabolic pathway thus provides a tool for finding the intermediate *before* the block and also allows for the testing of suspected intermediates that come *after* the block.

In a similar manner, a cell mutant that lacks the genetic information for producing a specific enzyme may serve as a test organism for studying the steps of the pathway involving that enzyme.

Carbohydrate Metabolism

The steps in the catabolism of sugars are known in great detail. For the most part, cells decompose carbohydrates by similar metabolic pathways whether they are plant cells, animal cells, or bacterial cells. There are alternative methods of oxidizing carbohydrates, but the central pathway found in most cells is that outlined in Figure 10–2. Polysaccharides such as starch and glycogen are cleaved into their component saccharide units by the addition of inorganic phosphate (phosphorolysis) through the action of phosphorylase enzymes (reaction 10–6). Oligosaccharides are hydrolyzed first into their component monosaccharide units, and then the monosaccharides such as glucose are phosphorylated by enzymatic reaction with ATP (reactions 10–8 and 10–9). The phosphorylated sugars from each of these sources are sequentially subjected to the enzymatic reactions outlined in reactions 10–10 through 10–18, pro-

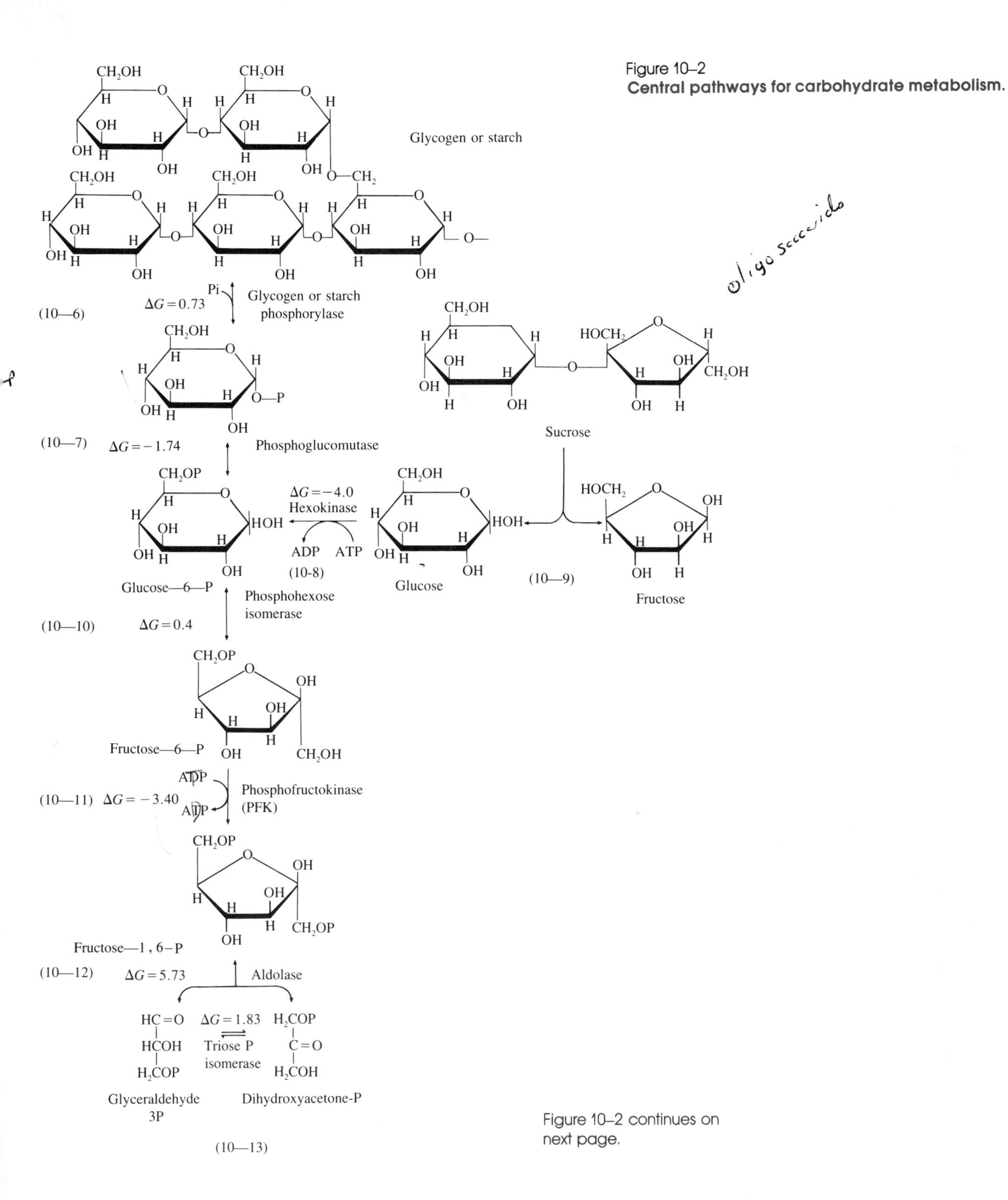

Figure 10–2
Central pathways for carbohydrate metabolism.

Figure 10–2 continues on next page.

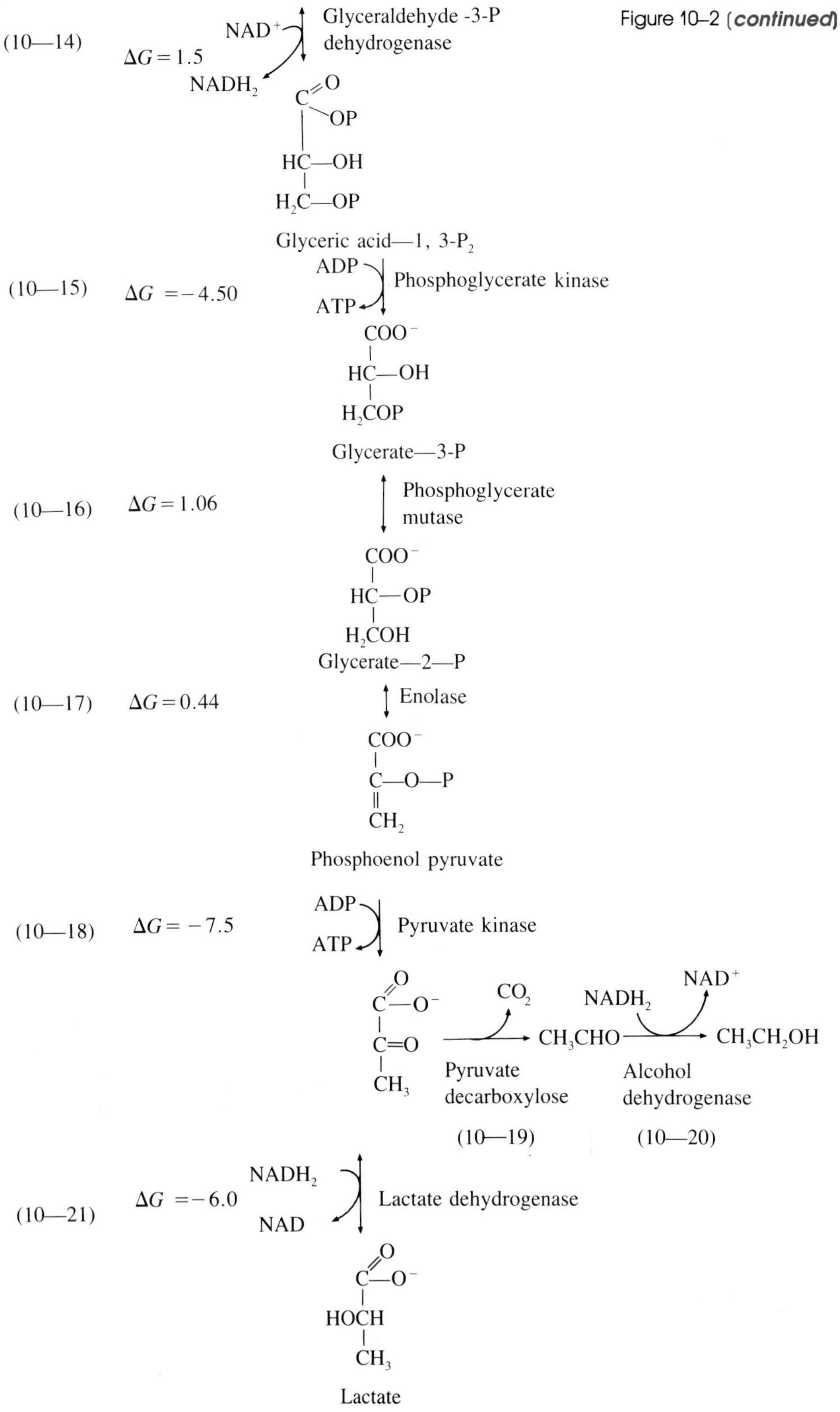

Figure 10–2 (*continued*).

(10—14) $\Delta G = 1.5$

NAD^+ → $NADH_2$ — Glyceraldehyde-3-P dehydrogenase

$$\begin{array}{l} C(=O)OP \\ | \\ HC—OH \\ | \\ H_2C—OP \end{array}$$

Glyceric acid—1, 3-P_2

(10—15) $\Delta G = -4.50$

ADP → ATP — Phosphoglycerate kinase

$$\begin{array}{l} COO^- \\ | \\ HC—OH \\ | \\ H_2COP \end{array}$$

Glycerate—3-P

(10—16) $\Delta G = 1.06$

Phosphoglycerate mutase

$$\begin{array}{l} COO^- \\ | \\ HC—OP \\ | \\ H_2COH \end{array}$$

Glycerate—2—P

(10—17) $\Delta G = 0.44$

Enolase

$$\begin{array}{l} COO^- \\ | \\ C—O—P \\ \| \\ CH_2 \end{array}$$

Phosphoenol pyruvate

(10—18) $\Delta G = -7.5$

ADP → ATP — Pyruvate kinase

$$\begin{array}{l} C(=O)—O^- \\ | \\ C=O \\ | \\ CH_3 \end{array}$$

→ (CO_2) Pyruvate decarboxylose (10—19) → CH_3CHO → ($NADH_2$ → NAD^+) Alcohol dehydrogenase (10—20) → CH_3CH_2OH

(10—21) $\Delta G = -6.0$

$NADH_2$ → NAD — Lactate dehydrogenase

$$\begin{array}{l} C(=O)—O^- \\ | \\ HOCH \\ | \\ CH_3 \end{array}$$

Lactate

ducing the intermediate pyruvate. This central pathway is frequently referred to as **glycolysis.**

Glycolysis

The major features of glycolysis are as follows:

1. The saccharides are phosphorylated. In the case of monosaccharides such as glucose, fructose, mannose, and the like, 2 moles of ATP per mole of monosaccharide are utilized. Glycogen and starch require only 1 mole of ATP per mole of glucose-equivalent, since inorganic phosphate is acquired during phosphorolysis.

2. The six-carbon sugar diphosphate is split by aldolase (10–12), producing 2 three-carbon units, glyceraldehyde-3-phosphate and dihydroxyacetone phosphate; the latter subsequently forms a second mole of glyceraldehyde-3-phosphate (10–13).

3. A major oxidation and phosphorylation of the substrate is catalyzed by glyceraldehyde-3-phosphate dehydrogenase. Two moles of hydrogen are removed per mole of substrate and reduce 2 moles of the coenzyme NAD. In the same reaction inorganic phosphate is bound to the acid.

4. In the final reactions of glycolysis the intermediates are dephosphorylated by reaction with ADP. For each mole of monosaccharide oxidized to pyruvate, 2 moles of ATP are consumed and 4 moles of ATP are produced, resulting in a net production of 2 moles of ATP per mole of monosaccharide. Note that in the case of glycogen or starch, there is a net production of 3 moles of ATP per mole of glucose-equivalent.

Pyruvate does not accumulate in very large amounts in cells. Instead, it is converted into other products. The enzymes that react with pyruvate vary with the type of organism and with the nature of the environment. The more common fates of pyruvate are (1) its fermentation to ethyl alcohol and carbon dioxide in cells such as yeast, (2) its anaerobic conversion into lactate in cells such as muscle, and (3) its conversion into acetate in the mitochondria of most organisms living under aerobic conditions.

Anaerobic Respiration and Fermentation

In glycolysis there is no specific requirement for oxygen. Oxidation reactions do occur, such as the removal of two hydrogens from glyceraldehyde-3-phosphate, and NAD is reduced to $NADH_2$, but oxygen per se is not consumed. In the absence of oxygen, pyruvate may be reduced to a variety of different compounds. Alcoholic fermentation (reactions 10–19 and 10–20) is a common pathway in microorganisms and is of industrial importance. In these last two steps, 1 mole of CO_2 is given off per mole of pyruvate (i.e., 2 moles of CO_2 per mole of monosaccharide) and $NADH_2$ is reoxidized to NAD in the final step, producing ethanol. The stoichiometry and cyclic action of NAD are important; in the oxidation of each mole of glyceraldehyde-3-phosphate (reaction 10–14), a mole of NAD is reduced to $NADH_2$; a mole of $NADH_2$ is reoxidized to NAD during the conversion of a mole of acetaldehyde to ethanol (reaction 10–20). The levels of NAD and $NADH_2$ in the cell are relatively small. If the NAD reduced in the earlier reaction was not reoxidized, this central pathway would soon be blocked at the glyceraldehyde-3-phosphate step by the lack of sufficient NAD. ATP and ADP are also cycled between the ATP-requiring reactions in the early steps of glycolysis and the ATP-producing reactions in the later steps. Cells contain pools of adenylates (ATP, ADP, and AMP), and these compounds are drawn from the pools when needed and then readded to the pools. Although the stoichiometry indicates that two ATP molecules are consumed for every four molecules of ATP produced in fermentation or glycolysis, the earlier reactions may proceed by drawing ATP from the pool, while the latter reactions return ATP to the pool.

Another common fate of pyruvate that occurs in the absence of oxygen is its conversion to lactate. This is a normal process in muscle cells when they are deprived of oxygen and in many plant and bacterial cells living under anaerobic conditions (without air or, specifically, without oxygen). In these instances, a single enzyme and reaction reduce pyruvate to lactate (reaction 10–21). The $NADH_2$ is reoxidized to NAD in equimolar amounts to that consumed in the earlier glyceraldehyde oxidation step.

Oxidation of Pyruvate

Except in the case of the *strict anaerobes* (organisms that cannot live in the presence of oxygen), when oxygen is present, most cells will oxidize pyruvate rather than reduce it to lactate, ethanol, or other compounds.

Energetically, there is an advantage for the cell to oxidize the pyruvate to CO_2 and water rather than to reduce it to lactate. In the production of lactate, there is an overall change in standard free energy ($\Delta G'$) of -47.0 kcal per mole of glucose and a net production of 2 moles of ATP. However, in the breakdown of a mole of glucose through glycolysis and further oxidation to CO_2 and water, the $\Delta G' = -680.0$ kcal per mole of glucose, and a net of 36 to 38 moles of ATP are produced.

All the enzymes concerned with the oxidation of pyruvate are found in the mitochondria, either in the central matrix or associated with the mitochondrial membranes. The stages of pyruvate oxidation—(1) formation of acetyl−CoA, (2) the tricarboxylic acid cycle (Krebs cycle) reactions, and (3) electron transport and oxidative phosphorylation—are so closely associated with the structure of the mitochondrion that their detailed description will be deferred to Chapter 16. The reactions are summarized in Figure 16–16.

Other Pathways of Carbohydrate Catabolism

Phosphogluconate Pathway

This alternative oxidative pathway, also called the **pentose phosphate pathway** or the **hexose monophosphate shunt,** occurs in plants and in most animal tissues. However, its activity in comparison with glycolysis is usually lower and varies considerably from tissue to tissue. The pathway serves most cells as the primary means of (1) converting hexoses into those pentoses necessary for the synthesis of nucleotides and nucleic acids, (2) degrading pentoses so that they may be catabolized in the glycolytic pathway, and (3) generating reduced pyridine nucleotides ($NADPH_2$) in the cytosol for synthetic reactions such as fatty acid synthesis, steroid synthesis, and amino acid synthesis. In animal tissues, the latter functions occur extensively in the liver, in the mammary glands, and in the cortex of the adrenal glands. It has been demonstrated that 20% of the hexose metabolized by the mammary gland occurs via the phosphogluconate pathway. In heart or skeletal muscle, little synthesis of fatty acids (or other related substances) occurs, and correspondingly little phosphogluconate pathway activity is observed.

Figure 10–3 illustrates the major reaction steps of the phosphogluconate pathway. Reactions 10–22 and 10–24 describe the two oxidative steps that produce the reduced $NADPH_2$ required in the synthetic reactions. Step 10–24 is also the decarboxylation step that degrades the hexose to a pentose. Through the action of an isomerase and an epimerase, three pentose phosphates may be formed: ribose-5-phosphate, ribulose-5-phosphate, and xylulose-5-phosphate. These compounds may be incorporated into nucleic acids or they may be acted upon by a set of transaldolases and transketolases. The transketolases catalyze the transfer of 2-carbon moieties from compounds such as the pentose phosphates above to other compounds, so that ribulose-5-phosphate could have a 2-carbon portion of the molecule transferred to xylulose-5-phosphate, and sedoheptulose-7-phosphate and glyceraldehyde-3-phosphate would result. In a somewhat similar action, transaldolases catalyze the transfer of 3-carbon moieties; for example, sedoheptulose-7-phosphate could lose a 3-carbon portion to glyceraldehyde-3-phosphate, thus forming erythrose-4-phosphate and fructose-6-phosphate. Because of the action of these enzymes 3-, 4-, 5-, 6-, and 7-carbon sugar phosphates are interconvertible, and the intermediates of the phosphogluconate pathway may all be channeled into glycolysis for subsequent breakdown. Further conversions may occur in the mitochondria in the tricarboxylic acid enzymes and terminal respiration compounds to form CO_2 and H_2O. Figure 10–4 shows the major reaction steps for the routing of hexoses through the phosphogluconate pathway and into the glycolytic pathway.

The enzymes and reactions of the phosphogluconate pathway are also found in the stroma of chloroplasts and are used there to regenerate pentoses in photosynthesis (Chapter 17).

The Glyoxylate Pathway

The glyoxylate pathway is essentially a bypass of the CO_2-evolving steps of the tricarboxylic acid cycle. The enzymes for this bypass are commonly found in plants and are generally absent in the tissues of animals. The enzymes are localized in organelles called glyoxysomes. Figure 10–5 summarizes the key reactions of the bypass. Two moles of acetate are required for each turn of the pathway rather than the 1 mole required in the tricarboxylic acid cycle. The first mole of acetate (acetyl CoA) condenses with oxaloacetate to form citrate and then isocitrate. The inducible enzyme isocitric lyase then catalyzes the conversion to succinate and glyoxylate. The glyoxylate then reacts

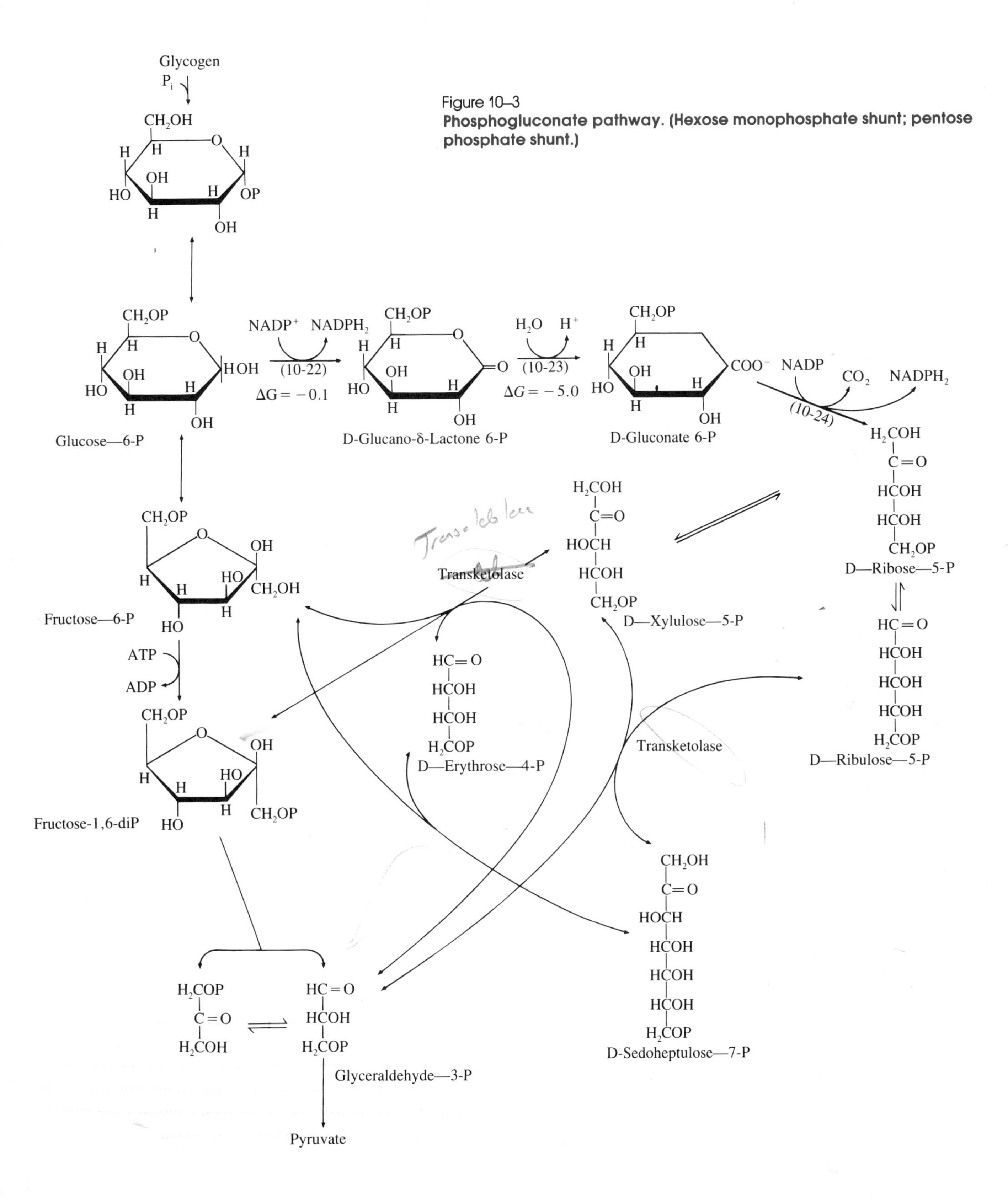

Figure 10–3
Phosphogluconate pathway. (Hexose monophosphate shunt; pentose phosphate shunt.)

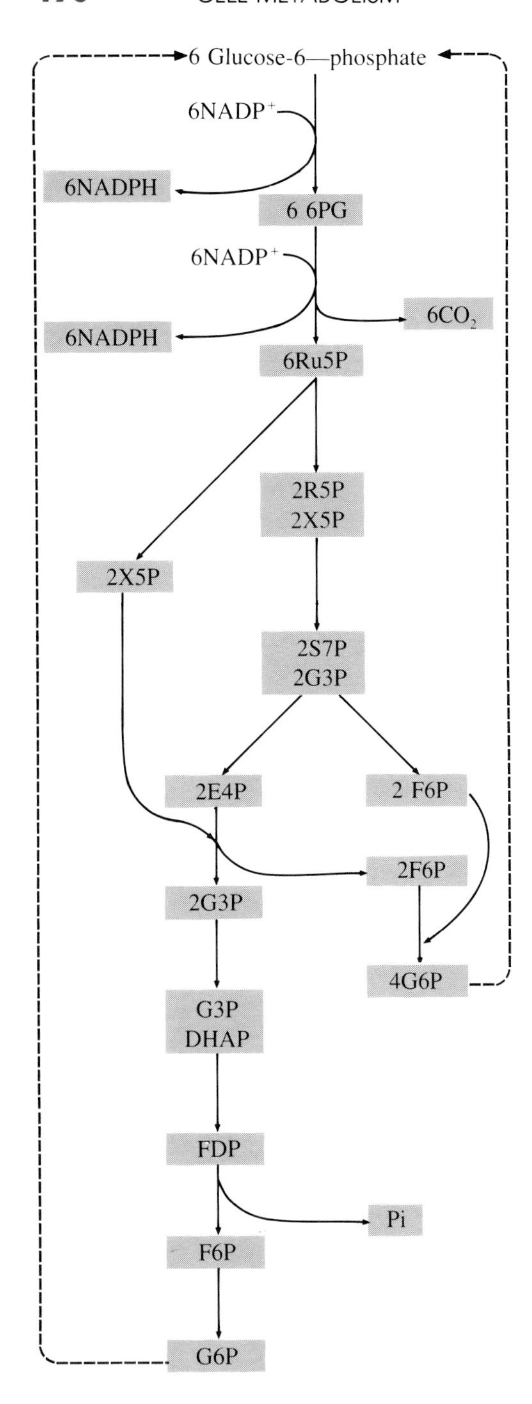

Figure 10–4
Balance sheet and diagram illustrating how the phosphogluconate intermediates formed from 6 moles of glucose-6-phosphate can be recycled back to glycolysis or 5 moles of glucose-6-phosphate. The other mole-equivalent becomes CO_2.

key:
6PG = 6—phosphogluconate
Ru5P = ribulose 5—phosphate
R5P = ribose 5—phosphate
X5P = xylulose 5—phosphate
S7P = sedoheptulose 7—phosphate
G3P = glyceraldehyde 3—phosphate
G6P = glucose 6—phosphate
E4P = erthrose 4—phosphate

with a second mole of acetate (acetyl CoA) to produce malate, which is converted back to oxaloacetate.

This bypass set of reactions enables plants and microbes to transform acetyl CoA derived from the breakdown of fatty acids and convert it into carbohydrate. The excess succinate produced by the glyoxylate pathway can be channeled back up the glycolytic pathway for the formation of sugar and subsequently polysaccharides. To achieve the reversal of the glycolytic pathway, certain bypass reactions are needed to make the synthesis of sugar energetically possible.

Gluconeogenesis

Glucose can be synthesized by reversing most of the reaction steps of glycolysis. Different initial reactions are utilized in different types of cells and tissues in order to get various substrates started back up the pathway. Also, the carbohydrates finally produced vary with conditions and with types of cells. Photosynthetic organisms can reduce CO_2 through a reversal of the glycolytic reactions after an initial fixation with ribulose-diphosphate (Chapter 17). Liver cells can regenerate glucose from lactate by reversing the glycolytic reactions. Almost all cells can transaminate or deaminate key amino acids into tricarboxylic acid cycle intermediates and, by conversion into phosphoenol pyruvate, reverse glycolysis to produce glucose. Plant and bac-

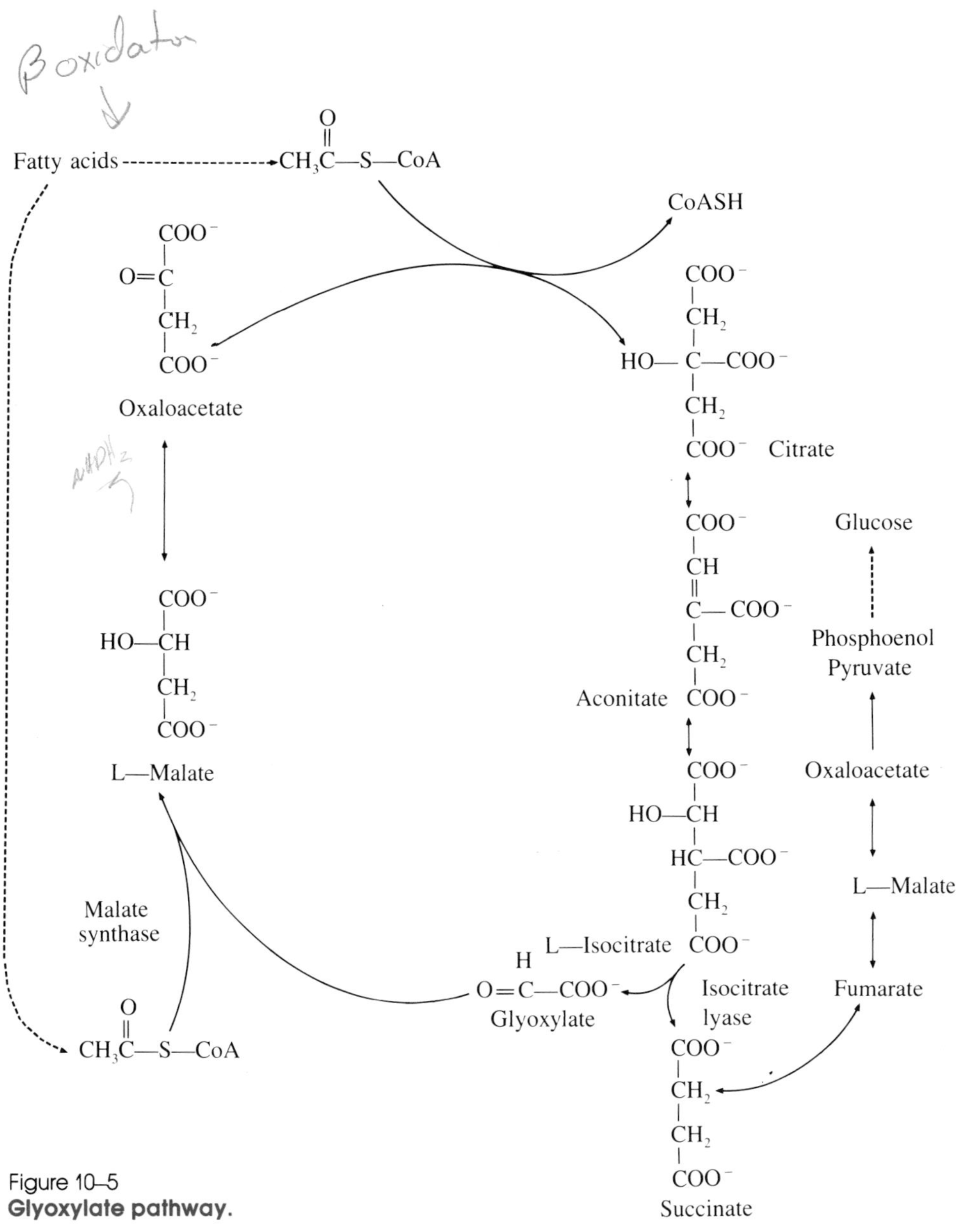

Fatty acids ⟶ $CH_3C(=O)$—S—CoA

CoASH

Oxaloacetate (COO^-—C(=O)—CH_2—COO^-)

Citrate (COO^-—CH_2—C(OH)(COO^-)—CH_2—COO^-)

Aconitate (COO^-—CH=C(COO^-)—CH_2—COO^-)

L—Isocitrate (COO^-—CH(OH)—HC(COO^-)—CH_2—COO^-)

Isocitrate lyase

Glyoxylate (H—C(=O)—COO^-)

Succinate (COO^-—CH_2—CH_2—COO^-)

Fumarate

L—Malate

Oxaloacetate

Phosphoenol Pyruvate

Glucose

Malate synthase

$CH_3C(=O)$—S—CoA

L—Malate (COO^-—CH(OH)—CH_2—COO^-)

Figure 10–5
Glyoxylate pathway.

terial cells as described above can oxidize fatty acids to acetyl CoA and, via the glyoxylate cycle, form tricarboxylic acid intermediates which can then be converted to phosphoenol pyruvate and thereby start reverse glycolysis. Figure 10–6 diagrams the major initial reactions and features of gluconeogenesis. In the reversal of glycolysis, there are three reactions that occur as alternates to those found in the catabolic sequence:

1. Pyruvate is not directly converted to phosphoenol pyruvate (the energetics of that particular reaction are not favorable; $\Delta G^\circ = +7.5$ kcal/mole); instead, the pyruvate is converted first to tricarboxylic acid cycle intermediates within the mitochondria, which then pass into the cytosol, where phosphoenolpyruvate carboxykinase, together with GTP, converts these to phosphoenolpyruvate ($\Delta G^\circ = +1.0$ kcal/mole).

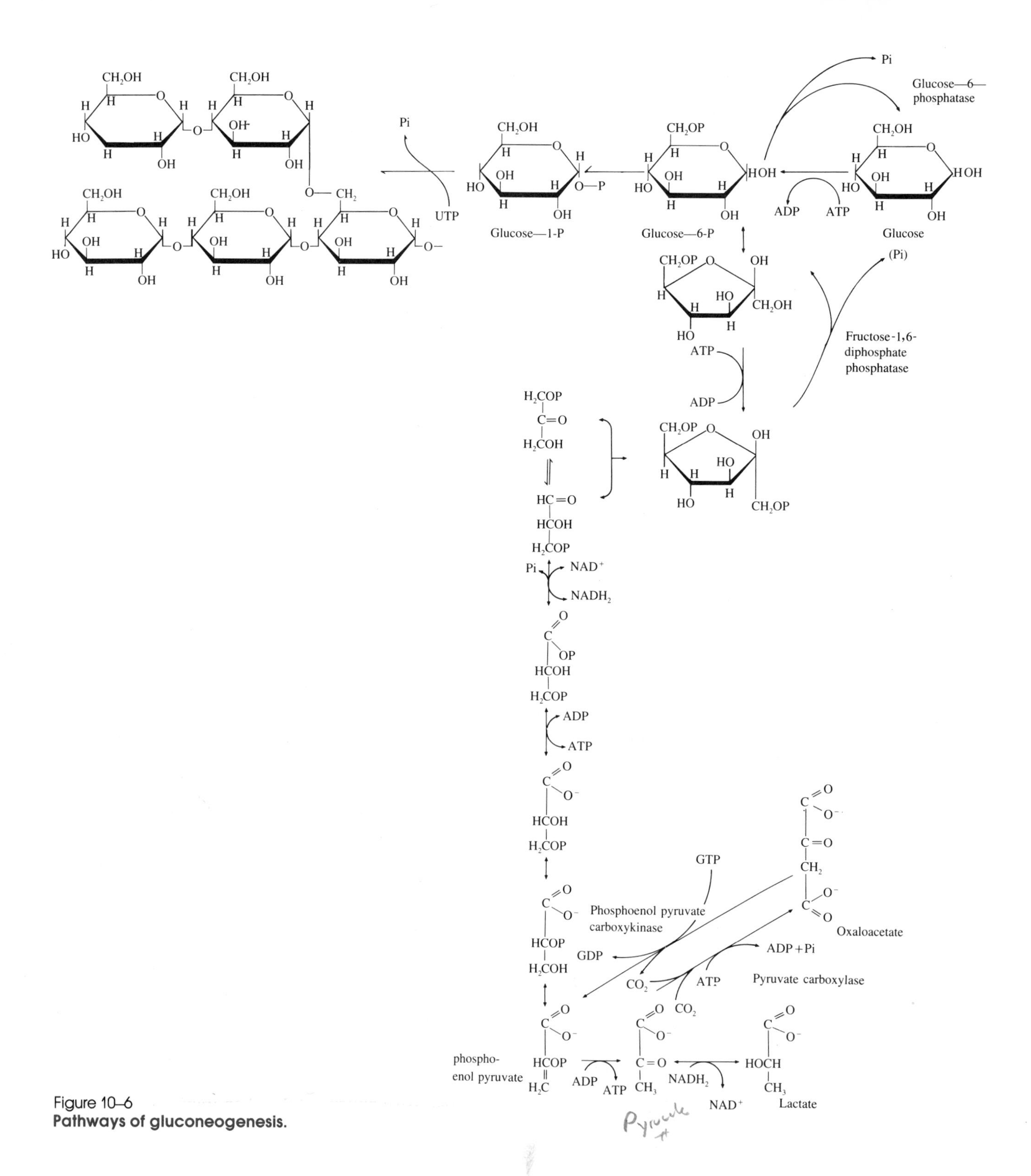

Figure 10–6
Pathways of gluconeogenesis.

2. The dephosphorylation of fructose-1,6-diphosphate is not coupled to ATP by the catabolic sequence enzyme (phosphofructokinase) but instead is acted upon by fructose diphosphatase to yield fructose-6-phosphate and inorganic phosphate.

3. In some tissues such as liver and kidney the glucose-6-phosphate may be broken down to glucose and inorganic phosphate by glucose-6-phosphatase rather than by the kinase.

Synthesis of Glycogen and Starch

The syntheses of glycogen and starch follow basically the same reaction steps beginning with glucose-6-phosphate; however, different enzymes are involved, and usually different nucleoside triphosphate sugars are formed. In both pathways (Fig. 10–7), glucose-6-phosphate is first converted to glucose-1-phosphate, which then reacts with UTP in the case of glycogen formation in animals and ATP in starch formation in plants. The resulting pyridine diphosphate-glucose component then becomes the glucosyl donor reacting with the preexisting polysaccharide chain (usually called the "primer" molecule). The glycogen or starch molecule is thereby lengthed by one glucose unit at a time through an $\alpha,1 \longrightarrow 4$ glycosidic linkage. A separate branching enzyme catalyzes the $\alpha,1 \longrightarrow 6$ linkages necessary to form the branches that occur in the glycosidic chains of glycogen and starch.

Lipid Metabolism

Triglycerides

The synthesis of fatty acids and their ultimate incorporation into triglycerides follows a pathway that is significantly different from that which results in the catabolic breakdown of these compounds. Basically, the linear fatty acid molecule is built up two carbon units at a time by the consumption of acetate in the form of acetyl CoA (Fig. 10–8). There are six enzymatic steps to add each 2-carbon unit to the chain:

1. The **priming reaction,** in which acetyl CoA is bound first to the nonenzymatic *acyl carrier protein* (ACP) and then transferred to the enzyme ACP-acyltransferase

2. The **malonyl reaction,** in which a second acetyl CoA converted to malonyl CoA by incorporation of HCO_3^- is bound to the acyl carrier protein

3. The **condensation reaction,** in which the malonyl group loses CO_2, becoming an acetyl group which then attaches to the acetyl group on the enzyme

4. A **first reduction step** involving $NADPH_2$

5. A **dehydration step**

6. A **second reduction step** involving $NADPH_2$

The resulting fatty acid, which is four carbons in length, can now act as the primer and repeat the sequence. Each cycle through the sequence adds two carbon units to the growing chain length. Desaturation of the fatty acids occurs by additional enzymatic steps. The acetyl CoA for the synthesis is primarily generated in the mitochondria by decarboxylation of pyruvate, by oxidative degradation of some amino acids, or by oxidation of other fatty acids. The mitochondrial acetyl CoA must be converted into other molecules to escape from the mitochondria into the cytosol, where acetyl CoA is reformed and incorporated into the fatty acids. The triglycerides are formed by the condensation of the fatty acyl CoA molecules onto dihydroxyacetone-phosphate (produced by the reactions of glycolysis) or glycerol-3-phosphate (formed through the reduction of dihydroxyacetone phosphate by $NADH_2$).

Degradation of triglycerides occurs initially in the cytosol by hydrolysis to glycerol and fatty acids; the glycerol then enters the glycolytic pathway. Fatty acids must be activated and transported into the matrix of the mitochondria, where they undergo the degradative and oxidative steps outlined in Figure 10–9. The enzymatic sequence of steps 3 to 6 (Fig. 10–9) repeats, removing two carbon units at a time until breakdown is complete and only acetyl CoA or malonyl CoA remains. This process is known as β-*oxidation,* since it is the bond at the second carbon atom that is cleaved. Other oxidations are known, for example, α-oxidation occurs in some germinating seeds producing CO_2. ω-oxidation occurs in the liver of mammals, but its role is not understood.

The synthesis and degradation of most other lipids such as phospholipids and sterols are known in detail but will not be considered here. Synthesis of cholesterol occurs by the progressive buildup of 2-carbon fragments (i.e., acetyl CoA units) and involves about 25 enzymatic steps.

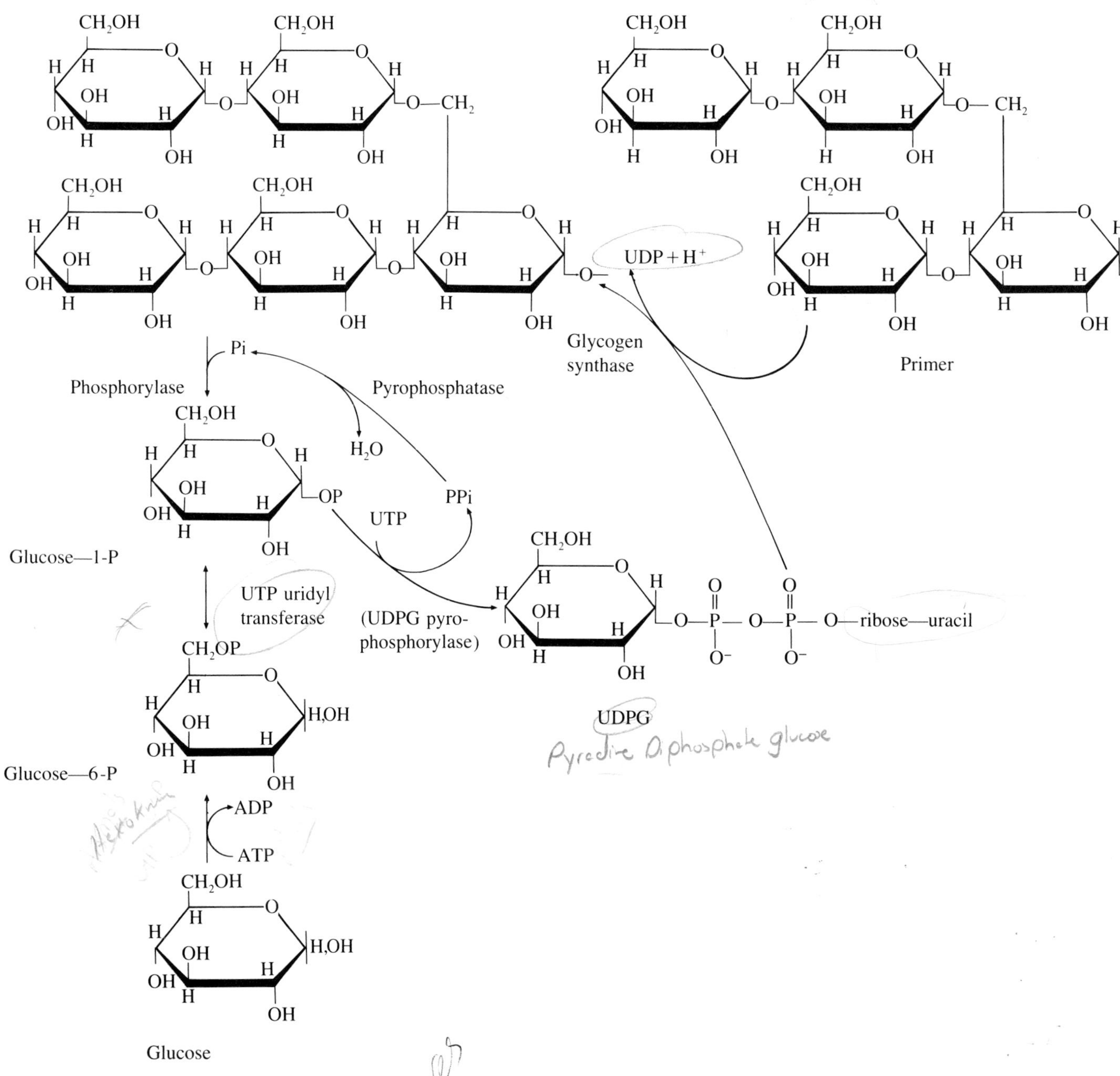

Figure 10–7
Pathways for glycogen and starch syntheses.

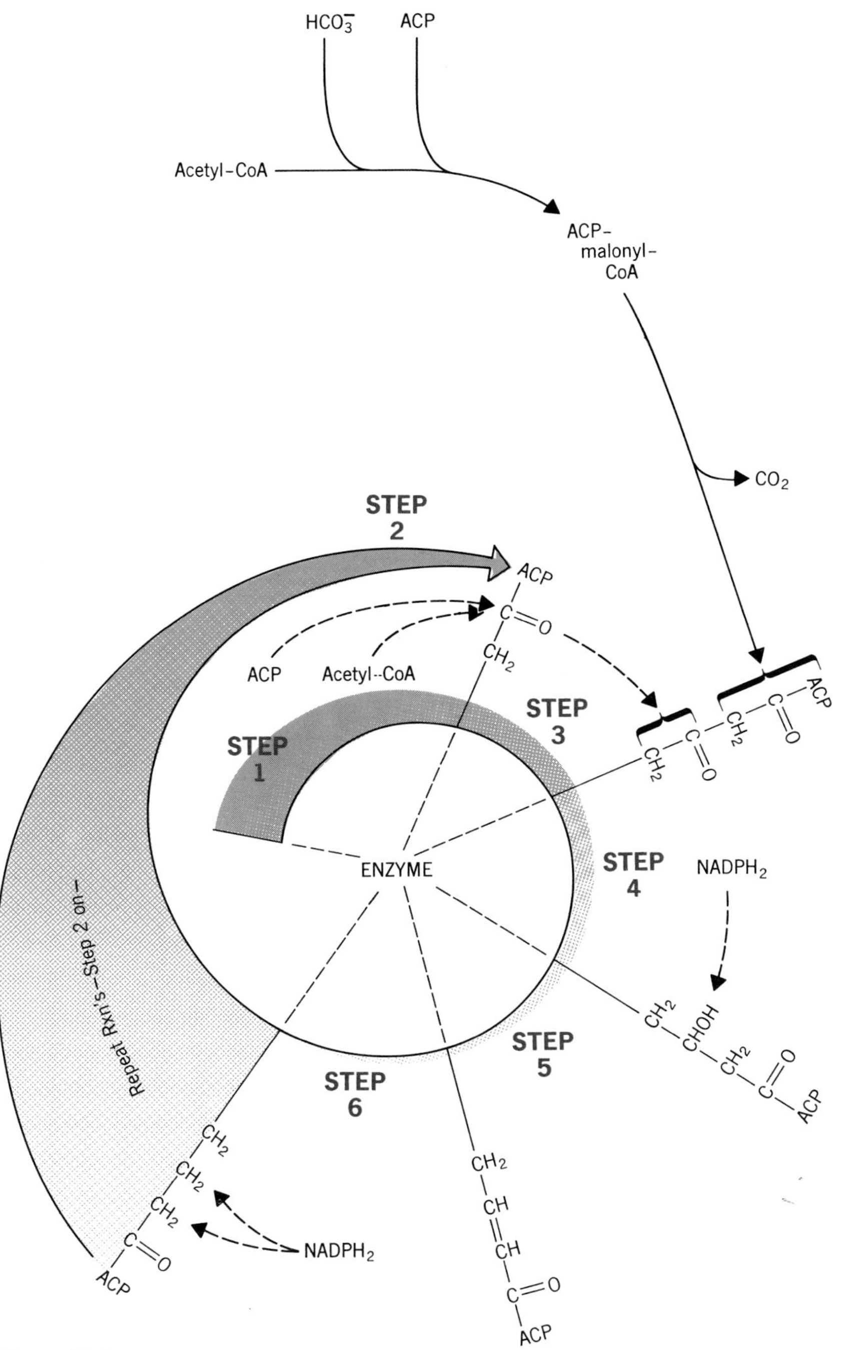

Figure 10–8
Fatty acid biosynthesis (see text).

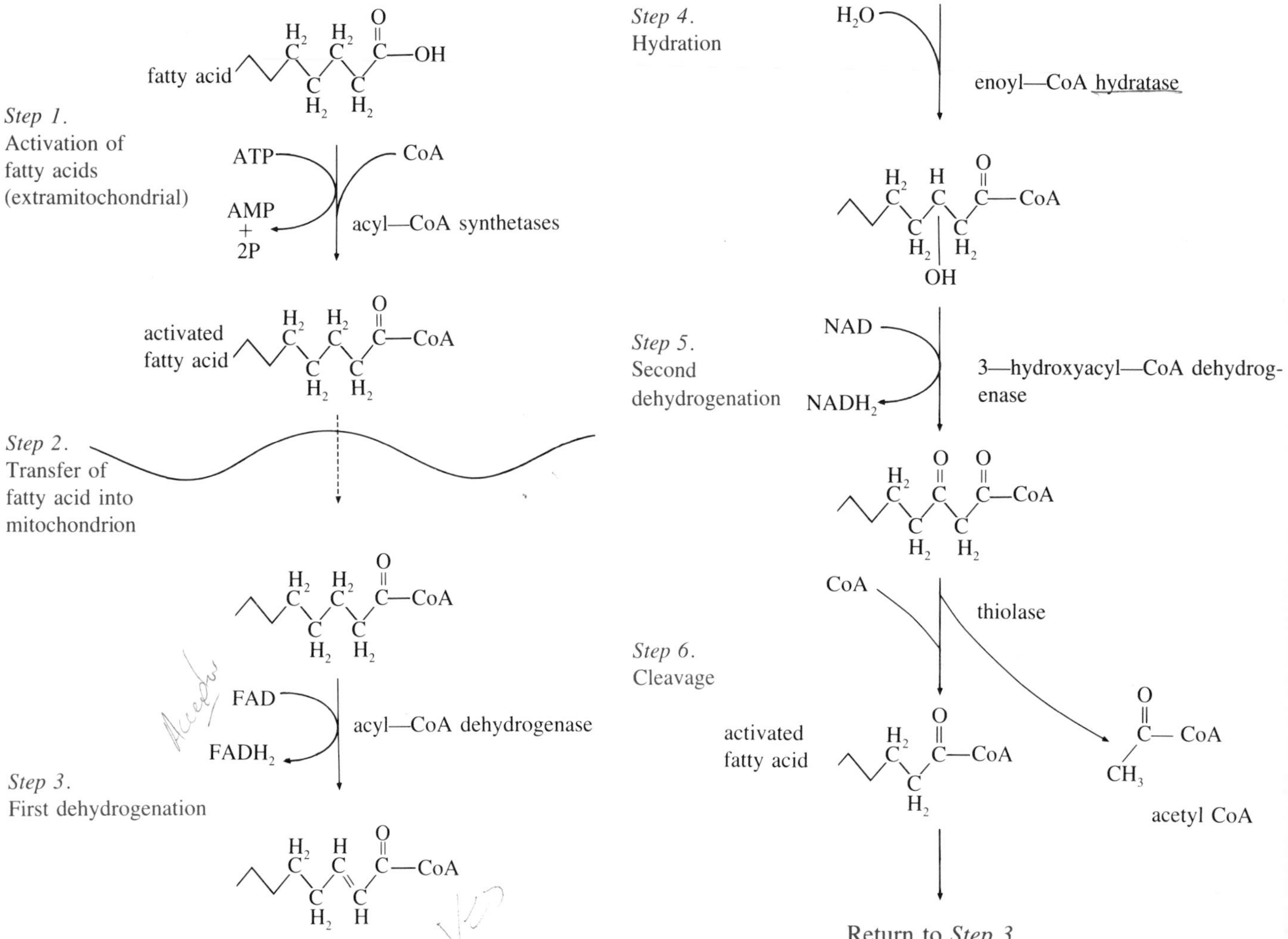

Step 1. Activation of fatty acids (extramitochondrial)

fatty acid; ATP; CoA; AMP + 2P; acyl—CoA synthetases; activated fatty acid

Step 2. Transfer of fatty acid into mitochondrion

Step 3. First dehydrogenation

FAD; $FADH_2$; acyl—CoA dehydrogenase

Step 4. Hydration

H_2O; enoyl—CoA hydratase

Step 5. Second dehydrogenation

NAD; $NADH_2$; 3—hydroxyacyl—CoA dehydrogenase

Step 6. Cleavage

CoA; thiolase; activated fatty acid; acetyl CoA

Return to Step 3.

Figure 10–9
Oxidation of fatty acids.

Nitrogen Metabolism

Cells require a variety of nitrogen compounds for survival. Central to the requirement for nitrogen is the formation of amino acids. Amino acids are necessary not only for the synthesis of proteins but also as the primary source of nitrogen in the synthesis of the nucleotide building blocks of nucleic acids. Not all cells or organisms are able to synthesize all the amino acids necessary to their existence. For example, the cells of humans beings are able to make only 10 of the 20 amino acids required for human protein synthesis. The other 10 are referred to as the *essential amino acids* and are obtained from plant or microbial sources in the diet. Nitrogen for the synthesis of the nonessential amino acids of humans and other higher animals is obtained from ammonium ions, since the enzyme systems for utilizing nitrate, nitrite, or atmospheric nitrogen are not present. However, many microbes, leguminous plants, and a few other plants can use atmospheric nitrogen. Most higher plants can make the amino acids needed for protein synthesis by using ammonia, nitrate, and nitrite as sources of nitrogen.

Each of the 20 different amino acids required for protein synthesis is synthesized by a different system of enzymes. Several utilize intermediates of the glycolytic and tricarboxylic acid cycle pathways. The decomposition of amino acids involves pathways different from the synthetic pathways but the products after deamination are usually further metabolized by the glycolytic or the tricarboxylic acid cycle reactions.

Cancer Cell Metabolism

Cancer cells have a distinct type of metabolism, one that is different from most normal tissues. While cancer cells possess all the enzymes necessary for glycolysis, the tricarboxylic acid cycle reactions, and terminal respiration, the rates of utilization of these pathways are distinctive. Less oxygen is consumed by cancerous tissue than by normal tissue, but significantly more glucose is consumed. Instead of converting most of the glucose to CO_2 and H_2O through terminal respiration as normal tissue does, cancerous tissue converts large amounts of glucose to lactate, even though oxygen is available. This phenomenon is called **aerobic** glycolysis. The lactate produced by cancerous tissue can be converted back into glucose in the liver, but this resynthesis process generally consumes six molecules of ATP per mole lactate (or glucose) compared to a net generation of only two molecules per glucose during its breakdown in the cancerous cells. In effect the cancerous cell tends to be a "parasite" of the other, normal tissues of the organism.

Functions of Metabolic Pathways

The sequences of metabolic reactions called metabolic pathways serve a number of functions. First, and most obvious, is the formation of an end product needed by the cell or organism. This product may be used as part of the structure of the cell, or in some cases when secreted from the cell, it may be incorporated into an extracellular structural part of an organism. The proteins in the matrix of cartilage and the pectins of the middle lamellae of plant cell walls are examples of such secretions. The end product may also function in the cell as a regulatory agent for other reactions. Enzymes and hormones are such end products. End products may also become storage or reserve compounds such as starch, glycogen, and certain lipids.

Second, a metabolic pathway may function to provide energy-rich compounds such as ATP, GTP, and UTP for other energy-requiring reactions. Metabolic pathways rarely provide these compounds as end products of reaction sequences but instead produce them at one or more of the intermediate steps of a pathway. For example, in glycolysis, ATP is produced at two intermediate steps (Fig. 10–2; reactions 10–15 and 10–18). Other energy-related compounds such as $NADH_2$ and $NADPH_2$ are also products of intermediate reactions and may be utilized directly in other reactions or oxidized to provide energy in the form of ATP.

Third, intermediates in metabolic pathways may be drawn upon or utilized as substrates for other metabolic sequences. For example, the tricarboxylic acid cycle (Krebs cycle) is a sequence of reactions that effectively oxidizes acetyl CoA to CO_2, $NADH_2$, and GTP, but many of the intermediates are drawn off and utilized for other purposes. Acetyl CoA itself is used in lipid syntheses, and α-ketoglutarate and oxaloacetate are used in amino acid syntheses. Metabolic reactions that serve in both a catabolic (energy-producing) and an anabolic (biosynthetic) function are called **amphibolic** pathways. Amphibolic pathways, which may have their intermediates diverted to other reaction sequences, frequently include specialized reactions that replenish these intermediates. The specialized enzymatic reactions are called **anapleurotic** reactions. The importance of anapleurotic reactions is apparent in the tricarboxylic acid cycle. To oxidize 1 mole of acetyl CoA, 1 mole of oxaloacetate is required. At the end of the sequence, 1 mole of oxaloacetate is produced and is therefore available to react with additional acetyl CoA. If, however, some of the intermediates of the cycle are diverted into side reactions to form amino acids or other compounds 1 mole of acetyl CoA will not result in the entire production of 1 mole of oxaloacetate, and the capability of oxidizing further acetyl CoA will be reduced. Anapleurotic reactions, which regenerate intermediates of the cycle from external compounds, could reestablish the full capacity of the cycle. The "Wood-Werkman" reaction in bacteria (carbon dioxide is bound to pyruvate to form oxaloacetate) is an example of such an anapleurotic mechanism. The glyoxylate pathway described earlier is also an anapleurotic mechanism, effectively forming the tricarboxylic acid cycle intermediates, succinate and malate.

Calculations of Energy Change

In most instances, there is sufficient knowledge about the major metabolic pathways of the cell so that the specific enzymatic reactions in which ATP (GTP or other related compounds) is formed or consumed are known. By inspection of a metabolic chart, such as that in Figure 10–2, one should be able to calculate the number of moles of ATP consumed or produced and the *net* change for any reactant-to-product sequence. For example, if 1 mole of sucrose is oxidized to pyruvate, 4 moles of ATP would be consumed,

1 mole in reaction 10-8a
1 mole in reaction 10-8b
2 moles in reaction 10-11
Total = *4* moles (consumed)

and 8 moles of ATP would be produced;

4 moles in reaction 10-15
4 moles in reaction 10-18
Total = 8 moles (produced)

Therefore, in the glycolytic oxidation of 1 mole of sucrose, there is a net production of 4 moles of ATP. These "paper calculations" provide the theoretical amounts of ATP expected from the glycolytic sequence of reactions. Experimentally, these numbers are approached but not always obtained, for laboratory conditions do not always provide advantageous conditions, intermediates may be drawn into other reactions (especially in whole cell preparations), cofactors may not be in the proper concentration, and so on.

The efficiency of a metabolic pathway is usually determined by an analysis of the change in free energy, ΔG°. For example, in the conversion of glucose to lactate (Fig. 10–2), the enzymatic sequence may be considered to be composed of the exergonic reaction

$$\underset{\text{(Glucose)}}{C_6H_{12}O_6} \longrightarrow \underset{\text{(Lactate)}}{2\ C_3H_6O_3}$$

and the endergonic reaction

$$2\ ADP + 2\ P_i \longrightarrow 2\ ATP + H_2O$$

The standard free energy change (ΔG°) for the catabolic reactions is determined by adding the values for each of the reactions.

Note. The ΔG° values given for a reaction or shown in a metabolic reaction chart are for the reaction *as written*, including the *direction* written; for example,

$$\text{Glycerate-3-P} \xrightarrow[\Delta G^{\circ} = 1.06]{} \text{glycerate-2-P}$$

and

$$\text{Glycerate-2-P} \xrightarrow[\Delta G^{\circ} = -1.06]{} \text{glycerate-3-P}$$

If both an exergonic reaction and an endergonic reaction are written together, for example

$$\text{Phosphoenol pyruvate} \xrightarrow[\Delta G^{\circ} = -5.72]{\text{ADP} \quad \text{ATP}} \text{pyruvate}$$

then the ΔG° given is the algebraic sum of the two separate reactions; that is

$$\text{Phosphoenol pyruvate} \xrightarrow[\Delta G^{\circ} = -13.02]{} \text{pyruvate} + P_i$$

and

$$ADP + P_i \xrightarrow[\Delta G^{\circ} = 7.30]{} ATP$$

In the case of the catabolic reactions of Figure 10–2 to form lactate, a summation would be:

Reaction Number	ΔG° *(kcal mole^{-1})*
10-8*	+ 3.3
10-10	+ 0.4
10-11*	+ 3.9
10-12	+ 5.73
10-13	+ 1.83
10-14	+ 3.0, i.e. (+1.5 × 2)
10-15*	−23.6, i.e. (−11.8 × 2)
10-16	+ 2.12, i.e. (+1.06 × 2)
10-17	+ 0.88, i.e. (+0.44 × 2)
10-18*	−29.6, i.e. (−14.8 × 2)
10-21	−12.0, i.e. (−6.0 × 2)
Total	−44.04 kcal mole^{-1}

*Excluding the ΔG° attributed to ATP ⟶ ADP + P_i or ADP + P_i ⟶ ATP.

A summation of the net ADP/ATP reactions would be

Reaction Number	ΔG° *(kcal mole^{-1})*
10-8	− 7.3
10-11	− 7.3
10-15	+ 14.6, i.e. (7.3 × 2)
10-18	+ 14.6, i.e. (7.3 × 2)
Total	+ 14.6 kcal mole^{-1}

The reactions of glycolysis and the fermentation of glucose to lactate thus produce much more free energy than they consume in the formation of ATP. The efficiency of these pathways would therefore be

$$14.6/44.04 \times 100 \text{ or } 33.1\%$$

Attempts to measure the energetics of glycolysis and fermentation in intact cells (i.e., in vivo) have produced striking results. In red blood cells (which are ideal in such studies, since this cell derives most of its energy from glycolysis and from fermentation to lactate), the efficiency is about 53%; this is much higher than that expected from the calculations above. To determine this in vivo efficiency, the steady-state concentrations of all the glycolytic intermediates are measured, and from these values, the actual equilibrium constants are calculated and the ΔG values determined. The ΔG values (rather than the ΔG° values) reveal the greater efficiency of red blood cells. Skeletal muscle cells also reveal an efficiency level higher than that anticipated from calculations and summations of the type carried out above. Therefore while ΔG° values provide figures for easy comparison under defined circumstances (i.e., pH 7.0; 25°C), the differences in substrate concentration, pH, and other factors may bring about extreme variations in efficiency of enzymatic conditions under natural conditions, that is in vivo.

Summary

Cells break down compounds by sequences of enzyme reactions **(catabolism)** to obtain energy or to form smaller molecules that can be used to build other, larger molecules. Sequences of enzyme reactions also provide the mechanism for the **synthesis** of new molecules **(anabolism).** These reaction sequences, called **metabolic pathways,** are frequently associated with specific organelles. The enzymes that catalyze the reactions are often compartmentalized within, between, or on membranes of the organelles.

In general, carbohydrates, lipids, and proteins can be catabolized to yield energy and a pool of small molecules, and these may either be used in the synthesis of new macromolecules or be excreted from the cell. The products of catabolism in one organelle may be transported to other organelles for further catabolism or anabolism. Reaction sequences and their intracellular location have been determined by **marker** and **radioactive tracer techniques,** studies of **enzyme activity** and **inhibition,** and through the use of **mutant** cells.

Some of the more common pathways in cells are listed below:

1. Glycolysis—the catabolism of monosaccharides to pyruvate

2. Fermentation—the catabolism of monosaccharides to products such as ethanol and CO_2 in the absence of air (anaerobic conditions)

3. Oxidation of pyruvate (Krebs cycle)—catabolism to CO_2 and water in the presence of oxygen

4. Phosphogluconate pathway (pentose phosphate or hexose monophosphate shunt)—pathway for formation and/or catabolism of pentoses

5. Glyoxylate pathway—conversion of acetyl CoA to carbohydrate

6. Gluconeogenesis—synthesis of glucose from simple acids

7. Glycogen and starch synthesis

8. Fatty acid synthesis—pathway for formation of fatty acids from acetyl CoA

9. β-oxidation—pathway for breakdown of fatty acids

10. Protein synthesis—mechanism for formation of proteins from amino acids

References and Suggested Reading

Articles and Reviews

Krebs, H. A., The history of the tricarboxylic acid cycle, *Perspect. Biol. Med. 14,* 154 (1970).

Villar-Palasi, C., and Larner, J., Glycogen metabolism and glycolytic enzymes, *Annu. Rev. Biochem. 39,* 639 (1970).

Books, Monographs, and Symposia

Atkinson, D. E., *Cellular Energy Metabolism and Its Regulation,* Academic Press, Inc., New York, 1977.

Axelrod, B., Other pathways of carbohydrate metabolism, in *Metabolic Pathways* (D. M. Greenberg, ed.), Vol. 1, Academic Press, New York, 1967, p. 272.

Dagley, S., and Nicholson, D. E., *Metabolic Pathways,* John Wiley & Sons, Inc., New York, 1970.

Green, D. E., and Goldberger, R. F., *Molecular Insights into the Living Process,* Academic Press, Inc., New York, 1967.

Hinkle, P. C., and McCarty, R. E., How cells make ATP. *Sci. Am. 238*(3), 104 (Mar. 1978).

Larner, J., *Intermediary Metabolism and Its Regulation,* Prentice-Hall, Englewood Cliffs, N.J., 1971.

Lehninger, A. L., *Biochemistry,* Worth Publishers, New York, 1975.

Masoro, E. J., *Physiological Chemistry of Lipids in Mammals,* W. B. Saunders Co., Philadelphia, 1968.

Meister, A., *Biochemistry of the Amino Acids,* Academic Press, Inc., New York, 1965.

Stanbury, J. O., Wyngaarden, J. B., and Fredrickson, D. S., *The Metabolic Basis of Inherited Disease,* McGraw-Hill Book Co., New York, 1972.

Chapter 11
METABOLIC REGULATION

The diverse metabolic reactions and reaction sequences in cells have been briefly described and outlined in the preceding three chapters. From these descriptions, it is clear that a substrate can be enzymatically converted into a great variety of intermediates and products. Although there are a number of essentially unidirectional reactions, the metabolic network of reversible and cyclical reaction sequences provides for the possible conversion of almost any metabolite into any other metabolite. In a superficial comparison, one could visualize the metabolic pathways as a branching and connecting network of water pipes in which the water can be caused to flow between any two points in the network under suitable conditions. For example, simply an *excess* of water (metabolite) in one part of the system could cause flow to the other parts of the system. In vivo and in vitro experiments have indicated that the ''flow'' of metabolites through metabolic pathways is not as free and uncontrolled as this pipeline-network analogy. Metabolic conversions are controlled or regulated by a variety of mechanisms that channel metabolites into needed compounds or into stable reserve products and prevent energetically wasteful conversions.

Cells have evolved a diverse set of regulatory mechanisms. Individual reactions may be controlled by one or more processes from simple mass action to complex hormonally controlled enzyme systems. The more common of these processes is described in this chapter.

Regulation by Mass Action

For any reversible reaction, such as

$$[A] + [B] \rightleftharpoons [C] + [D] \qquad (11\text{–}1)$$

in which A and B are reactants and C and D are products, we can write an expression that indicates the ratio of the concentrations of products $[C]$ and $[D]$, and reactants $[A]$ and $[B]$ at equilibrium (see also Chapter 8). This ratio,

$$K_{eq} = \frac{[C]\ [D]}{[A]\ [B]}$$

forms a constant for the reaction and is fixed for a particular temperature. If the concentration of any one of the components of the reaction is altered, the concentration of at least one other component must change to maintain the equilibrium as expressed by the K_{eq}. A constant greater than 1 indicates that the equilibrium of the reaction lies to the right in reaction 11–1, and if the ratio is less than 1, the equilibrium is to the left.

In a sequence of reactions such as that shown in Figure 11–1, one reaction may be affected by the next reaction because the products of the first are used in the second and so forth. Thus the K_{eq} for an overall sequence may not be the sum of the individual reactions. (The important relationship between the K_{eq} and free energy, ΔG, was discussed in Chapter 9.) Even though the intermediate reaction in the sequence in Figure 11–1 has a K_{eq} less than 1 (with an equilibrium that lies to the left), the sequence may still proceed when glucose and ATP are introduced, since the formation of glucose-6-P will necessitate a conversion to fructose-6-P to maintain the K_{eq}.

A limited degree of regulation of alternative metabolic pathways is also achieved by the law of mass action at branch points in a pathway. Although the equilibrium is to the right in each of the reactions in Figure 11–2, the equilibrium is "further" to the right in the reaction forming fructose-6-P, and more product would be formed from reactant along that branch than in the branch that forms glucose-1-P. However, in glucose metabolism, both products are part of a continuing sequence of reactions, and subsequent reactions (not shown in Figure 11–2; see Fig. 10–1) would affect the net K_{eq} and could alter the overall direction of conversion.

Regulation by Enzyme Activity

Regulation of metabolism is most commonly controlled at the cellular level by altering the *activities* of the enzymes or by altering the *number* of enzyme molecules present. The regulation by altered activity can be brought about by (1) changes in substrate concentration, (2) allosteric effectors such as AMP or glucose-6-phosphate, (3) irreversible covalent bond modifications such as the hydrolytic activation of pancreatic zymogens, (4) reversible covalent bond modifications such as the phosphorylase enzyme activation, and (5) noncovalent modifications brought about by the action of one enzyme on another.

Substrate Concentration Effectors

As described in Chapter 8, the activity of an enzyme (i.e., the rate of product formation) increases as a hyperbolic function as the substrate concentration is raised (Fig. 11–

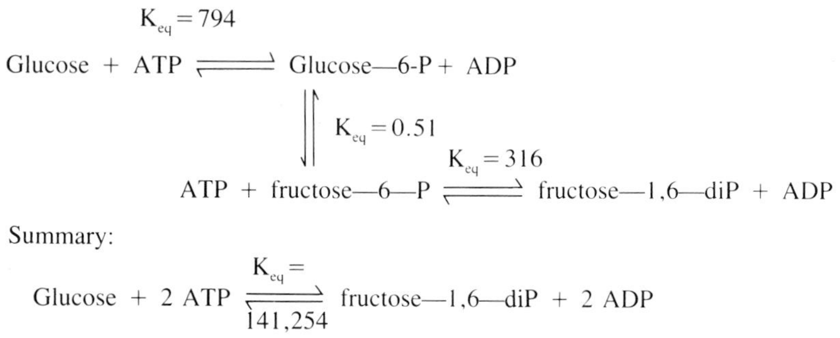

$$\text{Glucose} + \text{ATP} \underset{}{\overset{K_{eq} = 794}{\rightleftharpoons}} \text{Glucose—6-P} + \text{ADP}$$

$$\downarrow\uparrow \; K_{eq} = 0.51$$

$$\text{ATP} + \text{fructose—6—P} \overset{K_{eq} = 316}{\rightleftharpoons} \text{fructose—1,6—diP} + \text{ADP}$$

Summary:

$$\text{Glucose} + 2\ \text{ATP} \underset{141{,}254}{\overset{K_{eq} =}{\rightleftharpoons}} \text{fructose—1,6—diP} + 2\ \text{ADP}$$

Figure 11–1
Example of linear metabolic pathway.

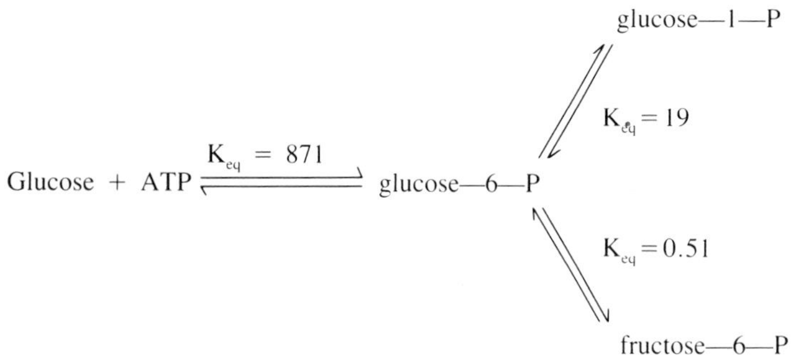

$$\text{Glucose} + \text{ATP} \overset{K_{eq} = 871}{\rightleftharpoons} \text{glucose—6—P}$$

$$\text{glucose—6—P} \overset{K_{eq} = 19}{\rightleftharpoons} \text{glucose—1—P}$$

$$\text{glucose—6—P} \overset{K_{eq} = 0.51}{\rightleftharpoons} \text{fructose—6—P}$$

Figure 11–2
Example of branching metabolic pathway.

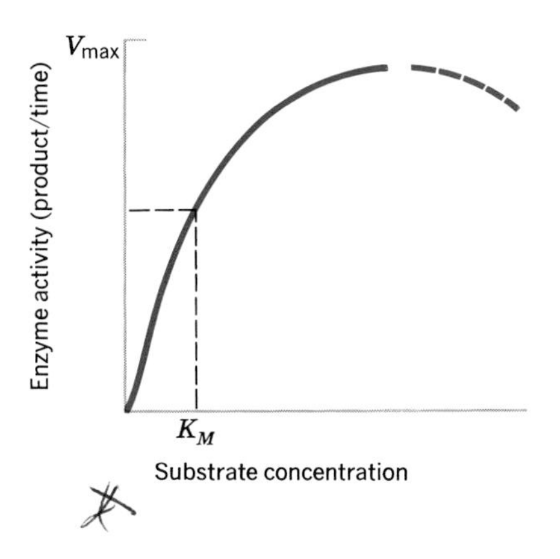

Figure 11–3
Hyperbolic effect of substrate concentration on enzyme activity.

3) until a maximum reaction velocity (V_{max}) is achieved. However, excessively high substrate concentrations may actually reduce enzyme activity. Each enzyme, subject to experimental conditions, has a characteristic maximum velocity, as exemplified by the maximum velocities of the glycolytic enzymes in brain tissue shown in Table 11–1. If the substrate in this tissue were in excess, then one would expect that aldolase with the lowest maximum velocity would be the rate-limiting reaction in the sequence. In vivo studies have indicated that enzymes are rarely saturated by substrate. In experiments with mouse brain tissue it has been shown that hexokinase phosphoglucoisomerase and aldolase, under normal conditions, function at a substrate concentration somewhat equal to or greater than the K_M (Michaelis-Menten constant). Small changes in substrate concentration do not significantly alter the rate of metabolism through this part of the glycolytic pathway. Of greater regulatory importance here would be the amount of enzyme present. The last six enzymes in mouse brain glycolysis function at substrate levels significantly below the K_M. Small changes in the substrate concentration at these levels directly and significantly alter the rate of enzyme activity. Therefore, under anoxic conditions the rate of production of lactate will not appreciably increase by supplementing the tissue with glucose. However, supplements of compounds such as glycerol that enter the glycolytic pathway below the level of aldolase will cause an increase in lactate production.

The limiting factor for the rate of an enzyme-catalyzed reaction in vivo is the *affinity* between the enzyme and the substrate. At branch points along metabolic pathways, two enzymes compete for the same substrate. The enzyme with the lower K_M value will react more rapidly with the substrate at low substrate concentrations, while the enzyme with the higher K_M value will be more active at high substrate concentrations (Fig. 11–4). Thus, when present at low levels, a substrate may be channeled primarily into one pathway, while the major direction of metabolism may shift to other pathways at higher substrate levels.

Allosteric Effectors

The regulatory effects of substrate concentrations and the mass action factor generally influence all reactions in metabolic pathways; therefore, these mechanisms are not very specific. There are, however, a number of mechanisms by which specific reactions in a pathway can be regulated. In one such mechanism, enzymes catalyzing specific reactions of the pathway are influenced by the type and amount of certain *regulatory metabolites* present. These enzymes are called **allosteric** enzymes because their catalytic activity is modified by the noncovalent binding of specific metabolites to a site on the protein *other than the active site* (see Chapter 8). Some of the more common regulatory metabolites are listed in Table 11–2. The binding of the regulatory metab-

Figure 11–4
Kinetics of two enzymes with different K_M at varying substrate concentrations.

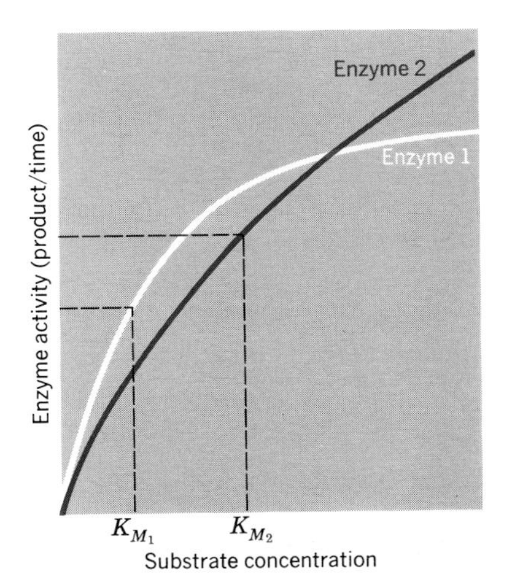

Table 11–1
Maximum Activity of Glycolytic Enzymes in Mouse Brain Tissue under Anoxic Conditions

Enzyme (in Order of Glycolysis)	V_{max} (mmoles/kg/min)
Hexokinase	15.2
Phosphoglucoisomerase	154.0
Phosphofructokinase	26.7
Aldolase	7.6
Glyceraldehyde-3-phosphate dehydrogenase	96.0
Phosphoglycerate kinase	750.0
Phosphoglycerate mutase	145.0
Enolase	36.0
Pyruvate kinase	95.0
Lactic dehydrogenase	129.0

Source Copyright © American Society of Biological Chemists, Inc., *J. Biol. Chem 239,* 31 (1964).

olite to the allosteric enzyme may (1) cause either an inhibitory or a stimulatory response in enzyme activity, and (2) cause a change in the level of activity of the pathway that is proportional to the concentration of the regulatory metabolite.

In general, inhibitory allosteric effects are caused by the accumulation of the product of a reaction sequence, with the resultant binding of the product to an enzyme at or near the beginning of the sequence. Where branches occur in a pathway, the end product usually affects one of the enzymes at the branch point (Fig. 11–5). This type of inhibition is called **end-product inhibition** or **feedback inhibition.** The end-product metabolite is called an **effector** (or **modulator**), and if its effect is inhibitory, it is termed a **negative effector. Positive effectors** also occur; generally, these are either the original substrate of a metabolic pathway or an early intermediate in the sequence and serve to increase the activity of an enzyme further along the metabolic pathway. This regulatory mechanism is called **feedforward stimulation,** in contrast to feedback inhibition. An allosteric enzyme may be affected by one modulator, in which case it is said to be **monovalent,** or by two or more modulators, in which case it is **polyvalent.** The kinetics of allosteric enzyme reactions are discussed in Chapter 8.

The regulation of amino acid synthesis in *Escherichia coli* provides a clear example of control of divergent metabolic pathways by feedback inhibition. An outline of the metabolic pathways for the synthesis of three amino acids is shown in Figure 11–6. Lysine, methionine, and threonine are each synthesized from aspartate, and each may be utilized in protein synthesis. Without metabolic controls the consumption or utilization of any one of these amino acids would stimulate the pathways and cause unneeded synthesis of the unused amino acids as well as the one utilized. Such an unregulated system would consume vital resources and energy; both factors could have survival implications to the organism and evolutionary consequences to the species. However, in *E. coli,* the allosteric regulatory mechanisms are most effective. The accumulation of each amino acid provides a feedback inhibition of the first enzyme in the specific branch of the pathway leading to the synthesis of that amino acid. In Figure 11–6, this negative effect is shown by the dotted lines. Moreover, an additional level of regulation is achieved through effects on the enzyme *aspartokinase,* which catalyzes the phosphorylation of aspartate (Fig. 11–6). This enzyme exists in three forms or isozymes, as described later in this chapter and in Chapter 8. The presence of the three isozymes is symbolized in Figure 11–6 by using three separate arrows to show the

Figure 11–5
Feedback inhibition mechanisms (- - -) for linear and branched pathways.

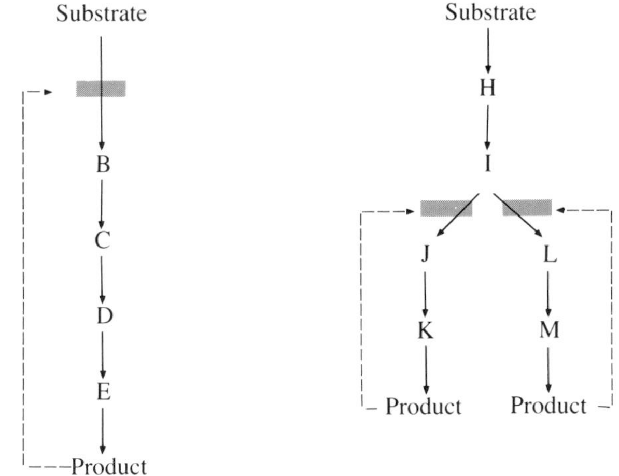

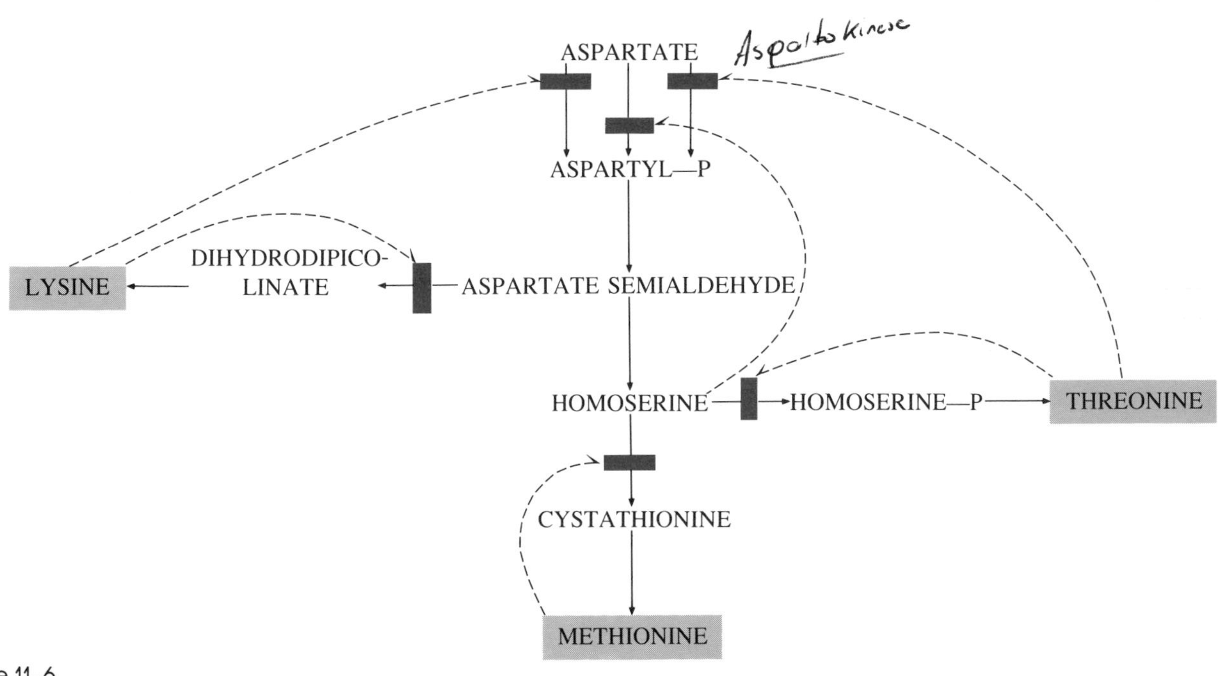

Figure 11–6
Allosteric feedback control of amino acid synthesis (⇢▮ = negative feedback).

conversion of aspartate to aspartylphosphate. One of the isozymes is specifically and completely inhibited by threonine; the second (which is present only in small amounts) is specifically inhibited by homoserine; the third isozyme is specifically inhibited by lysine. In addition, the latter isozyme is *repressed* by lysine. (**Repression** is a regulatory mechanism that reduces the *number* of enzyme molecules in the cell and is discussed later in the chapter.)

Covalent Bond Modification of Enzyme Activity

A number of enzymes are synthesized in what is called an *inactive* (or **zymogen**) form and must be covalently modified to become active. The modification may be irreversible, as is the case with the hydrolytic modification of zymogens such as pepsinogen (to form pepsin) and trypsinogen (to form trypsin), which are discussed in Chapter 8. Other enzymes may be reversibly covalently activated and deactivated. The reversible activation and deactivation of glutamine synthetase is a well-studied example (Figure 11–7). This enzyme catalyzes the conversion of glutamate to glutamine,

$$\text{Glutamate} + NH_3 + \text{ATP} \longrightarrow \text{glutamine} + \text{ADP} + \text{phosphate}$$

and the transfer of a glutamyl group to hydroxylamine,

$$\text{Glutamine} + NH_2OH \longrightarrow \gamma\text{-glutamyl-NHOH} + NH_3$$

Interestingly, it was found that glutamine synthetase was inactivated when treated with ammonia for 10 minutes. Inactivation was found to be the result of covalent adenylation, as shown in Figure 11–7. The active and inactive forms of the enzyme are spectroscopically different, have different metal specificities (Mg^{++} and Mn^{++}), and are themselves acted upon by two different enzymes (adenylating and deadenylating) which change their activities. The latter enzymes are further regulated by glutamine and ATP.

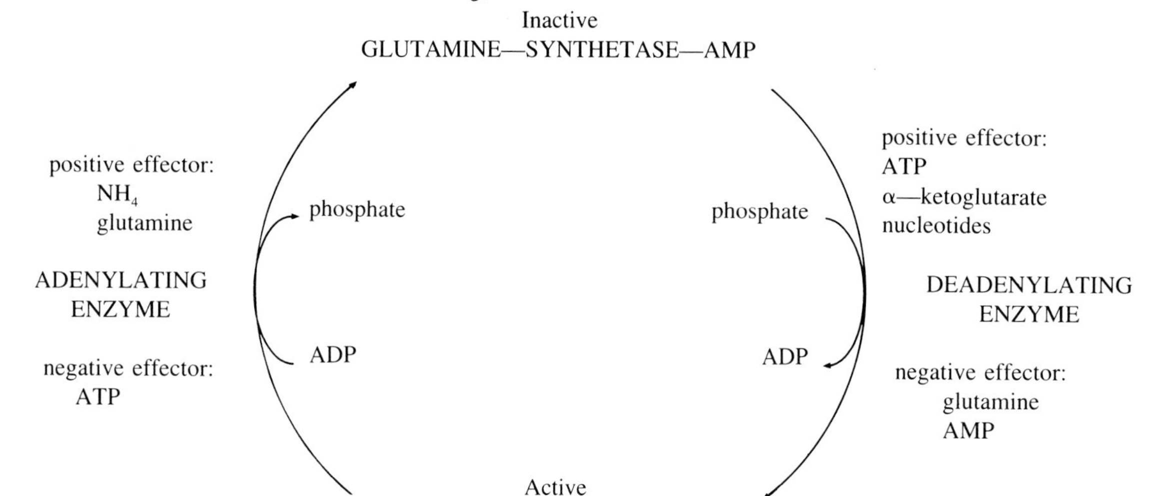

Figure 11–7
Regulation of glutamine synthetase by activation and inactivation.

Regulation by Number of Enzyme Molecules

Isozymes (Isoenzymes)

In a number of instances, enzymes that catalyze specific reactions have been found to exist in multiple forms in tissues and organisms. These isozymes or isoenzymes are coded for by different genes and therefore have different amino acid complements. Isozymes can usually be separated from one another by gel electrophoresis (Chapter 13) of the tissue extracts. One of the most exhaustively studied of the known isozymes is *lactic dehydrogenase* which catalyzes one of the terminal reactions in glycolysis; that is,

$$\text{Pyruvate} + NADH_2 \rightleftharpoons \text{lactate} + NAD$$

In rats and in a number of other vertebrates, this enzyme is present in five forms. Each of the five forms has a molecular weight of about 134,000 daltons, consists of four polypeptide chains of about 33,500 daltons each, and catalyzes the same reaction. Each of the four polypeptides may be of two types, usually referred to as *M* and *H*. In rat skeletal muscle tissue, the predominant form of the isozyme contains four polypeptides of the *M* type; in contrast, in rat heart muscle, the predominant form contains four polypeptides of the *H* type. The other isozyme forms are made up of three *H* and one *M*, two *H* and two *M*, and one *H* and three *M* polypeptides (usually abbreviated H_3M, H_2M_2, and HM_3). Some of these forms predominate in other tissues, although each tissue has some of each isozyme (Fig. 11–8).

The M_4 isozyme is prevalent in embryonic tissue and in skeletal muscle tissue. It has a low K_M for pyruvate and high V_{max} for converting pyruvate to lactate. It is well adapted to these tissues, which are frequently deprived of

Table 11–2
Common Regulatory Metabolites

Glucose-6-phosphate
Fructose-1,6-diphosphate
1,3-diphosphoglycerate
Citrate
Acetyl CoA
AMP, ADP, ATP
Cyclic AMP, cyclic GMP
GTP, UTP, etc.
NAD, $NADH_2$
Fatty acids
Amino acids

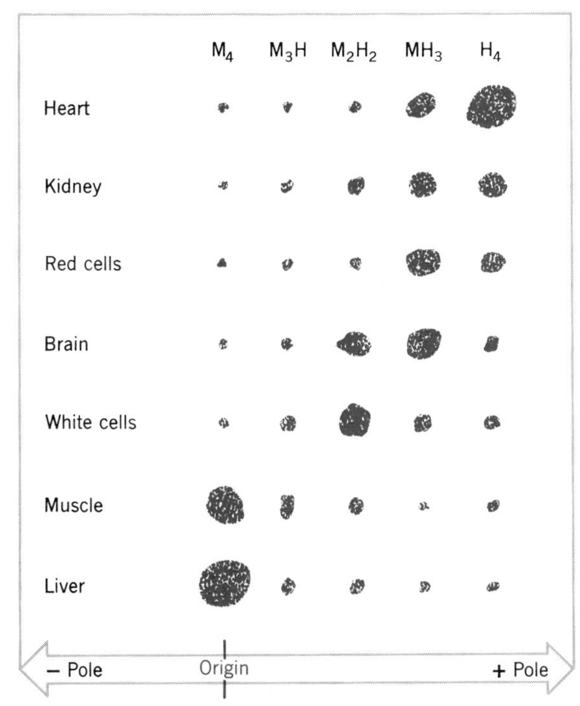

Figure 11–8
Lactic dehydrogenase isozymes. The spots show the relative amounts of each isozyme as they might appear when separated by gel electrophoresis.

oxygen and must depend upon the breakdown of glucose to lactate for energy. The H_4 isozyme is likewise beneficial to the regulation of metabolism in heart muscle. This isozyme has a high K_M and a low V_{max} for pyruvate-to-lactate conversions and is inhibited by excess pyruvate. Heart muscle is primarily aerobic converting pyruvate to CO_2 and H_2O rather than to lactate. The lactic dehydrogenase enzyme is most active during emergency conditions when the oxygen supply is low.

Regulation of Enzyme Synthesis

Protein (and therefore enzyme) biosynthesis and gene expression are discussed in detail in Chapters 2, 4, 20, and 21. The control of an enzyme-catalyzed reaction through the regulation of the number of enzyme molecules available in the cell may be achieved at the various biochemical steps leading from transcription of the DNA to mRNA to translation of the mRNA into polypeptides. Two levels of control based on these steps are usually recognized; these are **transcriptional control mechanisms** and **translational control mechanisms.** Most of the present understanding of these control mechanisms stems from work with procaryotic organisms, for the regulation of gene expression in eucaryotic cells is much more complex.

Constitutive and Induced Enzymes

Work with microorganisms has indicated that enzymes fall into two categories with respect to their occurrence and number in cells. Those that appear to always be present and that occur in relatively constant concentrations are called **constitutive enzymes.** The enzymes of the glycolytic pathway in microbes are usually constitutive. The second type of enzyme may be found lacking in cells or be present only in small amounts, whereas upon introduction of a specific metabolite, usually a substrate, these enzymes quickly increase in concentration. Since their synthesis appears to be induced by the presence of the substrate, they are called **inducible enzymes.**

One of the first thoroughly studied inducible enzymes was β-*galactosidase*. Wild-type *E. coli* cells normally metabolize glucose and will metabolize only glucose even if lactose is also present. The enzymes for glucose metabolism are all constitutive and are thus present, while the enzyme needed to initiate lactose metabolism, β-galactosidase, is present in only minor amounts—according to one study, no more than five copies per cell. If wild-type *E. coli* cells are placed in a growth medium containing only lactose as the carbon source, they are at first unable to utilize this disaccharide. Soon, the cells respond by synthesizing β-galactosidase and the lactose is thus hydrolyzed to glucose and galactose and the resulting sugars metabolized by glycolysis. A number of β-galactosides besides lactose are able to act as inducers; these include methyl β-galactoside and allolactose. Actually, the application of any of these inducers initiates the synthesis of not one but three *enzymes* in *E. coli*: (1) β-*galactoside permease,* an enzyme formed in the plasma membrane that promotes the rate of transfer of β-galactosides across the membrane even against a concentration gradient; (2) β-*thiogalactoside acetyltransferase,* used in the metabolism of galactosides; and (3) β-galactosidase, the key enzyme for initiating lactose decomposition. When, as in this case, induction can be brought about by a single agent and result in the appearance of several enzymes, the process is known as **co-ordinate induction.**

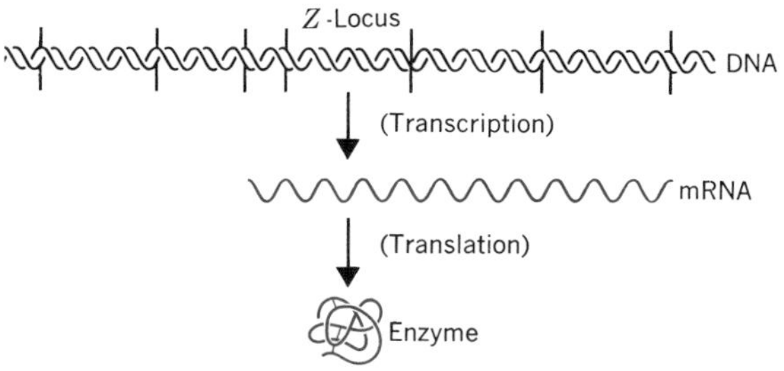

Enzyme Repression

The presence of a specific substance may inhibit the synthesis of an enzyme or sequence of enzymes in a metabolic pathway; this process is called **enzyme repression** (or in the case of the repression of a sequence of enzymes, **coordinate repression**). *E. coli* has all the enzyme systems necessary to synthesize the 20 amino acids from organic acids and NH_4^+, if no other nitrogen source is present. However, if one of the amino acids is introduced exogenously, the synthesis of the enzymes in the pathway leading to that amino acid will be inhibited, and the number of these enzymes quickly becomes reduced.

Catabolic Repression

Catabolic repression is a specific type of repression of enzyme synthesis in which a catabolite, usually glucose, functions to repress the formation of enzymes that would allow the decomposition of *other* substrates. For example, glucose represses the formation of β-galactosidase even when lactose (an inducer of this enzyme) is present. Glucose is even known to repress the formation of constitutive enzymes. A common phenomenon in microbes is the suppression of aerobic respiration and electron transport at *high* glucose concentrations, *even in the presence of ample oxygen*. Under these conditions, the cells utilize the glycolytic and fermentative pathways.

Repressors and Transcription

The relationships between induction and repression of enzymes in microbes were clarified by the studies of Monod and Jacob and their colleagues at the Pasteur Institute in France. These investigators showed that the mechanism of regulation is tied to the transcription of the DNA code into mRNA. Today, it is clear that portions of DNA are coded with specific information on the sequencing of amino acids to form a specific enzyme (or other protein). These segments of the DNA are termed the *structural genes* and designated the **z-locus.** Their encoded information is *transcribed* first into mRNA molecules, and these are then *translated* into polypeptide chains:

Z-Locus
DNA
(Transcription)
mRNA
(Translation)
Enzyme

Because they found mutants of *E. coli* that contained β-galactosidase even in the absence of an inducer (just as though β-galactosidase was a constitutive enzyme), Monod and Jacob concluded that a locus of the DNA other than the structural gene (i.e., the i-locus) must be responsible for inhibition in the absence of inducer. This **i-locus** is called the **regulatory gene.** When this gene undergoes mutation, it can no longer inhibit the structural gene, and therefore the structural gene is expressed as what appears to be a constitutive enzyme. Mutants of this kind are called **constitutive mutants.**

Pardee, Monod, and Jacob found that if the DNA containing a normal i-locus and z-locus is introduced into a cell containing a mutated i-locus, the cell behaves as a normal cell with inducible β-galactosidase. In other words, the normal regulatory gene (i-locus) regulated the structural genes (z-loci) of both normal and mutated DNA. Thus, they proposed the existence of a product of the regulatory gene, a **repressor** substance, which diffuses to other sites in the cell. The product of the regulatory gene, which has been found to be a small polypeptide, was postulated to diffuse to a site called the o-locus, which is next to or near the structural gene.

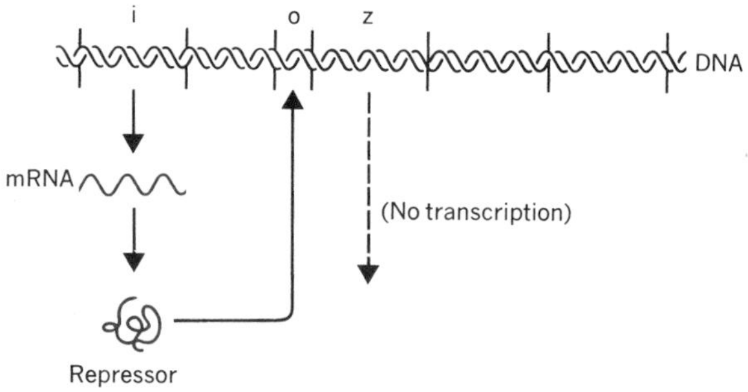

i
o
z
DNA
mRNA
(No transcription)
Repressor

The existence of this o-locus has been substantiated by the finding of mutants defective in DNA at this point. The o-locus or **operator** controls the transcription of the structural gene. In wild-type cells in the absence of an inducer, the regulatory gene is transcribed and translated into a repressor protein that diffuses to the operator, where it binds and inhibits the operator. With the operator locus inhibited, transcription of the structural gene does not occur. It is suggested that when an inducer such as lactose is added, it combines with the repressor protein to form a **repressor-inducer complex** that can no longer bind to and inhibit the operator locus:

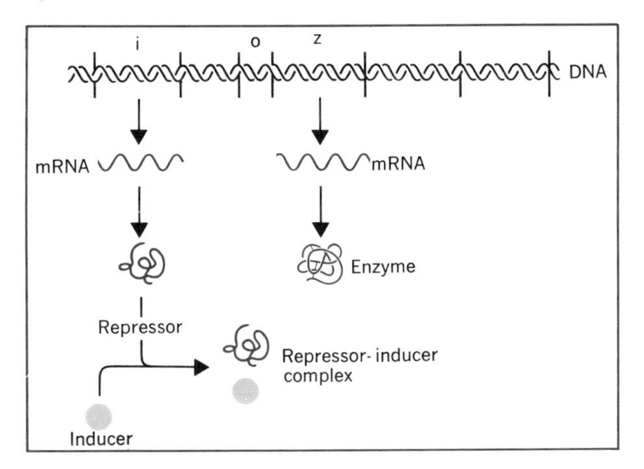

Binding of the inducer to the repressor is reversible so that when the inducer is removed the repressor is again free to bind to the operator locus.

The Operon

Jacob and Monod expanded their proposals for the regulation of enzyme synthesis to include coordinate induction. In the combined *Operon model* all enzymes induced by the same inducer have their structural genes (z, y, a) aligned sequentially along the DNA molecule following the operator locus.

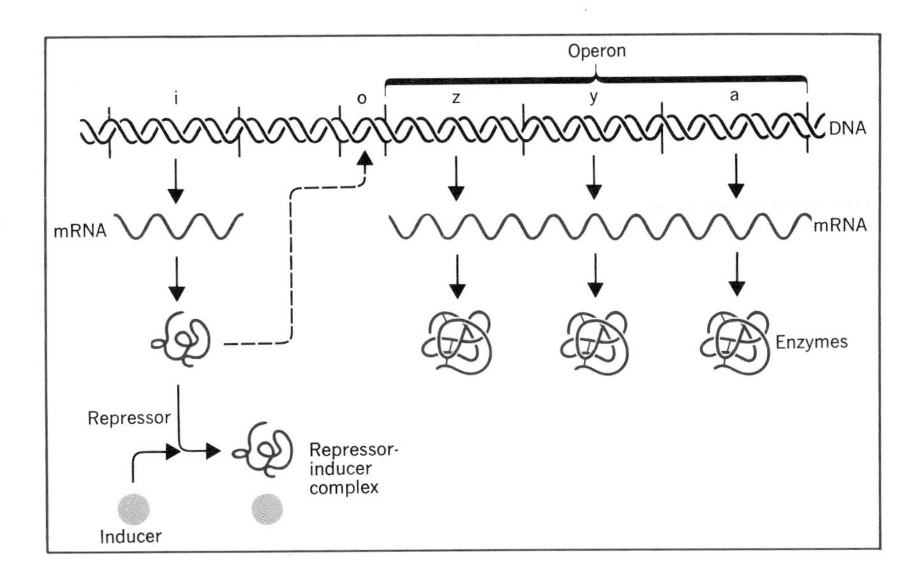

All of the structural genes in the operon would be controlled by the single operator, and the operator in turn would be affected by the single repressor protein formed from the regulatory gene (i-locus). The messenger RNA transcribed from the structural genes of the operon may code for several enzymes and is said to be polycistronic.

A variation of this model was also proposed by Monod and Jacob to account for enzyme repression. In this variation, it is proposed that the repressor protein is not effective alone but has to bind to a repressing metabolite **(corepressor)** to be active. This repressor-corepressor complex would in turn bind to the operator locus.

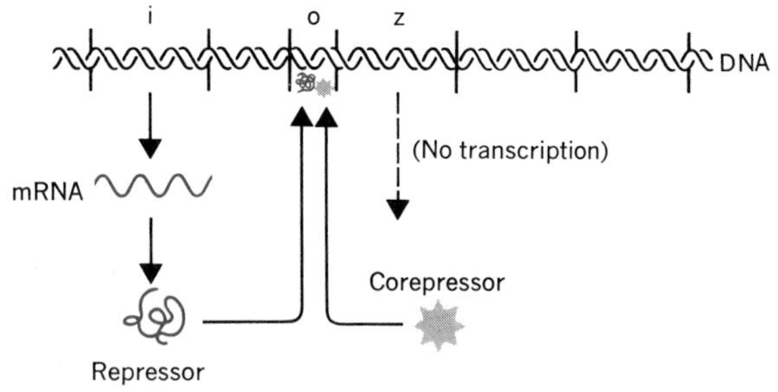

i o z
DNA
(No transcription)
mRNA
Corepressor
Repressor

Today a number of Operons are known in addition to the lactose or *lac operon* studied by Jacob, Monod, and their colleagues. Some of the better studied operons are listed in Table 11–3.

Subsequent Modifications of the Jacob-Monod Operon model

Another locus active in the Jacob-Monod model was found in the 1960s. This locus, called the **promotor locus,** is the point of recognition by DNA-directed RNA polymerase (Chapter 20) and is used to establish where transcription is to begin. This *p-locus* is very small but also contains the binding site of *cyclic AMP receptor protein (CRP)* or sometimes *catabolite gene-activator protein (CAP).*

Cyclic AMP (i.e., cAMP) is present only in low levels in bacteria. The level varies for some unknown reason according to the concentration of glucose; when glucose is present in high concentration, cAMP is low, but cAMP increases as glucose is depleted. Cyclic AMP first combines with CRP and then interacts with the promotor locus or possibly RNA polymerase. This event initiates transcription.

Table 11–3
Some More Common Operons in Procaryotes

Operon	Number of Enzymes	Function
lac	3	Hydrolysis and transport of galactose
his	10	Histidine synthesis
gal	3	Galactose ⟶ UDP-glucose
leu	4	Leucine synthesis
trp	4	Tryptophan synthesis
ara	4	Metabolism and transport of arabinose
pyr	5	Aspartate ⟶ UMP

For example, in the case of the lac operon, when the glucose level is high (even if lactose is present), cAMP is low, and therefore the cAMP-CRP complex is not available to bind to the promotor and start transcription. However, in the absence of glucose and in the presence of lactose, the repressor is bound by the lactose, and the cAMP is available to bind with CRP so that transcription is initiated. This interaction is shown in Figures 11–9 and 11–10.

Translational Control

Regulatory mechanisms usually function at the beginning of a sequence of reactions rather than at the end. Allosteric feedback control is effected by regulating the first enzyme in a pathway. Likewise, control of the number of enzyme molecules is most commonly brought about by transcriptional control. These regulatory mechanisms save the cell energy and prevent the needless buildup of products that will not be used. Therefore, *translational control* mechanisms are rare, and at the present time there is little evidence that this level is used for control. Some indirect evidence suggests the possible translational regulation of enzyme synthesis when the quantities in the cell of enzymes from the same operon are studied. For example, since the three structural genes in the lac operon are controlled by the same operator, one would expect that equal quantities of the three enzymes would be produced. However, many more copies of some enzymes are produced than the others. Presumably, the ribosome leaves the mRNA before completing the translation of *all* the messages in the polycistronic mRNA.

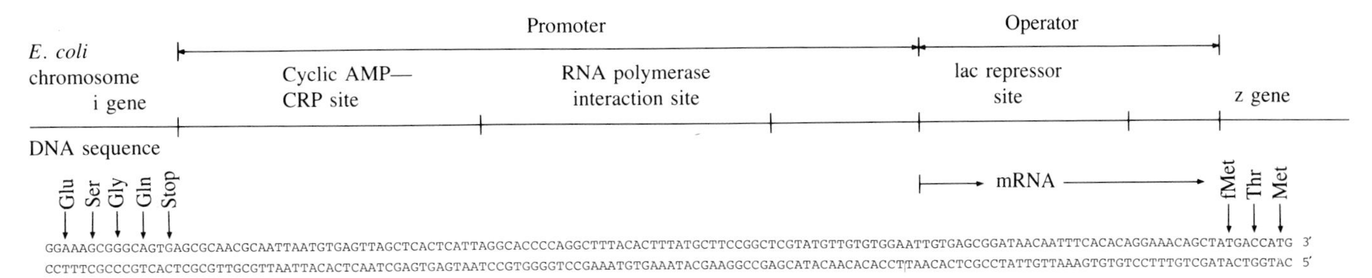

Figure 11–9
The molecular structure and relative location of the promotor locus. (Redrawn from *Science 187,* 32, 1975.)

The economical impact of control mechanisms is reflected also in the synthesis of ribosomal RNA. When *E. coli* cells are grown in a medium deficient in amino acids, the cells are unable to synthesize proteins. As a consequence, the cells also stop making ribosomal RNA and ribosomes, a mechanism called **stringent control.** Mutants do exist, however, that maintain an increased rate of rRNA synthesis; these are called **relaxed mutants.** Paper chromatographic analysis of nucleotide extracts of the stringent and relaxed mutants reveal two spots in the chromatograms of stringent cells and none for the relaxed mutants. Initially these spots were called ''magic spot I'' and ''magic spot II,'' but subsequently they were identified as guanosine-5′-diphosphate-2′-(or 3′)-diphosphate (ppGpp) and guanosine-5′-triphosphate-2′-(or 3′)-diphosphate (pppGpp), respectively. The studies of Cashel, Gallant, and their colleagues indicate that these nucleotides are used as messengers to turn off ribosomal RNA synthesis when amino acids for protein synthesis are lacking.

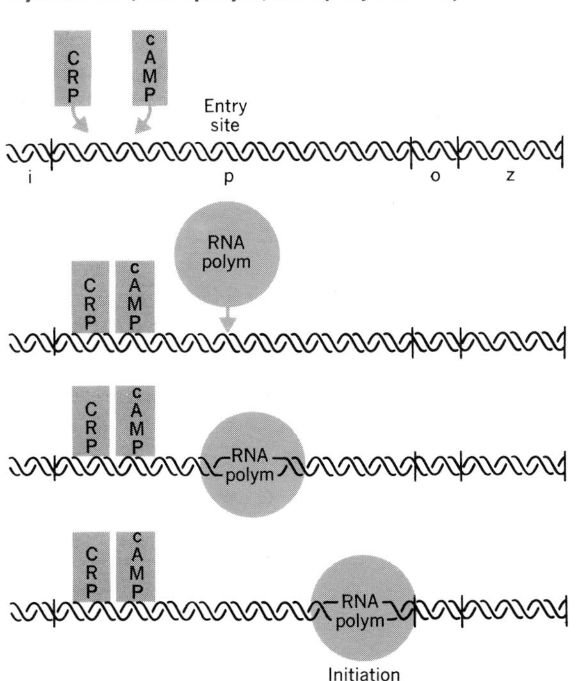

Figure 11–10
Model of the transcription of the *Lac* operon. (See text for description) (CRP, cyclic AMP receptor protein; cAMP, cyclic AMP; RNA polym, RNA polymerase).

Regulation of Enzyme Production in Eucaryotic Cells

Eucaryotic cells are much more complex than procaryotic cells. Eucaryotic cells usually contain several chromosomes instead of the single circular DNA molecule of procaryotes. At times, eucaryotic cell chromosomes are diploid, triploid, or even polyploid. The DNA in the chromosomes may be highly folded or compacted and the chromosomes physically separated from the cytoplasmic ribosomes by the perforated nuclear envelope. These factors undoubtedly make control of enzyme synthesis much more complex, even though there is little doubt that transcription and translation occur by basically the same mechanism as in procaryotes. However, groups of enzymes that would form an inducible component such as an operon are not found adjacent to one another on a chromosome and may, in fact, be distributed among different chromosomes.

Enzyme induction has been observed in eucaryotic cells of more primitive organisms such as yeast and *Neurospora* as well as vertebrates. The induction process is slower, and the change in concentration of enzymes is not as great. In yeast, the induction of β-galactosidase takes several minutes rather than seconds as occurs in *E. coli;* also, the increase in activity is 10-fold rather than 1000-fold. The enzyme *tryptophan 2,3-oxygenase* can be induced in the liver cells of vertebrates but requires many hours.

Enzyme Induction by Hormones

Enzyme induction in procaryotic cells is usually brought about by a potential metabolite (such as the induction of β-galactosidase by lactose). This form of induction occurs in eucaryotic cells as well; however, higher animals also have a highly developed control system initiated by *hormones*. Active in the system of hormone-induced enzymes are the sex hormones (androgens, estrogens, and progesterone), the glucocorticoids (adrenocortical steroids regulating metabolism), and thyroxin.

Estrogen acting on chick oviduct cells has been shown to bind to specific receptor proteins in the cytoplasm. This hormone-receptor complex then becomes incorporated in the chromatin, where presumably it acts as an inducer. Increased rates of transcription of specific genes occur, leading to higher rates of synthesis of ovalbumin and lysozyme.

The glucocorticoids have multiple effects on most animal cells. A synthetic analogue, dexamethasone, along with the naturally occurring hormones, has been found to promote adaptation to stress, promote development of various organs, and induce enzymes in the liver that are involved in gluconeogenesis. Thyroxin has also been shown to be capable of enzyme induction.

Repression in Eucaryotes

In higher animals and plants the differentiation of cells and tissues can be very extensive, especially when compared with the simplicity of development of a procaryotic cell. Interestingly, it has been shown that most differentiated cells of higher plants and animals contain *complete* genomes. In effect, the totipotency indicates that large segments of the genome must be repressed. In some cases the repression is reversible. For example, differentiated carrot root cells can be isolated and grown in tissue culture to produce a new set of differentiated cells including vascular tissue, storage tissue, epidermal tissue and so forth. However, in many differentiated tissues such as brain and muscle, the portion of the genome that would allow dedifferentiation or differentiation into other tissue types is permanently repressed.

The mechanism of eucaryotic repression is not yet understood. For many years, other components of the chromatin (Table 11–4) have been proposed as potential repressors, but their properties do not lend themselves to a full explanation. The histones have been studied in this context since the 1940s when Stedman and Stedman first proposed that they might act as gene repressors. However, there are only five major classes of histones in eucaryotic cells, and they occur in about equal amounts, with little variation between tissues of an organism or between species. It would appear that histone function is rather basic and the same among all organisms; if histones do act as repressors, then their similarities are difficult to reconcile with the fact that eucaryotic cells have 40,000 to 100,000 different genes, most of which are repressed in different combinations in different tissues. Yet, the histones do change in significant ways during the cell cycle (Chapter 2), and their potential regulatory role is still actively being studied.

Electrophoretic analysis indicates that the nonhistone proteins present in chromatin occur in much greater variety than the histones. However, their diversity is not sufficient to support a contention that they play a role in repression.

Model of Gene Regulation in Eucaryotes

Although several models have been proposed to explain gene regulation in eucaryotes, none has been substantiated with evidence that would give them the certainty which

Table 11–4
Relative Concentrations of Calf-Thymus Chromatin Components

DNA	100
Histone	114
Nonhistone protein	33
RNA	7

surrounds the Jacob-Monod operon model for procaryotic cells. One model for eucaryotes has attracted attention and does explain a number of observations about the eucaryotic nucleus, one of which is the finding that it contains a number of *heterogeneous nuclear RNA (hnRNA)* molecules. These molecules are much longer than mRNA, are not transported out of the nucleus, and have a high turn-over rate. The model was proposed by R. Britten and E. Davidson. It suggests that the various DNA strands in the eucaryotic nucleus contain a number of *sensor sites* that can recognize various agents such as metabolic inducers (substrates), hormone-receptor inducers, or regulatory nucleotides (ppGpp). When the sensor binds the inducer, it causes an adjacent *integrator gene* to transcribe a specific *activator RNA* molecule. The activator RNA can attach to appropriate *receptor sites* on the same or different chromosomes. The binding of the activator RNA to a receptor site initiates transcription of the adjacent structural gene into mRNA, which then diffuses out of the nucleus to cytoplasmic ribosomes where translation into protein occurs.

The model has a number of more elaborate modifications (Fig. 11–11) that help to explain the many variations in gene expression and differentiation in eucaryotes. As diagrammed in the upper portion of Figure 11–11, a number of different sensors (S_1, S_2, S_3) upon binding different inducers (I_1, I_2, I_3) cause their companion integrators to form different activator RNA molecules (a, b, c). The presence of *multiple* receptor sites for each structural gene would imply that different combinations of structural genes would be available for transcription, depending upon binding of the various activator RNAs to their respective receptor sites.

In the model shown in the upper half of Figure 11–11, each structural gene has two or more receptor sites, and the activation of any one would trigger transcription. An alternative proposal for transcription of structural genes in various combinations by single inducers is shown in the lower half of the figure. In this variation, the sensors initiate activator RNA synthesis in a number of adjacent integrators. Each activator RNA then associates with *one* receptor. The models have a number of interesting possibilities and also help to explain the observation that large portions of the DNA (40% in calf-thymus cells) are made up of repeating nucleotide sequences too small to be structural genes.

Compartmentalization

A final regulatory mechanism that is most evident in eucaryotes is the physical separation of groups of enzymes by cellular membrane boundaries. Selected groups of enzymes are *compartmentalized* in organelles. For example, the enzymes of the tricarboxylic acid cycle are physically separated from those of glycolysis by their confinement within the mitochondria. The enzymes of the ''dark reactions'' of photosynthesis (which function in basically the same manner as many of those of glycolysis) are physically isolated in the stroma of the chloroplast and are not associated with the enzymes of glycolysis that occur in the cytosol.

Many of these ''isolated'' enzyme sequences use substrates and/or cofactors produced by enzymes confined in other parts of the cell. Regulation of the transport of these compounds across the membranes from one cell compartment to another affords yet another level of control of metabolism.

Summary

The direction and rate of metabolic pathways and individual enzymatic reactions may be controlled by one or more **regulatory mechanisms:**

1. **Regulation by mass action:** The rate and direction of

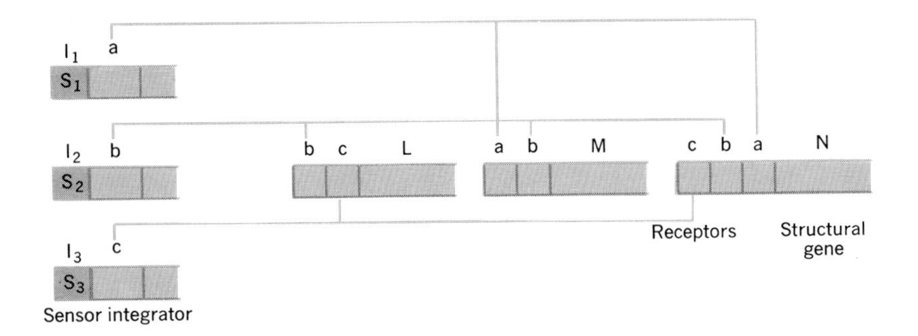

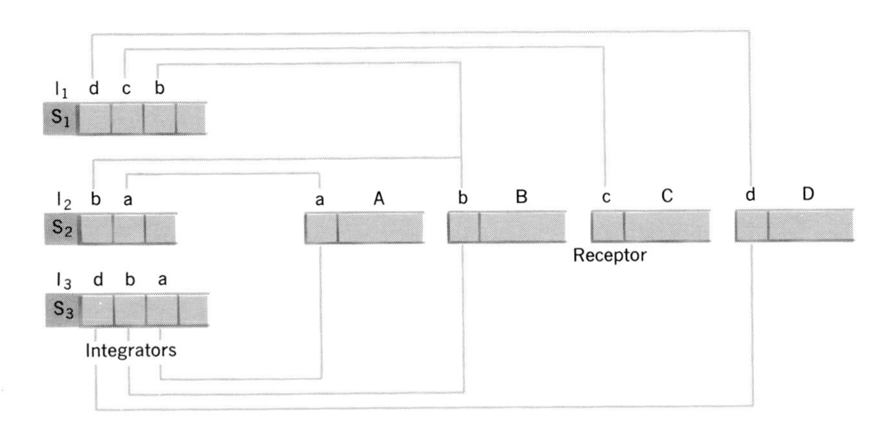

Figure 11–11
The Britton-Davidson model of gene regulation in eucaryotes (see text for description).

a reaction is changed by addition of substrates or removal of products.

2. Enzyme activity: The rate is altered by inhibition or activation of an enzyme.

3. Allosteric effectors: Metabolites that attach to enzymes at a site other than that occupied by the substrate may stimulate or inhibit activity.

4. Bond modification: Zymogens are activated by altering covalent bonds.

5. Isozymes: Metabolic regulation by means of multiple forms of an enzyme. Different forms have different activities under varying conditions.

6. Enzyme induction: Regulation at transcriptional and translational levels of enzyme synthesis.

(a) Transcriptional: Control through a repressor-inducer complex affecting an operator gene segment of DNA and controlling the expression of structural genes **(operon theory).**

(b) Translational: Rare, possibly a mechanism for controlling amounts of different enzymes regulated by the same operon.

7. Hormones: Hormone-stimulated induction of enzymes.

8. Compartmentalization: Isolation of enzymes and therefore metabolic pathways in organelles. Changes in permeability of membranes may separate substrates from enzymes.

References and Suggested Reading

Articles and Reviews

Maniatis, T., and Ptashne, M., A DNA operator-repressor system. *Sci. Am. 234*(1), 64 (June 1976).

O'Malley, B. W., and Schrader, W. T., The receptors of steroid hormones. *Sci. Am. 234*(4), 32 (February 1976).

Books, Monographs, and Symposia

Atkinson, D. E., *Cellular Energy Metabolism and Its Regulation,* Academic Press, New York, 1977.

Conn, E. E., and Stumpf, P. K., *Outline of Biochemistry,* John Wiley & Sons, Inc., New York, 1972.

Larner, J., *Intermediary Metabolism and Its Regulation,* Prentice-Hall, Inc., Englewood Cliffs, N.J., 1971.

Lehninger, A., *Biochemistry,* Worth Publishers, Inc., New York, 1975.

Vogel, H. J. (editor), *Metabolic Pathways,* vol. 5, *Metabolic Regulation,* Academic Press, Inc., New York, 1971.

Whelan, W. J., *Biochemistry of Carbohydrates,* Butterworth and Co. Ltd, London, 1975.

Part 4
TOOLS AND METHODS OF CELL BIOLOGY

Chapter 12
FRACTIONATION OF TISSUES AND CELLS

Much of our current knowledge concerning the structure, chemical composition, and function of the various cell organelles has been obtained following the isolation of these components from cells using various **fractionation** procedures. Such studies involve three major phases. In the first, the tissue or cell suspension is *disrupted* in order to release the cell components. The second phase involves the *sorting* of the cell components into *fractions,* such that the members of any single fraction are the same but differ from the members of any other fraction. The final phase consists of an *examination* and *analysis* of the separated cell fractions.

Methods for Disrupting Tissues and Cells

Various methods have been devised for disrupting tissues and suspensions of cells, but the method of choice is usually the procedure that causes *minimal damage* to the cell constituents. Most physical procedures are based on the effects of *shearing* forces, and because the separated parts of the cells undergo rapid degradation at room temperature, these procedures are usually carried out at low temperature and in cold buffer solution. Among the older methods used is the grinding of the sample with a *mortar* and *pestle,* often with the aid of abrasive materials such as sand, alumina, or ground glass. This procedure has several disadvantages, including the loss of some cell constituents by adsorption onto the abrasive and the necessity of removing the abrasive material either before or during the fractionation procedure.

Shearing forces adequate to disrupt most cells and tissues may also be obtained using a blender (such as a Waring Blender or an Omni-Mixer) in which steel blades rapidly rotate through the cell or tissue suspension. The product of this and similar techniques is called an **homogenate.** Tissues may also be homogenized by placing them, along with cold buffer solution, in a cylindrical glass tube fitted with a glass or Teflon plunger. As the plunger is driven down the tube (generally by hand), the tissue and buffer are forced upward through the narrow space between the wall of the tube and the plunger. The shearing forces so generated are usually sufficient to disperse the tissue after several up-and-down strokes. This is the method of choice for the disruption of soft tissues, such as liver, brain, and kidney.

One of the more rigorous methods for disrupting cells is by the use of a *pressure cell,* which consists of a steel cylinder and close-fitting steel piston. The piston is pushed into one end of the cylinder using a press or hydraulic jack, and the sample is forced out of the cylinder through a narrow opening at the other end. The size of this opening may be accurately controlled by a needle valve. Bacteria and other microorganisms enclosed within a tough cell wall are frequently disrupted using this approach.

Cells may also be disrupted by *insonation* (ultrasound) using a sonifier. In this procedure, the probe of the sonifer is immersed in the cell suspension and caused to vibrate in the fluid (usually at about 20,000 cycles per second). These ultrasonic vibrations produce a number of effects in the fluid that collectively cause the disruption of the cells. Shock waves (alternate compressions and rarefactions) arising from the tip of the probe create turbulent flow of the fluid in which the cells are suspended and may disrupt the cells. The shock waves also cause *cavitation* of the fluid; that is, microscopic bubbles are formed in the fluid near the tip of the probe, and these rapidly stream away from the probe along with some of the fluid. The friction created between this stream and the suspended cells also contributes to their disruption. Some of these bubbles disintegrate into smaller (microscopic) bubbles that travel away from the probe like miniature projectiles; when these impinge upon the cells, the shear force created may disrupt them.

Cells can also be disrupted by chemical means. For example, enzymes that specifically degrade the components of the cell wall or cell membrane may be added to the tissue or cell suspension. Alternately, proteolytic or lipolytic agents that dissolve the membrane may be used. Some cells are sufficiently fragile that they are readily disrupted by successive freezing and thawing. Erythrocytes (red blood cells) and certain other cells may be broken by the *osmotic pressure* created within them when they are placed in distilled water or hypotonic solution.

After cells have been disrupted, the goal is to separate and isolate the structures that have been released. Since cellular organelles and other constituents vary in size, shape, and density, they settle through the liquid in which they are suspended at different rates. Consequently, disrupted cells are more often fractionated by some form of **centrifugation** than by any other method. Indeed, centrifugation has become one of the most widely employed procedures in cellular research and one of the most important tools of the cell biologist. In the following section, we consider the principles and applications of centrifugation.

Centrifugation

Theory of Centrifugation

If a container is filled with a liquid suspension of particles of varying size and density, the particles will gradually settle to the bottom of the container under the influence of gravity. The rate at which settling occurs can be greatly increased by increasing the gravitational effect upon the particles. This is the underlying principle for isolating large or dense particles by centrifugation. A tube containing the suspension of particles (e.g., a tissue homogenate) is placed in the rotor of a centrifuge and then is rotated at high speed. The resulting acceleration greatly increases the gravitational pull on the suspended particles, causing their more rapid sedimentation to the bottom of the tube along paths that are *perpendicular* to the axis of rotation (i.e., along radii of the circle being swept out by the rotating tube). This relationship is shown in Figure 12–1.

If a force F is applied to a particle of mass m, the particle will be accelerated in a linear direction such that

$$F = (m)(a) \tag{12–1}$$

where a is the rate of linear acceleration. However, for *angular* acceleration (as occurs during centrifugation)

$$a = \frac{v^2}{x} \tag{12–2}$$

where v is the velocity of rotation (i.e., the circumferential distance traversed per unit of time) and x is the radius of

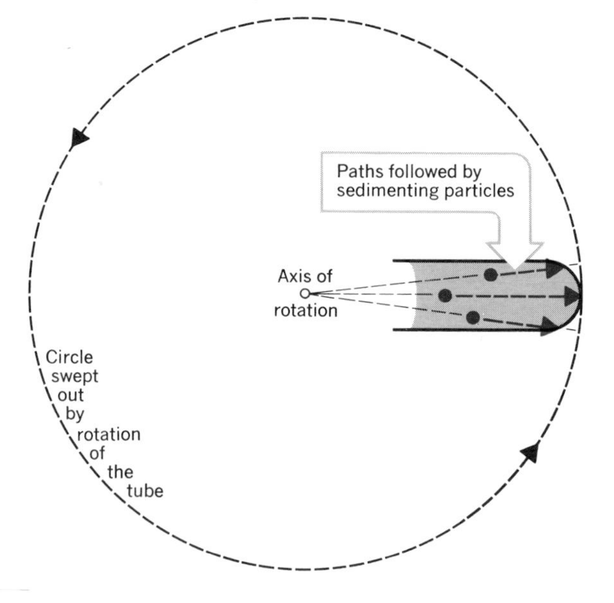

Figure 12–1
Paths followed by sedimenting particles during centrifugation. The tube is being viewed from a point above the horizontal plane of rotation.

the circle being swept out. Since for circular motion

$$v = (\omega)(x) \qquad (12\text{–}3)$$

where ω (omega) is the angular velocity in radians per second (one unit of circumference contains 2π radians), then

$$F = (m)(a) = m\frac{(v^2)}{x} = m\frac{[(\omega)(x)]^2}{x} = m\omega^2x \qquad (12\text{–}4)$$

Usually, the value given for the force applied to particles during centrifugation is a relative one; that is, it is compared with the force that the earth's gravitational pull would have on the same particles. It is called **relative centrifugal force** or RCF. The force of the earth's gravity upon a mass m is given by

$$F = mg \qquad (12\text{–}5)$$

where g is acceleration due to gravity and equals 980 cm/sec. Therefore,

$$\text{RCF} = \frac{F_{\text{centrifugation}}}{F_{\text{gravity}}} = \frac{m\omega^2x}{mg} = \frac{\omega^2x}{g} \qquad (12\text{–}6)$$

Since g is a constant, then

$$\frac{\omega^2}{g} = \frac{\left[\frac{(2\pi)(\text{rpm})}{60}\right]^2}{980} \qquad (12\text{–}7)$$

In equation 12–7, the term $(2\pi/60)^2/980$ is also a constant and is equal to 1.119×10^{-5}. Thus the equation for relative centrifugal force simplifies to

$$\text{RCF} = 1.119 \times 10^{-5}(\text{rpm})^2(x) \qquad (12\text{–}8)$$

Consequently, in order to determine the relative centrifugal force in effect during centrifugation it is necessary to measure the revolutions per minute (rpm) and the distance between the sample and the axis of rotation. Since RCF is the ratio of two forces, it has no units. However, it is customary to follow the numerical value of the RCF with the symbol g. This indicates that the RCF being applied is some multiple of the earth's gravitational force.

Sample Problem. What relative centrifugal force is applied when red blood cells are sedimented at 1000 rpm in a rotor of *maximum radius* equal to 10 cm?

$$\text{RCF}_{\text{max}} = 1.119 \times 10^{-5}(1000)^2(10)$$
$$= 1.119 \times 10^2 \text{ or } 112\ g$$

The sedimentation of particles by centrifugation is, in effect, a method for concentrating them; therefore, one of the major physical forces opposing such concentration is **diffusion.** In the case of the sedimentation of cells or subcellular particles such as nuclei or mitochondria, the effects of diffusion are essentially nil. However, when centrifugation is employed for the sedimentation of much smaller particles (such as cellular proteins, nucleic acids or polysaccharides), the effects of diffusion become significant.

Sedimentation Rate and Coefficient

The RCF or "g force" applied to particles during centrifugation may readily be calculated using equation 12–8 and is independent of the physical properties of the particles being sedimented. However, a particle's **sedimentation rate** at a specified RCF depends upon the properties of the particle itself. Also, since the RCF varies directly with x, it is clear that the sedimentation rate *changes* with changing distance from the axis of rotation. (For particles settling under the influence of the earth's gravity alone, the sedimentation rate becomes constant.) The instantaneous sedimentation rate of a particle during centrifugation is determined by three forces: (1) F_C (i.e., the centrifugal force), (2) F_B, the **bouyant force** of the medium, and (3) F_f, the frictional resistance to the particle's movement. For a spherical particle, P of volume V and density ρ_P sedimenting through a liquid medium of density ρ_M,

$$F_C = m_P\omega^2x \qquad (12\text{–}9)$$

The bouyant force is given by

$$F_B = (m_P/\rho_P)\rho_M\omega^2x \qquad (12\text{–}10)$$

where (m_P/ρ_P) is the volume of the particle and (m_P/ρ_P) (ρ_M) is the mass of the liquid medium displaced by the particle. The frictional force resisting the sedimentation of a sphere through a liquid at 1 cm/sec is given by **Stokes law,** according to which

$$f = 6\pi\eta r \tag{12–11}$$

In this equation, η is the viscosity of the medium and r is the radius of the spherical particle. For a particle sedimenting at a rate other than 1 cm/sec, the frictional resistance would be

$$F_f = f(dx/dt) = 6\pi\eta\, r(dx/dt), \tag{12–12}$$

where dx/dt is the instantaneous sedimentation rate.

If the sedimentation of a particle during centrifugation is viewed as a series of tiny incremental sedimentation rate increases, then the sedimentation rate *between* increases results from the balance of F_C, F_B, and F_f. That is, at any instant the sedimentation rate results from the fact that

$$F_C = F_B + F_f$$

Accordingly,

$$m_P\omega^2 x = [(m_P/\rho_P)\rho_M\omega^2 x] + [6\pi\eta r(dx/dt)] \tag{12–14}$$

Substituting the volume of a sphere (i.e., $\frac{4}{3}\pi r^3$) times its density for its mass in equation 12–14, we obtain

$$(\tfrac{4}{3}\pi r^3)(\rho_P)(\omega^2 x) = (\tfrac{4}{3}\pi r^3)(\rho_M)\omega^2 x + 6\pi\eta r(dx/dt) \tag{12–15}$$

By factoring and transposing, we then obtain

$$\tfrac{4}{3}\pi r^3(\rho_P\text{-}\rho_M)\omega^2 x = 6\pi\eta r(dx/dt) \tag{12–16}$$

Now, the **sedimentation coefficient,** s, is the rate at which a particle sediments under conditions of unit acceleration (i.e., when $\omega^2 x = 1$). For conditions other than unit acceleration, s is therefore given by

$$s = (dx/dt)/\omega^2 x \tag{12–17}$$

By transposing terms in equation 12–16 and solving for s as defined in equation 12–17, we obtain

$$s = \frac{\frac{4}{3}\pi r^3\,(\rho_P\text{-}\rho_M)}{6\,\pi\,\eta\,r} = \frac{2r^2\,(\rho_P\text{-}\rho_M)}{9\,\eta} \tag{12–18}$$

Equation 12–18 is very important to the understanding of particle sedimentation and should be carefully examined, for the equation shows that those properties of a particle that determine its rate of sedimentation during centrifugation are **radius** and **effective density** (i.e., the difference between the density of the particle and the density of the liquid through which the particle is sedimenting). Since sedimentation is a function of the square of the particle's radius but only a first-order function of the particle's density, it is clear that particle *size* is "more important" than particle density in determining sedimentation rate.

From equation 12–18, it may also be noted that two particles having different sizes or different densities may have similar sedimentation coefficients. On the other hand, two particles with either similar size or similar density may have different sedimentation coefficients.

The Analytical Ultracentrifuge

The sedimentation coefficient of a particle may be experimentally determined in an instrument known as an **analytical ultracentrifuge.** This instrument is equipped with an optical system (called Schlieren optics) that permits visual observation of the sedimenting particles as one or more moving boundaries formed between regions of the particle suspension having different refractive indexes. The measured *rate* at which these boundaries move under specified conditions is used to determine the sedimentation coefficient, and the *number* of boundaries formed is an index of the heterogeneity of the sample.

The first analytical ultracentrifuges were designed and built by the Nobel Prize-winning biochemist T. Svedberg in the 1920s. Following Svedberg's pioneering work, the analytical ultracentrifuge was advanced to its present status as one of the more important tools of the molecular biologist principally through contributions of E. G. Pickels and J. W. Beams.

The *rotor* that spins in the ultracentrifuge typically contains two compartments. Into one compartment is placed a "reference cell" containing the sample-free solvent, while the other receives the cell containing the sample to be analyzed. The interior of each cell is sector shaped and bounded above and below by parallel quartz windows to permit light from below the rotor to pass through the reference and sample during rotation (Fig. 12–2). As the particles sediment, boundaries are formed at the trailing edges of each particulate species. When these boundaries pass in front of the optical system, the resulting change in refractive index

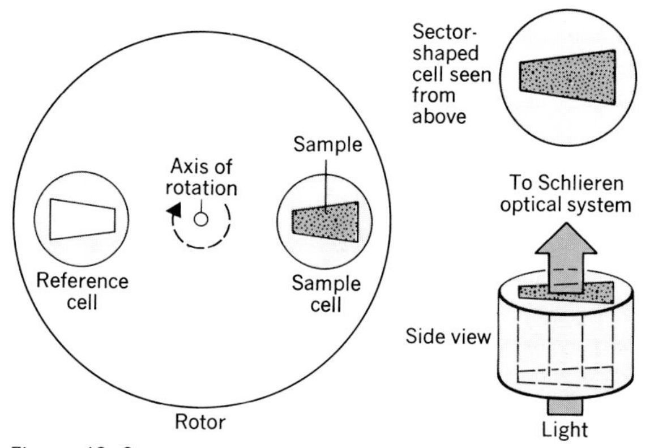

Figure 12–2
The analytical ultracentrifuge rotor and sector-shaped reference and sample cells.

is measured and recorded. Changes in the distance between each boundary and the axis of rotation are measured as a function of time. These measurements are then used to calculate the respective sedimentation coefficients of the particles in the suspension.

The sedimentation coefficients of many cellular macromolecules such as proteins, polysaccharides, and nucleic acids fall in the range 1×10^{-13} seconds to 200×10^{-13} seconds (i.e., the dimensions of s are seconds). For convenience, a unit called the *Svedberg unit* (after T. Svedberg) and abbreviated S is used to describe sedimentation coefficients and is equal to the constant 10^{-13} seconds. Thus, most cellular proteins have sedimentation coefficients between 1 and 200 S. The sedimentation coefficients of a number of cell constituents are listed in Table 12–1.

Calculation of the sedimentation coefficient from data collected with the analytical ultracentrifuge involves the following amplification of the relationship of equation 12–17. If a boundary is x_1 centimeters from the axis of rotation at time t_1 and x_2 centimeters at time t_2, then equation 12–17 may be solved by integration as follows. First, by transposition,

$$s\,(dt) = \frac{1}{\omega^2}\,(dx/x) \qquad (12\text{–}19)$$

Integrating between the limits set above, we obtain

$$s \int_{t_1}^{t_2} dt = \frac{1}{\omega^2} \int_{x_1}^{x_2} dx/x \qquad (12\text{–}20)$$

and

$$s\,(t_2 - t_1) = \frac{1}{\omega^2}\,(\ln x_2 - \ln x_1) = \frac{1}{\omega^2}\left(\ln \frac{x_2}{x_1}\right) \qquad (12\text{–}21)$$

Therefore,

$$s = \frac{1}{\omega^2\,(t_2 - t_1)} \ln \frac{x_2}{x_1} \qquad (12\text{–}22)$$

Sample Problem. A solution containing a single particulate species is accelerated in the analytical ultracentrifuge at 60,000 rpm. At time t_1, the boundary between the trailing edge of the sedimenting particles and the axis of rotation is 6.0 cm, while at t_2, 60 minutes later, the boundary is 6.8 cm from the axis. What is the sedimentation coefficient of the particles?

$$\omega = \frac{60{,}000\,(2\pi)}{60} \text{ radians per second}$$

$$= 6{,}280 \text{ radians per second}$$

$$s = \frac{1}{(6{,}280)^2(60)\,(60)}\,2.3 \log_{10}(6.8/6.0)$$

$$\frac{1}{1.42 \times 10^{11}}\,2.3\,(0.053)$$

$$= 8.6 \times 10^{-13} \text{ or } 8.6\,S$$

As its name implies, the analytical ultracentrifuge is an analytical instrument and does not physically separate the

Table 12–1
Some Representative Sedimentation Coefficients

Particle	Sedimentation Coefficient *(s)*
Proteins	
Cytochrome *c*	1.7
Hemoglobin	4.1
Fibrinogen	7.6
Hemocyanin	59
Cell organelles	
Nucleosomes	11
Ribosomes	70 to 80
Membrane fragments	10^2 to 10^4
Plasma membranes	up to 10^5
Lysosomes	4×10^3 to 2×10^4
Peroxisomes	4×10^3
Mitochondria	1×10^4 to 7×10^4
Chloroplasts	10^5 to 10^6
Nuclei	10^6 to 10^7

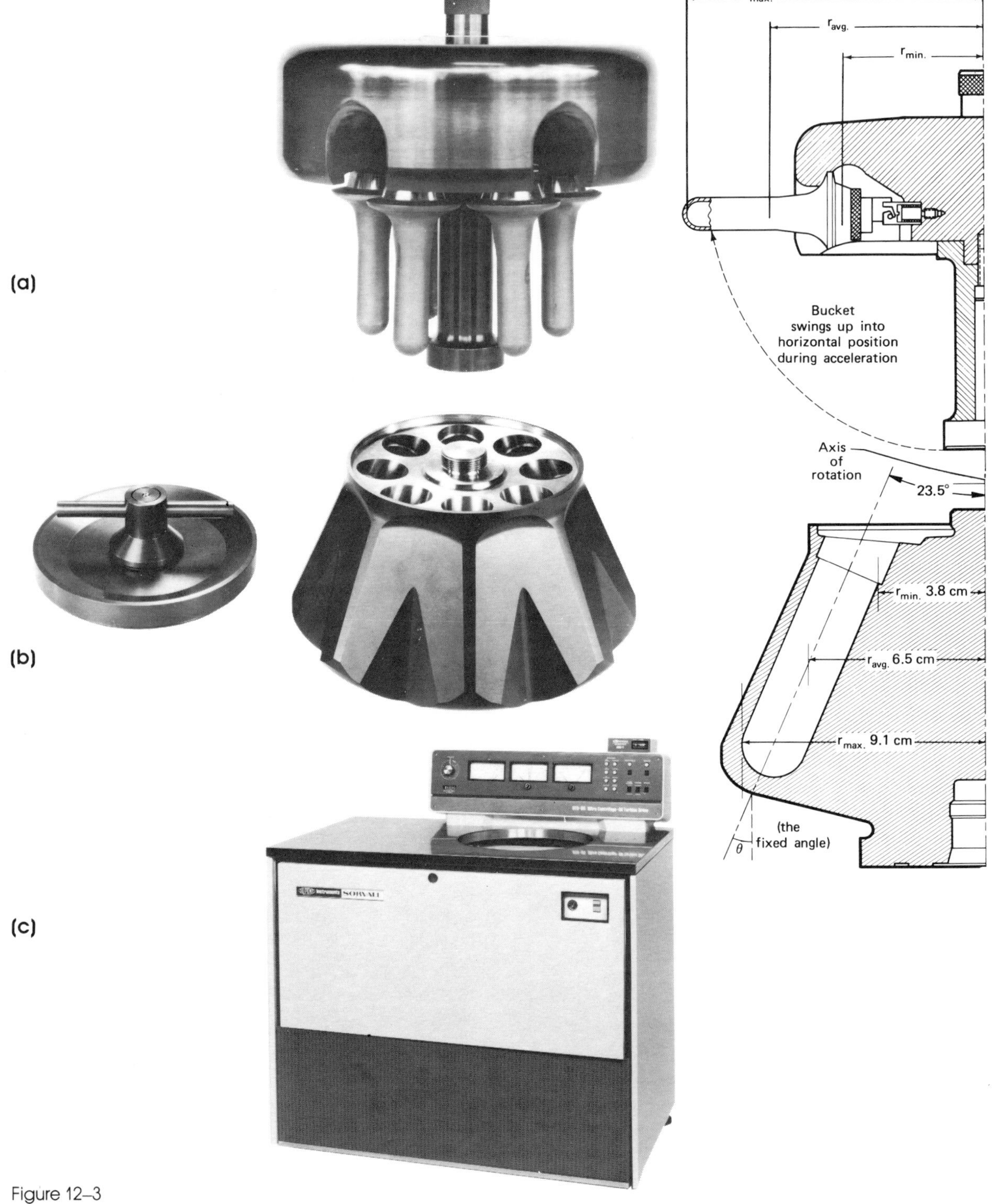

Figure 12–3
Swinging-bucket *(a)* and fixed angle *(b)* rotors. *(c)* A modern, oil-turbine driven ultracentrifuge. (Photos courtesy of DuPont Instruments.)

multiple components of a mixture from one another (except for the component that sediments most slowly). Furthermore, the amount of material that may be studied is quite limited as a result of the small size of the sample cell. The counterpart to analytical centrifugation is **preparative centrifugation,** which provides for the isolation of cell components for further analysis. Two basic types of centrifuge rotors are regularly employed for conventional preparative centrifugation; these are the *swinging-bucket* rotor and the *fixed angle* rotor (Fig. 12–3).

The swinging-bucket rotor consists of a series of metal buckets (usually three to six) in high-speed rotors (but many more in low-speed rotors) attached to the central *harness* of the rotor. The samples to be centrifuged (previously placed in the appropriate centrifuge tubes) are inserted into the buckets which swing upward from a vertical position to a horizontal position during acceleration of the rotor (Fig. 12–3). During deceleration, the buckets return to the vertical position and the centrifuge tubes are removed. In a fixed angle rotor, the centrifuge tubes are maintained at a constant angle (usually 15° to 45° from verticality) throughout the entire centrifugation (Fig. 12–3). As a result, the radially sedimenting particles quickly strike the tube wall, where convection currents rapidly carry them to the bottom of the tube.

Differential Centrifugation

During centrifugation, particles sediment through the medium in which they are suspended at rates related to their size, shape, and density. Differences in the sedimentation coefficients of the various subcellular particles provide the means for their effective separation. **Differential centrifugation,** a technique introduced to cellular research in the early 1940s by the noted biologist Albert Claude, is one of the classical procedures for isolating subcellular particles and involves the stepwise removal of classes of particles at increasing RCF.

The material to be fractionated is subjected first to low-speed centrifugation in order to sediment the largest (or densest) particles present. Following this, the unsedimented material (called the **supernatant**) is centrifuged at a higher speed to sediment particles of somewhat smaller size (and/or lower density). The sequence is repeated several times until all particles have been sedimented and the sediments then used for further experimentation and analysis. The procedure regularly employed for the differential fractionation of liver tissue may serve as a convenient example of this method and is shown diagrammatically in Figure 12–4.

The removed liver tissue is homogenized in cold buffer and centrifuged for 10 minutes at 700 g. This is usually sufficient to sediment all the cell nuclei to the bottom of the centrifuge tube, thereby providing the **nuclear fraction.** Depending upon the effectiveness of the homogenization procedure, some unbroken cells and large cell fragments may also be recovered in this fraction. The overlying supernatant (called the *nuclear supernatant*) is removed and transferred to another tube for a second centrifugation at 20,000 g for 15 minutes. This sediments nearly all the mitochondria (i.e., the *mitochondrial fraction*). Again, the supernatant (i.e., *mitochondrial supernatant)* is removed and is subjected to a third centrifugation at 105,000 g for 60 minutes. This causes the sedimentation of a fraction called *microsomes,* which includes ribosomes and small fragments of intracellular membranes. The *microsomal supernatant* is referred to as the *soluble phase* of the cells, or *cytosol,* and includes soluble proteins, soluble nucleic acids, soluble polysaccharides, lipid droplets, and other small particles. In the procedure just described the liver tissue is separated into four major fractions.

Differential centrifugation has several major disadvantages. Since the homogenate is initially distributed uniformly throughout the centrifuge tube, the first particles sedimented will necessarily be contaminated with all other constituents of the homogenate. This effect is shown in Figure 12–5. As the smaller particles are sedimented, they in turn will be contaminated by even smaller particles. In fact, the only particles to be obtained in relatively pure form will be those that sediment most slowly. In the example given above, the initial nuclear fraction would contain some mitochondria, microsomes, and soluble components; the mitochondrial fraction obtained in the second centrifugation would be contaminated with microsomes and soluble components, and so on. These disadvantages may not be serious in some instances, since the major subcellular particles of a tissue homogenate have sedimentation values that differ from one another by one or more orders of magnitude. However, serious difficulties arise if the particles to be separated have similar sedimentation coefficients. This problem is illustrated in Fig. 12–4f in which the mitochondrial fraction also contains the cells' lysosomes.

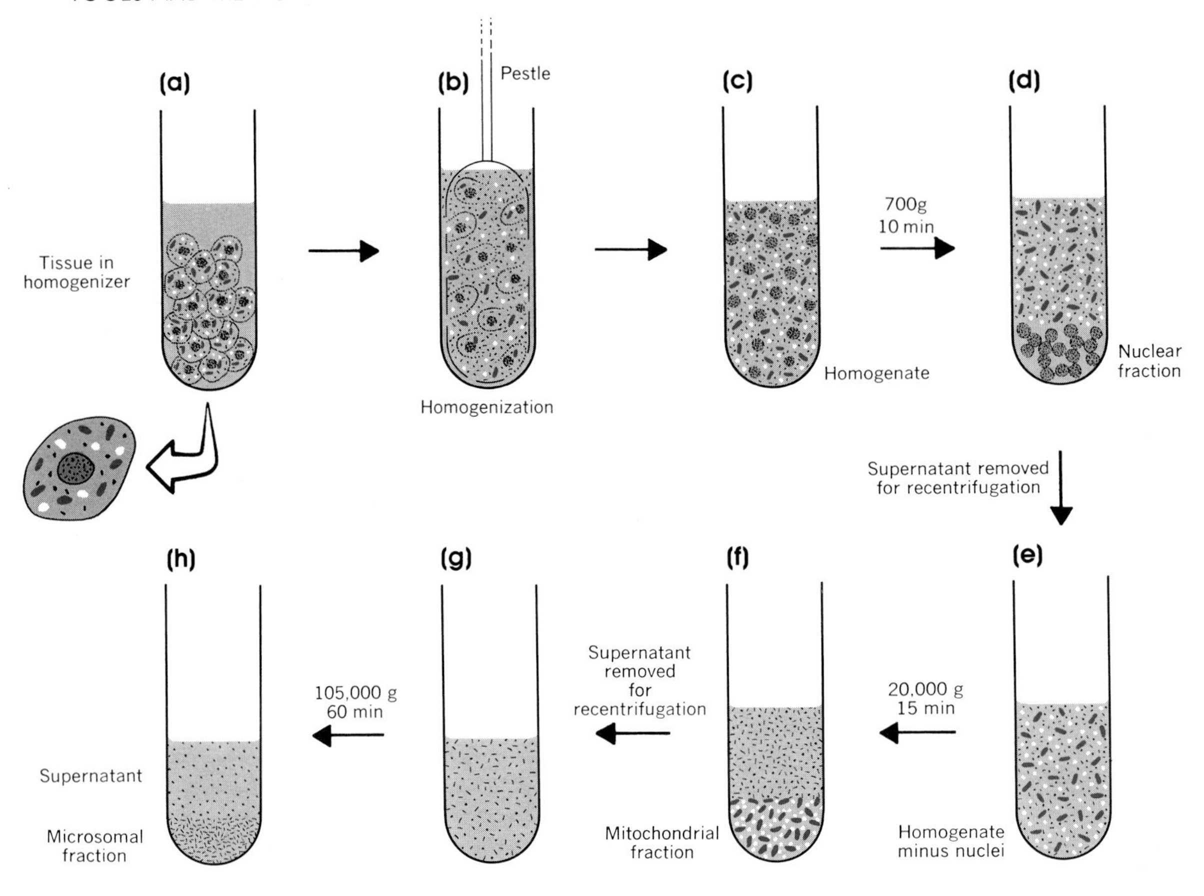

Figure 12–4
Fractionation of liver tissue by differential centrifugation. Nuclei (pale color), mitochondria (solid color), lysosomes (white), microsomes (black) and cytosol (grey).

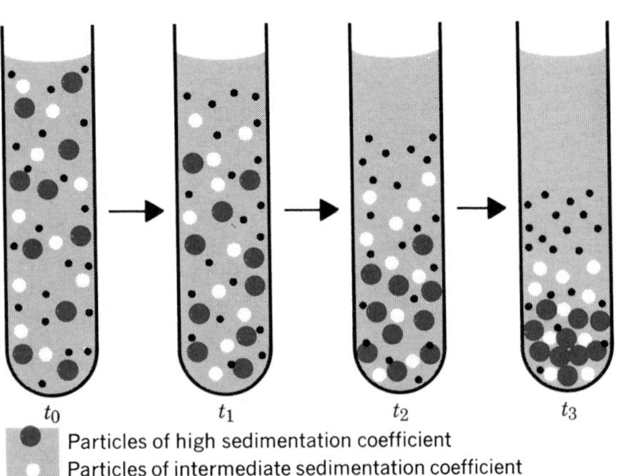

Figure 12–5
Effects of differential centrifugation. At time t_0, all particles are uniformly distributed through centrifuge tube. During centrifugation (times t_1, t_2, and t_3), all particles sediment according to their sedimentation coefficients. Although particles of highest sedimentation coefficient quickly reach the bottom of the tube, cross-contamination by smaller particles is unavoidable.

Density Gradient Centrifugation

The resolution achieved during centrifugation can be greatly improved if the mixture of particles is confined at the outset to a narrow zone at the top of the centrifuge tube and the particles then permitted to sediment from this position. Initial stability under these conditions can only be obtained if the particles are layered onto a **density gradient,** that is, a column of fluid of increasing density. The technique, known as density gradient centrifugation, was introduced to cellular research in 1951 by M. K. Brakke.

Density gradients may be *stepwise (discontinuous)* or *continuous*. A step gradient is prepared by successively layering solutions of decreasing density in the centrifuge tube. Continuous density gradients are prepared by mixing dense and light solutions in varying proportions at a controlled rate and delivering the mixture to the tube in a continuous stream. A simple procedure for producing a ''linear'' density gradient is shown in Figure 12–6*a*. In this procedure, the light solution (in the right cylinder) flows into and is mixed with the dense solution (in the left cylinder). The mixture is then delivered to the centrifuge tube. Using this method, a density gradient is formed that decreases linearly as a function of volume between the limiting densities originally present in the two cylinders. Gradients with other shapes (hyperbolic, logarithmic, etc.) may be prepared using containers of various noncylindrical shapes or gradient-generating devices specifically manufactured for this purpose. Solutes used to provide solutions for density gradients are selected on the basis of their solubility in water

Figure 12–6
(a) **Illustration shown production of a linear density gradient in a centrifuge tube using a simple, two-chamber mixing device.** ***(b)*** **A commercial gradient maker (courtesy of DuPont Instruments).**

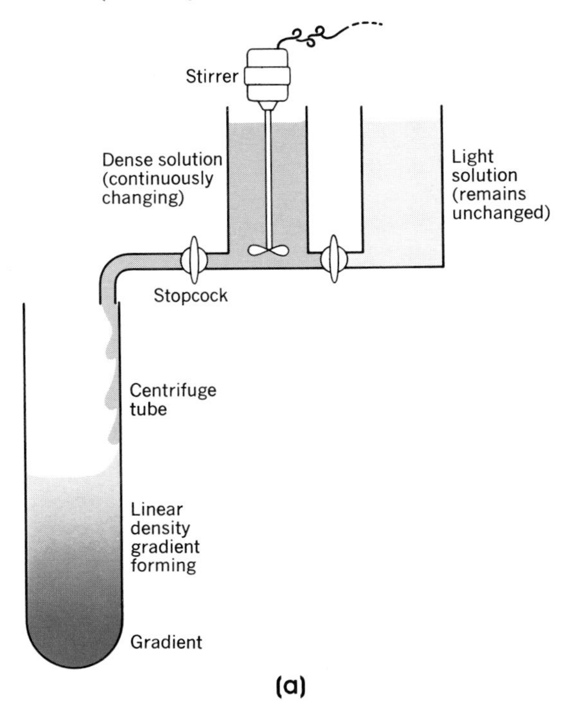

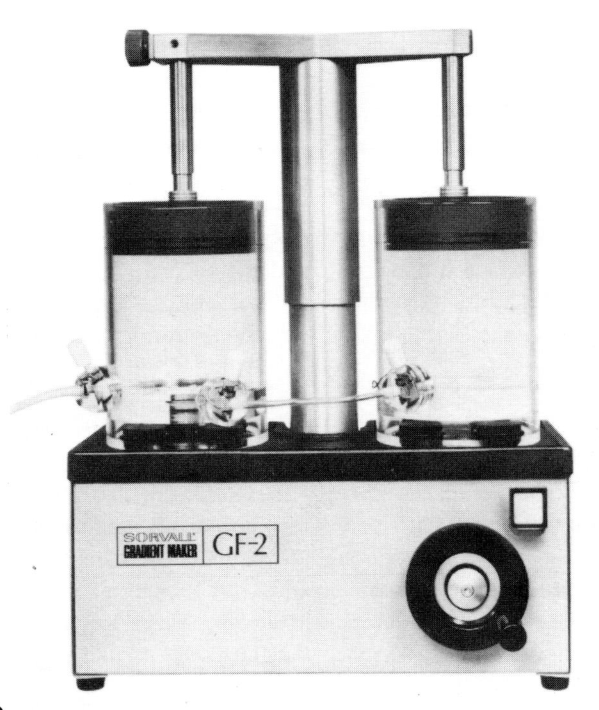

and their compatibility with cells and subcellular organelles. Sucrose, Ficoll (a copolymer of sucrose and epichlorohydrin), and dextran (a linear α, 1 $\longrightarrow$ 6 polymer of glucose) are the most popular, since they have minimal detrimental effects on organelles and provide densities up to about 1.3 g/cm^3. Cesium chloride and other dense salts are frequently used when the limiting density must be considerably higher (i.e., up to 1.9 g/cm^3).

The rate at which particles sediment through a density gradient is given by equation 12–18. Since the quantity $(\rho_P - \rho_M)$ changes as the particles sediment deeper into the gradient, s also changes. If a particle reaches a position in the gradient where the particle's density and the gradient's density are the same (i.e., $\rho_P = \rho_M$), then s becomes zero, and no further sedimentation of the particle occurs. If a particle suspension is overlaid with a graident whose maximum density is greater than the particles (i.e., $\rho_M > \rho_P$), then during centrifugation the particles will *rise* through the gradient, for when ρ_M is greater than ρ_P, s becomes a *negative* term.

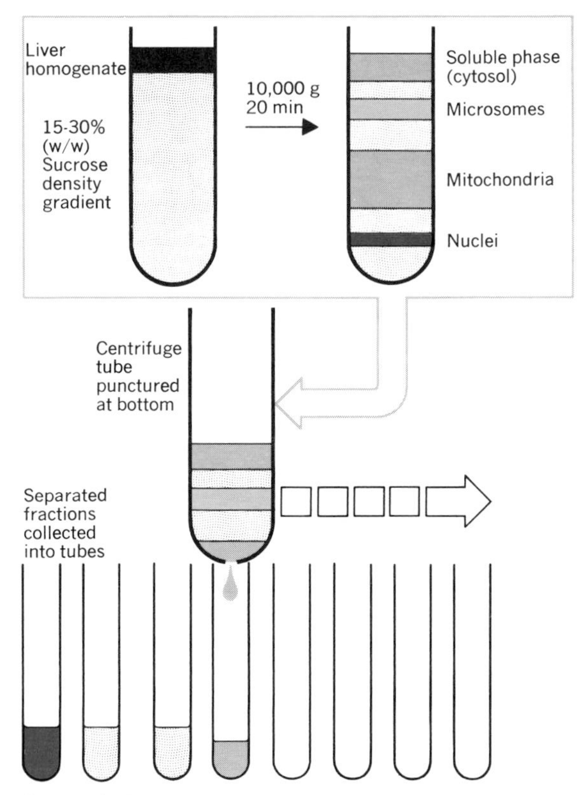

Figure 12–7
Rate centrifugation of a liver tissue homogenate. The homogenate is layered onto a 15 to 30% (w/w) sucrose density gradient which is then centrifuged at 10,000 *g* for 20 min. The homogenate is separated into four major fractions: nuclear, mitochondrial, microsomal, and soluble (or cytosol). Following centrifugation, the separated fractions may be collected by puncturing the bottom of the centrifuge tube and permitting the gradient to drip into a series of test tubes.

Rate sedimentation. When the densest region of a density gradient (i.e., at the bottom of the centrifuge tube) is less dense than the particles being sedimented, all particles will eventually reach the bottom of the tube. However, if the duration of centrifugation is carefully limited, this will not occur; instead, the particles will be distributed through the density gradient *in order of their sedimentation coefficient*. Fractionations carried out in this manner are called **rate** separations and are based upon the combined contributions of particle size and particle density. The rate fractionation of a liver tissue homogenate is depicted in Figure 12–7. The four major cell fractions (nuclear, mitochondrial, microsomal, and soluble) are separated during a single centrifugation.

Isopycnic Sedimentation. When the densest region of a density gradient is denser than the particles to be sedimented, no particles will reach the bottom of the tube no matter how long centrifugation is carried out. Instead, the particles will sediment through the gradient until they reach their **isodense** or **isopycnic** position. Fractionations carried out on the basis of particle density alone are called isopycnic separations. It is to be noted that this implies that small, dense particles with low sedimentation coefficients sediment further through a steep density gradient than large, light particles with high sedimentation coefficients (Fig. 12–8).

In rate separations, both particle size and particle density determine the final positions of particles in the density gradient. However, of the two parameters, size is more important, since s is a second-order function of particle radius but a first-order function of particle density (examine equation 12–18). In isopycnic separations, particle density *alone* determines final position in the density gradient.

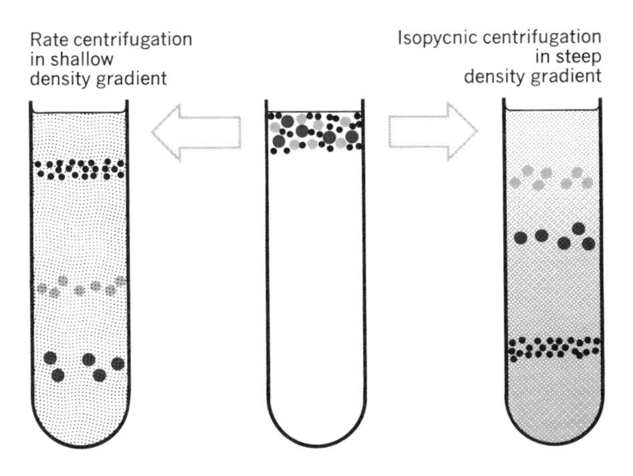

Figure 12–8
Comparison of *rate* and *isopycnic* centrifugation. During rate centrifugation in a shallow density gradient, the particles become distributed in order of sedimentation coefficient. Consequently, a large but light particle may sediment further through the gradient in a short period of time than a small dense particle. During isopycnic centrifugation, particles become distributed through the steep gradient in order of density and regardless of particle size.

Maximum resolution of particle mixtures by density gradient centrifugation is achieved when both rate and isopycnic centrifugations of the sample are carried out in sequence. First, using rate centrifugation, the particle mixture is separated into fractions of similar particle sedimentation coefficient; then, collected fractions are subfractionated into classes of equal density by isopycnic centrifugation. This technique is known as two-dimensional centrifugation and provides for the separation of cell components of similar sedimentation coefficient or similar density. (It is extremely unlikely that two different particles will have similar sedimentation coefficients *and* similar densities.) Figure 12–9 compares the sedimentation coefficients and densities of a variety of cellular components. After centrifugation is completed, the separated zones may be recovered from the density gradient by puncturing the bottom of the centrifuge tube and allowing the gradient to elute (Fig. 12–7).

In certain instances, it is not necessary to layer the sample to be fractionated onto a preformed density gradient. Instead, the sample and the density gradient solutions are mixed and placed in the centrifuge tube. During centrifugation at high speed, the density gradient forms automatically within the tube, and the particles migrate (upward and/or downward) to their isopycnic positions. This procedure is known as **equilibrium isopycnic centrifugation** and is regularly carried out in using solutions of CsC1 or other heavy salts (Fig. 12–10).

Although density gradient centrifugation is usually carried out using swinging-bucket rotors, isopycnic (but not rate) separations can also be achieved in fixed-angle rotors (Fig. 12–11*a*). When the gradient with sample layered above is placed in the inclined position in the rotor, the sample zone and equal density planes within the gradient form ellipses. During acceleration of the rotor, each isodense elliptical layer becomes a small section of a **paraboloid of revolution** with focal point on the axis of rotation. The family of paraboloids becomes increasingly steep as rotor acceleration continues and eventually approaches verticality; the phenomenon is called **gradient reorientation.** Particles in the gradient sediment radially until they encounter the sloping tube wall; from there, they are carried down through the gradient by centrifugal force and convection until they reach the isodense region of the gradient, where they form layers. Rotor deceleration reorients the gradient again so that the separated particles now form a series of horizontal layers.

A recent and rather novel modification of the fixed-angle approach to isopycnic centrifugation involves the use of *vertical rotors* in which the fixed angle is reduced to 0° (i.e., the centrifuge tubes are maintained in the vertical position throughout; Fig. 12–11*b*). Here, too, the gradient is reoriented during rotor acceleration and deceleration.

Zonal Centrifugation

Density gradient centrifugation using tubes is the most widely employed technique for separating cells and cell organelles and for isolating cellular macromolecules. However, although it is one of the cell biologist's most valuable tools, it is not without disadvantages, since the amount of material that can be fractionated in a single tube is so small. When very large quantities of sample must be fractionated (in order to isolate sparse organelles such as lysosomes or

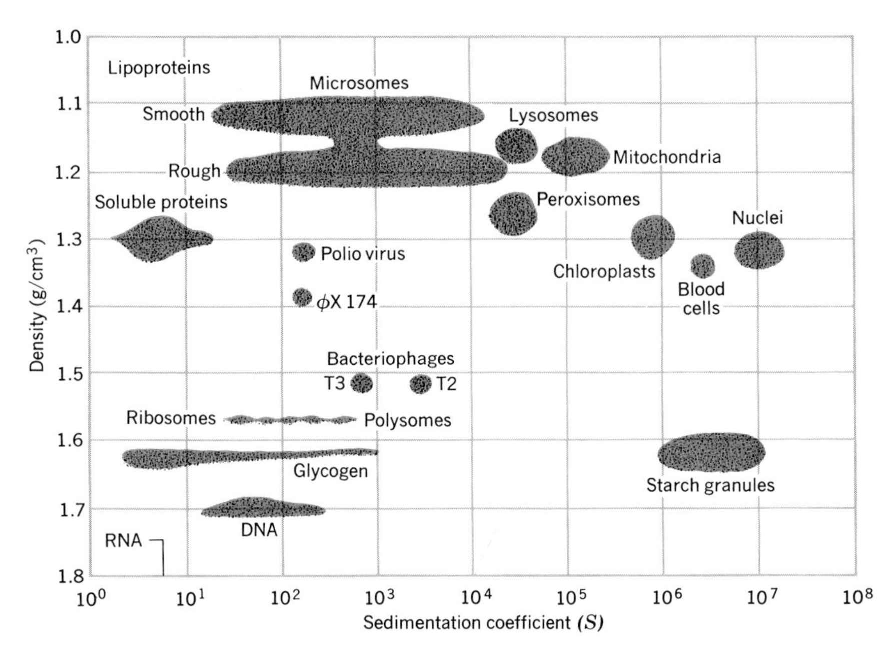

Figure 12–9
Densities and sedimentation coefficients of cells, subcellular particles, and viruses.

peroxisomes), the numbers of tubes and gradients needed can be overwhelming. The lack of sector shape in a centrifuge tube causes "wall effects" (Fig. 12–12), including clumping, premature sedimentation down the tube wall, and convection disturbances. These problems served as the impetus for the development in the 1960s of **zonal rotors,** a technological advancement pioneered by N. G. Anderson.

A zonal rotor consists of a large cylindrical chamber subdivided into a number of sector-shaped compartments by vertical *septa* (or vanes) that radiate from the axial *core* to the rotor wall. The entire chamber is used during centrifugation and is loaded with a single density gradient, each sector-shaped compartment serving as a large centrifuge tube. The large chamber capacity of these rotors (typically 1 to 2 liters) eliminates the need for multiple runs and multiple density gradients. Two basic forms of the zonal rotor are regularly used for tissue fractionation; these are (1) **dynamically unloaded (rotating-seal)** rotors and (2) **reorienting gradient** ("reograd") rotors. The similarity in appearance of these rotors (Fig. 12–13) is misleading, since the basic principles of operation considerably differ.

Figure 12–10
Equilibrium isopycnic centrifugation. During centrifugation, the solution in which the particles are initially suspended automatically forms a density gradient. Particles, which are initially distributed uniformly through the centrifuge tube, sediment or float to their isopycnic positions.

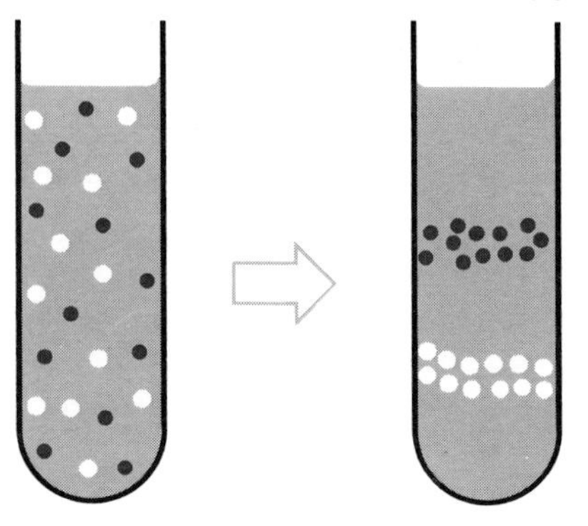

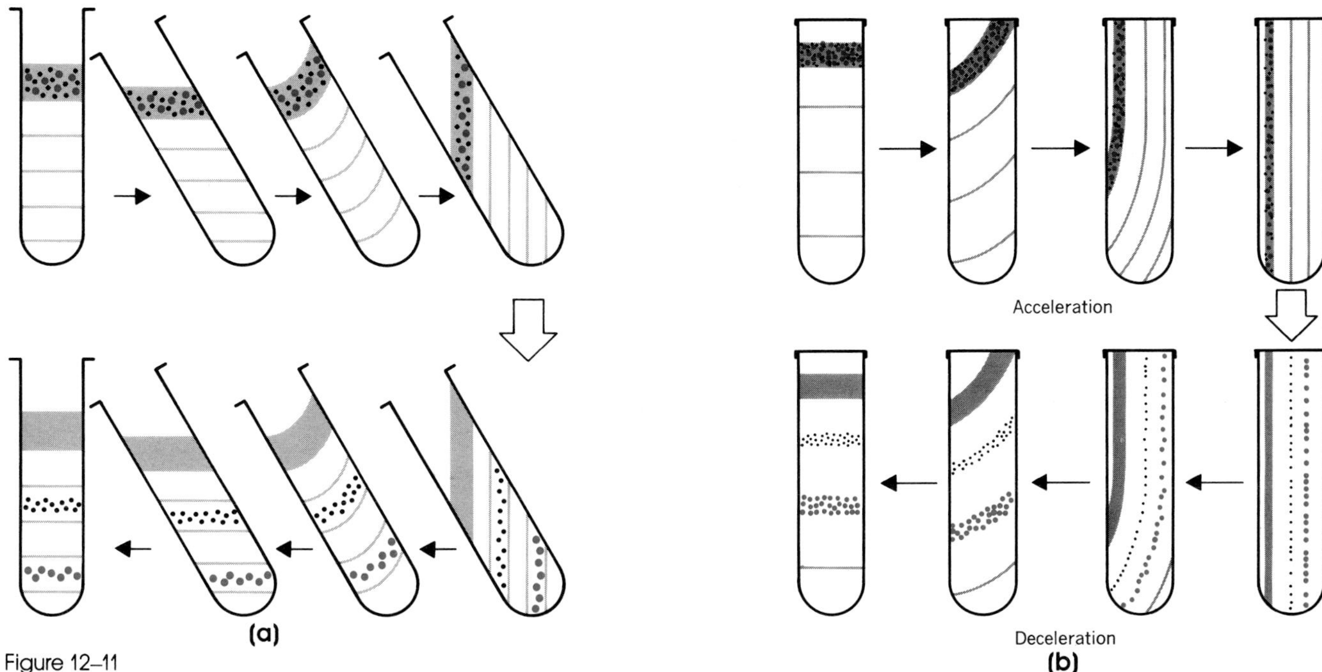

Figure 12–11
(a) Isopycnic centrifugation using a fixed-angle rotor. *(b)* Isopycnic centrifugation in a "vertical rotor." In both cases the axis of rotation is to the left of the tube, and the density gradient is caused to reorient between the vertical and radial position during rotor acceleration and deceleration.

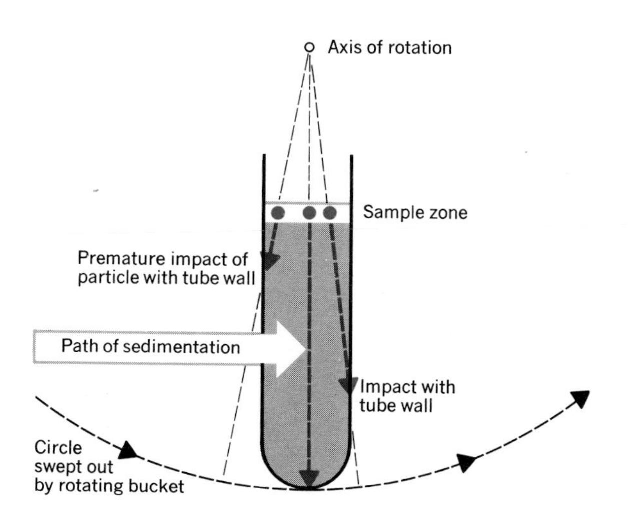

Figure 12–12
Wall effects. The path of sedimentation in the centrifuge tube is perpendicular to the axis of rotation. Consequently, particles initially near the center of the sample zone will sediment radially and contact the tube at its base. However, particles initially located further toward the periphery of the starting zone will strike the wall of the tube before reaching the base. This can result in particle clumping, premature sedimentation down the wall, convective disturbances, and other anomalous effects.

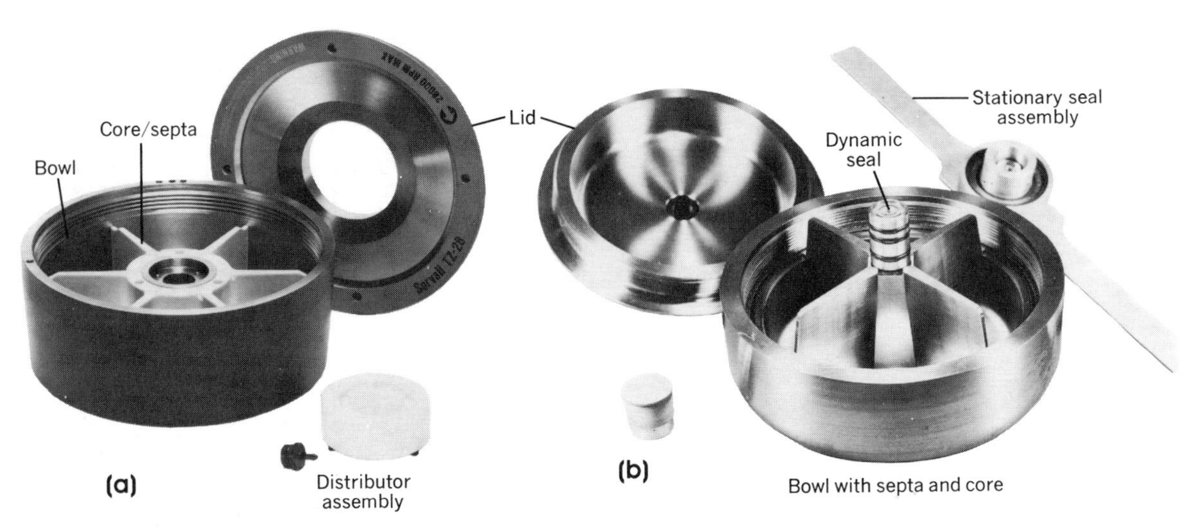

Figure 12–13
***(a)* Reorienting gradient zonal rotor (TZ-28) and *(b)* dynamically unloaded zonal rotor (B-29). The bowl of the TZ-28 rotor is unloaded through lines in the distributor, core, and septa after the rotor has decelerated to rest and the density gradient reoriented. The B-29 zonal rotor is unloaded while the rotor is spinning; this is achieved using a static and rotating seal assembly that provides continuous communication with the core and wall of the rotor bowl through lines in the core and septa.**

Dynamically Unloaded Zonal Rotors

Operation of a dynamically unloaded rotor is depicted schematically in Figure 12–14. Two fluid lines connect the center and edge of the rotor chamber with a rotating-seal assembly that permits loading the density gradient (and sample) *while the rotor is spinning*. The center fluid lines open into the rotor chamber through the core section, while the edge lines pass radially through each of the septa and open at the rotor wall.

The zonal rotor is filled with the density gradient while rotating at low speed. The light end of the gradient is loaded first through the edge lines and is followed by the denser mixture (Fig. 12–14*a*). The dense end of the gradient gradually displaces the lighter fluid toward the core of the rotor. Addition of a dense "cushion" forces some of the light end of the gradient out of the rotor (Fig. 12–14*b*). The sample to be fractionated is introduced through the center lines, thereby displacing some of the cushion out of the edge lines (Fig. 12–14*c*). Additional light fluid (called *overlay*) is then pumped into the center line to push the sample clear of the core region (Fig. 12–14*d*). Now, the upper (stationary) portion of the seal assembly is removed and the rotor is accelerated to a higher speed for separation of the particles in the sample (Fig. 12–14*e*). The separated particles form a series of concentric cylindrical zones in the rotor bowl. Following particle separation, the rotor is decelerated to a lower speed, the static seal reinserted, and the entire gradient displaced through the center lines by pumping dense fluid through the edge line (Fig. 12–14*f*). The eluting gradient may be monitored and collected in tubes for subsequent analysis.

Reograd Zonal Rotors

The operation of reograd zonal rotors (developed in the 1960s by P. Sheeler and J. R. Wells) is depicted in Figure 12–15. Unlike dynamically unloaded zonal rotors, the gradient and sample can be loaded into the reograd rotor either at rest through the septa lines (which communicate with the bowl floor) or while spinning using the core lines. Unloading is *always* carried out with the rotor at rest. If the rotor is loaded at rest, then the gradient is reoriented from the vertical to the radial position during acceleration (Fig. 12–

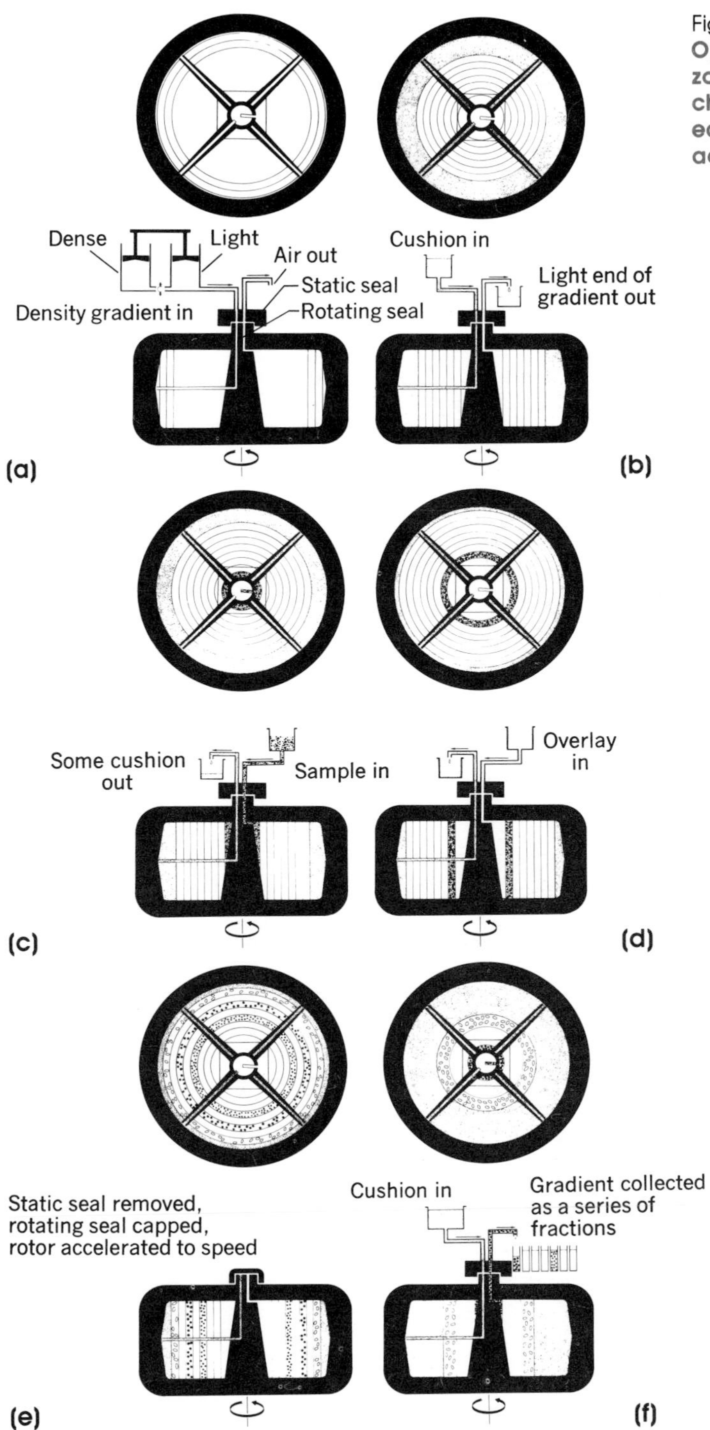

Figure 12–14
Operation of a dynamically unloaded zonal rotor. Lines drawn inside the rotor chambers depict hypothetical planes of equal gradient density. See text for additional details.

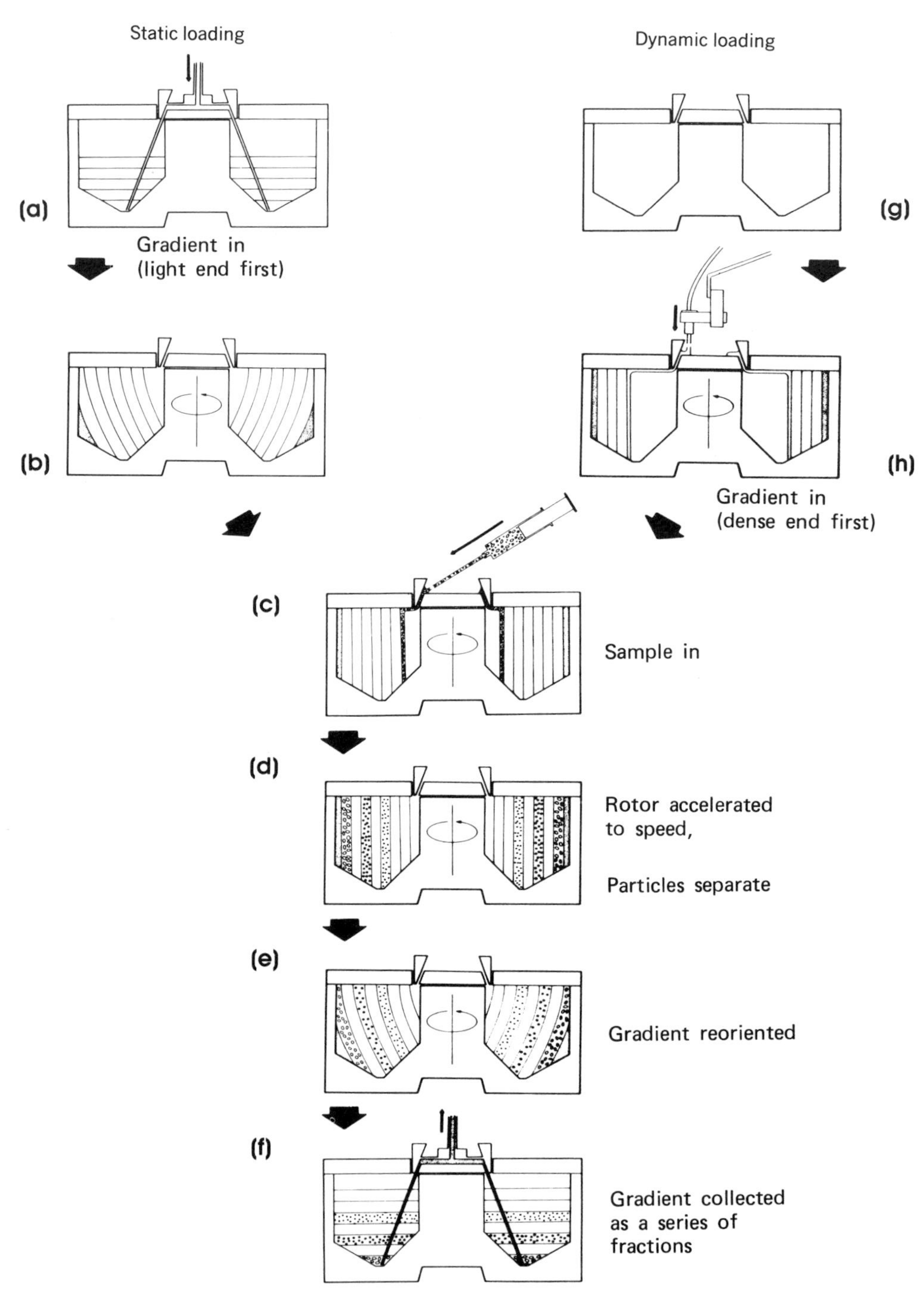

Figure 12–15
Operation of a reograd zonal rotor. Lines drawn inside the rotor chambers depict hypothetical planes of equal gradient density. See text for additional details.

15*a,b*). During centrifugation, different particles in the sample sediment to form a family of concentric cylindrical zones in the radial gradient (Fig. 12-15*d*), but during rotor deceleration, they are reoriented to form horizontal layers (Fig. 12–15*e,f*). The density gradient and entrained particles are then withdrawn from the stationary rotor through the septa lines and collected as fractions for further analysis.

Because the chambers of zonal rotors are sector shaped, detrimental wall effects are minimized or eliminated entirely. Zonal rotors have greater capacities than most swinging-bucket rotors of comparable speeds, and therefore, larger quantities of material can be fractionated during a single centrifugation. In order to fractionate a comparable amount of tissue or cells using conventional rotors, several successive centrifugations would be required and separate density gradients would necessarily be prepared for each tube.

Examination and Analysis of Separated Cell Fractions

The final phase of any study involving the fractionation of cells or tissues is the *examination* and *analysis* of the separated fractions so that the morphology, chemical composition and organization, or function of the isolated components may be revealed or better understood. A host of procedures can be employed in such studies, but they will not be pursued here, since they are more properly considered individually in conjunction with closely allied topics in other chapters of the text. For example, the isolation of the plasma membrane-containing fraction (or fractions) from a tissue homogenate might be followed by morphologic or biochemical analyses involving perhaps electron microscopy, chemical characterization and isolation of the membrane constituents, determination of component function in membrane transport, and so on. These topics are dealt with in Chapters 1, 13, and 15, which specifically consider the principles of electron microscopy, chemical fractionation procedures, and plasma membrane structure and function. Indeed, a considerable portion of this book is devoted to the presentation of the results of examination and analysis of isolated cell components and to the integration of those results into a functional whole.

Methods for Separating Whole Cells

Throughout this chapter, we have been concerned with methods used to separate and isolate particles released from disrupted cells and tissues. Until quite recently, little attention was directed to a related problem—namely, how to separate *whole viable* cells from one another when the tissue or culture being studied was heterogeneously composed. For example, an organ such as the liver is composed of many different types of cells, including hepatic cells, Kupfer cells, connective tissue cells, smooth muscle cells, blood cells, and so on. Therefore, an homogenate of liver tissue contains subcellular particles from diverse kinds of cells. Even a culture of the same cell type may be heterogeneous with regard to cell ages (see Chapter 2) and therefore be representative or a broad spectrum of morphological characteristics or physiologic activity. Although of equal importance, the development of methods for separating different types of cells present in a tissue has lagged behind technological advances in the area of subcellular fractionation. In the following section, we will examine some of the problems associated with whole-cell separations and some of the more important methods that have evolved to effect such separations.

Tissue Disaggregation

If the tissue to be fractionated consists of suspensions of individual cells (e.g., cultures of microorganisms, some tissue cultures, blood cells, certain tumors), the problem of whole-cell separation is far less difficult than when the cells comprise a solid tissue (such as liver, kidney, brain, etc.). It is therefore not surprising that, to date, most efforts directed toward whole-cell separations have involved natural suspensions of cells as the starting material. Some success has been obtained with solid tissues by employing chemical agents that induce tissue disaggregation—primarily digestive enzymes and chelating agents. These materials weaken the connections between neighboring cells, making it possible to mechanically disperse the tissue into individual cells without appreciable cell breakage. The tissue may be so treated after its removal from the animal, although *perfusion* of the organ with a solution of the disaggregating agent prior to its excision is more often preferred.

Once the tissue has been reduced to a suspension of

individual cells, fractionation into subpopulations follows. If the objective is to isolate a particular subpopulation for further study, then the remaining cells in the suspension may be *selectively* destroyed or removed by chemical means. For example, the leukocytes of blood may be separated from the erythrocytes present by selective destruction of the erythrocytes by *osmotic* or *chemical lysis*. Purification of a particular subpopulation of cells may also be achieved by taking advantage of *differential cell agglutinability* in the mixed population. Simply *freezing* and *thawing* a suspension of cells may differentially lyse specific subpopulations. In general, chemical procedures cause some changes in *all* the cells in the mixed population, so that the method of choice is more generally one that achieves a separation by mild physical means. Among the most popular of the latter methods are **adherence and filtration, conventional and zonal centrifugation, centrifugal elutriation, unit gravity separation, countercurrent distribution, electrophoresis,** and **fluorescence-activated cell sorting.**

Adherence and Filtration

Separations of cells using differences in adherence phenomena or filtration properties are among the oldest physical procedures used. Some cells readily adhere to glass beads, nylon wool, glass wool, and so on and may be separated from nonadhering cells by passing the cell suspension through a hollow glass column packed with these materials. Success has also been obtained by coating glass or plastic beads with antibodies, antigens, or haptens so that cells will be differentially adsorbed to the beads on the basis of chemical interactions between the cell membrane and the coating material. Sieves of varying pore diameter can also be used to separate populations of cells on the basis of differences in cell diameter.

Conventional and Zonal Centrifugation

Because of their relatively large size (i.e., in comparison with organelles and macromolecules), whole cells sediment quite rapidly. Consequently, attempts to fractionate suspensions of cells using centrifugation involve rotation at low rpm (i.e., small RCF) for short periods of time (typically less than $500 \times g$ for a few minutes). As with subcellular centrifugal fractionations, greatest resolution is obtained using density gradients in which the mixture of cells in the starting zone is separated into subpopulations on the basis of differences in average cell size and/or density. Most cells behave like miniature osmometers, so that strict attention must be paid to the selection of gradient solute. Sucrose, salts, and other small molecules are rarely used to prepare density gradients for cell separations because of their deleterious osmotic effects. Large, impermeable, and biologically inert polymers such as dextran and Ficoll are the more frequent choices.

Significant wall effects accompany particle sedimentation in fixed-angle rotors; therefore, conventional approaches to the centrifugal separation of cells involve swinging-bucket rotors. Since some minimum rpm (and therefore minimum RCF) must be attained before the buckets reach the horizontal position, special swinging-bucket rotors in which the minimum and maximum radii are particularly small (i.e., the buckets are close to the axis of rotation) are used in order to maintain the low centrifugal force required. Although offering the advantage of greatly increased sample size, dynamically unloaded zonal rotors are rarely used because of the continuing cell sedimentation that occurs during the extended dynamic unloading period. Reograd zonal rotors can be used, however, since gradient unloading is carried out at $1 \times g$.

Centrifugal Elutriation (Counter-Streaming Centrifugation)

Centrifugal elutriation is an ingenious technique pioneered in the 1940s by P. E. Lindahl and brought to its present state of the art principally through the work of C. R. McEwen. In centrifugal elutriation, a suspension of cells is pumped through a marginally located entry port (Fig. 12–16) into a specially designed rotor chamber and is followed by a continuous supply of suspending medium. Centrifugal sedimentation of the cells is opposed by the centripetal flow of the suspending medium. Both effects vary in magnitude across the radial dimension of the rotor chamber, since (1) the centrifugal force increases with distance from the rotor axis, and (2) the rate of centripetal liquid flow varies according to the cross-sectional area of the chamber (the area increases exponentially as the liquid travels toward the rotor axis). Depending upon its initial position in the chamber and its sedimentation coefficient, each cell will either sediment radially under the centrifugal force or be carried centripetally by the liquid flow. As a result, the cells migrate

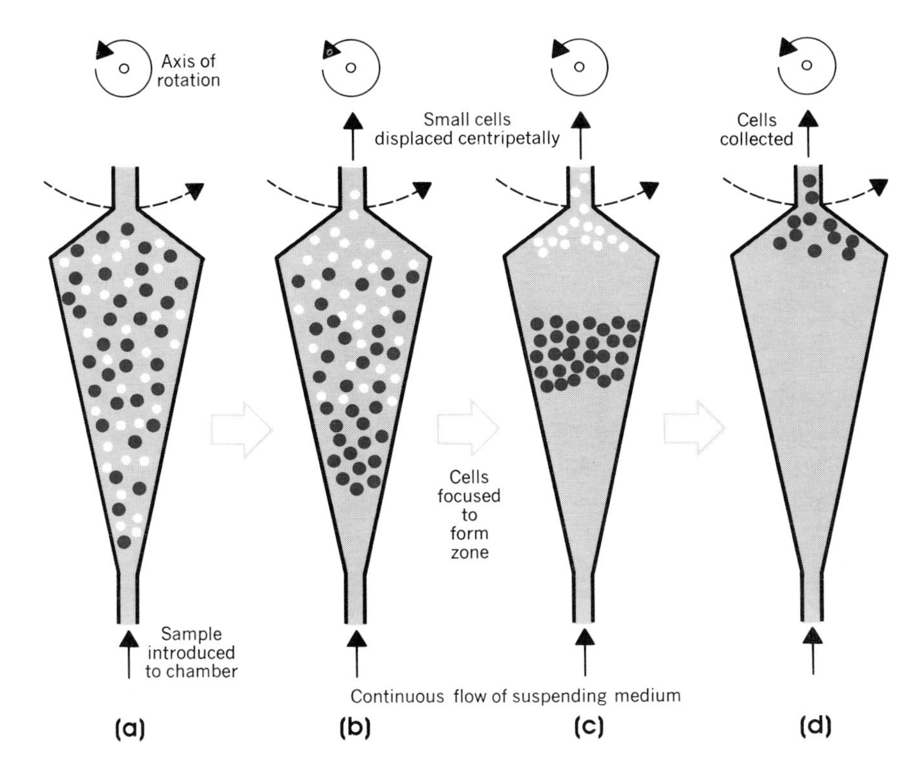

Figure 12–16
Centrifugal elutriation. *(a)* Sample containing mixture of cells introduced to chamber while rotor is spinning. *(b)* Combined effects of centrifugal force and centripetal liquid flow displace cells. *(c)* Small cells swept from rotor, while larger cells are focused into zone. *(d)* Cells collected by displacement.

through the chamber to positions where these two forces cancel one another. Some cells (e.g., the smallest ones) may be swept from the rotor chamber entirely, while others form a zone within the rotor chamber and can be collected for further study. Centrifugal elutriation has been successfully applied to separations of blood cells, algae, yeasts, and other cells in culture.

Unit Gravity Separation

The separation of particles on the basis of sedimentation rate differences may not necessitate centrifugation if the particles are sufficiently large. For example, whole cells sediment fairly quickly even at 1 *g* (i.e., at unit gravity). Unit gravity procedures have been used effectively to separate different types of blood cells, tissue culture cells, populations of microorganisms, and so on into subpopulations. The separation is achieved by layering the mixture of cells onto the top of a stationary density gradient and allowing the cells to settle through the gradient for some period of time. The gradient and separated cell populations are then collected as a series of fractions. Devices used to separate cells in density gradients at unit gravity are called "sta-put" devices and vary from simple cylindrical chambers to more elaborate apparatus having moving, conical end caps. The principle is illustrated in Figure 12–17.

Not only is it possible to separate heterogeneous mixtures of cells using this simple approach, but a population of a single cell type (e.g., a cell culture) can be fractionated according to cell age when age and size are related. In this way, events that occur during the *cell cycle* (see Chapter 2) can be studied by examining the cells in different collected fractions.

A number of methods for separating and isolating the molecular constituents of cells are considered in Chapter 13. Some of these methods have been appropriately modified and applied with varying degrees of success to separations of different kinds of cells that make up a tissue; **countercurrent distribution** and **electrophoresis** will be mentioned briefly here.

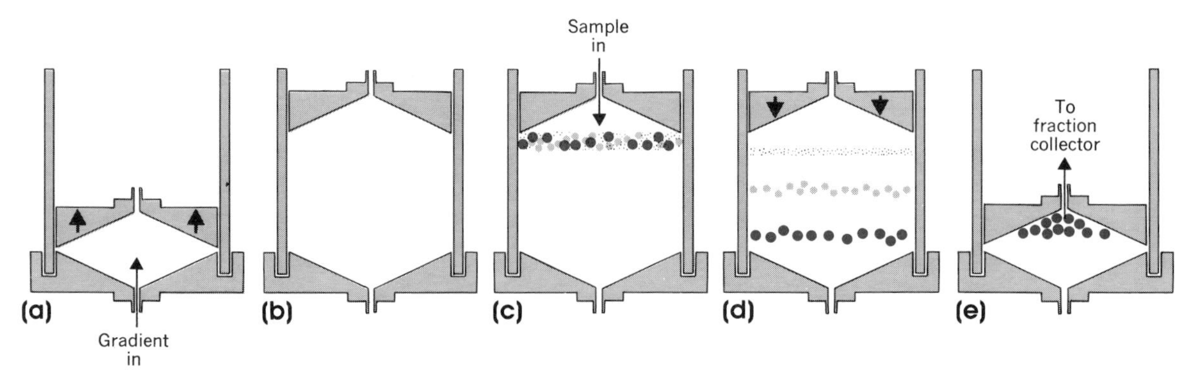

Figure 12–17
Separation of cells at unit gravity using a "Sta-Put" device. The fractionation may be carried out in a cylindrical chamber containing two funnel-like end caps. As the density gradient is pumped into the device *(a)*, the upper end cap is displaced vertically *(b)*. The sample is introduced through the upper end cap *(c)* and allowed to settle through the density gradient *(d)*. The separated particles are then collected either by withdrawal through the lower end cap or by downward displacement of the upper end cap as shown *(e)*.

Countercurrent Distribution

Cells may be separated from one another on the basis of differences in their partition between two immiscible liquids. Naturally, if the cells are to be separated without undue damage, the milieu selected must be compatible with the cells with respect to ionic composition, concentration, osmotic pressure, and so on. This demand significantly restricts the selection of liquids, especially in comparison with the range of choices available when countercurrent distribution is employed for molecular separations. Greatest success has been obtained with phases consisting of polyethylene glycol and aqueous solutions of dextran (polyglucose in which most glycosidic linkages are $1 \longrightarrow 6$). The technique has been especially fruitful in separations of different microorganisms and different blood cells.

Electrophoresis

Electrophoresis is one of the most popular methods used for separating different molecular species, especially proteins (see Chapter 13). However, electrophoresis can be used to separate whole cells. Cell separations using this technique are based on the fact that the plasma membranes of cells contain charged groups (e.g., proteins, sialic acid, short carbohydrate chains) that impart a net electrostatic charge to the cell surface. Different types of cells possess different net charges so that when they are placed in the appropriate conductive medium and subjected to an electrical current, they will migrate through the medium at different rates. Hence, they become separated into subpopulations that can be collected for further study. Various chemical substances may be applied to the cells in order to selectively alter their normal surface charge distribution and assist in their electrophoretic separation. Electrophoresis has been applied successfully to separations of microorganisms, blood cells, ascites tumor cells, HeLa cells and other cell cultures.

Fluorescence-Activated Cell Sorting

Fluorescent dyes such as *fluorescein* can react with and bind to the surfaces of cells; the type and quantity of dye bound varies for different kinds of cells. This differential property has been used for years to *visually* distinguish different types of cells in a mixed population and very recently has been employed in an elegant instrument that physically separates the cells. The instrument, known as a **fluorescence-activated cell sorter** and depicted diagrammatically in Figure 12–18, has a complex history, but its development may be credited to the combined contributions of M. J. Fulwyler, L. A. Herzenberg, R. G. Sweet, W. A. Bonner, and H. R. Hulett. The fluorescein-treated suspension of cells is mixed with electrolyte solution ("sheath fluid") and forced downward through a tiny nozzle vibrating at 40,000 cycles per second. The vibrations of the nozzle break the emerging stream into uniform droplets approximately equal in number to the frequency of vibration (i.e., 40,000 droplets per second). The population density of the original cell suspension and the flow rate are adjusted so that each droplet contains no more than one cell (indeed, most droplets contain no cells).

Figure 12–18
Major components of the fluorescence-activated cell sorter.

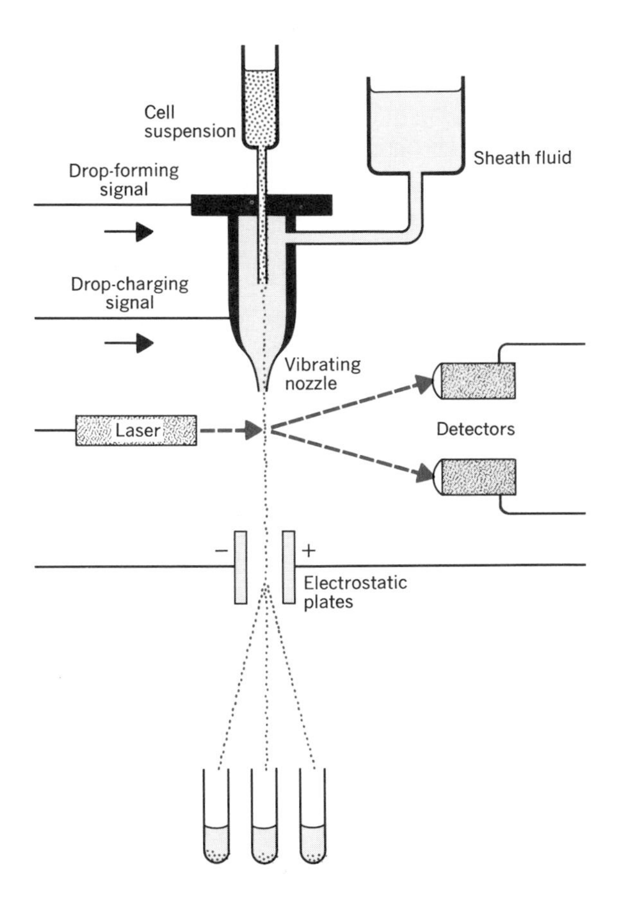

Just prior to droplet formation, the stream is illuminated with an argon-ion laser beam that excites the fluorescent material in the cell surfaces. Two detectors respectively measure the amount of fluorescent light and the volume of the cell and trigger an electrical pulse that charges each cell-containing droplet (and some empty droplets too) as it is formed. The amount and sign of the electrostatic charge borne by the droplet depend upon the size of the entrained cell and the number of fluorescein molecules bound to its surface. These charge parameters can be selected by the operator and effectively divide the droplets into *three* classes: positively charged, negatively charged, and uncharged. The droplets then pass between two electrostatic plates; the charged droplets are appropriately deflected as they pass through the field between these plates, while uncharged droplets continue on their original course. Finally, the three droplet streams are collected in reservoirs. The left and right streams contain different populations of cells, while the undeflected center stream consists primarily of empty droplets, unwanted cells, and debris. The fluorescence-activated cell sorter can separate about 5000 cells per second.

Harvesting Cells and Subcellular Components: Continuous-Flow Centrifugation

This discussion of methods for the differential isolation and separation of cells and subcellular components would not be complete without a description of the use of centrifugation for *harvesting* large quantities of particulate material from large-volume suspensions. The technique is generally known as **continuous-flow** centrifugation.

The most common application of continuous-flow centrifugation is the harvesting of bacteria, algae, protozoa, and other cells grown in multiliter cultures as a preliminary to chemical, physiological, or morphological analysis. However, the technique is also frequently employed (1) to collect cell-free culture media prior to the isolation and assay of cellular excretion products such as enzymes, vitamins, hormones, and growth substances; (2) to separate blood plasma from whole blood; (3) to remove the larger subcellular components such as nuclei, chloroplasts, and mitochondria from large volumes of tissue homogenates; and (4) to collect precipitates from large volumes of aqueous suspensions.

During continuous-flow centrifugation, the suspension of particles is introduced into the spinning centrifuge rotor as a continuous, uninterrupted stream. As the suspension passes through the rotor, particles are sedimented out of the stream and are trapped and concentrated within specific rotor chambers, while the clarified supernatant leaves the rotor and is collected separately. Continuous-flow centrifuge rotors thereby eliminate the need for a series of batch separations when very large volumes of particle suspensions must be processed. When processing is completed, the rotor is simply decelerated, opened, and the trapped cells or other particles removed.

Although cells or other particles present in multiliter volumes can be harvested using conventional swinging-bucket and fixed-angle rotors, this approach is far less efficient. Even the largest conventional rotors generally accommodate less than 6 liters of suspension, so that a succession of spins is necessary when larger volumes of material must be handled. Equally important, the increased size and weight of these rotors restricts their maximum operating speeds and may necessitate extended centrifugation time in order to ensure total particle ''cleanout.'' Since at any instant, continuous-flow rotors contain only a small fraction of the total volume of material to be centrifuged, they may be quite small. Thus, in addition to eliminating the need for successive runs, continuous-flow rotors can be operated at much higher speeds (hence greater RCF), providing more rapid and efficient particle cleanout.

The RCF exerted on the particles as they enter the collection chambers of the rotor depends on rotor speed and causes the particles to sediment at specific rates. If the rate of particle sedimentation is *greater* than the rate at which the surrounding liquid moves toward a centripetal exit port in the chamber, then the particles become trapped in the rotor. However, if the sedimentation rate is *less* than the rate of centripetal flow, then the particles are carried toward the exit ports and out of the rotor. Usually flow rate and rotor speed are selected to provide maximum cleanout of the particle suspension. However, for heterogeneous populations of particles, the flow rate and rotor speed can often be adjusted so that a *differential fractionation* of the particles is achieved. That is, depending on their sizes, shapes, and densities, some particles will be trapped in the rotor, while others are conducted out of the rotor with the supernatant.

Summary

Tissue fractionation begins with the disruption of the cells and the preparation of a subcellular particle suspension or homogenate. Cells are most frequently disrupted using the **shear** forces generated by special **grinders, blenders, pressure cells,** or **insonators.** Chemical procedures involving lytic agents or osmotic pressure are also used in certain instances.

Once the tissue homogenate is prepared, the method of choice for separating subcelluar organelles and particles is **centrifugation.** Centrifugal fractionation may involve one or a combination of different approaches. In **differential** centrifugation, gross differences in the **sedimentation rates** of certain subcellular particles are used to produce a series of particulate sediments at successively higher ''*g* forces.'' Different families of particles may also be separated on the basis of size and/or density differences using the **density gradient** approach. A variety of centrifugal devices are used to effect the purification of subcellular particles including conventional swinging-bucket and fixed-angle rotors and the more sophisticated vertical and zonal rotors.

Heterogeneous mixtures of very large particles, such as whole cells, may be fractionated by centrifugal **elutriation** or simply by **unit gravity** sedimentation. Electronic sorting, countercurrent distribution, and electrophoresis are alternative methods. When particularly large quantities of cells or particles are to be harvested, **continuous-flow** centrifugal procedures are generally employed.

References and Suggested Reading

Articles and Reviews

Brakke, M. K., The origins of density gradient centrifugation. *Fractions No. 1,* 1 (1979).

de Duve, C., Tissue fractionation—past and present. *J. Cell Biol. 50,* 20 (1971).

Grabske, R. J., Separating cell populations by elutriation. *Fractions No. 1,* 1 (1978).

Herzenberg, L. A., Sweet, R. G., and Herzenberg, L. A. Fluorescence-activated cell sorting. *Sci. Am. 234,* 108 (1976).

Horan, P. K., and Wheeless, L. L., Quantitative single cell analysis and sorting. *Science 198,* 149 (1977).

Pretlow, T. G., Weir, E. E., and Zettergren, J. G. Problems connected with the separation of different kinds of cells. *Int. Rev. Exp. Pathol. 14,* 91 (1975).

Sheeler, P., Reorienting density gradient zonal centrifugation. *Am. Lab. 3,* 19, (1971).

Shortman, K., Physical procedures for the separation of animal cells. *Annu. Rev. Biophys. Bioengineering 1,* 93 (1972).

Books, Monographs and Symposia

Anderson, N. G. (editor), *The Development of Zonal Centrifuges and Ancillary Systems for Tissue Fractionation and Analysis* (Natl. Cancer Inst. Monograph 21), U.S. Dept. Health, Education and Welfare, U.S. Govt. Printing Office, Washington, D.C., 1966.

Cutts, J. H. *Cell Separation. Methods in Hematology*. Academic Press, Inc., New York, 1970.

Dorvyl, G. (editor), *European Symposium of Zonal Centrifugation in Density Gradient,* Editions Cite Novelle, Paris, 1973.

Hinton, R., and Dobrota, M., *Density Gradient Centrifugation,* North-Holland Publishing Co., Amsterdam, 1976.

Reid E. (editor), *Separations with Zonal Rotors.* University of Surrey Printers, Guildford, U. K., 1971.

Reid, E. (editor), *Methodological Developments in Biochemistry,* Vol. 3, *Advances with Zonal Rotors,* Longman Group, Ltd., London, 1973.

Reid, E. (editor), *Methodological Developments in Biochemistry,* Vol. 4, *Subcellular Studies,* Longman Group, Ltd., London, 1974.

Trautman, R. ''Ultracentrifugation,'' in *Instrumental Methods of Experimental Biology* (D. W. Newman, ed.), Macmillan, New York, 1964.

Chapter 13
THE ISOLATION AND CHARACTERIZATION OF CELLULAR MACROMOLECULES

Over the years, a number of sophisticated analytical and preparative techniques have been developed for separating, analyzing, and isolating the various macromolecular constituents of cells and tissues. Various forms of electrophoresis, chromatography, gel filtration, and ultracentrifugation are now in routine use in most laboratories engaged in molecular biological studies and have greatly increased our understanding of the chemistry and properties of the cellular macromolecules. Most of the methods used to separate and isolate different members of a class of macromolecules simultaneously provide information concerning their chemistry because parameters such as molecular size, density, net and absolute charge, differential solubility, and so forth are used as the basis for the separation.

Nearly all the techniques used for separating and isolating macromolecules require that these substances initially be in a dissolved state. Consequently, macromolecules in extracellular fluids such as plasma, lymph, and digestive and hormonal secretions are most easily isolated and have been the subject of the most intensive studies to date (e.g., albumin, globulins, pepsin, chymotrypsin, insulin, ACTH, etc.). However, if the component to be isolated is normally a constituent of the soluble cell phase or cytosol, then centrifugal isolation of this phase from disrupted cells quickly provides the starting material. Greater difficulty is encountered when macromolecules are to be isolated from particulate cell components such as nuclei, mitochondria, cell membranes, and ribosomes where the molecule may be an integral part of the organelle's structure.

In Chapter 12, a number of physical methods were described for disrupting cells. Often, more rigorous or more extensive applications of the same procedures to whole cells or to isolated organelles will also free some of the constituent macromolecules. For example, extended insonation or homogenization of mitochondrial suspensions renders many of the mitochondrial enzymes and other constituents soluble, and nucleic acids may be released from isolated nuclei under similar conditions. Chemical procedures may also be employed to extract or solubilize the desired class of macromolecules. Lipids are often extracted from whole cells or isolated organelles using fat solvents such as chloroform-methanol mixtures or acetone. Proteins may be extracted from membranous elements using dissociating agents (such as urea and mercaptoethanol), chelating agents (such as ethylenediamine tetraacetic acid, EDTA) or organic detergents (such as sodium deoxycholate, sodium lauryl sulfate and Triton X-100). Whatever the method used, once a soluble mixture of macromolecules is obtained, these may then be separated from one another and isolated using one or a combination of methods.

In this chapter, a number of methods either regularly employed today by cell biologists for isolating and characterizing the macromolecules (proteins, carbohydrates, lipids, and nucleic acids) of cells or of significant historical interest will be considered. Most methods in routine use (see Table 13–1) rely upon differences in molecular size, shape, electrostatic charge, solubility, or biological activity to effect the separation.

Salting In and Salting Out

Among the oldest methods for fractionating and isolating mixtures of macromolecules (especially proteins) are chemical procedures that differentially alter a molecule's solubility. Among other factors, a protein is maintained in the dissolved (i.e., solubilized) state by the interaction of its charged groups with water (which is partially polar) and with salt ions in the solution. The salt concentration (or **ionic strength**) of the protein solution significantly and differentially influences the solubility of the proteins present. For example, many proteins are insoluble in pure (i.e., salt-free) water, whereas addition of small quantities of salt renders these molecules soluble. This effect is believed to be due to an interaction between the salt ions and certain charged groups of the protein that otherwise react with each other, resulting in insolubility. Even proteins that are soluble in distilled water may be dissolved in much greater quantities by the parallel addition of small amounts of salts. This phenomenon is known as **salting in.**

If the salt concentration of a protein solution is successively increased, a point is eventually reached at which some of the proteins begin to precipitate. Further addition of salt results in greater precipitation. Precipitation occurs because spheres of hydration formed around the salt ions effectively "remove" the water molecules necessary to hydrate certain surface charges of the protein. Protein—protein interactions begin to dominate over protein—solvent interactions with the resulting precipitation of the proteins.

Table 13–1
Methods for the Isolation and Characterization of Cellular Macromolecules

Method	Principal Impelling Factor(s)	Principal Retarding or Opposing Factor(s)	Separation Depends Primarily upon
Countercurrent distribution	Mechanical	Solubility	Differential partition
Dialysis/ultrafiltration	Osmotic effects, concentration gradients, hydrodynamic force	Molecular sieve effects	Molecular size
Ultracentrifugation	Centrifugal force	Friction, bouyancy, diffusion	Molecular size, shape, effective density
Electrophoresis			
Moving-boundary	Electrostatic force	Friction, diffusion	Molecular ionic properties
Zone	Electrostatic force	Friction, diffusion, molecular sieve effects	Molecular ionic properties
Discontinuous	Electrostatic force, Kohlrausch function	Friction, diffusion, molecular sieve effects	Molecular ionic properties and molecular size
Immunoelectrophoresis	Electrostatic force	Diffusion, molecular sieve effects	Molecular ionic properties, biological activity
Isoelectric focusing	Electrostatic force	Diffusion	Molecular ionic properties
Paper chromatography	Hydrodynamic force	Association/dissociation effects, diffusion	Adsorption/partition differences
Thin-layer chromatography	Hydrodynamic force	Association/dissociation effects, diffusion	Adsorption/partition differences
Ion-exchange chromatography	Hydrodynamic force	Electrostatic forces	Molecular ionic properties
Affinity chromatography	Hydrodynamic force	Molecular affinity	Biological activity
Gel filtration	Hydrodynamic force	Molecular sieve effects	Molecular size
Gas chromatography	Gas pressure	Diffusion	Adsorption/partition differences

In effect, the proteins are being "squeezed out of solution." The phenomenon is called **salting out.** The effect of salt concentration on the solubility of hemoglobin is shown in Figure 13–1.

Because different proteins are salted out at different salt concentrations, solutions containing mixtures of different proteins may be fractionated using this approach. Among the most popular salts used is ammonium sulfate, $(NH_4)_2SO_4$, because of its high solubility in water and because it does not irreversibly denature most proteins. The ammonium sulfate concentration of the protein solution is serially increased, each increment precipitating another group of proteins which are then removed (usually by filtration or centrifugation) before the next addition of salt.

Isoelectric Precipitation

The distribution of polar groups in proteins and nucleic acids is influenced by pH. For a given protein, a pH may be identified at which there are equal numbers of positive and negative charges, and this is known as the **isoelectric point** (or **isoelectric pH**). Most proteins are *least* soluble and many proteins are *insoluble* at the isoelectric point, so that their removal from solution is most easily achieved by first adjusting the pH of the solution to the isoelectric point. Electrostatic charge distribution and isoelectric properties of proteins and other macromolecules are considered in detail later in the chapter in connection with electrophoresis.

Dialysis and Ultrafiltration

Semipermeable membranes, such as those prepared from *cellophane* or *collodion* (cellulose nitrate), may be used to separate solutes on the basis of molecular weight differences. In **dialysis,** the solute mixture is placed in a bag formed from tubular sheets of the semipermeable membrane, and the bag is immersed in an aqueous medium (usually distilled water). Molecules larger than the pores of the membrane are confined to the tubing, while smaller molecules diffuse into the surrounding liquid (Fig. 13–2). Semipermeable membranes can be treated chemically or physically in order to alter the sizes of the pores so that solutes of varying molecular weight are rendered permeable. Generally, dialysis is used with unmodified membranes to quickly separate low-molecular-weight solutes (i.e., molecular weight of less than 5000) such as salts, sugars, and amino acids from proteins and polysaccharides present

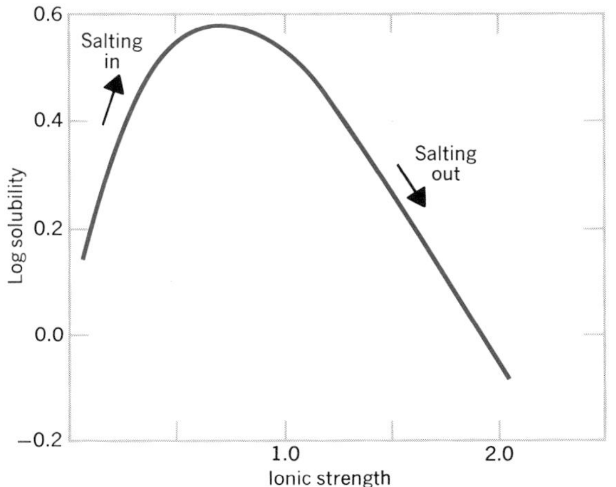

Figure 13–1
Salting in and salting out of carbonmonoxyhemoglobin using K_2SO_4. Ionic strength is given by the equation $I = \frac{\Sigma M_n Z^2_n}{2}$ where *M* is the molar concentration of each ionic species and *Z* is the electrostatic charge of each ion.

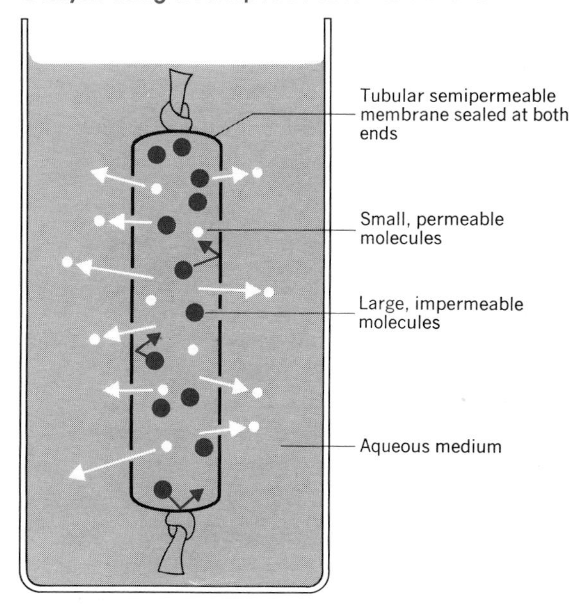

Figure 13–2
Dialysis using a semipermeable membrane.

in the solution. However, high-molecular-weight solutes do *differentially* penetrate membranes having large pore sizes (Table 13–2).

In **ultrafiltration,** force is used to drive the smaller molecules *along with solvent* through the semipermeable membrane. As a result, not only are permeable and impermeable molecules separated but the impermeable species is also simultaneously *concentrated* (Fig. 13–3).

Ultracentrifugation

Centrifugation can be employed not only for the separation of cells, subcellular organelles, and other particulate constituents of cells but also for molecular separations. Since its initial development in the 1920s by Svedberg, the **analytical ultracentrifuge** (described in Chapter 12) has been used repeatedly to evaluate the heterogeneity or purity of molecular constituents extracted from cells and to estimate molecular sizes on the basis of sedimentation rate. Physical separations (as opposed to analytical studies) became possible with the development of ultracentrifuges and preparative rotors capable of generating an RCF in excess of 500,000 *g*. Using forces of this magnitude, true separations and isolations of molecular constituents of cells in rate or isopycnic gradients have become routine.

Figure 13–3
Ultrafiltration through a semipermeable membrane using hydrodynamic force.

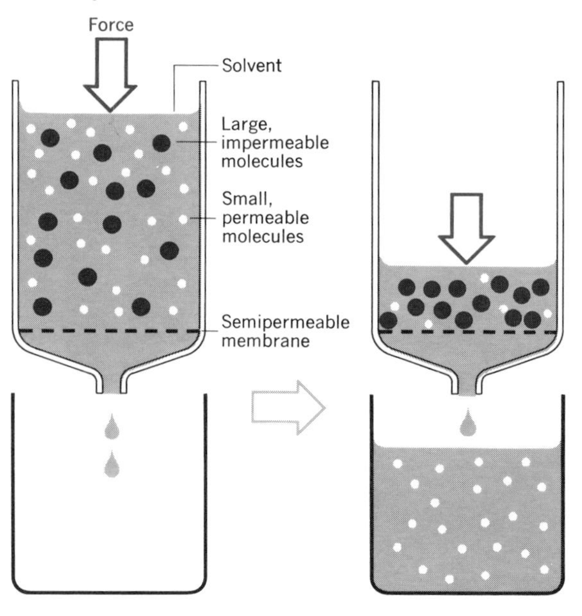

Table 13–2
Relationship Between Molecular Weight and Permeability to Modified Cellophane Membranes

Solute	Molecular Weight	$T_{1/2}^{a}$ (min.)
Tryptophan	204	4
Bacitracin	1,422	15
Cytochrome *c*	12,000	60
Ribonuclease	13,600	120
Lysozyme	14,000	138
Trypsin	20,000	240
Chymotrypsin	25,000	300
Pepsin	35,500	4,800

[a] $T_{1/2}$ is the amount of time required for one-half of the solute to permeate the membrane.

The principles of analytical and preparative centrifugation were considered at length in Chapter 12 and will not be pursued here. However, it should be noted that certain problems are encountered during the fractionation of macromolecules in density gradients that do not arise during particle or organelle fractionations. Most important among these problems is **diffusion.** Macromolecules (or, for that matter, any solute) in solution exhibit a net migration *away* from any region in which they are concentrated and *toward* regions containing a lower concentration of the solute. This movement, called diffusion, results in the *uniform* distribution of molecules throughout the space that they occupy.

The rate at which diffusion occurs is given by Ficke's law:

$$\frac{d(A)}{dt} = -D\,(A) \qquad (13\text{-}1)$$

in which $d(A)/dt$ is the rate of change in the localized concentration of molecular species A as diffusion ensues, and D is a constant (called the **diffusion constant**) that characterizes the behavior of molecule A in a given solvent. It should be noted that the rate of change is proportional to the concentration of A, so that the tendency for diffusion increases as the concentration of the diffusing species increases. It follows that the banding of a macromolecule in a density gradient is an attempt to concentrate that molecule and is increasingly opposed by the diffusion of that molecule. For large cellular particles or organelles, the effects of diffusion are negligible because the diffusion constants of large particles are so small. However, the diffusion constants of proteins, nucleic acids, and other macromolecules are considerably greater; consequently, during centrifugation, an equilibrium between sedimentation and diffusion is eventually reached, with the result that no further concentration of a given molecular species by isopycnic banding is possible. In effect, diffusion places a limit upon the resolution attainable when separating macromolecular mixtures in density gradients by ultracentrifugation. It is clear from Table 13–1 that diffusion also influences the effectiveness of other methods of macromolecular fractionation.

Electrophoresis

The term **electrophoresis** originally described the migration of charged particles through a liquid or semisolid medium under the influence of an electrical potential. More recently, the word has come to refer to any technique by which molecules are separated from one another in electrical potential gradients on the basis of differences in their net charges regardless of the conducting medium. The method is employed most often for the separation of different proteins, since the side chains of many of the constituent amino acids exist in a dissociated form and contribute some number of positive and negative charges to the macromolecule (see Chapter 4). However, other molecules such as carbohydrates can also be separated by electrophoresis after being complexed with inorganic ions or other charged groups.

The fundamental principles of electrophoresis are quite simple. If two electrodes are inserted into a solution containing an electrolyte and a suspension of macromolecules of varying net charge, the macromolecules will be accelerated toward the electrode of opposite sign with a force proportional to the magnitude of the charge on the macromolecule and the strength of the applied electrical potential. Since each of the migrating particles will encounter frictional resistance to its movement, each will soon attain some maximum velocity of migration. Therefore, if a mixture of these macromolecules is exposed to a constant field strength, the maximum velocities attained will be different for particles of differing net charge. In addition to net charge and field strength, several other factors influence the rate of electrophoretic migration of proteins, including molecular size and shape, pH, and the nature of the medium through which migration occurs. These factors will be considered later.

The rate at which a protein migrates toward one or the other electrodes during electrophoresis is called its **electrophoretic mobility** (usually expressed in square centimeters per second per volt) and is dependent on the relative numbers of positively charged and negatively charged amino acid side chains. Whether or not a particular amino acid side chain carries a charge is, in turn, determined in part by the pH of the protein solution. As the pH is lowered (i.e., the concentration of H^+ is increased), negatively charged groups such as the secondary COO^- of aspartic and glutamic acid and the O^- of tyrosine become protonated, thereby neutralizing these negative charges. At the same time, some secondary amino groups such as those of lysine and arginine may accept additional protons, thereby increasing the number of positive charges associated with the protein. In contrast, as the pH is raised (i.e., the concentration of OH^- is increased), protons are dissociated from these side chains and make the protein more negative. These relationships are shown in Figure 13–4. Therefore, the numbers and types of charges associated with the amino acid side chains of a protein are determined by pH. At low pH, proteins tend to carry more positive than negative side chains and, therefore, possess a net positive charge and migrate toward the cathode (the negative electrode) during electrophoresis. At high pH, negatively charged side chains predominate, and the protein migrates toward the anode.

It follows from the above discussion that for every protein there will be a pH at which the numbers of positive and negative charges will be equal. If electrophoresis is carried out at this pH, no migration occurs, since the protein has no net charge. Above this pH, the net charge on the protein becomes increasingly negative, and its electrophoretic mobility toward the anode increases. Below this pH, electrophoretic mobility toward the cathode increases. The relationships between pH and electrophoretic mobility for the proteins egg albumin and plasma beta lactoglobulin are shown in Figure 13–5. The pH at which a protein possesses no electrophoretic mobility is called the **isoelectric point** and is a characteristic of each protein (Table 13–3). Since pH markedly influences electrophoretic mobility, it is important to maintain a constant electrolyte pH during elec-

Aspartic acid residue

$$HC—CH_2—COO^{\ominus} \underset{OH^{\ominus}}{\overset{H^{\oplus}}{\rightleftharpoons}} HC—CH_2—COOH$$

Lysine residue

$$HC—(CH_2)_4—NH_2 \underset{OH^{\ominus}}{\overset{H^{\oplus}}{\rightleftharpoons}} HC—(CH_2)_4—NH_3^{\oplus}$$

$$HC—CH_2—C_6H_4—O^{\ominus} \underset{OH^{\ominus}}{\overset{H^{\oplus}}{\rightleftharpoons}} HC—CH_2—C_6H_4—OH$$

Tyrosine residue

Figure 13–4
Influence of pH on the charges of amino acid side chains. As the pH is lowered by addition of H^+ negative side chains of certain amino acids accept H^+ and are neutralized while amino groups of other side chains are protonated and contribute additional positive charges to the protein. In contrast, when the pH is raised, protons are dissociated from certain side chains, increasing the number of negative charges on the protein.

Figure 13–5
For each protein, a characteristic curve relates electrophoretic mobility to pH. The curves for beta lactoglobulin of blood plasma and albumin of egg white are shown.
The respective isoelectric points of these two proteins (indicated by arrows) are 5.1 and 4.6.

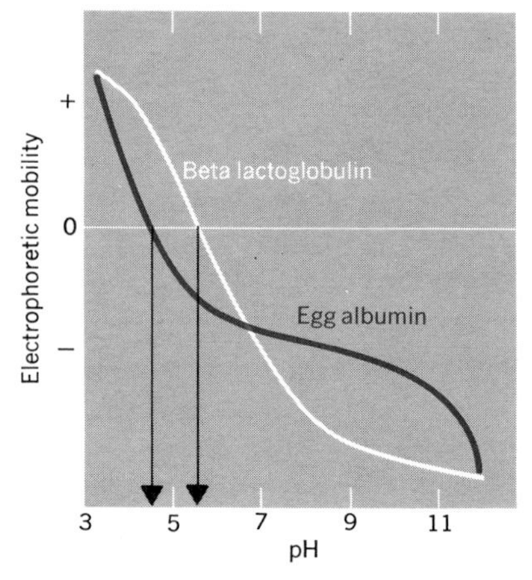

Table 13–3
Isoelectric Points of Some Proteins

Protein	Isoelectric pH
Lysozyme	11.0
Cytochrome *c*	10.6
Ribonuclease	9.5
Normal human hemoglobin	7.1
Myoglobin	7.0
Horse hemoglobin	6.9
Transferrin	5.9
Fibrinogen	5.8
Insulin	5.4
Beta lactoglobulin	5.1
Urease	5.0
Plasma albumin	4.8
Egg albumin	4.6
Haptoglobin	4.1
Pepsin	1.0

trophoresis. This is accomplished by including buffers in the electrolyte solutions.

Moving-Boundary Electrophoresis

Electrophoresis of proteins was introduced in its original form by Arne Tisulius in 1937 and called ''moving-boundary'' electrophoresis. The protein mixture, dissolved in a buffer solution that served as the electrolyte and maintained the desired pH, was placed in a glass tube connected to electrodes. When an electrical potential was applied across the tube, the protein molecules migrated toward one or the other electrodes. Since different proteins migrate at different rates, a number of interfaces or boundaries formed between the leading (and trailing) edge of each protein type and the remaining mixture. A Schlieren optical system recorded the number of boundaries formed and their rates of migration. Moving-boundary electrophoresis was used for the analysis but not for the fractionation of complex mixtures of proteins, since the proteins in the mixture were not really separated from each other.

Zone Electrophoresis

Zone electrophoresis offers a number of important advantages over moving-boundary electrophoresis. In zone electrophoresis, the physical separation of the different proteins in a sample is actually realized. For this reason, the technique can be *both* analytical and preparative. Generally, zone electrophoresis also yields greater resolution of the protein components of a mixture. In zone electrophoresis, the proteins are separated in a semisolid or porous supporting medium such as filter paper or various types of gels (polyacrylamide, cellulose acetate, hydrolyzed starch, etc.); these usually take the form of narrow sheets or slabs.

The principles of zone electrophoresis may be illustrated by considering the filter paper technique as an example. The strip of filter paper is saturated with the buffer/electrolyte solution to be employed and is tautly suspended between two (inner) baths (Fig. 13–6). The inner baths communicate with two additional (outer) baths containing the electrodes; the connections may be achieved using baffles or wicks saturated with the buffer electrolyte. (The electrodes are not inserted directly into the inner baths because electrolysis would dramatically alter the pH of the bath and also the filter paper strip. By confining the electrolysis to the outer baths, the pH of the inner baths and the filter paper remains constant.) The mixture of proteins to be separated is applied as a narrow zone perpendicular to the long axis of the filter paper.

When an electrical potential is applied, the proteins in the sample zone migrate through the filter paper toward the appropriate electrode. In so doing, they form a number of discrete zones distributed along the length of the paper (Fig. 13–6). Following separation, the individual zones or bands may be visualized using special stains. Alternatively, the

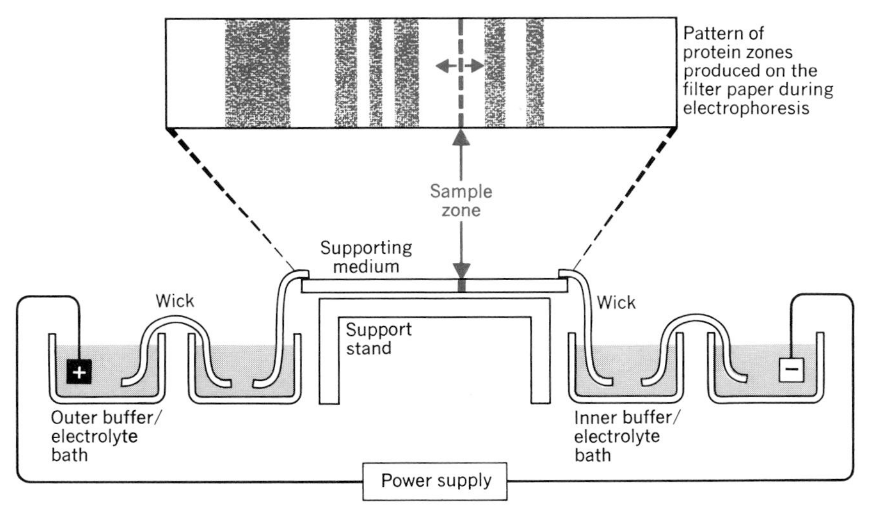

Figure 13–6
Essential components of a zone electrophoresis assembly.

paper strip may be cut into a number of sections and the proteins in each eluted and collected for further study.

Since the medium through which the proteins migrate is porous, the size of the protein influences its electrophoretic mobility—smaller proteins migrating more rapidly than larger proteins of equal net charge. Consequently, the sieving effect of the supporting medium results in a separation based upon both molecular size and charge. Polyacrylamide gels are especially effective for zone electrophoretic separations, since the sieving effect may be varied by changing the concentration of acrylamide in the gel (i.e., the greater the acrylamide concentration, the smaller the pore size in the gel).

An interesting recent modification of zone electrophoresis is "diagonal electrophoresis" or *two-dimensional* electrophoresis. In this approach, the sample is separated first on a rectangular sheet of filter paper or polyacrylamide gel at high pH, the supporting medium then rotated 90° and equilibrated at low pH, and the separated zones electrophoresed again.

The result is a better resolved distribution of the components in the original sample, since it is highly unlikely that two components would have the same electrophoretic mobility at grossly different pH's.

Disc or Discontinuous Electrophoresis

Discontinuous electrophoresis is a specialized form of zone electrophoresis developed in the early 1960s by L. Ornstein and B. Davis. In view of its widespread use and importance as a research tool, it will be considered separately. The procedure provides the highest degree of resolution so far attainable by electrophoretic methods, and this is a consequence of the extreme thinness of starting zones that may be achieved. Disc electrophoresis is carried out in either cylindrical blocks or sheets of polyacrylamide gel, the pore sizes of which may be accurately controlled by regulating the concentration of polyacrylamide in the gel. Consequently, the gel acts like a molecular sieve, and separation is based on both the charge and the size of the macromolecule.

In order to separate two proteins by conventional methods of zone electrophoresis, it is necessary to permit migration to continue until one of the proteins has traveled at least one starting zone thickness further than the other. For example, suppose that the sample zone was 1 mm thick and contained two different proteins. If one of the proteins had no electrophoretic mobility at all under the pH conditions employed, then the other would have to migrate at least 1 mm in order for separation of the two proteins to occur. If both proteins migrate in the field, then an even greater distance must be traversed by the protein of higher electrophoretic mobility. Since the sharpness of the zone occupied by each protein decreases with time owing to diffusion in the gel, it is most desirable to begin with the narrowest starting zone possible and to minimize the duration of electrophoresis. In disc electrophoresis, the sample zone undergoes a preliminary concentration during which its thickness is reduced from one or more centimeters to just a few microns; as a result, high resolution is achieved in very short runs. The preliminary concentration also makes it possible to analyze samples too small or too dilute to be studied using other methods.

The concentration of the starting zone during disc electrophoresis is based on a phenomenon first noted and described by F. Kohlrausch in 1897. He showed that if two solutions of ions having significantly different electrophoretic mobilities are layered over one another (slow ion above, fast ion below) and subjected to an electrical field, the boundary between the two ionic species would be sharply maintained as it migrates downward. That is, under these conditions the ions of lower electrophoretic mobility migrate at the same rates as the faster ions. This may be explained as follows: The velocity at which an ion migrates in an electrical field is determined by the product of its electrophoretic mobility and the applied voltage gradient (volts/centimeter). Therefore, an ion of low mobility can migrate as rapidly as an ion of high mobility if their mobility–voltage products are equal. In the example cited above, at the instant voltage is applied, the trailing edge of the fast ions at the boundary moves away from the leading edge of the slow ions, resulting in the temporary formation of a zone of lower conductivity and increased voltage. The increased voltage in this zone accelerates the slow ions to keep up with the fast ions, thereby creating a steady state in which the mobility–voltage products of the two ions remain constant.

Taking this a step further, consider the situation in which a region below the fast ion–slow ion boundary contains a number of ions of intermediate electrophoretic mobility (Fig. 13–7*a*). When an electrical potential is applied, the increased voltage gradient behind the downward-moving boundary accelerates both the ions of low mobility and

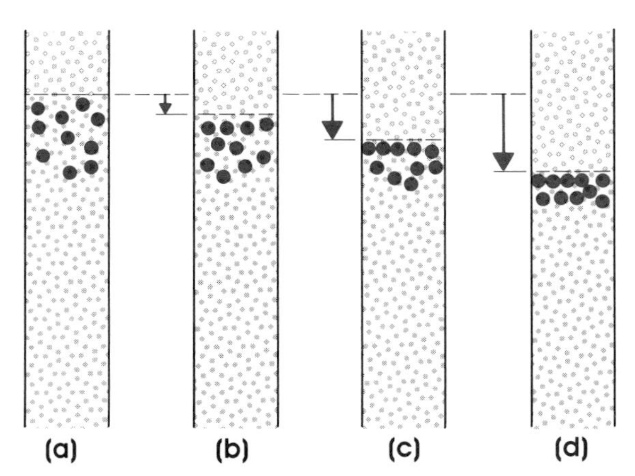

Figure 13–7
Concentration of ions of intermediate electrophoretic mobility (color) at the boundary between fast (small, closed circles) and slow (small, open circles) ions. *(a)* Before an electrical potential is applied, the ions of intermediate mobility occupy a region below the slow ion—fast ion boundary. *(b)* When an electrical potential is applied, the boundary moves downward and an increased voltage gradient is formed behind. *(c)* and *d)* The increased voltage gradient accelerates both the slow ions and the ions of intermediate mobility, causing the latter to be concentrated into a narrow zone between the trailing edge of the fast ions and the leading edge of the slow ions. This phenomenon underlies the formation of extremely thin starting zones during disc electrophoresis of proteins.

those of intermediate mobility. The mobility–voltage product of the latter, however, will be somewhat greater, so that these ions will be swept up by the boundary and form a zone of decreasing thickness between the fast and slow ions (Figs. 13–7 *b, c, d*). If instead of a single ionic species of intermediate mobility, a mixture is placed in the region below the slow ion–fast ion boundary, these would be concentrated into narrow zones stacked one above the other in order of decreasing mobility. In disc electrophoresis, conditions are chosen such that the proteins in the sample have mobilities intermediate to two especially selected fast and slow ions.

Disc electrophoresis gels are cylindrical blocks or flat sheets of polyacrylamide suspended between two reservoirs of buffer/electrolyte (Fig. 13–8). The polyacrylamide gel is divided into three regions called the **sample gel, stacking gel,** and **separating gel.** The sample gel contains the mixture of proteins to be separated and is prepared using low concentrations of polyacrylamide so that pore sizes are large and do not affect protein migration. The stacking gel is similar to the sample gel but lacks the proteins. The sample gel, stacking gel, and reservoirs have the same pH (usually 8.3). The separating gel differs from the other two regions in that it has a higher pH (usually 9.5) and is prepared with greater concentrations of polyacrylamide; this results in smaller pore sizes and provides the sieving effect. All three regions of the polyacrylamide gel contain the fast (leading) ion, whereas the buffer/electrolyte contains the slow (trailing) ion. In most instances, Cl^- serves as the fast ion and gylcine (NH_2-CH_2-COO^-) as the slow ion; other amino acids or weak acids may also serve as the slow ion. At pH 8.3, nearly all proteins have an electrophoretic mobility between that of Cl^- and glycine.

Figure 13–8
Essential components of discontinuous electrophoresis.

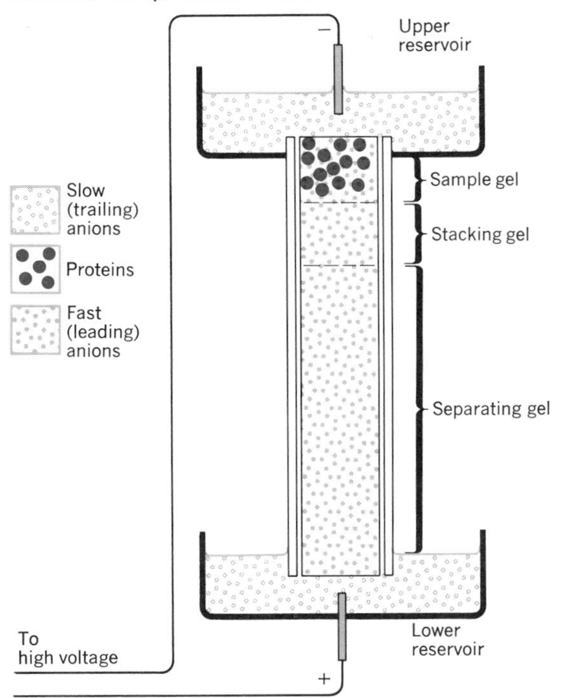

When an electrical potential is applied, the voltage gradient causes the chloride ions to migrate down from the top of the gel. The increased voltage gradient created immediately behind the trailing edge of the chloride ions accelerates the slower glycine ions to keep pace. As the Cl^- -glycine boundary moves downward, it overtakes the more slowly migrating proteins, and they, too, are accelerated by the increased voltage gradient behind the boundary. As a result, by the time the trailing edge of the chloride ions

Figure 13–9
Electrophoretic pattern of human plasma proteins obtained by discontinuous electrophoresis. Five samples (varying in concentration) have been separated in a single slab of polyacrylamide gel. More than 25 different proteins are resolved.

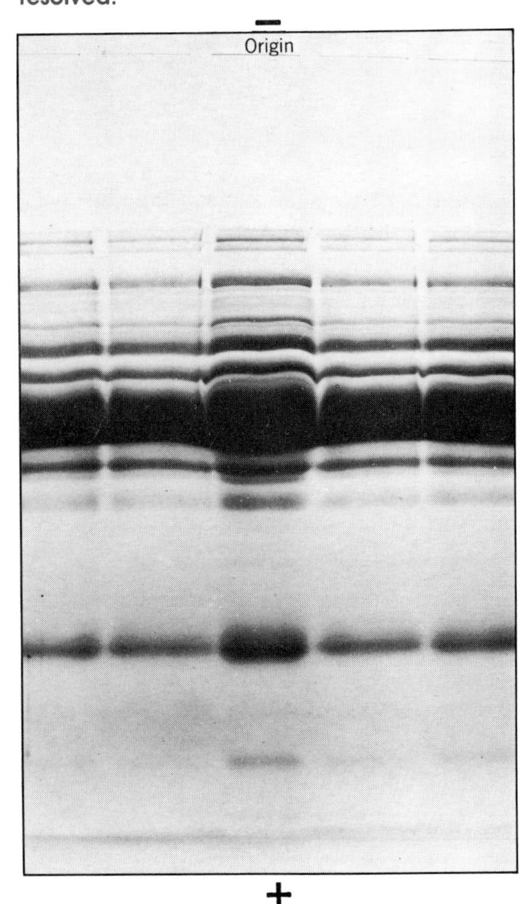

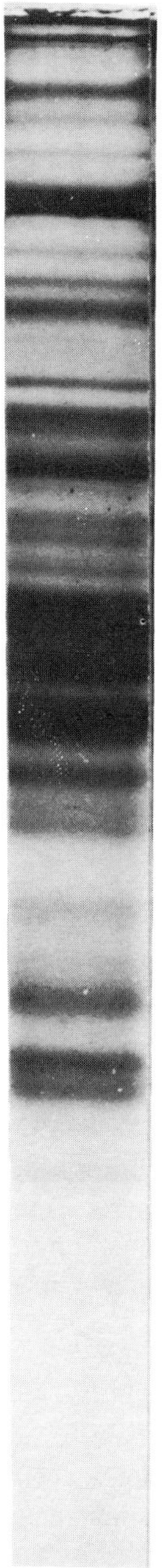

Figure 13–10
Discontinuous electrophoretic separation of proteins extracted from isolated plasma membranes of liver cells.

reaches the end of the stacking gel, all the proteins in the original sample have been concentrated into a series of contiguous thin zones.

Once the separating gel is reached, the change in pH dramatically alters the electrophoretic mobility of the trailing ion. In the case of glycine, the degree of ionization is very low at pH 8.3 but is several times greater at pH 9.5. As a result, the glycine ions now overtake each of the protein zones and catch up with the trailing edge of the Cl^- (the mobility of Cl^- is unaffected by the pH change), and the new interface then migrates rapidly through the remainder of the separating gel. The proteins, on the other hand, are physically retarded by the smaller pores in the separating gel, and since each protein zone is now in a uniform voltage gradient, these separate from one another strictly on the basis of net charge and size differences (just as in ordinary zone electrophoresis).

It is apparent from the foregoing discussion that disc electrophoresis differs markedly from ordinary zone electrophoresis and is a clever and unusual technique providing extraordinary resolution (Fig. 13–9). The method has been widely accepted since its introduction and has been applied in diverse cell studies, especially in the analysis of the protein components of isolated cell organelles. An example of this is given in Figure 13–10, which shows more than 30 different protein components separated from the plasma membranes of liver cells.

Immunoelectrophoresis

Immunoelectrophoresis, a technique developed in the 1950s by C. A. Williams and P. Grabar, combines the principles of electrophoresis and immunochemistry in order to separate and identify antigens, antibodies, and other proteins. In this technique, a portion of the sample of proteins (antigens) to be analyzed is first injected into an experimental animal (usually a horse or cow). After a period of time during which the animal produces antibodies against the injected proteins, the immunoglobulin-rich blood serum is collected and prepared as the **antiserum.** The remaining portion of the sample to be analyzed is then subjected to conventional zone electrophoresis in a rectangular block of agar or other supporting media. This distributes the various proteins of the original sample as a linear series of zones (Fig. 13–11 *a*). A trough cut in the agar block parallel to the direction of electrophoretic migration is then filled with the anti-

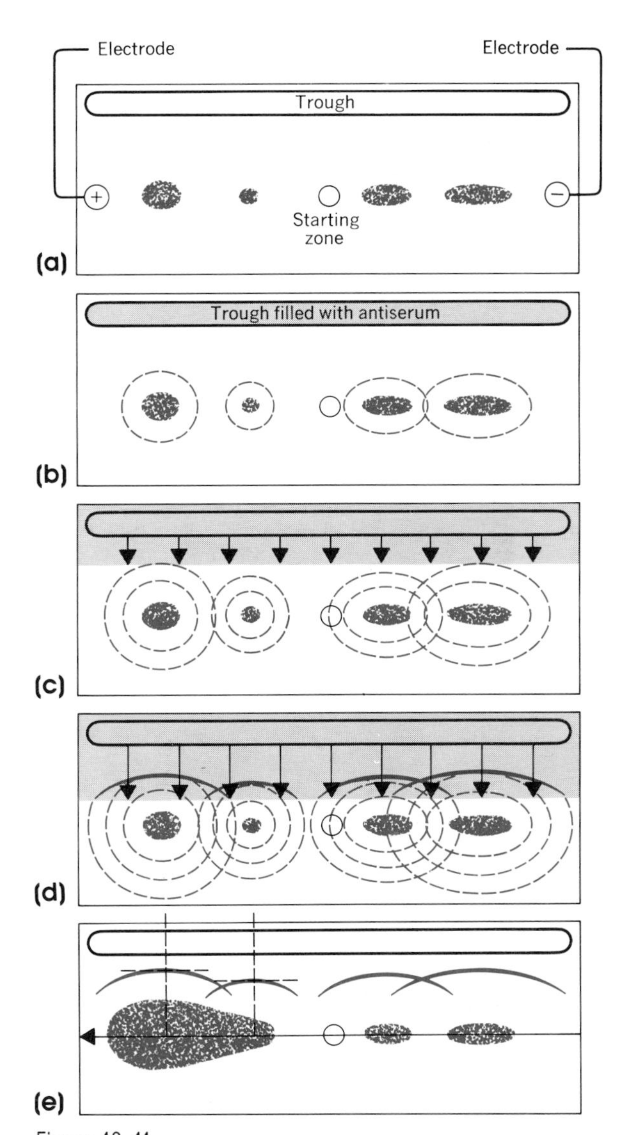

Figure 13–11
Immunoelectrophoresis. *(a)* Mixture of proteins separated by zone electrophoresis. (*b* and *c*) Trough filled with antiserum; diffusion of protein zones and antibodies. (*d*) Antibody-antigen reaction forms lines of precipitate (*e*). By drawing a line perpendicular from the widest point of the arc to the path of electrophoretic migration, the center of concentration of each protein can be determined.

serum. The protein zones in the agar block diffuse radially and eventually encounter the front of inwardly diffusing antibody, where the resulting antibody–antigen reaction produces a fine, arc-shaped line of precipitation (Fig. 13–11*d*). Each line of precipitation corresponds to a *pair* of precipitants (a protein in the sample plus an antibody from the antiserum); crossed arcs indicate that all the reactants are immunochemically different proteins.

Isoelectric Focusing

Isoelectric focusing is a recently developed form of electrophoresis used to determine the isoelectric pH's of proteins and to separate proteins from one another on the basis of differences in their isoelectric points. The method is extremely sensitive; proteins having isoelectric pH's that differ by as little as 0.02 units may be resolved. In isoelectric focusing, a glass column is filled with a solution of synthetic charged ampholytes (aliphatic aminocarboxylic acids of various molecular weights) having a broad and continuous range of isoionic points. The ends of the column contain the electrodes and electrode solutions: a strong organic base for the cathode (usually ethanolamine) and a strong acid for the anode (usually phosphoric acid). The distribution of constituents in the column is stabilized by the presence of a sucrose density gradient that also prevents convection during electrophoresis. The mixture of proteins to be separated may initially be confined to a specific region of the column or uniformly distributed through the density gradient (Fig. 13–12).

When a potential is applied to the electrodes, the ampholytes migrate in the column to positions where they are equally attracted by both electrodes. Thus, they arrange themselves in order of their isoelectric points, the most acidic ampholytes being located near the anode and the most basic near the cathode. An examination of the pH in different regions of the column at the time would reveal that a pH gradient has been established in which pH increases in progressing from the anode to the cathode. Depending upon the pH of the surrounding solution and the nature of the amino acid side chains present in the molecule, each protein assumes a characteristic net charge. Proteins close to the anode will bear a net positive charge and be repelled by that electrode, while the same proteins near the cathode carry net negative charges and are repelled by that electrode. As a result, most proteins will migrate away from the electrodes and toward the center of the column. As this occurs, the proteins pass through the pH gradient, and the charges on their amino acid side chains are altered. Proteins migrating away from the anode become less and less positive as they pass through regions of increasing pH, whereas proteins migrating away from the cathode become less negative (see Fig. 13–12). Eventually, the net charge of each protein becomes zero as that region of the pH gradient corresponding to the isoelectric pH is reached, and at that time the electrophoretic migration of the protein ceases. Consequently, the proteins become distributed through the gradient as a series of narrow zones in order of their isoelectric points (see Fig. 13–12). Once the migration of all proteins terminates, the contents of the column may be drained and the separated proteins collected as a series of

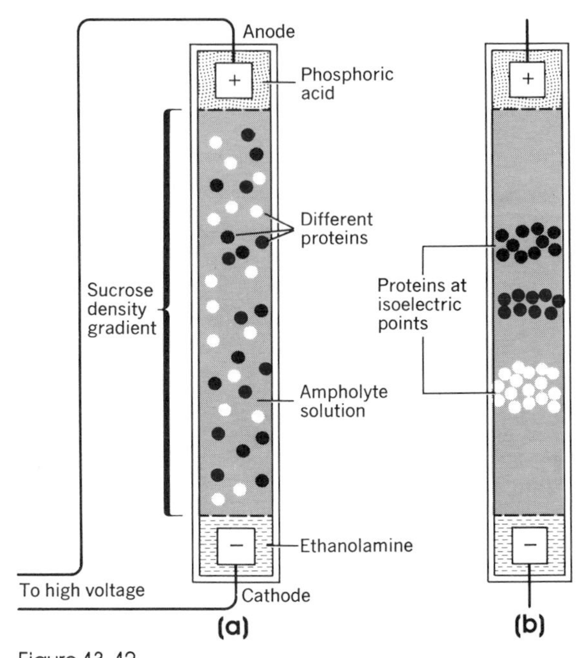

Figure 13–12
Essential features of isoelectric focusing. In (*a*) three proteins having different isoelectric points are shown distributed through the ampholyte-sucrose density gradient. When an electrical potential is applied, the ampholytes migrate to their isoionic points establishing a pH gradient. This is followed (*b*) by electrophoresis of proteins to their isoelectric positions in the gradient, forming a series of narrow zones.

fractions and further studied. The isoelectric point of each protein is determined from the pH profile of the collected gradient.

A description of all electrophoretic methods currently in use is beyond the scope of this discussion. However, it should be recognized that other forms of electrophoresis such as **density gradient** electrophoresis and **continous-flow** electrophoresis are also regularly employed. In density gradient electrophoresis, the separation of proteins is carried out in a glass column filled with buffer and stabilized by a density gradient. Continous-flow electrophoresis provides for the continuous application of small volumes of sample onto the supporting medium, coupled with the continuous removal of the proteins as separation is achieved. With this technique it is possible to process very large volumes of starting material.

Although all methods of electrophoresis have proven extremely valuable for the separation and identification of proteins, certain limitations of the technique should be noted. The resolution of a protein mixture into a number of discrete zones does not guarantee that all the different proteins present in the original sample have been separated, for two or more different proteins having similar sizes may also have the same net charge under a given set of conditions and display the same electrophoretic mobilities (see Fig. 13–5). These proteins would not be resolved by zone electrophoresis. By the same token, two different proteins may have the same isoelectric pH and would not be separated by isoelectric focusing. Therefore, in order to evaluate the effectiveness of a separation and to determine the purity of the separated fractions, it is necessary to carry out electrophoresis under a variety of pH conditions or with a variety of supporting media or even to apply a combination of altogether different methods in the analysis.

Countercurrent Distribution

A brief, preliminary description of countercurrent distribution was given in Chapter 12 in connection with the separation of whole cells in mixed populations (e.g., blood). However, the technique lends itself more directly to the separation of various molecular species in solution; and because the physicochemical principles in effect during countercurrent distribution are the foundation of many other separation procedures (i.e., paper chromatography, thin-layer chromatography, ion-exchange chromatography, affinity chromatography, gas chromatography, etc.), this method will be treated in some detail.

It has been known for a century or more that many solutes differentially distribute or *partition* themselves between the separated phases of two immiscible liquids. Solutes having gross differences in their physical properties may therefore be separated by one or a few simple extraction procedures (e.g., using separatory funnels). Reasoning that the separation of a complex mixture of solutes on the basis of their partition differences could be made considerably more effective using a *cascade* of several dozen, several hundred, or even several thousand discrete extraction steps, L. C. Craig in the late 1940s and early 1950s developed a number of special instruments to achieve this. These instruments and their more modern counterparts are generally known as Craig apparatuses and the technique as *countercurrent distribution*.

The fundamental type of Craig apparatus consists of a series or *train* of chambers or ''cells'' divided into upper and lower compartments that house the respective immiscible liquid phases (e.g., butanol/water, phenol/water, propanol/water, etc.). The upper phase is usually the *mobile* phase, and the lower phase is the *stationary* phase. Equilibration of a dissolved solute between the phases within a chamber is achieved by agitation, so that one phase becomes finely dispersed in the other (thereby greatly increasing the interfacial surface area between the two phases). Once equilibration is achieved, the phases are allowed to separate. The mobile phase is then transferred to the next chamber of the train and mixed with a fresh volume of stationary phase. Simultaneously, a fresh volume of mobile phase is added to the first chamber. Agitation, equilibration, separation, and transfer are then serially repeated for the entire train of chambers.

A solute dissolved in a given pair of immiscible liquids will partition itself in a characteristic way—a specific solute concentration ending up in each phase. The *partition coefficient*, P, is given by

$$P = \frac{\text{concentration of solute in upper phase}}{\text{concentration of solute in lower phase}} = \frac{(S)_U}{(S)_L} \quad (13\text{–}2)$$

For example, if a solute has a partition coefficient of 1.0 in a butanol/water mixture, it will be *equally* divided between the butanol and the water phases once equilibration and separation have been achieved.

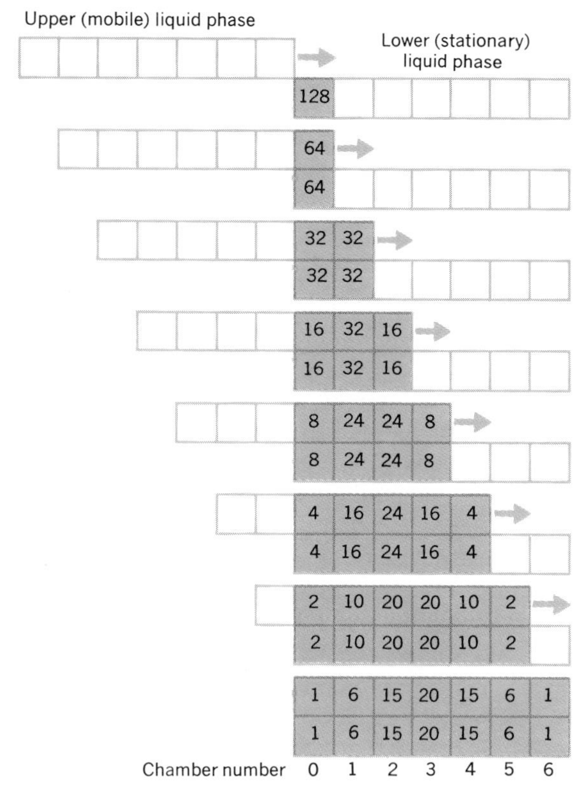

Figure 13–13
Countercurrent distribution of 128 molecules of a hypothetical solute having a *P* value of 1.0 in a train of seven chambers. At the outset, 128 molecules are dissolved in the lower half of chamber 0 and then equilibrated with an equal volume of mobile phase; this results in the equal partition of the solute in the two solvents. The upper phase of chamber 0 is then transferred and mixed with the lower phase of chamber 1, while fresh upper phase is added to chamber 0. Equilibration produces a new partition in chambers 0 and 1. The process is repeated until all seven chambers (i.e., C_0 through C_6) of the train have been used yielding a final distribution of the solute. In practice, from several dozen to several thousand partition steps may be employed.

Figure 13–13 dipicts the countercurrent distribution of 128 molecules of a hypothetical solute having $P = 1.0$ when carried through a train consisting of seven sets of chambers. In this example, the peak of the solute distribution in the train occurs in chamber number 3, which contains 40 molecules (or 31%) of the original solute. For a train consisting of $(C + 1)$ chambers (numbered C_0, C_1, C_2, C_3, etc.), each containing phases of equal volume, the fraction F of solute in compartment C_n at the conclusion of the countercurrent distribution is given by

$$F = \frac{C!P^n}{n!\,(C - n)!\,(P + 1)^C} \tag{13–3}$$

where n is the number of transfers needed to reach chamber C_n (and is equal to the chamber number). For the example given in Figure 13–13, the fraction of solute in chamber C_3 would be $6!(1.0)^3/3!(6–3)!(2.0)^6$, which equals 31%, as already noted. Figure 13–14 tabulates the final distributions of a mixture of five solutes having different P values through a train of seven chambers and illustrates how the peak concentrations of each solute would be found in different chambers in the train. Although it may be noted that in this example, considerable overlap exists among the chambers, it should be borne in mind that the number of chambers that make up the train has deliberately been kept low in order to simplify the example. Usually, a train consists of at least several dozen chambers. Accordingly, Figure 13–15 shows the resolution attainable for solutes having P values of 0.5 and 2.0 when the train consists of 30 chambers. It can be shown that the best resolution of two solutes, A and B, is achieved when $P_A \times P_B = 1.0$. The number of chambers occupied by a particular solute as a function of the total number

Figure 13–14
Percentage distributions in each chamber of a seven-chamber train of five solutes having various *P* values. Attention is drawn to those chambers containing the peak concentration of each solute.

P \ *C*	0	1	2	3	4	5	6
0.2	33	**40**	20	5	1	0	0
0.5	9	26	**33**	22	8	2	0
1.0	2	9	23	**31**	23	9	2
2.0	0	2	8	22	**33**	26	9
3.0	0	0	3	13	30	**36**	18

Rounded off to the nearest whole percent.

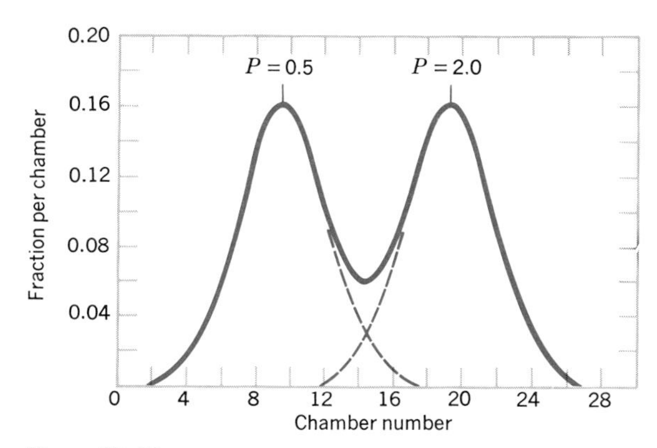

Figure 13–15
Countercurrent distribution of solutes having *P* values of 0.5 and 2.0 in a train consisting of 30 chambers.

of chambers present in the train rapidly diminishes as n increases. Hence, it is clear why trains in which hundreds or even thousands of transfers are possible may be used to provide maximum resolution of the solutes present in the original mixture.

The principles and mathematical relationships given above for a series of discrete partition transfers during countercurrent distribution apply equally well to seemingly quite distinct physical separation methods such as the various forms of chromatography described later in the chapter. In chromatography, the partition steps occur on numerous particles of tightly hydrated media such as starch, agarose, cellulose, silica, and paper fibers. This tightly bound solvent serves as the stationary phase, while a mobile phase containing solvent and the solute mixture to be fractionated flows by. This *continuous* process may be thought of as an infinite sequence of discrete, microscopic partition steps. Each microscopic step would be analogous to the stages shown in Figure 13-13, the upper mobile phase representing a continuously flowing solvent, and the lower stationary phase, the stationary supporting medium. Although equilibrium is not truly achieved at each "step," the total number of steps is so great that the solute mixture is effectively fractionated.

Countercurrent distribution is particularly effective when dealing with molecules having molecular weights below 5000 (peptides, oligonucleotides, phospholipids, etc.), but the technique is also used successfully with larger solutes. The first protein isolated and purified using countercurrent distribution was the hormone insulin, which has a relatively low molecular weight (i.e., 6500). However, ribonuclease (molecular weight of 13,600), lysozyme (mol. wt. of 14,000), albumin (mol. wt. of 68,000), and the α and β chains (mol. wt. of 16,000) of human hemoglobin have also been isolated using this approach. Little success is obtained with polysaccharides because of the problem of finding an organic phase providing useful partition coefficient differences. Ribonucleic acids are readily separated using countercurrent distribution; indeed, R. W. Holley's pioneering analysis of the primary structure of transfer RNA (Chapter 21) was carried out using alanyl-tRNA isolated and purified by countercurrent distribution. Lipids, especially steroids, are well suited for separation using this approach because of their greater solubility in organic solvents and the wide choice of solvent combinations that can be used.

Paper Chromatography

Paper chromatography is a technique in which a mixture of solutes is separated into discrete zones on a sheet of filter paper on the basis of (1) differences in solute partition between a stationary aqueous phase tightly bound to the cellulose fibers of the paper and a mobile organic liquid phase passing through the sheet by capillary action (i.e., liquid–liquid partition) and (2) differences in solute *adsorption* to the cellulose fibers and dissolution in the mobile liquid (i.e., solid–liquid partition). Although the separation is based upon a combination of both phenomena, liquid–liquid partition differences are the more significant. The principles of the technique were set down in the 1940s by R. Consden, A. H. Gordon, A. J. Martin, and R. L. Synge and are not unlike those that are in effect during countercurrent distribution.

In practice, a rectangular sheet of filter paper is saturated with the aqueous phase and allowed to air dry, and the sample is applied near the end of the sheet as a narrow zone. The sheet is then suspended in a closed chamber in which the air has been saturated with the vapors of the mobile organic phase, the edge of the paper immersed in a bath containing the mobile phase (Fig. 13-16*a*). Capillary action causes the mobile phase to slowly percolate through the paper from one end (the end containing the mixture to be separated) to the other. Movement of the liquid may be downward (descending chromatography) or upward (ascending chromatography). The solute mixture differentially partitions itself between the flowing solvent and the stationary phase time and time again as the solvent front advances toward the edge of the paper. Usually, the solvent front is allowed to migrate through the paper until it has almost reached the other end, at which time the sheet is removed and dried and the solute zones located by the appropriate chemical or physical means. However, in certain instances where the solutes trail far behind the solvent front, it is desirable to allow the solvent to run off the edge of the paper sheet (descending chromatography only) so that maximum resolution of the solutes is achieved.

The rate of movement of a solute during paper chromatography is usually expressed as a dimensionless term R_f, where R_f is the *ratio* of the distance traveled by the solute to the distance traveled by the solvent front. Naturally, the R_f can only be calculated in those instances when the solvent is not allowed to leave the end of the paper sheet.

Greater resolution of the solutes may be obtained using **two-dimensional paper chromatography** (Fig. 13-16*b*): after chromatography using a particular solvent system in one direction along the paper sheet, the sheet is dried and rotated 90°, and another solvent system is used to chro-

Figure 13–16
Paper chromatography. (*a*) Essential features of the ascending (left) and descending (right) chromatographic apparatus. (*b*) Stages in *two-dimensional* chromatography.

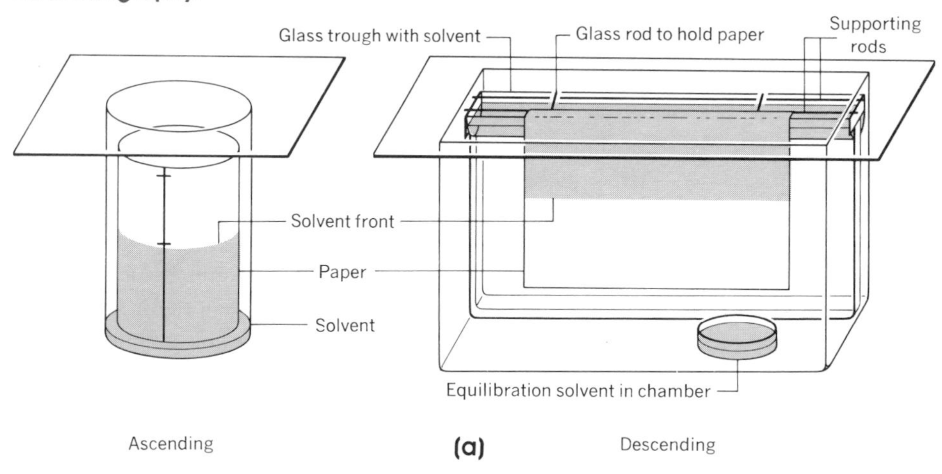

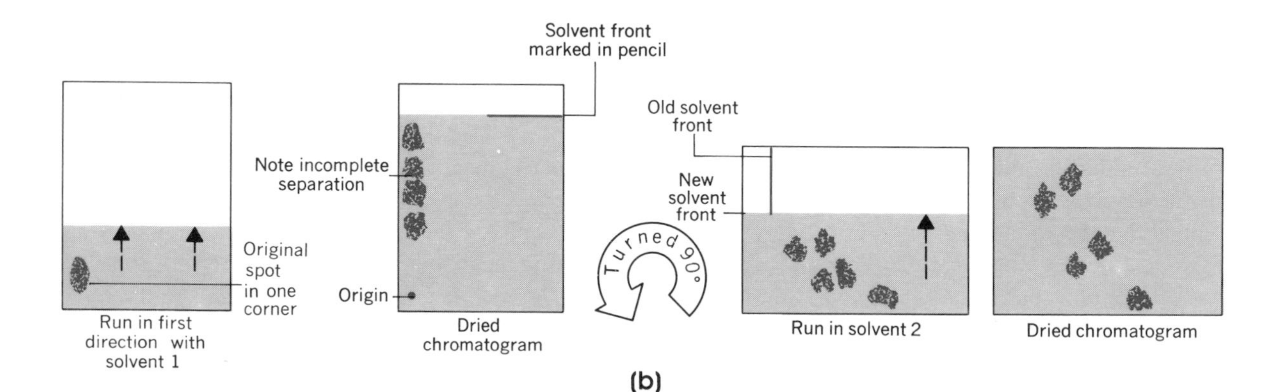

matograph the solutes a second time. In this manner, solutes not fully separated by partition in the first solvent may be completely separated using the second solvent (and vice versa). Paper chromatography can also be combined with zone electrophoresis to provide two-dimensional analysis of a solute mixture, a technique known as "fingerprinting" (see Chapter 21).

Thin-Layer Chromatography

Thin-layer chromatography (abbreviated TLC) is an especially valuable method for rapidly separating unsaturated and saturated fatty acids, triglycerides, phospholipids, steroids, peptides, nucleotides, and numerous other biological substances. In effect, TLC is a modification of paper chromatography in which the sheets of filter paper are replaced by glass plates covered with a thin, uniform layer of adsorbent. The essential features of the technique may be described as follows. An aqueous slurry of the selected adsorbent is uniformly spread over a glass plate to produce a thin layer and is then dried. Following this, the sample (usually prepared in a volatile solvent) is applied near one end of the long axis of the plate as a spot or thin line. When the sample has dried, the plate is supported vertically so that the end near the sample zone is immersed in a tray containing a shallow layer of the eluting solvent (usually an organic solvent of low polarity). Capillary action causes the solvent to ascend slowly through the layer of adsorbent, and as in paper chromatography, the solutes become distributed along the plate on the basis of differential partition between the stationary and mobile phases (Fig. 13-17).

Table 13-4
Adsorbents Used for Thin-Layer Chromatography

Adsorbent	Materials Separated
Silica gel	Amino acids, polypeptides, fatty acids, steroids, phospholipids, glycolipids, plasma lipids
Alumina	Amino acids, steroids, vitamins
Kieselguhr	Oligosaccharides, amino acids, fatty acids, triglycerides, steroids
Celite	Steroids
Cellulose powder	Amino acids, nucleotides
Hydroxylapatite	Polypeptides, proteins
Polyethylenimine	Nucleotides, oligonucleotides

Although ascending TLC is the most common, descending and horizontal separations may also be carried out. TLC separations are very rapid, rarely exceeding 20 to 30 minutes.

After the separation has been achieved the glass plate is removed from the tray of solvent and allowed to dry. Zones containing colored substances can be detected directly, and others may be identified if they contain compounds that fluoresce when exposed to ultraviolet light. Many zones may also be rendered visible by spraying the plate with certain reagent dyes or stains. Adsorbent may also be scraped off various regions of the plate and the separated molecules eluted from the adsorbent particles. Table 13-4 lists some frequently used adsorbents and their applications.

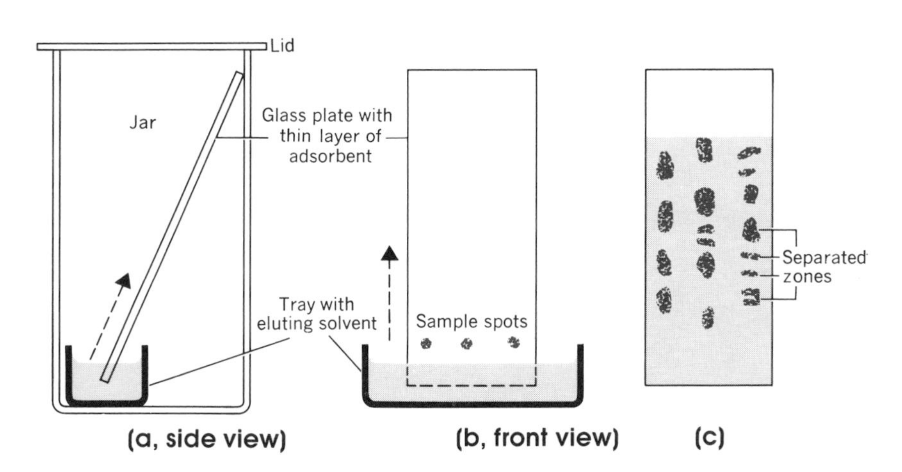

Figure 13–17
Thin-layer chromatography. (*a* and *b*) Glass plate containing thin layer of adsorbent and samples is immersed in tray containing shallow layer of eluting solvent. Capillary action causes solvent to ascend through adsorbent (arrows) which separates sample spots into series of zones (*c*).

Ion-Exchange Column Chromatography

Proteins and other macromolecules may also be separated by the technique known as ion-exchange chromatography. The separation is carried out in tall glass columns packed with grains of an ion-exchange **resin** (polymers to which numerous ionizable groups have been chemically added). Resins bearing negative charges are called **cation exchangers** and positively charged resins **anion exchangers.** As a solution of ions is passed through the column, the ions compete with each other for the charged sites on the resin. Consequently, the rate of movement of any ion through the column depends on its affinity for the resin sites, its degree of ionization, and the nature and concentration of competing ions in the solution. The differential rates of movement of ions through the column is the basis for protein and nucleic acid separations, since these molecules possess a variety of positively and negatively charged groups. Some ion exchangers used for protein and nucleic acid separations are listed in Table 13-5.

Among the resins most widely used for protein separations are diethylaminoethyl cellulose (DEAE-cellulose, an anion exchanger) and carboxymethylcellulose (CM-cellulose, a cation exchanger) developed by E. A. Peterson and H. A. Sober. These exchangers, like many others, are produced by reacting uncharged, high-molecular-weight polymers with ionizable compounds (Fig. 13-18). It is generally believed that the interaction between a resin and a protein involves the formation of a number of electrostatic bonds between the charged sites on the resin particle and oppositely charged dissociated side chains of certain amino acids. Since a number of bonds are formed, proteins are more firmly bound than singly charged substances. For example, at the appropriate alkaline pH, DEAE-cellulose could form a number of bonds with the negatively charged side chains of aspartic acid, glutamic acid, and tyrosine residues present in a protein. This also implies that the affinity of a protein for the resin can be altered by a change in pH. A reduction in pH sequentially suppresses the formation of bonds between the protein and DEAE-cellulose as hydrogen ions bind to the negative side chains of the protein and displace the resin. A similar result would be effected by the addition of salt ions which would compete with the protein for the resin sites (Fig. 13-19).

The separation of proteins by ion-exchange chromatography is carried out as follows. The resin is suspended in a buffer or salt solution (called the starting solvent), and the resulting slurry is used to fill the chromatographic column. At this time, the charges on the resin are neutralized by ions in the solvent. The sample (also dissolved in the starting solvent) is applied at the top of the column as a narrow zone. Depending upon the pH and salt concentration of the solvent, certain proteins present in the sample will form electrostatic bonds with sites on the resin (displacing solvent ions formerly bound at those sites), while the remaining proteins remain in the solvent. If a volume of the starting solvent is now passed through the column, the unbound proteins will be carried away with the flow and elute at the base of the column. In order to displace other proteins from the sample zone, it is necessary to change to another

Table 13-5
Some Commonly Used Ion Exchangers

	Polymer	Functional (Ionic) Group
Anion exchangers		
Diethylaminoethylcellulose	Cellulose	$-O-(CH_2)_2-N^+H-(C_2H_5)_2$
Dowex-2	Polystyrene-divinyl-benzene	$-N^+-(CH_3)_2C_2H_5OH$
Polylysine-Kieselguhr (PLK)	Polylysine	$-(CH_2)_4-NH_3^+$
Amberlite IRA-400	Polystyrene-divinyl-benzene	$-N^+(CH_3)_3$
Cation exchangers		
Carboxymethylcellulose	Cellulose	$-O-CH_2-COO^-$
Hydroxylapatite	—	$-PO_4\equiv$
Amberlite XE-64	Polymethacrylate	$-COO^-$
Sephadex SE	Dextran	$-(CH_2)_2-SO_3^-$

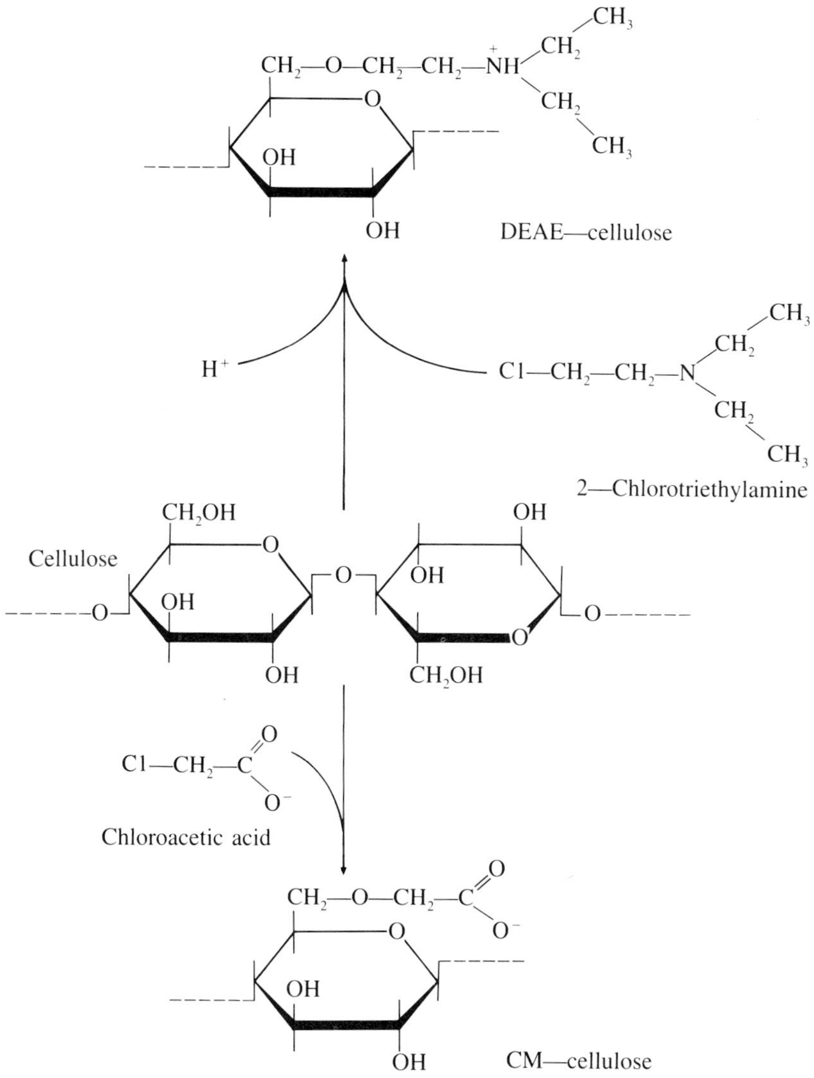

Figure 13–18
Synthesis of DEAE-cellulose and CM-cellulose.

solvent whose pH or ionic strength will alter the degree of ionization of the amino acid side chains or compete more effectively with the proteins for the charged sites of the resin. In the case of proteins adsorbed to DEAE-cellulose or other anion exchangers, this can be accomplished by decreasing the pH and increasing the salt concentration of the solvent. If the second solvent differs only slightly from the first, only a small group of proteins will be released from the sample zone. During their descent through the column, transient bonds will be formed with the resin many times as the proteins and solvent ions compete for the resin sites. Depending upon the relative affinities of the proteins for the resin and the solvent ions, different proteins may reach the bottom of the column at different times. Thus, changing the solvent not only releases an additional group of proteins but may also separate them into a series of zones during their passage through the column. The sequential addition of a series of solvents of different pH and salt

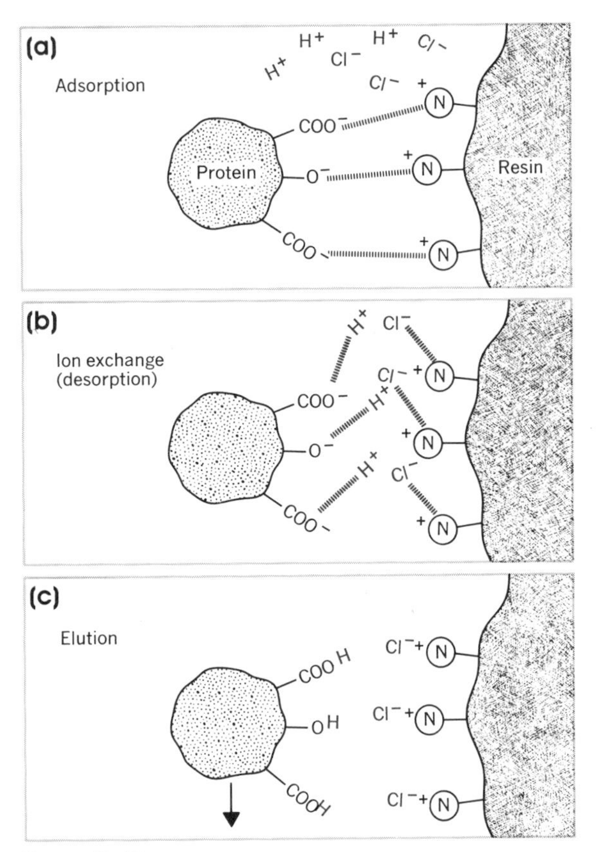

Figure 13–19
Desorption of protein from an anion-exchange resin by reducing the pH of the solvent. In (*a*), a theoretical protein molecule is adsorped to the resin by forming electrostatic bonds between certain dissociated aminio acid side chains and oppositely charged sites of resin. (N^+ represents any of several quaternary nitrogen groups associated with anion-exchangers; see Table 13–5). As the pH of the solvent flowing through the column is reduced, hydrogen ions compete more effectively for the dissociated amino acid side chains and break the bonds with the resin (*b*). Anions in the solvent (in this case Cl^-) bind to the resin, while the released protein is carried away with the flow of solvent (*c*). A similar effect is produced by increasing the salt concentration of the solvent. Generally, altering the pH and increasing the salt concentration are carried out simultaneously so that a combination of interactions result in protein desorption.

concentration, called **stepwise elution,** eventually displaces and separates all the proteins in the sample zone. Alternatively, the proteins may be eluted by passing a solvent of continuously changing pH and ionic strength through the column—a method known as **gradient elution.** The effluent from the column is collected as a series of fractions (Fig. 13–20).

Figure 13–20
Diagram of components used in ion-exchange column chromatography. Gradient elution method is shown. The solvent is pumped through the optical unit of an ultraviolet light absorption monitor before entering the column. Flow out of the column is also monitored and the two absorbances compared in order to identify that portion of the effluent containing proteins. The absorbances are plotted on a strip-chart recorder. Peaks in the tracing (called a chromatogram) correspond to protein zones emerging from the column.

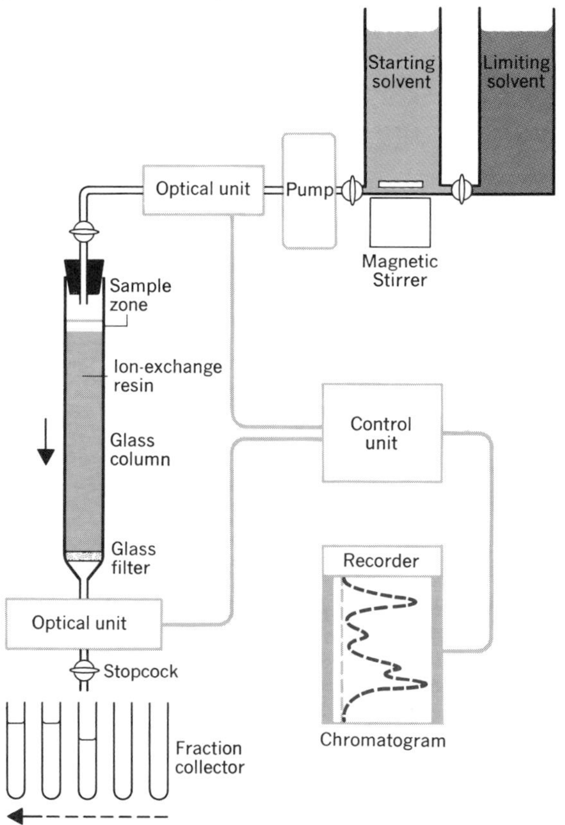

The conditions required for protein desorption depend on the number of bonds formed between the protein and resin and on the nature of the ionized amino acid side chains. Therefore, two proteins having the same *net* surface charge density (and, consequently, similar electrophoretic mobilities) might be desorbed under quite different conditions and emerge from the column at different times if their *absolute* surface charges differ. For example, under a given set of conditions, one protein may bear eight negative and five positive amino acid side chains and have a net charge of -3; another may have four negative and one positive side chain and be more weakly bound to the resin (desorbing earlier) but would have a similar electrophoretic mobility (i.e., its net charge would also be -3). For the same reason, two proteins desorbed under the same conditions and emerging from the column together may have quite different electrophoretic mobilities. Consequently, the rechromatography of a given protein fraction or its further examination by electrophoresis is usually recommended in order to evaluate the purity of isolated components.

Although the discussion of ion-exchange chromatography has centered around protein separations, it should be recalled that mixtures of different nucleic acids may also be resolved using this technique. Polylysine-kieselguhr columns have been particularly successful for chromatographic separations of both ribonucleic acids and deoxyribonucleic acids.

Affinity Chromatography

Affinity chromatography is a novel form of column chromatography in which the molecules (principally proteins and nucleic acids) to be isolated from the sample under study are retarded in their passage through the column by their specific *biological* reaction with the column matrix. It is the biological nature of the interaction between the sample and the column matrix that distinguishes this form of chromatography from others. The specificity may take the form of an antigen–antibody reaction, the complex of an enzyme with its substrate or inhibitor, hydrogen bonding between complementary polynucleotides, and so on.

The column is packed with porous gel particles (usually *agarose,* a linear polymer of the monosaccharide *galactose*) to which ligands having a high affinity for specific biological components have been covalently coupled. When the sample is applied to the column, only the constituents having a high affinity for the ligand are bound, while other components are rapidly eluted. The ultimate desorption and isolation of the bound species is achieved by significantly altering the pH and/or ionic strength of the eluent (Fig. 13–21).

Affinity chromatography has been used with great success for the isolation of specific immunoglobulins from antisera. This is achieved by first coupling the gel with specific antigenic materials. When antisera is applied to the column, specific immunoglobulins react with the antigen-coated gel

Figure 13–21
Affinity chromatography (see text for discussion).

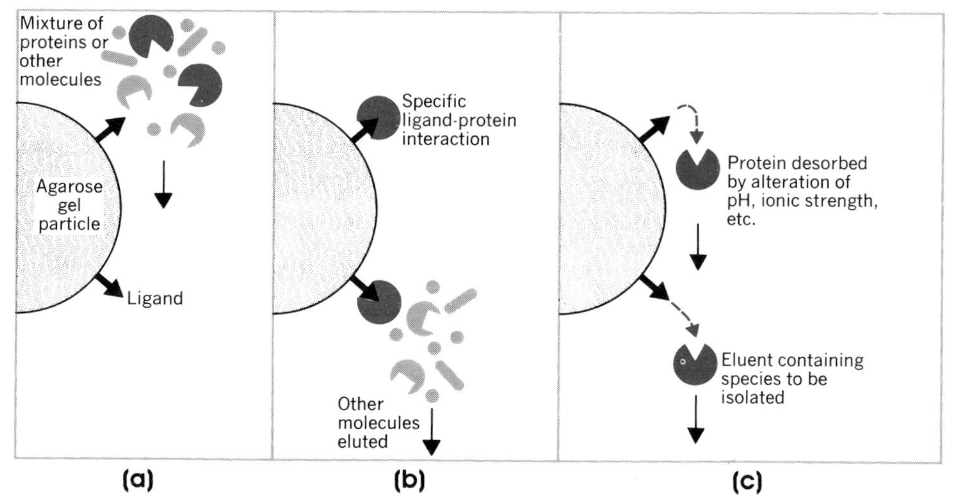

particles and are retained by the column, while other species are eluted. Gels to which the hormone insulin has been covalently bound have been used to isolate insulin-binding *receptor sites* of cell membranes. Specific cellular enzymes may be isolated by passing a tissue extract through a column in which the gel particles have previously been derivitized with the enzyme's substrate or specific cofactor or competitive inhibitor. Messenger RNA can be isolated from tissue homogenates using agarose impregnated with polyuridylic acid (polyU); mRNA's affinity for the column matrix results from the occurrence of sequences of polyA in many eucaryotic and procaryotic mRNAs (Chapter 21).

Gel Filtration (Molecular Sieving)

Gel filtration is a method for separating molecules on the basis of size (i.e., molecular weight) differences. The procedure is carried out using glass columns packed with **nonionic,** porous gel particles. Since the gels do not possess ionic groups, separation of a mixture of macromolecules does not involve the temporary formation of bonds with the gel. The most commonly used gels are **cross-linked dextrans** (produced by reacting dextran with epichlorohydrin) in which the degree of cross-linkage determines the average pore size of the gel. Gel particles may be produced with a variety of pore sizes.

In gel filtration, the dry gel particles are first swollen by hydration in water or buffer and then packed into a glass column. The volume of the column is effectively divided into two phases: the solvent surrounding the gel particles (solvent phase) and the solvent within the gel particles (gel phase). Whether or not a given solute molecule can pass between phases depends on the molecular weight of the solute and on the pore sizes of the gel. Any solute molecules larger than the largest pore size cannot pass into the gel phase and are said to be above the **exclusion limit** of the gel. Smaller molecules are able to penetrate the gel phase to varying degrees.

If a mixture of molecules of various sizes is placed at the top of the column and is followed by the passage of solvent through the column, molecules above the gel exclusion limit will pass between the gel particles and emerge from the bottom of the column most rapidly. Molecules below the exclusion limit will pass through both the solvent and gel phases and emerge later. Therefore, molecules in the original sample are quickly separated into two different populations—molecules above and those below the gel exclusion limit (Fig. 13–22). However, if two molecules are below the gel exclusion limit but one is larger than the other, the larger molecule will pass through the column more rapidly, since it will spend less time in the gel phase than the smaller molecule (i. e., the chances that a sufficiently large gel pore will be encountered during descent through the column are less for larger molecules below the exclusion limit than for smaller molecules). Hence, even molecules below the gel exclusion limit undergo a separation and emerge from the column in order of decreasing size.

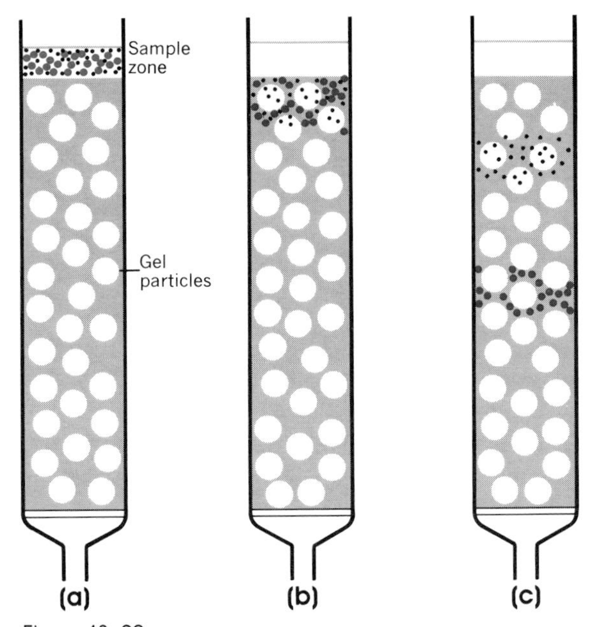

Figure 13–22
Separation of molecules by gel filtration. (*a*) Glass column packed with gel particles and overlaid with sample zone containing molecules of various sizes (small and large dots). (*b*) Molecules with molecular weights in excess of the gel exclusion limit percolate between the gel particles, while smaller molecules may enter the gel phase. (*c*) As a result, molecules reach the bottom of the column in order of decreasing size.

Cross-linked dextran and cross-linked agarose gels (known commercially as Sephadex and Sepharose) may be obtained with a variety of exclusion limits (Table 13–6) and are used routinely for protein, nucleic acid, polysaccharide, and lipid

Table 13–6
Exclusion Limits and Fractionation Ranges of Cross-Linked Dextrans and Cross-Linked Agarose

Type	Exclusion Limit (Molecular Weight Units)	Usable Fractionation Range (Molecular Weight Units)
Cross-linked dextrans		
G-10	700	0–700
G-15	1,500	0–1,500
G-25	5,000	1,000–5,000
G-50	30,000	1,500–30,000
G-75	40,000	3,000–40,000
G-100	150,000	4,000–150,000
G-150	400,000	5,000–400,000
G-200	800,000	5,000–800,000
Cross-linked agarose		
2-B	40,000,000	100,000–40,000,000
4-B	20,000,000	100,000–20,000,000
6-B	4,000,000	100,000– 4,000,000

separations. In most cases, a single solvent may be used to effect the separation, since molecules are not bound to the gel.

In addition to its usefulness for separating particle mixtures, gel filtration has been used to estimate molecular weights. As in ion-exchange chromatography and electrophoresis, the determination of the purity of separated components may require additional analyses, since different molecules may have the same size and pass through the column at the same rate.

Gas Chromatography

Gas chromatography is a special form of column chromatography in which a gas is used as the mobile phase (instead of a liquid) and either a liquid or a solid is used as the stationary phase. When a liquid is used as the stationary phase, the technique is called **gas-liquid chromatography** (GLC), and separations are based primarily on differences in the partition of the molecules in the sample between the stationary liquid and the moving gas. In **gas-solid chromatography** (GSC), separations result from the differential adsorption of sample molecules to the stationary phase as they are carried through the column by the gas. Of the two methods, GLC is, by far, the method most often employed.

The basic components of the gas chromatograph are the source of gas, the sample introduction chamber, the chromatographic column, the detector, and the recorder (Fig. 13–23). The gas (usually nitrogen, carbon dioxide, helium, or argon) is contained within a high-pressure cylinder connected to the column through metal tubing. A valve, pressure gauge, and flowmeter are used to accurately regulate the flow of gas. The sample is introduced into the flow of gas using a microsyringe and needle inserted through a self-sealing diaphragm in the sample chamber. The chamber itself is enclosed within a heating block so that the sample (if it is not already in a gaseous form) will immediately be vaporized upon introduction into the chamber and will be swept into the column by the gas. Gas chromatograph columns are made of glass, copper, or stainless steel tubing and are also enclosed in an oven; since they may be several feet long, they are often twisted to form a spiral.

The selection of packing material for the column depends upon whether the separation is to be based upon partition or adsorption. For GLC, the column is packed with an inert solid such as kieselguhr which is impregnated and lightly coated with a liquid of low volatility (so that it will not be eluted from the column at the operating temperature used). For GSC, the stationary phase is an adsorbent such as charcoal or silica gel. Different molecules in the sample will be carried through the column at different rates, depending on their adsorption or partition characteristics, and will emerge from the end of the column at different times. Located near the exit of the column and also housed within an oven is the detector that monitors the composition of the emerging gas and relays electrical signals proportional to the amounts of separated components to a strip-chart recorder. The separated components are thus recorded as a series of peaks in the chromatogram tracing. The most common form of detector is the *flame ionization* chamber in which the components are successively mixed with hydrogen and air and burned in a high-voltage field. Migration of the ionized fragments in this field creates a current registered by the recorder. Most chromatographic separations are analytical, and the technique is so sensitive that minute quantities of sample are required. However, separated components may also be collected as a condensate in tubes as they emerge from the heated column and are rapidly cooled.

Gas chromatography may be used to separate lipids, oligosaccharides, and amino acids after their preliminary con-

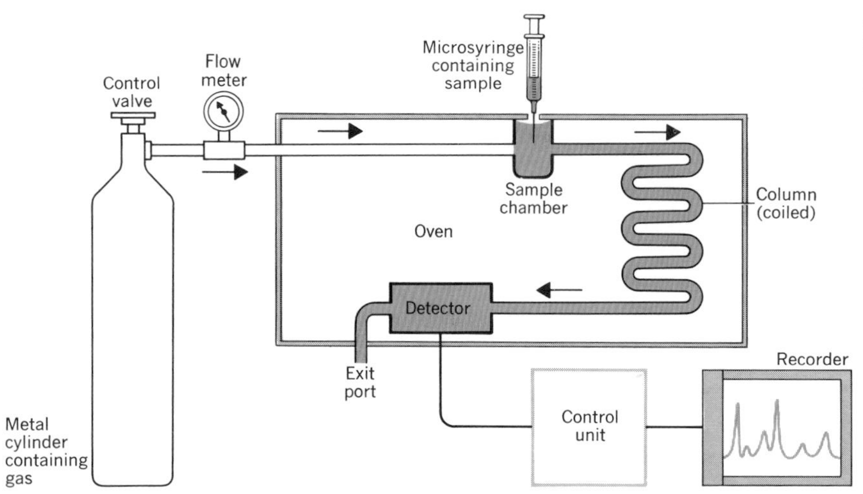

Figure 13–3
Basic components of a gas chromatograph.

version to volatile derivatives (many lipids may be chromatographed without conversion). Gas chromatography has been used with great success for the separation of different fatty acids. Depending upon whether a nonpolar or a polar liquid phase is used, fatty acids may be separated according to boiling point and size or degree of saturation.

The methods described in this chapter for separating, isolating, and studying macromolecules and their constituents are widely used in cellular research. Students interested in pursuing this subject can find a more comprehensive discussion of these methods in the books and articles listed at the end of the chapter, together with descriptions of other separation methods that are used less often but are also important.

Summary

Over the years, a variety of analytical and preparative techniques have been developed for separating, analyzing, and isolating the macromolecular constituents of cells and tissues. Older methods such as **salting in, salting out,** and **isoelectric precipitation** relied on differences in the solubility of the molecular species under investigation. Differences in molecular size are used to achieve separations in such diverse approaches as **dialysis, ultrafiltration, gel filtration** (molecular sieving), and **ultracentrifugation.** Differences in the net electrostatic charges of molecules are used in various types of **electrophoretic** separations, including **moving-boundary** electrophoresis, **zone** electrophoresis, and **disc** electrophoresis. Biological activities of the molecules are used to advantage in such techniques as **immunoelectrophoresis** and **affinity chromatography.** Many **chromatographic** techniques utilize **partition** differences to achieve separations, the most popular formats being **thin-layer** chromatography, **anion** and **cation exchange** chromatography, and **gas** chromatography. Because such molecular parameters as size, charge, and solubility are used as the basis for effecting a separation, information concerning the unique physical and chemical properties of the species under investigation is simultaneously acquired.

References and Suggested Reading

Articles and Reviews

Cooper, A. R., and Matzinger, D. P., Aqueous gel permeation chromatography. *Am. Lab.*, p. 13 (Jan. 1977).

Gray, G. W., Electrophoresis. *Sci. Am. 185*(6), 45 (Dec. 1951).

Haglund, H., Isoelectric focusing in natural pH gradients—a technique of growing importance for fractionation and characterization of proteins. *Sci. Tools 14,* 17 (1967).

Laurent, T. C., and Killander, J., A theory of gel filtration and its experimental verification. *J. Chromatography 14,* 317 (1964).

Peterson, E. A., and Sober, H. A., Chromatography of proteins. I. Cellulose ion-exchange adsorbents. *J. Am. Chem. Soc. 78,* 751 (1956).

Sober, H. A., Gutter, F. J., Wyckoff, M. M., and Peterson, E. A., Chromatography of proteins. II. Fractionation of serum proteins on anion-exchange cellulose. *J. Am. Chem. Soc. 78,* 756 (1956).

Vesterberg, O., Isoelectric focusing. *Am. Lab.,* p. 13 (June 1978).

Williams, C. A., Immunoelectrophoresis. *Sci. Am. 202*(3), 130 (March 1960).

Books, Monographs and Symposia

Bobbit, J. M., *Thin-Layer Chromatography,* Reinhold Publishing Co., New York, 1963.

Dean, J. A., *Chemical Separation Methods,* Van Nostrand and Reinhold Co., New York, 1969.

Determann, H., *Gel Chromatography,* Springer-Verlag, Inc., New York, 1968.

Krugers, J., and Keulemans, A. I. M., *Practical Instrumental Analysis,* Elsevier Publishing Co., Amsterdam, 1965.

Lehninger, A. L., *Biochemistry* (2nd ed.), Worth Publishers, Inc., New York, 1975.

Lowe, C. R., and Dean, P. D. G., *Affinity Chromatography,* John Wiley & Sons, Ltd., London, 1974.

Mangold, H. K., and Schmid, H. H. O., Thin-layer chromatography (TLC), in *Methods of Biochemical Analysis,* Vol. XII (D. Glick, ed.), John Wiley & Sons (Interscience), New York, 1964.

Morris, C. J. O. R., and Morris, P., *Separation Methods in Biochemistry,* Pitman and Sons, Ltd., London, 1964.

Newman, D. W., *Instrumental Methods of Experimental Biology,* MacMillan Co., New York, 1964.

Porath, J., Cross-linked dextrans as molecular sieves in *Advances in Protein Chemistry,* Vol. XVII (C. B. Anfinsen, ed.), Academic Press, Inc., New York, 1962.

Randerath, K., *Thin-Layer Chromatography,* Academic Press, Inc., New York, 1963.

Stock, R., and Rice, C. B. F., *Chromatographic Methods,* Chapman and Hall, Ltd., London, 1967.

Whipple, H. E. (editor), Gel electrophoresis, in *Annals of the New York Academy of Sciences,* Vol. CXXI, New York Academy of Sciences, New York, 1964.

Chapter 14
THE UTILIZATION OF RADIOACTIVE ISOTOPES AS TRACERS IN CELL BIOLOGY

Isotopes are chemical elements that have the same **atomic number** (i.e., the number of protons in the nucleus of the atom) but different **atomic masses** (i.e., the sum of the number of protons and neutrons in the nucleus). Certain isotopes are unstable and undergo spontaneous disintegrations **(transmutations)** accompanied by the emission of **particulate** and sometimes also **electromagnetic** radiations. These atoms are said to be radioactive and are called **radioisotopes** or **radionuclides;** their presence may readily be detected by instruments sensitive to their radiations. Generally, an organism cannot distinguish between the stable and radioactive forms of the same element so that both are metabolized in an identical manner. It is for this reason that radioisotopes have proven extremely useful to biologists, since these elements may conveniently be employed as **tracers.** That is, the fate of a given element (or molecule) in an organism (or even in an individual cell) may be studied by introducing the radioactive form of that element and following the uptake and subsequent localization of the radioactivity. Some of the radioisotopes frequently employed as tracers are listed in Table 14–1.

If the radioisotope is initially a part of a larger molecule, then the fate of all or part of that molecule may similarly be followed. The use of radioisotopically labeled compounds is particularly desirable when the compound to be administered is a normal constituent of the cell or organism and would be impossible to distinguish from stable molecules already present. Many organic and inorganic compounds of biological interest may now be obtained that have one or more specific atomic positions occupied by radioisotopes. Because of the extremely high sensitivities of most radiation detectors, the quantities of radioisotopes employed in tracer studies are small enough to preclude significant damage to cell constituents by the radiation.

Advantages of the Radioisotope Technique

Results obtained from experiments involving the use of radioisotopes are quantitative, since the amount of radioactivity present and available for detection is directly proportional to the radioisotope content. Moreover, numerous biological studies carried out routinely using radioisotopes can only be performed with great difficulty or are virtually impossible without them. Some examples may be cited to illustrate the value of the radioisotope technique.

Determination of Molecular Fluxes Under Conditions of Zero Net Exchange

The movement of a particular ionic or molecular species between different tissues of an organism or between a cell and its surroundings is often in **dynamic equilibrium.** That is, although a continuous exchange of a given substance

Table 14–1
Some Radioisotopes Frequently Used as Tracers

Isotope	Radiation Emitted: Beta Particles	Radiation Emitted: Gamma Rays	Half-life	Used as a Tracer of
^{3}H	Yes	No	12.3 years	Virtually any organic compound
^{14}C	Yes	No	5570 years	Virtually any organic compound
^{24}Na	Yes	Yes	15 hours	Salt metabolism, exchanges across membranes
^{32}P	Yes	No	14.3 days	Nucleic acid metabolism, phospholipid metabolism, salt metabolism
^{35}S	Yes	No	87.2 days	Protein metabolism
^{36}Cl	Yes	No	300,000 years	Salt metabolism
^{42}K	Yes	Yes	12.5 hours	Salt metabolism, exchanges across membranes
^{45}Ca	Yes	No	164 days	Salt metabolism, bone deposition
^{59}Fe	Yes	Yes	45.1 days	Heme synthesis, hemoglobin synthesis
^{131}I	Yes	Yes	8.1 days	Protein metabolism

between one region and another takes place, no *net* transfer of material occurs. Alternatively, the concentration of a given substance within a tissue or cell may remain fairly constant as a result of the balanced biosynthesis and degradation of that material. These situations cannot be easily detected or studied by routine chemical analyses. The continuous flux of Na^+ and K^+ across the plasma membrane of the erythrocyte is a typical example of this type of equilibrium. The concentration of K^+ within the mammalian erythrocyte is much higher than in the surrounding blood plasma, while the reverse is true for Na^+. These large concentration differences remain constant in spite of the continuous passage of Na^+ and K^+ across the plasma membrane in both directions. Consequently, this dynamic steady state is not revealed by chemical analysis and, prior to the utilization of radioisotopes of sodium and potassium to study this situation, the concentration differences were interpreted as the consequence of the impermeability of the erythrocyte membrane to Na^+ and K^+. When the radioisotopes of sodium and potassium, $^{24}_{11}Na$ and $^{42}_{19}K$, were used to label the plasma surrounding the red cells, the gradual passage of radioactivity into the erythrocytes was observed, eventually approaching an equilibrium identical to the concentration equilibrium determined chemically. These studies unequivocally demonstrated that the plasma membrane of the red cell is readily permeable to Na^+ and K^+ and that a continuous exchange of these two ions between the cell and the surrounding plasma takes place. Observations of this kind provided great impetus to the development of concepts concerning the continuous metabolic turnover of cellular materials present in constant concentrations and the *active transport* (Chapter 15) of materials across the cell membrane against concentration gradients.

In addition to providing qualitative information, radioisotopes may be used to accurately measure the *rates* at which the metabolic turnover of materials within a cell or tissue occurs, as well as the rates of exchange of materials across the cell membrane. Prior to the availability of radioisotopes, no satisfactory methods were available to measure these rates. In the case of the erythrocyte, it has been shown that about 2% of the cell's Na^+ and K^+ are exchanged with the plasma each hour.

Simplification of Chemical Analyses

Depending on the nature of the requisite chemical analyses, quantitative determinations of the distribution of a given element or compound in different tissues of the body or in different parts of an individual cell may be very difficult. This difficulty may be compounded if certain components to be analyzed contain only trace amounts of substance to be measured. Since radioisotope-containing compounds are not distinguished from their stable (nonradioactive) counterparts by an organism, the use of labeled compounds in studies of this sort can greatly simplify and reduce the number of analyses. Two requirements must be fulfilled: (1) the administration of the labeled compound must quickly be followed by its uniform distribution among stable equivalents already present in the system, and (2) once a uniform distribution is attained, the **specific activity** of the substance in question must be determined in one of the several components (i.e., tissues, cell fractions, etc.) to be analyzed. Specific activity is defined here as *the quantity of radioactive element or compound* (the units of measurement are described later) *per unit weight of total element or compound.* If the labeled compound does become uniformly distributed in proportion to the stable form already present in the system, then the specific activity will be equal throughout the system. Consequently, all subsequent measurements of the quantity of that compound may be made simply by measuring the amount of radioactivity present in the tissue or cell part to be analyzed. In nearly every instance, this is far easier than performing a series of quantitative chemical measurements of the element or compound in question.

The rates at which different elements or compounds are incorporated by individual cells or by the body tissues, together with a determination of the specific loci of the deposition, can also be conveniently determined using radioisotopes. For example, cells may be incubated in media containing labeled material or, in the case of measurements in many whole animals, the material may be introduced into the bloodstream. In these cases, the specific activity of the element or compound is determined before it is made available for incorporation and the subsequent rate of incorporation determined from either the rate of disappearance of radioactivity from the medium or bloodstream or from the rate of appearance of radioactivity in the cells or tissue. The specific locus of deposition or utilization within the cell may be determined from separate radioactivity measurements of **fractionated** material (see Chapter 12) or by **autoradiographic** procedures (to be described in detail later) combined with light or electron microscopy. Again, it should be noted that specific quantitative chemical analyses of the material being incorporated by the cells are unnecessary when the radioisotope technique is applied.

"Isotope Dilution" Methods

A common problem confronting physiologists and one that lends itself to the radioisotope technique is the determination of the *total* quantity or volume of a given material in the body or in a cell when quantitative isolation of that material for analysis is not possible. Determinations of total circulating blood volume, erythrocyte mass, chloride space, exchangeable sodium, and so on fall in this category. The manner in which such unknown quantities are measured using radioisotopes may be exemplified by considering the following generalized case. Suppose that material X occurs in a system (i.e., a tissue, cell, etc.) in unknown abundance and a labeled form of this material, X^*, is either available commercially or can be prepared experimentally having a known specific activity, SA, given in this instance by the equation

$$SA = \frac{(X^*)}{(X^* + X)} \tag{14–1}$$

A measured quantity of the labeled material is introduced into the system and permitted to equilibrate thoroughly with the stable material already present. Following this, the specific activity of a small quantity of the material isolated from the system is determined. Since the X^* originally introduced was uniformly mixed with X already present, its specific activity will have been reduced in direct proportion to the total amount of X present in the system. Consequently, the total amount of X originally present can be determined from the relationship

$$X_T = X^*(\frac{SA_1}{SA_2} - 1) \tag{14–2}$$

where

X_T = the total amount of X originally present in the system,
X^* = the quantity of labeled material added,
SA_1 = the original specific activity of the material added, and
SA_2 = the specific activity of the sample removed for analysis.

Precursor–Product Relationships

Radioisotopes have been widely applied to study precursor–product relationships, for in many cases the introduction of a labeled compound (i.e., **precursor**) into a system is soon followed by its chemical conversion into another form (i.e., **product**). For example, soon after ^{32}P-labeled phosphate is introduced into the bloodstream, radioactivity appears in a variety of tissue phospholipids. In a similar manner, the introduction of radioactive iron is soon followed by the appearance of radioactivity in liver ferritin and also in newly synthesized hemoglobin of red blood cells maturing in the bone marrow. In other words, radioisotopically labeled compounds may be used to determine which of a variety of alternative metabolic pathways is followed by a compound by examining the radioisotope content of the alternative metabolic products. If several different pathways are open to the precursor, then the extent to which each is followed under a variety of experimental conditions may be determined from the respective radioisotope contents of the products.

The formation of an end product often involves a number of intermediate precursors and products; that is,

A →	B →	C →	D
Precursor-1	Product-1 Precursor-2	Product-2 Precursor-3	End product

Before radioisotopes were available to study such metabolic pathways, it was often difficult to determine which compound was the immediate precursor of each product. However, if each intermediate has a single precursor (as above), then the precise sequence can be determined using labeled material by following the change in specific activity of each intermediate compound with time. This may be done because *the specific activities of an intermediate and its immediate precursor are equal when the intermediate's specific activity reaches a maximum value* (Fig. 14–1). By degrading the intermediate and final products of a series of chemical reactions and determining the distribution of radioisotope among the constituent atoms, it is often possible to determine the fate of each atom in the pathway.

Properties of Radioactive Isotopes

Most radiations emitted by radioisotopes are the result of changes in the arrangement of unstable atomic nuclei. Whether or not a given atomic nucleus is stable depends in turn upon the numbers of neutrons (N) and protons (Z) that it contains. The relationship between nuclear stability and the neutron: proton composition of the nucleus is shown graphically in Figure 14–2. It should be noted that for the lighter elements, nuclear stability exists when $N \cong Z$, whereas in the stable heavier elements, the number of neutrons exceeds the number of protons, with the neutron excess increasing with atomic number. Isotopes with N and Z numbers outside of the stable region shown in Figure 14–2 undergo spontaneous changes in which nuclear neutrons and protons are interconverted. These nuclear transmutations are accompanied by the emission of particulate and electromagnetic radiation.

Figure 14–1
Relationship between the specific activities of precursors and products in the chemical pathway A → B → C → D.

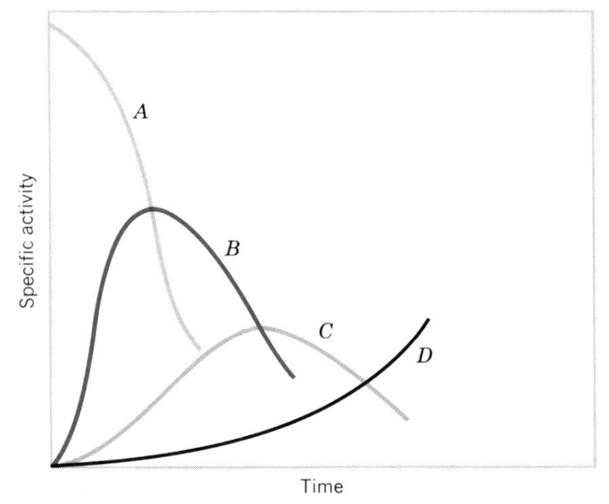

Figure 14–2
Relationship between stability of atomic nucleus and the number of neutrons and protons present.

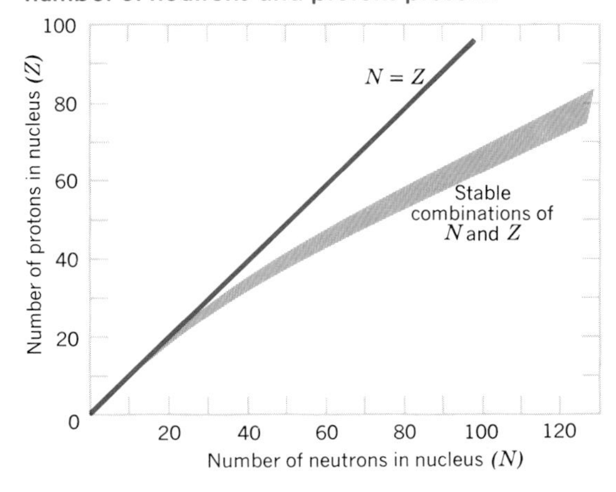

Types of Radiation Emitted by Radioisotopes

The most common types of nuclear radiations are **alpha particles,** positive and negative **beta particles,** and **gamma rays.** Alpha particles, which consist of two protons and two neutrons and are therefore identical to helium nuclei, are emitted by radioisotopes of high atomic number such as uranium, polonium, thorium, and radium; that is,

$$\underset{\text{Atomic number}}{\overset{\text{Atomic mass}}{{}^{238}_{92}\text{U}}} \longrightarrow {}^{234}_{90}\text{Th} + \underset{\text{(Alpha particle)}}{{}^{4}_{2}\alpha}$$

and

$${}^{210}_{84}\text{Po} \longrightarrow {}^{206}_{82}\text{Pb} + {}^{4}_{2}\alpha$$

Alpha-emitting radioisotopes are rarely used as tracers in biological studies.

Positive beta particles, also called **positrons,** are emitted from nuclei in which the N:Z ratio is below that which is stable. The nuclear transmutation involves the conversion of a proton into a neutron, positron and neutrino, the latter two being ejected from the nucleus. The positron possesses a unit positive charge and is equal in mass to an electron, while the neutrino has neither mass nor charge. The isotope ${}^{11}_{6}\text{C}$ *decays* by positron emission:

$${}^{11}_{6}\text{C} \longrightarrow {}^{11}_{5}\text{B} + \underset{\text{(Positron)}}{{}^{0}_{(+1)}\text{e}} + \text{Neutrino}$$

In view of the similarity between beta particles and electrons, the symbol "e" is often used to describe the beta particle. Very few positron-emitted radioisotopes are employed as biological tracers.

Nearly all radioisotopes used as tracers by biologists emit negative beta particles **(negatrons).** It has become customary to drop the term "negative" so that the expression "beta particle" is understood to imply negative beta particle. Although technically incorrect, this terminology is widely used and will be employed here. Beta particles are emitted from nuclei in which the N:Z ratio is above that which is stable. The nuclear change involves the conversion of a neutron to a proton with the resulting ejection of a beta particle and neutrino. ^{14}C and ^{33}P may serve as examples of this form of decay:

$${}^{14}_{6}\text{C} \longrightarrow {}^{14}_{7}\text{N} + \underset{\text{(Beta particle)}}{{}^{0}_{(-1)}\text{e}} + \text{Neutrino}$$

and

$${}^{32}_{15}\text{P} \longrightarrow {}^{32}_{16}\text{S} + {}^{0}_{(-1)}\text{e} + \text{Neutrino}$$

The emission of beta particles from the nuclei of ^{3}H, ^{14}C, ^{32}P, ^{35}S, ^{36}C1, and ^{45}Ca atoms changes the N:Z ratio to a stable value and reduces the energy content of the nucleus to the ground state. The energy lost by the nucleus in this process is distributed between the beta particle and neutrino. However, for many radioisotopes, including ^{24}Na, ^{59}Fe, and ^{131}I, intranuclear changes resulting in the emission of beta particles and neutrinos produce nuclei with stable N and Z combinations but with energy levels still above the ground state. In these instances, the excess energy is eliminated and the ground state is attained by the emission of one or more gamma rays; for example,

$${}^{59}_{26}\text{Fe} \longrightarrow {}^{59}_{27}\text{Co} + {}^{0}_{(-1)}\text{e} + \text{Neutrino} + \text{Gamma ray(s)}$$

Unlike alpha and beta radiation, gamma radiation is not particulate but is **electromagnetic.**

Energy of Radiation and Its Interaction with Matter

The energy of radiation is measured in **electron volts** (abbreviated ev), one electron volt being the kinetic energy acquired by an electron in a potential difference of one volt. Beta particles and gamma rays emitted by radioisotopes often used as tracers have energies ranging from about 1×10^4 to 4×10^6 ev or 0.01 to 4.0 Mev (1 Mev equals 1 million electron volts). This energy range may be compared with the energy of chemical bonds, which is of the order of a few electron volts. The kinetic energy acquired

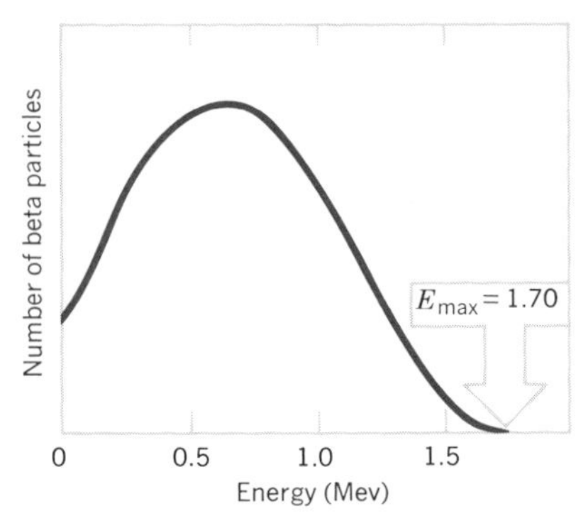

Figure 14–3
Energy spectrum of beta particles emitted by ^{32}P.

by a beta particle during transmutation of the parent nucleus can vary from 0 Mev up to some maximum value, E_{max}, which is a characteristic of the particular radioisotope. As an example, the energy spectrum for ^{32}P beta particles is shown in Figure 14–3. It should be noted that the maximum energy of a ^{32}P beta particle is 1.71 Mev but that the average energy is much less (actually, 0.70 Mev). Although the value for E_{max} varies, the *shape* of the energy curve is similar for all radioisotopes.

In order to reduce the energy level of the radioisotope nucleus to the ground state, a specific quantity of energy, Q, must be given off as radiation during the nuclear change. For isotopes emitting only beta particles, Q equals E_{max}, and the energy of the neutrino accounts for the difference between E_{max} and the actual kinetic energy acquired by the beta particle. For example, if the transmutation of a particular ^{32}P atom results in the emission of a 1.20 Mev beta particle, then the accompanying neutrino would have an energy of 0.51 Mev (i.e., 1.20 Mev + 0.51 Mev = 1.71 Mev = E_{max}). If the beta particle had escaped the nucleus with the maximum possible energy, then no neutrino would have been emitted. Table 14–2 lists the maximum beta particle energies of some radioisotopes.

In contrast to beta particles, gamma rays are emitted at specific energy levels. Thus, all gamma rays emitted by ^{42}K have an energy of 1.51 Mev (Table 14–2). The Q value of an isotope emitting both beta and gamma radiation is equal to the sum of the energies possessed by the beta particle, neutrino, and subsequent gamma ray(s). Occasionally, as in the case of ^{24}Na, the emission of a beta particle may be followed by the emission of two (or more) successive gamma rays in order to reduce the nuclear energy level to the ground state. It may also be noted from Table 14–2 that the decay of some radioactive isotopes (e.g., ^{24}K, ^{59}Fe, and ^{131}I) proceeds along two or more alternative pathways; that is, beta particles having more than one E_{max} may be emitted, and since the Q value is constant, these are necessarily followed by gamma rays of different energies.

Beta particles and gamma rays interact with matter by ionizing and exciting atoms in their path. In the case of beta particles, ionization results from the repulsion of orbital electrons by the negatively charged particle. As a result, beta particles tracking through matter produce a *wake* of electrons and positively charged ions (called **ion pairs**). The number of ion pairs produced per unit path length, called the **specific ionization,** is not constant but increases as the beta particle slows down. Also, the path of the beta particle is not linear but is quite erratic as a result of its repulsion by the orbital electrons of atoms with which it interacts. High-energy beta particles may traverse the linear equivalent of 1 to 2 m in air. Since gamma radiation is electromagnetic, its probability of interacting with matter is less than that for beta particles. Consequently, gamma rays have a much lower specific ionization and a much longer and also linear path length.

Half-Life

The number of atoms in a sample of radioisotope that disintegrate during a given time interval decreases logarithmically with time and is unaffected by chemical and physical factors that normally alter the rates of chemical processes (i.e., temperature, concentration, pressure, etc.). Radioactive decay is therefore a classical example of first-order reaction. A convenient term used to described the rate of decay of a radioisotope is the **physical half-life,** T_p—that is, the amount of time required to reduce the amount of radioactive material to one-half its previous value. Each radioactive isotope decays at a characteristic rate and therefore has a unique half-life (see Table 14–1). For example, the amount of radioactivity arising from a sample of ^{59}Fe is reduced to one-half its original value in 45.1 days, to one-fourth in 90.2 days, to one-eighth in 135.3 days, and so on. The amount of decay occurring in the course of a tracer experiment must be taken into account when radioisotopes of short physical half-life such as ^{23}Na, ^{32}P, ^{36}Cl,

Table 14–2
Energies of Beta Particles and Gamma Rays Emitted by Radioisotopes Frequently Used as Tracers

Isotope	ENERGY (Mev) Beta Particle	(E_{max})	Gamma Ray	Q (Mev)
^{3}H	0.0176			0.0176
^{14}C	0.154			0.154
^{24}Na	1.39	plus	1.37 & 2.75	5.51
^{32}P	1.71			1.71
^{35}S	0.167			0.167
^{36}Cl	0.714			0.714
^{42}K	3.60			3.60
	or 2.10	plus	1.50	3.60
^{45}Ca	0.254			0.254
^{59}Fe	0.27	plus	1.29	1.56
	or 0.46	plus	1.10	1.56
^{60}Co	0.31	plus	1.17 & 1.33	2.81
^{131}I	0.61	plus	0.36	0.97
			or 0.28 & 0.08	
	or 0.34	plus	0.63	0.97

^{42}K, and ^{131}I are used. Of course, this is not a problem in experiments involving ^{3}H and ^{14}C (Table 14–1).

When radioisotopes are used in in vivo experiments of extended duration, the turnover rate of the element in the body (or in the cell) must also be considered, for the rate of decrease of radioactivity will be a function of *both* radioactive decay and metabolic turnover. In these instances, a more useful term is the **effective half-life,** T_e, which is the amount of time required to reduce the radioisotope content of the body (or cell) to one-half its original value by the combined effects of decay and turnover; it is determined using the relationship

$$T_e = \frac{T_b \times T_p}{T_b + T_p} \tag{14–3}$$

where

T_e = effective half-life,
T_p = physical half-life, and
T_b = **biological half-life** and is defined as the normal amount of time required for the turnover of one-half of the body content of a given element (radioactive or nonradioactive).

The physical, biological, and effective half-lives of several elements are compared in Table 14–3. It should be noted that T_e can never be greater than T_p and that the slower rate of turnover of an element, the closer T_e approaches T_p.

On the basis of their biological and physical half-lives, their loci of deposition within the body, the types of radiation emitted, and the energy of the radiation, radioisotopes may be classified according to their degree of hazard to the investigator if accidentally ingested (Table 14–4).

Table 14–3
Physical Biological and Effective Half-Lives of Some Radioisotopes

Element	Radio-isotope	Half-life Physical	Biological	Effective
Hydrogen	^{3}H	12.3 yr	19 days	19 days
Carbon	^{14}C	5,570 yr	180 days	180 days
Phosphorous	^{32}P	14.3 days	3 yr	14.1 days
Calcium	^{45}Ca	164	73 yr	163 days
Iron	^{59}Fe	45.1 days	3.4 yr	27.1 days
Cobalt	^{60}Co	5.3 yr	8.1 days	8.0 days
Iodine	^{131}I	8.1 days	156 days	7.7 days
Radium	^{226}Ra	1,620 yr	104 days	104 days

Table 14–4
Relative Internal Hazard of Selected Beta- and Gamma-Emitting Isotopes

Degree of Hazard	Radioisotope
Slight hazard	^{24}Na ^{42}K ^{64}Cu
Moderately dangerous	^{3}H ^{14}C ^{22}Na ^{32}P ^{35}S ^{36}Cl ^{59}Fe ^{60}Co
Very dangerous	^{45}Ca ^{55}Fe ^{90}Sr

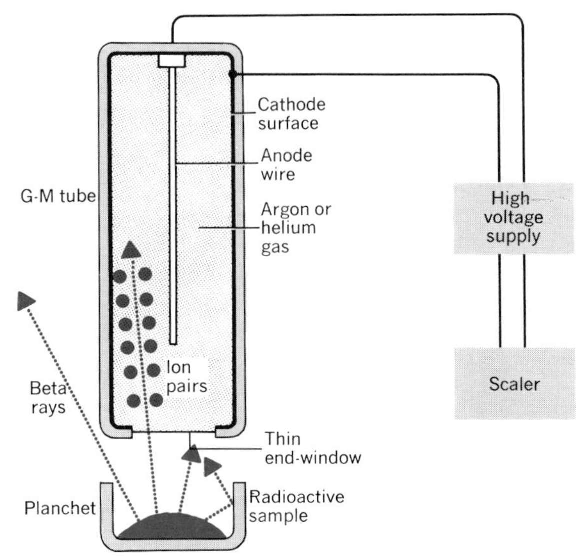

Figure 14–4
Basic components of a Geiger-Müller counter.

Detection and Measurement of Radiation

The selection of instruments for the detection and measurement of radioisotopes is based primarily on the type and energy of the emitted radiation. The most commonly used detectors are: (1) **Geiger-Müller counters,** which are employed primarily with isotopes emitting beta particles of intermediate or high energy (E_{max} above 0.2 Mev) and which may also be used at low efficiency for the measurement of gamma radiation; (2) **solid scintillation counters,** which are generally employed with gamma-ray-emitting isotopes; and (3) **liquid scintillation counters,** which are used with isotopes emitting low-energy beta particles (E_{max} below 0.2 Mev).

Geiger-Müller Counters

The most widely used instrument for the detection and measurement of radiation is the Geiger-Müller (or G-M) counter. The detector itself, called a G-M tube, consists of a cylinder several inches long containing two electrodes and filled with a readily ionizable inert gas such as helium or argon. The insulated metallic internal surface of the cylinder serves as the cathode, and a narrow wire passing down the center of the tube serves as the anode (Fig. 14–4). One end of the G-M tube is covered by a thin material such as mylar plastic or mica and is called the **end-window.** The anode and cathode terminals at the other end of the tube are connected to a source of high voltage and a **scaler,** a device that simply counts electrical pulses.

When a radioactive sample (usually deposited on a small metal disk called a *planchet*) is placed near the end-window, radiation enters the G-M tube, ionizing some of the gas molecules and forming a number of ion-pairs (i. e., positively charged argon or helium atoms and electrons). If a sufficiently high electrical potential is applied to the electrodes, the ion-pairs will migrate toward the appropriate electrode. During this migration, the ions collide with and ionize additional gas molecules, so that the passage of a single beta particle or gamma ray through the gas results in a large number of ions being collected at the electrodes. These events produce an electrical pulse that is recorded by the scaler as a **count.** Ideally, each ionizing ray entering the G-M tube is registered as a count and the amount of radioactivity expressed as *counts per minute* (cpm).

Since radiation is emitted in all directions from a radioactive source, it is apparent that only a small percentage of the rays arising from the sample are directed toward the end-window. Therefore, even if all the rays entering the G-M tube are detected and counted, the cpm recorded for the sample is only a fraction of the true *rate* of disintegration (i. e., *disintegrations per minute, dpm*) of the isotope. This does not pose a serious problem when the *relative* isotope contents of a number of samples are to be determined (this is generally the case) and if constant *geometric* conditions are maintained for each sample (i. e., distance of the sample from the end-window, volume of sample, etc.).

For some radioisotopes such as ^{3}H, ^{14}C, ^{35}S, ^{45}Ca, and others of low E_{max}, much or all of the energy of the emitted beta particles may be dissipated before the ray enters the ionizing gas. For example, the energy may be expended within the sample itself (called **self-absorption**), in the air between the sample and the end-window, or in the material of the end-window. Even with radioisotopes emitting beta particles of high E_{max} (Table 14–2), beta particles in the low region of the energy spectrum may go undetected. Because of the low specific ionization of gamma rays and the low density of the gas in the G-M tube, gamma rays may pass through the tube without causing ionizations and therefore go undetected. For these reasons, Geiger-Müller counters are not the most suitable instruments for the detection and measurement of radioisotopes emitting gamma rays or beta particles of low E_{max}.

Even when no radioactive sample is placed below the end-window of the G-M tube, a small count is recorded. This is known as the *background* count and results from cosmic radiation (primarily gamma rays), naturally occurring radioisotopes in laboratory materials (such as ^{40}K in glass and naturally occurring ^{14}C and ^{3}H in organic compounds), radioactive samples left in the vicinity of the detector, and electronic "noise" within the components of the counting system. Therefore, the background count must always be subtracted from the count obtained for a radioactive sample. The magnitude of the background count may be reduced by placing lead shielding around the detector so that much of the cosmic radiation and radiations from other sources are absorbed before reaching the detector.

The total amount of radioisotope present in a sample at any instant in time may be determined from its rate of disintegration, dpm; the basic unit of measurement is the *curie* and is defined as that quantity of radioisotope undergoing 2.22×10^{12} dpm. (Note that the curie content of a radioactive sample decreases exponentially with time at a rate determined by the physical half-life of the radioisotope.) In most tracer experiments, the quantity of radioisotope used is generally at the millicurie or microcurie level. The curie content of a labeled compound is generally provided at the time of purchase so that the efficiency of the counting system may be determined by comparing the recorded cpm of an aliquot of the isotope with its known dpm. Generally, this value is 10% or less for G-M counters but is much higher in solid and liquid scintillation systems. Once the efficiency of the counting system is known, then the specific activity of a radioactive sample (which we may now define as the *number of curies per unit mass of element*) collected during the course of a tracer experiment may be calculated from its observed counting rate and composition.

Automated G-M counters of modest efficiency are available in which fresh ionizing gas is continuously supplied to the detector and which also permit the planchet containing the sample to seal to the end of the detector. This eliminates the end-window together with air and end-window absorption. These "windowless gas-flow counters" may also be equipped with automatic sample (i. e., planchet) changers. G-M counters are effectively employed in tracer experiments involving ^{24}Na, ^{32}P, ^{36}Cl, and other "hard beta" emitters but generally are not used with ^{3}H and ^{14}C, which emit "soft beta" rays.

Solid Scintillation Counters

Solid scintillation counters are used to detect and measure radioisotopes emitting gamma rays. The cylindrical detector (Fig. 14–5) contains a crystal (usually thallium-activated sodium iodide) and a photomultiplier tube encased in an aluminum housing and interfaced with a preamplifier, a source of high voltage, and a scaler. The radioactive sample to be counted is placed either against the end of the detector containing the crystal or, for greatly improved counting efficiency, into a well-shaped opening drilled into the crystal's surface (Fig. 14–5). Because of its high density, the crystal absorbs much of the energy of the gamma rays, causing excitation of orbital electrons of atoms composing the crystal. This is followed by the emission of flashes of light or *scintillations* proportional in number to the number and energy of the gamma rays exciting the crystal. The light photons are converted by the adjacent photomultiplier tube into electrical pulses of proportional magnitude and frequency that are relayed to the scaler. Since the magnitude of the electrical pulses produced is proportional to the energy of the gamma rays, and since gamma rays are monoenergetic, the inclusion of the appropriate circuitry in the counting system (i. e., a pulse height analyzer, see Chapter 2) allows different gamma-ray-emitting isotopes to be easily distinguished.

In contrast to G-M counters, few or no problems involving self-absorption and end-window absorption are incurred when solid scintillation methods are used with gamma-ray-emitting isotopes. However, the use of constant geometry and lead shielding around the detector to reduce the magnitude of the background count is important.

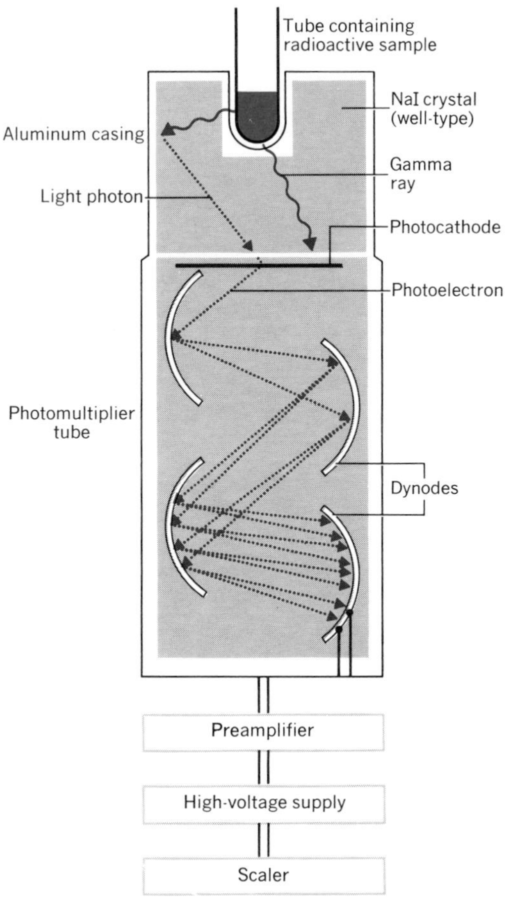

Figure 14–5
Basic components of a solid scintillation counter.

Liquid Scintillation Counters

In view of the ubiquitous occurrence of hydrogen- and carbon-containing compounds in cells, ^{3}H and ^{14}C are used more extensively than any other radioisotopes in tracer experiments. The energies of the beta particles emitted by these isotopes are very low (especially tritium), so that it is only with great difficulty that they can be accurately measured using gas ionization methods, even in windowless counters. The great demand for a sensitive and accurate method for measuring ^{3}H and ^{14}C, as well as other soft beta emitters, led in the 1950s to the development of the **liquid scintillation** technique.

Before considering this important technique in detail, the basic mechanism involved in the detection process will briefly be summarized. The radioactive sample to be counted is mixed in a glass or plastic vial with a ''scintillation fluid.'' Certain molecules dissolved in the fluid indirectly absorb the energy dissipated by the beta particles and respond by scintillating. The resulting photons of light are detected by photomultiplier tubes that relay proportional electrical pulses to the scaling units. Certain advantages of the procedure are immediately apparent, for if the sample is thoroughly mixed with the scintillation fluid, no self-absorption occurs. Moreover, since the technique is employed almost exclusively with weak beta-emitting isotopes, the energy of *all* beta rays is dissipated within the fluid, making it possible, at least in theory, to detect most if not all of the nuclear disintegrations.

The scintillation fluid contains three major components: (1) an organic *solvent* constituting the bulk of the mass (typically toluene, xylene, or dioxane); (2) a *primary solute* or *fluor* [e.g., 2, 5-diphenyloxazole (''PPO'') or 2-phenyl 5(4-biphenylyl)-1,3,4-oxadiazole (''PBO'')] and (3) a *secondary solute* or *fluor* [e.g., p-*bis*-[2-(5-phenyloxazolyl)]-benzene (''POPOP'') or 2,5-*bis* [5′-tert-butylbenzoxazolyl-(2′)]thiophene (''BBOT'')]. The energy of the beta particle is transferred to the solvent, causing ionization or excitation of solvent molecules. The latter effect is more important, since in returning to the nonexcited ground state, energy is transferred to the primary solute molecules, causing their excitation. In returning to the ground state, the excited primary solute emits photons of near ultraviolet light. The photocathodes of most photomultiplier tubes are not sensitive to photons having wavelengths in this region, and it is the role of the secondary solute to absorb the light energy from the primary solute and reemit it as light of longer wavelength. (Secondary solutes are sometimes called ''wave shifters.'') For example, the light emitted by PPO has a wavelength maximum at 380 nm, whereas POPOP emits light having a wavelength maximum at 420 nm and is close to the maximum sensitivity region of most photomultiplier tubes.

The size of the electrical pulse produced by the photomultiplier tube is proportional to the energy dissipated by the beta particle in the liquid scintillation fluid. Consequently, a pulse height spectrum is produced having a profile almost identical to that of the beta particle energy spectrum. In the case of ^{3}H and ^{14}C, the sizes of many of the pulses produced are so low that spurious low-amplitude pulses arising spontaneously within the photomultiplier tube itself (i.e., "noise") result in a false high count. To minimize this problem, in most liquid scintillation counters the sample vial is placed between the faces of *two* photomultiplier tubes connected through a pulse height discriminator to a *coincidence circuit* (Fig. 14–6) which permits a count to be registered only when *simultaneous* pulses are produced by the tubes. Therefore, a noise signal from one photomultiplier tube is not recorded unless another noise signal is generated by the other tube at the same time (actually, within 10^{-7} seconds of each other). Since noise signals are random, coincidental noise pulses are few in number.

As in G-M and solid scintillation counting, the background count must be subtracted from each recorded count. In liquid scintillation counting, the background count is derived from a number of sources including coincidental tube noise, naturally occurring radioisotopes in the vial and scintillation fluid (most glass contains small amounts of ^{40}K and organic solvents contain small amounts of ^{3}H and ^{14}C), cosmic radiation, and Cerenkov radiation.

Most liquid scintillation counters contain pulse height discriminators that provide for the rejection of pulses above and below a selected size range. This feature is particularly valuable in "double-label" experiments (experiments in which two different radioisotopes are used). Since the spectrum of pulse heights produced is related to the beta particle energy spectrum of the radioisotope, the discriminator settings may be adjusted to distinguish pulses arising from each of the isotopes present in the sample. This is particularly easy in the case of samples containing both ^{3}H and ^{14}C, since their beta particle energy spectra differ so markedly (Fig. 14–7).

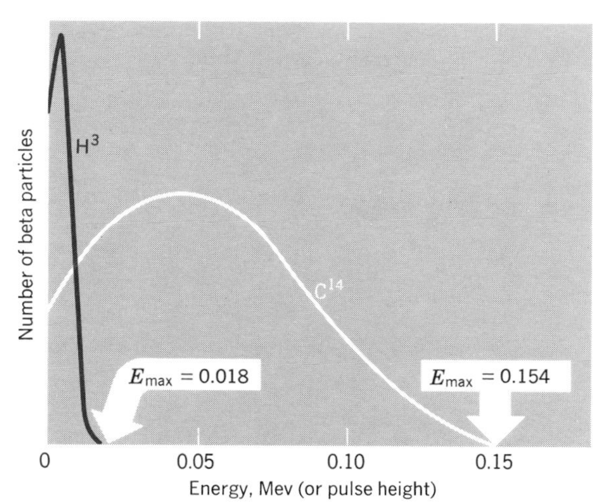

Figure 14–7
Energy or pulse height distributions of ^{3}H and ^{14}C. The energy and pulse height distributions differ so markedly that pulses arising from ^{14}C only in a sample containing both isotopes may readily be obtained by adjusting the pulse height position of the discriminator.

Figure 14–6
Basic components of a liquid scintillation counter.

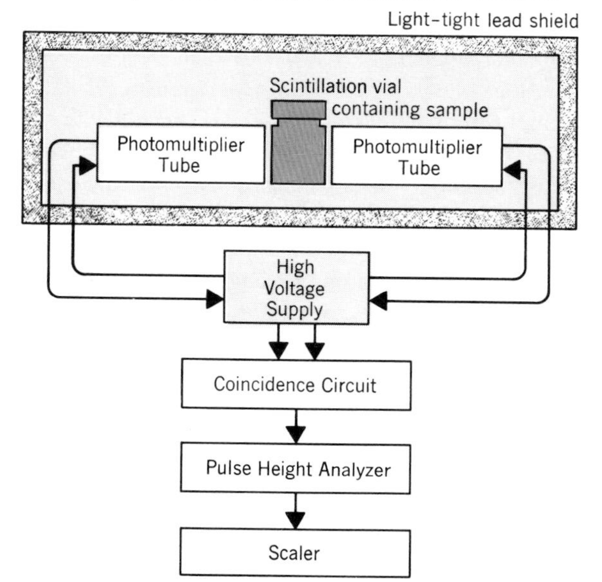

Autoradiography

Any discussion of the uses of radioisotopes as tracers in cell biology would be incomplete without at least a brief description of **autoradiography.** The materials and equipment used in this technique differ significantly from those employed in the methods previously described. In autoradiography, the biological sample containing the radioisotope is placed in close contact with a sheet or film of photographic emulsion. Rays emitted by the radioisotope enter the photographic emulsion and expose it in a manner similar to visible light. After some period of time (usually several days to several weeks), the film is developed and the location of the radioisotope in the original sample determined from the exposure spots on the film. Unlike the methods described earlier, autoradiography is generally not employed as a quantitative technique (although it can be under certain conditions); instead, it is used to determine the specific region of **localization** of a radioactive tracer. The method is most often used with histological sections to determine the precise location of a labeled compound in a tissue or in a cell and may be applied either at the light microscope or electron microscope level. Autoradiography has also been successfully employed in conjunction with electrophoresis, chromatography (i.e., thin-layer chromatography, paper chromatography, etc.), and other molecular fractionation methods (see Chapter 13) for identifying zones containing labeled compounds (Fig. 14–8).

Figure 14–8

Autoradiogram of an electrophoresis gel. ^{14}C-labeled proteins from whole T4 phage and from virus subfractions were separated in polyacrylamide and the gel covered with a sheet of film to expose the protein bands. (*A*) Whole phage; (*B*) heads; (*C*) necks + tails; (*D*) extended tails; (*E*) tails. Courtesy of Drs. D. Coombs and F. A. Eiserling.)

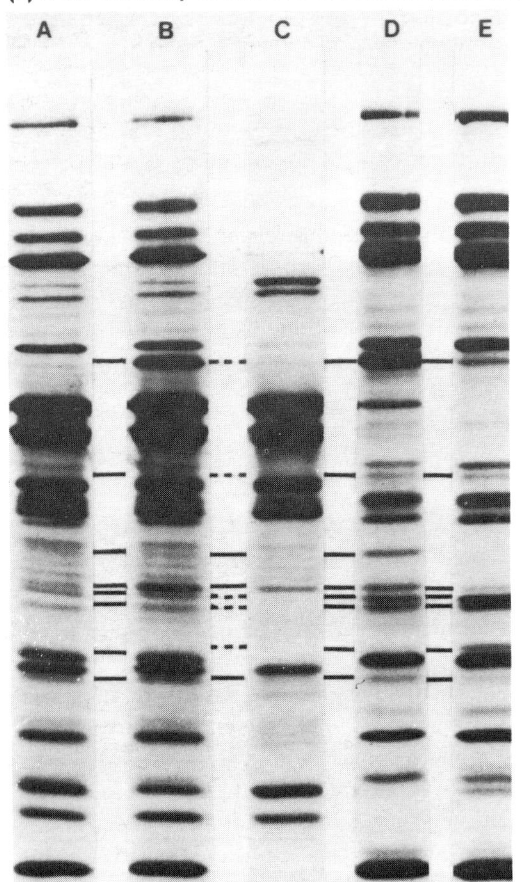

Since the main goal of autoradiography is to determine the precise location of the tracer, the degree of resolution obtained in the autoradiograph is of primary importance. Resolution depends on (1) the type of photographic emulsion used, (2) the distance between the radioactive sample and the emulsion, and (3) the nature of the emitted radiation. A variety of photographic emulsions are available varying in sensitivity and resolution according to the size and concentration of their silver halide grains—the least sensitive films generally offering the highest degree of resolution. Emulsions can also be obtained that are particularly sensitive to specific types of radiation. Since radiation is emitted in all directions from the radioactive source, the greater the distance between the film and the source, the more diffuse the resulting image. It is therefore very important to use very thin samples and place them in very close contact with the film. The highest degree of resolution is obtained with radioisotopes that emit rays of short path length and that have a high specific ionization. Alpha particles provide high resolution, but radioisotopes emitting this type of radiation are rarely useful in biological experiments. Autoradiographs become increasingly diffuse as the E_{max} of a beta emitter increases, but ^{3}H, ^{14}C, ^{35}S, and ^{45}Ca do yield good resolution. Gamma rays are inefficient as a result of their extremely high ranges and low specific ionizations. It should also be noted that some problems may be encountered when radioisotopes with short half-lives are used, since the exposure time required may be quite lengthy (i.e., weeks or even months).

Autoradiography may be employed with large but thin slices of tissues or organs (gross autoradiography) or with smaller pieces sectioned and prepared for light or electron microscopy. For example, the deposition of calcium in bone has been studied using ^{45}Ca by cutting thin, flat, longitudinal slices through bone and placing them against large sheets of photographic film. When used in conjunction with light microscopy, the paraffin sections are first mounted on slides which are then coated with the photographic emulsion and stored in a lighttight, usually lead-lined box (in order to

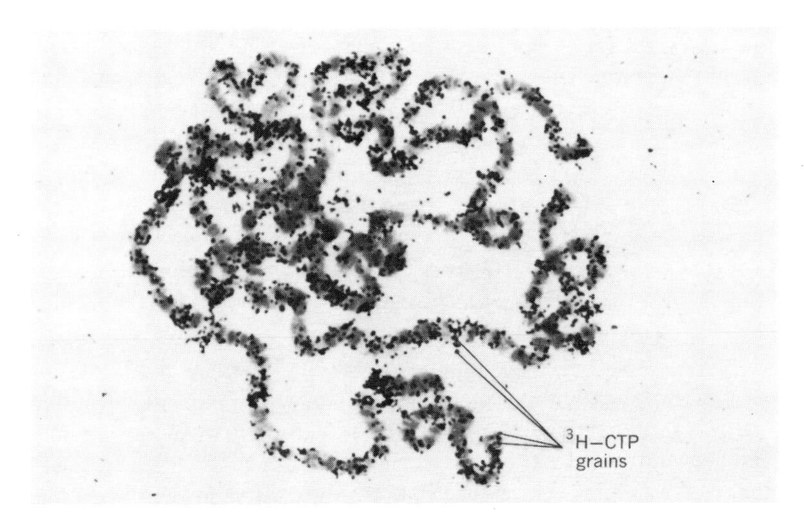

Figure 14–9
Autoradiogram of chromosomes from the polytene nucleus of Drosophila salivary gland cell. The dark exposure spots or "grains" reveal the incorporation of ^{3}H-labeled cytidine triphosphate into actively transcribing regions of the chromosomes. The heavily labeled regions were specifically induced to transcribe by incubating the salivary gland at 37°C in medium containing the tracer. (Courtesy of Dr. J. Lee Compton.)

minimize background exposure that results from the effects of cosmic radiation and other sources of radiation). Several days or weeks later, after photochemical development, the sections are conventionally stained to better visualize the biological material and the distribution of radioisotope determined microscopically from the location of dark exposure spots (called ''grains'') on the section (Fig. 14–9).

Autoradiography may also be used with thin sections of tissues prepared for electron microscopy (Fig. 14–10). This procedure involves preliminary examination and photography of the thin section followed by coating the grid with a very thin layer of photographic emulsion containing silver halide crystals of particularly small size. After several days, the grid is developed and again examined and photographed with the electron microscope in order to identify those regions of the original section that contain clusters of metalic silver grains and therefore contain the radioactive tracer.

Figure 14–10
Autoradiogram of thin section through dividing *E. coli* cells. The bacteria were cultured on a medium containing ^{3}H-labeled thymidine which was incorporated into the DNA of the nucleoids. (Courtesy of Dr. F. A. Eiserling.)

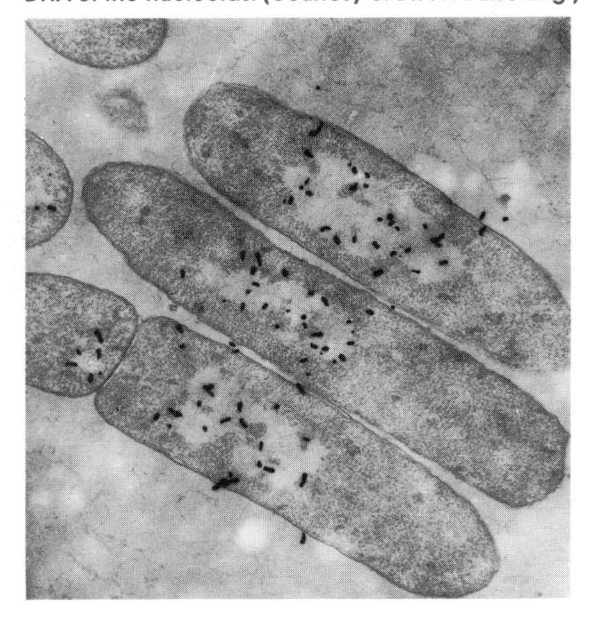

Summary

Certain atomic **isotopes** are unstable and emit particulate (and sometimes electromagnetic) radiations. A number of radioactive isotopes, especially ^{3}H, ^{14}C, ^{32}P, ^{35}S, and ^{59}Fe, are particularly valuable to cell biologists as **tracers** of metabolism and function. With radioisotopes, it is possible to follow ionic or molecular flux under conditions of zero net exchange, to simplify chemical analyses, to measure pool sizes, and to determine precursor–product relationships.

The most common types of radiations are **alpha particles, beta particles,** and **gamma rays.** Only negative beta-particle-emitting isotopes and gamma-ray-emitting isotopes are normally used by biologists. The energy acquired by a beta particle during **nuclear transmutation** may vary over a specific range, whereas gamma rays have discrete energy values. Beta particles and gamma rays interact with matter by ionizing or exciting atoms in their path, and it is this property that is used in instruments designed to detect and measure the quantity of radioisotope present in a sample. The simplest and most common detector is the **Geiger-Müller counter** and is used principally with tracers emitting beta particles of intermediate or high energy. **Solid scintillation counters** are used with isotopes emitting gamma rays, and **liquid scintillation counters,** with isotopes emitting weak beta rays. ^{3}H and ^{14}C are used as tracers more often than any other isotopes, and since these emit very weak beta particles, the liquid scintillation approach is a particularly important technique in cell biology. In the typical liquid scintillation counter, the sample is mixed with a scintillation fluid in a glass vial. The energies of the beta rays emitted by the sample are ultimately converted to flashes of light which are detected by photomultiplier tubes and counted.

Autoradiography is a special modification of the radioisotopic tracer technique most often used in conjunction with light and/or electron microscopy. Biological samples (usually tissue sections) containing the tracer are placed in contact with special photographic films or emulsions. The emitted rays expose the film, producing **grains,** the numbers and distribution of which within a cell or tissue provide quantitative and qualitative information about the movement and/or localization of metabolic intermediates or products.

References and Suggested Reading

Article

Joftes, D. L., Radioautography, principles and procedures. *J. Nucl. Med. 4,* 143 (1963).

Books, Monographs, and Symposia

Chase, G. D., and Rabinowitz, J. L., *Principles of Radioisotope Methodology* (3rd ed.), Burgess Publishing Co., Minneapolis, 1967.

Finlayson, J. S., *Basic Biochemical Calculations: Related Procedures and Principles,* Addison-Wesley Publishing Co., Reading, Mass., 1969.

Friedlander, G., Kennedy, J. W., and Miller, J. M., *Nuclear Radiochemistry* (2nd ed.), John Wiley & Sons, Inc., New York, 1964.

Glasstone, S., *Sourcebook on Atomic Energy* (3rd ed.), Van Nostrand Co., Princeton, N.J., 1967.

Gude, W. D., *Autoradiographic Techniques,* Prentice-Hall, Inc., Englewood Cliffs, N.J., 1968.

Overman, R. T., and Clark, H. M., *Radioisotope Techniques,* McGraw-Hill Book Co., Inc., New York, 1960.

Sheppard, C. W., *Basic Principles of the Tracer Method,* John Wiley & Sons, Inc., New York, 1962.

Wang, C. H., and Willis, D. L., *Radiotracer Methodology in Biological Science,* Prentice-Hall Inc., Englewood Cliffs, N.J., 1965.

Part 5
STRUCTURE AND FUNCTION OF THE MAJOR CELL ORGANELLES

Chapter 15
THE PLASMA MEMBRANE

The plasma membrane delimits the cell, physically separating the cytoplasm from the surrounding cellular environment. This implies that all substances either entering or exiting the cell must pass *through* the plasma membrane. Only rarely does the plasma membrane play a passive role in the exchange of molecules between the cell and its surroundings. Instead, the flux of substances may actually be facilitated by continuous molecular changes within the membrane. In many instances, transport through the membrane is achieved by the active participation of carrier molecules within the membrane and incurs the expenditure of large amounts of chemical energy. The cellular ingestion (or excretion) of some materials is associated with gross movements and separations of fragments of the membrane from the main body. Stages of this activity can be seen and studied with the electron microscope (sometimes also the light microscope).

In this chapter, we consider the structure and chemical organization of the plasma membrane and the mechanisms by which the transport of materials across the membrane may be achieved. As will become evident in subsequent chapters, much of the information presented here can be directly extrapolated to the membranes that encase cell organelles as well as to other cytomembranes.

Early Studies on the Chemical Organization of the Plasma Membrane

Among all animal and plant cells, none has been more extensively studied than the mammalian erythrocyte or red blood cell. The erythrocyte has long been the favorite of investigators studying the plasma membrane because relatively pure membrane preparations are so easily obtained. The mature erythrocyte contains no nucleus, mitochondria, ribosomes, or other organelles and no intracytoplasmic membranes; instead, this highly specialized cell consists essentially of a concentrated (semicrystalline) solution of hemoglobin encased in a membrane. Because of the cell's simplicity, its membranes are easily separated from other cytoplasmic constituents (primarily hemoglobin) by centrifugation following osmotic cell lysis.

Results obtained using erythrocytes have frequently been extrapolated to all cells. This is unfortunate because the erythrocyte is not a typical cell, and it is therefore unlikely that the chemical composition and organization of its limiting membrane are representative. Indeed, with increased interest in studying the plasma membranes of other cells and with the advent of methods for isolating these membranes, our knowledge of the plasma membrane has expanded rapidly in recent years. During this time, it has become increasingly obvious that many of the properties of the erythrocyte membrane are unique. For historical perspective, however, we will begin by considering the early studies of the red blood cell membrane.

Existence of Lipid in the Membrane

As early as 1899, E. Overton recognized that the boundary of animal and plant cells was "impregnated" by lipid material. Overton's conclusions were based on exhaustive studies of the rates of penetration of more than 500 different chemical compounds into animal and plant cells. In general, compounds soluble in organic solvents entered the cells more rapidly than compounds soluble in water. These differences were attributed to the "selective solubility" of the membrane; that is, lipid soluble materials would pass into the cell by dissolving in the corresponding lipid elements that made up the membrane. Overton suggested that cholesterol and lecithins might be among the lipid constituents of the plasma membrane, a suggestion that was later substantiated chemically. The pioneering studies of Overton at the turn of the century set the stage for Gorter, Grendel, Cole, Danielli, Harvey, Davson, and others who attempted to determine the specific manner in which the lipid might be organized within the membrane.

The Langmuir Trough

One of the most valuable instruments used to study the behavior of lipid films is the **Langmuir trough** (Fig. 15–1). If lipid containing hydrophilic groups (such as the carboxyl groups of fatty acids or the phosphate groups of phospholipids) is dissolved in a highly volatile solvent and several drops are then carefully applied to the surface of water, the lipid spreads out to form a thin, monomolecular film in which the hydrophilic parts of each molecule project into the water surface while the hydrophobic parts are directed up, away from the water (Fig. 15–2). In 1917, I. Langmuir introduced a clever technique for measuring the specific minimum surface area occupied by a monomolecular film of lipid and the force necessary to compress all the lipid molecules into this area. His device, known as the Langmuir trough or Langmuir film balance, has been used extensively over the past several decades in connection with physical measurements of membrane lipids.

Figure 15–1
The Langmuir trough.

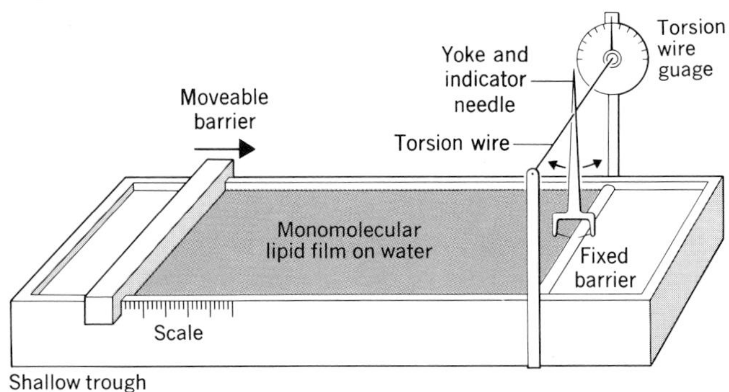

Figure 15–2
Formation of a monomolecular lipid film on water. Phosphatidyl ethanolamine (*a*) represents a typical lipid molecule possessing polar (hydrophilic) and non-polar (hydrophobic) regions. These regions of the phospholipid molecule are depicted diagrammatically in (*b*). When spread on water, the hydrophilic parts of each lipid project into the water surface, while the hydrophobic parts are directed up, away from the water (*c*).

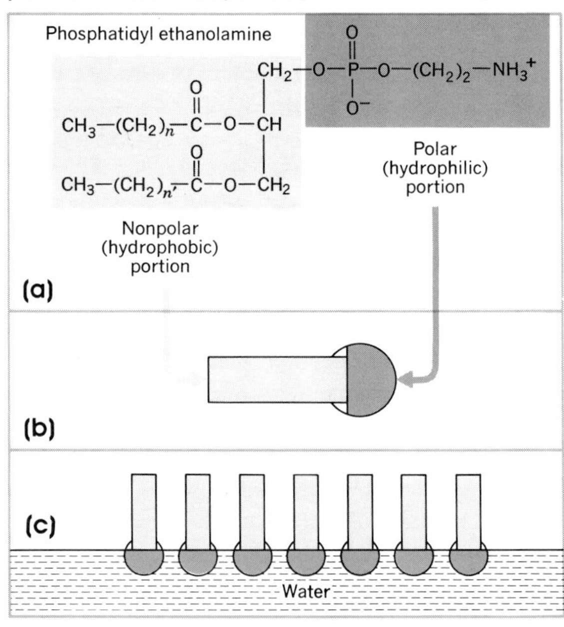

The apparatus (shown in Fig. 15–1) consists of a shallow trough coated with a nonwettable material so that it can be filled with water to a level slightly higher than its edges. A bar is placed across the width of the tray and is used to sweep dirt and dust from the water. A second bar, called the "fixed barrier" (which, in fact, is slightly movable), floats on the water surface and is connected to a torsion balance. After the surface of the water is swept clean, a third bar, called the "movable barrier," is placed across the clean region of the trough, and a known quantity of dissolved lipid is added to the space between it and the fixed barrier. The lipid spreads out over the water surface, forming a layer one molecule thick. The movable barrier is then slowly pushed toward the fixed barrier, thereby compressing the lipid until it forms a continuous and rectangular monomolecular film. The force exerted by the film on the fixed barrier can be measured using the torsion wire gauge. From the area occupied by the lipid and the amount initially added to the trough, it is possible to calculate the surface area occupied by a single lipid molecule. Langmuir himself employed this device to study the behavior of the surface films formed by a variety of organic compounds, and for this work, together with his electronics innovations, he received the Nobel Prize in 1932. Others have applied the same technique in specific studies of membrane lipids.

Gorter and Grendel's Bimolecular Lipid Leaflet Model

In 1925, E. Gorter and F. Grendel published the results of their studies on the organization of lipid in the membrane of the red blood cell. Their studies were carried out using blood from a variety of mammals, including dogs, sheep, rabbits, guinea pigs, goats, and humans, and all yielded essentially the same results. The lipid present in accurately measured quantities of washed red blood cells was extracted with acetone and the acetone was then evaporated, leaving the lipid as a residue. This residue was redissolved in benzene and spread in a Langmuir trough to form a tightly packed monomolecular layer. The surface area occupied by the extracted lipid was then measured.

Gorter and Grendel determined the numbers of red cells present in each sample of blood analyzed and estimated the *total* surface area of the cells by multiplying the cell number by the average surface area per cell. (The surface area of the erythrocyte was estimated using the relationship proposed by Knoll that for red blood cells the surface area $= 2d^2$, d being the diameter of the cell determined microscopically. This estimate was subsequently shown to be in error.) By dividing the total surface area occupied by a monomolecular layer of membrane lipid extracted from these cells by the total cell surface area, the number of lipid layers present in the membrane was obtained. The value varied between 1.8 and 2.2, leading Gorter and Grendel to propose that the cell membrane was formed by a *bimolecular* lipid sheet. They further suggested that the polar ends of the lipid molecules of one layer were directed outward (from the cell) toward the surrounding plasma, while the polar ends of the lipid molecules forming the other layer were directed inward toward the cell hemoglobin. Thus the nonpolar and extremely hydrophobic ends of the lipid molecules in each layer would face one another (Fig. 15–3).

In the past 10 years, the results of Gorter and Grendel have been reexamined under improved conditions by a number of investigators. The validity of Gorter and Grendel's bimolecular lipid leaflet model depends on the assumptions that (1) all the erythrocyte lipids are in the plasma membrane, (2) all of this lipid was extracted using their acetone procedure, and (3) the average surface area of the cells was accurately estimated. The first assumption has been verified—all the erythrocyte lipid *is* in the plasma membrane. However, it is now clear that Gorter and Grendel extracted only 70 to 80% of the total lipid. This error would seriously alter the predicted ratio of *lipid film area to cell surface area* if it were not for the fact that Gorter and Grendel also

Figure 15–3
Bimolecular lipid leaflet model for the structure of the red blood cell membrane proposed by Gorter and Grendel.

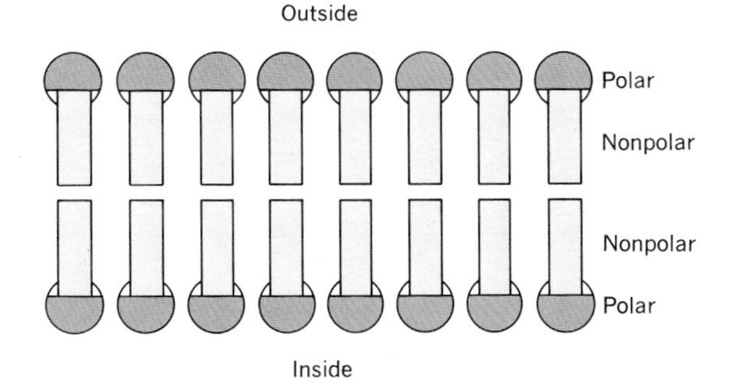

underestimated the red blood cell surface area by a comparable amount. The two errors cancelled each other out, so that the ratio of 2:1 is still obtained.

The Danielli-Davson Membrane Model

A consistent observation made for cell membranes that was not explained by the bimolecular lipid leaflet model was the very low surface tension of the cell membrane. In 1935, J. F. Danielli and E. N. Harvey proposed that oil droplets and other lipid inclusions in cells were bounded at their surfaces by an organized layer of the lipid and a layer of protein. It was postulated that the protein, which consisted of a monomolecular layer of hydrated molecules, faced the aqueous cytoplasm and simultaneously interacted with the polar portions of the lipid layer. The nonpolar portions of the lipid layer faced the hydrophobic oil phase of the droplet interior (Fig. 15–4). In this structure, the natural surface activity of the protein would account for the low interfacial tension of the droplet membrane. Shortly thereafter, Danielli and H. Davson suggested that the plasma membrane itself might be composed of two such lipid-protein bilayers—one facing the interior of the cell and the other facing the external milieu. This arrangement is shown in Figure 15–5. Danielli and Davson proposed that such a membrane would exhibit **selective permeability,** being capable of distinguishing between molecules of different size and solubility properties and also between ions of different charge.

By the early 1950s, several modifications were made in the Danielli-Davson membrane model. Using the polarizing microscope, F. O. Schmitt showed that in the erythrocyte membrane, lipid molecules were oriented with their long axis perpendicular to the membrane surface; in contrast, the protein molecules were oriented tangentially. Accordingly, layers of polypeptides arranged in the pleated sheet configuration or as alpha helices were substituted for the hydrated globular molecules of the original model. It was also suggested that glycoproteins might be adsorbed to the outer membrane surface, thereby accounting for the antigenic properties of cell membranes. Pores in the membrane were presumed to be formed by periodic continuities (bridges) between the outer and inner protein layers. The modified Danielli-Davson membrane model is shown in Figure 15–6. In this arrangement, the association between the surface proteins and the bimolecular lipid leaflet would be maintained primarily by electrostatic interactions between the polar ends of each lipid molecule and charged amino acid side chains of the polypeptide layers. Either electrostatic or van der Waals bonds could bind other groups to the outer protein surface.

It should be kept in mind that the model shown in Figure 15–6 was formulated *before* the plasma membrane was first seen, since the use of the electron microscope to study the organization of the plasma membrane began around 1957. Nevertheless, even the thickness of the membrane was estimated to be about 100 Å, based on the known lengths of extended phospholipid molecules (about 35 Å) and the thickness of *several* layers of pleated sheet polypeptide or a *single* layer of the alpha-helix form (10–15 Å).

Robertson's Unit Membrane

In the late 1950s, electron microscopy provided additional information about the structure of the plasma membrane. J. D. Robertson was a pioneer in this area, showing that membranes fixed with osmium tetroxide revealed a char-

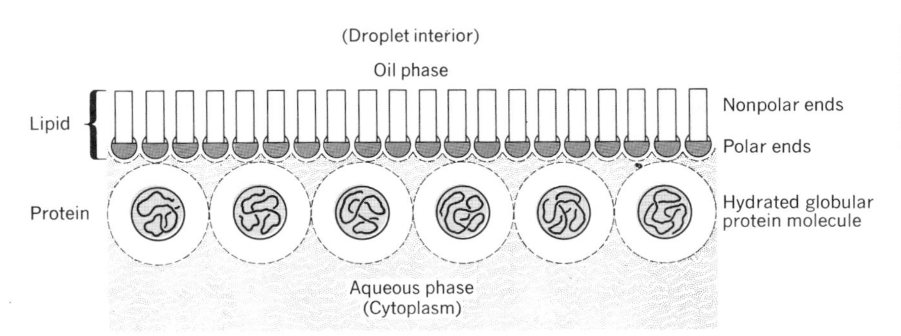

Figure 15–4
Danielli and Harvey's 1935 model of the protein-lipid bilayer formed at the interface between a cell oil droplet and aqueous cytoplasm.

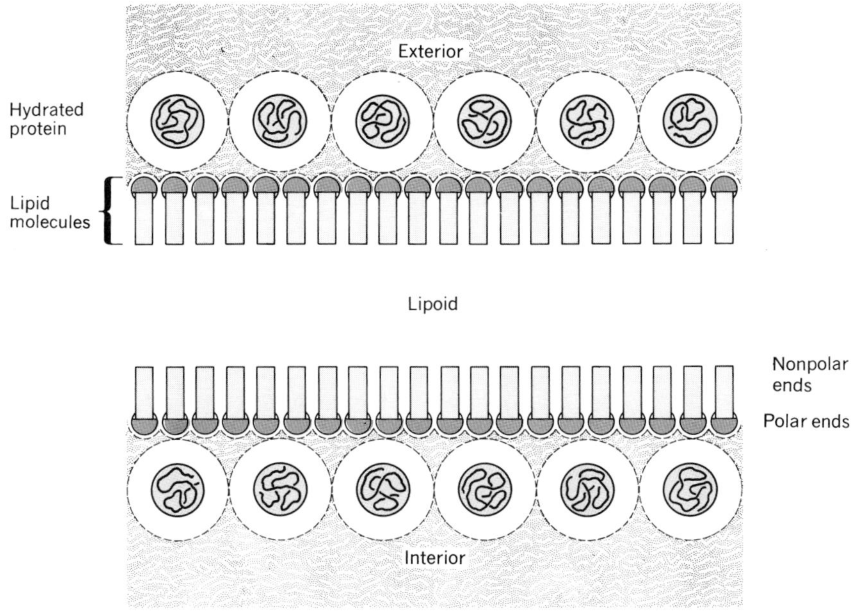

Figure 15–5
Danielli-Davson membrane model (1935).

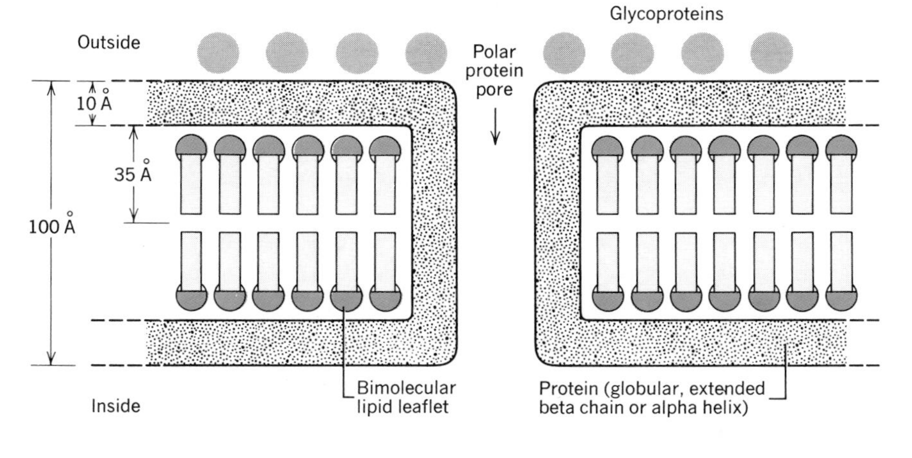

Figure 15–6
Modified Danielli-Davson membrane model (1950).

acteristic *trilaminar* appearance consisting of two parallel outer dark (osmiophilic) layers and a central light (osmiophobic) layer (Fig. 15–7). The osmiophilic layers typically measured 20–25 Å in thickness and the osmiophobic layer measured, 25–35 Å, yielding a *total* thickness of 65–85 Å. This value compared favorably with the thickness predicted on the basis of chemical studies. However, the thickness of the osmiophilic outer layer was much greater than that predicted for the outer protein coats, while the thickness of the central osmiophobic layer was too small to be the bimolecular lipid leaflet (see Fig. 15–6). To account for these apparent discrepancies, Robertson suggested that the dark layers might be produced by osmium ions binding to *both* the polar amino acid side chains of the protein and the polar ends of the phospholipid molecules, while the light central layer represented only the nonpolar fatty acid chains of each phospholipid that would not bind osmium. In some cells, the outer dark line was thicker than the internal dark

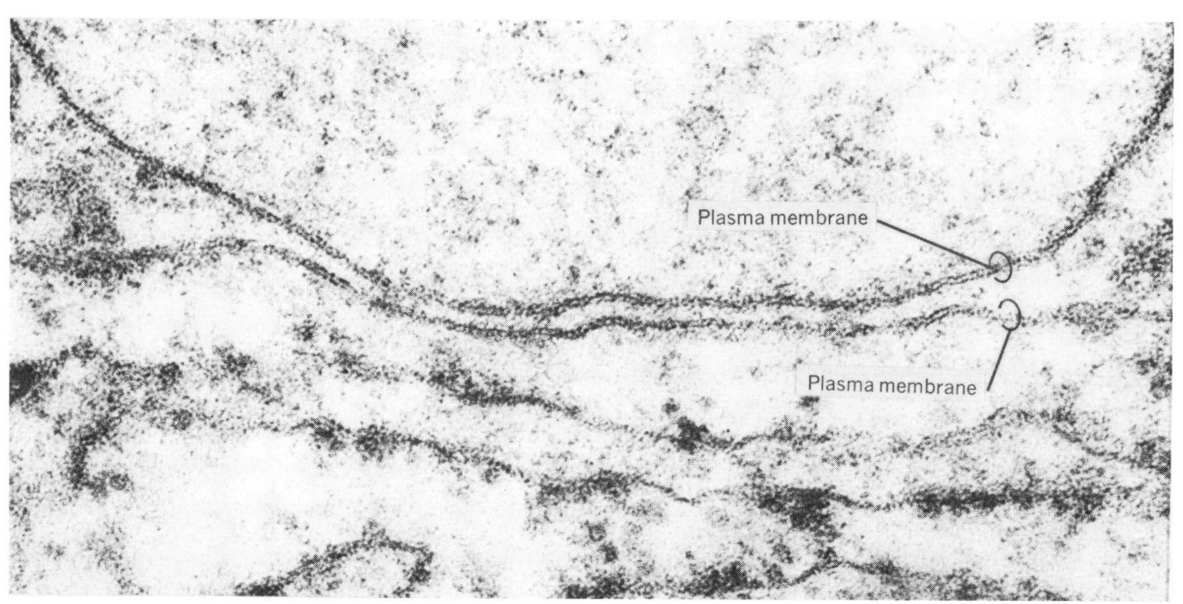

Figure 15–7
Trilaminar appearance of the plasma membrane. (Electron photomicrograph courtesy of R. Chao.)

line, and this was presumed to be due to the binding of additional osmium ions by adsorbed glycoproteins or other osmiophilic molecules.

Robertson and others demonstrated that the trilaminar pattern was characteristic of many other cellular membranes, including the endoplasmic reticulum, the membranes of mitochondria, chloroplasts, and Golgi bodies. In view of the underlying unity in the appearance of nearly all cell membranes studied, Robertson proposed his now famous *unit membrane* model. According to Robertson, the unit membrane consisted of a bimolecular lipid leaflet sandwiched between outer and inner layers of protein organized in the pleated sheet configuration. Such an arrangement was presumed to be basically the same in all cell membranes. While acknowledging specific chemical differences between membranes (i.e., the particular molecular species that make up each membrane differ), Robertson proposed that the pattern of molecular organization was fundamentally the same.

Robertson extended his unit membrane model to include the notion that continuity exists between the membranes of the nuclear envelope and the plasma membrane via the endoplasmic reticulum. Furthermore, he suggested that vesicular organelles including mitochondria might arise from this continuous membrane system and that these are subsequently pinched off to form separate cytoplasmic structures. While there is some supportive evidence for this in the case of certain cell structures including lysosomes and pinocytic vesicles, this most attractive notion still remains speculative.

From its inception, Robertson's unit membrane model was highly controversial and almost continuously argued and questioned. Nonetheless, the model held a cornerstone position in membrane biology until the late 1960s, when a number of new findings made the unit membrane no longer tenable. Although there can be no doubt about the similar electron-microscopic appearance of nearly all membranes, so strict a chemical interpretation to account for the uniformity is no longer supportable. Among the more important observations and viewpoints that argue against a generalized unit membrane model are the following:

1. Cellular membranes vary in biological functions, in kinds and relative amounts of phospholipids, and in quantitative lipid-to-protein ratios. Consider, for example, the plasma membrane of a parenchymal cell in the liver. The membrane forms junctions with at least three different neighboring structures (Fig. 15–8): (a) one face of the liver cell plasma membrane forms a junction with the membranes of the capillary sinuses, and across this face pass various substances exchanged between the bloodstream and the liver cell cytosol (e.g., sugars, amino acids, insulin, etc.). The membrane in this region would be expected to contain specific transferases involved in capillary exchange; (b) a second face of the liver cell is in contact with the *bile canaliculi* into which bile salts and bile pigments are continuously discharged; here too the organization and composition of the plasma membrane would be expected to be rather specific; (c) a third junction is formed with the plasma membranes of neighboring cells, and it is across this face that intercellular exchange takes place. W. H. Evans and others have clearly shown that the different faces of the liver cell plasma membrane are clearly distinguishable in enzyme content and molecular composition. The epithelial cells that line the small intestine also reveal a differential distribution of membrane proteins. The portion of the membrane facing the intestinal lumen is rich in glycoproteins, while the opposite membrane face contains *sodium pumps* (discussed later). The membrane uniformity apparent during electron microscopy of cells belies the chemical heterogeneity that actually exists.

Figure 15–8
Three faces of the liver cell. (A) Juxtaposition of plasma membrane and sinusoid capillary membrane. (B) Juxtaposition with neighboring cell. (C) Juxtaposition with bile canaliculus. Each face contains specific membrane constituents and properties.

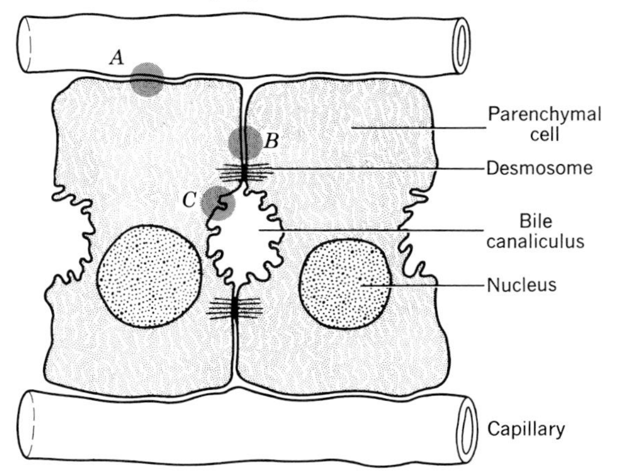

2. There is no rigorous evidence unequivocally relating the trilaminar electron-microscopic appearance of membranes to a specific arrangement of protein and lipid. Indeed, the chemistry of the interaction between protein and lipid and electron-dense stains such as osmium tetroxide is unclear. For example, the trilaminar appearance of some cell membranes is not altered by the preliminary extraction of the membrane lipids. Moreover, it has been shown chemically that osmium can react with the unsaturated fatty acid side chains of phospholipids.

3. There is now ample evidence to substantiate that the protein and lipid composition of the plasma membrane does not remain constant. Instead, the composition changes as protein and lipid molecules are added to the membrane, removed from the membrane, and redistributed through the membrane.

For all intent and purposes, the only aspect of Robertson's unit membrane model that remains undisputed is the existence in the membrane of a bimolecular leaflet of lipids. Yet, even this should more properly be credited to the earlier studies of Gorter and Grendel. In the time since their original studies were conducted, the quantitative distributions of membrane lipids in erythrocytes have been reevaluated using newer and improved techniques that guarantee total lipid extraction from the membranes and more accurate measurements of cell surface area. Some of the new methods account also for the specific physical dimensions of various lipid molecules known to be present in the membrane (fatty acids, phospholipids, neutral lipids, etc.) in making calculations of the lipid surface area. For the most part, these studies indicate that indeed there is enough membrane lipid present to form a layer two molecules thick.

Other sources of support for the presence of the lipid bilayer include results of x-ray diffraction analysis of dispersions of isolated cell membranes and electron spin resonance (ESR) studies which indicate that the lipid molecules are oriented with their long axes perpendicular to the plane of the membrane. Attempts have been made to reconstruct lipid bilayers using molecular models for the various membrane lipids and by considering the attractive and repulsive forces resulting from van der Waals interactions, hydrogen bonds, and electrostatic interactions. These studies indicate that a stable bimolecular lipid leaflet can be formed by a tail-to-tail interdigitation of the lipids.

The dimensions, electron-microscopic appearance, permeability, surface tension, and electrical capacitance of artificially created lipid bilayers (i.e., *liposomes*) are very similar to those of natural biological membranes. Finally, freeze-fracture techniques (see Chapter 1) used in transmission electron microscopy produce images of cell membranes, suggesting a natural plane of cleavage at the center of the membrane. A plane susceptible to such cleavage would be provided by the space at the center of the bimolecular lipid leaflet. This will be pursued in the next section.

The Fluid-Mosaic Model of Membrane Structure

At the present time, the most widely accepted model of membrane structure is the **fluid-mosaic model** (an expression introduced by S. J. Singer and G. Nicolson to describe both the properties and organization of the membrane). According to this model (Fig. 15–9), the membrane contains a bimolecular lipid layer, the surface of which is interrupted by proteins. Some proteins are attached at the polar surface of the lipid (i.e., the *peripheral,* or *extrinsic,* proteins), while others penetrate the bilayer or span the membrane entirely (i.e., the *integral,* or *intrinsic,* proteins). The peripheral proteins and those parts of the integral proteins that occur on the outer membrane surface frequently contain chains of sugars (i.e., they are glycoproteins). The sugar chains are believed to be involved in a variety of physiological phenomena including the adhesion of cells to their neighbors. Membrane lipid is primarily phospholipid, although quantities of neutral lipids may also be present. Some of the lipid at the outer surface is complexed with carbohydrate to form glycolipid.

Freeze-Fractured Membranes

The fluid mosaic model of membrane structure is beautifully supported by the visual evidence provided when freeze-fractured membranes are examined with the transmission electron microscope. D. Branton, who pioneered this field, showed that membranes rapidly frozen at the

Figure 15–9
The *fluid mosaic* model of membrane structure.

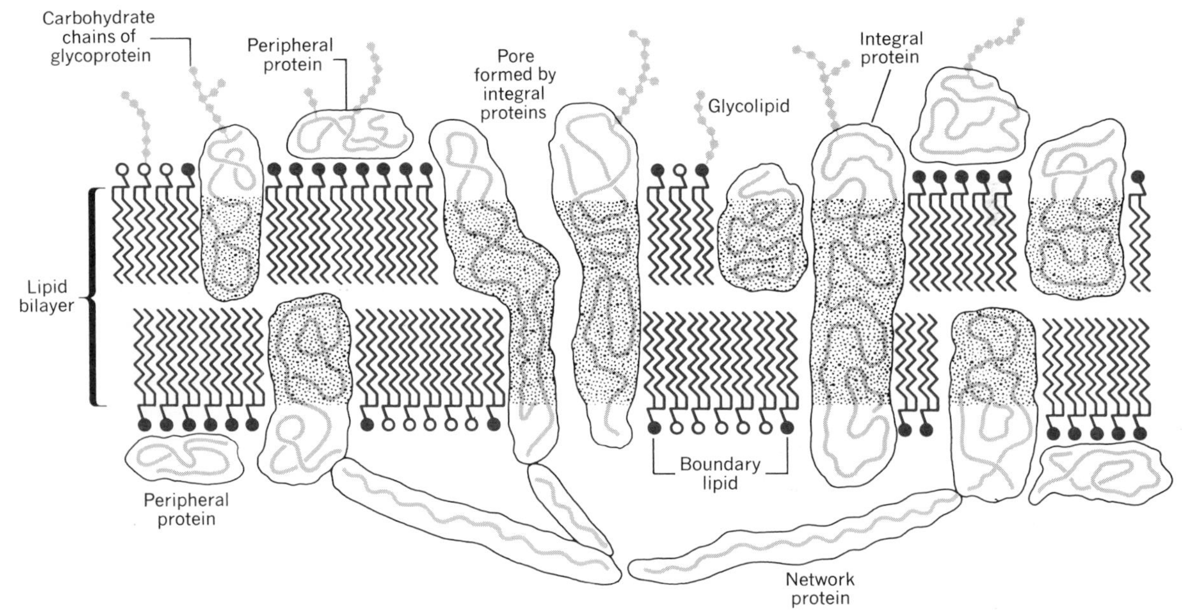

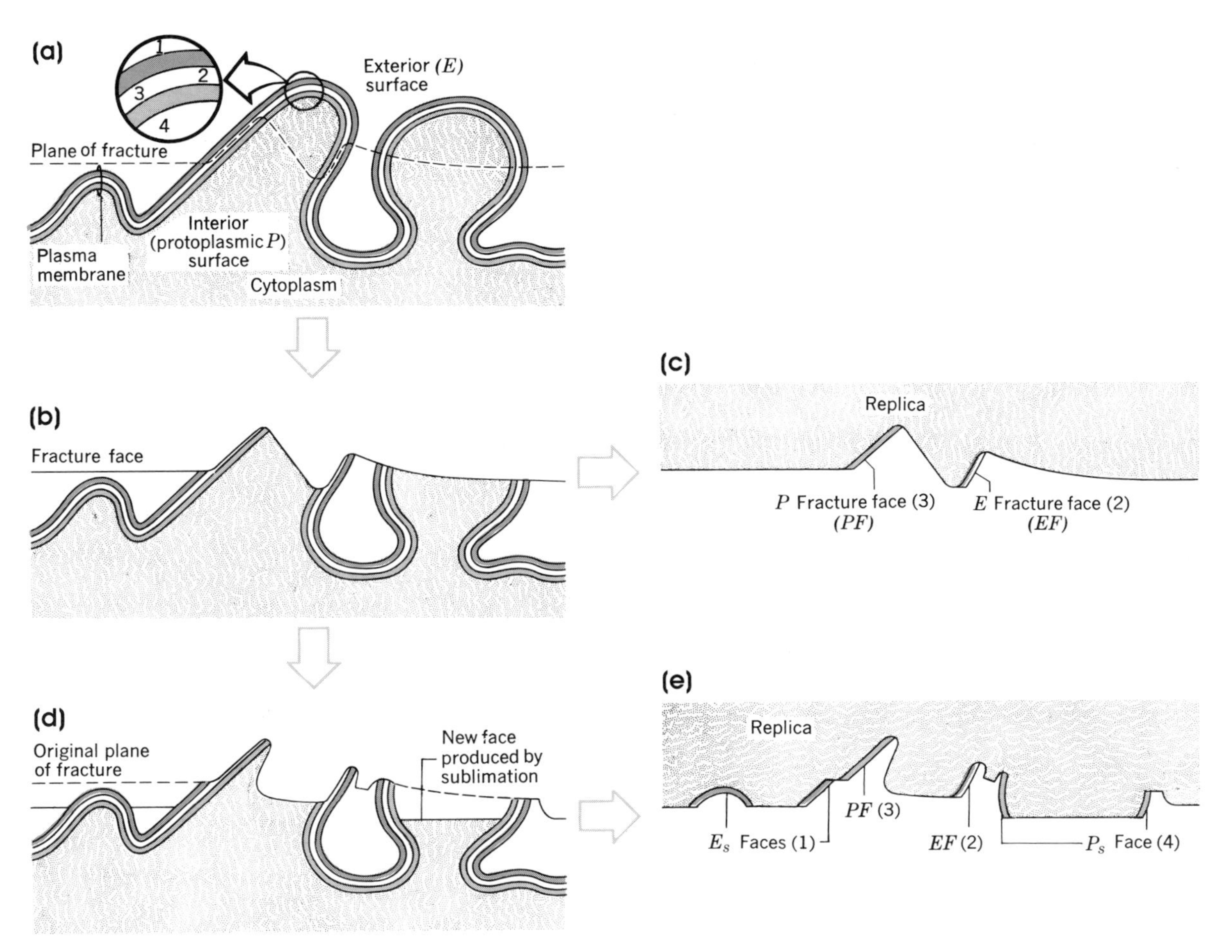

Figure 15–10
Membrane faces exposed by freeze-fracturing alone and by freeze-fracturing followed by sublimation. (*a*) The fracture passes along the lipid bilayer for some distance following its intersection with the membrane. (*b*) This exposes the PF and EF fracture faces seen in the replica (*c*). Alternatively, the surface of the specimen is lowered by sublimation (*d*), exposing the outer and inner surfaces (i.e., the E_s and P_s faces) as well as the fracture faces (*e*).

temperature of liquid nitrogen and cut or chipped with a microtome blade readily fracture along specific planes. When the plane of the fracture intersects the plane of the membrane, the membrane is split along the center of the lipid bilayer, producing two "half-membranes" called the *E* half and the *P* half. The *E* half is that portion of the membrane that faced the cell *exterior*, while the *P* half corresponds to the portion that faced the protoplasm (cytosol). One side of each half-membrane is the original membrane surface, called the *E* and *P* faces, while the other side is the newly exposed **fracture face,** called the *E* fracture face (EF) and *P* fracture face (PF). The fracture faces are extremely delicate and are not examined directly. Instead, a thin film of platinum and carbon is evaporated onto the surface of the fracture faces to produce a *replica* (see Chapter 1) which is then examined by transmission electron microscopy.

In many instances, before the replica is made, water (as well as other volatile materials) on or near the fracture surfaces is eliminated by sublimation (i.e., by carefully

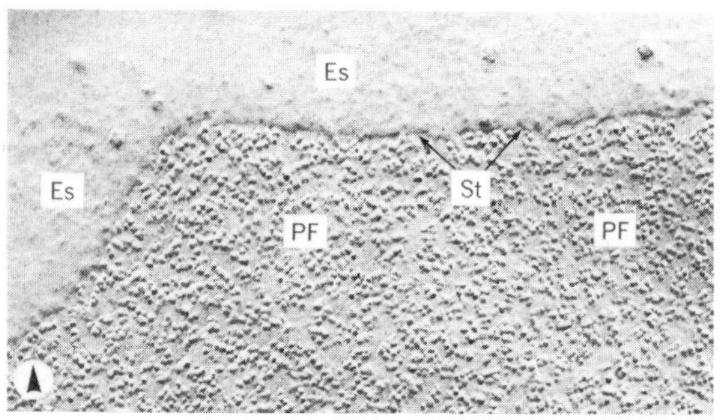

Figure 15–11
Freeze-fractured and sublimated plasma membrane of an erythrocyte. The fracture face of the protoplasmic half of the membrane (i.e., PF) is covered with particles believed to be integral proteins. Sublimation reveals the outer surface of the exterior half of the membrane (i.e., E_s). The fracture face is seen to be separated by a step (St) from the outer surface of the membrane. Circled arrow indicates direction of shadowing. Magnification, 44,500 X. (Photomicrograph courtesy of Dr. C. Stolinski.)

raising the temperature of the sample). This step, which used to be called "freeze etching," exposes additional surface features of the fracture face (see Fig. 15–10 and especially Fig. 1–11).

Electron micrographs of freeze-fractured cells show the membranes to be covered by numerous small particles (Fig. 15–11). There is convincing evidence that the particles are membrane proteins (e.g., they disappear when the membranes are first treated with proteolytic enzymes). This suggests that the plane of fracture passes around the protein molecules rather than through them. This relationship is depicted in Figure 15–12. The relatively uniform background apparent in the fracture face in Figure 15–11 corresponds to the surface of one-half of the lipid bilayer.

Membrane Proteins

Peripheral (Extrinsic) Proteins

Peripheral or extrinsic membrane proteins are generally loosely attached to the membrane and are more readily removed than are the integral proteins. Peripheral proteins are rich in amino acids with hydrophilic side chains that permit interaction with the surrounding water and with the polar surface of the lipid bilayer. Peripheral proteins on the cell's exterior membrane surface often contain chains of sugars.

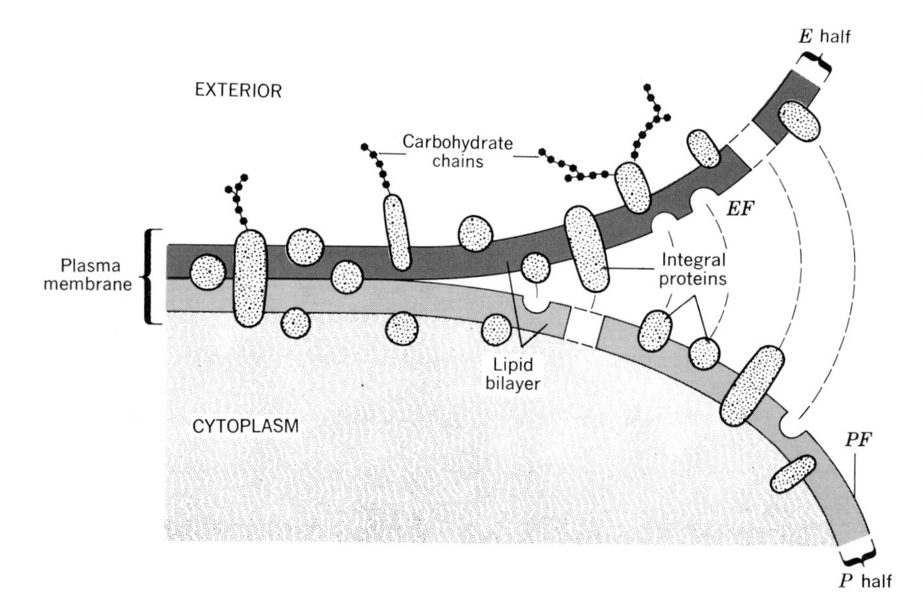

Figure 15–12
Freeze-fracturing of the plasma membrane. The fracture plane occurs at the center of the lipid bilayer and passes over (or under) the integral membrane proteins.

Integral (Intrinsic) Proteins

Integral or intrinsic membrane proteins contain both hydrophilic and hydrophobic regions. Those portions of the protein that are buried in the lipid bilayer are rich in amino acids with hydrophobic side chains. The latter are believed to form hydrophobic bonds with the fatty acid tails of the membrane phospholipids. Portions of integral proteins that project outward from the lipid bilayer are rich in hydrophilic amino acids; those projecting from the outer membrane surface may contain carbohydrate chains.

Integral Proteins That Span the Membrane. It was M. Bretscher who first demonstrated the existence of integral proteins that span the entire membrane. In a series of elegant experiments, Bretscher showed that radioactive ligands specific for membrane proteins of the erythrocyte were bound in smaller quantities to intact cells than to disrupted cells. Disruption of the cells was shown to expose portions of the membrane proteins previously facing the cell interior, thereby allowing additional radioactive ligand to associate with the protein.

T. L. Steck developed a technique for converting fragments of disrupted erythrocyte membranes into small vesicles that were either "right side out" (i.e., the external face of the membrane also formed the external face of the vesicle) or "inside out" (Fig. 15–13). When proteolytic enzymes were added to separate suspension of each type of vesicle, certain of their membrane proteins were found to be equally susceptible to digestion and could therefore be enzymatically attacked from either membrane surface. These proteins clearly spanned the membrane. Other proteins were susceptible to enzymatic digestion only when present in right-side-out or inside-out vesicles, indicating their differential distribution in the membrane's outer and inner surfaces.

Integral proteins that span the entire membrane contain outer regions that are hydrophilic and a central region that is hydrophobic. Carbohydrate associated with the hydrophilic region facing the cell's surroundings is believed to play a role in maintaining the orientation of the protein within the membrane. The hydrophilic sugars, together with the hydrophilic side chains of amino acids in the outer region of the protein, effectively prevent reorientation of the protein in the direction of the hydrocarbon core of the lipid bilayer.

Figure 15–13
Experiments of T. L. Steck showing that proteins of the erythrocyte membrane may be confined to one or the other surface or may span the membrane. Membranes were fragmented and allowed to form suspensions of vesicles. When the vesicles were "right side out," proteolytic enzymes added to the suspension destroyed those proteins normally present at the outer surface of intact membranes but had no effect on proteins normally present at the inner surface. In contrast, when "inside out" vesicles were similarly treated, proteins normally present at the inner surface were destroyed. Proteins spanning the membrane were destroyed in both cases.

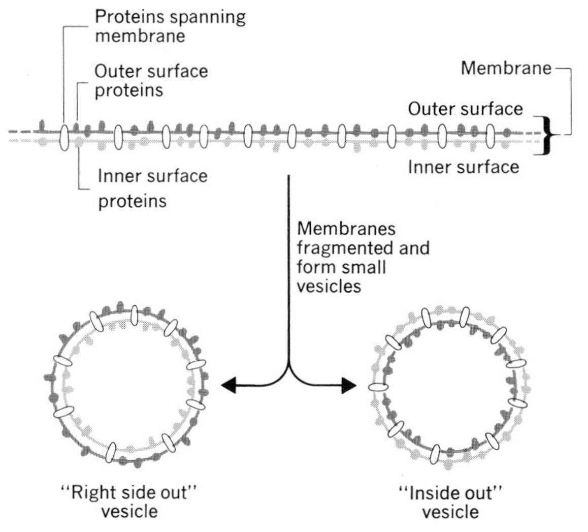

Asymmetric Distribution of Membrane Proteins

The outer and inner regions of the cell membrane do not contain either the same types or equal amounts of the various peripheral and integral proteins. For example, the outer half of the erythrocyte membrane contains far less protein than does the inner half. In addition, various membrane proteins may be present in significantly different quantities; the membranes of some cells contain a hundred times as many molecules of one protein species as another. Moreover, regardless of absolute quantity, all copies of a given membrane protein species have exactly the same orientation in the membrane. This is in stark contrast with the more uniform distribution of the various membrane lipids. The differential distribution of proteins in the various regions of the plasma membrane within a single cell was described earlier in connection with liver cells and intestinal epithelium. This irregular distribution of membrane proteins is known as *membrane asymmetry*. Not only plasma membranes but also membranes of the endoplasmic reticulum and vesicular organelles (e.g., mitochondria) are asymmetric.

Mobility of Membrane Proteins

When cells are grown in culture, there is an occasional fusion of one cell with another to form a larger cell. The frequency of cell fusion can be greatly increased by adding *Sendai virus* to the cell culture. In the presence of this virus, even different strains of cells can be induced to fuse, producing *hybrid* cells or **heterokaryons.** D. Frye and M. Edidin utilized this phenomenon to demonstrate that membrane proteins may not maintain fixed positions in the membrane but may move about laterally through the bilayer. Frye and Edidin induced the fusion of human and mouse cells to form heterokaryons and, using fluorescent antibody labels, followed the distribution of human and mouse membrane proteins in the heterokaryon during the time interval that followed fusion. At the onset of fusion, human and mouse membrane proteins were respectively restricted to their "halves" of the hybrid cell, but in less than an hour both protein types became uniformly distributed through the membrane (Fig. 15–14). The distribution of the membrane proteins was not dependent on the availability of ATP and was not prevented by metabolic inhibitors, indicating that lateral movement of proteins in the membrane occurred by diffusion.

Not all membrane proteins are capable of lateral diffusion. G. Nicolson and others have obtained evidence suggesting that some integral proteins are restrained within the membrane by a *network* of protein lying just under the membrane's inner surface (Fig. 15–9). This network may, in turn, be associated with a system of microfilaments and microtubules in the cytosol (i.e., the "cytoskeleton").

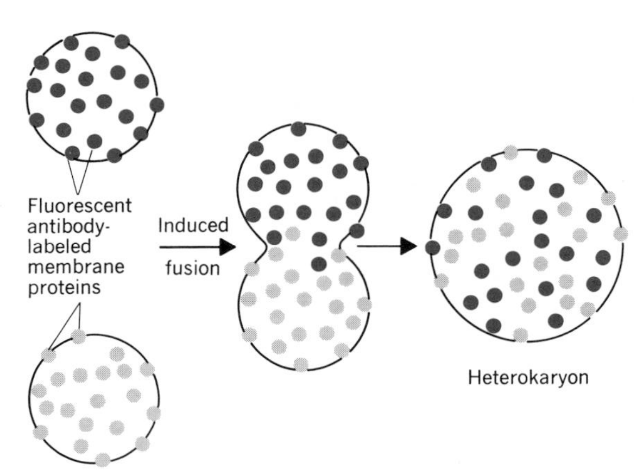

Figure 15–14
Movement of proteins in the plasma membrane. When fluorescent antibody labeling of the membrane proteins of different cells is followed by *Sendai virus* induced cell fusion, the proteins are soon observed to distribute throughout the membrane of the heterokaryon.

Enzymatic Properties of Membrane Proteins

Membrane proteins have been shown to possess enzymatic activity. Table 15–1 lists some of the enzymes that are now recognized as constituents of the plasma membrane. To this list of proteins must be added *receptor* proteins (such as the insulin-binding sites of the liver cell membrane) and *structural* proteins. Recent evidence suggests that glycosyl trasferases may also be present on the membrane's outer surface, where they add sugars to the ends of oligosaccharides associated with membrane protein or lipid.

Isolation and Characterization of Membrane Proteins

Because of the relative ease with which they may be purified, the plasma membranes of erythrocytes provided much of the early information on the chemistry of proteins (and lipids) present in membranes. Now, however, plasma membranes can be obtained from many cell types in a reasonably uncontaminated state using various forms of density gradient centrifugation. Nonetheless, the individual protein constituents of the membrane are not so easily extricated for individual study because of their high degree of insolubility. Varying degrees of success in extracting proteins from the plasma membrane have been achieved using sodium dodecyl sulfate (SDS) and Triton X-100 (two organic detergents) and concentrated solutions of urea, *n*-butanol, and ethylene diamine tetraacetic acid (EDTA). These chemicals have a disaggregating effect on membranes, causing the release of many of the membrane proteins by disso-

Table 15–1
Enzymes Present in the Plasma Membrane

Adenosine triphosphatase (Mg^{++} stimulated)
Adenosine triphosphatase (Mg^{++}, Na^{+}, K^{+} stimulated)
Adenosine triphosphatase (Mg^{++}, Ca^{++} stimulated)
Nucleoside diphosphate phosphatase
Nucleoside triphosphate pyrophosphatase
5′ Nucleotidase
Adenylcyclase
Protein kinase
Acetylphosphatase (K^{+} stimulated)
Alkaline nitrophenylphosphatase
Acid nitrophenylphosphatase
NAD pyrophosphatase
Alkaline glycerophosphatase
Alkaline phosphodiesterase
NAD glycohydrolase
Cholesterol esterase
Phosphatidyl inositol kinase
Diglyceride kinase
Phosphatidate phosphatase
Sphingomyelinase
Monoglyceride lipase
Triglyceride lipase
Acetylcoenzyme A synthetase
Invertase
Maltase
Isomaltase
Lactase
Trehalase
Furanase
Cellobiase
UDP glycosidase
Nitrophenyl glycosidase
Collagen glycosyl transferase
Leucyl β-naphthyl amidase
NADH dehydrogenase
Nucleoside kinase
Triosephosphate dehydrogenase
Phosphoribose isomerase
Xanthine oxidase

ciating the bonds that link the proteins together or to other membrane constituents. Often, the removal of these agents from a preparation of solubilized membrane proteins is quickly followed by the reassociation or reaggregation of the proteins to form an intractable matrix.

Once solubilized, the membrane proteins can be separated into discrete classes using electrophoresis, chromatography, or other procedures (see Fig. 13–10). This generally demands that the dissociating agents be present in the separating medium (e.g., the electrophoresis gel, the column eluent, etc.); otherwise, application of the membrane extract to the medium is followed by membrane protein reaggregation into insoluble complexes that will not separate into distinct fractions. For example, the separation of liver plasma membrane proteins shown in Figure 13–10 is achieved only if the electrophoresis gel contains SDS and Triton X-100. The solubility problem has been the greatest barrier to progress in isolating and fully characterizing the proteins of membranes.

Many of the plasma membrane enzymes listed in Table 15–1 have not actually been isolated from the membranes, since removal and isolation of the enzyme is *not* a prerequisite for establishing its presence. Instead, the enzyme activity can be measured directly in the (unsolubilized) membrane preparation.

Membrane Lipids

Much more is known about the specific lipid composition of cell membranes, because the lipids are more readily extracted from the membranes using a variety of organic solvents. Once extracted from isolated membranes, the lipids may be separated and identified using chromatographic or other procedures. Nearly all the membranes studied so far appear to contain the same types of lipid molecules. Phospholipids such as phosphatidyl ethanolamine, phosphatidyl serine, phosphatidyl inositol, phosphatidyl choline (lecithin), and sphingomyelin are the most common constituents, but cholesterol may also be present. The chemical structures of these lipids may be found in Chapter 6. Table 15–2 lists the most common lipids found in a variety of cell membranes and also shows their protein-to-lipid weight ratios; the latter varies considerably.

Mobility of Membrane Lipids

Lipids exhibit a higher degree of mobility in the membrane than do proteins, although lateral mobility is very much greater than transverse (''flip-flop'') mobility. A single lipid molecule may move several microns laterally through the membrane in just 1 or 2 seconds! Lipid molecules that are in direct contact with membrane proteins are not as mobile

Table 15–2
Lipids Present in the Plasma Membrane

Plasma Membrane of	Major Lipids Present	Protein/Lipid (wt/wt)
Liver cell	Cholesterol, phosphatidyl choline, phosphatidyl ethanolamine, phosphatidyl serine, sphingomyelin	1.0–1.4
Intestinal epithelial cell	Cholesterol, phosphatidyl choline, phosphatidyl ethanolamine, phosphatidyl serine, sphingomyelin	4.6
Erythrocyte	Phosphatidyl inositol, cholesterol, phosphatidyl choline, phosphatidyl ethanolamine, phosphatidyl serine, sphingomyelin	1.6–1.8
Myelin	Cholesterol, cerebrosides, phosphatidyl ethanolamine, phosphatidyl choline,	0.25
Gram-positive bacteria	Diphosphatidyl glycerol, phosphatidyl glycerol, phosphatidyl ethanolamine	2.0–4.0

as lipid molecules that are solely in contact with one another; such immobilized lipid is called **boundary lipid** (Fig. 15–9).

The mobility of lipid and protein molecules in the plasma membrane attests to the membrane's fluidity. C. F. Fox and H. M. McConnell have shown that the degree of fluidity is dependent, in turn, on the types and the balance of fatty acid side chains of phospholipids in the membrane.

Fatty acid side chains of membrane phospholipids can be either *saturated* or *unsaturated*. In saturated side chains, all the carbon-carbon bonds are single, with the remaining carbon bonds carrying hydrogen atoms; in unsaturated side chains, one or more pairs of neighboring carbon atoms are linked by double bonds (see Chapter 6). In phospholipid layers consisting exclusively of saturated fatty acids, the side chains are aligned next to one another in an ordered, crystalline array; the result is a relatively rigid structure (Fig. 15–15). In phospholipid layers consisting of a mixture of saturated and unsaturated fatty acid side chains, the packing of neighboring molecules is less orderly (and therefore more fluid). The double bonds of the unsaturated side chains produce bends in the hydrocarbon chain, and these give rise to structural deformations that prevent formation of the more rigid crystalline structure. The greater the number of double

Figure 15–15
Disruption of the orderly stacking of phospholipids having saturated fatty acid tails by unsaturated fatty acids having one or more double bonds.

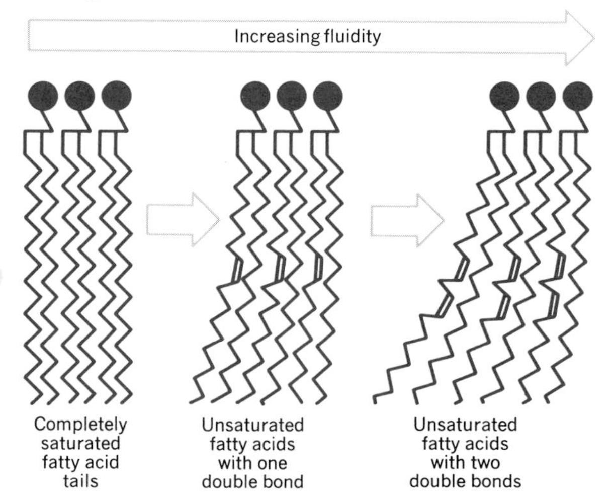

bonds, the more disordered (and fluid) is the lipid bilayer (Fig. 15–15).

The rigidity of lipid layers is also affected by temperature. Almost everyone is familiar with the ''melting'' of fats and waxes at elevated temperatures. In order to maintain membrane fluidity, cells living at low temperatures have higher proportions of unsaturated fatty acids in their membranes than do cells at higher temperatures. Evidence also exists, suggesting that cells can alter the balance of saturated and unsaturated fatty acids in their membranes as an adjustment to changing temperature or the demands for altered membrane fluidity.

Lipid Asymmetry

The various membrane lipids are not equally distributed in both monolayers, although the asymmetry is not nearly as marked as in the case of protein. The distribution of lipids in the erythrocyte membrane is shown in Table 15–3 and reveals that the choline phosphatides are primarily in the outer monolayer and the aminophosphatides are in the inner monolayer. Cholesterol is believed to be present in both surfaces in large amounts. Although lipid asymmetry is a general property of membranes, the type of asymmetry varies considerably from one membrane to another. Asymmetry, once established, is most likely maintained because of the high activation energy that would be required to move the polar groups through the hydrophobic center of the bilayer.

Membrane Carbohydrate

It has already been noted that carbohydrate is present in the plasma membrane as short, sometimes branched chains of sugars attached either to exterior peripheral proteins (forming glycoproteins) or to the polar ends of phospholipid molecules in the outer lipid layer (forming glycolipid). No membrane carbohydrate is located at the interior surface.

Table 15–3
Distribution of Lipids in the Erythrocyte Membrane

	Interior Lipid Monolayer (%)	Exterior Lipid Monolayer (%)
Total	50	50
Sphingomyelin	6	20
Phosphatidyl choline	9	23
Phosphatidyl ethanolamine	25	6
Phosphatidyl serine	10	0
Phosphatidyl inositol	0	0

The oligosaccharide chains of the membrane are formed by various combinations of six principal sugars: *D-galactose, D-mannose, L-fucose, N-acetylneuraminic acid* (also called *sialic acid*), *N-acetyl-D-glucosamine* and *N-acetyl-D-galactosamine* (see Chapters 4 and 5 for chemical structures). All of these may be derived from glucose.

Possible Functions of Membrane Carbohydrate

Several roles have been suggested for the carbohydrate present on the outer surface of the plasma membrane. One possibility is that because they are highly hydrophilic, the sugars help to orient the glycoproteins (and glycolipids) in the membrane so that they are kept in contact with the external aqueous environment and are unlikely either to rotate toward the interior or to diffuse transversely.

Certain plasma transport proteins, hormones, and enzymes are glycoproteins, and in these molecules, carbohydrate is important to physiological activity. It would therefore not be inappropriate to expect that in certain glycoproteins of the plasma membrane the carbohydrate moiety is basic to either enzymic or some other activity.

Surface carbohydrate is clearly responsible for the various human blood types (e.g., ABO types, MN types, etc.) and other tissue types. That is, the sugar sequence and the arrangement of the sugar chains in the membranes of blood cells of an individual with type A blood differ from those of an individual with type B blood, and so on. The carbohydrate is responsible for cell type specificity and is therefore fundamental to the specific antigenic properties of cell membranes. These antigenic properties are linked in some manner to the body's *immune system* and the ability of that system to distinguish between cells that should be present in the organism (i.e., native cells) and foreign cells. Foreign cells (such as bacteria, other microorganisms, transplanted tissue, or transfused blood) may be recognized as foreign because their membrane glycoproteins contain different carbohydrate markers than those present in the individual's own tissues. Such a situation triggers the immune response. In contrast, an individual's own cell membrane carbohydrate organization is recognized as being native (referred to as "recognition of self") and does not normally trigger an immunological response. Of course, neither does blood transfusion or tissue transplantation if the carbohydrate organization in the membranes of the "donor's" and "recipient's" cells is the same. Cell-specific membrane carbohydrate organization is considered further below in connection with the actions of *lectins* and *antibodies*.

Oppenheimer, Roseman, Roth, and others have clearly implicated surface carbohydrate in the adhesion of a cell to its neighbors in a tissue; presumably, the carbohydrate acts as an adhesive maintaining the integrity of the tissue by linking neighboring cells together. **Contact inhibition,** the phenomenon in which cells grown in culture stop dividing when they touch one another (thereby limiting the growth of the population), may possibly be attributable to a mechanism triggered by interaction of carbohydrates on neighboring cells.

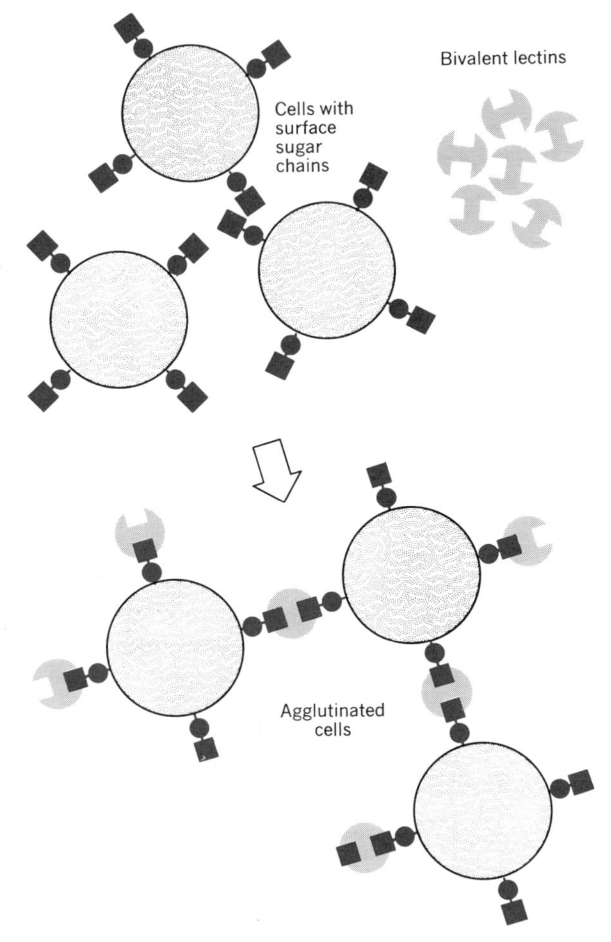

Figure 15–16
Agglutination of cells by lectins.

Lectins, Antibodies, Antigens and the Plasma Membrane

Lectins

Lectins are a special class of proteins (found principally in plants, especially legumes, and also in some invertebrates) that have a high affinity for sugars and combine with them in much the same manner as an enzyme combines with its substrate or an antibody combines with an antigen. Because the interaction of the lectin with sugar is specific (see Table 15–4), lectins can be used to map the distribution of sugars on the cell surface.

Following their discovery in plants some 90 years ago by H. Stillmark, the lectins were for some time called **phytohemagglutinins** because of their ability to cause the agglutination of red blood cells. However, lectins will agglutinate many kinds of cells, including bacteria. Lectin molecules contain two or more sugar-binding sites, and when large numbers of lectins bind *simultaneously* to sugars on the surfaces of separate cells (thereby cross-linking the cells), the result is agglutination (Fig. 15–16). It should be noted that binding of the lectin to the cell surface sugars can occur *without* ensuing agglutination *if no cross-linking takes place*. The presence and extent of cross-linking is dependent on the balance of lectin concentration and the numbers of surface sugars. In this respect, the lectin-sugar interaction is much like that of an antibody-antigen reaction (see below). However, unlike antibodies, which chemically are very similar proteins (Chapter 4), lectins are of diverse structure, organization, and size.

Lectins will bind free sugars as well as sugars attached to cell membranes. Consequently, lectin-induced cell agglutination can be blocked by preliminary addition of the appropriate free sugar to a suspension of cells.

Lectins have been used to verify that the plasma membranes of malignant cells and normal cells differ. Malignant cells are much more readily agglutinated by lectins than the normal cells from which they are derived; that is, the malignant cells can be caused to agglutinate at much lower lectin concentrations than are required to agglutinate normal

Table 15–4
Some Lectins and Their Sugar Specificities

Lectin	Sugar Specificity
Concanavalin A (Con A) (from jack beans)	D-Mannose
Wheat germ agglutinin (WGA)	*N*-Acetyl-*D*-glucosamine
Ricinis communis agglutinin (RCA)	D-Galactose *N*-Acetyl-D-galactosamine
Soybean agglutinin (SBA)	*N*-Acetyl-D-galactosamine
Lima bean lectin	*N*-Acetyl-D-galactosamine
Limulus lectin	*N*-Acetylneuraminic acid

cells. It has been found that the increased agglutinability of the malignant cells results from increased glycoprotein mobility in the lipid bilayer of the plasma membrane. Since the malignant cell membrane is more fluid, lectins are able to cluster the glycoproteins in the membrane (i.e., draw them together) and thereby make it possible to form greater numbers of cross-bridges.

How lectins can be used to determine the composition of the sugar chains of surface carbohydrate is illustrated in Figure 15–17. The lectins bound by unmodified cell membranes establish the choice of terminal sugars; these may then be cleaved from the carbohydrate using the specific enzyme. The newly exposed terminal sugars are now examined for their lectin-binding characteristics, following which another enzymatic sugar removal is carried out. Repetition of this sequence of treatments progressively reveals the order of sugars.

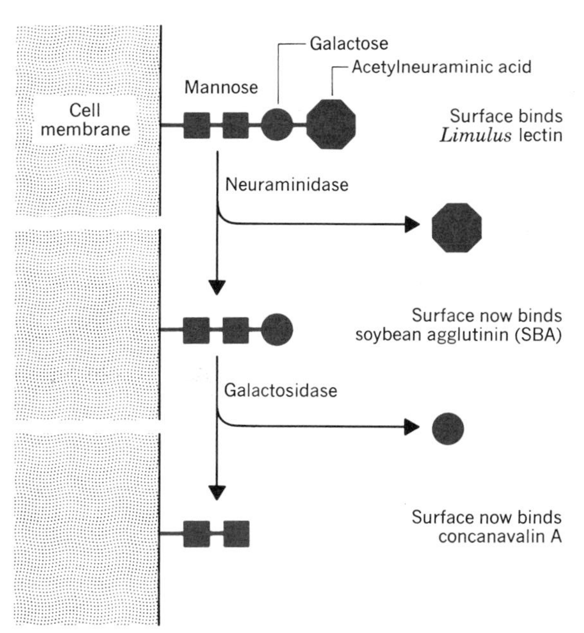

Figure 15–17
Structure of surface carbohydrate is established by successive treatments with lectins and terminal sugar-cleaving enzymes. In this illustration, binding of *Limulus* lectin by the cells indicates the presence of terminal acetylneuraminic acid groups. Their removal using *neuraminidase* is followed by binding of soybean agglutinin, indicating that the acetylneuraminic acid was linked either to galactose or acetylglucosamine. Which of these two alternative exists is revealed by sensitivity to the specific enzyme added in the next round. In the illustration, it is *galactosidase* that now allows concanavalin A binding, thereby indicating that the terminus was galactose. Removal of galactose and binding of Con A indicates that the next sugar is mannose.

Antigens and Antibodies

An antigen may be defined as any molecule that has the capacity to stimulate antibody production by the immune system of higher animals. Typically, antigens are glycoproteins in the membranes of cells or in other particles foreign to the animal. For example, the antigens present in the membranes of bacterial cells or in viruses act to stimulate antibody production by the immune system of the infected animal. The antibodies or immunoglobulins (see Chapter 4) produced in response to the presence of the antigen combine with the antigen to form a complex, and this is followed by a series of reactions in which the antigen-bearing agents (e.g., the bacteria) are destroyed.

Antibodies are synthesized by lymphocytes, a subpopulation of white blood cells produced either in the bone marrow (B-lymphocytes) or in the thymus gland (T-lymphocytes). The reaction between antibody and antigen is very specific, a particular antibody combining with only one type of antigen. An enormous variety of B- and T-lymphocytes are present in the body's tissues, each capable of manufacturing only a single antibody type (and therefore capable of reacting with a single type of antigen or foreign cell).

Some of the antibodies manufactured by a lymphocyte are maintained in its plasma membrane. In the presence of the corresponding antigen, a reaction takes place on the surface of the lymphocytes that acts as a stimulus to the production and secretion of additional quantities of antibody. The surface reactions also trigger lymphocyte proliferation, so that even larger quantities of antibody became available. Following the initial reaction on the lymphocyte cell surface, subsequent antigen-antibody reactions may not involve lymphocytes directly. Instead, the secreted antibodies react with either free antigens or more likely with antigens in the invading cells' membranes. The involvement of either lymphocyte membranes or foreign cell membranes (or both) in the antigen-antibody reaction makes these molecules especially valuable tools for studying the properties of the cell surface.

It should be clear that the surface antibodies of the lymphocytes of one animal can serve as antigens if these lymphocytes are transferred to the bloodstream of another animal. The transferred lymphocytes will be treated much like any other foreign cell and serve to stimulate the production of **anti-immunoglobulin antibodies** (AIA). AIA has been especially useful in probing the distribution of glycoproteins in the cell membrane.

M. C. Raff and S. dePetris prepared AIA which they then coupled with **ferritin.** Ferritin is a liver protein rich in iron and readily discernible as dark spots by transmission electron microscopy (i.e., the iron renders the ferritin electron-dense). When ferritin-coupled AIA is added to suspensions of lymphocytes, it combines with their surface immunoglobulins, thereby assisting in their identification. Raff and dePetris showed that when lymphocytes are incubated at 4°C with ferritin-coupled AIA, the electron-dense spots are distributed all over the membrane surface, whereas at 20°C, the ferritin-coupled AIA was clustered together at one pole of the cell. As in the case with lectins described earlier, the AIA was able to *cluster* the surface immunoglobulin at 20°C but not at 4°C because at the higher temperature the plasma membranes of the cell were more fluid.

Origins of Plasma Membrane Protein and Lipid Asymmetry

Illustrated in Figure 15–18 is a mechanism recently proposed by J. E. Rothman and J. Lenard to account for the specific asymmetric distribution of proteins in the plasma membrane. Peripheral and integral proteins that face the cell exterior are synthesized by ribosomes that attach to the inner surface of the membrane. During synthesis, the hydrophilic portion of the protein (which ultimately is to be in contact with the polar exterior) is discharged through a polar, protein-lined pore in the hydrocarbon core of the membrane. If the protein being synthesized is an integral protein, the nonpolar region that is to be anchored in the lipid bilayer is synthesized next, accompanied by the loss of the polar pore. For proteins that span the membrane, synthesis is completed with the production of the interior hydrophilic segment. This mechanism is not unlike *vectorial synthesis* of proteins by ribosomes of the rough en-

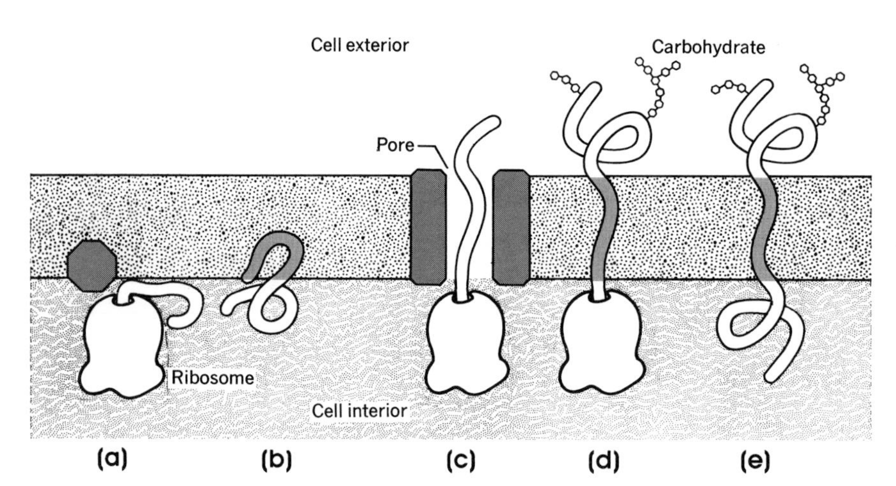

ıure 15–18
Origin of membrane protein asymmetry. (*a* and *b*) Synthesis of interior integral protein. (*c*, *d*, and *e*) Synthesis of integral protein spanning the membrane. Colored regions of polypeptide chains represent hydrophobic sections. Geometric areas of the membrane represent a ribosome attachment site and a channel for protein extrusion. See text for details.

doplasmic reticulum (see Chapter 21), except that the protein is left integrated in the membrane rather than extruded through it completely.

Membrane proteins located on or in the internal surface may be synthesized either by cytoplasmic ribosomes or ribosomes attached to the plasma membrane. Peripheral interior membrane proteins synthesized by cytoplasmic ribosomes would diffuse to the inner membrane surface where polar interactions would cause their spontaneous attachment. Such proteins could not pass through to the exterior surface because they could not traverse the hydrophobic membrane core. Integral interior membrane proteins synthesized by ribosomes attached to the plasma membrane would insert in the membrane once the hydrophobic region of the protein was completed (Fig. 15–18).

The distribution of lipid in the plasma membrane is not absolutely asymmetric in that nearly every type of lipid is present in some amount in both monolayers. The membrane phospholipids are thought to be synthesized from cytoplasmic precursors by enzymes that are part of the interior half of the membrane. Following synthesis, the phospholipid becomes, at least temporarily, part of the interior monolayer but may be enzymatically translocated to the outer monolayer. Since outer monolayer lipids are derived from the inner monolayer, an absolute asymmetry is precluded.

Special Cell Surface Properties Revealed by Erythrocytes

In view of the fact that the mature erythrocyte lacks organelles, this cell has always been a popular source of plasma membranes. Indeed, the pioneering studies of Gorter and Grendel, which were the first to indicate the existence of the lipid bilayer, were carried out using erythrocytes. It is now clear that the erythrocyte membrane may possess a number of unusual properties in addition to others that are believed to be widespread. Notwithstanding its apparent chemical and structural specializations, the erythrocyte is being studied more extensively today than at any time previously.

Like the plasma membranes of other cells, the red blood cell membrane is asymmetric. The lipid asymmetry of the erythrocyte membrane has already been described (Table 15–3). With regard to protein asymmetry, the peripheral proteins account for about 40% of all membrane proteins *but are restricted to the interior surface*. The most abundant of these proteins and the first to be isolated is a molecule called **spectrin.** Spectrin is believed to be an important component of a weblike network of proteins on the interior membrane surface.

There are two major integral proteins, and both apparently span the lipid bilayer. One of these, called **glycophorin-A,** has been fully sequenced and reveals several very interesting properties. Glycophorin-A consists of a single chain of 131 amino acids; 16 short carbohydrate chains are linked to residues near the N-terminus of the polypeptide (primarily to serine and threonine side chains), the carbohydrate accounting for about 60% of the total mass of the glycoprotein. The N-terminal region of glycophorin-A is thought to project beyond the exterior membrane surface. The C-terminal end of glycophorin-A is rich in acidic amino acids, especially glutamic acid, and is believed to project into the cell interior. A segment of about 20 amino acids in the middle of the polypeptide consists exclusively of nonpolar and hydrophobic amino acids and apparently is that portion of glycophorin-A that spans the lipid bilayer. The C-terminal ends of the glycophorin-A molecules are thought to interact internally with the spectrin-containing network. Experimental manipulation of the network brings about a corresponding rearrangement of glycophorin-A in the membrane.

The differentiation of the erythrocyte in the bone marrow is accompanied by a major structural reorganization of the cell in which organelles like the nucleus, mitochondria, intracellular membranes, ribosomes, and so forth are progressively lost. Mature erythrocytes also fail to reveal microfilaments and microtubules, which in other cells are believed to from a *cytoskeleton* radiating through the cell with anchor points at the inner surface of the plasma membrane. In effect, the erythrocyte may be described as a membranous bag of hemoglobin. Despite its seeming simplicity, the erythrocyte has a characteristic shape. In humans (and in most other mammals), the cell takes the form of a *biconcave disk* having a diameter of about 8 μ (Figure 15–19; see also Fig. 1–15*b*), although changes in shape are readily induced by variations in osmotic pressure (see later in chapter). The biconcave shape of the erythrocyte is important in its biological function, since such a shape maximizes oxygen diffusion from the cell into the tissues and promotes efficient stacking of cells (*rouleaux* formation) as they circulate through the narrow capillary passageways.

The characteristic shape of the erythrocyte was for some

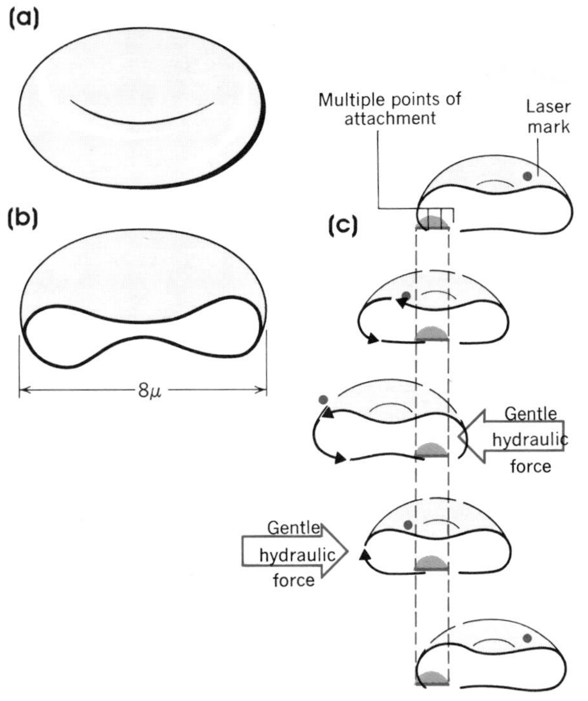

Figure 15–19
(*a*) Erythrocyte. (*b*) Cross section showing biconcavity. (*c*) Effect of lateral force on erythrocyte shape.

time believed to be due to the differential lateral distribution of lipid (and perhaps also protein) in the membrane. However, it is now known that the erythrocyte membrane is uniformly organized laterally throughout its surface. Despite the existence of a positive hydrostatic pressure internally and the absence of internal structure, the cell does not normally assume the spherical shape expected. The biconcave shape must be related in some manner to the properties and lateral arrangement of neighboring lipid and protein molecules in the cell's membrane or to constraints resulting from the organization of the paracrystalline hemoglobin content.

In some rather startling experiments bearing on this problem, B. Bull and J. D. Brailsford have shown that when an erythrocyte is attached by a portion of its undersurface to a glass slide and a laser used to make a visible mark on the membrane's surface so that membrane movement can be followed, slight lateral displacement of the cell using hydraulic force is accompanied by the membrane *rolling* in the direction of the force, much like a tractor track does (Fig. 15–19*c*). The laser mark travels over the cell surface, following the contours of the biconcave shape. In other words, though rolling laterally, the biconcave shape of the cell is maintained with the vertical biconcavity moving parallel to the glass surface.

Erythrocytes traveling through capillaries and other blood vessels are often arranged as stacks or **rouleaux,** with their biconcave faces juxtaposed. Such an orderly procession makes it possible for so large a number of cells to pass through the body tissues in short periods of time. P. B. Canham has shown that when a cell in a rouleaux is struck by a laser beam of sufficient energy, the cell membrane is disrupted and the cell lyses (bursts). However, several cells in the rouleaux on either side of the target cell (and not directly affected by the laser) also are seen to undergo gradual lysis. This "contagious lysis" apparently results from the fact that the membranes of erythrocytes in rouleaux transiently adhere to one another. Sudden movements of the membranes of one cell (as during lysis) create sufficient shear forces at contact points with neighboring cells that their membranes are also affected. The nature of membrane interaction between neighboring erythrocytes in a rouleaux is unknown, but it is clear that much remains to be learned about the "simple" red blood cell.

Intercellular Junctions and Other Specializations of the Plasma Membrane

The plasma membranes of neighboring cells in a tissue frequently exhibit specialized junctional regions believed to play roles in cell-to-cell adhesion and in intercellular transport. The most common of these junctions are (1) **tight junctions** *(zonula occludens),* (2) **intermediate junctions** or **belt desmosomes** (also called *terminal bars* or *zonula adherens)*), (3) **spot desmosomes** *(macula adherens),* (4) **gap junctions** *(nexuses),* and (5) **plasmodesmata** (Figs. 15–20, 15–21, 15–22, and 15–23).

In tight junctions, the plasma membranes of the neighboring cells fuse at one or more points. The tight junctions generally occur in the same circumferential region of the cell, so that they give rise to *belts* of fusion points with neighboring cells. The belt obliterates the intercellular space and acts as a barrier to the flow of materials between the cell surfaces. Internally, the belts of tight junctions are

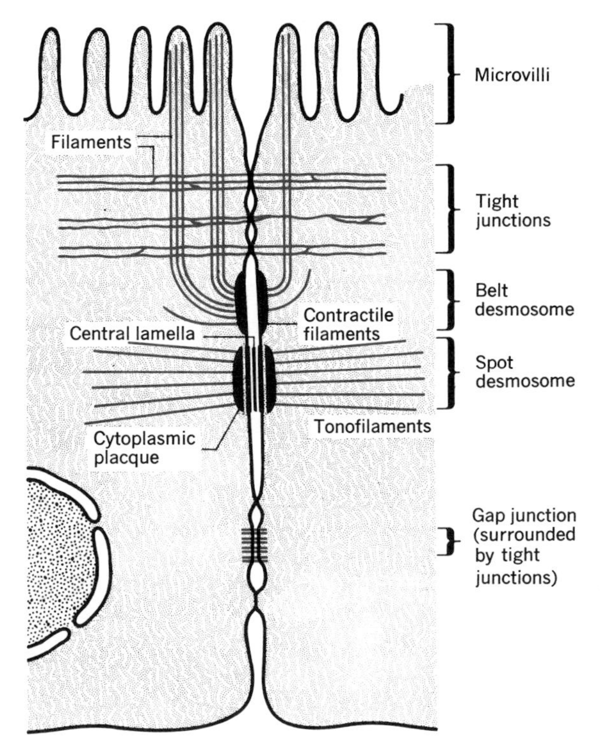

Figure 15–20
Specializations of the plasma membrane.

reinforced by a network of fine filaments radiating into the cytoplasm. The tight junctions between cells are formed by two interdigitating rows of membrane particles (probably integral proteins), one row contributed by each cell. The number of rows of particles (usually called *sealing strands*) and the extent to which they interconnect to form a network vary from one type of tissue to another. It is thought that sealing strands also act to deter the movements of other proteins within the membrane. In this way, the differential distribution of certain membrane proteins is maintained, providing for the functional specialization of different faces of the cell (see earlier).

Intermediate junctions, or belt desmosomes, are girdles of contractile filaments (i.e., they contain actin) attached to the interior surface of the plasma membrane. The girdle of contractile filaments interweaves with another web of filaments that extends to the microvilli (Fig. 15–20).

Unlike tight junctions and belt desmosomes, spot desmosomes (macula adherens) do not form a belt around the cell. Instead they are discrete, buttonlike attachment points scattered over the opposing membrane surfaces. In the region of the spot desmosome the adjacent cell membranes are strictly parallel, somewhat thicker, and separated by an intracellular space of about 300 Å. This gap characteristically is granular in appearance and contains a central dense band (the **central lamella**). The cytoplasm adjacent to the plasma membrane is divided into two regions: a lucid zone that lies immediately next to the membrane and a neighboring dense band called the cytoplasmic placque. Microfilaments called **tonofilaments** arise in this region, radiate into the cell, and may be linked to other spot desmosomes. Spot desmosomes are believed to be the strongest points of attachment between neighboring cells.

Epithelium is a good example of a tissue in which neighboring cells contain the junctions just described. Beginning at the free surface of the epithelium (i.e., the *apical* surface of the cells) and "descending" through toward the basement membrane (i.e., the *basal* surface), the junctions occur in a characteristic order: a belt of tight junctions is followed by a belt of intermediate junctions; spot desmosomes are most abundant near the basal ends of the cells.

Gap junctions are the most complex modification of adjacent plasma membranes. In the gap junction, the distance between opposing plasma membranes is reduced to about 30 Å and the space penetrated by a number of parallel cylindrical structures that run from the cytoplasmic surface of one plasma membrane to the other (Figs. 15–20 and 15–22). The cylinders are each formed from 12 protein subunits, 6 associated with each cell membrane. The channel in the center of the cylinder is about 20 Å in diameter and permits the intercellular exchange of cytoplasmic constituents having molecular weights under 1000 daltons (e.g., ions, amino acids, sugars, nucleotides, vitamins, certain hormones, etc.). A gap junction is depicted schematically in Figure 15–22. In freeze-fracture studies, gap junctions are revealed as patches of closely packed particles on the protoplasmic fracture face of the plasma membrane (Fig. 15–21*d*). These particles are probably the cylinders described above. The observations by M. Ravazzola and L. Orci that in some tissues the sealing strands characteristic of tight junctions occur at the periphery of gap junctions have lead these researchers to suggest that the gap junctions are thereby enclosed in a compartment of controlled chemical composition (compare the gap junction of Fig. 15–20 with the freeze-fracture view of Fig. 15–21*d*).

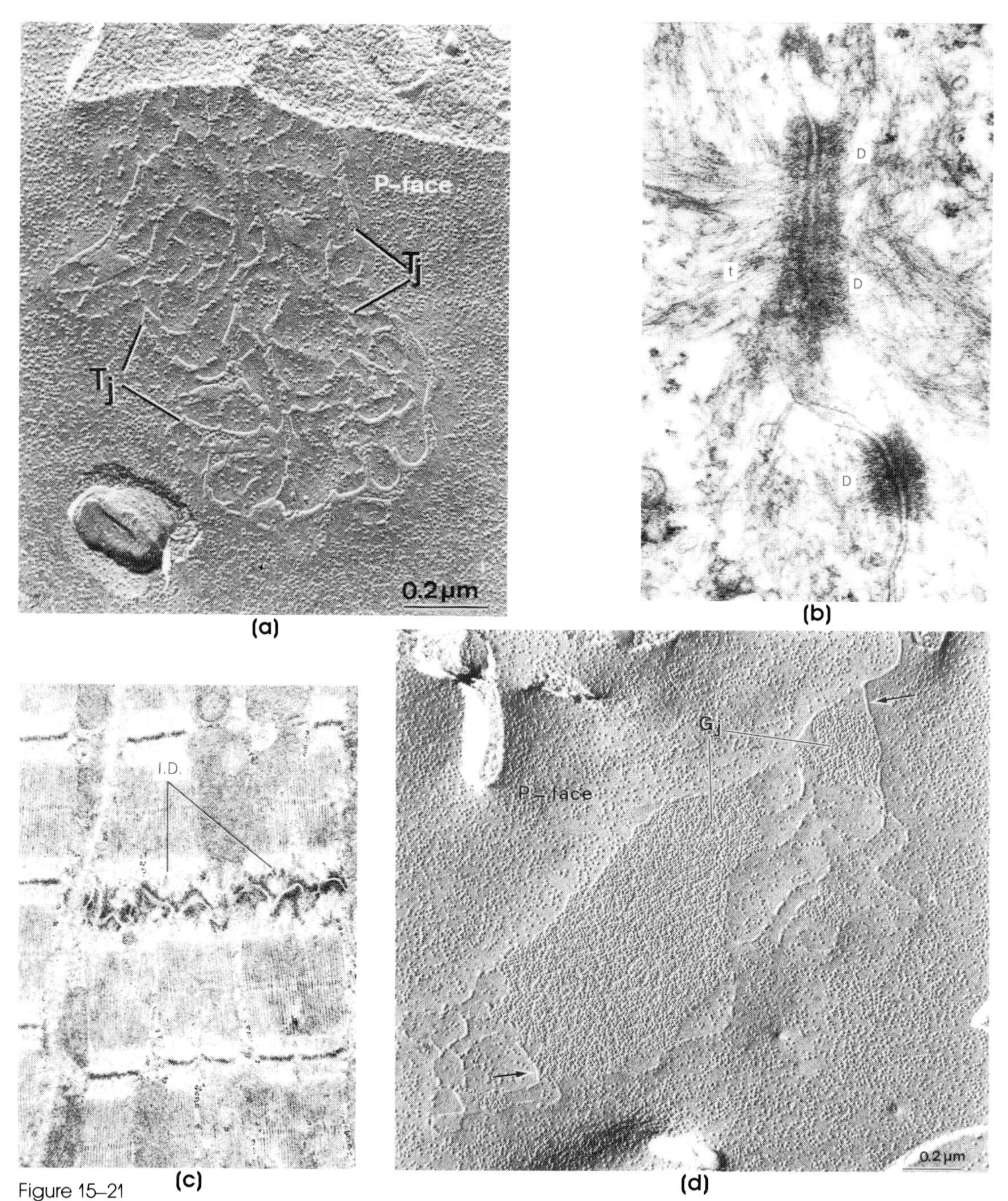

Figure 15–21
Examples of intercellular junctional complexes. (*a*) Freeze-fracture view of tight junctions (Tj) formed by *sealing strands* of membrane proteins; the *P* fracture face (PF) of one of the two cell membranes is seen in this view. (*b*) Desmosomes (D) seen in thin section; note the tonofilaments (t). (*c*) Thin section of cardiac muscle revealing the array of desmosomes that form the *intercalated disks* (I.D.) linking neighboring cells. (*d*) Freeze-fracture view of two gap junctions (Gj). Note the sealing strands (arrows) that border the tightly packed clusters of particles comprising the junctions. Magnifications: (*a*) 60,000 X; (*b*) 22,000 X; (*c*) 7,200 X; (*d*) 42,500 X. (*a* and *d*, courtesy of Dr. L. Orci; *b* and *c*, courtesy of R. Chao.)

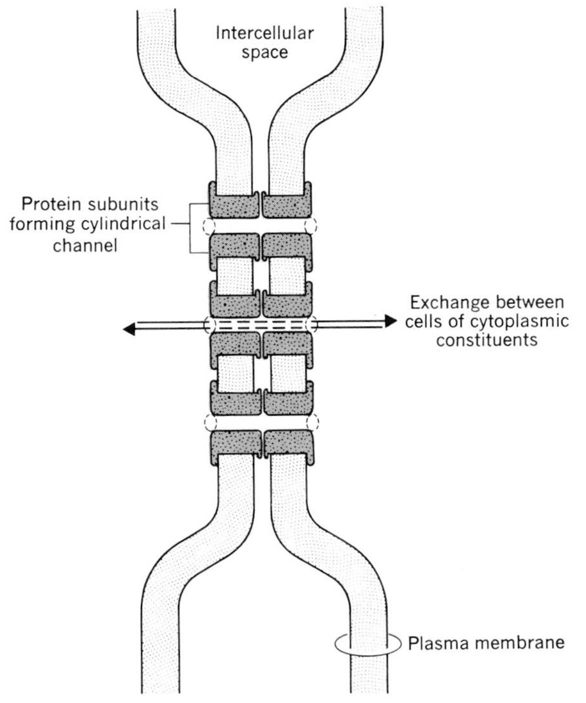

Figure 15–22
The gap junction (nexus).

Extensive regions of the plasma membranes of some cells are modified to form numerous elongate outfoldings called **microvilli** (Fig. 15–20). The microvilli greatly increase the surface area of the membrane and are especially abundant in but not restricted to cells having a transport activity, such as absorptive epithelium or glandular epithelium.

In plant tissues, the cytoplasm of neighboring cells may be connected through numerous narrow channels that penetrate the fibrous cell wall separating the cells. The channels, called **plasmodesmata,** are formed by extensions of the plasma membranes of the cells (Fig. 15–23) and are much larger than the channels of the nexus. Plasmodesmata provide for the direct exchange of materials between neighboring cells in the tissue.

Passive Movements of Materials Through Cell Membranes

In this section, we consider some of the fundamental principles that govern the passive movements of water, ions, and various other molecules through cell membranes. The term *passive* is intended to denote that the movement of the substance through the membrane is not associated with any chemical or metabolic activities in the membrane. Passage through the membrane in such instances is regulated by such factors as the **concentration gradient** across the membrane and the chemical and physical relationships between the membrane and substances inside and outside the cell. Later, we will direct our attention to movements through

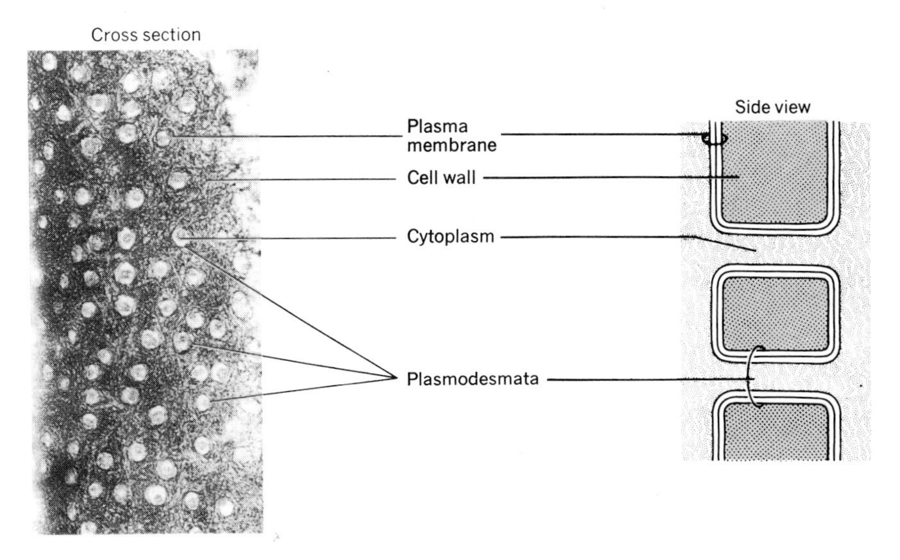

Figure 15–23
Plasmodesmata.

the membrane that are accompanied by chemical changes, metabolic activity, or gross molecular rearrangements within the membrane itself.

Osmosis and Diffusion Across Membranes

Substances that are able to pass through membranes are said to be **permeable** to the membrane. Nearly all plasma membranes are permeable to water. If water (or some other solvent) is the only substance that can pass through the membrane, the membrane is said to be **semipermeable.** Membranes that display a gradation of permeability to water and dissolved solutes (i.e., membranes that permit water to pass through more readily than salts, sugars, etc.) are said to be **selectively permeable.**

Water molecules are continuously moving into and out of the cell through the plasma membrane. Such movements are generally not discernible as changes in cell size or shape because the flux in each direction is the same. When the concentrations of solutes inside and outside the cell differ, the water flux in one direction may be greater than in the other direction, and the cell may swell or shrink. Water moves from a region of *low* solute concentration to one of *higher* solute concentration in order to establish a concentration equilibrium. The movement of water (or some other solvent) in response to such a solute *concentration gradient* is known as **osmosis.**

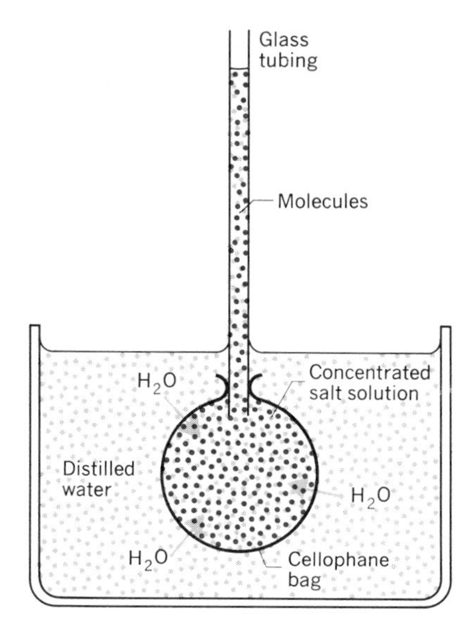

Figure 15–24
Osmosis of water into a cellophane bag containing a concentrated salt solution. Because the solute concentration inside the bag is greater than outside the bag (i.e., the solute concentration outside the bag is zero), water enters through the cellophane membrane by osmosis. The influx of water causes the water level in the attached glass tubing to rise.

Osmosis may readily be demonstrated using an *artificial* membrane such as *cellophane*, which is permeable to water and small molecules such as salts, sugars, and amino acids but is impermeable to larger molecules such as proteins. If a cellophane bag filled with a concentrated salt solution is connected to a length of vertical glass tubing and is then immersed in a container of distilled water, water will pass into the bag by osmosis. The entry of water into the cellophane bag will cause the water level of the glass tubing to rise (Fig. 15–24). Salts permeate cellophane membranes more slowly than water, so that some time elapses before the salt molecules pass out of the bag into the surrounding water. The movement of solute molecules (in this case, salt molecules) from a region of *high* concentration (inside the bag) to one of *lower* concentration (outside the bag) occurs by the process of **diffusion.** In the case illustrated in Figure 15–24, the initial movement of water into the cellophane bag is followed by the outward diffusion of salt from the bag. As the solute concentration inside the bag decreases, the liquid level in the glass tubing again falls as water molecules leave the bag by osmosis. The movements of salt and water molecules continues until the salt concentrations inside and outside the bag are equal.

Consider a case in which the cellophane bag is filled with a solution containing an impermeable solute. As in the previous instance, water will enter the bag by osmosis, causing the liquid level in the glass tubing to rise. Since the solute is impermeable, it cannot diffuse from the bag, and a concentration equilibrium across the membrane cannot be achieved. Consequnetly, water will continue to enter the bag and rise in the glass tubing until a height is reached at which the pressure at the base of the water column is just great enough to prevent any further water influx. In theory, the water will remain at this level indefinitely. The pressure that is created inside the bag by the impermeable solute and that supports the column of water is called **osmotic pressure.** Its value can be approximated by measuring the height of the water column. Usually, osmotic pressure is expressed in *millimeters of mercury* (i.e., mm Hg) rather than in inches

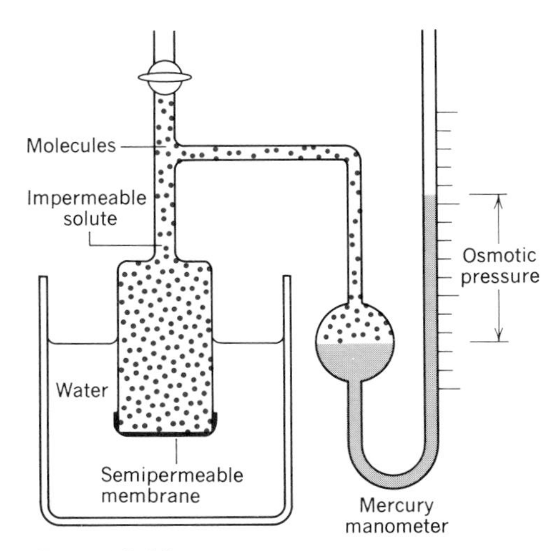

Figure 15–25
An osmometer.

of water. Devices used to measure osmotic pressure are called **osmometers** (Fig. 15–25).

Osmosis and Diffusion Across Cell Membranes

Cellular phenomena associated with osmosis and diffusion across the plasma membrane are readily demonstrated using red blood cells, sea urchin eggs, or certain plant cells. The *plasma* in which the red blood cells are normally suspended contains the same concentration of *impermeable* salt (0.15 *M* NaCl) as the erythrocyte cytoplasm (0.15 *M* KCl). Normal plasma is said to be **isotonic** to the red cell. If the plasma is diluted with water, its salt concentration will decrease, and the plasma will become **hypotonic** to the red blood cell. Any suspending medium containing an impermeable solute concentration that is *lower* than the corresponding solute concentration in the cells suspended in that medium is considered hypotonic. In the case of hypotonic plasma, water will enter the red cells by osmosis, causing the cells to swell. The same effect can be produced by placing red blood cells in any hypotonic solution. Just how much water will enter the cell depends upon how hypotonic the suspending medium is. For example, if the red cells are suspended in plasma containing *one-half* the normal salt concentration, water will enter the cells until they swell to twice their original volume. This will reduce the internal salt concentration to one-half its former value, bringing the internal and external salt concentrations into equilibrium (Fig. 15–26).

If the cells are suspended in a solution of even greater hypotonicity, then proportionately more water will have to enter the cell to reduce the internal salt concentration to that outside the cell. Obviously, cells can tolerate only a certain amount of swelling before the membrane ruptures, spilling the cell contents into the surrounding medium; this is called **osmotic lysis.** Red blood cells lyse when suspended in very dilute salt solutions or in distilled water (Fig. 15–26). In the specific case of the red blood cell, this phenomenon is called hemolysis, since hemoglobin from the red blood cell is released into the suspending medium. Other animal cells behave in a similar manner in appropriately hypotonic media.

Figure 15–26
Behavior of red blood cells in hypotonic solutions.

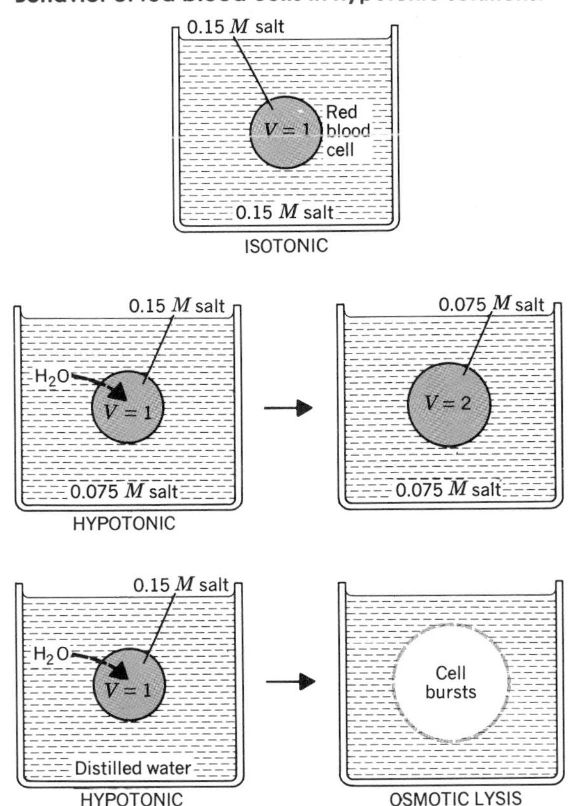

Plant cells generally do not lyse even when placed in distilled water because cell swelling is limited by the rather inflexible cellulose cell wall. In hypotonic solutions, plant cells swell as water enters the cytoplasmic vacuoles by osmosis. This forces the cytoplasm to the margins of the cell wall. Under these conditions, the plant tissue becomes **turgid** (Fig. 15–27).

Equal concentrations of impermeable salts and nondissociating (nonionizing) molecules (such as sucrose and other sugars) do *not* exhibit the same osmotic effects. For example, 0.15 M NaCl exerts twice the osmotic pressure as does 0.15 M sucrose. This is because 0.15 M NaCl undergoes dissociation in water to produce twice the number of particles (i.e., ions) per cubic centimeter as 0.15 M sucrose. Thus, 0.15 M NaCl, 0.15 M KCl, 0.10 M $CaCl_2$, and 0.3 M sucrose would all exert the *same* osmotic pressure, since they produce the same particle concentrations in water.

Figure 15–27
Behavior of plant cells in hypotonic solutions.

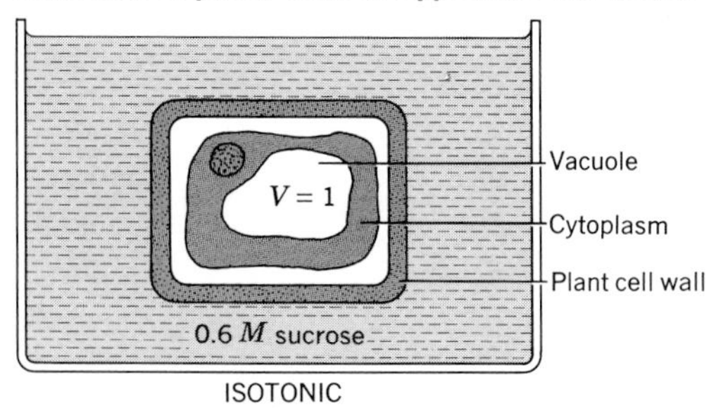

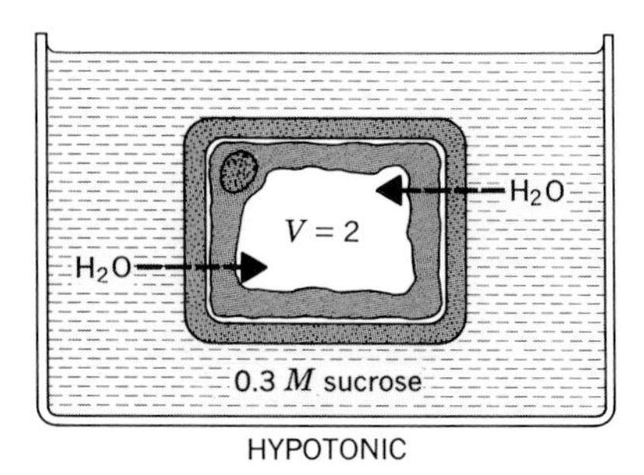

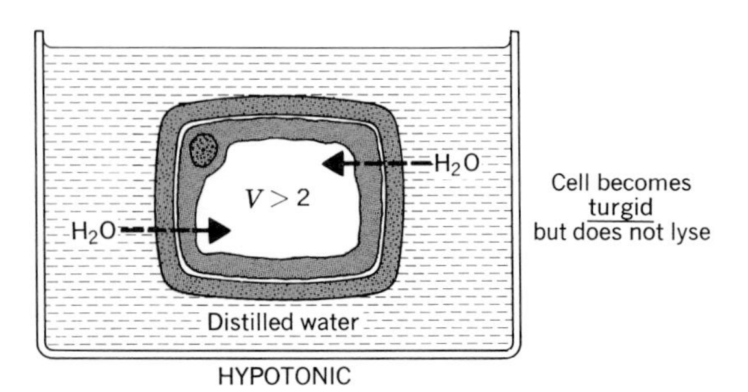

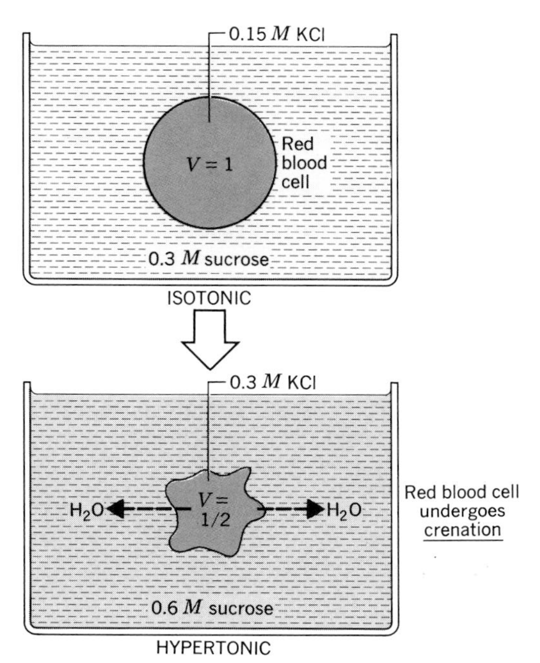

Figure 15–28
Crenation of red blood cells in hypertonic solution.

Sucrose is impermeable to most cell membranes, including the membrane of the red blood cell. Therefore, both 0.15 M NaCl and 0.3 M sucrose are isotonic to red blood cells. Solutions that contain higher concentrations of impermeable solute than are found inside cells are said to be **hypertonic.** What happens to red blood cells placed in hypertonic sucrose is depicted in Figure 15–28. Water moves by osmosis from the cells into the medium until the concentrations of solute inside and outside the cell are the same. The shrinkage associated with such water loss is called **crenation.** Other animal cells behave in a similar manner.

When plant cells are placed in hypertonic media, they undergo **plasmolysis;** that is, water passes from the cyto-

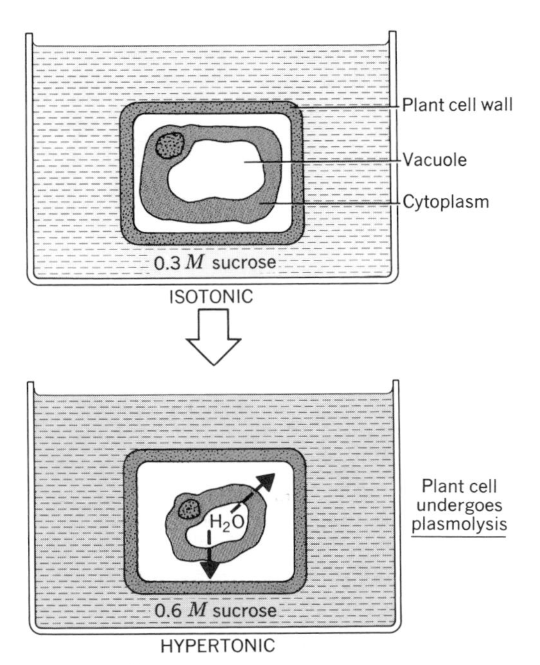

Figure 15–29
Plasmolysis of plant cells in hypertonic solution.

plasmic vacuoles into the space between the cell wall and cell membrane (Fig. 15–29).

So far, we have considered only the movement of water into and out of the cell. When the concentration of a permeable solute is greater outside a cell than inside, the solute molecules diffuse into the cell until the external and internal concentrations are in equilibrium. In most instances, the combined volume of all the cells present in a cell suspension is only a small fraction of the total volume of the suspending medium. Therefore, the quantity of solute entering (or leaving) the cells has a negligible effect on the solute concentration of the medium.

Consider what happens when red blood cells are placed in an isotonic saline solution (e.g., 0.15 *M* NaCl) containing 0.1 *M* urea. Urea is permeable to the red blood cell membrane, and since red blood cells normally contain little or no urea, urea molecules will diffuse into the cells until the intracellular urea concentration reaches 0.1 *M*. Even after a concentration equilibrium is achieved, urea molecules will continue to pass through the cell membrane into and out of the cell. However, the migration in each direction will be the same, so that no *net* change in the urea concentration of the cell occurs. If the cells are transferred to an isotonic saline solution lacking urea, urea molecules will diffuse back out of the cell. Many substances enter and leave cells by diffusion through the plasma membrane.

The Permeability Constant. The quantity of solute diffusing from one region to another depends on the concentration difference between the two regions—that is, the **concentration gradient.** Concentration gradients often exist between the internal and external environments of a cell. If the solute is permeable to the cell membrane, diffusion in the direction of the concentration gradient into (or out of) the cell ensues.

The **diffusion rate** for a solute is given by **Fick's equation:**

$$\frac{dS}{dt} = D\,A\,\frac{C_1 - C_2}{x} \qquad (15\text{–}1)$$

where dS/dt = the number of moles of solute, S, diffusing from region 1 to region 2 in the time interval dt.

D = the **diffusion coefficient** of the solute. Each solute has a specific diffusion coefficient; the units of diffusion coefficient are moles per unit cross-sectional area per unit concentration gradient per unit time. This reduces to the dimensions square centimeters per second. The diffusion coefficient of a solute is determined in part by the solute molecule's size and shape.

A = the cross-sectional area through which diffusion occurs.

C_1 and C_2 = the concentrations of S in regions 1 and 2, respectively.

x = the distance between regions 1 and 2

$\frac{C_1 - C_2}{x}$ = the concentration gradient

In the case of diffusion into or out of the cell, we may let

x = the membrane thickness,

$C_1 = C_{out}$ = the concentration of S outside of the cell,

$C_2 = C_{in}$ = the concentration of S inside of the cell.

A new term, the **permeability constant K,** which describes the diffusion of a particular substance across a particular membrane, may be substituted for D/x in equation 15–1 to yield

$$\frac{dS}{dt} = K\,A\,(C_{out} - C_{in}) \qquad (15\text{–}2)$$

The dimensions of K are centimeter per second. Since the total amount of substance S in a cell of volume V is VC_{in}, substitution of VC_{in} for S in equation 15–2 yields

$$\frac{dC_{in}}{dt} = K\,\frac{A}{V}\,(C_{out} - C_{in}) \qquad (15\text{–}3)$$

Thus, it can be seen that other things being equal, the rate of change of the internal solute concentration (i.e., dC_{in}/dt) depends upon the surface area to volume ratio of the cell (i.e., A/V). Since the ratio A/V is at a minimum for spherical objects, the rate at which solute is exchanged by diffusion between a cell and its environment is increased by deviation from spherical shape.

If the external cell volume is sufficiently large in comparison with the cell volume (this generally is the case), then C_{out} will not change significantly with time as a result of diffusion. That is, during the time interval between t_0 and t_1, C_{out} will remain constant. Therefore, equation 15–3 may be solved by integration as follows:

$$\int_{C_{out} - C_{in} \text{ at } t_1}^{C_{out} - C_{in} \text{ at } t_0} \frac{dC_{in}}{C_{out} - C_{in}} = \int_{t_1}^{t_0} K\,\frac{A}{V}\,dt \qquad (15\text{–}4)$$

$$\ln\,(C_{out} - C_{in} \text{ at } t_0) - \ln\,(C_{out} - C_{in} \text{ at } t_1) = K\,\frac{A}{V}\,\Delta t \qquad (15\text{–}5)$$

where Δt equals the time interval between t_0 and t_1. Solving equation 15–5 for K, we obtain

$$K = \frac{V}{A\,\Delta t}\,\ln\,\frac{C_{out} - C_{in} \text{ at } t_0}{C_{out} - C_{in} \text{ at } t_1} \qquad (15\text{–}6)$$

C_{out}, C_{in}, A, V, and Δt may be determined experimentally, so that equation 15–6 can be employed to find the permeability constant. Evaluation of the permeability constants for a variety of substances entering a cell permits much more to be learned about the chemical nature and behavior of the cell membrane. The permeability constants of glycol, urea, glycerol, and sucrose for the membranes of the red blood cell, the bacterium *Beggiatoa,* and the alga *Chara* are compared in Table 15–5. The marked differences that can be seen in these values reflect the variation that must exist in the composition and organization of the plasma membranes of these cells.

Table 15–5
Some Permeability Constants

Substance	Permeability Constant (cm/sec × 10^5)		
	Red Blood Cell	*Beggiatoa*	*Chara*
Glycol	0.21	1.39	1.2
Urea	7.8	1.58	0.11
Glycerol	0.0017	1.06	0.021
Sucrose	[a]	0.14	0.0008

[a] *Sucrose is impermeable to the red blood cell membrane.*

Yes

Factors Influencing Permeability. A number of different molecular parameters influence the permeability of a substance to the cell membrane. Important among these are **distribution coefficient, molecular size,** and **charge.** The distribution coefficient of a substance relates its solubility in oil to its solubility in water. In general, the more *nonpolar* a substance, the more soluble it is in oil and the less soluble it is in water. In contrast, the more polar a substance is, the less soluble it is in oil and the more soluble it is in water.

One of the most extensive studies of the relationship between partition coefficient and membrane permeability was that carried out in the 1930s by R. Collander and H. Bärlund using the unicellular alga, *Chara*. They found that, in general, the higher the distribution coefficient of a compound, the greater its permeability to the cell membrane. Overton had made similar but less quantitative observations may years earlier, and this led him to propose that the cell membrane was composed of lipid. The relationship between lipid solubility and membrane permeability is the basis for the proposal that lipid-soluble substances readily pass into or out of the cell by dissolving through lipid regions of the cell membrane.

For chemically related substances, permeability increases with increasing distribution coefficient *regardless of molecular size*. However, in general, where two molecules have simple partition coefficients, the smaller molecule is more permeable than the larger molecule. The fact that water has a very high permeability constant in spite of its low lipid solubility led to the suggestion that there are *hydrophilic* pores in the cell membrane through which small polar molecules may more readily diffuse. The permeability constant for water is about 1×10^{-4} cm/sec. This value is only 0.001% of the rate at which water molecules diffuse across a water layer the thickness of the cell membrane. This implies that if they are present, hydrophilic pores must cover only a small percentage of the surface area of the cell. In the case of the red blood cell, A. K. Solomon has estimated that 0.06% of the surface area is occupied by such pores.

In addition to the distribution coefficient, the *effective* size of a molecule influences its permeability to the cell membrane. In general, small molecules are more permeable than large molecules. Substances of very high molecular weight (such as starch, glycogen, and many proteins) are usually unable to permeate the membrane at all. This is not to imply that cells cannot incorporate or eliminate very large molecules. Indeed, they do; however, the process involves mechanisms other than diffusion (see later).

Electrolytes enter cells more slowly than do *nonelectrolytes* of similar effective molecular size, and *strong* electrolytes enter more slowly than *weak* electrolytes. Since pH influences the degree of ionization of many electrolytes, permeability to electrolytes varies with the pH. In general, monovalent ions (e.g., Na^+, K^+, I^-, Cl^-) permeate membranes more readily than divalent ions (e.g., Ca^{++}, Mg^{++}, SO_4^{--}), which in turn are more permeable than trivalent ions (e.g., Fe^{+++}). The relative permeability of the membrane to ions of the same valency number but different sign (i.e., K^+ vs. Cl^-) depends on the particular cell considered.

Not all ions of the same valency and sign are equally permeable. Most differences can be explained in terms of the relative effective sizes of the ions in an aqueous medium. The charge on the ion attracts neighboring water molecules, causing them to align themselves around the ion to form **spheres of hydration.** The effective size of the ion is therefore determined by the number of these hydration spheres (Fig. 15–30). Ions of low molecular or atomic weight attract more water molecules than do ions of higher molecular weight because there are fewer electron shells around the atomic nuclei to neutralize the ionic charge. Accordingly, the effective size of Li^+ (atomic number = 3) is greater than Na^+ (atomic number = 11), which is greater than K^+ (atomic number = 19), and so on. Therefore, K^+ is more permeable than Na^+, and Na^+ is more permeable than Li^+. The spheres of hydration about an ion do not totally neutralize its charge.

Ions may enter or leave a cell through pores in the cell membrane. Cations would be attracted to and pass more readily through pores lined by negative charges (e.g., negatively charged R groups of amino acids of membrane pro-

Li+

Spheres of hydration

Lithium ion in water (atomic number = 3)

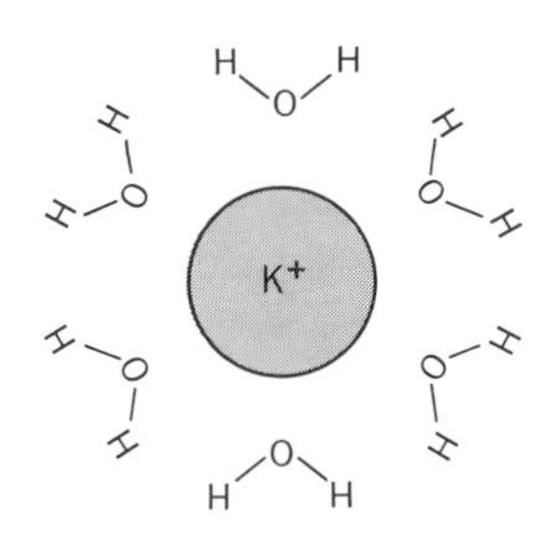

Potassium ion in water (atomic number = 19)

Figure 15–30
Spheres of hydration formed around lithium and potassium ions in water. Ions of lower atomic or molecular weight attract and orient more water molecules than ions of higher atomic or molecular weight. Therefore, their effective size is greater.

teins, etc.). Anions would enter or leave the cell through positively charged pores.

The Gibbs-Donnan Effect

Proteins behave like ions because the R groups of their amino acids may bear positive or negative charges. The net ionic charge of the protein molecule depends on the *relative* numbers of positive and negative side groups. Most soluble proteins behave like *anions* because they possess more negative than positive sites. Unlike many other ions, proteins are generally too large to permeate cell membranes. The **Gibbs-Donnan effect** (after J. W. Gibbs and F. G. Donnan) describes the effect that protein ions have on the equilibrium distributions of small ions across a selectively permeable membrane. When a selectively permeable membrane separates an electrolyte solution containing proteins from one that lacks proteins, the concentrations reached at equilibrium for each permeable ionic species will *not* be the same on both sides of the membrane. Instead, the concentration of small ions having the *same* sign as the protein will be lower on the side of the membrane containing the protein (i.e., inside the cell), while the concentration of small ions of *opposite* sign will be higher. The Gibbs-Donnan effect does not influence the equilibrium distributions of nonionized substances such as glucose, urea, and the like.

The Gibbs-Donnan effect may be illustrated as follows. Suppose that two chambers of *equal* and *fixed* volumes are separated by a membrane permeable to water and small ions but impermeable to proteins. Into chamber 1 is placed a sodium proteinate (NaPr) solution of concentration C_1, and into chamber 2, a NaCl solution of concentration C_2. Since the chloride ions are permeable and present at a greater concentration in chamber 2 than in chamber 1 (i.e., chamber 1 initially has no Cl^-), Cl^- will diffuse into chamber 1. In order to preserve the electrical neutrality of each chamber, the diffusion of Cl^- into chamber 1 must be accompanied by the diffusion of an equivalent amount of Na^+. Therefore, chamber 2 will be left with equal but reduced concentrations of Na^+ and Cl^-. According to the law of mass action, **at equilibrium the products of the diffusable ion concentrations in each chamber will be equal.** Therefore, if we let X equal the concentration of Cl^- diffusing from chamber 2 into chamber 1 (X will of course also equal the concentration of Na^+ accompanying Cl^-), then at equilibrium

$$(C_1 + X)(X) = (C_2 - X)(C_2 - X)$$

$$C_1X + X^2 = (C_2)^2 - 2C_2X + X^2$$

$$C_1X = (C_2)^2 - 2C_2X \qquad (15\text{–}7)$$

and

$$X = \frac{(C_2)^2}{C_1 + 2C_2} \qquad (15\text{–}8)$$

Equation 15–8 is called the Gibbs-Donnan equilibrium equation and may be used to determine the equilibrium distribution of ions between the protein-containing and protein-free portions of the two-chamber system.

Let us consider a specific case in which chamber 1 initially contains 0.01 *M* sodium proteinate (i.e., 0.01 *M* Na^+ and 0.01 *M* Pr^-) and chamber 2 contains 0.03 *M* NaCl (i.e., 0.03 *M* Na^+ and 0.03 *M* Cl^-). The concentration of Cl^- (and Na^+) diffusing across the membrane and into chamber 1 from chamber 2 may be determined using equation 15–8 as follows:

$$X = \frac{(0.03)^2}{0.01 + 2(0.03)}$$

$$= 0.013$$

Therefore, the equilibrium distribution of ions would be (see also Fig. 15–31):

Chamber 1	*Chamber 2*
Na^+ 0.023*M*	Na^+ 0.017 *M*
Cl^- 0.013 *M*	Cl^- 0.017 *M*
Pr^- 0.010 *M*	Pr^- 0.000 *M*

Note that (1) an electrical balance exists *within* each chamber, (2) the osmotic pressure is greater in chamber 1 than chamber 2 as a result of the higher total concentration of particles (no osmosis occurs because the chamber volumes are fixed), and (3) there are more sodium ions but fewer chloride ions in chamber 1 than in chamber 2.

By extrapolating the observations made for the artificial two-chamber system just described to cells (i.e., chamber 1) and their surrounding milieu (i.e., chamber 2), it may be seen that the Gibbs-Donnan effect would result in an unequal distribution of permeable ions across the cell membrane. That is, a *stable* ionic concentration gradient would be established. Just such a phenomenon accounts in part for the resting electrochemical membrane potentials possessed by many cells (see Chapter 23).

In the cases we have considered so far, the chamber volumes have been equal and fixed. This, of course, is not necessarily the case for a cell and its surroundings. The flexibility of the plasma membrane permits notable fluctuations in cell shape and

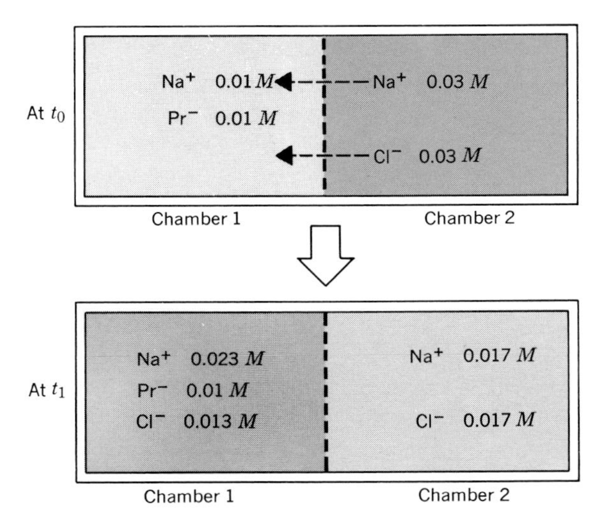

Figure 15–31
Gibbs-Donnan effect: Initial (t_0) and equilibrium (t_1) distributions of ions across a selectively permeable membrane separating two chambers of equal and fixed volume. When one ion (P_r^-) is impermeable. At equilibrium the sum of anions equals the sum of cations on each side of membrane.

volume. The greater osmotic pressure created inside the cell by the Gibbs-Donnan effect causes water to pass into the cell by osmosis, increasing the cell's volume and lowering its ion concentration. Since this disturbs the Gibbs-Donnan equilibrium previously established, a new equilibrium is achieved by the passage of additional permeable ions into the cell. Osmosis halts when the hydrostatic pressure created by water inside the cell balances the osmotic pressure generated by the dissolved cell solute.

Facilitated (Mediated) Diffusion through the Cell Membrane

A variety of compounds including sugars and amino acids pass through the plasma membrane and into the cell at a much higher rate than would be expected on the basis of their size, charge, distribution coefficient, or magnitude of the concentration gradient. The increased rate of transport through the membrane is believed to be facilitated by specific membrane *carrier* substances and is called **facilitated** (or **mediated**) **diffusion.**

During facilitated diffusion, the rate at which the solute permeates the membrane increases with increasing solute concentration *up to a limit*. Above this limiting concentration, no increase in the rate of transport across the membrane is observed. In other words, facilitated diffusion exhibits **saturation kinetics** (Fig. 15–32) and is therefore similar to the relationship between reaction rate and substrate concentration in enzyme-catalyzed reactions (see Chapter 8). Other characteristics of facilitated diffusion are also similar to enzyme catalysis. Transport is *specific;* for example, in the erythrocyte, the inward diffusion of glucose, but not fructose or lactose, is facilitated. The rate of solute permeation can also be affected by the presence of structurally similar chemical compounds, much as in *competitive enzyme inhibition*. Facilitated diffusion exhibits *pH dependency*. Although facilitated diffusion results in a more rapid attainment of a concentration equilibrium across the membrane than passive diffusion, *the normal equilibrium concentrations are not altered*. Substances are *not* transported through the membrane *against a concentration gradient*. Although facilitated diffusion is not affected by chemicals that act as metabolic inhibitors, it is affected by *enzyme inhibitors* such as sulfhydryl blocking agents.

Facilitated diffusion is believed to result from the interaction of solute with specific membrane molecules, presumably proteins, thereby forming a **carrier-solute complex.** The complex undergoes translation or rotational diffusion within the membrane in such a way that the solute now faces the other membrane surface and is released from

Figure 15–32
Kinetics of facilitated diffusion. Note similarity to kinetics of enzyme-catalyzed reactions.

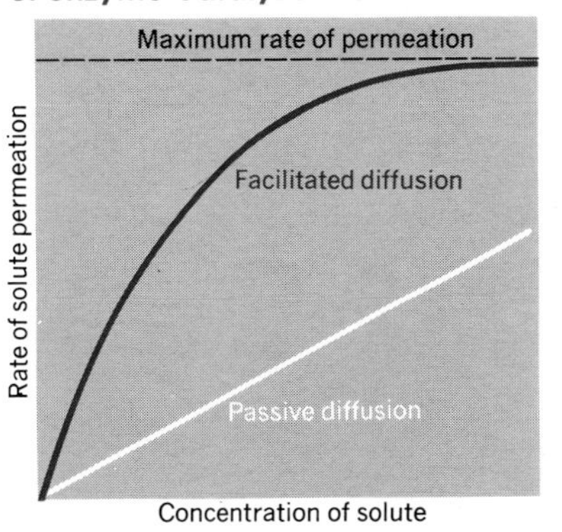

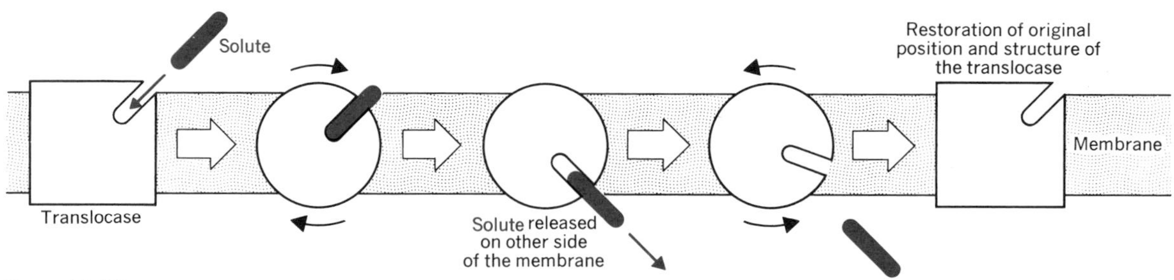

Figure 15–33
Stages of facilitated diffusion.

the carrier (Fig. 15–33). An alternative suggestion is that the carrier may be a small molecule with the solute-carrier complex formed by an enzyme-catalyzed reaction within the membrane. Once formed, the solute-carrier complex diffuses to the other side of the membrane where the solute is released in a second reaction.

A good and well-defined example of facilitated diffusion occurs in the bacterium *E. coli*. The sugar *galactose* does not permeate wild-type *E. coli* cells and cannot be hydrolyzed by cytoplasmic extracts of these cells. However, when *E. coli* is in a medium containing galactose, the enzyme *beta-galactosidase*, which hydrolyzes galactose, soon appears in the cell cytoplasm. The presence of the substrate in the growth medium is said to *induce* the formation of the enzyme by the cells. Such cells are able to take up galactose from the medium as well as metabolize it. Certain mutants of *E. coli* can be induced in this way to form beta-galactosidase even though galactose remains impermeable and cannot be incorporated by the cells. Finally, other *E. coli* mutants can be found that are able to incorporate galactose but cannot metabolize it once inside the cell.

It is now clear from these studies that in wild-type cells the presence of galactose in the growth medium induces *both* the formation of the hydrolytic enzyme *and* a carrier system—called a **permease** or **translocase.** The sets of genes controlling the formation of the permease and the hydrolytic enzyme are coordinately induced in the presence of galactose. The loss or alteration of either set of genes (as in the *E. coli* mutants) results in a corresponding inability to induce the formation of the enzyme or the permease.

Active Transport

During diffusion (passive or facilitated), substances pass through the plasma membrane until some sort of equilibrium is achieved. The equilibrium may be of the Gibbs-Donnan variety or may be a simple concentration equilibrium. Both involve an interplay between the concentrations of soluble solute inside and outside the cell. Cells can also accumulate solutes in quantities far in excess of that expected by any of the above mechanisms if the solute is rendered insoluble once it has entered the cell, since insoluble materials do not contribute to concentration gradients. Alternatively, once inside the cell, a solute may enter a metabolic pathway and be chemically altered, thereby reducing the concentration of that particular solute and allowing additional solute permeation. In all the cases we have so far considered, solute permeation of the membrane hinges on the presence of a concentration gradient, with the solute moving in the direction of the gradient.

Substances can also move through the plasma membrane into or out of the cell *against a concentration gradient*. This requires the expenditure of energy on the part of the cell and is called *active transport*. Active transport ceases when cells are (1) cooled to very low temperatures (such as 2–4°C), (2) treated with metabolic poisons such as cyanide or iodoacetic acid, or (3) deprived of a source of energy. The best understood and most exhaustively studied instances of active transport are those that involve the movements of sodium and potassium ions across the plasma membranes of erythrocytes, nerve cells, and *Nitella* cells and that result in an ionic concentration gradient across the cell membrane. The mechanism that establishes and maintains these gradients appears to be basically similar in all of these cells and can be illustrated with the erythrocyte.

The Na^+/K^+ Exchange Pump

The cytoplasm of the erythrocyte contains 0.150 M K^+, whereas the surrounding blood plasma contains only 0.005 M K^+. In contrast, the erythrocyte contains only 0.030 M Na^+ while the plasma contains 0.144 M Na^+.

Hence, marked K^+ and Na^+ concentration gradients exist across the cell membrane. In Chapter 14 we noted that tracer studies utilizing radioactive isotopes of Na and K clearly established that these ions are permeable to the erythrocyte membrane and are constantly diffusing through it. Yet, in spite of this permeability, Na^+ and K^+ concentration gradients across the membrane are maintained. The gradients are maintained because sodium ions diffusing into the cell from the plasma under the influence of the concentration gradient are transported outward again, and potassium ions diffusing out of the cell are replaced by the inward transport of K^+ from the plasma. That these movements involve active metabolic processes is clearly demonstrated when the temperature of a blood sample is reduced from 37°C (normal blood temperature) to 4°C, when cyanide is added to the blood, or when plasma glucose consumed during erythrocyte metabolism is not resupplied to the blood sample. Under these conditions, cell metabolism is interrupted and is followed by the inward diffusion of Na^+ and the outward diffusion of K^+ until the ionic concentrations on both sides of the erythrocyte membrane are in a passive equilibrium.

It is apparent from these and other studies that energy-producing reactions in the cell are in some way coupled to the active transport mechanism. Since many of the substances actively transported into or out of the cell are freely permeable, interference with energy-producing cell reactions results in the diffusion of these materials in the direction opposite to that of active transport.

In the case of red blood cells and nerve cells, the active transport of Na^+ and K^+ appears to be linked. That is, the mechanism responsible for the outward transport of Na^+ simultaneously transports K^+ inward. An enzyme isolated from nerve cell membranes and believed to be involved in Na^+ and K^+ transport has been shown to have two sites that bind one or more of each of these cations. The enzyme is believed to be an integral protein spanning the lipid bilayer.

Active transport of Na^+ and K^+ through the membranes of nerve cells and erythrocytes requires ATP, and ATP cannot be replaced by other nucleoside triphosphates such as GTP, UTP, and ITP. ATP is converted to ADP during active transport by a membrane-bound Na^+- and K^+-stimulated ATPase. This enzyme and that involved in the transport of Na^+ and K^+ may be one and the same. The membranes of cells from many other mammalian tissues seem to possess a similar ATPase activity.

Two K^+ and three Na^+ are transported through the membrane for each molecule of ATP dephosphorylated. Transport of Na^+ and K^+ through the plasma membrane is believed to occur in the following stages (see Fig. 15–34). Three sodium ions and one molecule of ATP inside the cell are bound to specific sites on the enzyme carrier, while two potassium ions are bound to a site on the same enzyme facing the exterior of the cell. Binding of the substrates results in and is followed by a change in the tertiary structure of the carrier molecule such that the bound sodium and potassium ions are "translocated" across the membrane. It is presumed that at some stage during this process, ATP is split, releasing ADP. Translocation is followed by an alteration of the binding sites such that the sodium ions are "released" outside the cell, while the potassium ions are released inside the cell. Once the ions are released, the carrier undergoes another change in structure, priming it for another round of the transport cycle. The stage, called

Figure 15–34
The Na^+/K^+ exchange pump. Changes in the tertiary structure of the enzyme carrier are shown as transitions of geometric shape (square to circle, etc.).

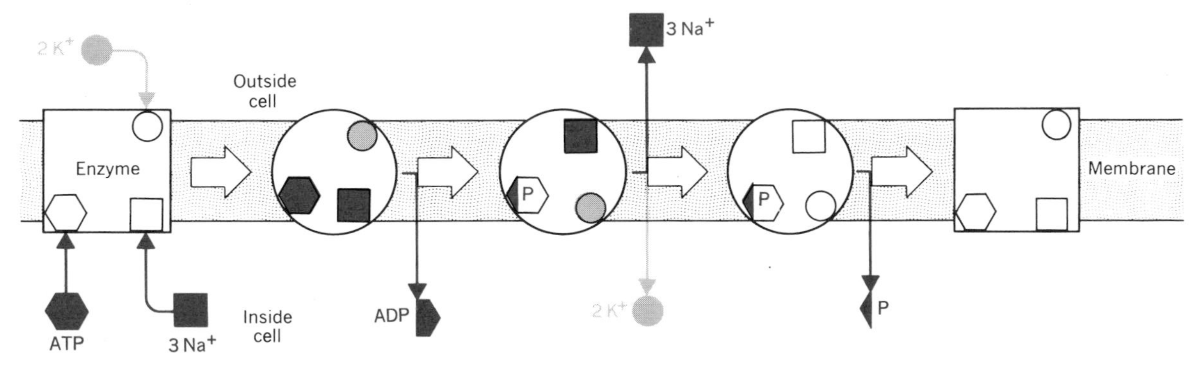

''recovery,'' is accompanied by the release of inorganic phosphate. Although this model is widely accepted, it has also been suggested that the enzyme site that binds Na^+ on the inside of the cell binds K^+ on the outside following translocation, while the site that initially binds K^+ on the outside binds Na^+ on the inside following translocation. In this manner, the recovery phase would result in an additional movement of ions through the membrane and would be more efficient.

Cotransport: The Electrogenic Pump

Amino acids, sugars, and other metabolites are also actively transported through the plasma membrane into the cell. In many cells, the transport of these metabolites is coupled to the movements of sodium, as shown in Figure 15–35. The Na^+/K^+ exchange pump creates a steep concentration gradient across the plasma membrane favoring the inward diffusion of Na^+. Carrier proteins in the membranes bind both Na^+ and the metabolite, following which a change in carrier orientation brings both substrates to the cell interior, where they are released. However, release of the Na^+ is followed by its active extrusion back through the membrane. The latter event is coupled to ATP hydrolysis and results in the maintenance of the steep Na^+ gradient. Thus the steep Na^+ gradient acts as the driving force for the inward transport of metabolites and the simultaneous movements of Na^+ together with metabolites constitute **cotransport.** Steep ionic gradients that serve to drive transmembrane movements of solutes are called *electrogenic pumps*. As will be seen in Chapter 16, the potential energy of an electrogenic pump is coupled to ATP synthesis in mitochondria.

"Simple" Active Transport

The passage of some substances through membranes against a concentration gradient is unidirectional but not coupled to ionic movements even though ATP is consumed in the process. Such movement is called **simple active transport.** The carrier enzyme cyclically binds the solute at one membrane surface and releases it at the other. The cycle is accompanied at some point by the hydrolysis of ATP.

Bulk Transport Into and Out Of Cells

Before we consider several mechanisms that result in **bulk transport** into and out of cells, it is important to reconsider what is meant by ''inside the cell'' and ''outside the cell'' (see also Chapter 1). For convenience, we have so far depicted the plasma membrane as a more-or-less continuous, smooth sheeting enclosing the cell. In reality, of course, this is an oversimplification, for in most cells the membrane exhibits numerous outfoldings and infoldings. Outfoldings of the plasma membrane cover microvilli, cilia, flagella, and other cytoplasmic extensions. Infoldings of the membrane form small pockets and narrow channels that descend into the cytoplasm. Some of these infoldings may join the network of cisternae that forms the endoplasmic reticulum and that furrows the cytoplasm. This implies that the *lumenal phase* of the endoplasmic reticulum (Chapter 1) may

Figure 15–35
Cotransport. The coupled transport of Na^+ plus metabolite (in this case, sugar). Changes in the tertiary structure of the enzyme carrier are shown as transitions of geometric shape (square to circle, etc.).

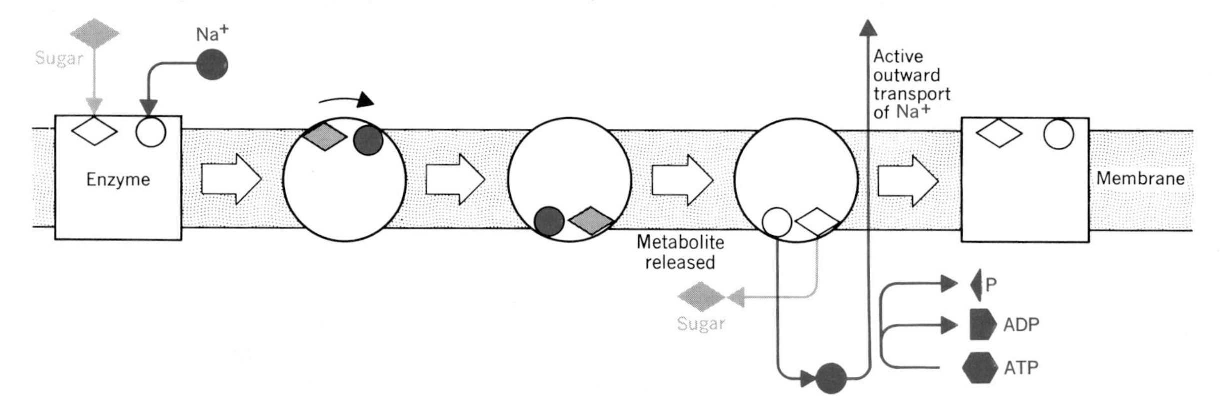

be in direct continuity with the surrounding cell environment and that the movement of materials between the lumenal phase and the cell surroundings does not require passage across any membranes. Consequently, materials within the cisternae of the endoplasmic reticulum (i.e., in the lumenal phase) may be regarded as "outside" the cell. In order to get "inside" the cell, substances in the lumenal phase must pass through the membranous walls of the endoplasmic reticulum and into the cytosol (or hyaloplasm).

Many intracellular vesicles appear to be derived from the endoplasmic reticulum by being "pinched off" from the latter. The contents of these vesicles are "outside" of the cell and separated from the cytosol by membranes. Some vesicles are derived from the invagination and pinching off of parts of the plasma membrane; the contents of these vesicles are also to be considered as "outside" the cell. These relationships are depicted in Figure 15–36 in which the external (or cisternal) and internal (or cytosol) halves of the membrane are distinguished. Particles within such a vesicle are in contact with the external face of the membrane.

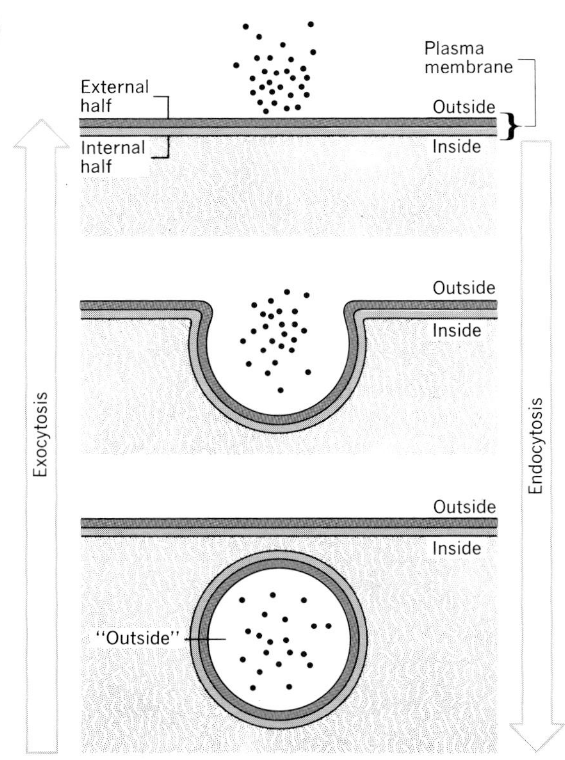

Figure 15–36
Relationship of external and internal halves of the plasma membrane to the inner and outer faces of vesicles in exocytosis and endocytosis.

From the preceding discussion, it should now be clear that while substances may be directly exchanged between the cell surroundings and the cytosol through the plasma membrane, many exchanges also occur across membranes of cytoplasmic vesicles and endoplasmic reticulum. These exchanges are mediated by the same mechanisms described above in connection with movements directly through the plasma membrane (i.e., diffusion, facilitated diffusion, active transport, etc.).

The formation of cytoplasmic vesicles from the plasma membrane and the consequent entrapment within these vesicles of materials formerly in the cell surroundings is called **endocytosis.** Several different kinds of endocytosis have been described including **pinocytosis, rhopheocytosis,** and **phagocytosis.** Movements of materials from the cell into its surroundings by the fusion of cytoplasmic vesicles with the plasma membrane constitute **exocytosis.** As seen in Figure 15–36, endocytosis and exocytosis are variations of the same fundamental phenomenon, differing from one another in the *direction* of bulk solute movement.

Endocytosis

Pinocytosis. Using time-lapse photography to study tissue culture cells, W. H. Lewis in 1931 described a phenomenon in which small amounts of culture medium were trapped in invaginations of the plasma membrane and then pinched off to form small cytoplasmic vesicles. Since the entire process appeared much like some form of organized cell drinking, Lewis termed the phenomenon **pinocytosis** (*pinos* means "I drink" in Greek). Lewis' observations with tissue culture cells were confirmed in 1934 by S. O. Mast and W. L. Doyle studying amoebae in which pinocytosis is readily observed with the light microscope. Using electron microscopy, it became clear in the 1950's that pinocytosis is a common phenomenon occurring at intervals in many different kinds of cells including leukocytes, kidney cells, intestinal epithelium, liver macrophages, and plant root cells.

Pinocytosis is induced by the presence of appropriate concentrations of proteins, amino acids, or certain ions in the medium surrounding the cell. The first step in the process involves a simple binding of the inducer substance to

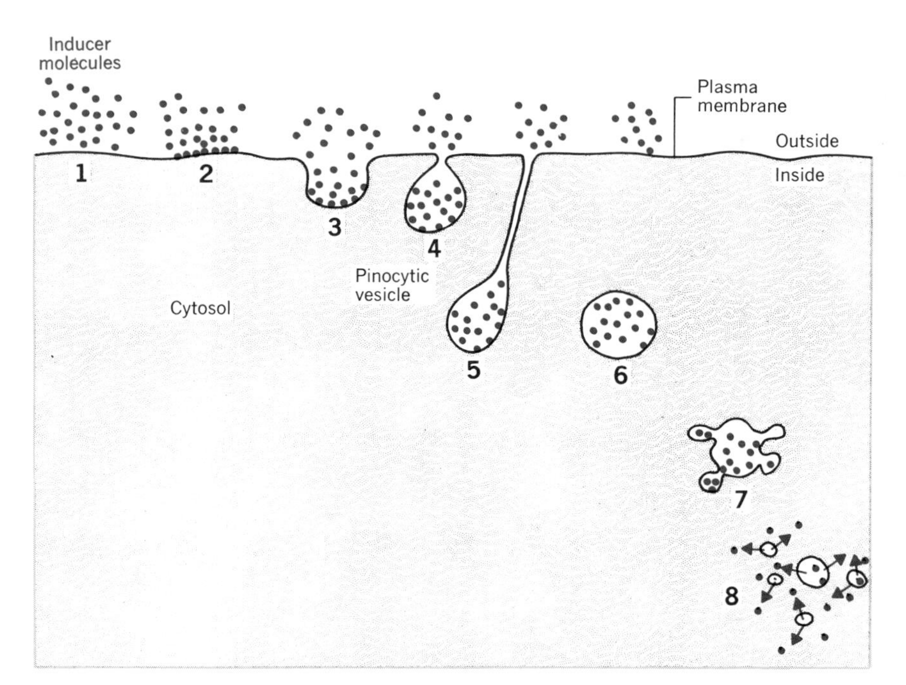

Figure 15–37
Stages of pinocytosis: 1–2, binding of inducer molecules to plasma membrane; 3–5, invagination of the membrane; 6–8, detachment from plasma membrane and fragmentation into smaller vesicles.

specific receptor sites on the cell membrane. This is followed by invagination of the membrane to form either small pinocytic vesicles or channels (Fig. 15–37). Although binding of the inducer is not inhibited by cyanide or low temperature, the formation of pinocytic vesicles is, and it is therefore dependent on cell metabolism. Formation of pinocytic and other endocytic vesicles is believed to involve the contractions of intracellular microfilaments whose ends are anchored in the plasma membrane.

Pinocytic vesicles (which usually are less than 1 nm in diameter) detach from the cell membrane and migrate toward the interior of the cell. In the cell interior, these may fragment into smaller vesicles or coalesce to form larger ones. Unless the vesicles are "tagged" by inducing pinocytosis in the presence of radioactive tracers, they soon become indistinguishable from other vacuoles in the cell.

Although pinocytosis is induced by the presence of specific substances in the cell surroundings, other materials are also enclosed by the pinocytic vesicles including water, salts, and so on. These substances, together with the inducer molecules, may enter the cytosol from the vesicle by diffusion, active transport, or related transport mechanisms.

Rhopheocytosis. Rhopheocytosis is a bulk transport mechanism in which small quantities of cytoplasm, together with their inclusions, are transferred from one cell to another. Rhopheocytosis was first demonstrated in bone marrow tissues by M. Bessis. In the marrow, maturing red blood cells (erythroblasts) are attached to reticuloendothelial cells to form large numbers of *erythroblast islands* (Fig. 15–38). The reticuloendothelial cells of the marrow contain large quantities of iron derived from the breakdown of hemoglobin from old red blood cells. In the reticuloendothelial cells, hemoglobin iron is converted to *ferritin*—a high-molecular-weight protein containing up to 23% iron by weight. Because of their high density, clusters of ferritin in reticuloendothelial cells are readily identified with the electron microscope, which has therefore been employed to study the fate of these molecules. During the maturation of red blood cells, ferritin granules, together with small amounts of cytoplasm, are transferred from reticuloendothelial cells to erythroblasts by the simultaneous invagination and evagination of their adjacent plasma membranes (Fig. 15–38). Once inside the erythroblast, ferritin iron (and iron derived directly from the circulating blood plasma)

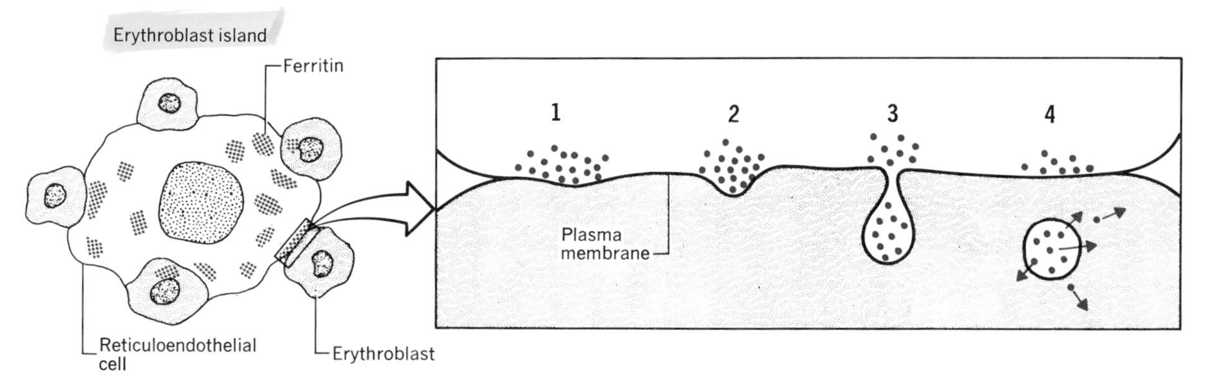

Figure 15–38
Rhopheocytosis in the bone marrow: stages 1–4 show successive evagination and invagination of adjacent plasma membranes transferring ferritin and some cytoplasm from the reticuloendothelial cell to the erythroblast.

passes into the cytosol and is used in the synthesis of new hemoglobin molecules.

Phagocytosis. Phagocytosis which was first described by E. Metchnikoff in the late nineteenth century is similar to pinocytosis but involves the engulfment of much larger quantities of particulate material. For example, entire ciliates, rotifers or other microscopic organisms may be phagocytosed by an amoeba and enclosed within one or more vacuoles called *phagosomes, food vacuoles,* or *food cups* (see Fig. 15–39). During phagocytosis, the "prey" may be temporarily immobilized by secretions from the phagocytic cell. The phagocytosis of ciliates by amoebae is characterized by the flowing of the amoeba's cytoplasm into footlike projections (*pseudopodia*) that gradually encircle and fully encapsulate the ciliate. Using a similar mechanism, certain white blood cells phagocytose hundreds of bacteria. The removal and destruction of old red blood cells in the liver, spleen, and bone marrow by reticuloendothelial cells in these organs also occur by phagocytosis. Following phagocytosis, the phagosomes fuse with *primary lysosomes* in the cell. The hydrolytic enzymes from these lysosomes digest the engulfed material, converting it to a form that may be transported across the vacuolar membranes (see Chapter 19).

Exocytosis

Exocytosis is the mechanism by which large quantities of material enclosed within a cell vacuole are transferred to the cell surroundings by fusion of the vacuole with the plasma membrane. In a sense, the process is the reverse of pinocytosis or phagocytosis, the contents of the vacuole being emptied into the extracellular space. The best understood form of exocytosis is secretion. When the secretory vesicle touches the plasma membrane, lipids in both membranes are moved aside, making the membranes more fluid. After the vesicle's contents have been discharged to the outside, the vesicle membrane is incorporated into the plasma membrane (Fig. 15–40). The electron photomicrographs of Figure 15–41 vividly depict this phenomenon. Exocytosis is also discussed in Chapters 18 and 19, which deal with the structure and functions of Golgi bodies and lysosomes.

Summary

The plasma membrane delimits the cell and actively participates in the movement of materials between the cell and its surroundings. Early studies on the chemistry of the plasma membrane focused upon the content and organization of its lipids resulting in the **bimolecular lipid leaflet model.** This concept was subsequently modified to account for the membrane proteins, resulting first in the **unit membrane model** and more recently in the **fluid mosaic model.** The latter model appears to be more consistent with bio-

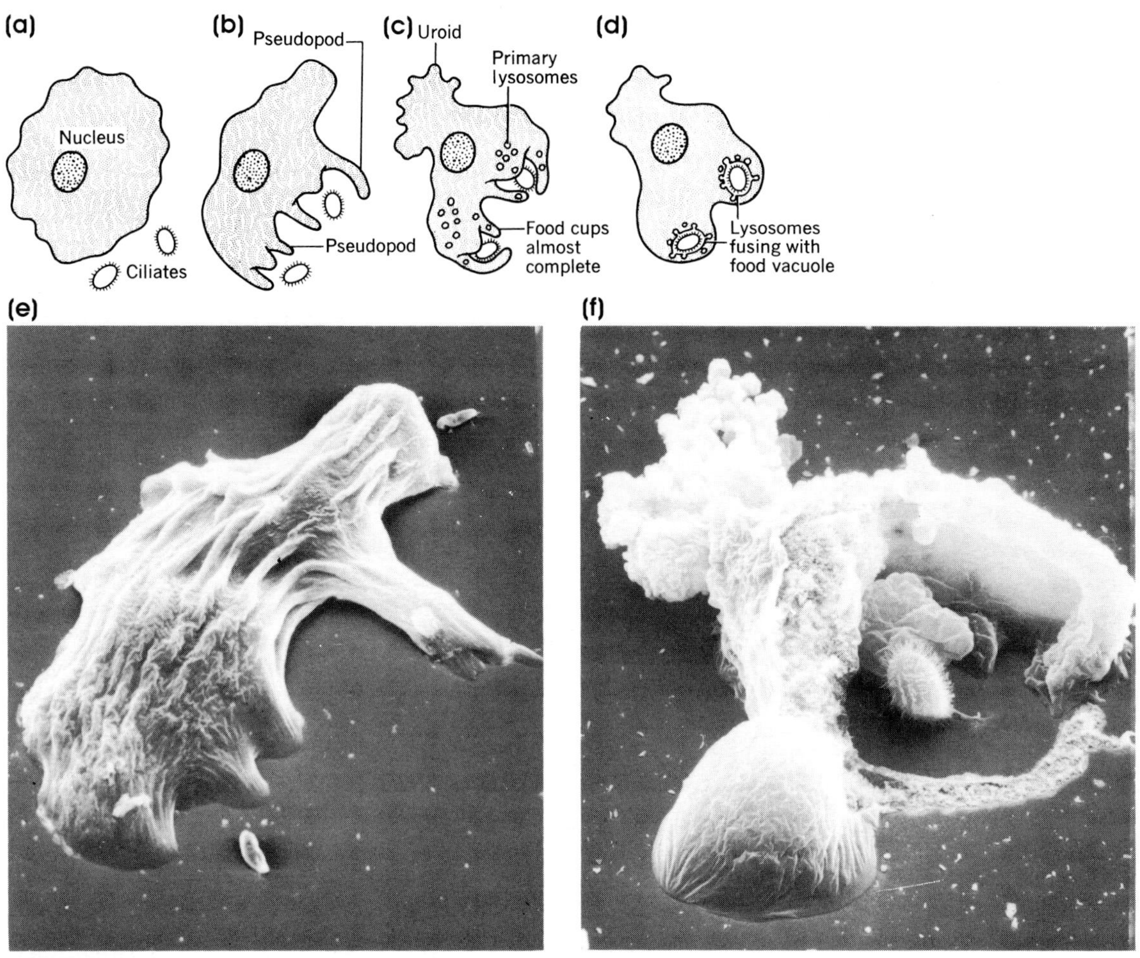

Figure 15–39
Phagocytosis. In the presence of phagocytosable material (*a*), the cell forms pseudopodia (*b*) that entrap the prey in vacuoles (*c* and *d*). The fusion of primary lysosomes with these vacuoles is followed by digestion of the prey. Scanning electron micrographs are an amoeba prior to (e) and during (f) phagocytosis and correspond to stages *b* and *d* of the diagrams. (Photomicrographs kindly provided by Dr. K. W. Jeon.)

chemical studies and electron-microscopic analyses of plasma membranes. Both the proteins and lipids of the membrane are asymmetrically distributed across the inner and outer halves. In some cells, the proteins in the inner half of the membrane may be anchored in position by a **cytoskeletal network.** Carbohydrate chains associated with protein (and lipid) on the outer membrane surface play roles in maintaining membrane **organization,** in transmembrane **transport,** in cell-to-cell **adhesion,** and in providing membrane **antigenic** properties. In tissues, the plasma membranes of neighboring cells exhibit specialized junctional regions including **tight junctions, desmosomes, gap junctions,** and **plasmodesmata.** Such junctions are important in maintaining tissue integrity and in regulating the passage of substances across a tissue and from cell to cell.

The movement of materials across the cell membrane

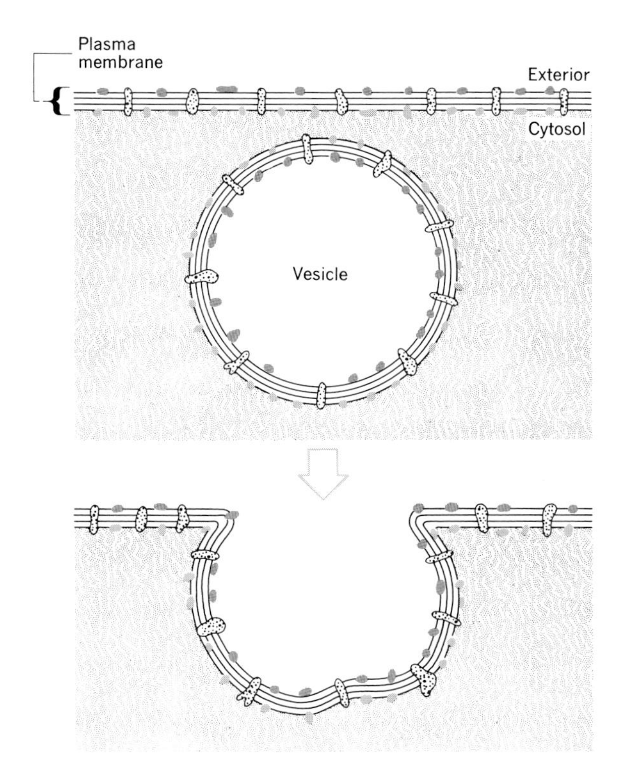

Figure 15–40
Fusion of exocytic vesicle with the plasma membrane.

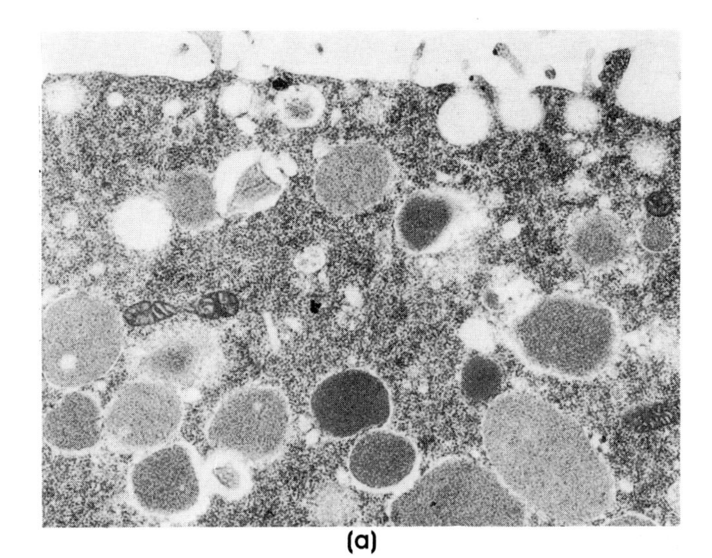

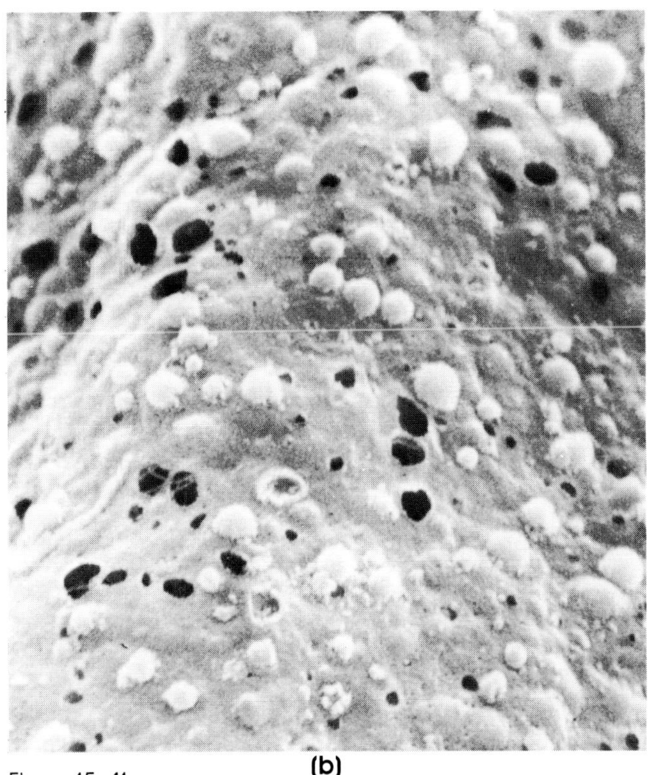

Figure 15–41
Exocytosis is captured in progress in these electron photomicrographs of the surface of an egg cell. In (*a*), several cortical vesicles are seen at various stages of fusion with the membrane (top of photo). The scanning photomicrograph (*b*) shows another form of exocytosis called "blebbing" (i.e., formation of bubble-like out pocketings of the plasma membrane). Magnifications: (*a*), 14,000 X; (*b*), 7000 X. (Courtesy of Dr. E. G. Pollock.)

may be achieved by **passive** and **active** mechanisms. The passive movements of solutes (by **diffusion**) and water (by **osmosis**) across the membrane occur principally as the result of **concentration gradients.** In some cases, diffusion through the membrane is **facilitated** by carrier molecules processing the properties of enzymes. Substances can also move through the plasma membrane against a concentration gradient; this consumes cellular energy (i.e., ATP is hydrolyzed in the process) and is termed **active transport.** For individual ions and small molecules, active transport is effected by membrane-associated **enzymes** acting as pumps; however, transport into and out of cells (called **endocytosis** and **exocytosis**) can occur in *bulk* through gross movements of the plasma membrane and takes the form of **pinocytosis, rhopheocytosis,** and **phagocytosis.**

References and Suggested Reading

Articles and Reviews

Branton, D., Membrane structure, in *Annual Review of Plant Physiology*, Vol. 20 (L. Machlis, W. R. Briggs, and R. B. Park, eds.) Annual Reviews Inc. Palo Alto, Calif. (1969), p 209.

Capaldi, R. A. A dynamic model of cell membranes. *Sci. Am. 230* (3) 26 (March 1974).

Collander, R., The permeability of plant protoplasts to small molecules. *Physiol. Plantarum 2*, 300 (1949).

Fox, C. F., The structure of cell membranes. *Sci. Am. 226* (2), 30 (Feb. 1972).

Gorter, E., and Grendel, F., On bimolecular layers of lipoids in the chromocytes of the blood. *J. Exp. Med. 41*, 439 (1925).

Hendler, R. W., Biological membrane ultrastructure. *Physiol. Rev. 51*, 66 (1971).

Kaplan, D. M., and Criddle, R. S., Membrane structural proteins. *Physiol. Rev. 51*, 249 (1971).

Keynes, R. D., Ion channels in the nerve-cell membrane. *Sci. Am. 240* (3), 126 (Mar. 1979).

Korn, E. D., Structure of biological membranes. *Science 153*, 1491 (1966).

Korn, E. D., Cell membranes: structure and synthesis, in *Annual Review of Biochemistry*, (E. E. Snell, ed.) Annual Reviews Inc., Palo Alto, Calif., 1969, p. 263.

Lodish, H. F., and Rothman, J. E., The assembly of cell membranes. *Sci. Am. 240* (1), 48 (Jan. 1979).

Luria, S. E., Colicins and the energetics of cell membranes. *Sci. Am. 233* (6), 30 (Dec. 1975).

Marchesi, V., Furthmayr, H., and Tomita, M., The red cell membrane. in *Annual Review of Biochemistry*, Vol. 45 (E. E. Snell et al., eds.), Annual Reviews Inc., Palo Alto, Calif., 1976.

Paul, S. M., and Skolnick, P., The red cell as a fluid droplet: tank tread-like motion of the human erythrocyte membranes in shear flow. *Science 202*, 894 (1978).

Pollock, E. G., Fine structural analysis of animal cell surfaces: membranes and surface topography. *Am. Zool. 18*, 25 (1978).

Raff, M. C., Cell surface immunology. *Sci. Am. 234* (5), 30 (May 1976).

Ravazzola, M., and Orci, L., Intercellular junctions in the rat parathyroid gland: A freeze-fracture study. *Rev. Biol. Cellulaire 28*, 137 (1977).

Robertson, J. D., The membrane of the living cell. *Sci. Am. 206* (4), 65 (Apr. 1962).

Rothman, J. E., and Lenard, J., Membrane asymmetry. *Science 195*, 743 (1977).

Satir, B., The final steps in secretion. *Sci. Am. 233* (4), 28 (Oct. 1975).

Sharon, N., Lectins. *Sci. Am. 236* (6), 108 (June 1977).

Staehelin, L. A., and Hull, B. E., Junctions between living cells. *Sci. Am. 238* (5), 141 (May 1978).

Stolinski, C., Freeze-fracture replication in biological research: development, current practice and future prospects. *Micron 8*, 87 (1977).

Books, Monographs, and Symposia

Bessis, M., Weed, R. I., and Leblond, P. F. (eds.), *Red Cell Shape*, Springer-Verlag, New York, 1973.

Branton, D., and Park, R. B. (eds.), *Papers on Biological Membrane Structure*, Little, Brown and Co., Boston, 1968.

Gomperts, B. D., *The Plasma Membrane*, Academic Press, London, 1977.

Haggis, G. H., *Introduction to Molecular Biology*, John Wiley & Sons, Inc., New York, 1964.

Hendler, R. W., *Protein Biosynthesis and Membrane Biochemistry*, John Wiley & Sons, Inc., New York, 1968.

Stein, W. D., *The Movement of Molecules Across Cell Membranes*, Academic Press, New York, 1967.

Weiss, L., *The Cell Periphery, Metastasis and Other Contact Phenomena*, North Holland Publishing Co., Amsterdam, 1967.

Chapter 16
THE MITOCHONDRION

Most of the energy-requiring, or **endergonic,** reactions carried out in cells either directly or indirectly consume *adenosine triphosphate* (ATP). This "energy-rich" substance is converted in the process usually to adenosine diphosphate (ADP) and occasionally to adenosine monophosphate (AMP) (Fig. 16–1). Cells have evolved three major ways of producing this vital source of chemical energy: (1) ATP is produced in the cytosol during the chain of exergonic reactions called **glycolysis** in which sugars are catabolized; (2) ATP may be produced within the **chloroplasts** of certain plant cells, utilizing the energy of sunlight; and (3) ATP may be produced within the **mitochondria** present in virtually all plant and animal cells by the **oxidation** of a variety of elementary substrates. It is for this reason that mitochondria are often referred to as the cell's "powerhouses."

ATP production in chloroplasts is considered in detail in Chapter 17 which also deals with photosynthesis, and further discussion of that subject is deferred until then. ATP synthesis in glycolysis occurs by **substrate-level phosphorylations** in which the phosphate is enzymatically transferred directly from the substrate to ADP to form ATP (see Chapters 9 and 10). Some ATP is generated in mitochondria by substrate-level phosphorylations but the greater amount is formed by special **electron transport system** (ETS) oxidation-reduction reactions. Although much of the present chapter will be devoted to the manner in which mitochondria function in ATP production, it is to be noted that mitochondria also play several other important roles; for example, in eucaryotic cells the mitochondria are also responsible for the oxidation of fatty acids and other lipids and are one of the sites for fatty acid chain lengthening. Some subunits of cytochromes are also synthesized in mitochondria, and final assembly occurs there.

In recent years, it has been clearly established that mitochondria contain their own genetic apparatus, as well as the machinery for the synthesis of an array of enzymatic and structural mitochondrial proteins, and that mitochondria are capable of semiautonomous proliferation within the cell. These subjects are dealt with here and also in Chapters 4 and 20.

Discovery of Mitochondria

Mitochondria were first observed and isolated from cells about 100 years ago when Köllicker mechanically teased these organelles from insect striated muscle tissue and studied their osmotic behavior in various salt solutions. Köllicker, whose work extended over several decades beginning around 1850, concluded that these "granules" were independent structures not directly connected to the interior structure of the cell. In 1890, Altmann defined stains specific for these granules and named them "bioblasts"; this term was superseded by Benda who introduced the expression "mitochondrion" (Greek: *mito-* = "thread" + *chon-*

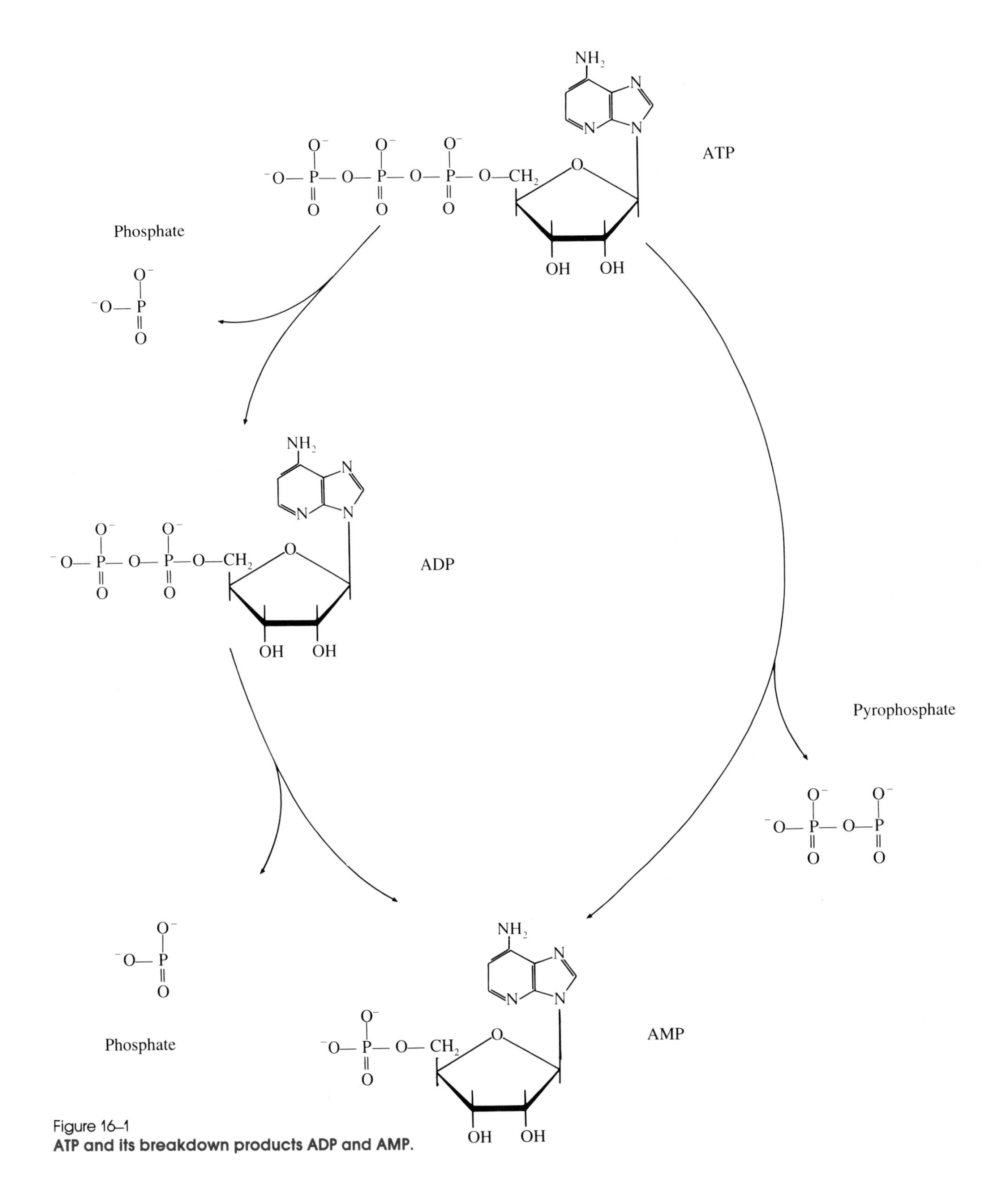

Figure 16–1
ATP and its breakdown products ADP and AMP.

drion = "granule") because of the threadlike appearance of these granules under the light microscope. In 1900, Michaelis introduced the use of the supravital dye Janus green B to specifically stain mitochondria to the exclusion of other cellular components and showed that oxidative reactions in the mitochondria caused color changes in the dye. Janus green B is still frequently employed as a cytological marker for mitochondria.

In 1910, Warburg showed that the "large granule" fraction isolated by low-speed centrifugation from tissues disrupted by grinding contained enzymes catalyzing oxidative cellular reactions, and Kingsbury in 1912 suggested that the mitochondria were the specific loci of the oxidation. Warburg's findings were confirmed in the 1930s by Claude, Bensley, and Hoerr, who employed more sophisticated methods to obtain a mitochondrial preparation essentially free of contaminating cell structures. They disrupted liver tissue using procedures almost identical to those currently used to prepare a conventional "homogenate" (see Chapter 12) and isolated the mitochondria by repeated differential centrifugation; the final mitochondrial preparation was then examined biochemically. At about the same time, Sir Hans Krebs elucidated the various reactions of the tricarboxylic acid (TCA) cycle (or Krebs cycle; see later) and by 1950, Lehninger, Green, Kennedy, Hogeboom, and others had clearly shown that these reactions, as well as those of fatty acid oxidation and oxidative phosphorylation (i.e., ATP generation), were properties of mitochondria.

Structure of the Mitochondrion

The size, shape, and structural organization of mitochondria, as well as the number of these organelles per cell and their intracellular location, vary considerably depending on the organism, tissue, and physiological state of the cell examined.

Some cells, usually unicellular organisms, contain a single mitochondrion. Figure 16–2 shows a photomicrograph of the single mitochondrion in the motile swarm spore of *Blastocladiella emersonia,* a fungus, and a model of the single mitochondrion of *Polytomella agilis,* an alga. At the other end of the scale are cells such as *Chaos chaos,* an amoeba, which contains several hundred thousand mitochondria. Cells of higher animals also contain various numbers of mitochondria. Sperm cells have fewer than 100 mitochondria. Kidney cells generally contain less than 1000, while liver cells can contain several thousand. As described in Chapter 1, procaryotic cells such as bacteria and blue-green algae do not contain mitochondria. The functions associated with mitochondria are carried out in the cytosol or associated with the cell (plasma) membrane in these organisms.

The distribution of the mitochondria in the cell may vary, although in most cells the distribution appears to be at random. In *Blastocladiella* (Fig. 16–2), the single mitochondrion is at the base of the flagellum. A concentration of mitochondria also appears in metabolically active areas of cells (Fig. 16–3*a*). In epithelial cells lining the lumen of the small intestine, the mitochondria occur in greater numbers near the surface of the cell adjacent to the lumen (where active absorption of digestive products is occurring). In general, when mitochondria are present in greater numbers in one part of the cell than another, it is usually near a site where significant ATP utilization is occurring. For example, in muscle tissue the mitochondria are aligned in rows parallel to the contractile fibrils (Fig. 16–3*b*). In many plant cells, cyclosis, the active streaming of the cytoplasm about the cells, tends to distribute the mitochondria in a uniform manner.

The number and distribution of mitochondria in a cell are closely related to the activity of the cell and its organelles. Cells that are actively growing, producing especially large amounts of some product such as digestive enzymes, actively transporting materials into the cell, or undergoing movement may actually increase the number of mitochondria during periods of activity and reduce these numbers during periods of quiescence. Yeast that are grown anaerobically produce cells in successive generations with fewer and fewer mitochondria per cell. However, cells that have been grown in this manner without oxygen will rapidly produce greater numbers of mitochondria per cell when oxygen and appropriate nutrients are added to the culture. An increased rate of cell growth and division is also observed as greater numbers of mitochondria are able to produce more ATP to facilitate absorption of nutrients and synthesis of cell constituents.

The origins of mitochondria appear to be somewhat diverse. By light microscopy and cinematography, they have been observed dividing by a fissionlike process. The mitochondrion appears to stretch out in a tubelike manner, constrict near the center to form a dumbbell-shaped struc-

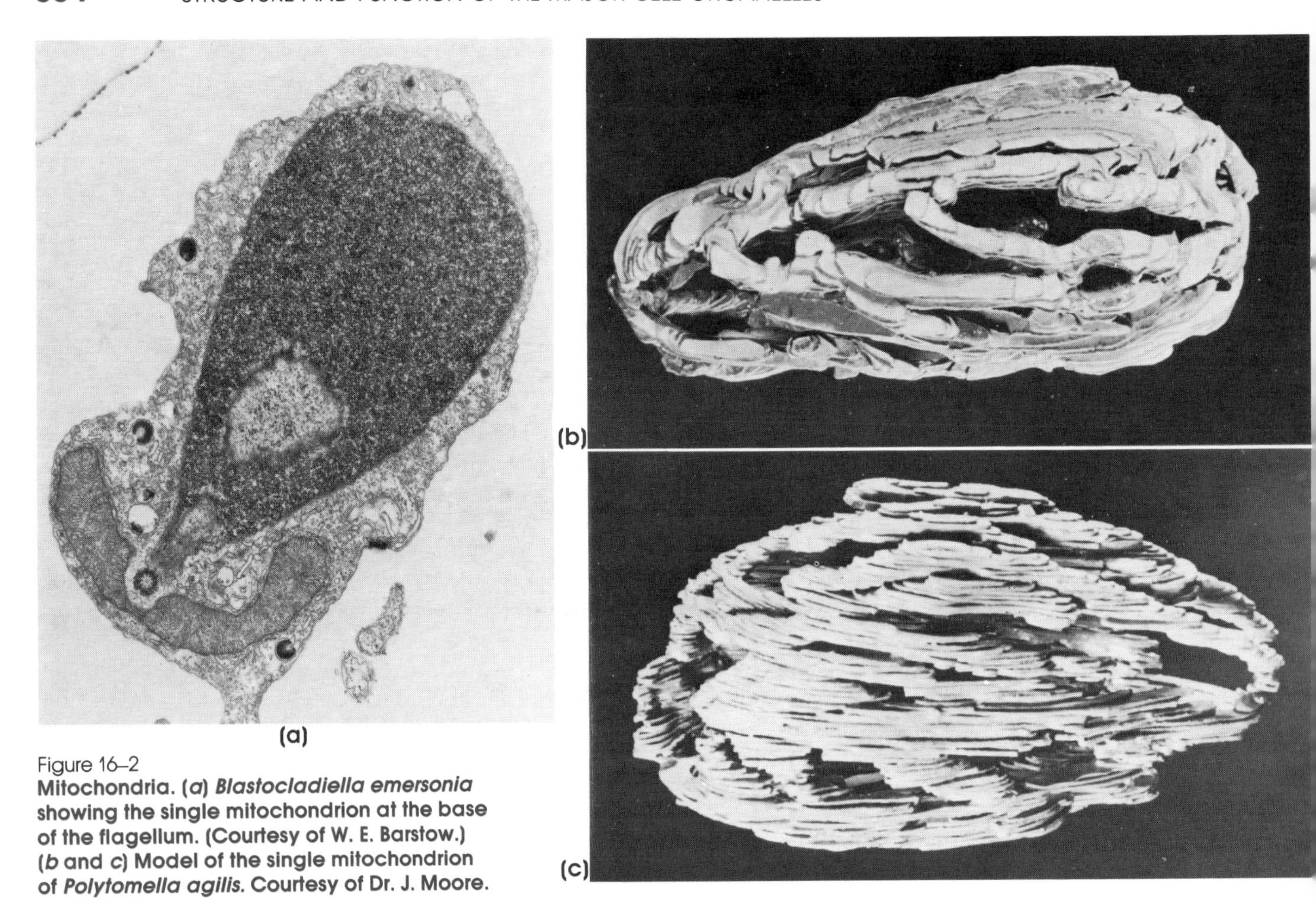

Figure 16–2
Mitochondria. (*a*) *Blastocladiella emersonia* showing the single mitochondrion at the base of the flagellum. (Courtesy of W. E. Barstow.) (*b* and *c*) Model of the single mitochondrion of *Polytomella agilis*. Courtesy of Dr. J. Moore.

ture, and then pinch off to form two halves that round up (Fig. 16–4). Mitochondria have also been observed fusing together. There is also evidence to suggest that they may arise from microbodies and from a number of different membranes. D. B. Roodyn and D. Wilkie in 1968 obtained electron photomicrographs that appear to indicate that the mitochondria could arise from the endoplasmic reticulum, the plasma membrane, and the nuclear envelope. During cell division, there does not appear to be a specific method for partitioning the mitochondria into the two developing daughter cells. In most cells the cytoplasm becomes very turbulent during cell division, and the random distribution of the organelles is enhanced. In some instances, the mitochondria appear to stick to the spindle fibers. In certain yeasts during cell division, one developing daughter cell appears to be ''blown out'' from the parent cell, a process known as **budding.** Rather than an equal division of the cytoplasm, little parent cytoplasm enters the bud along with the divided nucleus. Occasionally, one of these buds will fail to obtain a mitochondrion. The subsequent growth of this ''deficient'' bud cell is much slower than normal cells, since the cell must essentially rely on the glycolytic process in the cytosol for ATP synthesis. Fletcher and Sanadi in 1961 and Wilson and Dove in 1965 studied the turnover rate for mitochondria by using labeled compounds. The first workers labeled proteins and lipids and followed the dis-

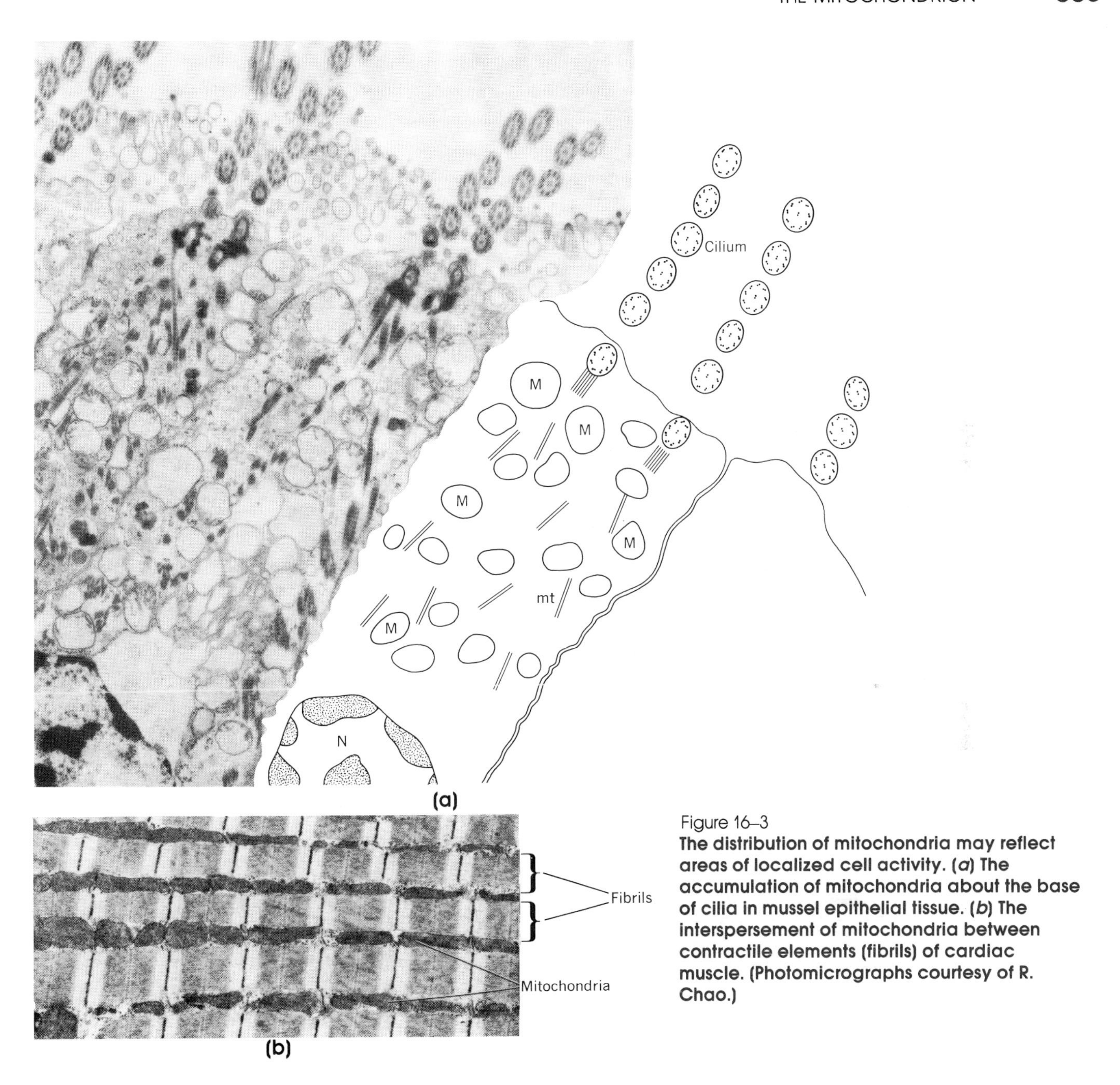

Figure 16–3
The distribution of mitochondria may reflect areas of localized cell activity. (*a*) The accumulation of mitochondria about the base of cilia in mussel epithelial tissue. (*b*) The interspersement of mitochondria between contractile elements (fibrils) of cardiac muscle. (Photomicrographs courtesy of R. Chao.)

appearance of the label with time. The latter workers utilized a technique with tritiated water. From both studies, it would appear that there is a complete turnover of mitochondria every 20 days.

There have been some interesting theories proposed for the evolutionary origin of mitochondria in eucaryotic cells. Mitochondria possess some DNA and ribosomes of their own and have a number of functions independent of the

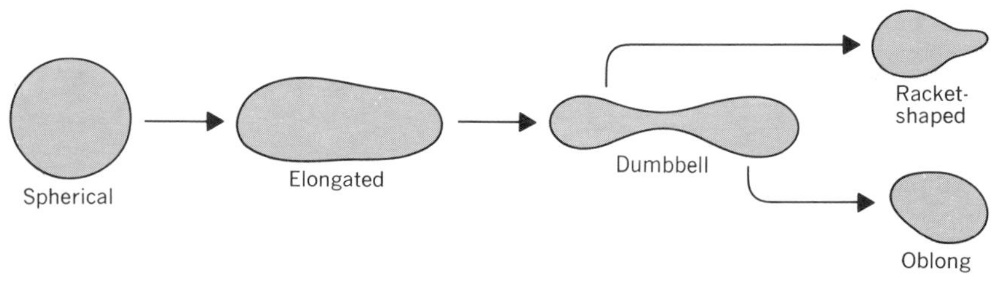

Figure 16–4
Stages in the division of a mitochondrion.

nuclear and ribosomal processes of the cell. (Protein synthesis and ribosome formation in mitochondria are discussed in Chapter 21.) The idea that mitochondrial ancestors could have been procaryotic cells is supported by the fact that mitochondrial DNA, like procaryotic DNA, is a single molecule in the shape of a circle (i.e., no free ends). Likewise, mitochondrial ribosomes are made up of subunits having molecular weights and sedimentation coefficients more similar to procaryotic ribosomes than to eucaryotic ribosomes. It has been proposed that one aerobic procaryote invaded or infected another procaryote and that, through time, the infective procaryote evolved morphologically and physiologically into the mitochondrion. This fascinating theory of the evolution of mitochondria must be evaluated in the light of evidence that indicates that most mitochondrial proteins are encoded by nuclear DNA and are synthesized by ribosomes in the cytosol or on the endoplasmic reticulum. It would therefore appear that there is a greater transfer of information from DNA in the cell nucleus to the mitochondrion than arises from DNA found in the mitochondrion itself.

The size and shape of mitochondria, like the number in a cell, vary according to the organism's tissues and the physiological state of the cell. Most common mitochondria are between 0.5 and 1.0 μ in diameter and may be up to 7 μ in length. Usually, the smaller the number of mitochondria per cell, the larger the individual organelles are likely to be. The mitochondria may be spherical, ovoid, cylindrical, dumbbell-shaped, racket-shaped, or irregular with branches (Fig. 16–5). In contrast to the gross structure, the internal structure of the mitochondrion is unique.

The light microscope reveals little about the structure of mitochondria, since these organelles are so small. However, with the electron microscope, the mitochondrion is revealed as a highly complex organelle. Regardless of source, all mitochondria exhibit features in common, and we may therefore describe a "generalized" organelle (Fig. 16–6). The mitochondrion contains two distinct membranes, one inside the other, called the **outer** and **inner membranes** (Fig. 16–7). The inner membrane separates the internal volume into two phases: the **matrix,** which is a gel-like fluid volume enclosed by the inner membrane, and the fluid **intermembrane space** between the two membranes. The

Figure 16–5
Electron photomicrographs of *Fucus* mitochondria showing various morphologies. (Courtesy of Dr. E. G. Pollock.)

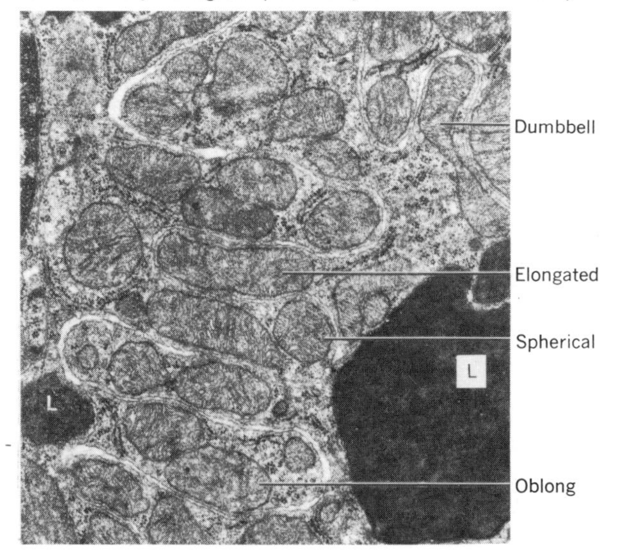

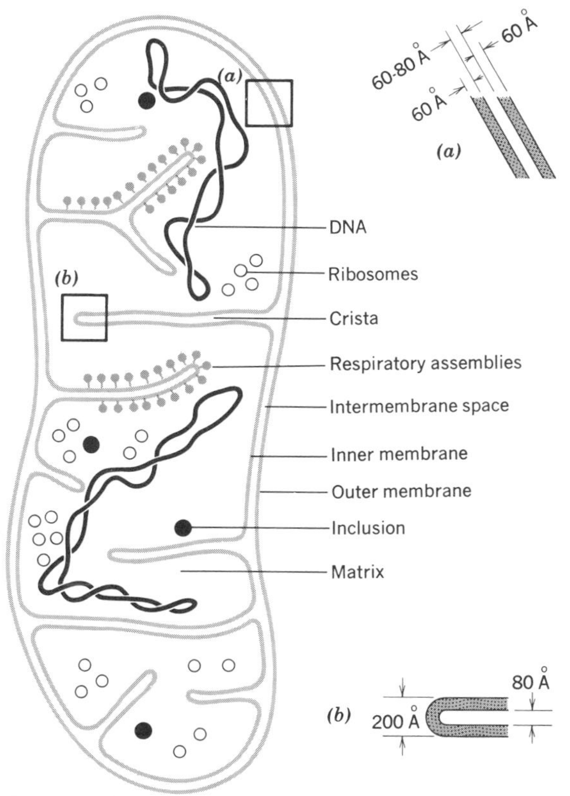

Figure 16–6
The generalized mitochondrion.

outer and inner membranes themselves, as well as the matrix and intermembrane space, contain a variety of enzymes (Table 16–1). The matrix contains a number of the enzymes of the Krebs cycle (tricarboxylic acid cycle, or TCA cycle) as well as salts and water. Suspended in the matrix are circular strands of DNA (Fig. 16–8) and ribosomes. A number of other inclusions have been described for mitochondria from diverse tissues. These include filaments and tubules, what appear to be crystalline proteins, and a number of small, nonribosomal granules. The granules vary in density, and little is known about their function except that some of them appear to strongly bind divalent cations.

The intermembrane space has been identified as containing a couple of enzymes (Table 16–1), but generally it appears to be devoid of inclusions. Several types of connections through this space that join the membranes have been reported.

The inner and outer membranes are distinctly different both in structure and function. Although determination of

Figure 16–7
Transmission EM of inner and outer membranes of the mitochrondrion.

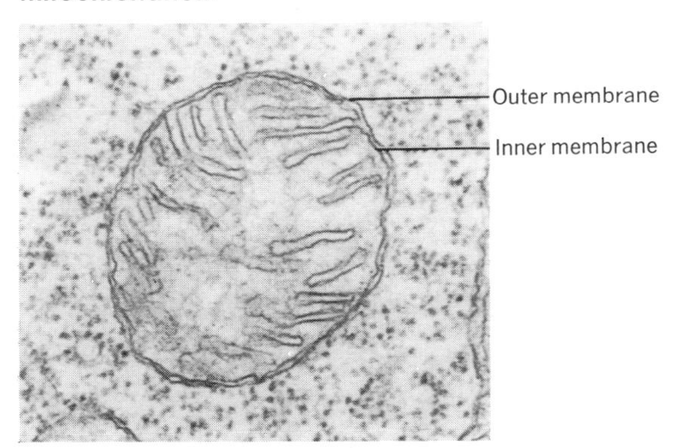

Figure 16–8
DNA in the mitochondrion. M, mitochondrion; N, nucleus. (Courtesy of Dr. E. G. Pollock.)

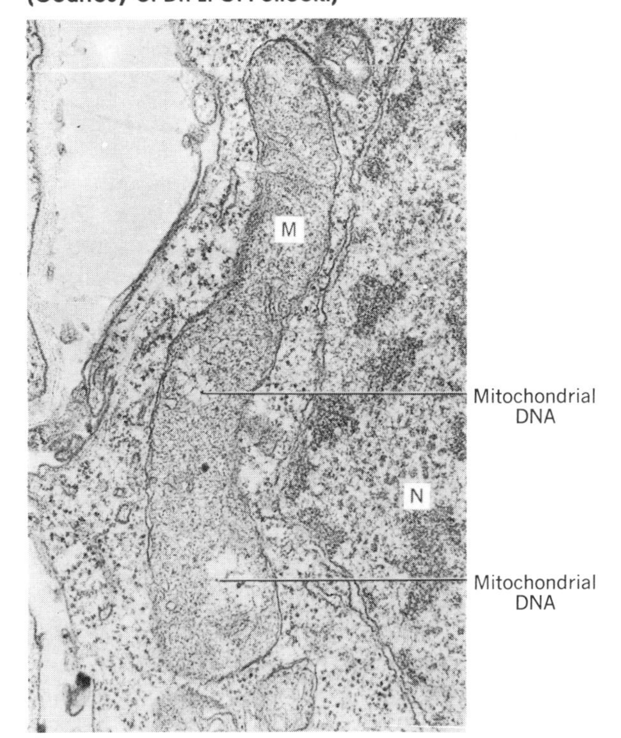

Table 16–1
Location of some Mitochondrial Enzymes

Location / Enzyme
Outer membrane:
Monoamine oxidase
Fatty acid thiokinases
Kynurenine hydroxylase
Rotenone-insensitive cytochrome *c* reductase
Space between the membranes:
Adenylate kinase
Nucleoside diphosphokinase
Inner membrane:
Respiratory chain enzymes
ATP-synthesizing enzymes
α-Keto acid dehydrogenases
Succinate dehydrogenase
D-β-Hydroxybutyrate dehydrogenase
Carnitine fatty acyl transferase
Matrix:
Citrate synthase
Isocitrate dehydrogenase
Fumarase
Malate dehydrogenase
Aconitase
Glutamate dehydrogenase
Fatty acid oxidation enzymes

Source. Courtesy A. Lehninger, *Biochemistry*, 2nd ed. Worth Publishing Co., New York, 1975, p. 512 (copyright © Worth Publishing Co.).

the thickness of the membranes is difficult from electron photomicrographs because various fixatives cause different amounts of swelling, the inner membrane appears to be somewhat thicker (60–80 Å) than the outer membrane (about 60 Å). The inner membrane has a greater surface area because it is folded or extended into the matrix. These projections are called **cristae** and vary in number and shape. With distinct exceptions, it would appear that the cristae in higher animal cells are broad folds that sometimes almost bridge the matrix. Usually the cristae lie almost parallel to one another across the short axis of the mitochondrion, but they may run longitudinally in some organelles or form a branching network. In protozoa and many plants, the cristae form a set of tubes that project into the matrix from all sides, sometimes twisting in different directions (Fig. 16–9). The number of cristae may increase or decrease depending upon aerobic activity. Active aerobic tissues producing great amounts of ATP generally have mitochondria with extensive cristae.

The outer and inner mitochondrial membranes have been analyzed by transmission electron microscopy using conventional thin-section techniques, negative staining procedures, and freeze-cleavage methods. Each method has provided some insight into the structure of these membranes. Since both outer and inner membranes (during pioneering electron microscopy) appeared to be double layers in high-resolution photomicrographs, they were presumed to fit nicely into the *unit membrane* structural model (see Chapter 15). However, the apparent thickness of the membranes, the finding that the double-layer appearance remained after extraction of lipids, and the greater than average amount of protein in these membranes made this simple explanation unlikely. A simple interpretation based upon current evidence would indicate that the mitochondrial membranes are composed of globular proteins contained within a lipid framework (Fig. 16–10). The evidence from freeze cleavage techniques is in agreement with this concept. Figure 16–11 is a detailed diagram of the layers of the outer and inner membranes showing the granulation ascribed to the protein units of the membrane.

Although the general model for the structure of the two membranes is about the same, there are some notable differences. The inner membrane can be stripped of 60% of its protein by treatment with 7% acetic acid. This would indicate that a large proportion of the inner membrane protein is extrinsic. Further, the matrix side of the inner membrane contains many spheres larger (80–90 Å), and more

Figure 16–9
Various arrangements of cristae within the mitochondrion.

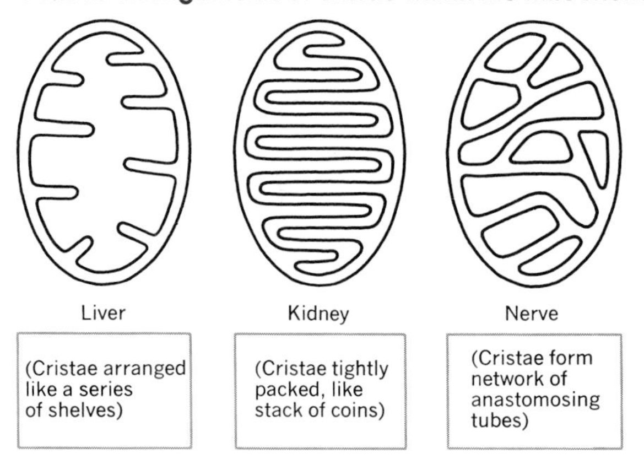

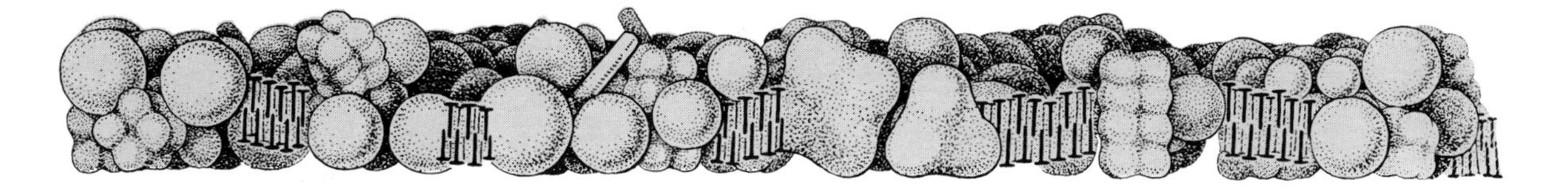

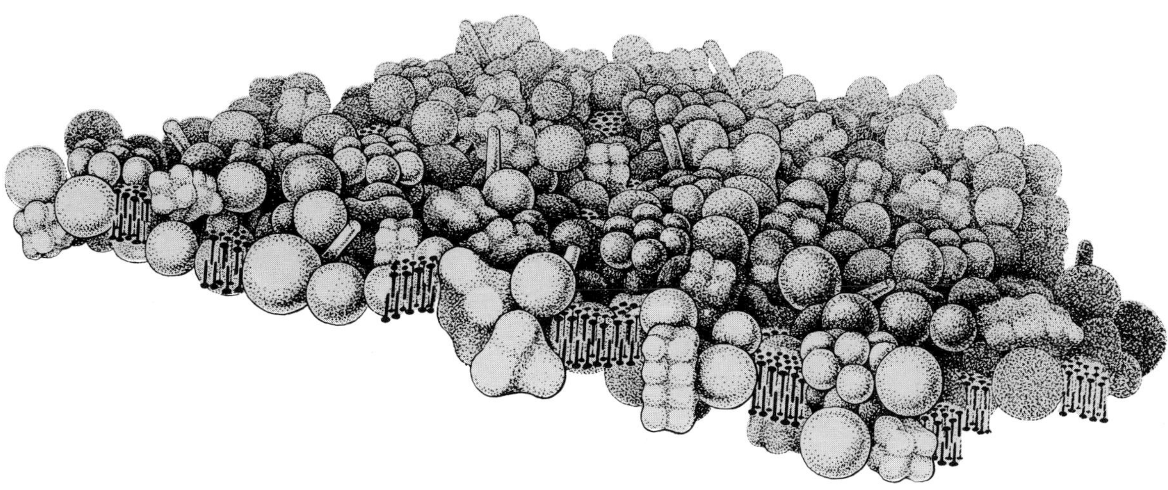

Figure 16–10
Sjostrand and Barajas model of the mitochondrion membrane. In this model, mitochondrial membrane enzymes and other proteins are depicted as globular bodies of various shapes. The lipid bilayer is shown as rows of "T-shaped structures. (Courtesy of F. Sjostrand, copyright Academic Press, *J. Ultrast. Res. 32,* 298, 1970.)

distinct than any of the granules on any of the other three membrane surfaces. These **inner membrane spheres,** first described by Fernández-Morán in 1962, have been revealed in good detail by negative staining (Fig. 16–12). They appear to be spaced regularly across the membrane and borne on stalks. It is these inner membrane spheres that have been identified as the primary site of oxidative phosphorylation and electron-transport-system generation of ATP.

In addition to their structural differences, the outer and inner membranes differ significantly in their permeability. The outer membrane appears to be very porous and freely permeable to a wide variety of substances up to a molecular weight of about 10,000. When isolated, the intramembranous fluid therefore tends to reflect the water-soluble small-molecular-weight components of the cytosol or suspending medium. The inner membrane has a very limited permeability especially to substances with molecular weights above 100 to 150.

The difference in permeability between the two membranes is utilized to separate the outer membrane from the inner. Although several variations on the method exist, the basic technique is to disrupt the cell and isolate intact mitochondria by differential, density gradient, or zonal centrifugation. The mitochondria are then placed in a hypotonic solution that causes them to swell. Usually, a phosphate buffer is used for this purpose. As the mitochondria swell, the outer membrane ruptures and fragments; the inner membrane also swells, causing the loss of cristae but normally

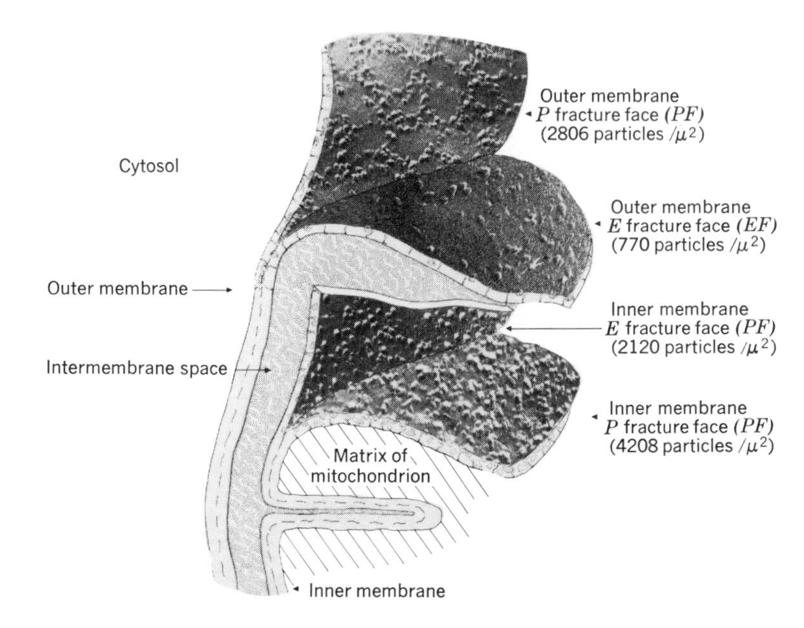

Figure 16–11
Diagram of the fracture faces of the outer and inner membranes of the mitochondrion. (Courtesy of L. Packer, copyright New York Academy of Science, *Ann. N.Y. Acad. Sci. 227*, 167, 1974.)

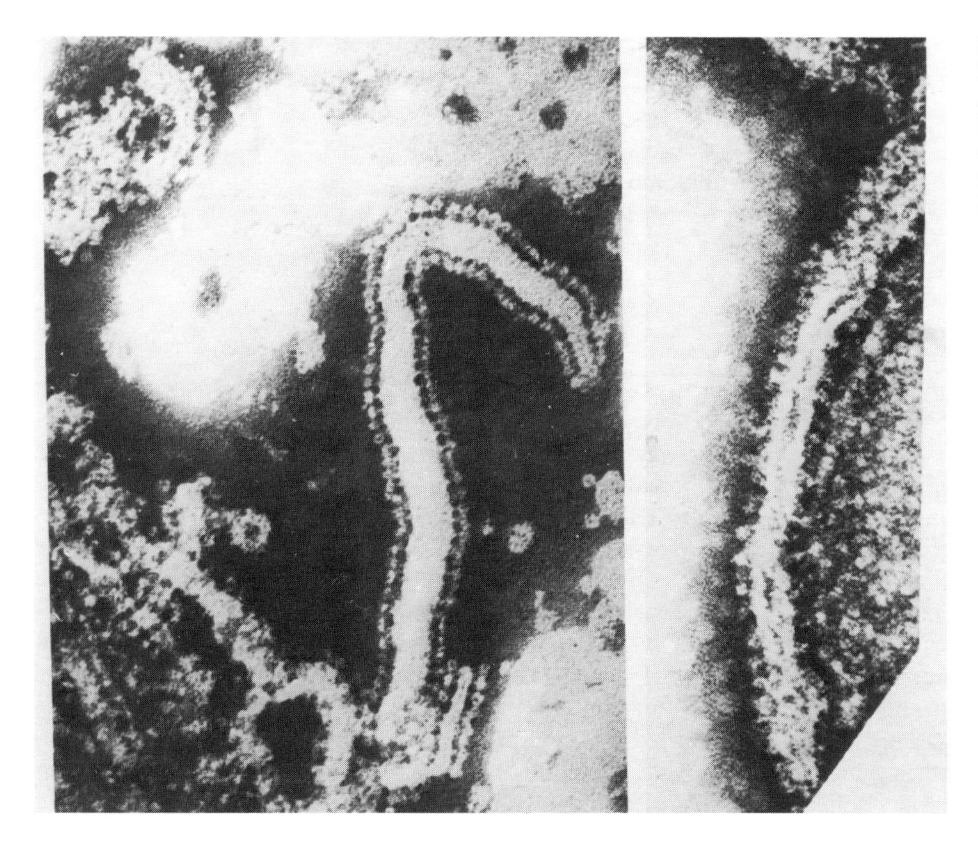

Figure 16–12
Electron photomicrographs of thin sections of mitochondria showing the inner membrane with spheres borne on stalks on the inner surface. (Courtesy B. Chance, University of Pennsylvania.)

does not break. The mitochondria are then placed in a hypertonic solution, causing the inner membrane matrix to shrink. The hypertonic solution is frequently a sucrose solution containing ATP and Mg^{++}. Since the inner membrane appears to be attached to the outer membrane at several locations (probably by proteinaceous connecting strands), some workers use digitonin, EDTA, or sonication to aid the separation. Resuspension in an isotonic solution allows the inner membrane matrix to reestablish typical morphology. Separation of the membrane fractions can be done by differential centrifugation. The results of a typical separation are shown in Figure 16–13.

The differences and relationships between the membranes are also apparent when the metabolic state of the mitochondria is altered. These reversible **conformational states,** illustrated in Figure 16–14, were originally proposed by C. R. Hackenbrock. The **orthodox conformation** is the normal state and is transformed to the **condensed conformation** by binding of ADP to ADP-ATP translocase molecules on the inner membrane. The **swollen** and **contracted conformations** produced by ion shifts illustrate drastic osmotic changes.

The unique structure of the mitochondrion has prompted many investigators to try to correlate structure with func-

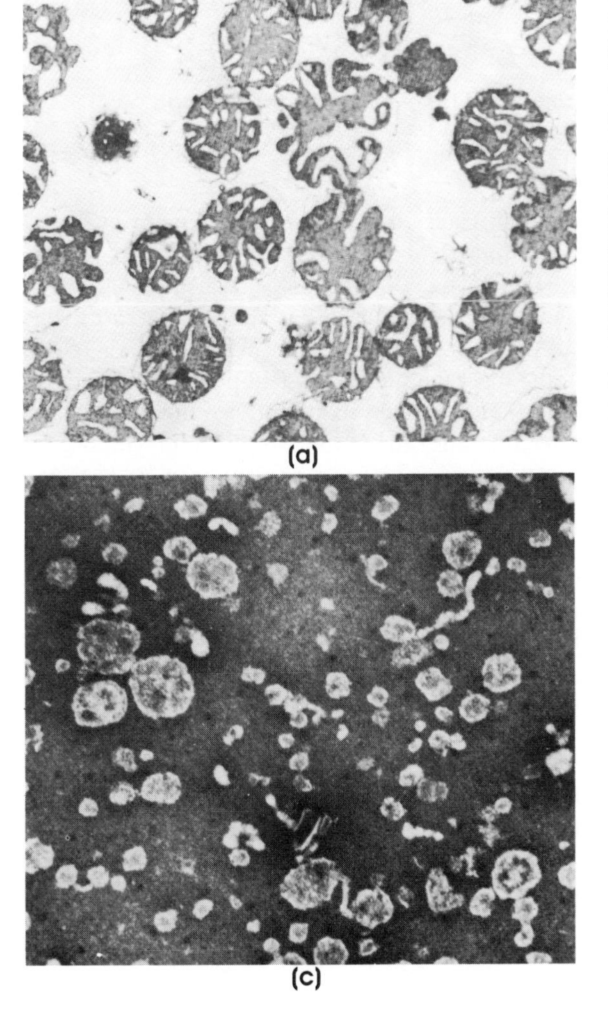

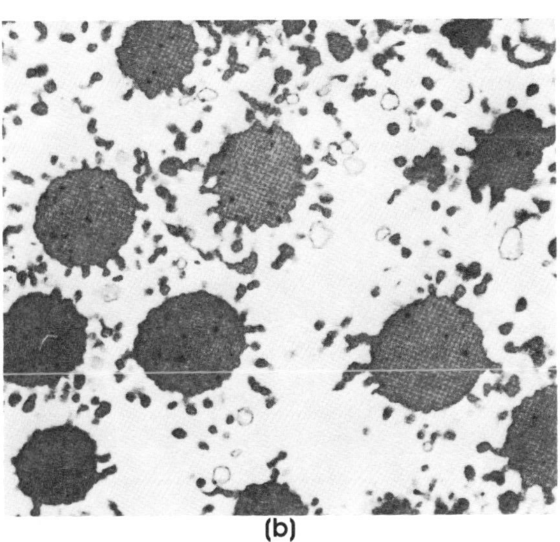

Figure 16–13
Mitochondrial preparations from rat liver. (*a*) Freshly isolated whole mitochondria. (*b*) Inner membrane and matrix isolated after digitonin treatment and centrifugation. (*c*) Outer membrane preparation isolated after digitonin treatment and centrifugation. (Courtesy of C. A. Schnaitman; copyright The Rockefeller Press, *J. Cell Biol. 38,* 170, 1968.)

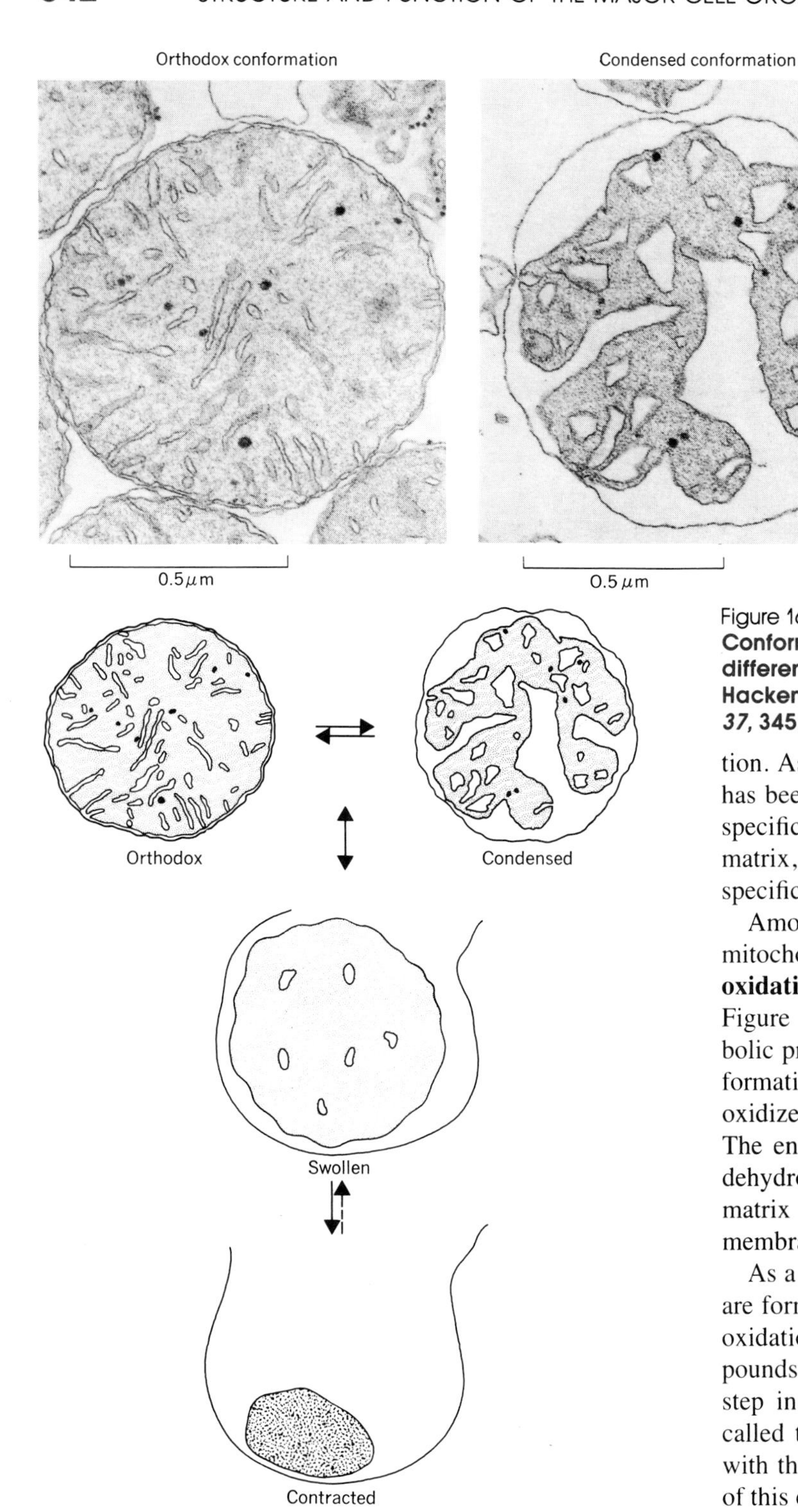

Figure 16–14
Conformational states of mitochondria induced by different physiological states. (Courtesy of C. R. Hackenbrock; copyright The Rockefeller Press, *J. Cell Biol.* *37*, 345, 1968.)

tion. As listed in Table 16–1, the location of many enzymes has been determined, and it is generally possible to assign specific metabolic functions to the intermembrane space, matrix, and inner membrane spheres. In some cases, the specific surface of the membrane can be assigned a function.

Among the most thoroughly studied processes unique to mitochondria are **substrate oxidation, respiratory chain oxidation-reductions,** and **oxidative phosphorylation.** In Figure 16–15, these major events are diagrammed. Metabolic products of reactions in the cytosol (such as pyruvate formation during glycolysis) enter the mitochondrion to be oxidized by the Krebs or tricarboxylic acid cycle enzymes. The enzymes for each of these reactions (except succinic dehydrogenase; see later) are believed to be localized in the matrix or on or near the matrix facing surface of the inner membrane.

As a result of the Krebs cycle oxidations, CO_2 and water are formed as waste products, and a number of specialized oxidation-reduction compounds are reduced. These compounds (e.g., $NADH_2$, see Chapter 10) act as the initial step in a sequence of oxidation and reduction reactions, called the respiratory chain, that is specifically associated with the inner membrane of the mitochondrion. The result of this chain of reactions is the reduction of O_2 to form H_2O

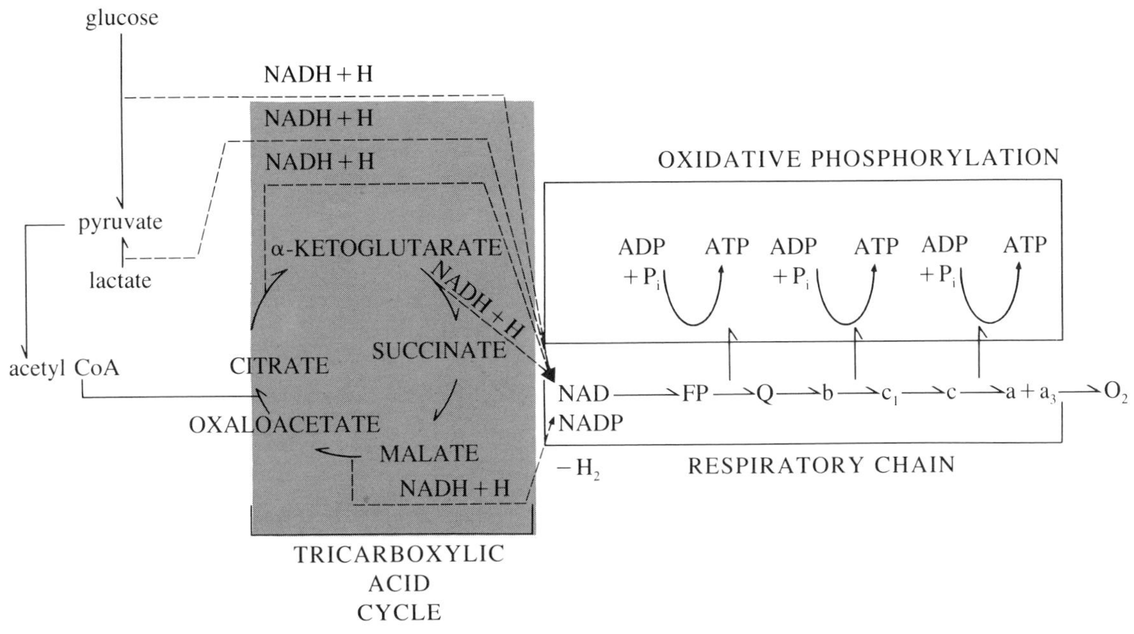

Figure 16–15
Relationship between substrate oxidation, respiratory chain oxidation-reductions, and oxidative phosphorylations.

as a "waste product" and the induction of a third major process called **oxidative phosphorylation.** Oxidative phosphorylation, the net result of which is the production of ATP, is intimately associated with the inner membrane spheres.

In effect, pyruvate and other small molecules of cytosol metabolism must diffuse through the very permeable outer membrane and across the intermembrane space. It is upon entering the inner membrane that the three major reaction sequences (Krebs cycle, respiratory chain oxidations-reductions, and oxidative phosphorylation) begin. During glycolysis in the cytosol, reduced pyridine nucleotides (e.g., $NADH_2$) are produced, as are others (e.g., $NADPH_2$) in the hexose monophosphate shunt and other metabolic pathways. These reduced compounds may also pass through the outer mitochondrial membrane or transfer their reductive capacity to the respiratory chain compounds and thereby initiate the latter two major reaction sequences: respiratory chain oxidations-reductions and oxidative phosphorylation. Each of these three major sequences is described in more detail below.

Tricarboxylic Acid Reactions

The **tricarboxylic acid cycle** is frequently called the Krebs cycle because the major steps of the cyclic reactions were first proposed in 1937 by H. A. Krebs. At that time, radioactively labeled compounds were not available for experimentation, the cellular site of the oxidation reactions was not determined, and even the compound that initiates the cycle (i.e., citric acid) was not known with certainty. Krebs' experiments and his analysis of the data were outstanding. It was not until 1948, when E. P. Kennedy and A. L. Lehninger homogenized rat liver and fractionated and assayed the enzyme activity of its components, that it was finally shown that the Krebs cycle reactions were localized in the mitochondria.

The Krebs cycle reactions are utilized by the cell to further metabolize a number of products of other reactions in the cytosol. These can be products as diverse as amino acids, fatty acids, and pyruvate. However, the greatest metabolic merit of the cycle is its oxidation of the pyruvate—the product of carbohydrate metabolism in the cy-

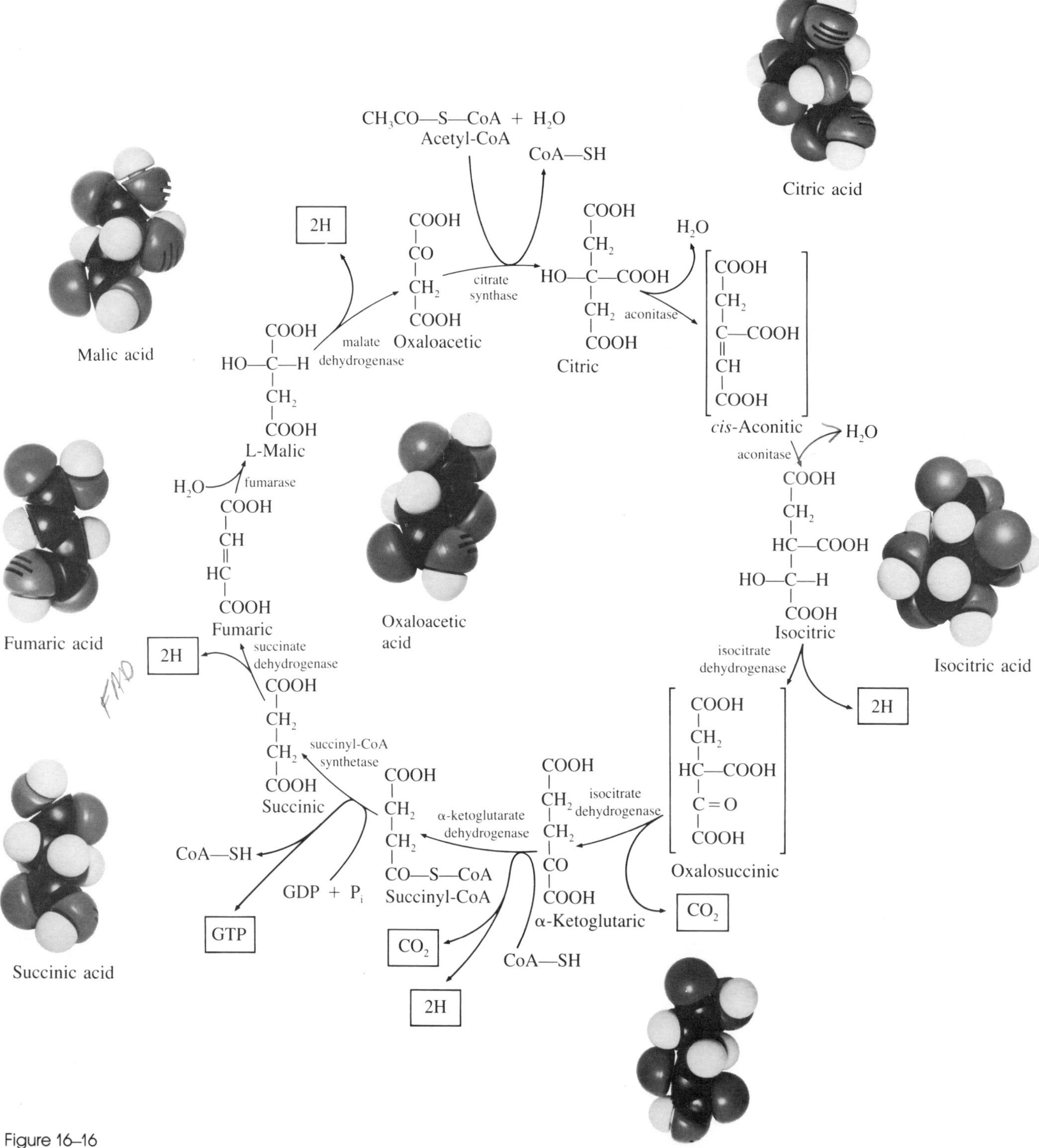

Figure 16–16
The Krebs cycle (tricarboxylic acid cycle). Photos by Kathy Bendo.

tosol. The oxidation of this compound in the TCA cycle provides the reducing power to subsequently make significant amounts of ATP. The reactions are summarized in Figure 16–16 and are described below.

The oxidation of pyruvate is brought about by a complicated set of reactions catalyzed by the **pyruvate dehydrogenase complex.** This multienzyme complex is located in the matrix of the mitochondrion. The first step is the decarboxylation of the pyruvate to yield CO_2 and an α-hydroxyethyl unit attached through thiamine pyrophosphate (TPP) (a coenzyme) to that enzyme (E). That is,

$$\begin{array}{c} E_1\text{—TPP} \\ + \\ CH_3COCOOH \\ \downarrow \searrow CO_2 \\ E_1\text{—TPP—CHOH—CH}_3 \end{array} \qquad (16\text{–}1)$$

The hydroxyethyl group is then dehydrogenated (oxidized) to form an acetyl group, which is then transferred to one of the sulfur atoms of lipoic acid.

$$\begin{array}{c} E_1\text{—TPP—CHOH—CH}_3 \\ + \\ E_2\text{(lipoyl, S—S ring)} \\ \downarrow \\ E_1\text{—TPP} \\ + \\ E_2\text{(lipoyl: S—C(=O)—CH}_3\text{, SH)} \end{array} \qquad (16\text{–}2)$$

The acetyl group is then enzymatically transferred to coenzyme A (Fig. 3–8) forming *acetyl CoA,* which separates from the enzyme complex. One hydrogen from the lipoyl group and one from coenzyme A are transferred to a flavin adenine dinucleotide (FAD) coenzyme (Fig. 3–8) on the enzyme complex.

$$\begin{array}{c} E_2\text{(lipoyl: S—C(=O)—CH}_3\text{, SH)} \\ + \\ \text{CoA—SH} \\ + \\ E_3\text{—FAD} \\ \downarrow \\ E_2\text{(lipoyl, S—S ring)} \\ + \\ E_3\text{—FADH}_2 \\ + \\ CH_3CO\text{—S—CoA} \end{array}$$

The resulting $FADH_2$ complex is then reoxidized by the transfer of the hydrogens to NAD^+.

$$FADH_2 + NAD^+ \longrightarrow FAD + NADH + H^+ \qquad (16\text{–}3)$$

The $NADH + H^+$ is then freed of the enzyme complex. Thus the overall reaction is

$$\begin{array}{c} \text{Pyruvate} + NAD^+ + \text{CoA} \\ \downarrow \\ \text{acetyl-CoA} + NADH + H^+ + CO_2 \end{array} \qquad (16\text{–}4)$$

The $\Delta G^{\circ\prime}$ (see Chapter 9) is -8.0 kcal mole^{-1}, indicating that the reaction is highly exergonic, with the equilibrium lying well to the right. In most animal tissues, the reaction is considered nonreversible.

The first step of the Krebs cycle is a *condensation* reaction between acetyl CoA and oxaloacetate, a product of the cycle itself. The reaction is catalyzed by citrate synthetase, with citrate produced and CoA freed for reutilization in the prior reaction.

$$\text{Acetyl-CoA} + \text{oxaloacetate} + H_2O \longrightarrow \text{citrate} + \text{CoA} \qquad (16\text{–}5)$$

The $\Delta G^{\circ\prime}$ is -7.7 kcal mole^{-1}. This reaction is the "pacemaker" or primary rate-limiting reaction of the cycle. The rate is controlled by the availability of acetyl CoA and oxaloacetate. Succinyl CoA, the product of a later step in the cycle, is a competitive inhibitor of this reaction, since it competes with acetyl CoA for the active site on the enzyme.

The enzymatic conversion of citrate to isocitrate is a two-step conversion in which the intermediate, *cis*-aconitate remains attached to the enzyme *aconitase* and therefore is frequently not shown in the Krebs cycle. The equilibrium of this reaction lies toward citrate, $\Delta G^{\circ\prime}$ is 1.59 kcal mole^{-1}, but the isocitrate is rapidly oxidized in the next step, thus shifting the direction of the reaction by removal of the product.

$$\text{Citrate} \xrightarrow{H_2O} \left[\textit{cis}\text{-aconitic acid}\right] \xrightarrow{H_2O} \text{isocitrate} \qquad (16\text{–}6)$$

The oxidation of isocitrate to α-ketoglutarate is also a two-step process, with the intermediate remaining attached to the enzyme *isocitric dehydrogenase*. The first part of the reaction is an oxidation with the two hydrogens being transferred to NAD$^+$. Actually, there are two isocitric dehydrogenases in the mitochondrial matrix. One is linked to NAD$^+$ and the other to NADP$^+$. An NADP$^+$-linked isocitric dehydrogenase is also found in the cytosol. However, it appears that the NAD$^+$-isocitric dehydrogenase is the most active form in Krebs cycle reactions. This form of the enzyme is allosteric and specifically stimulated by ADP. When large amounts of ATP are consumed in the cell, ADP would be produced, acting as a stimulus for the Krebs cycle. Alternately, when ATP accumulated there would be little ADP and the TCA cycle reactions would slow down owing to inactivity of ADP-dependent NAD$^+$-isocitric dehydrogenase. This point of rate control is considered secondary however to the citrate synthetase step.

The second part of the reaction catalyzed by isocitric dehydrogenase is the decarboxylation of the β-carboxyl group. The $\Delta G^{\circ\prime}$ for the entire reaction is -5.0 kcal mole^{-1}.

$$\text{Isocitrate} \xrightarrow{NAD^+ \rightarrow NADH + H^+} \text{Oxalosuccinate} \xrightarrow{\rightarrow CO_2} \alpha\text{-Ketoglutarate} \qquad (16\text{–}7)$$

The oxidation of α-ketoglutarate to succinyl CoA is very similar in reaction sequences and enzyme complex to the conversion of pyruvate to acetyl CoA. The enzyme complex is called the **α-ketoglutarate dehydrogenase complex.** In this sequence of reactions, CO_2 is removed by first complexing with thiamine pyrophosphate. During addition of CoA, two hydrogen atoms removed from lipoic acid and coenzyme A reduce NAD$^+$ to NADH + H$^+$. The $\Delta G^{\circ\prime}$ is -8.0 kcal mole^{-1}.

$$\alpha\text{-Ketoglutarate} + NAD^+ + \text{CoA} \rightleftharpoons \text{succinyl-CoA} + CO_2 + NADH + H^+ \qquad (16\text{–}8)$$

The removal of the CoA is coupled with a substrate level phosphorylation of GDP catalyzed by **succinyl CoA synthetase.**

$$\text{Succinyl CoA} + P_i + \text{GDP} \longrightarrow \text{succinate} + \text{GTP} + \text{CoA} \qquad (16\text{–}9)$$

The phosphate is actually attached to the enzyme-succinyl-CoA complex first and then transferred to GDP. In *E. coli*, the phosphate is transferred to ADP; in animal and most plant tissues GTP is formed and then the GTP donates the phosphate to ADP to form ATP.

$$\text{GTP} + \text{ADP} \rightleftharpoons \text{ATP} + \text{GDP} \qquad (16\text{–}10)$$

The ΔG° is -0.7 kcal mole^{-1}.

Succinic dehydrogenase catalyzes the oxidation of succinate to fumarate. This is the one enzyme of TCA cycle reactions that has been shown to be firmly bound to the inner surface of the inner membrane and not associated with the matrix as are the other enzymes. This enzyme contains a flavin adenine dinucleotide (FAD) as a coenzyme, and it is this coenzyme that accepts the two hydrogens removed from the succinate during oxidation.

$$\text{Succinate} + \text{FAD} \longrightarrow \text{fumarate} + \text{FADH}_2 \quad (16\text{–}11)$$

The $\Delta G^{\circ\prime}$ is 0.

Fumarate is converted to malate by *fumarase*. The reaction has a $\Delta G^{\circ\prime}$ = -0.88 kcal mole^{-1}.

$$\text{Fumarate} + \text{H}_2\text{O} \longrightarrow \text{malate} \quad (16\text{–}12)$$

Malate is oxidized by *malate dehydrogenase*, which is an NAD^+-containing enzyme. Although the $\Delta G^{\circ\prime}$ is 7.1 kcal mole^{-1} and indicates that the isolated reaction

$$\text{Malate} + \text{NAD}^+ \longrightarrow \text{oxaloacetate} + \text{NADH} + \text{H}^+ \quad (16\text{–}13)$$

is endergonic, the products of the reaction are both readily removed in vivo and the reaction is therefore driven in the direction as written. Malic dehydrogenase is found not only in the matrix of mitochondria but also in the cytosol. The oxaloacetate produced by the reaction initiates the cycle again by combining with acetyl CoA to form citrate.

Summary of the TCA Cycle

It is possible to account for all the atoms entering the cycle. There are two acetyl carbons in acetyl CoA, and during the Krebs cycle, one is converted to CO_2 at the isocitric dehydrogenase step (16–7) and the other at the α-ketoglutarate dehydrogenase step (16–8). Although these are not the same carbon atoms that entered the cycle, a balance is achieved—two carbons enter the cycle and two leave the cycle.

The same accounting of carbons can be made when starting with glucose. This monosaccharide is broken down during glycolysis in the cytosol (Chapter 10) into two molecules of pyruvate. Each pyruvate molecule loses one carbon to CO_2 as it enters the Krebs cycle at the pyruvate dehydrogenase step (16–1, 16–2) and another at each of the two steps described above in the Krebs cycle:

$$\underset{\text{(6 carbons)}}{\text{Glucose}} \longrightarrow \underset{\text{(3 carbons each)}}{\text{2 pyruvate}} \longrightarrow 6\ \text{CO}_2 \quad (16\text{–}14)$$

A balance can also be made of the hydrogen atoms during glycolysis and Krebs cycle reactions. One glucose ($C_6H_{12}O_6$) and 6 H_2O molecules contribute a total of 24 hydrogens. During glycolysis, 4 hydrogens are removed at the glyceraldehyde-3-phosphate dehydrogenase step (10–14) to form 2 $NADH + H^+$ in the cytosol. Again, the two pyruvates from glucose each yield two more hydrogens (2 $NADH + H^+$) at the pyruvate dehydrogenase step (16–1, 16–2). Four more are removed at the isocitric dehydrogenase step (16–7) to form $NADH + H^+$, four more at the α-ketoglutarate dehydrogenase step (16–8), four more at the succinate dehydrogenase step (16–11), and finally four more at the malate dehydrogenase step (16–13). This makes a total of 24 hydrogens. Again, although these are not exactly the same hydrogen atoms that were present in the original glucose and water molecules, a balance is achieved.

In a similar way, the oxygens also balance. Glucose contributes 6 oxygens, and the 6 waters added bring the total to 12 oxygens entering the system. Twelve oxygens are removed from the system in the 6 CO_2 molecules produced. A summary of all of these balances is apparent in the overall equation for the respiration of glucose through the Krebs cycle reactions:

$$\text{C}_6\text{H}_{12}\text{O}_6 + 6\ \text{H}_2\text{O} + \left\{\begin{array}{c}10\ \text{NAD}^+\\ 2\ \text{FAD}\end{array}\right\} \longrightarrow 6\ \text{CO}_2 + \left\{\begin{array}{c}10\ \text{NADH}+\text{H}^+\\ 2\ \text{FADH}_2\end{array}\right\} \quad (16\text{–}15)$$

The important part of the metabolism of the Krebs cycle reactions is the production of tremendous chemical reducing power in the matrix of the mitochondrion by the accumulation of $NADH + H^+$ and $FADH_2$. These compounds now enter the electron transport system reactions, and their potential drives oxidative phosphorylation thereby producing ATP.

Electron Transport System

The reduced $NADH + H^+$ and $FADH_2$ that accumulate as the TCA cycle operates form an enormous potential energy pool. These compounds will be reoxidized and will transfer their reducing capabilities through a sequence of compounds to oxygen where the acceptance of electrons and H^+ will form water. The sequence of compounds through which the electrons and H^+ are passed is called the **electron transport system.** Each transfer of electrons (or H^+) from one compound to the next results in the oxidation (electron loss) of the donor molecule and the reduction (electron acceptance) of the acceptor molecule. Such a transfer occurred in reaction 16–3 in the matrix during pyruvate dehydrogenase action. The enzyme-linked FAD (E-FAD) accepted the electrons from the substrate and became reduced, that is, E-$FADH_2$. Subsequently, the E-$FADH_2$ transferred hydrogens to NAD, which in turn became reduced while the FAD was reoxidized.

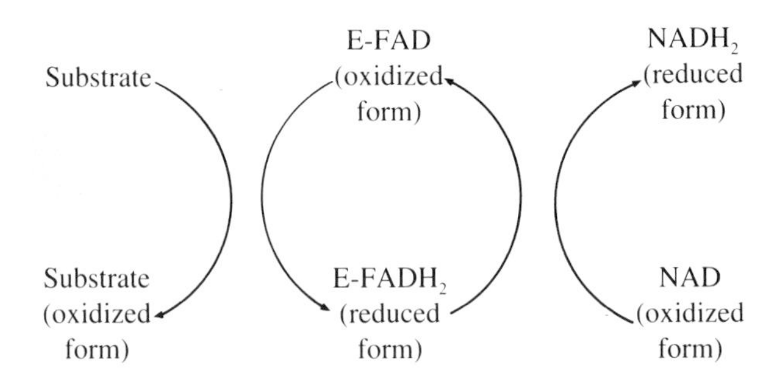

In the electron transport system the electrons are shuttled between about eight intermediate compounds before they eventually reach oxygen to form water. Although it is possible to reoxidize NADH + H^+ directly by adding oxygen, in the mitochondria this is prevented. The enzymes of the electron transport system appear to be arranged physically in the inner membrane in such a manner that the transfer of the electrons *must* proceed through the specified series of compounds. With each transfer, there is an energy change. The energy released from electron transfer is utilized to synthesize ATP—a process termed oxidative phosphorylation and described in the next section.

Oxidation-Reduction Reactions

In Chapter 3 the ability of acids and bases to lose or gain protons was described. In acid-base systems, one compound acts as the proton (H^+) donor and the other as the proton acceptor. The proton donor is called the acid, and the proton acceptor is the base. The two form a pair that is called a conjugate acid-base pair. In a similar manner, oxidizing and reducing agents function as pairs. In this case, they are called **redox pairs** or **redox couples.** The member of the pair that donates the electron is called the **reducing agent,** or **reductant,** and the electron acceptor is called the **oxidizing agent,** or **oxidant.**

$$\underset{\text{(reducing agent)}}{\text{Electron donor}} \longrightarrow e + \underset{\text{(oxidizing agent)}}{\text{electron acceptor}}$$

The terms donating and accepting fail to convey the forces or energy involved. In effect, the reducing agent has a certain ability (power) to hold electrons, as does the oxidizing agent. In a redox couple, one member attracts electrons more strongly than the other, and in effect the oxidizing agent can pull the electrons away from the reducing agent. This power to gain (or lose) electrons can be measured and is called the **oxidation-reduction potential** or **redox potential.** (Redox potentials are described in Chapter 9.) It is frequently expressed in volts. For convenience, the potential of most redox couples is standardized against a reference, which usually is a hydrogen electrode.

The measurements are made using 1.0 *M* concentrations of oxidant and reductant at 25°C and at pH 7.0. The hydrogen electrode is equilibrated with H_2 gas at 1 atmosphere and $[H^+]$ is 1.0 *M* at 25°C. When the pH is adjusted to 7.0 (i.e., $[H^+] = 10^{-7}M$), the redox potential of the reference hydrogen electrode indicates -0.42 volts. The redox potentials of electron transport system compounds are given in Table 16–2. The greater the E_0' value, the more strongly the oxidant binds electrons. However, an oxidant having a lower E_0' than another compound could become the reductant or electron donor to that compound.

The Henderson-Hasselbach equation given in Chapter 3 describes the relationship between pH and the dissociation constant. In a similar way, the **Nernst equation** shows for a redox pair the relationship between the standard redox potential (E'_0), the observed potential at any concentration, and the concentration ratio of the oxidant and reductant:

$$E_h = E'_0 + \frac{2.3\,RT}{n\mathcal{F}} \log \frac{[\text{oxidant}]}{[\text{reductant}]}$$

where E'_0 is the standard redox potential, E_h is the observed electrode potential, *R* is the *gas constant* (8.31 joule

Table 16–2
Oxidation-Reduction Potentials of the Electron Transport System. (Values are based on two-electron transfers at pH = 7.0 and 25–30°C)

Electrode equation	E'_o V
$2H + 2e^- \rightleftharpoons H_2$	−0.421
$NAD^+ + 2H^+ + 2e^- \rightleftharpoons NADH + H^+$	−0.320
$NADP^+ + 2H^+ + 2e^- \rightleftharpoons NADPH + H^+$	−0.324
Ubiquinone + $2H^+ + 2e^- \rightleftharpoons$ ubiquinol	+0.10
2 cytochrome $b_{K(red)} + 2e^- \rightleftharpoons$ 2 cytochrome $b_{K(ox)}$	+0.030
2 cytochrome $c_{red} + 2e^- \rightleftharpoons$ 2 cytochrome c_{ox}	+0.254
2 cytochrome $a_{3(red)} + 2e^- \rightleftharpoons$ 2 cytochrome $a_{3(ox)}$	+0.385
$\frac{1}{2}O_2 + 2H^+ + 2e^- \rightleftharpoons H_2O$	+0.816

deg^{-1} $mole^{-1}$), T is the absolute temperature, n is the number of electrons transferred, and $\mathfrak{F}$ is the faraday (96,406 joule $volt^{-1}$). When two electrons are transferred at a time (which is usual in biological systems) and the constants are combined, the Nernst equation reduces to

$$E_h = E'_o + 0.03 \log \frac{[\text{oxidant}]}{[\text{reductant}]}$$

Classes of Electron-Transport-System Compounds

There are five groups of compounds associated with the electron transport system (ETS). Three of these groups consist of enzymes whose coenzyme fraction or prosthetic group is known to be responsible for the transfer. They are (1) *pyridine-linked dehydrogenases,* which have either NAD^+ or $NADP^+$ as coenzymes; (2) *flavin-linked dehydrogenases,* which are linked to flavin adenine dinucleotides (FAD) or flavin mononucleotides (FMN); and (3) the *cytochromes,* which contain iron-porphyrin prosthetic groups. Forming the fourth category is *coenzyme Q* or *ubiquinone,* a lipid-soluble coenzyme functioning in electron transport. The fifth group is composed of iron-sulfur proteins.

Pyridine-linked dehydrogenases require as their coenzyme either NAD^+ or $NADP^+$. The structures of these two coenzymes are shown in Figure 3–6. As shown, both compounds can accept two electrons at a time; one as a hydride ion (H^-) and the other as in a hydrogen atom. There are about 200 dehydrogenases that employ NAD^+ or $NADP^+$ as coenzymes. Although the NAD^+ and $NADP^+$ dehydrogenases are found both in the cytosol and in the mitochondria and are known to transfer electrons between compounds in both places, it appears that only the NAD^+-linked compounds are involved in the electron transport system.

Flavin-linked dehydrogenases require either FAD (Fig. 3–8) or FMN. Both are prosthetic groups and can accept two hydrogen atoms on the isoalloxazine ring. (Prosthetic groups are firmly bound to the protein and are not considered free carriers as are the coenzymes.) Flavin-linked enzymes are involved in a number of enzyme systems, the more common of which are associated with fatty acid oxidation, amino acid oxidation, and Krebs cycle activity (pyruvate dehydrogenase and succinic dehydrogenase). It is not uncommon to have flavin prosthetic groups and NAD^+ coenzymes linked to the same protein in dehydrogenases.

The **cytochromes** are proteins containing iron-porphyrin (or *heme*) groups (Fig. 16–17). Most cytochromes are found in mitochondria, although some function in the endoplasmic reticulum and in chloroplasts. There are a large number of cytochromes in cells; in mitochondria there appear to be at least four different ones associated with the inner membrane. These are identified as cytochrome b, c_1, a, and a_3. Some occur in two or more forms, but all transfer electrons by reversible valence changes of the iron atom ($Fe^{+++} \rightleftharpoons Fe^{++}$). In cytochromes b, c_1, c, and a, the manner of binding of iron in the porphyrin ring, and its association with the protein prevents the iron ligands from forming with oxygen and therefore these reduced cytochromes cannot be

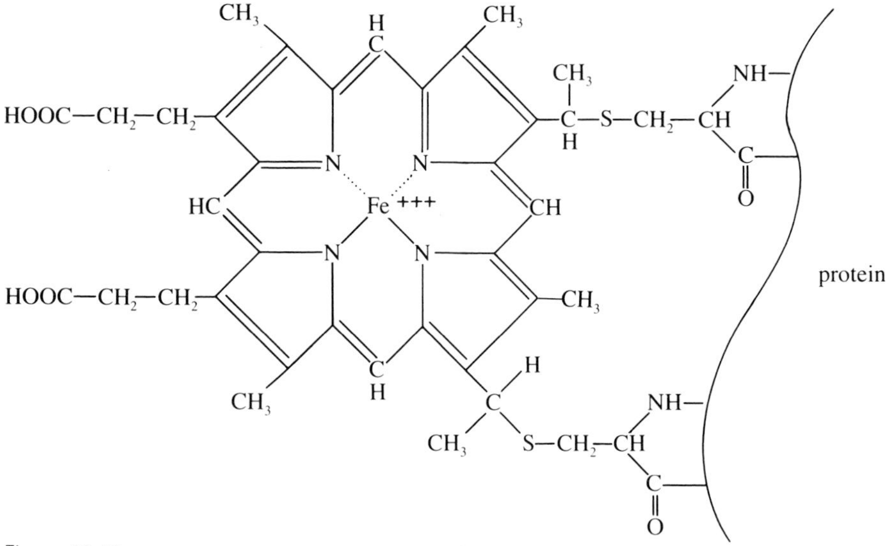

Figure 16–18
Ubiquinone.

directly oxidized by molecular oxygen. Cytochrome a_3, which is the terminal carrier in the electron transport system and which together with cytochrome *a*, forms the *cytochrome oxidase* complex, is an exception and can be directly oxidized by oxygen. In addition to cytochromes *a* and a_3, cytochrome oxidase contains copper. Electrons received from cytochrome *c* are first picked up by cytochrome *a* and then transferred to a_3. It is believed that their electrons are finally transferred from a_3 to oxygen by a $Cu^{++} \longrightarrow Cu^{+}$ intermediation.

Ubiquinones were so named because of their occurrence in so many different organisms and because of their structure (Fig. 16–18). They are found in several different forms and are related to the *plastoquinones* of chloroplasts. The form present in mitochondria is often called coenzyme Q (CoQ) and will accept two hydrogen atoms at a time.

Electron Transport Pathway

The chain of electron transfers shown in Figure 16–19 has taken almost half a century to work out. During the early 1900s, workers discovered a number of dehydrogenases that remove hydrogen from substrates. Perhaps most notable of these early workers was T. Thunberg. In 1913, O. Warburg discovered that cyanide inhibits oxygen consumption but does not interfere with dehydrogenases. He proposed the existence of iron-containing "respiratory enzymes," now recognized as the cytochromes. The flavoproteins were identified by A. Szent-Gyorgyi as the intermediates between dehydrogenases and the respiratory enzymes. R. A. Morton added the information about ubiquinone and a number of workers, especially Keilin, Green, Okunuki, Singer, King,

Figure 16–17
Cytochrome *c*.

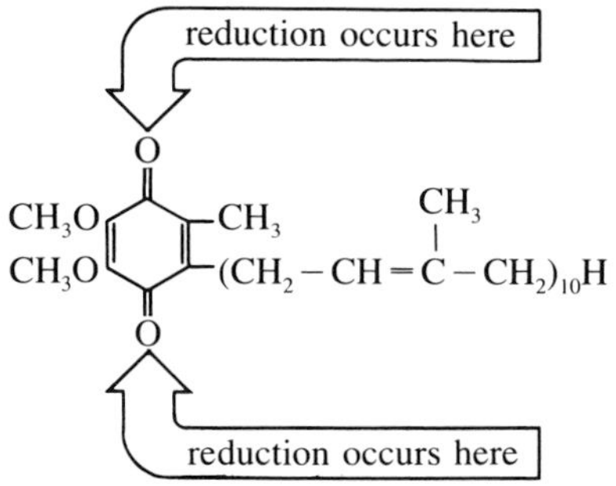

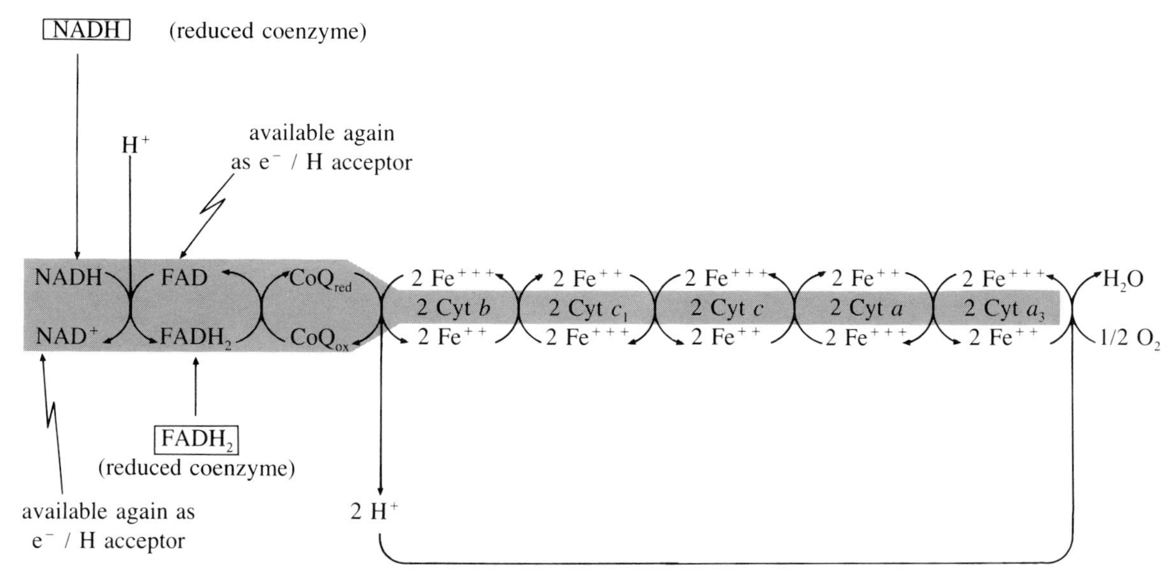

Figure 16–19
Pathway of oxidation-reduction.

Chance, Williams, and Racker, added the details on compounds, structures, and sequences.

Today we visualize the electron transport chain as a sequence of compounds through which the hydrogens or electrons are passed in order, two at a time. The fact that the electrons do not jump or "short-circuit" to stronger oxidizing agents is probably the result of the physical positioning of the coenzymes in the inner membrane and the nonreactivity of some of the enzymes. However, hydrogen or electrons may enter the chain at various points.

Most electrons are removed from the metabolic substrates in the cytosol or matrix of the mitochondria by the NAD^+- (or $NADP^+$-) linked dehydrogenases. The reduced $NADH + H^+$ acts as the collection mechanism for carrying these electrons to the NAD^+-flavoprotein-linked dehydrogenase in the inner membrane of the mitochondrion. As shown in Figure 16–19, the reduced $NADH + H^+$ is then oxidized by FAD, which now being reduced is in turn reoxidized by CoQ, which is subsequently reoxidized by cytochrome *b*. The reduced iron in cytochrome *b* is then reoxidized by cytochrome *c*, and finally the cytochrome oxidase accepts the electrons from cytochrome *c*, passing them from coenzyme *a* to a_3 and subsequently to oxygen. The acceptance of the electrons by oxygen is simultaneous with the incorporation of $2H^+$ to form water. Each of the coenzymes in the chain has been reoxidized and is available to be reduced again. Pyruvate dehydrogenase, malic dehydrogenase, isocitrate dehydrogenase, and α-ketoglutarate dehydrogenase are examples of enzymes initiating the electron transfer chain through NAD^+. Succinate dehydrogenase oxidizes succinate, and the electrons removed are bound to FAD on the succinate dehydrogenase. This FAD_{suc} is reoxidized by giving up its electrons to CoQ, thereby bypassing the NAD^+ step. Fatty acyl-CoA and glycerol phosphate are oxidized in a similar manner with entry of electrons directly from flavoprotein to CoQ.

Balance of Electrons from Glycolysis and TCA Cycle Metabolism

As was pointed out in the earlier section on TCA metabolism, 10 molecules of $NADH + H^+$ become reduced and 2 molecules of $FADH_2$ become reduced when 1 molecule of glucose is metabolized via glycolysis and the TCA cycle. Each of these 12 compounds passes the pair of electrons through the electron transport system and finally to oxygen. For each pair of electrons, $\frac{1}{2}0_2$ is absorbed and 1 H_2O is formed. Therefore, for 12 electron pairs 6 0_2 are consumed and 12 H_2O are produced. You will recall that of the 12 waters, 6 are consumed in the TCA cycle, so that in the overall oxidation of glucose there is a net production of only 6 H_2O.

$$C_6H_{12}O_6 + 6\ O_2 \longrightarrow 6\ CO_2 + 6\ H_2O$$

The Energetics of Electron Transport

The standard free energy change, $\Delta G°$ (see Chapter 9), can be calculated from the redox potentials by the following equation:

$$\Delta G°' = n\mathcal{F}\,\Delta E'_0$$

where n = the number of electrons transferred at a time (usually 2), $\mathfrak{F}$ = the number of faradays (23,062 cal V^{-1}), and $\Delta E'_0$ is the change in standard redox potential (the E'_0 of the accepting redox couple less the E'_0 of the donating couple). The $\Delta G°'$ for the entire chain from NAD to oxygen would be -52.7 kcal $mole^{-1}$. The stepwise sequences are shown in Figure 16–20. The fate of the energy released by the transfer of electrons is discussed later in the chapter under oxidative phosphorylation. It is interesting to note at this point that there is enough energy to produce several moles of ATP at a standard free energy of formation of 7.3 kcal $mole^{-1}$. Generally, only three are made, corresponding to the three steps producing sufficient energy for the reaction (Fig. 16–20).

Transport of Protons

Transport of H^+ closely parallels electron transport. Some of the electron carriers actually carry electrons in the form of complete hydrogen atoms, as was noted earlier. These are NAD^+, $NADP^+$, FAD, FMN, and ubiquinone. FAD (FMN) carries a pair of hydrogens, while NAD^+ ($NADP^+$) carries a hydrogen plus one electron (hydronium ion): the other H^+, however, associates closely with the NADH. The cytochromes, on the other hand, only pick up and release electrons. It appears that each cytochrome carries only one electron at a time; therefore, there is speculation that the cytochromes work in pairs. The H^+ thus dissociated at the ubiquinone/cytochrome *b* step and not utilized until the oxygen step therefore accumulates and may have a functional significance in ATP synthesis as described later.

Electron Transport Inhibitors

Inhibitors of the electron transport system are categorized into three main groups: (1) One group includes those inhibitors that act on NAD^+ ($NADP^+$) dehydrogenase and

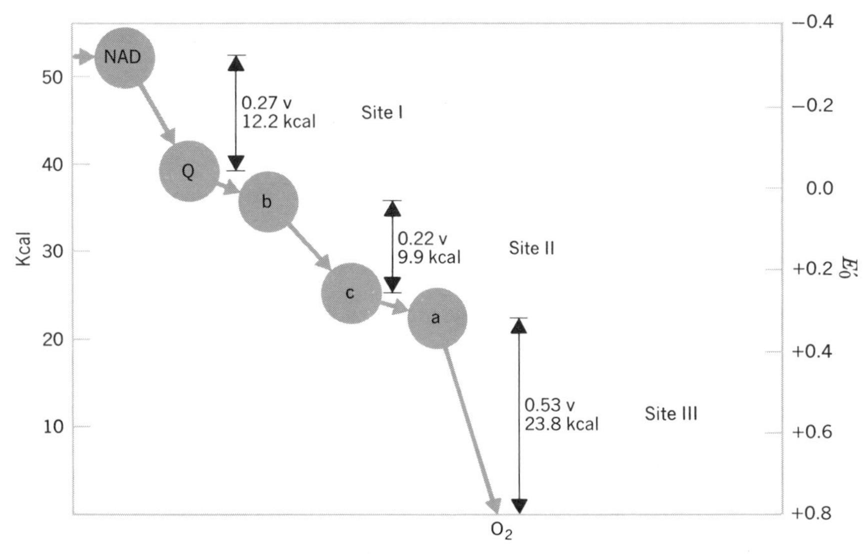

Figure 16–20
Changes in free energy as electrons pass through the electron transport chain of enzymes. The three changes generating sufficient energy to drive the synthesis of ATP are indicated as sites I, II, and III.

ubiquinone. Examples are *rotenone* (a plant substance once used by some South American Indians as a fish poison and occasionally used today as an insecticide), *amytal,* and *piericidin* (an antibiotic similar in structure to and competitive with ubiquinone). (2) The second group includes those that block electron transport between cytochrome *b* and *c;* an example is *antimycin A* an antibiotic from *Streptomyces griseus*. (3) The third group includes those inhibitors that block electron transport from cytochrome oxidase ($a + a_3$) to oxygen and includes cyanide, hydrogen sulfide, and carbon monoxide.

NAD $[P]^+$ Transhydrogenases

The electron transport chain usually accepts $NADH + H^+$ as the initial donor of hydrogen. In many animal tissues and in microorganisms, hydrogens from $NADPH + H^+$ are also accepted. The entry of these hydrogens is apparently brought about by the enzyme $NAD[P]^+$ transhydrogenase, which is located in the mitochondrial membrane and which catalyzes the transfer of hydrogen to NAD^+.

$$(NADPH + H^+) + NAD^+ \longrightarrow (NADH + H^+) + NADP^+$$

Oxidative Phosphorylation

The linking of phosphorylation with oxidative metabolism was first proposed in the 1930s. The first evidence was that phosphate was removed from the medium when TCA cycle intermediates were metabolized by cells. Subsequently, it was shown that organic phosphates such as glucose-6-phosphate and fructose-6-phosphate accumulate. Shortly after, ATP was recognized as the primary product of oxidative phosphorylation. By quantitatively measuring the uptake of oxygen and the disappearance of phosphate, investigators found that calculations of the P/O ratio were consistently about 3. Since one oxygen is consumed for each oxidation in the TCA cycle, it was possible to calculate that 15 ATP are synthesized for each pyruvate oxidized through the cycle (i.e., five oxidation steps; each releases two hydrogens that pass through the electron transport system to oxygen to form H_2O). In 1948, Kennedy and Lehninger established that these oxidations and phosphorylations occurred exclusively in the mitochondria. By 1951, Lehninger was able to show that the Krebs cycle oxidations could be bypassed entirely by adding $NADH + H^+$. For each $NADH + H^+$ added, three phosphates were consumed and one oxygen used ($P/O = 3$). It was therefore concluded that the electron transport chain was important for ATP synthesis.

The exergonic reactions of electron transfer from the NAD^+-dehydrogenases to cytochrome oxidase provide the energy to drive the endergonic oxidative phosphorylations. The exergonic reactions can be summarized as

$$NAD + H^+ + \tfrac{1}{2}O_2 \longrightarrow NAD^+ + H_2O \qquad (16\text{-}16)$$

which has a $\Delta G^{\circ\prime} = -52.7$ kcal $mole^{-1}$. The endergonic reactions are summarized as

$$3\ ADP + 3\ P_i \longrightarrow 3\ ATP$$

with a $\Delta G^{0\prime} = +21.9$ kcal for 3 moles ($= 3 \times 7.3$ kcal $mole^{-1}$). The efficiency of the coupling of the two systems is about 42% (i.e., $21.9 \times 10^2/52.7$).

The sites of the coupling are known. Figure 16–20 shows the major energy changes along the electron transfer chain. The step from NAD to Q represents the reactions of the dehydrogenases that link NAD^+ and FAD to CoQ. This reaction sequence (*site I*) provides enough energy (12.2 kcal) to drive the first endergonic formation of ATP. The step from CoQ to cytochrome *b* does not provide enough energy, but the following step from cytochrome *b* to cytochrome *c* does yield sufficient energy and is identified as the second coupling site (*site II*) of the ETS and oxidative phosphorylation. The transition from cytochrome *c* to cytochrome *a* is too low, but the final transition from cytochrome *a* to oxygen is highly exergonic and is easily identified as the third site (*site III*). Thus, a pair of electrons passing through the oxidation-reduction reactions generates the energy for the formation of ATP at sites I, II, and III, thereby forming three ATPs per pair of electrons. However, not all electrons from oxidized substrates pass into the electron transport system at the initial NAD^+ acceptor. Some electrons, like those from succinate, pass from the coenzyme FAD of succinate dehydrogenase directly to CoQ. When this happens, site I is skipped and, as could be predicted, only two ATPs are made per pair of electrons.

Molecular Events in Oxidative Phosphorylation

Although countless experiments have been run by accomplished investigators, the actual mechanism whereby energy is transferred from the electron transport system to oxidative phosphorylation is far from clear. It would appear that there are yet to be discovered coupling factors and even enzymes involved in the process. What is known is a collection of observations that includes (1) mechanisms of controlling the rate of electron transport and oxidative phosphorylation, (2) inhibitors that uncouple electron transport from oxidative phosphorylation, (3) partial oxidative phosphorylation reactions, and (4) coupling factors.

In intact mitochondria, maximal electron transport occurs only if adequate ADP and inorganic phosphate are present. Once the ADP has been converted to ATP, the rate of oxygen uptake decreases to a very low level called *state 4 respiration* (Fig. 16–21). If ADP is added, the rate temporarily increases to *state 3 respiration* and then returns to state 4 after the ADP has been consumed. Although it would appear that the exergonic ETS reactions should drive the oxidative phosphorylations, this evidence indicates some type of control mechanism obviously operating at the molecular level that allows for *ADP-acceptor control*. Interestingly, the configuration of mitochondria changes between state 4 and state 3 respiration. During state 4 respiration, the mitochondrion assumes the orthodox state (Fig. 16–14) but changes to the condensed state when added ADP induces state 3.

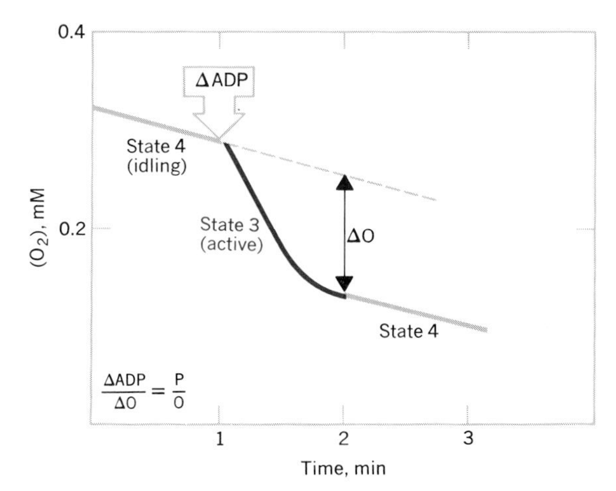

Figure 16–21
Changes in oxygen consumption and physiological state with the addition of ADP. Initially the rate of oxygen consumption is low owing to the lack of ADP (most of that available has been converted to ATP) and the idling state is called state 4 respiration. Addition of ADP induces a more active state 3 respiration, which continues until available ADP is used up.

There are a number of inhibitors that uncouple oxidative phosphorylation from the ETS. Several common ones are 2,4-dinitrophenol, dicumarol, and the salicylanilides. The presence of these compounds does two interesting things. First, they cause the speedup of electron transport and oxygen consumption without synthesis of ATP, even in the absence of ADP. In other words, ADP-acceptor rate control is uncoupled and so is energy transfer. Second, the hydrolysis of ATP occurs—the *opposite* of the normal event in mitochondria.

Oligomycin is an inhibitor of oxidative phosphorylation that is not considered an uncoupler. It can inhibit both ATP synthesis and oxygen consumption but does not prevent electron transfer. A group of agents called ionophores also prevent oxidative phosphorylation by preventing transfer of energy, but they work only in the presence of certain monovalent cations such as K^+ and Na^+.

Some of the reactions of oxidative phosphorylation are known. An *ATPase* is known to function in the inner membrane and is stimulated by 2,4-dinitrophenol but is inhibited by oligomycin. Another reaction is the exchange of inorganic phosphate with the terminal phosphate on ATP, a reaction called *phosphate-ATP exchange*. This reaction is inhibited by both 2,4-dinitrophenol and oligomycin. An exchange of atoms also occurs between the oxygens of water and those of inorganic phosphate, called *phosphate-water exchange*. The terminal phosphate of ATP can also exchange with ADP, a process inhibited by 2,4-dinitrophenol and oligomycin and called the *ADP-ATP exchange* reaction.

Some of the more interesting work that ties together structure and function in mitochondria involves the disruption and fractionation of mitochondria and the reconstruction of small functional units. In this type of study, the inner membrane is separated from the outer membrane, and the inner membrane is disrupted by sonication or detergents (Fig. 16–22) to form submitochondrial vesicles. These vesicles apparently form from the cristal membranes by rounding up; the inner membrane spheres are on the outside of the vesicles. Intact submitochondrial vesicles have functioning ETS and oxidative phosphorylation reactions. If the membranes do not round up to form a closed vesicle, these functions appear to be lost.

If the submitochondrial vesicles are treated with urea or

trypsin, the inner membrane spheres are removed and can be separated from the vesicles, which lose their oxidative phosphorylation function. However, if the spheres are added back to the vesicles, function is regained. The spheres were initially called *coupling factor one* (F_1). Subsequently, it was learned that this factor contains ATPase. The enzyme has been purified and studied. It has a molecular weight of about 360,000, a diameter of 9 nm, and requires Mg^{++}. In the mitochondrion, it probably catalyzes the synthesis of ATP from ADP and inorganic phosphate, rather than the reverse, and as such would be better termed ATP-synthetase. The particle's functioning is not inhibited by oligomycin. However, another factor has been isolated which when present with F_1 renders the ATPase sensitive to oligomycin. This latter factor is called F_0 or *oligomycin-sensitivity-conferring factor (OSCF)*. F_0 is also a large protein; it has been speculated that this factor could constitute the stalk that holds the sphere to the inner membrane (Fig. 16–12).

These assorted bits of information about oxidative phosphorylation add little to our understanding of the process. How they all fit together is open to speculation. Three hypotheses currently having wide support are the **chemical-coupling hypothesis,** the **conformational-coupling hypothesis,** and the **chemiosmotic-coupling hypothesis.** To some extent, there are similarities between each of them.

The Chemical-Coupling Hypothesis

This model was the first proposed to explain the mechanism of coupling the exergonic reactions of electron transport with the endergonic oxidative phosphorylation reactions. The mechanism is a carryover of the basic manner in which energy in substrates is coupled to ATP snythesis for substrate-level phosphorylation. In this mechanism an intermediate compound is formed during electron transport that has a strongly negative standard free energy of hydrolysis. The intermediate then reacts directly with ADP, switching its high-energy bond to form ATP. The reactions appear to be thermodynamically possible, and an enzyme complex could link the set of reactions. However, in the 25 years since the hypothesis was made, there has been no substantial evidence for the existence of the proposed intermediate. In the meantime, evidence has grown that an intact inner membrane is a prerequisite for oxidative phosphorylation. The requirement of an intact membrane may be an indication that the intermediate is located in the lipid phase and easily hydrolyzed on exposure to water, as could occur on disruption of the membrane.

The Conformational-Coupling Hypothesis

In this hypothesis, it is proposed that the energy yielded by electron-transport changes the conformation of the carrier

Figure 16–22
The disruption of mitochrondria and the reconstitution of vesicles from the inner membrane with reattached spheres.

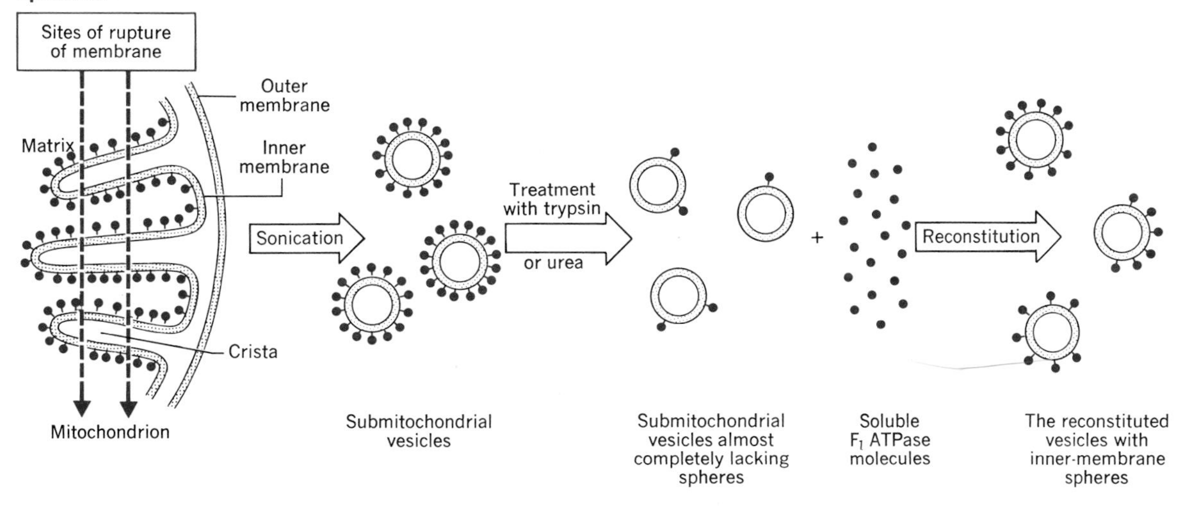

protein or a coupling factor such as ATPase. The conformational change resulting from the energy absorption would shift the number and/or location of weak bonds in the protein. The protein would then take on an energized configuration and on returning to its normal state would directly transfer the energy into forming ATP.

Little evidence has been presented thus far to support this model. However, the fact that a somewhat similar mechanism exists in muscle fiber contraction lends support to the possibility of the operation of the model.

The Chemiosmotic-Coupling Hypothesis

This hypothesis incorporates into its mechanism the need for an intact inner membrane but does not require the presence of a high-energy intermediate. The basis of the model is a shift in or creation of a gradient of hydrogen ions across the inner membrane as electron transport occurs. The energy of electron transport establishes an active transport pump that moves H^+ from the matrix across the inner membrane into the intermembrane area of intact mitochondria or into the vesicles of fragmented submitochondrial preparations. An intact membrane bound vesicle is needed for the establishment of the electrochemical gradient. The gradient, in turn, provides the energy to drive the synthesis of ATP.

It is proposed that the molecules involved in electron transport are not randomly arranged in or on the inner membrane but are spacially specifically oriented. As hydrogen atoms are removed from the matrix substrate, the electron transport molecules separate electrons from protons (H^+). The H^+ is released to the "outside" (i.e., the intermembrane space), and simultaneously the electrons are transferred to the next molecule. This causes the matrix to become especially alkaline (high OH^- content). As was discussed earlier in the chapter, some electron transport intermediates accept H^+ as well as electrons (e.g., NAD, FMN, CoQ, O_2), while others accept only electrons (e.g., cytochromes *b, c, a,* and a_3 and the little-known iron-sulfur compounds). As the electrons are passed from molecule to molecule along the membrane, H^+ is picked up from the matrix by those intermediates that accept H^+ as well as electrons and the H^+ is then released to the intermembrane space when the electrons are transferred to an intermediate not carrying H^+. In Figure 16–23, these events are shown with the electron transport intermediates in order. An additional step (X) has been proposed to account for observed measurements. In some models, the usual order (FAD — CoQ — cyto *b*) is reversed (FAD — cyto *b* — CoQ) but still accounts for the same observations. The vectorial flow of protons shown in Figure 16–23 creates a protonmotive force, which is the source of energy for phosphorylation.

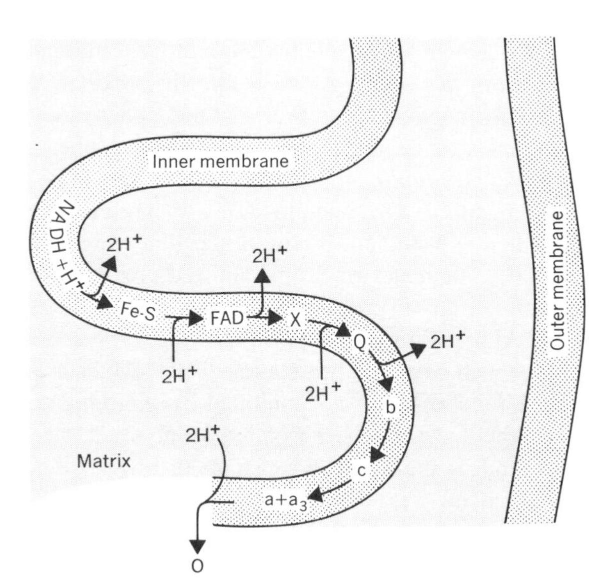

Figure 16–23
Illustration of the "pumping action" of H+ across the inner membrane by the electron transport system. See text for explanation.

The oxidative phosphorylation enzymes in the model are also spacially fixed. ATP synthetase binds ADP and inorganic phosphate and the energy of the gradient causes H^+ to dissociate from the ADP and enter the *alkaline sink* in the matrix (Fig. 16–24). (This alkaline sink was formed by the loss of H^+ across the inner membrane.) At the same time, the OH^- of inorganic phosphate is directed to the acid (H^+) intermembrane space. The net effect of oxidative phosphorylation is thus observed, namely the combination of ADP and P to form ATP. As oxidative phosphorylation occurs, the alkaline sink is neutralized as is the acid pool. Continued electron transport is necessary to maintain the gradient required to drive the phosphorylation.

There is good evidence to support this model. In addition to the intact membrane requirement discussed above, it has been observed that the inner membrane is impermeable to H^+ movement in the direction leading to the matrix but that 2,4-dinitrophenol (an uncoupling agent) results in the reverse permeability of H^+. Electron transport has been

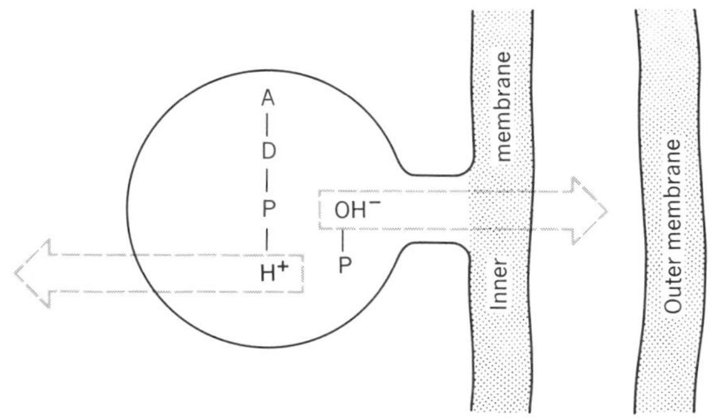

Figure 16–24
Illustration of the attraction of H^+ to the matrix alkaline sink and Oh^- to the acid intermembrane space resulting in phosphorylation of ADP. See text for explanation.

shown to be accompanied by the outward pumping of H^+ and ATP synthesis by the inward movement of H^+. Although the model seems fairly reasonable, there continues to be concern as to whether a sufficient gradient can be produced to drive phosphorylation. In addition, there is the possibility that the observed acid pool of the intermembrane space is an inconsequential result of another set of reactions and is not the driving force of oxidative phosphorylation. It could be that the high-energy intermediate of the chemical-coupling hypothesis removes H^+ from the matrix, making it alkaline, and OH^- from the intermembrane space, making it acidic. This process works in reverse but causes the same observed result.

Other Functions of Mitochondria

Mitochondria are generally described as the powerhouses of the cell, and as such, most interest is directed to the processes that evolve the most energy, namely the TCA cycle, electron transport, and oxidative phosphorylation. However, a great number of other reactions occur in mitochondria.

The Glyoxylate Cycle

This chain of reactions was described in Chapter 10. The enzymes are localized in the matrix, and some of them are the same as those of the TCA cycle. In effect, the glyoxylate cycle is a modified form of the Krebs cycle, but its function appears to be primarily associated with the conversion of acetate from fatty acid decomposition into oxaloacetate. Oxaloacetate is also an important intermediate in the conversion of fatty acids to carbohydrates. Animals lack certain enzymes of the glyoxylate cycle and are incapable of converting fatty acids to carbohydrates using this pathway. Both plants and microorganisms have functional glyoxylate systems. In higher plants, some of the enzymes of the system (e.g., isocitric lyase and malate synthetase) are localized in specific organelles called *glyoxysomes* (Chapter 18) as well as in mitochondria.

Fatty Acid Oxidation

Free fatty acids are rarely found in more than trace quantities in cells because they are highly toxic. The fatty acids associated with mono-, di-, and triglycerides and in phospholipids are generally hydrolyzed from the glycerol in the cytosol and immediately activated for transport into the matrix of the mitochondrion where they are then oxidized. These reactions are also described in Chapter 10.

Fatty Acid Chain Elongation

Fatty acids are generally synthesized on the smooth endoplasmic reticulum. However, there are a number of enzymes in mitochondria that catalyze the elongation of palmitic and other saturated fatty acids by successive additions of acetyl CoA to the carboxyl end. In smooth ER, both unsaturated and saturated fatty acids are elongated but by the addition of malonyl CoA rather than acetyl CoA. Synthesis of fatty acids is discussed in Chapter 10.

Superoxide Dismutase and Catalase

During electron transport, a number of toxic reductive products of oxygen are formed. The most common are superoxide (O_2^-) and hydrogen peroxide (H_2O_2). A protective enzyme has been identified in mitochondria called *superoxide dismutase,* which decomposes the O_2^- into hydrogen peroxide. The hydrogen peroxide, in turn, is decomposed by *catalase*.

$$2\,O_2^- + 2H^+ \xrightarrow{\text{dismutase}} H_2O_2 + O_2$$

$$H_2O_2 \xrightarrow{\text{catalase}} H_2O + \tfrac{1}{2}O_2$$

Amphibolic and Anapleurotic Reactions

The intermediates of the TCA cycle can act as precursors of a variety of anabolic products. The TCA cycle thus is amphibolic in that it can act in both a catabolic and anabolic manner. Most of the intermediates can be precursors for a variety of products of different metabolic pathways (Table 16–3).

These anabolic reactions can drain the intermediates away from the TCA cycle and thus deplete the oxaloacetate supply necessary to keep the cycle functioning. Compensatory reactions that supply the TCA cycle with intermediates are also well known (Table 16–4).

Permeability of the Inner Membrane

As was discussed earlier, the outer membrane is porous and most metabolites can pass through this membrane freely. The inner membrane is not freely permeable. Other than in special cases, sugars cannot pass through the membrane, ions such as Na^+, K^+, Cl^-, and Br^- are impermeable, and NAD^+, NADH, $NADP^+$, NADPH, AMP, CDP, GDP, CTP, GTP, CoA, and acetyl CoA collect in the matrix, unable to mix or exchange with pools of these molecules in the cytosol. There are, however, special transport systems in the inner membrane that are specific for select metabolites, and via these systems, ADP, ATP, P_i, pyruvate, and a number of TCA cycle intermediates are transported across. The transport systems are genetically determined and species-specific and in most cases are associated with membrane proteins called *carriers, translocases,* or *porters*. Table 16–5 lists a number of these systems and indicates their functions.

These systems can function as *passive carriers* facilitating the exchange of metabolites when there is a favorable concentration gradient, or they can function in an active manner if coupled to the energy-producing electron transport system. When electron transport is active, there is an accumulation of hydroxyl ions in the matrix, as shown in Figure 16–24. The accumulated hydroxyl ions can be transported outside via the phosphate carrier, which simultaneously carries inorganic phosphate into the matrix on an exchange basis (Fig. 16–25). The phosphate accumulated

Table 16–3
TCA Precursors of Anabolic Pathways

TCA Cycle Intermediate	Pathway or Anabolic Reaction
Citrate	Citrate + ATP + CoA → oxaloacetate + ADP + P_i + acetyl CoA → fatty acid biosynthesis
Isocitrate	Isocitrate → glyoxylate → malate → carbohydrate synthesis
α-Ketoglutarate	α-Ketoglutarate + alanine → pyruvate + glutamate (→ protein synthesis)
Succinyl CoA	Heme biosynthesis
Malate	Carbohydrate synthesis
Oxaloacetate	Oxaloacetate + alanine → pyruvate + asparate → protein synthesis

Table 16–4
Anapleurotic Reactions of the TCA Cycle

TCA Cycle Intermediate Produced	Non-TCA Cycle Reaction Generating the TCA Cycle Intermediate
α-Ketoglutarate	Glutamate + pyruvate → α-ketoglutarate + alanine (from protein breakdown)
Succinate	From glyoxylate cycle
Malate	Malic enzyme: pyruvate + CO_2 + NADPH + H^+ → malate + $NADP^+$
Oxaloacetate	Pyruvate carboxylase: pyruvate + CO_2 + ATP + H_2O → oxaloacetate + ADP + P_i
Oxaloacetate	Aspartate + pyruvate → oxaloacetate + alanine (from protein breakdown)

Table 16–5
Mitochondrial Membrane Transport Systems

System	Exchange
Dicarboxylate carrier	Exchange on mole-for-mole basis of malate, succinate, fumarate, and phosphate between matrix and cytosol.
Tricarboxylate carrier	Exchange on mole-for-mole basis citrate and isocitrate between matrix and cytosol. Exchange citrate or isocitrate for dicarboxylate.
Aspartate-glutamate carrier	Exchange aspartate for glutamate across membrane.
α-Ketoglutarate-malate carrier	Specifically exchange α-ketoglutarate for malate across membrane.
ADP-ATP carrier	Exchange of ADP for ATP.

can be partly consumed in ATP synthesis, with the ADP-ATP carrier functioning to bring ADP into the matrix, or the phosphate can be exchanged with external dicarboxylate or tricarboxylate required in TCA cycle reactions using the appropriate carboxylate carrier.

One of the clearest connections between active transport across the inner mitochondrial membrane and electron transport system has been provided by the studies of Ca^{++} transport. Ca^{++} uptake is accompanied by an equivalent uptake of phosphate. It has been shown that for each pair of electrons transported through the electron transport system from NADH + H^+ to oxygen, six Ca^{++} are transported across the inner membrane into the matrix. When the Ca^{++} is transported, there is no phosphorylation of ADP. Specific inhibitors indicate that the same sites responsible for phosphorylation along the electron transport chain are the energy-providing sites for the active transport of Ca^{++} which must function in connection with the phosphate carrier. Mn^{++} and Fe^{++} appear to be coupled to electron transport and phosphate carriers in a similar manner.

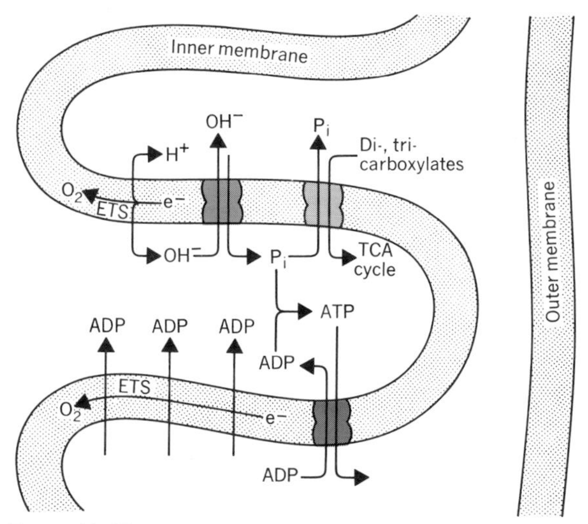

Figure 16–25
The function of carriers across the inner membrane. See text for explanation.

Cytosol-Matrix Exchange of NADH and NADPH

Reduced coenzymes such as NADH and NADPH do not permeate the inner membrane of the mitochondrion to any significant extent. However, reduced pyridine nucleotides are known to be produced in a number of reactions in the cytosol. The reduction of NAD at the glyceraldehyde step in glycolysis is an important example. The reoxidation of these compounds by the mitochondrion does occur. The mechanism involves a set of reactions called a **shuttle.** In the case of NAD, the shuttle involves glycerol phosphate dehydrogenase in the cytosol, glycerol phosphate dehydrogenase on the outer surface of the inner mitochondrial membrane, and the reduction of CoQ in the electron transport chain. Dihydroxyacetone phosphate and $NADH + H^+$ react with the cytosol enzyme to form glycerol phosphate, which diffuses through the porous outer membrane and to the outer surface of the inner membrane. There the glycerol phosphate reacts with the membrane dehydrogenase to form dihydroxyacetone phosphate, which returns to the cytosol. The membrane-bound dehydrogenase has a flavoprotein (FP) rather than NAD as a coenzyme, and the FP becomes reduced during the reaction. Subsequently, the electrons from the FPH_2 are transported directly into the electron transport system at the CoQ step. Because the NAD $\longrightarrow$ FAD step of electron transport is skipped, only two ATP are generated per pair of electrons that enter in this fashion from the cytosol.

Another shuttle, the **malate-aspartate shuttle,** can transport NAD across the inner membrane in either direction. $NADH + H^+$ transported in this manner into the matrix enters the electron transport chain in the usual manner, and as a result, three ATP are generated for each pair of electrons.

Total Energy Production from Catabolism of Glucose

The total energy produced from the complete oxidation of a molecule of glucose metabolized through glycolysis, the TCA-cycle, and electron transport is usually quoted as equivalent to a *gross* production of 40 ATP or a *net* production of 38 ATP. These ATP are produced by both substrate-level phosphorylations and electron transport coupled with oxidative phosphorylation. The accounting of the ATP is diagrammed in Figure 16–26.

Summary

Mitochondria are the ''powerhouses'' of the cell and the primary sites of cell **oxidations.** Within the membranes of these organelles, elementary substrates produced by the breakdown of carbohydrates, lipids, or nitrogen macromolecules in other cell locations are oxidized to CO_2 and water. The energy released from the exergonic oxidations serves to form ATP.

Mitochondria occur in almost every type of aerobic eucaryotic cell. Although their shape may vary, all mitochondria contain two structurally and functionally different membranes—an **inner** and an **outer membrane.** Between the two membranes is the fluid, **intermembrane space.** The inner membrane surrounds the innermost compartment or **matrix.** Projections of the inner membrane into the matrix are called **cristae.** New mitochondria form by pinching off from other mitochondria and may also arise from microbodies.

The oxidative and phosphorylation reactions occur in the

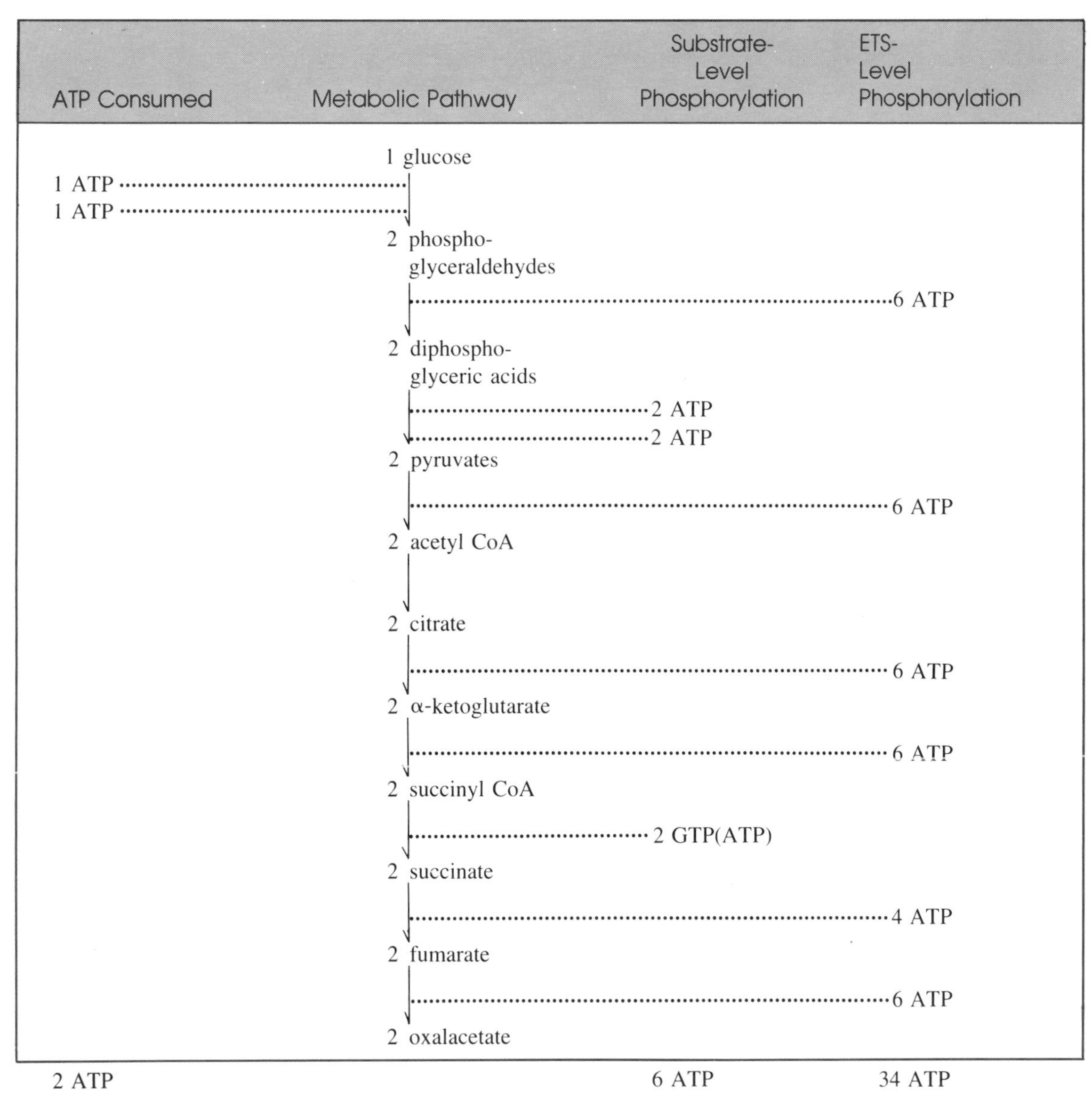

Figure 16–26
Balance sheet of ATP during glycolysis, TCA cycle, and electron transport.

inner membrane or in the matrix. The tricarboxylic acid or **Krebs cycle** reactions constitute the first phase of the oxidation of substrates such as acetate. In these reactions, the molecules are enzymatically degraded to CO_2 and water. Hydrogen and electrons from the substrates reduce NAD and FAD to NADH + H^+ and $FADH_2$. Some ATP is synthesized at a **substrate level** by these reactions. Most of the ATP generated in the mitochondrion is by the **elec-**

tron transport system (ETS) which oxidizes the NADH + H^+ and $FADH_2$ formed by the Krebs cycle. The ETS functions as a multistep series of oxidation-reduction reactions, transferring electrons through a set of intermediates associated with the **inner membrane** and ultimately reducing molecular oxygen to water. These intermediates include **pyridine** or **pyrimidine-linked dehydrogenases, flavin-linked dehydrogenases, cytochromes,** and **ubiquinones.**

Associated with the ETS is a mechanisms for formation of ATP called **oxidative phosphorylation.** For each pair of electrons transferred from NADH + H^+, three molecules of inorganic phosphate are added to ADP to form ATP (two ATP for $FADH_2$). Oxidative phosphorylation is a function of the inner membrane. Three hypotheses for the mechanism of phosphorylation are the **chemical-coupling hypothesis,** the **conformational-coupling hypothesis,** and the **chemiosmotic-coupling hypothesis.**

The mitochondrion is also the site of the **glyoxylate cycle** reactions, **fatty acid oxidation, fatty acid chain elongation, superoxide mutase,** and **catalase** reactions and a number of **amphibolic** and **anapleurotic** reactions.

References and Suggested Reading

Articles and Reviews

Cohen, S. S., Mitochondria and chloroplasts revisited. *Am. Scientist 61*, 437 (1973).

Fernández-Morán, H., Cell membrane ultrastructure. *Circulation 26*, 1039 (1962).

Fletcher, M. J., and Sanadi, D. R., Turnover of rat liver mitochondria. *Biochim. Biophys. Acta 51*, 356 (1961).

Hinkle, P. C., and McCarty, R. E., How cells make ATP. *Sci. Am. 238* (3) 104 (Mar. 1978).

Raven, P. H., A multiple origin for plastids and mitochondria. *Science 169*, 641 (1970).

Sjöstrand, F. S., The structure of mitochondrial membranes: a new concept. *J. Ultr. Res. 64*, 217 (1978).

Wilson, J. E., and Dove, J. L., Turnover of mitochondria in rat liver, kidney and heart. *J. Elisha Mitchell Sci. Soc. 81*, 21 (1965).

Books, Monographs, and Symposia

Lehninger, A. *Biochemistry*, 2nd ed., Worth Publishers, New York, 1975.

Racker, E., *Membranes of Mitochondria and Chloroplasts*, Van Nostrand Reinhold Co., New York, 1970.

Roodyn, D. B., and Wilkie, D., *The Biosynthesis of Mitochondria*, Methuen, London, 1968.

Tedeschi, H., *Mitochondria: Structure, Biogenesis and Transducing Functions*, Springer-Verlag, New York, 1976.

Chapter 17
THE CHLOROPLAST

Chloroplasts are organelles in plant cells that are responsible for the absorption of light energy, the synthesis of carbohydrates, and the evolution of molecular oxygen. Light energy captured by the chloroplast is converted into potential chemical energy in the form of carbohydrate and in this state starts the "energy chain" in nature. The process is called **photosynthesis.** The energy is first made available to satisfy the needs of the cell and whole plant carrying out photosynthesis; the energy is then passed on to the consumers of the plant and their subsequent predators. The oxygen evolved during the capture of light energy becomes the ultimate oxidizing agent for cellular metabolic reactions, as described in Chapter 10. Mitochondrial oxidations, as well as other oxidations in plant, animal, and microbial cells, depend on this primary source of oxygen. It is currently believed that the entire supply of oxygen in the atmosphere today was derived from and is presently maintained by photosynthesis.

A **chloroplast** is any membrane-encased organelle containing **chlorophyll** that belongs to a group of related organelles in plants called **plastids.** The plastids have a variety of morphological forms, carry out diverse functions, and store many different compounds. For example, the *amyloplast* is the starch-storing plastid of potato tubers; the *chromoplast* is the lycopene-containing plastid that gives the fruit of tomatoes its red color. Each of the diverse plastids is believed to arise from a common **proplastid** precursor. These organelles are found exclusively in the cells of green plants. All of the major groups of plants, with the exception of the fungi, contain chloroplasts. There may be a single chloroplast or dozens of chloroplasts in each cell. Most frequently, the simpler plants, such as the algae, contain single chloroplasts, while the higher plants, such as the cone-bearing and flowering plants, have many chloroplasts in each cell. For most laboratory studies of photosynthesis, the single-celled plant *Chlorella* is employed. This alga has one cup-shaped chloroplast that practically fills the cell. The organism can easily and conveniently be cultured with artificial lighting in the laboratory in solutions of inorganic salts. Because of the one-to-one relationship between cell and chloroplast and the ease with which the cells can be grown and enumerated, quantitative studies using chloroplast preparations of known content are possible.

For studies with chloroplasts of higher plants, leaves are generally used, and spinach and parsley leaves are probably the most popular source. In leaves, the chloroplasts are found in greatest numbers in two internal tissues, the *palisade parenchyma* and the *spongy parenchyma* mesophyll. Both tissues lie between an upper and lower epidermis (Fig. 17–1) and have thin cell walls that are easily broken. The number of chloroplasts in each cell varies within an organism with changing environmental conditions and varies greatly from one species to another. In spinach, there are

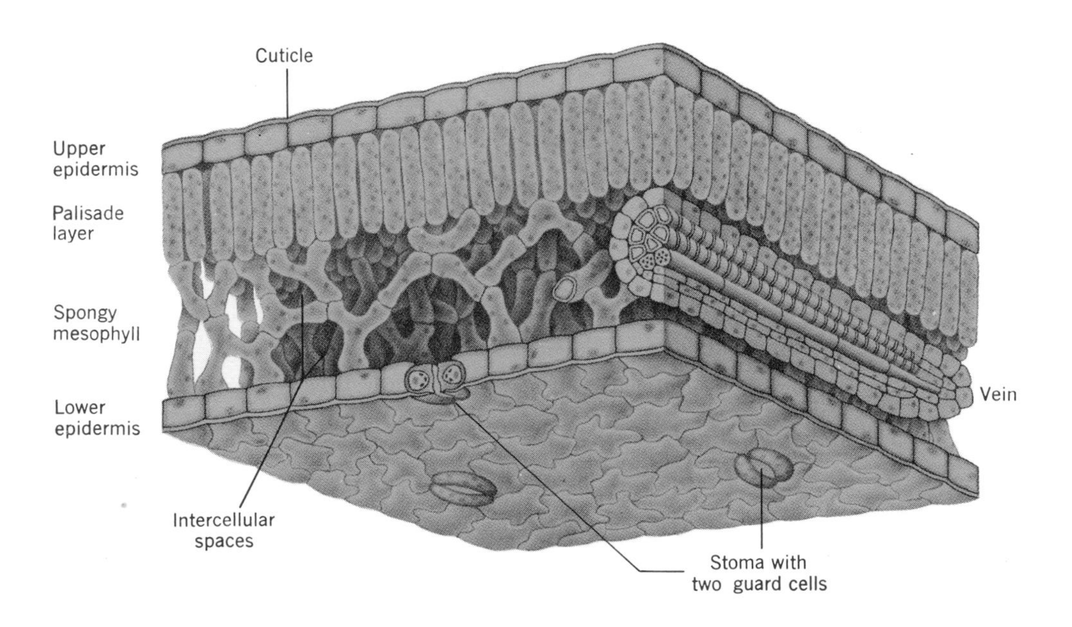

Figure 17–1
Diagram of a cross section of a leaf showing the position of the chloroplast-containing tissues, the palisade, and spongy parenchyma.

between 20 and 40 chloroplasts in each palisade parenchyma cell. In the palisade parenchyma, the chloroplasts lie along the side walls of the cell, the center of the cell being filled with large vacuoles. In the spongy parenchyma, the chloroplasts are more randomly distributed throughout the cytoplasm of the cell. In many genera, cytoplasmic streaming (i.e., *cyclosis*) moves the chloroplasts about the cell, and in a few instances, an active amoeboid-type of movement of the chloroplasts has been observed.

Chloroplasts are routinely isolated from plant tissues by differential centrifugation following the disruption of the cells. Leaves are homogenized in an ice-cold isotonic saline solution such as 0.35 *M* NaCl buffered at pH 8.0. The disruption is generally carried out with short spins in a Waring blender. After a preliminary filtration through nylon gauze (20 μm pore size) to remove the larger particles of debris, cell nuclei, tissue fragments, and unbroken cells, the chloroplasts are separated by centrifugation at 200 *g* for 1 minute. The chloroplast-rich pellet is then resuspended and centrifuged again at 2000 *g* for 45 seconds to resediment the chloroplasts. Chloroplast preparations obtained by this procedure are generally mixtures of intact and broken organelles. Since the chemical composition, rate of photosynthetic activity and other properties of intact chloroplasts differ significantly from those of damaged organelles, it is often desirable to separate the two populations. This may be accomplished by rate or isopycnic density gradient centrifugation of the chloroplast preparation using sucrose, Ficoll, or Ludox gradients.

Chloroplast size is quite variable. Although the average diameter of a chloroplast in higher plant cells is between 4 and 6 μm, the size may fluctuate according to the amount of available illumination. In sunlight, chlorophyll is more readily synthesized by the plant, and the chloroplasts increase in size; in the shade, chlorophyll synthesis declines, and there is a corresponding reduction in chloroplast size. Polyploid cells contain larger chloroplasts than comparable diploid cells. Changes in chloroplast size and shape are also observed after short-term exposure of plants to light. Short-term light exposure produces a small but measurable decrease in chloroplast volume. Presumably, this is due to a light-induced production of ATP, for the addition of ATP to chloroplasts in the dark causes a reduction in volume.

The shape of most chloroplasts in higher plants is spheroid, ovoid, or discoid (Fig. 17–2). Other irregular shapes sometimes occur but are more common in lower plants. For example, in algae, cup-shaped chloroplasts as well as spiral bands, star shapes, and digitate forms are observed. The shape and structure of chloroplasts can also be altered by the presence of starch granules. During periods of active photosynthesis, the sugars formed in the chloroplast are polymerized into starches that precipitate as small granules. The starch granules are usually ellipsoidal and up to 1.5 nm long.

Fine Structure of the Chloroplast

The chloroplasts of higher plants are composed of two membrane layers similar to those of mitochondria. Each membrane is about 50 Å thick. The *outer membrane,* which lacks folds or projections, serves to delimit the organelle and regulate the transport of materials between the cytoplasm and the interior of the organelle. The *inner membrane* parallels the outer membrane, but inward folds of this membrane are extensive. The inward growth of the inner membrane gives rise to a series of internal parallel membranes called **lamellae** (Fig. 17–3*b*). The interior of the chloroplasts in which the lamellae are suspended is a granular fluid that appears somewhat electron-dense in electron micrographs. This matrix is referred to as the **stroma.** The lamellae form a complex series of membranes through the stroma.

Most of the lamellae are organized to form disk-shaped sacs called **small thylakoids.** The small thylakoids are often arranged in stacks called **grana** (sing. **granum**) having a diameter of about 3000 to 6000 Å. Since these thylakoids are round, the grana appear much like a stack of coins (Figs. 17–3*a* and 17–3*b*). A typical chloroplast has between 40

Figure 17–2
Electron photomicrographs of chloroplasts of higher plants. (*a*) tobacco (Courtesy of P. Seyer and D. Marty; copyright 1975, North-Holland Publishing Co., *Cell Differentiation 4,* 191.) (*b*) sugar beet (Courtesy of W. Laetsch.)
Figure 17–2 continues on the next page.

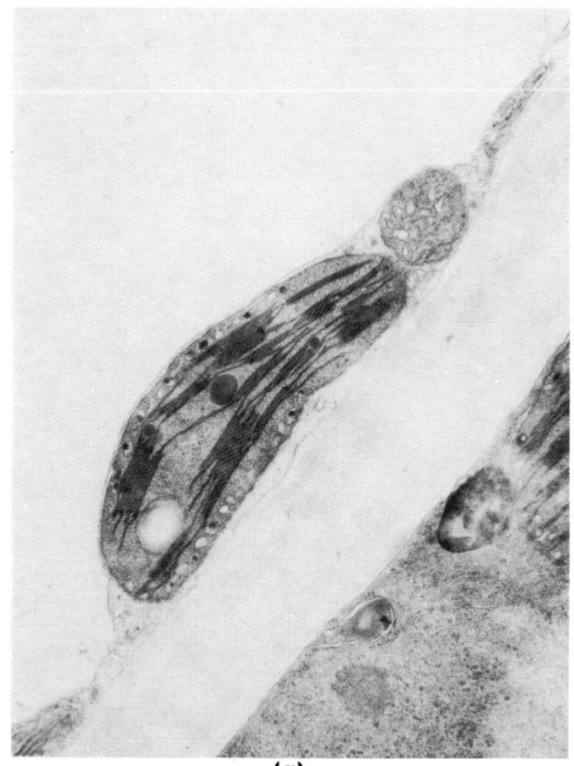
(a)

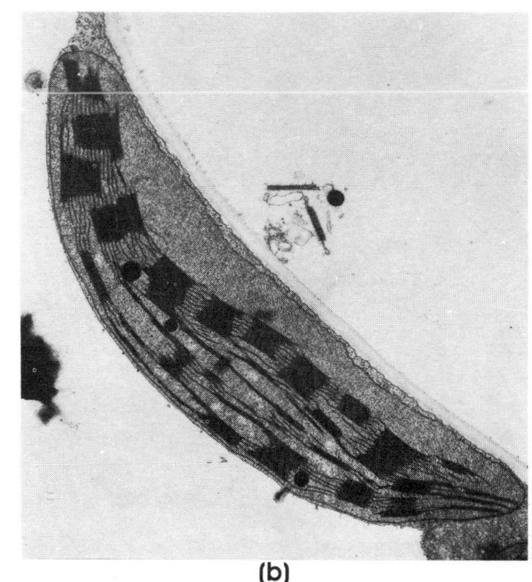
(b)

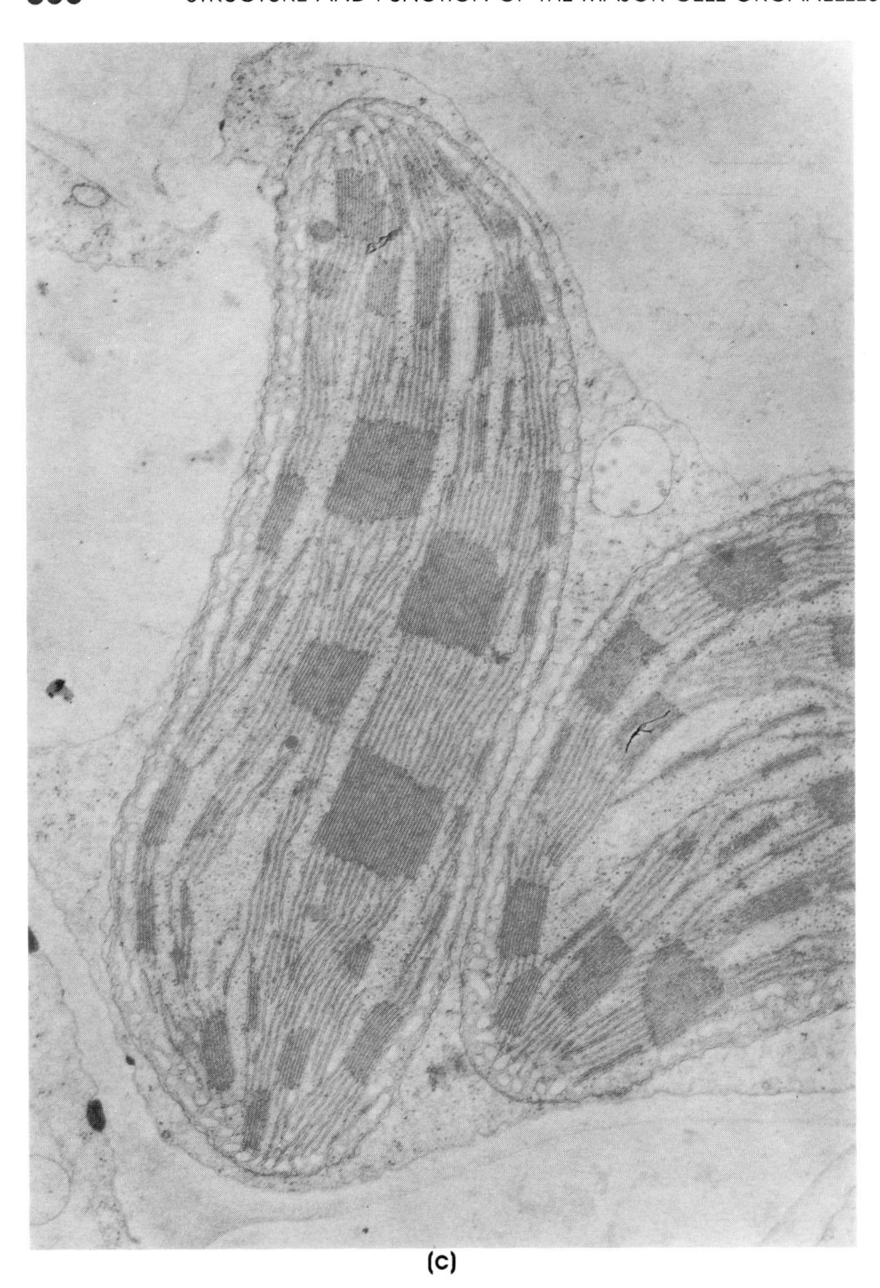

(c)

Figure 17–2 (*continued*)
Electron photomicrograph of chloroplasts of higher plants. (*c*) grass (Courtesy of W. Laetsch.)

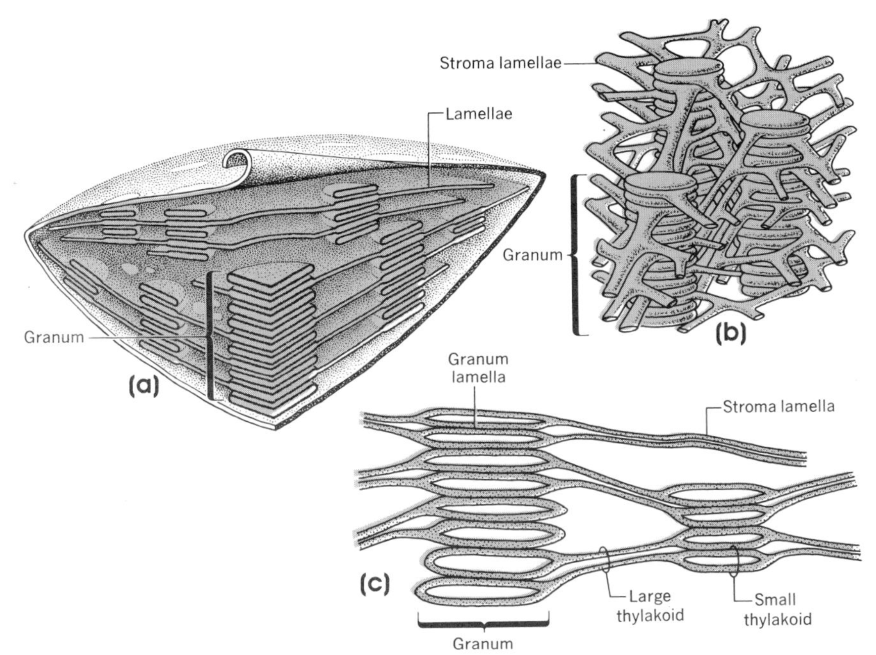

Figure 17–3
Diagram of the membranes of the chloroplast. (*a*) Cross section showing the position of lamellae and grana within the chloroplast. (*b* and *c*) Thylakoids of grana and stroma and relationship to lamellae.

and 60 grana, and each may be composed of 2 to 100 small thylakoids. Frequently, a small portion of the thylakoid extends radially into the stroma forming a branching tube, or **large thylakoid** that communicates with other small thylakoids and grana (Fig. 17–3*c*). Collectively, the branching and anastomosing network is called the **stroma lamellae.**

Structure of the Thylakoid

The adjacent membranes of neighboring thylakoids within each of the grana form thick layers called **grana lamellae.** Electron photomicrographs of grana lamellae fixed with glutaraldehyde and stained with osmium reveal a five-layered arrangement consisting of three dark 40-Å-thick osmiophilic layers enclosing two 17-Å-thick light osmiophobic spaces. Freeze-fracture techniques indicate that these membranes contain numerous particles. The particles, which are undoubtedly protein in composition, appear to be of two basic sizes (Fig. 17–4). The stroma lamellae contain only the smaller size particles. However, the grana lamellae contain both large and small size particles.

When the inner surface of the thylakoid of some plants such as spinach is examined by combined freeze-etch and shadowing techniques, a regular array of discrete internal units is observed (Fig. 17–5). These units have been termed **quantasomes.** Each is about half protein and half lipid and measures 185 x 155 x 100 Å. The quantasomes are occasionally arranged in a paracrystalline manner at the center of the thylakoid but are more random at the periphery. Each quantasome appears to be composed of four smaller subunits.

Stroma Structures

The granular stroma contains a variety of particles. The presence of starch granules was noted earlier. Electron micrographs also reveal a number of osmiophilic granules and groups of ellipsoidal structures called *stromacenters*. Of particular interest are the strands of DNA scattered through the stroma and ribosomelike particles. Some of these structures are shown in Figure 17–2, which also clearly depicts the relationship between the stroma and grana lamellae.

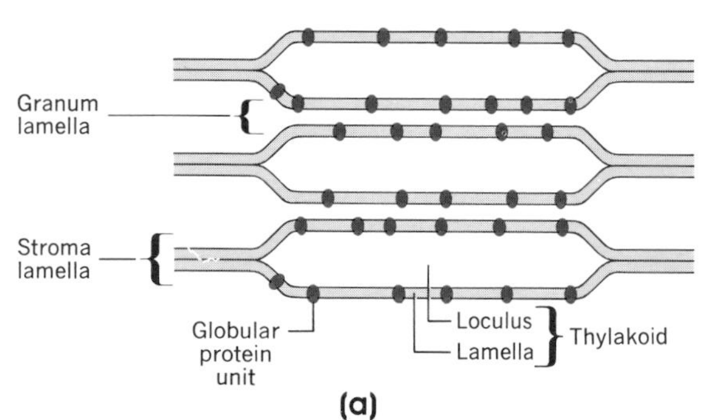

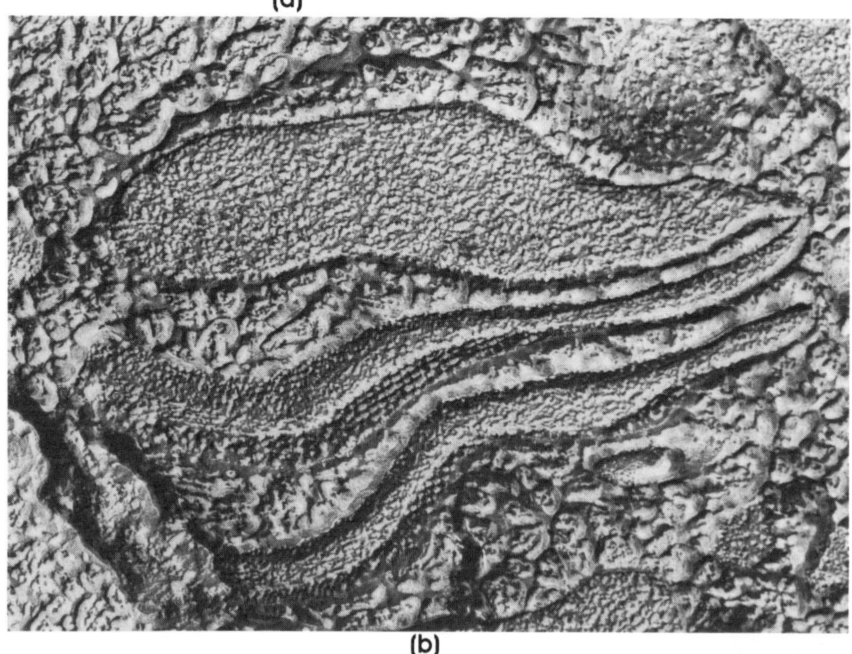

Figure 17–4
(*a*) Diagram of the stroma lamellae and grana lamellae showing layering and arrangement of large protein particles in membranes. (*b*) Electron photomicrograph of partly unfolded grana membrane that has been fractured and sublimed. The sublimation has revealed surfaces that show the large particles and small particles. (Electron photomicrograph courtesy of K. Muhlethaler; copyright Springer-Verlag, 1965, *Planta 67,* 305.)

Chemical Composition of Chloroplasts

The organic constituent present in greatest quantity in the chloroplast is protein. Up to 69% of the dry weight of the chloroplast may be protein (Table 17–1). For leaf cells, 75% of the total cell nitrogen is found within the chloroplasts. Both structural and soluble proteins have been identified, but only a few of these have been extracted and purified. A peptide analysis by SDS gel electrophoresis shows compositional differences between stroma and grana lamellae, but the differences are primarily quantitative rather than qualitative. The relative compositions of stroma lamellae and grana lamellae are shown in Table 17–2.

Essentially all the pigments and cytochromes are located in the lamellae. The stroma lacks these compounds but contains DNA and RNA which are not present in the lamellae. Most of the RNA is associated with the ribosomes of the stroma. The amount of DNA is low; estimates are 10^{-15} to 10^{-14} g per chloroplast or about 0.03% of its dry weight. However, this is enough to carry ample information for the synthesis of chloroplast proteins including many enzymes active in photosynthesis. The disposition of chloroplast DNA during chloroplast division is unclear.

Lipids and lipid-soluble pigments account for about 34% of the dry weight of the spinach chloroplast. An exceedingly large number of different lipid compounds have been identified. The more common lipids are the galactosyl digly-

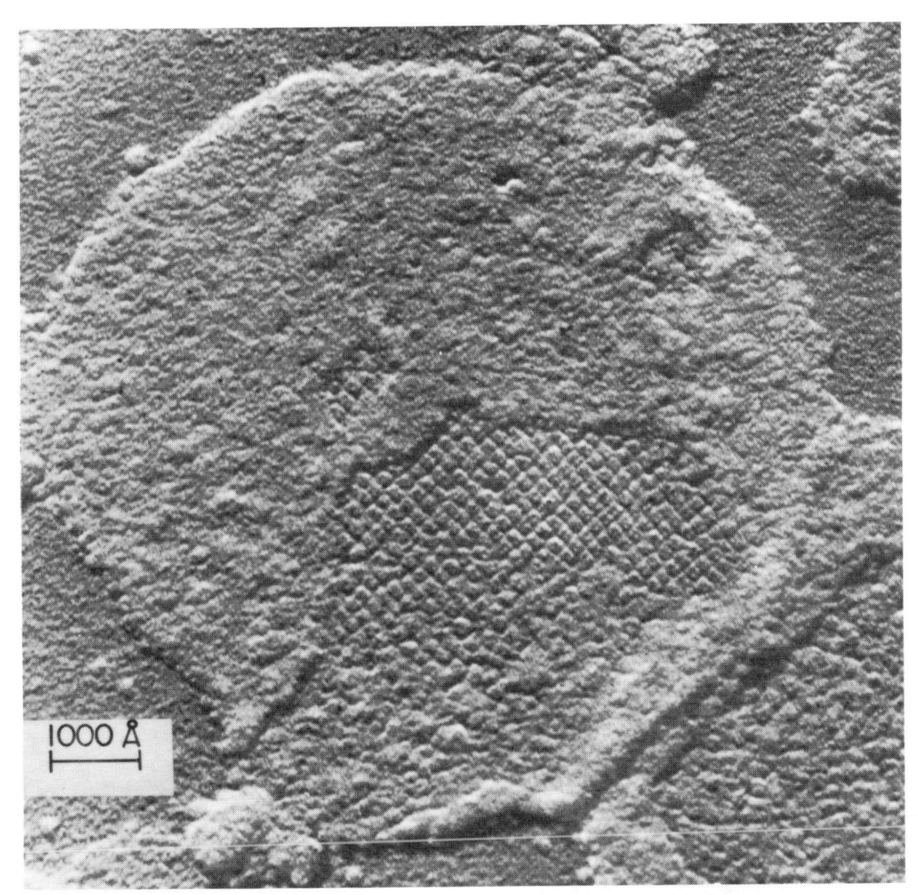

Figure 17–5
Quantasomes. (Electron photomicrograph courtesy of R. B. Park and J. Biggins; copyright 1964, American Association for the Advancement of Science, *Science 144,* 1009.)

cerides, phospholipids, quinones (including vitamin K), and sterols.

The Chlorophylls

The green pigments of the chloroplast and the main sources of the color of green plants are the chlorophylls. Although a large number of chemically distinct chlorophylls have been identified in a variety of different plants, the structures of these chlorophylls are basically the same. (The structures of chlorophylls *a* and *b* are given in Fig. 17–6.) It is customary to identify each chlorophyll by a different letter. All photosynthetic plants have been found to contain chlorophyll *a,* but the presence of the secondary chlorophylls *b, c,* or *d* depends on the type of plant. Higher plants usually have chlorophyll *b*. Photosynthetic bacteria contain a unique chlorophyll called *bacteriochlorophyll*. Together, chlorophylls *a* and *b* represent about 5% of the dry weight of the spinach chloroplast, with an *a:b* weight ratio of 2.05 to 3.52. In most plants, the *a:b* ratio varies according to the light intensity. For example, alpine plants have a ratio of 5.5. The ratio is much lower in shade plants.

Each chlorophyll has a characteristic light absorption spectrum. The absorption spectra of chlorophylls *a* and *b* are shown in Figure 17–7. Chlorophyll *a* has absorption maxima at 430 and 670 nm, whereas the absorption maxima

Table 17–1
Chemical Composition of Spinach Chloroplasts

Component	Percentage of Chloroplast Dry Weight	
	Chloroplasts Isolated in Water	Values Corrected for Loss of Soluble Protein
Total protein	50	69
Water-insoluble protein	50	31
Water-soluble protein	0	38
Total lipid	34	21
Chlorophyll	8	5
Carotenoids	1.1	0.7
Ribonucleic acids	—	1.0–7.5
Deoxyribonucleic acids	—	0.02–0.1
Carbohydrate (starch, etc.)	Variable	

Source: Modified from J. T. O. Kirk, and R. A. E. Tilney-Bassett, *The Plastids,* W. H. Freeman and Co., San Francisco, 1967.

Table 17–2
Major Components of Stroma and Grana Lamellae

	Stroma Lamellae	Grana Lamellae
Total chlorophyll	278[a]	401
Chlorophyll a	238	281
Chlorophyll b	40	130
P_{700}	2.5	0.6
β-Carotene	21	17
Lutein	10	29
Violaxanthin	15	20
Neoxanthin	8	16
Phospholipid	76	66
Monogalactosyl diglyceride	231	214
Digalactosyl diglyceride	172	185
Sulfolipid	65	59
Cyt *b* (total)	1.0	3.4
Cyt *f*	0.5	0.7
Manganese	0.3	3.2

[a]Values in micromoles of component per gram of membrane protein.

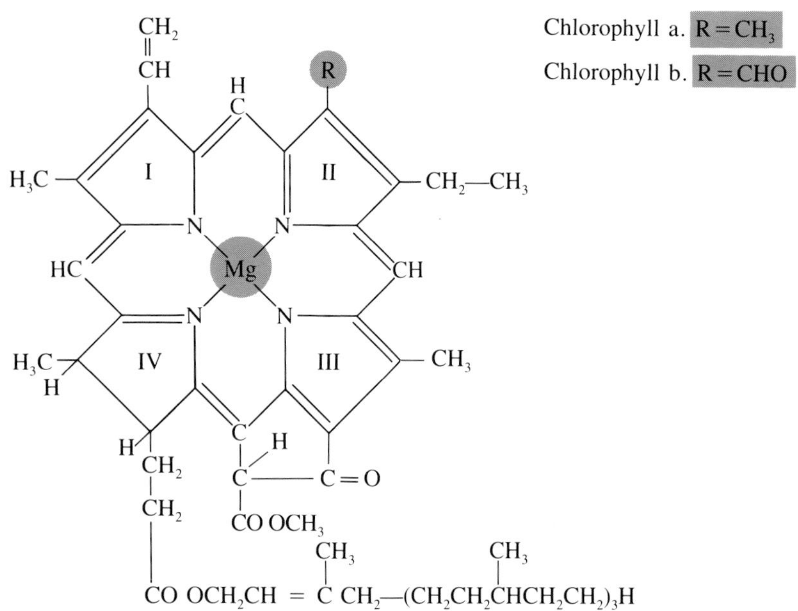

Figure 17–6
Structures of chlorophyll *a* and *b*.

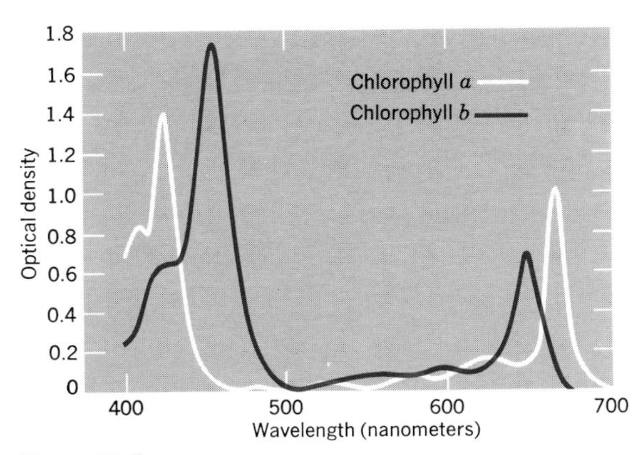

Figure 17–7
Absorption spectra of chlorophylls *a* and *b* in an ether solvent.

of chlorophyll *b* occur at 455 and 640 nm. Absorption maxima of other plant and bacterial pigments are indicated in Table 17–3. The absorption spectrum and maxima of plant pigments vary according to the solvent used for extraction. Experiments conducted in vivo indicate that the native absorption maximum for chlorophyll *a* occurs at 677 nm. Absorption maxima determined in vivo or in extracts are the result of mixtures of chlorophyll molecules of the same type. However, each molecule exhibits its own characteristic maxima, and these can differ somewhat from the positions of the average peaks. One very important form of chlorophyll *a* that is readily bleached by light has an absorption maximum at 700 nm. This form, which represents only about 0.1% of the total chlorophyll *a* molecules present in a sample, is called P_{700} or chlorophyll a_I. The roles of the various chlorophylls in photosynthesis are discussed later in the chapter.

The Carotenoids

The carotenoids are all long-chain isoprenoid compounds having an alternating series of double bonds. Although these compounds are synthesized only in plant tissue and participate in photosynthesis, they also serve as precursors of vitamin A in animal tissues. Most carotenoids are yellow, orange, or red. The formulas of α-, β, and γ carotene are shown in Figure 17–8, and their absorption maxima are given in Table 17–3. Most of these pigments are located in the chloroplast lamellae and are believed to function as accessory pigments for light absorption during photosynthesis.

Table 17–3
Absorption Maxima[a] of Plant and Bacterial Pigments

Pigment	Wavelength (nm)	Occurrence
Chlorophyll *a*	430, 670	All green plants
Chlorophyll *b*	455, 640	Higher plants; green algae
Chlorophyll *c*	445, 625	Diatoms; brown algae
Bacteriochlorophyll	365, 605, 770	Purple and green bacteria
α-Carotene	420, 440, 470	Leaves; some algae
β-Carotene	425, 450, 480	Some plants
γ-Carotene	440, 460, 495	Some plants
Luteol	425, 445, 475	Green leaves; red and brown algae
Violaxanthol	425, 450, 475	Some leaves
Fucoxanthol	425, 450, 475	Diatoms; brown algae
Phycoerythrins	490, 546, 576	Red and blue-green algae
Phycocyanins	618	Red and blue-green algae
Allophycoxanthin	654	Red and blue-green algae

[a]Absorption maxima vary according to the solvent in which the pigment is dissolved.
Source. From E. Rabinowitch and Govindjee, *Photosynthesis*, John Wiley & Sons, Inc., New York, 1969.

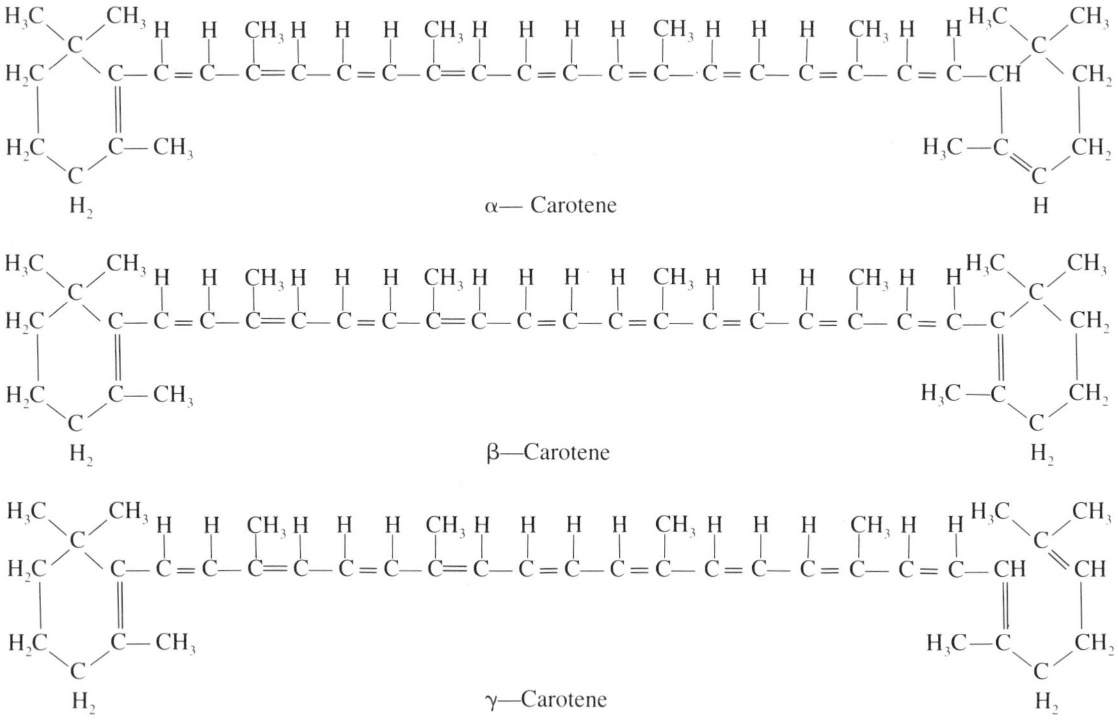

α— Carotene

β—Carotene

γ—Carotene

Figure 17–8
Fomulae of alpha, beta, and gamma carotene.

Location and Arrangement of the Pigment

Both the chlorophylls and the carotenoids are located almost exclusively in the chloroplast lamellae. The lamellae are about 52% protein and 48% lipid, and the two pigments reside primarily in the lipid component. Some lipid is also found in the osmiophilic granules of the stroma, but these are not believed to contain chlorophyll. Because each chlorophyll molecule has a hydrophilic portion (the tetrapyrrole) and a lipophilic portion (the phytl chain), the chlorophyll molecules are thought to be aligned in a specific manner within the lamallae. The pyrrole groups form weak bonds with the lamellar protein, while the phytyl chains extend into the lamellar lipid. The carotenoids are dissolved in the lipid adjacent to the chlorophyll molecules.

It has been calculated that each quantasome contains about 230 chlorophyll molecules and 48 carotenoid molecules. Such a unit would also contain 1 molecule of P_{700}. On the basis of these calculations, as well as physiological studies, it has been proposed that the quantasome is the fundamental photosynthetic unit.

The chloroplast stroma contains many of the enzymes associated with photosynthesis. Chloroplast protein synthesis also takes place in the stroma. DNA strands about 150 μ long have been isolated from the chloroplast along with ribosomes and polyribosomes. Chloroplast ribosomes belong to the 70 *S* class and contain 23 *S* and 16 *S* RNA (see Chapter 21); these ribosomes are smaller than those found in the cytoplasm of the plant cell.

Development of Chloroplasts

Chloroplasts develop by a pinching process (Fig. 17–9) or from proplastids in the cell. The cells of young shoots of higher plants may contain 20 to 40 of these small bodies. These proplastids, which are ovoid to round in shape and are surrounded by two membranes, can develop into a num-

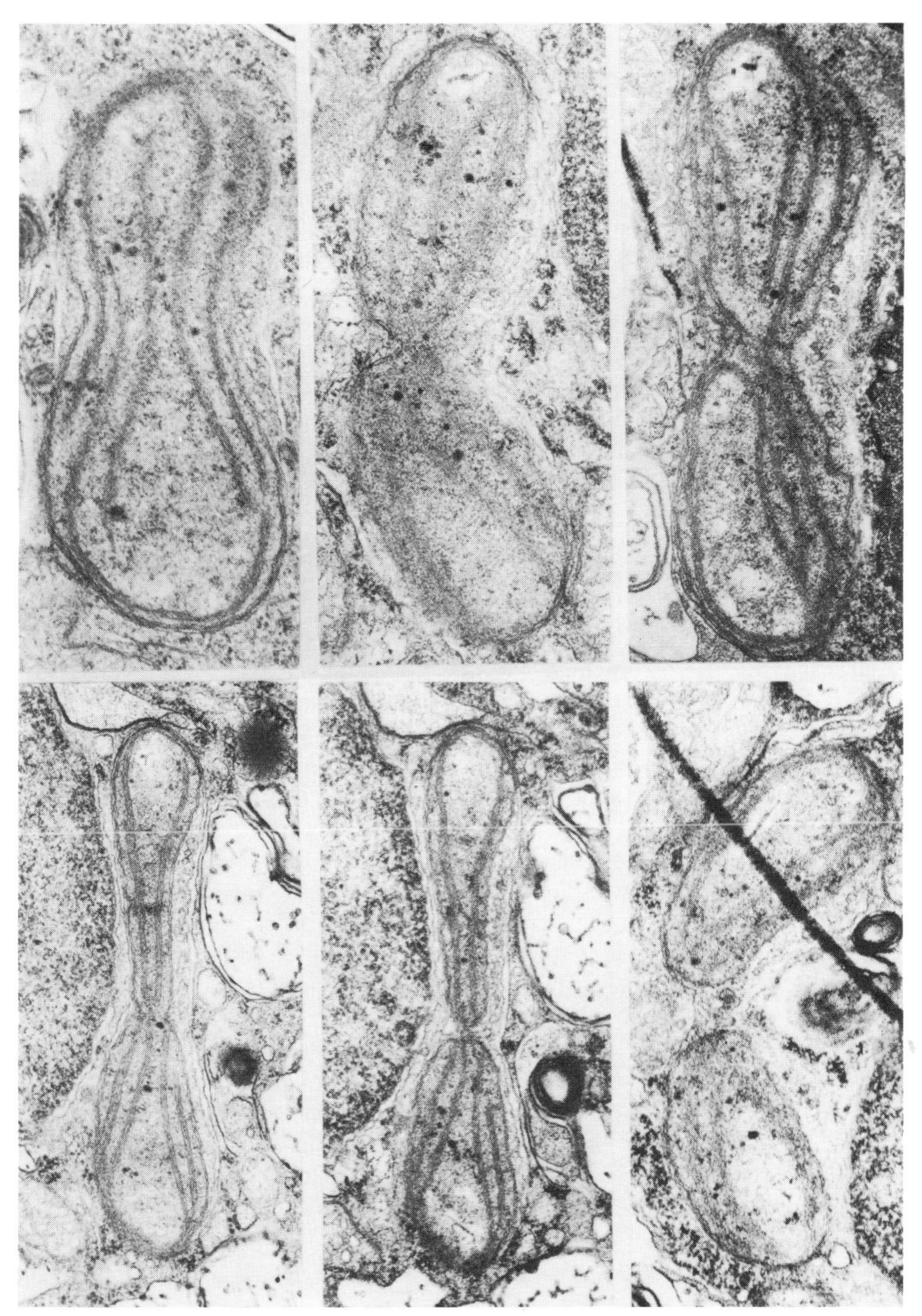

Figure 17–9
Sequence of chloroplast division stages in the brown alga, *Fucus.* (Photomicrographs courtesy of Dr. E. G. Pollock.)

ber of different kinds of plastids in addition to chloroplasts. As the chloroplast develops, the inner membrane gives rise to internal membranes, which then form the lamellae and thylakoids. Proplastids, as well as mature chloroplasts, increase in numbers by a form of division.

The control of development of chloroplasts is not fully understood. Clearly, the chloroplast contains information in the strands of DNA found with the organelle. Proteins are synthesized by the ribosomes in response to messages from this DNA. Whole chloroplasts may divide and the resultant halves grow. When a plant propagates by the production of seeds, the seeds rarely contain chloroplasts, but they do contain proplastids that develop into chloroplasts as the seed germinates and matures. *Euglena* cells treated with streptomycin lose their chloroplasts. Plants treated in this manner and then washed free of streptomycin are unable to resynthesize chloroplasts. Apparently, some organellar genetic and developmental mechanism is necessary. Information in the cell nucleus and cytoplasmic mechanisms alone cannot produce a chloroplast. However, there are examples of the genetic influence of the nuclear DNA on the morphology of the chloroplast.

The hypothesis that the evolutionary ancestors of chloroplasts and mitochondria were procaryotic organisms that "invaded" other organisms and established a symbiotic relationship with the host was discussed in Chapter 16. In the case of the chloroplast, the "invaded cell" could have been a heterotrophic eucaryotic cell. Today, it is possible to observe the inclusion of chloroplasts in animal cells. Nudibranchs, marine snails lacking a shell, feed on algae. As the cytoplasm of the algae cells passes through the gut of the nudibranch, whole chloroplasts are absorbed into the gut tissue and may persist for the life of the animal. When the animal is in light, oxygen evolution can be detected.

Photosynthesis also occurs in many procaryotic organisms such as the blue-green algae and photosynthetic bacteria. These procaryotes do not have true chloroplasts; instead, they have lamellated structures called **chromatophores** that carry out only the light-absorbing reactions of photosynthesis but not the carbohydrate-synthesizing reactions.

Photosynthesis—Historical Background

The overall reactions of photosynthesis may be summarized by the equation

$$6\ CO_2 + 12\ H_2O \xrightarrow[\text{chlorophyll}]{\text{light}} C_6H_{12}O_6 + 6\ O_2 + 6\ H_2O$$

The studies of many individuals over the past 300 years have led to our present understanding of this process. One of the first studies was that made in 1648 by Jan Baptista van Helmont (Dutch, 1577–1640). Van Helmont planted a 5-lb willow shoot in a large pot containing 200 lb of soil. The plant was regularly watered over a 5-year period and was then carefully removed and weighed. While the willow tree was found to weigh almost 170 lb, the original soil weighed only a few ounces less than the original 200 lb. Van Helmont concluded that the increase in weight of the willow tree was due primarily to the addition of water, for he was not aware of the role played by gases in the air in the growth of the plant. Today, we also recognize that the small quantity of material removed from the soil included minerals, nitrogen, phosphate, and calcium salts.

Joseph Priestley (English, 1733–1804) was the first to show that plants exchange gases with the atmosphere. In particular, Priestley found that oxygen was evolved by plants during photosynthesis. Late in the eighteenth century, Jan Ingenhousz (Dutch, 1730–1799) showed that oxygen was produced by the green parts of plants when exposed to light. Furthermore, he observed that the amount of oxygen produced varied according to the amount of light to which the plant was exposed. In 1804, Nicholas de Saussure (Swiss, 1767–1845) found that the increase in the carbon content of plants resulted from the accumulation of carbon from carbon dioxide in the air. De Saussure also showed that leaves respire in darkness—that is, they take in oxygen and release carbon dioxide.

The studies of Julius Sachs in the late nineteenth century showed that chlorophyll was confined to the chloroplasts and was not distributed throughout the plant cell. He also showed that sunlight causes chloroplasts to absorb carbon dioxide and that chlorophyll is formed in chloroplasts only in the presence of light. Sachs also noted that one of the products of photosynthesis was starch. It was not until 1918 that Wilstätter and Stoll isolated and characterized the green pigments chlorophyll *a* and *b*.

On the basis of these early findings, the reactions of photosynthesis were described by the equation

$$6\ CO_2 + 6\ H_2O \xrightarrow[\text{chlorophyll}]{\text{sunlight}} (C_6H_{12}O_6)_n + 6\ O_2$$

However, photosynthesis is not simply the fusing of carbon dioxide and water molecules through the use of light energy. Instead, two complex series of chemical reactions are involved. One set of reactions, called the **photochemical** or **light reactions,** occurs in the lamellae of the chloroplasts. In these reactions, light energy is absorbed and used to form ATP, and water molecules are split, releasing oxygen and hydrogen; the latter is used in the reduction NADP. The second set of reactions, called the **synthetic** or **dark reactions,** occurs in the stroma of the chloroplasts, and although these reactions do not require light, they do depend on the availability of ATP and $NADPH_2$ from the photochemical reactions in order to reduce the carbon dioxide to form sugars.

Photosynthesis—Photochemical (Light) Reactions

The Absorption of Light By Chlorophyll

The absorption of electromagnetic energy by any atom or molecule often involves a shift of electrons from one atomic orbital to another. Each electron possesses energy, and the amount is determined by the location of the electron orbital in space and the speed at which the electron moves. When an atom absorbs light energy, an electron is either raised to an orbital of higher energy level or accelerated in its orbit. In either case, certain discrete quantities of energy are required, for when light photons have either too much or too little energy, they are not absorbed. Electrons may orbit in pairs within an orbital, the members of the same orbital spinning in *opposite* directions. Most atoms at their lowest energy level (i.e., *ground state*) have all their electrons paired in this fashion and are said to be in the *singlet* (i.e., *S*) state. When a photon of light is absorbed and an electron is thereby raised to a higher, unoccupied orbital, it may continue to spin in the direction opposite its former partner (in which case, the atom is still in the *S* state), or it may spin in the same direction as its former partner (in which case, the atom is said to be in the *triplet* or *T* state).

In a molecule, some electrons orbit exclusively about specific atomic nuclei; others may be shared between two nuclei forming a bond (called localized or π^* electrons), or may orbit about several nuclei (called delocalized or π electrons). The absorption of a light photon may move a π electron to a π^* position.

Chlorophyll is normally a singlet in its ground state. Although the absorption of light causes some π electrons to be raised to a π^* orbital, chlorophyll remains in a singlet state. When red (680 nm) light is absorbed, an electron is raised to a higher, $S\pi^*$, orbital. Blue (430 nm) light possesses more energy per photon than red light, and its absorption raises an electron to a higher (but still $S\pi^*$) energy state. These transitions are summarized in Figure 17–10.

Once a molecule has absorbed light energy and is in an excited state, the ground state may be reestablished in three different ways: (1) the energy may be reemitted in the form of radiation of longer wavelength (i.e., fluorescence); (2) the energy may be converted to heat; and (3) the excited molecule may transfer its excess energy to another molecule. The transfer of energy from one molecule to another often involves the exchange of a high-energy electron for one of lower energy. Referring to Figure 17–10, the blue light absorbed by chlorophyll raises an electron to the $S^b\pi^*$ state. By losing energy in the form of heat, that electron could "drop" back to the $S^a\pi^*$ state. The electron could then drop back from the $S^a\pi^*$ state to the $S\pi$ state (i.e., the ground state) by the immediate loss of energy as heat or fluorescence. Another possibility also exists, for the electron could drop from the $S^a\pi^*$ state to a triplet (i.e., $T\pi^*$) state by heat loss and then to the ground state by either

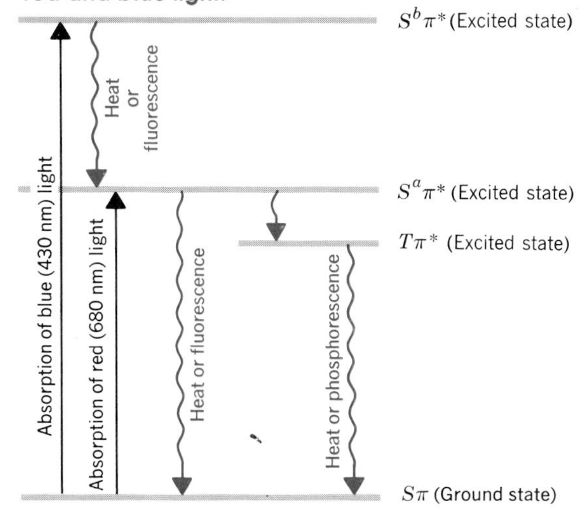

Figure 17–10
Main electronic states in the excitement of chlorophyll by red and blue light.

phosphorescence (delayed fluorescence) or heat loss. Concentrated solutions of extracted chlorophyll will strongly fluoresce red when placed in a beam of sunlight or white light. Finally, it should also be noted that if the chlorophyll molecule is sufficiently excited, it may not return to the ground state by heat loss or reradiation; instead, the excited electron may be transferred to another molecule, leaving the chlorophyll in a temporary, oxidized state.

Primary Photochemical Events in Photosynthesis

As shown in Figure 17–10, chlorophyll absorbs both blue and red light, and this raises electrons to $S^b\pi^*$ or $S^a\pi^*$ states. The return of an electron from the $S^b\pi^*$ state to the $S^a\pi^*$ state is extremely fast (about 10^{-12} seconds) and does not afford an opportunity for the energy to be lost by fluorescence or by transfer to another molecule. Consequently, the energy is lost as heat. The decay of an electron from the $S^a\pi^*$ state, whether brought there by red light absorption or by prior decay from the $S^b\pi^*$ state, *does* permit the transfer of energy to another molecule, and this is the event that initiates photosynthesis. Consequently, a photon of red light is just as effective in initiating photosynthesis as a photon of blue light, even though the former is much less energetic.

As we have already noted, the chloroplast contains many different pigment molecules (other chlorophylls, carotenoids, phycobilins, etc.) and the electrons of these molecules may be excited to various energy states by the absorption of light. As these excited accessory pigment molecules return to the ground state, the resulting energy is transferred to chlorophyll *a* molecules, causing their excitation. Since the chlorophyll *a* molecules present in a quantasome vary in their absorption maxima, varying quantities of energy are required to raise their electrons to the $S^a\pi^*$ state. The molecule that requires the least energy is postulated to be a pigment absorbing long, red wavelengths—namely the P_{700} molecule. It seems reasonable that light energy captured by the accessory pigments and transferred to chlorophyll *a* is in turn transferred from the latter to P_{700}. Since it appears that only one molecule of P_{700} is present in each quantasome, the quantasome may be the basic unit for the absorption of light energy. Each accessory pigment or chlorophyll *a* molecule can only pass its energy on to a pigment having an absorption maximum of longer wavelength because these require less energy to be activated to the $S^a\pi^*$ state. Since P_{700} has its absorption maximum at the longest wavelength, it serves as the final energy trap.

Two Photosystems

When some plants are exposed to light having a wavelength of 690 nm or longer, photosynthetic efficiency decreases. The effect is called the **red drop** (Fig. 17–11). Since the absorbed energy is funneled to P_{700}, it would be expected that the absorption of light by the accessory pigments, chlorophyll *a* or even P_{700}, should be equally efficient. The efficiency can be increased through the addition of shorter wavelength radiations. This *enhancement* phenomenon can increase the photosynthetic rate 30 to 40% above the rate obtained by either the short wavelength or long wavelength alone. The synergistic effect of the two different wavelengths led early investigators to conclude that two distinct photochemical reactions exist.

The photoreaction sequence involving pigments absorbing light above 690 nm is called **photosystem I.** If the photosynthetic rate of a plant is enchanced by shorter wavelength of light, the second series of photoreactions is called **photosystem II.** In higher plants, photosystem I is a unit

Figure 17–11
The photosynthetic activity or *action spectrum* (i.e., oxygen evolution) and *absorption spectrum* of the green alga *Ulva*. The difference between the two curves at wavelengths above 690 nm is called the *red drop*.

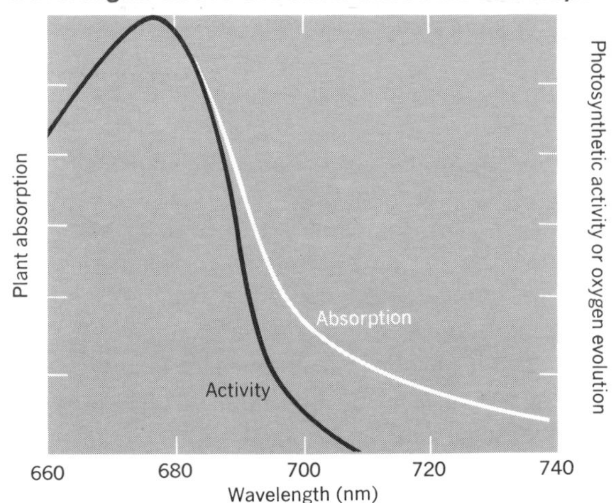

containing about 200 molecules of chlorophyll of mostly type *a*, 50 carotenoids, one chlorophyll P_{700} molecule, one cytochrome *f*, one plastocyanin, two cytochrome b_{563}, and one or two ferredoxin molecules. Photosystem II has about 200 molecules of chlorophyll of both *a* and *b* types, 50 carotenoids, a trapping chlorophyll (a primary electron donor), a primary electron acceptor that is believed to be a quinone, four plastoquinones, six Mn atoms, and two cytochrome b_{559} molecules.

Sequence of Energy (Electron) Flow

The absorption of light by chlorophyll alters the state of the orbiting electrons, and if the energy is not lost by reradiation or heat, the excited electrons can be transferred to another compound. Such an electron loss "bleaches" the chlorophyll and leaves it in an oxidized state. Oxidized P_{700} may be reduced by absorbing an electron from photosystem II, while the reduction of the oxidized chlorophyll of photosystem II is brought about by the oxidation of water.

In 1938, Robin Hill demonstrated that isolated chloroplasts exposed to light could evolve oxygen and reduce a variety of compounds without consuming carbon dioxide. Three years later, S. Ruben and M. Kamen, using the isotope ${}^{18}_{8}O$, were able to show that the oxygen liberated during whole plant photosynthesis was derived from water molecules. That is,

$$6\ CO_2 + 12\ H_2{}^{18}O \xrightarrow{\text{light}} C_6H_{12}O_6 + 6\ {}^{18}O_2 + 6\ H_2O$$

The method by which the water molecules are split is unknown, although four water molecules are required for the evolution of one oxygen molecule and four quanta of light are necessary:

$$4\ H_2O \xrightarrow{4\ h\nu} O_2 + 4\ (H^+ + e^-) + 2\ H_2O$$

Figure 17–12
Absorption of light energy, transfer of electrons between photosystems, and the regeneration of reduced chlorophylls.

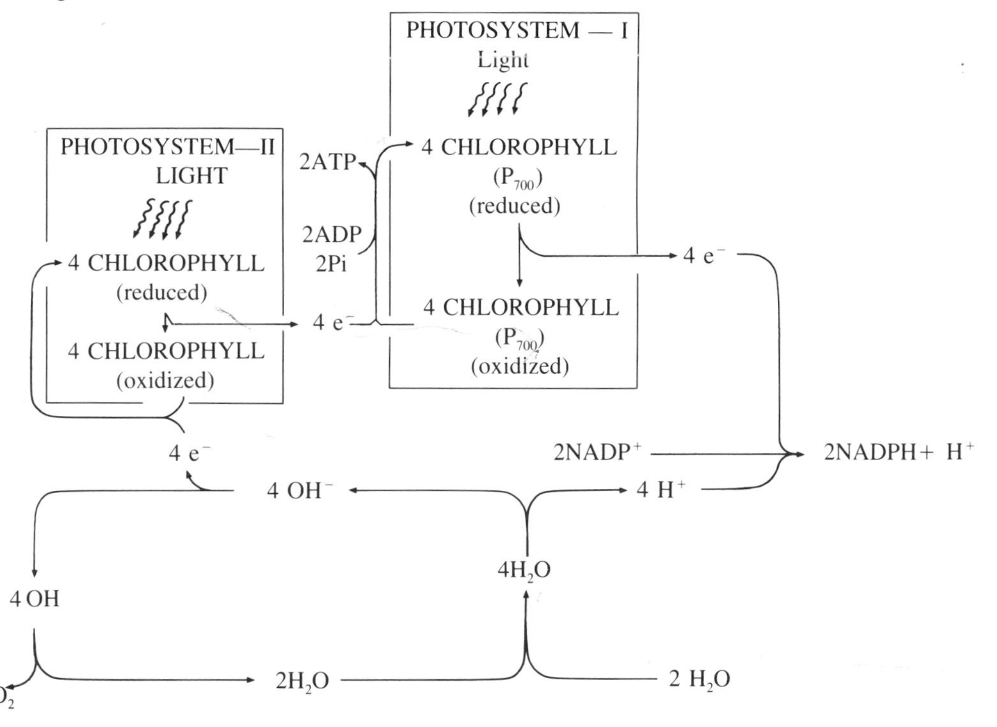

Protein-bound Mn and Cl may also be required. Some of these reactions are summarized in Figure 17–12.

Redox Reactions

Electrons are always transferred from activated chlorophyll, P_{700}, and other molecules in the chloroplast to more positive or oxidized molecules. Each molecule has a different amount of electrical potential called **redox potential.** The redox potential can be measured and the molecules listed in order of the magnitude of their potential. During electron transfer, molecules may accept electrons from less positive (more negative) molecules and may donate electrons to more positive (less negative) molecules. When a molecule accepts electrons, it is said to be *reduced,* and when electrons are given up, the molecule is said to be *oxidized*. The transfer of electrons following the absorption of light energy by chlorophyll therefore involves a sequence of oxidation-reduction reactions. This sequence is shown in Figure 17–13, which also identifies the intermediate electron acceptors and their redox potentials.

Photosystem I is located in the membranes of the grana thylakoids and the stroma thylakoids. Photosystem II is found only in the membranes of the grana thylakoids. Therefore, in the stroma thylakoids, photosystem I functions independently but may function in conjunction with photosystem II in the grana thylakoids. Absorption of light ultimately by P_{700} of photosystem I causes the ejection of electrons to ferredoxin (Fig. 17–13). Other molecules can be substituted as electron acceptors experimentally, but it would appear that the membrane-bound nature of ferredoxin and the positioning of P_{700} make this electron shift requisite in the chloroplast.

The ferredoxin appears to be reoxidized in the chloroplast by the transfer of electrons to either of two compounds, NADP or cytochrome *b*. The reduction of NADP requires a flavoprotein (ferredoxin-NADP reductase) and the absorption of H^+ from the reduction pool created by the splitting of H_2O, that is,

$$NADP^+ + (H^+ + e^-) + H \longrightarrow NADPH + H^+$$

The reduced $NADPH + H^+$ thus generated in the lamellae spills into the stroma, where it is reoxidized in the dark reactions (described later). The reduction of ferredoxin by cytochrome *b* initiates a sequence of redox reactions passing electrons on to cytochrome *f*, plastocyanin, and ultimately to P_{700}, completing a cycle. The function of this cyclic set of redox reactions is coupled to the phosphorylation of ADP (*cyclic photophosphorylation,* described below).

In the grana lamellae the P_{700} oxidized by the absorption of light can be reduced by the cyclic return of the electrons from ferredoxin or by the transfer of electrons from photosystem II. In photosystem II (Fig. 17–13), light energy is transferred to an ultimate trap chlorophyll molecule (or other ultimate electron donor) by the accessory pigments. The ejected electrons are absorbed by an electron acceptor, which is currently unidentified but could be a quinone. This photoevent initiates another set of redox reactions in which the electrons pass from the unidentified electron acceptor (quinone?) to plastoquinone to cytochrome b_{559} to the cytochrome *f* and plastocyanin of photosystem I and ultimately to P_{700}. The electrons in photosystem II do not cycle back to trap chlorophyll (or other electron donor), but their transfer through the redox reactions to P_{700} is coupled to phosphorylation of ADP (*noncyclic photophosphorylation,* described below).

The reduction of the oxidized trap chlorophyll (or other electron donor) is brought about by the oxidation of water via an enzyme system that is closely associated with the structure of the thylakoid and that also produces molecular oxygen. The enzyme system presumably processes four protons simultaneously to produce O_2.

$$2\,H_2O \longrightarrow O_2 + 4\,H^+ + 4\,e^-$$

Mn is a known cofactor in the system. The electrons produced act to reduce the light-oxidized trap chlorophyll, and the H^+ forms a pool available for the reduction of NADP.

Cyclic and Noncyclic Photophosphorylation

Phosphorylation of ADP occurs in the chloroplast during the light reactions, as described briefly above. This photophosphorylation occurs in the lamellae of the stroma and grana thylakoids as a part of both photosystems I and II. There is a clear similarity between the mechanism of photophosphorylation in the chloroplast and electron transport system phosphorylation in the mitochondrion (Chapter 16). In both systems, the mechanism is closely associated with a membrane and with a compartment separated from the rest of the organelle. In the chloroplast, the compartment

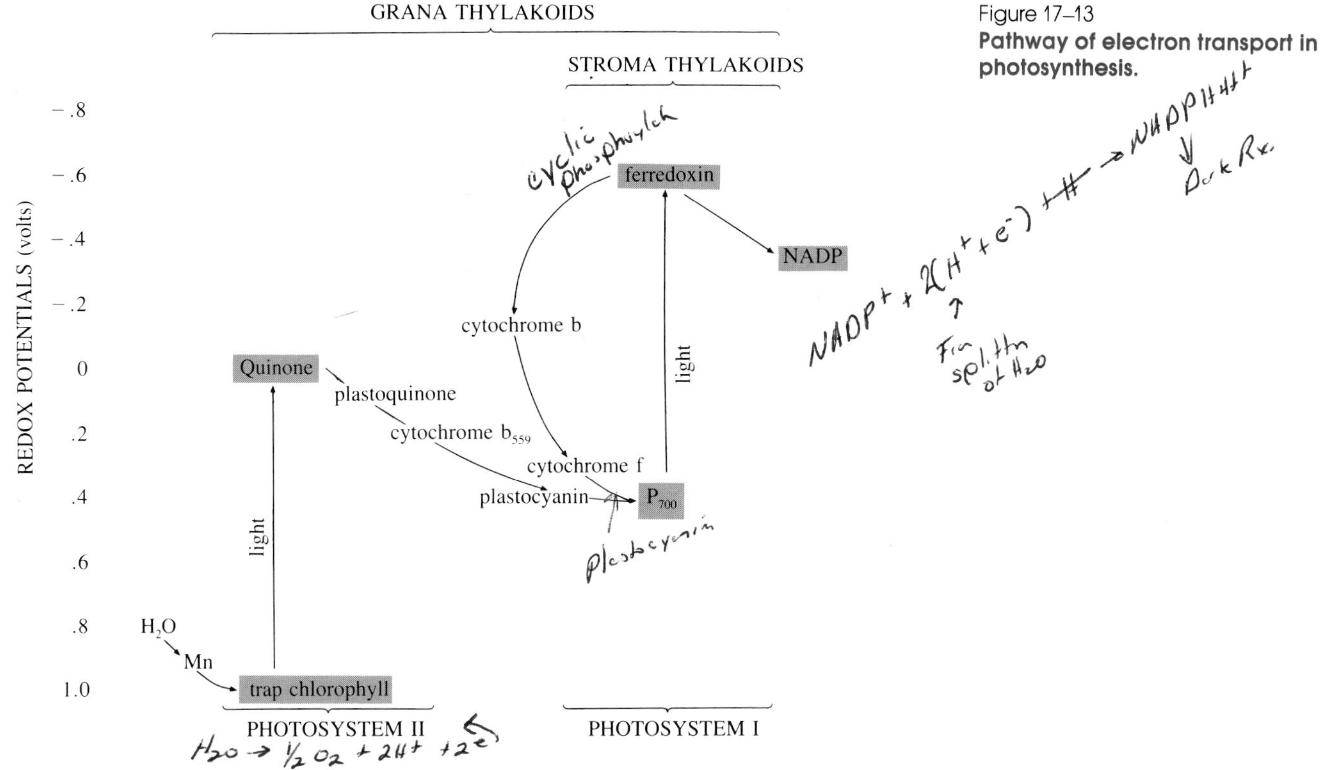

Figure 17–13
Pathway of electron transport in photosynthesis.

is the loculus of the thylakoid; in the mitochondrion, it is the matrix within the inner membrane. In both systems, phosphorylation occurs coupled to electron transport through a sequence of redox reactions. Several of the components of the redox reactions are very similar, for example, the quinones and cytochromes.

As described above, **cyclic photophosphorylation** occurs while electrons released from P_{700} by light are being shunted back to P_{700} through ferredoxin, cytochrome *b*, cytochrome *f*, and plastocyanin. The energy from these exergonic redox reactions is coupled to phosphorylation. **Noncyclic phosphorylation** occurs while electrons released from trap chlorophyll of photosystem II are being shuttled via plastoquinone, cytochrome *b*, cytochrome *f*, and plastocyanin to P_{700} of photosystem I. The energy from these noncyclic exergonic redox reactions is coupled to phosphorylation.

At the present time, the mechanism of coupling is unknown, although evidence to support a proton gradient across the thylakoid membrane is growing (Fig. 17–14). Chloroplasts illuminated in an unbuffered medium quickly cause the medium to become alkaline, implying that protons

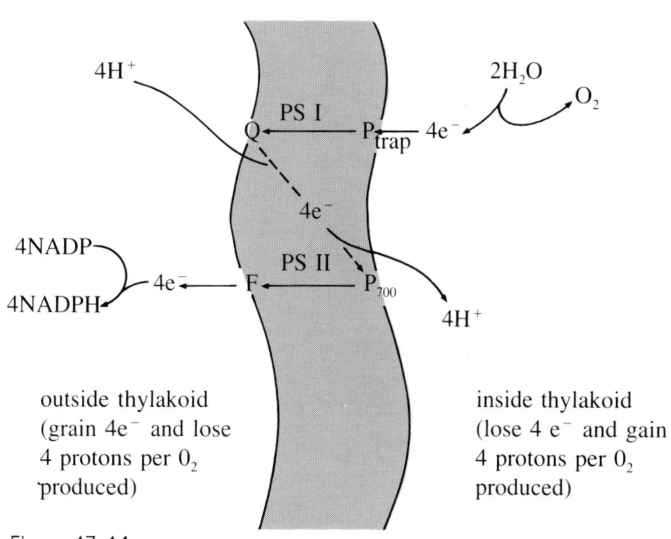

Figure 17–14
Model illustrating accumulation of protons inside thylakoids during illumination. Q, electron acceptor of photosystem I (PSI); P_{trap}, electron donor of photosystem II (PS II); F, electron acceptor of PS II.

are transported into the thylakoid sac. The system will reequilibrate in darkness. If the external medium is made alkaline in the dark, phosphorylation of ADP can be induced. Compounds that destroy the membrane, such as detergents, also cause leakage of the proton pool and prevent phosphorylation.

The standard free energy to generate ATP from ADP and P_i is about 10 kcal/mole, which is equivalent to a redox potential of about 0.43–0.61 volts. During the movement of a pair of electrons from photosystem II to I (noncyclic photophosphorylation), most measurements indicate that one ATP is produced. Recent measurements indicate that possibly more than one ATP is produced between O_2 evolution and NADP reduction, leading to strong speculation that the first ATP is produced during O_2 evolution (photolysis) and the second during electron transport between photosystems. For a proper balance between the light reactions and the dark reactions (discussed later), three ATP must be produced for each pair of electrons transported from H_2O to NADP ($NADPH + H^+$). The third ATP is probably generated by cyclic photophosphorylation.

Summary of the Light Reactions

Two photosystems function during the light reactions of photosynthesis. As each system absorbs four quanta of light energy, trap chlorophyll (system II) and P_{700} (system I) are activated, passing two pairs of electrons to acceptor molecules. Trap chlorophyll is then returned to its reduced state by the oxidation of four water molecules, and in the process one molecule of oxygen is evolved. P_{700} is returned to its reduced state by the flow of electrons from photosystem II through intermediate oxidation-reduction compounds. The transfer of two pairs of electrons from water through photosystem II to photosystem I causes the noncyclic photophosphorylation of two ADP molecules producing two ATP and two H_2O. The two pairs of electrons released from photosystem I reduce two NADP to two $NADPH + H^+$ Therefore, the net result is

$$2\,H_2O + 2\,NADP + (2\,ADP + 2\,P_i) \xrightarrow{\text{8 quanta}} O_2 + 2\,NADPH + H^+ + (2\,ATP + 2\,H_2O)$$

The two ATP and two $NADPH + H^+$ molecules produced by the light reactions occurring in the grana lamellae are used in the synthetic (dark) ractions that take place in the stroma. The latter reactions fix CO_2 into sugars. For each CO_2 molecule fixed, two $NADPH + H^+$ and three ATP molecules are required (see below). Where the third ATP molecule is formed is not yet clear; it may be obtained from cyclic photophosphorylation.

Photosynthesis—Synthetic (Dark) Reactions

The eludication of the sequence of chemical reactions that result in the incorporation of carbon dioxide into sugars and starches relied heavily on the use of radioactive isotopes. Using ${}^{14}_{6}C$-labeled carbon dioxide, it was possible to add ${}^{14}_{6}CO_2$ at known times to an actively photosynthesizing system, halt the process a short time later, and then identify the compounds into which the carbon dioxide became incorporated. Identification of the ${}^{14}_{6}C$-containing intermediates was carried out using combined paper chromatography (Chapter 13) and autoradiography (Chapter 14).

Ruben and co-workers first showed that the active form of CO_2 in the chloroplast was carbonic acid. Calvin and co-workers studied the sequence of reactions that follow the formation and entry of carbonic acid into the chloroplast. They added ${}^{14}_{6}CO_2$ to cultures of the alga *Chlorella* and allowed the cells to photosynthesize for given periods of time (usually between 2 and 60 seconds). The *Chlorella* were then killed and the soluble cell components extracted and concentrated. The extracts containing radioactive carbon were chromatographed on paper, and the spots containing radioactivity were identified.

When photosynthesis in the presence of ${}^{14}_{6}CO_2$ was allowed to proceed for only 2 seconds, the major labeled compound identified was 3-phosphoglyceric acid (PGA). After 7 seconds, sugar phosphates and diphosphates were found in addition to PGA. A 60-second exposure to ${}^{14}_{6}CO_2$ produced labeled phosphoenolpyruvic acid (PEP), carboxylic acids, and amino acids. Using many different time intervals, the entire sequence of reactions was uncovered, and it was found that many of the steps were the reverse of those in the glycolytic pathway (Fig. 17–15).

In the chloroplast stroma, CO_2 in the form of carbonic acid reacts with the sugar ribulose diphosphate (RuDP) to form an unstable 6-carbon compound that immediately

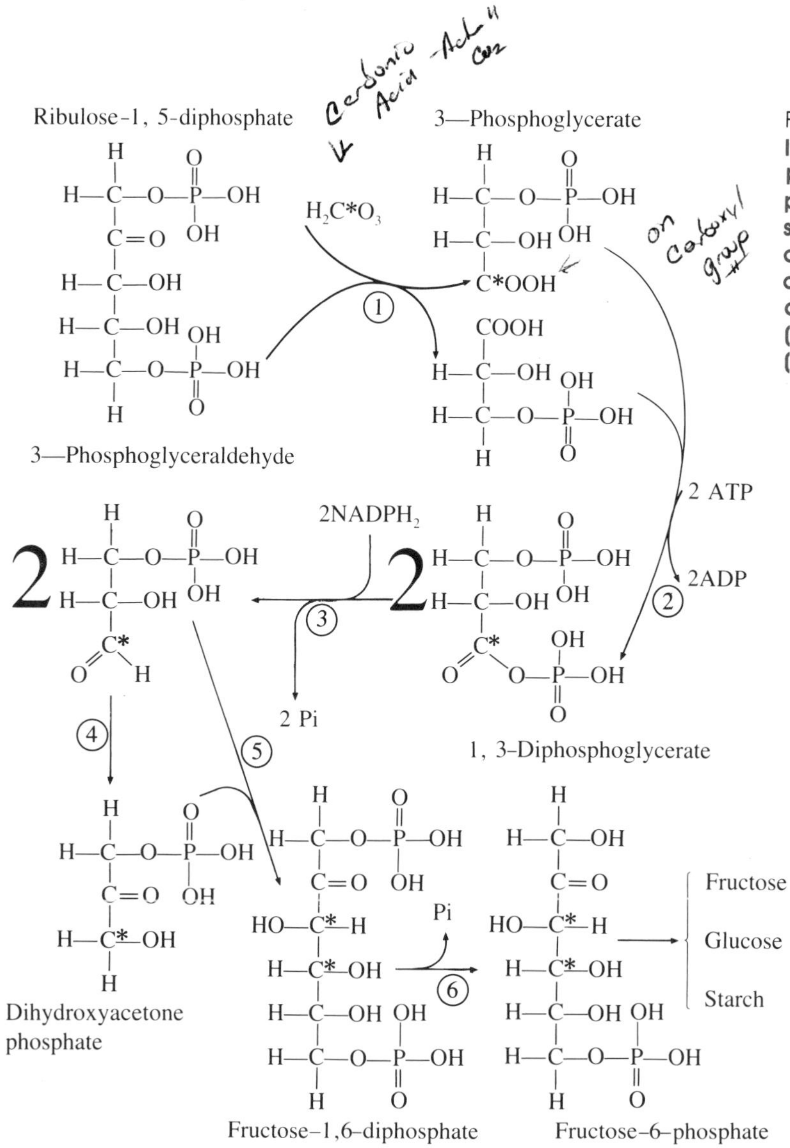

Figure 17–15
Initial steps in the fixation of CO_2 during photosynthesis. C* = $^{14}_{6}CO_2$ and traces the path of carbon through the reaction sequence. Enzymes: (1) carboxydismutase; (2) phosphoglyceric acid kinase; (3) triose phosphate dehydrogenase; (4) isomerase; (5) fructose diphosphate aldolase; (6) fructose diphosphatase.

splits to form two molecules of 3-phosphoglyceric acid (PGA). The enzyme catalyzing this reaction is carboxydismutase and the radioactive carbon of $^{14}_{6}CO_2$ is incorporated into the carboxyl group of PGA.

PGA is then reduced to 3-phosphoglyceraldehyde (PGAL) in two steps. First, each PGA is phosphorylated by ATP and then reduced by NADPH + H^+. The ATP and NADPH + H^+ were produced by photochemical reactions in the grana lamellae. Thus, for each molecule of CO_2 fixed and converted to PGAL, two ATP and two NADPH + H^+ molecules are required. These reactions are catalyzed by a kinase and dehydrogenase.

Some of the PGAL is isomerized by triose phosphate isomerase to form dihydroxyacetone phosphate (DHAP). The enzyme aldolase then condenses PGAL and DHAP to produce fructose-1,6-diphosphate (FDP). Fructose diphosphatase splits off the phosphate group of the first carbon atom, producing fructose-6-phosphate (F6P). F6P may then be converted to fructose, glucose, or starch.

F6P and PGAL are also used for the resynthesis of RuDP (Fig. 17–16). F6P and PGAL are converted to erythrose-4-phosphate (E4P) and xylulose-5-phosphate (X5P). An aldolase then catalyzes the condensation of E4P and DHAP to form sedoheptulose-1,7-diphosphate (SDP),

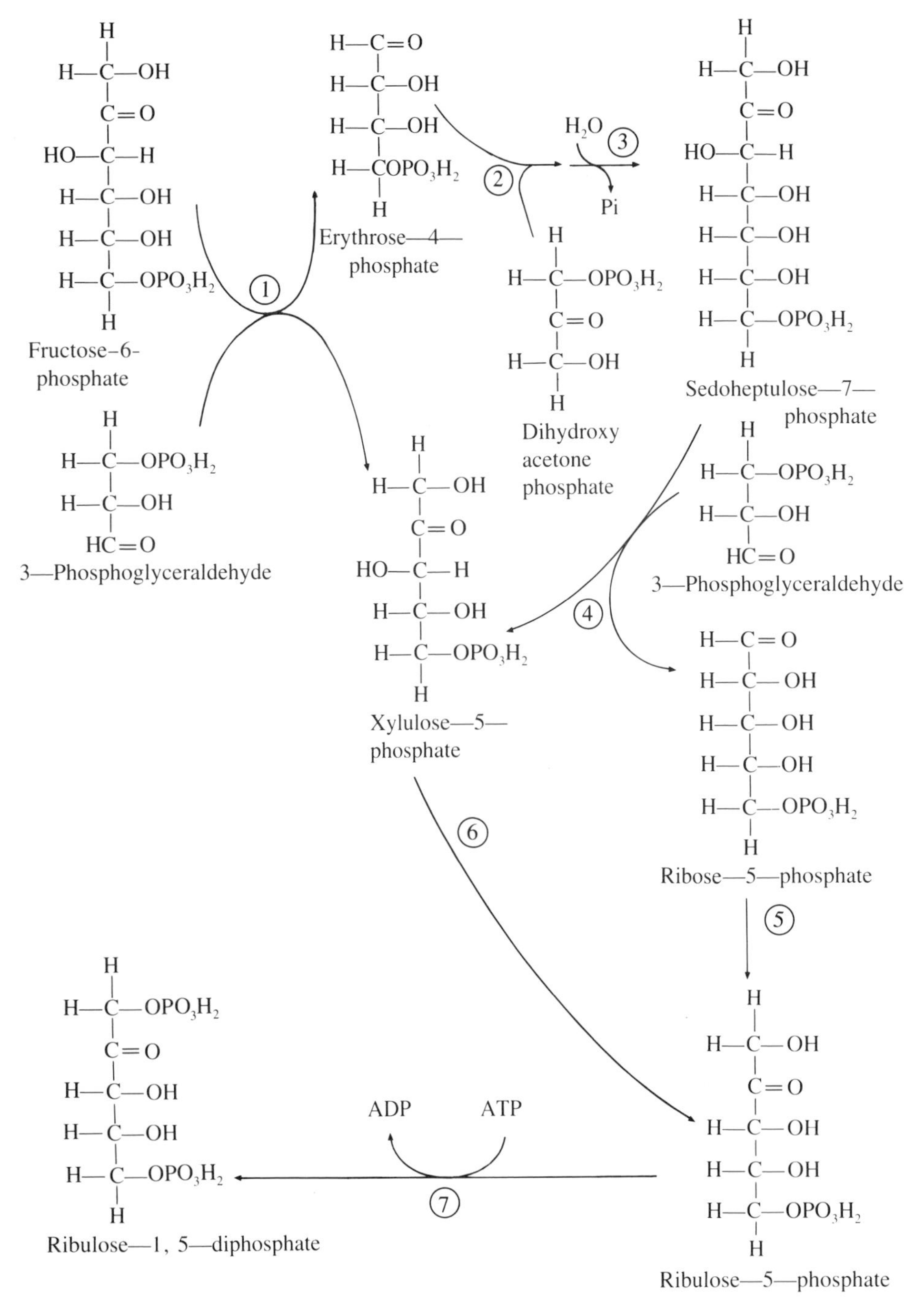

Figure 17–16
Resynthesis of ribulose-1, 5-diphosphate. Enzymes: (1) transketolase; (2) aldolase; (3) sedoheptulose-1, 7-diphosphatase; (4) transketolase; (5) isomerase; (6) epimerase; (7) kinase.

which is then converted to sedoheptulose-7-phosphate (S7P). S7P and PGAL also react to form ribose-5-phosphate and X5P. The R5P is then isomerized to form ribulose-5-phosphate (Ru5P). Ru5P is also formed from X5P. Finally, ATP phosphorylates Ru5P to form RuDP.

For each CO_2 fixed, one RuDP, two ATP, and two $NADPH + H^+$ are consumed and one phosphate sugar is produced. In actuality, all the RuDP is resynthesized. Fixation of three molecules of CO_2 results in the formation of six PGAL. Five of these are recycled to replenish the pool of RuDP. The extra PGAL is the photosynthetic product and is used for sugar and starch synthesis. Figure 17–17 summarizes all the steps and the requisite numbers of molecules participating in the dark reactions; the pathway is known as the **Calvin cycle.**

Since additional ATP is consumed in the formation of RuDP, a total of nine ATP and six $NADPH + H^+$ are required for the fixation of three CO_2 molecules. Since six CO_2 molecules are required to produce one 6-carbon sugar molecule, 18 ATP molecules are consumed in the fixation (or three ATP/CO_2).

It has been estimated that about 5 x 10^{16} g of carbon are fixed annually by photosynthesis and this corresponds to a storage of 4.8 x 10^{17} kcal of energy. Since about 6.7 x 10^{21} kcal of light energy fall on the earth each year, photosynthesis traps a mere 0.0072%.

Other CO_2-fixation Pathways

At the present time, three pathways for CO_2 fixation during photosynthesis in plants are recognized: the Calvin cycle or (C_3 pathway) just described, the Hatch-Slack (or C_4 pathway), and crassulacean acid metabolism (or CAM pathway). Species of higher plants can be characterized by the pathways they utilize. Although referred to as three "different" pathways, in effect, all these contain the Calvin

Figure 17–17
The path of carbon in photosynthesis. The molecule of PGAL in the box is the photosynthetic product and is employed for sugar and starch synthesis. Ru5P, ribulose-5-phosphate; RuDP, ribulose-1,5-diphosphate; PGA, 3-phosphoglyceric acid; PGAL, 3-phosphoglyceraldehyde; DHAP, dehydroxyacetone phosphate; F6P, fructose-6-phosphate; SDP, sedoheptulose-1,7-diphosphate, S7P, sedoheptulose-7-phosphate; E4P, erythrose-4-phosphate; X5P, xylulose-5-phosphate; R5P, ribose-5-phosphate; P_i, inorganic phosphate.

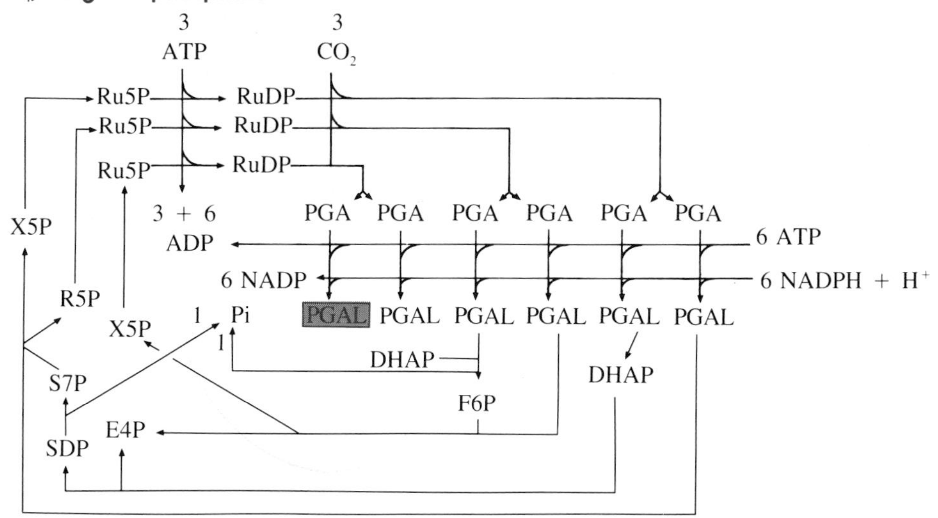

cycle (C_3) reactions, but two contain additional steps that provide CO_2 in a more efficient manner.

The Hatch-Slack (C_4 pathway) is common to corn, sugar cane, a number of other grasses, and several other plant species. How widespread this pathway is in the plant kingdom is not fully known, but it is believed to be substantial. One of the characteristics of plants with this photosynthetic mechanism is a conservation of water, and therefore, plants that grow in relatively arid environments could be expected to have evolved such a mechanism. The primary characteristics of "C_4 plants" are (1) high photosynthetic and growth rates, (2) low photorespiration rate, (3) an unusual leaf anatomy, (4) dimorphic chloroplasts, and (5) low rate of water loss.

Unlike the leaf anatomy described earlier in this chapter and typical for plants with Calvin cycle photosynthesis, C_4 plants have an internal arrangement of chloroplast-containing cells that are oriented around the veins (vascular bundles). These cells are arranged in basically two layers (Fig. 17–18). These cells, which form a layer immediately about the vein, are called the **bundle sheath cells.** They contain chloroplasts, but the chloroplasts may lack grana (as in

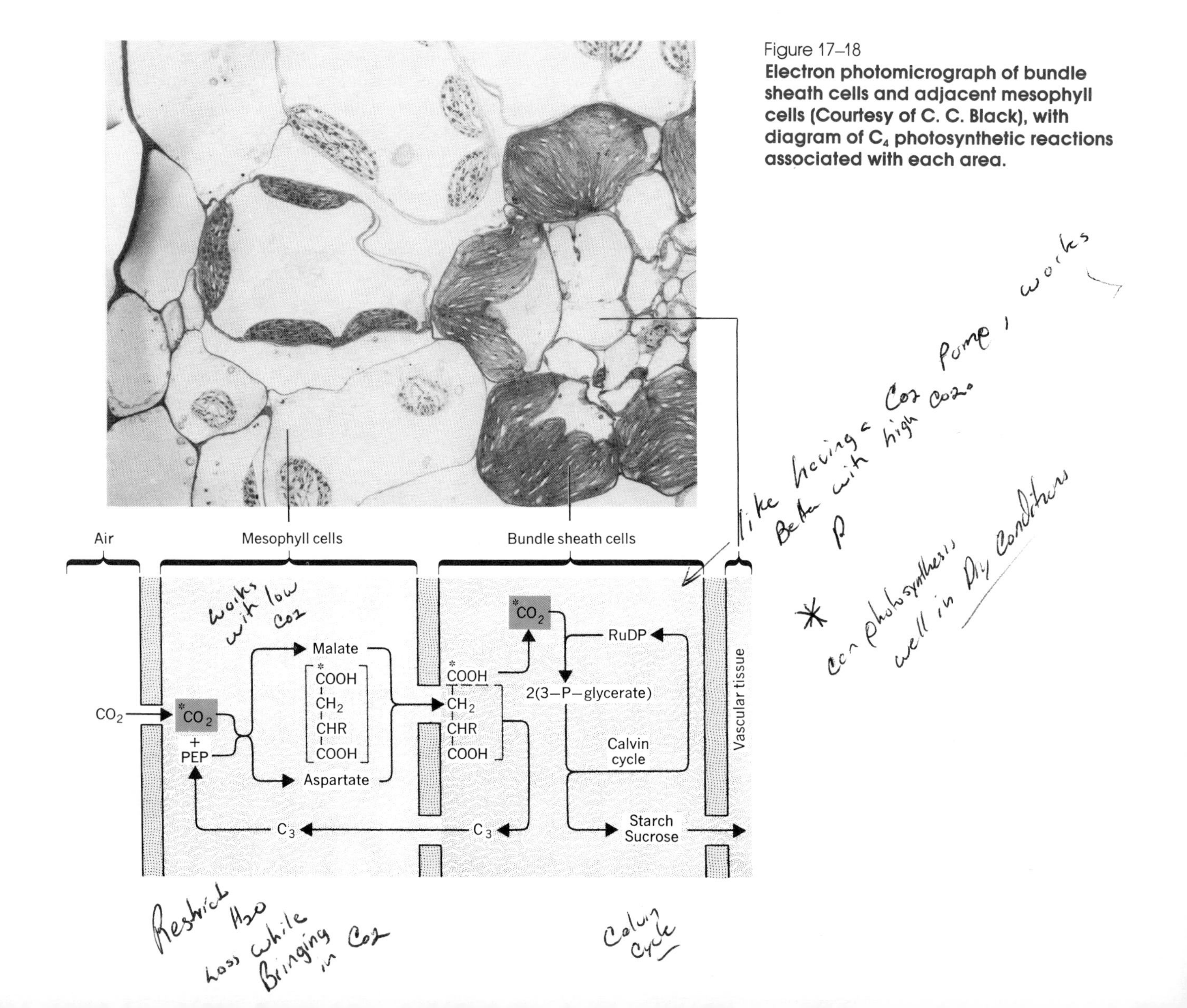

Figure 17–18
Electron photomicrograph of bundle sheath cells and adjacent mesophyll cells (Courtesy of C. C. Black), with diagram of C_4 photosynthetic reactions associated with each area.

sugar cane) or the grana may be very reduced in size. These bundle sheath cells also accumulate starch in the chloroplasts during active photosynthesis. The bundle sheath cells also contain numerous microbodies and many large mitochondria.

Surrounding the bundle sheath cells is another layer of photosynthetic tissue called the **mesophyll** layer. The cells of this layer contain chloroplasts with extensive grana, have few mitochondria and microbodies, and do not accumulate starch. It is believed that the two tissue layers work together in photosynthesis. C_4 plants vary as to the exact sequence of steps but in general the steps are as illustrated in Figure 17–18. CO_2 absorbed from the air spaces of the leaf into the mesophyll cells is fixed with phosphoenolpyruvate (PEP) in the chloroplast. Oxalacetate is the initial product that is subsequently converted to malate or aspartate. These dicarboxylic acids are then transported into the bundle sheath cells and decarboxylated releasing CO_2. The CO_2 is then refixed by the Calvin cycle reactions in the chloroplasts, as described earlier. The 3-carbon compounds formed after decarboxylating malate or aspartate are transported back into the mesophyll where they are reconverted into PEP. The sugars that accumulate from the Calvin cycle reactions are temporarily converted to starch under active photosynthetic conditions. At night or during darkness or dim light the starches are converted back to the sugars and transported out of the leaf by the vascular tissue of the veins (vascular bundles).

The presence of the C_4 pathway seems like a needless addition to the Calvin cycle system. However, there is evidence indicating that the mesophyll is able to build up a high concentration of potential CO_2 by this method, which could provide evolutionary advantages. In addition, the rapid and efficient fixing and storing of CO_2 decreases the leaf's need to have a large number of stomates (openings in the leaf epidermis that allow CO_2 and other gases to diffuse into the leaf). Open stomates, while allowing the passage of CO_2 into the plant, also allow water to escape from the plant—a disadvantage to plants in arid climates!

Crassulacean acid metabolism (CAM) is a special form of metabolism associated with photosynthesis that is carried out by members of the plant family *Crassulaceae* (succulent herbs such as sedum). Plants carrying out this form of metabolism have closed stomates during the daylight hours and therefore cannot absorb sufficient CO_2 for photosynthesis. But during the night (dark) hours, the stomates open

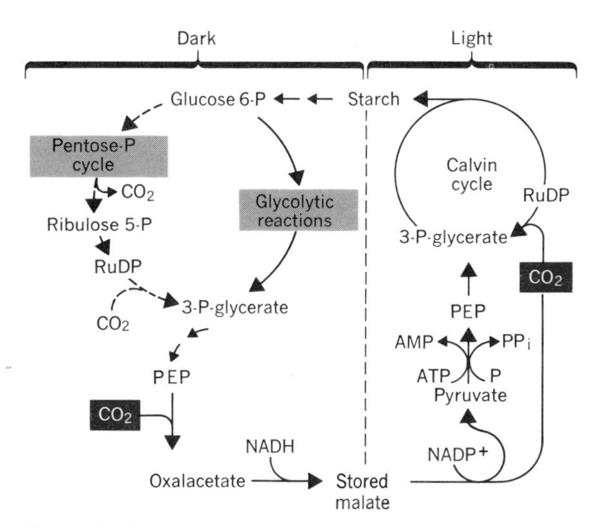

Figure 17–19
Diagram of the major reaction steps of the light and dark phases of crassulacean acid metabolism.

and the leaf cells can fix CO_2 in the dark by combining it with PEP to form oxalacetate. The oxalacetate is converted into malate for storage. During the following daylight hours, the malate is decarboxylated, and the CO_2 is utilized by the Calvin cycle reactions. The 3-carbon pyruvate remaining after decarboxylation is converted first into PEP and then into phosphoglyceric acid and is utilized by the Calvin cycle as well. The PEP required for the dark fixation of CO_2 is derived from some of the starch produced from the Calvin cycle products (Fig. 17–19)

Bacterial Photosynthesis

Photosynthesis in procaryotic organisms occurs in lamellar membrane systems called **chromatophores.** The chromatophores contain the pigments for the photochemical reactions but none of the subsequent biosynthetic enzymes. The pigment system includes the chlorophylls, carotenoids, and in some cases phycobilins. However, in bacteria, *bacteriochlorophyll* is the ultimate light-trapping molecule (not chlorophyll *a*).

The most important distinction between plant and bacterial photosynthesis is that water is not used as the reducing agent and oxygen is not an end product. The power to reduce CO_2 may come from molecular hydrogen, H_2S, or organic compounds. Two major groups of bacteria that carry out photosynthesis are the green and purple sulfur bacteria; these organisms utilize H_2S and produce sulfur and sulfate; that is

$$6\ CO_2 + 12\ H_2S \xrightarrow[\text{bacteriochlorophyll}]{\text{light energy}} C_6H_{12}O_6 + 6\ H_2O + 12\ S$$

During photosynthesis, sulfur accumulates as granules of elemental sulfur and may be further metabolized later.

Nonsulfur purple bacteria use organic compounds such as acetic acid as electron donors. The acetic acid is anaerobically oxidized via the Krebs cycle reactions (Chapter 10). Acetic acid can also be reduced to hydroxybutyric acid. Certain members of the sulfur and nonsulfur purple bacteria can use molecular hydrogen to reduce either CO_2 or acetic acid; that is,

$$6\ CO_2 + 12\ H_2 \longrightarrow C_6H_{12}O_6 + 6\ H_2O$$

and

$$2\ CH_3\text{-}C(=O)OH + H_2 \longrightarrow \underset{\text{Hydroxybutyric acid}}{CH_3\text{-}CH(OH)\text{-}CH_2\text{-}C(=O)OH} + 2\ H_2O.$$

Other Plastids

The chloroplast is only one of several different plastids found in plant cells. Other plastids such as *etioplasts, amyloplasts* and *chromoplasts* have a different structure and function. They are all called plastids because they appear to develop from a common structure or from one another.

Proplastids are small, generally colorless structures found in young or dividing cells. They have little internal structure but are delimited by a double membrane. Proplastids give rise to other types of plastids.

Etioplasts are prevalent in the leaves of plants grown in the dark. Their ellipsoidal and sometimes irregular structure is also bound by a double membrane. Internally, etioplasts contain a latticelike arrangement of tubules. Etioplasts are changed into chloroplasts upon exposure to light.

The outer membrane of the **amyloplast** encloses the stroma, containing one to eight starch granules. In certain plant tissues such as the potato tuber, the starch granules within the amyloplasts may become so large that they rupture the encasing membrane. Starch granules of amyloplasts are typically composed of concentric layers of starch.

Chromoplasts contain carotenoids and are responsible for imparting color (yellow, orange, red) to certain portions of plants such as flower petals, fruits, and some roots. The chromoplasts of carrots contain large quantities of lipid that reduces their overall density to less than 1 g/ml; consequently, during centrifugation of root homogenates, the amyloplasts do not sediment but rise to the surface of the tube. Chromoplast structure is quite diverse; they may be round, ellipsoidal, or even needle-shaped, and the carotenoids that they contain may be localized in droplets or in crystalline structures. The function of the chromoplasts is not clear, but in many cases such as flowers and fruit, the color that they produce probably plays a role in attracting insects and other animals for pollination or seed dispersal.

A number of other, less frequently occurring plastids have been described, including the oil-filled *elaioplasts,* the protein-containing *proteoplasts,* and the sterol-rich *sterinochloroplasts.*

Summary

Photosynthesis, like mitochondrial reactions, is concerned in eucaryotes with the formation of energetically important ATP, involves hydrogen and electron transport in compounds like NADPH + H^+ and cytochromes, and occurs in or between membranes. The two processes differ in that photosynthesis uses light rather than chemical substrates as the source of energy, CO_2 and water are consumed rather than produced, and O_2 and carbohydrate are produced rather than consumed. The overall reaction

$$6\ CO_2 + 12\ H_2O \xrightarrow{\text{light}} (C_6H_{12}O_6)_n + 6\ O_2 + 6\ H_2O$$

can be broken down into a light phase (photolysis) and a dark phase (CO_2 fixation and dark reactions). In the light phase, visible light is absorbed by chlorophyll or a variety of other pigments located in the membranous **thylakoids** of the chloroplast. The light energy excites the molecules, inducing them to reemit light or heat or transfer the energy to a **chlorophyll** P_{700} molecule or **trap chloroplyll.** This activation of trap chlorophyll induces the reactions of **photosystem II,** which generates ATP by the process called **noncyclic photophosphorylation** and terminates with the activation of P_{700}. Activation of P_{700} by photosystem II or absorption of light energy directly initiates **photosystem I.**

This photosystem can also result in the formation of ATP (by the process of **cyclic photophosphorylation**) or the reduction of NADP to NADPH + H^+. The ATP and NADPH + H^+ transported into the **stroma** of the chloroplast are consumed in the dark reactions.

During the dark reactions, CO_2 is fixed by binding to ribulose diphosphate and is subsequently reduced by NADPH + H^+. ATP acts as the source of energy for these endergonic reactions. The final product is carbohydrate, usually in the form of a sugar.

Chloroplasts vary in shape and are found only in eucaryotic plant cells. They form from proplastids, which may be responsible for the formation of other plastids, such as **chromoplasts** and **leukoplasts.** Plastids associated with **C_3 photosynthesis** have inner membranes arranged in layers and organized into **grana.** Plastids associated with **C_4 photosynthesis** may have a similar structure but frequently lack grana.

References and Suggested Reading

Articles and Reviews

Arnon, D. I., The role of light in photosynthesis. *Sci. Am. 203*(5), 104 (Nov. 1960).

Bassham, J. A., The path of carbon in photosynthesis. *Sci. Am. 206*(6), 88 (June 1962).

Calvin, M., The path of carbon in photosynthesis. *Sci. 135,* 879 (1962).

Calvin, M., and Androes, G. M., Primary quantum conversion in photosynthesis, *Science, 138,* 867 (1962).

Cheniae, G. M., Photosystem II and O_2 evolution. *Annu. Rev. Plant Physiol. 21,* 467 (1970).

Hatch, M. D., and Slack, C. R., Photosynthetic CO_2-fixation pathways. *Annu. Rev. Plant Physiol. 21,* 141 (1970).

Rabinowitch, E. I., and Govindjee, The role of chlorophyll in photosynthesis. *Sci. Am. 213*(1), 74 (July 1965).

Stainer, R. Y., Photosynthetic mechanisms in bacteria and plants; development of a unitary concept. *Bacteriol. Rev. 25,* 1 (1961).

von Wettstein, D., Genetics and submicroscopic cytology of plastids. *Hereditas 43,* 303 (1957).

Books, Monographs, and Symposia

Bassham, J. A., Photosynthesis: The path of carbon, in *Plant Biochemistry* (J. Bonner and J. E. Varner, eds.), Academic Press, New York, 1965.

Bonner, J. and Varner, J. E., *Plant Biochemistry,* 3rd ed. Academic Press, New York, 1976.

Clayton, R. K., *Molecular Physics in Photosynthesis,* Blaisdell Publishing Co., New York, 1965.

DeRobertis, E. D. P., Nowinski, W. W., and Saez, F. A., *Cell Biology,* 5th ed., W. B. Saunders Co., Philadelphia, 1970.

Hall, D. O., and Whatley, F. R., The Chloroplast, in *Enzyme Cytology* (D. B. Roodyn, ed.), Academic Press, London, 1967.

Hatch, M. D., Osmond, C. B., and Slatyer, R. O. *Photosynthesis and Photorespiration,* Wiley-Interscience, New York, 1971.

Kirk, J. T. O., and Tilney-Bassett, R. A. E., *The Plastids,* W. H. Freeman and Co., San Francisco, 1967.

Kok, B., Photosynthesis: The path of energy, in *Plant Biochemistry* (J. Bonner and J. E. Varner, eds.), Academic Press, New York, 1965.

Mahler, H. R., and Cordes, E. H., *Biological Chemistry,* Harper & Row Publishers, New York, 1966.

Mazia, D., and Tyler, A. (eds.). *General Physiology of Cell Specialization,* McGraw-Hill Book Co., New York, 1963.

Nobel, P. S., *Plant Cell Physiology,* W. H. Freeman and Co., San Francisco, 1970.

Novikoff, A. B., and Holtzman, E., *Cells and Organelles,* Holt, Rinehart and Winston, Inc., New York, 1970.

Park, R. B., The chloroplast, in *Plant Biochemistry* (J. Bonner and J. E. Varner, eds.), Academic Press, New York, 1970.

Price, C. A., *Molecular Approaches to Plant Physiology,* McGraw-Hill Book Co., New York, 1970.

Rabinowitch, E., and Govindjee, *Photosynthesis,* John Wiley & Sons, Inc., New York, 1969.

Chapter 18
THE GOLGI APPARATUS

The **Golgi apparatus** (or Golgi body) is another organelle along with the lysosomes, peroxysomes, and other vesicles that make up part of the **endomembrane system** of eucaryotic cells. The term ''endomembrane system'' is used to refer to the collection of cytoplasmic and vesicular membranes believed to arise from the same common source. The common source is either the outer nuclear envelope or the endoplasmic reticulum. The mitochondria and chloroplasts are not included in this class because they are known to arise de novo or by division of preexisting organelles of the same type.

The Golgi apparatus was first described by C. Golgi in 1898 (Fig. 18–1). Although he referred to the structures as the ''internal reticular apparatus,'' they soon were identified with his name and became one of the more controversial cell structures. Part of the controversy arose from the fact that the Golgi apparatus is not easily seen and is variable in structure and position within the cell. Golgi worked out a method for differentiating the structure from the rest of the cell by adding silver stains, but since the structure was not consistent from cell to cell, many investigators interpreted the variability as being a direct result of the artificial application of chemicals and stains—in other words, *artifacts*. The controversy was not resolved until the structure was seen and described by electron microscopy in the 1950s.

Figure 18–1
The "internal reticular apparatus" of the cell as first depicted by Camillo Golgi in 1898.

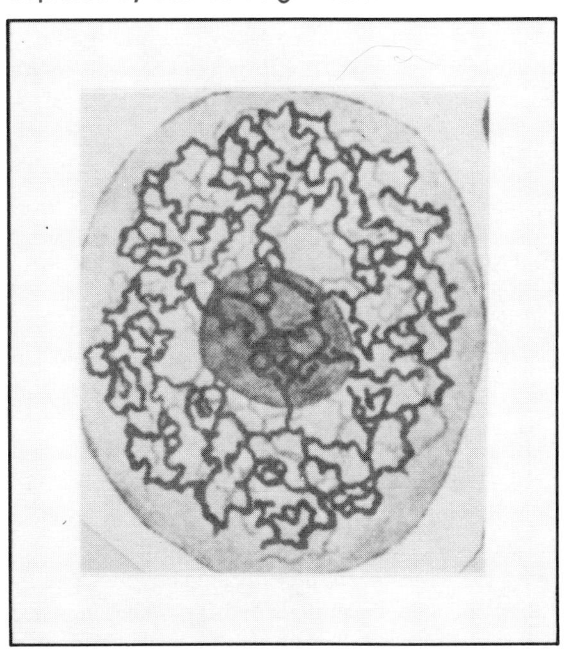

Structure of the Golgi Apparatus

The Golgi apparatus is a complex of membrane-lined vesicles called **cisternae.** A cisterna is a fluid-filled sac or container, and the cisternae of the Golgi apparatus vary from large flattened vesicles to branching and anastomosing vesicles to individual spherical vesicles (Fig. 18–2). In most cells, the Golgi apparatus is composed of *layers* of large flattened vesicles. These appear to be arranged in order, the small vesicles lying nearest the nuclear envelope or endoplasmic reticulum—the origin of the vesicles—and the large ones lying at the opposite side of the structure. The stacks of vesicles nearest the nucleus comprise the "forming face" of the Golgi apparatus while the vesicles on the side closer to the plasma membrane constitute the "maturing face." The membranes bounding the vesicles are smooth, with no evidence of particles such as ribosomes. Three different levels of organization of the Golgi structures may be described: the *cisternae,* the *dictyosome,* and the whole *apparatus*. Changes in the Golgi structures are interpreted from electron photomicrographs that show only static conditions. The sequence of changes is a judgment based upon changes in the activity of the cell.

The cisternae initially are relatively small spherical vesicles. Some of them subsequently become flattened sacs or tubular elements. Those that become flattened may develop pores about the periphery, called **fenestrae,** and may form tubes that extend outward (Fig. 18–3). The ends of the tubular extensions may fuse with each other or enlarge. As these ends mature, they frequently detach and become secretory vesicles (discussed later). In some cases, the mature edges of the cisternae become coated with electron-dense material.

The **dictyosome** is a collection of cisternae that forms a stack and acts as a unit (Figs. 18–4 and 18–5). The cisternae are formed at one end of the stack, progressively mature, and are released as vesicles on the opposite side of the stack. The size and number of dictyosomes vary from one type of cell to another and according to the metabolic activity of the cell. Some cells have been reported to have only a single dictyosome; others may have hundreds. Since one of the functions of the Golgi apparatus is secretion, as one might expect, the size and occasionally the number of dictyosomes increases when the cells are actively secreting material. In plant cells, the number of dictyosomes increases during cell division when these organelles secrete material for the cell plate, which then develops into the cell wall separating the two new cells. In goblet cells of intestinal epithelium there is a single dictyosome, but its size increases significantly during periods of digestion.

The location of the dictyosomes in the cell is as variable as their number and size. Most often, the distribution appears to be at random; however, in some cases, the location is related to a specific cell function. For example, the dictyosomes occur in greatest numbers near the site of cell plate formation in dividing plant cells; the dictyosomes in goblet cells of the intestinal epithelium are located next to the region of the cell where mucigen granules are stored prior to secretion (Fig. 18–6).

The Golgi **apparatus,** when used in the sense of hierarchy of structure, refers to the association of dictyosomes which together make up a cluster or distinct functional group. In some cases, the term "apparatus" refers to a single dictyosome together with the many surrounding small vesicles. In other cases, when there is evidence to suggest that they are associated in function, the term may apply to the entire collection of dictyosomes in the cell.

Origin of Golgi Structures

There are three proposed sources of new cell dictyosomes: (1) from vesicles arising from the endoplasmic reticulum or nuclear envelope, (2) from other structures or vesicles in the cytoplasm, and (3) by the division of preexisting dictyosomes. There is good electron-microscopic evidence to support the concept that vesicles arise either from the outer membrane of the nuclear envelope (Fig. 18–7*a*) or the endoplasmic reticulum, then migrate to the dictyosome, and become cisternae of that structure, thereby contributing to its growth. However, evidence is less than sufficient to indicate that a *complete* dictyosome can be formed from vesicles arising from such a source.

Aggregations of small vesicles occur in areas of the cytoplasm called **zones of exclusion,** which are free of ribosomes. These zones are frequently surrounded by endoplasmic reticulum membranes or are next to the nuclear envelope. Small dictyosomes, which are presumably early stages of formation, are also found in these zones of exclusion. Dormant seeds of higher plants generally lack a Golgi apparatus but do have zones of exclusion with aggregations of these small vesicles. Photomicrographs of early stages of germination suggest progressive development of dictyosomes in these zones of exclusion (Fig. 18–7*b*), and the development of the dictyosome coincides with the disappearance of the aggregations of vesicles.

In frog oocytes it is also possible to follow the devel-

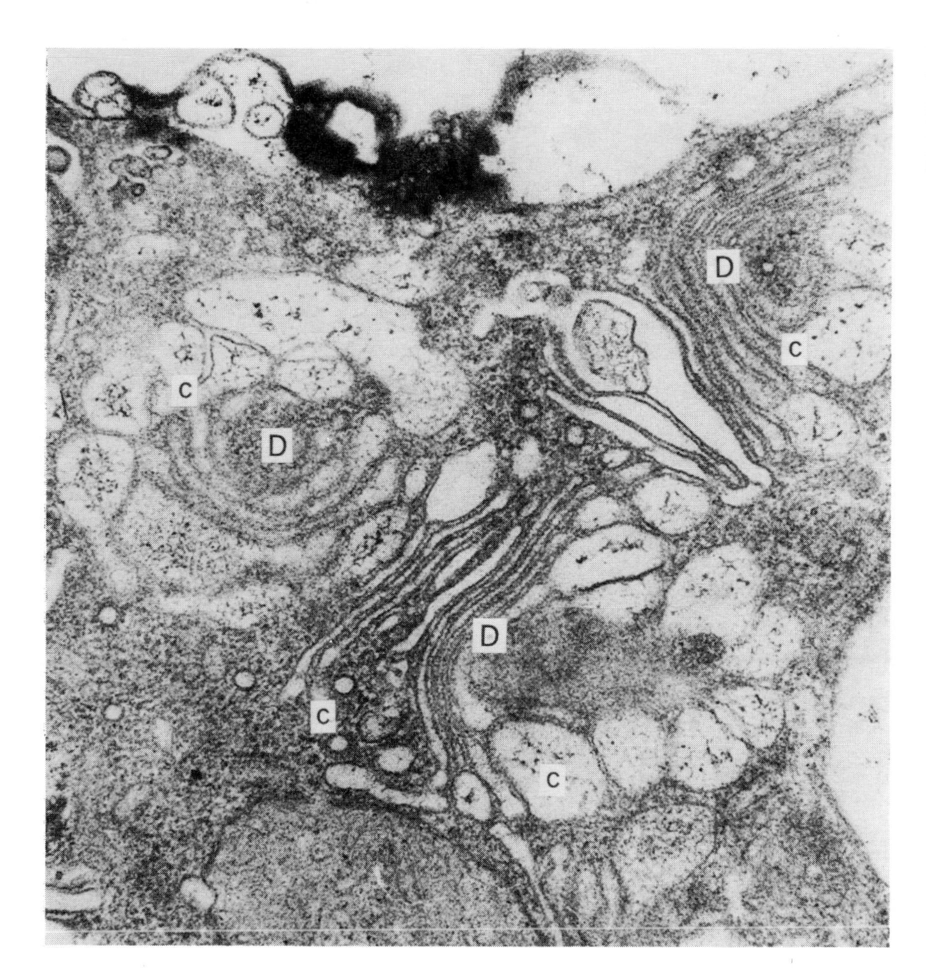

Figure 18–2
Various shapes of cisternae (c) in a dictyosome (D). (Electron photomicrograph courtesy of Dr. E. G. Pollock.)

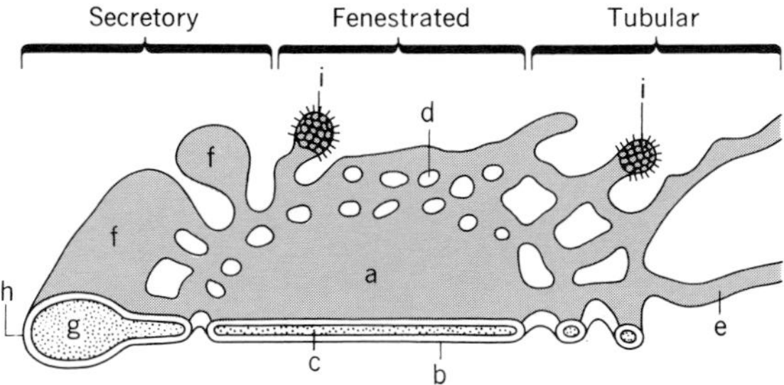

Figure 18–3
Composite diagram of dictyosome cisternae: (*a*) Central, plate-like region. (*b*) membrane; (*c*) lumen; (*d*) fenestrae (perforations); (*e*) peripheral tubules; (*f*) secretory vesicles; (*g*) vesicle lumen; (*h*) vesicle membrane; (*i*) coated vesicles (600 to 750 Å diameter). (Diagram courtesy J. Morré et al, in *Origin and Continuity of Cell Organelles* (Reinert and Urspring, eds.), copyright 1971 Springer-Verlag.)

opment of dictyosomes from vesicles in the zones of exclusion (Fig. 18–8). However, R. Ward and E. Ward have interpreted their electron photomicrographs as indicating that the vesicles arise de novo from fine fibers within the zone rather than from small vesicles that are pinched off from the endoplasmic reticulum. The early stages are probably dependent upon protein synthesis. G. Werz has shown that actinomycin D (an inhibitor of protein synthesis) prevents formation of dictyosome prestages in *Acetabularia*.

During cell division in both plants and animals, the number of dictyosomes increases, for the number of dictyosomes in each daughter cell just after division is about the same as the number in the parent cell prior to division. Direct evidence for the division of the dictyosome is lacking. Dictyosomes do persist throughout mitosis and can be observed. Photographs taken at successive stages indicate that more dictyosomes are present in metaphase and anaphase than during other phases of the cell cycle, but the source of added organelles is not clear. Many workers have observed "paired dictyosomes," but whether or not the pairing results from the fission of a larger dictyosome is not

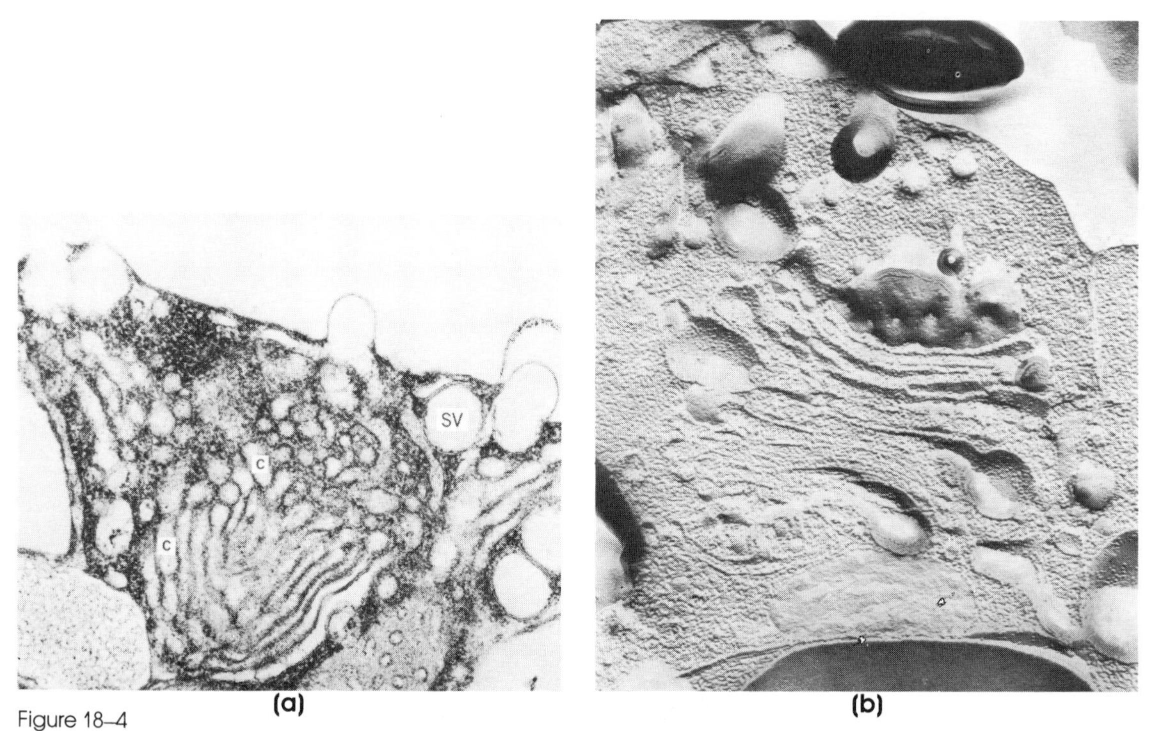

Figure 18–4

Golgi apparatus. (*a*) Transmission electron photomicrograph showing large number of cisternae (c) and secretory vesicles (SV). (*b*) Freeze-fractured preparation showing arrangement of cisternae. (Electron photomicrographs courtesy of Dr. E. G. Pollock.)

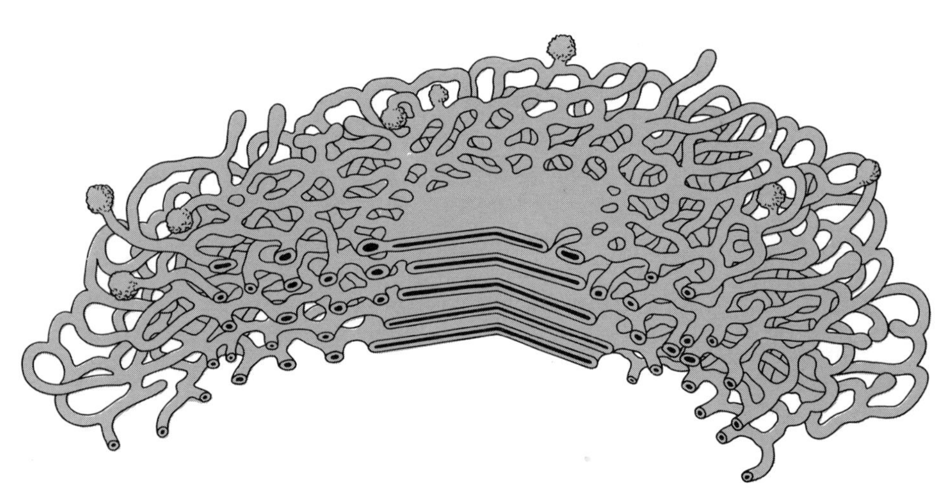

Figure 18–5

Three-dimension drawing of dictyosome with highly fenestrated and tubular cisternae arranged in layers.

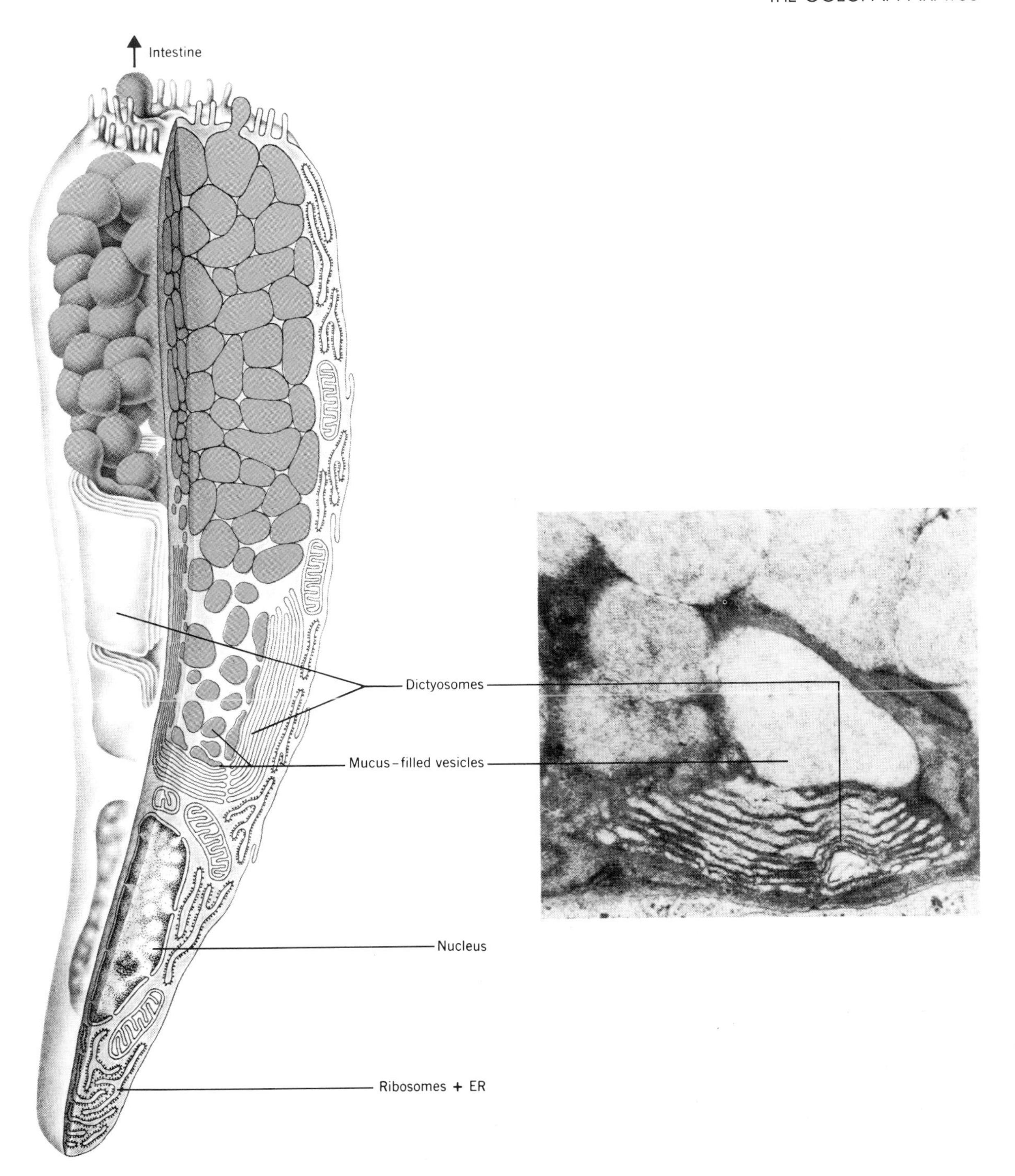

Figure 18–6
Goblet cell of the intestinal epithelium showing the location of the dictyosomes and mucigen granules.

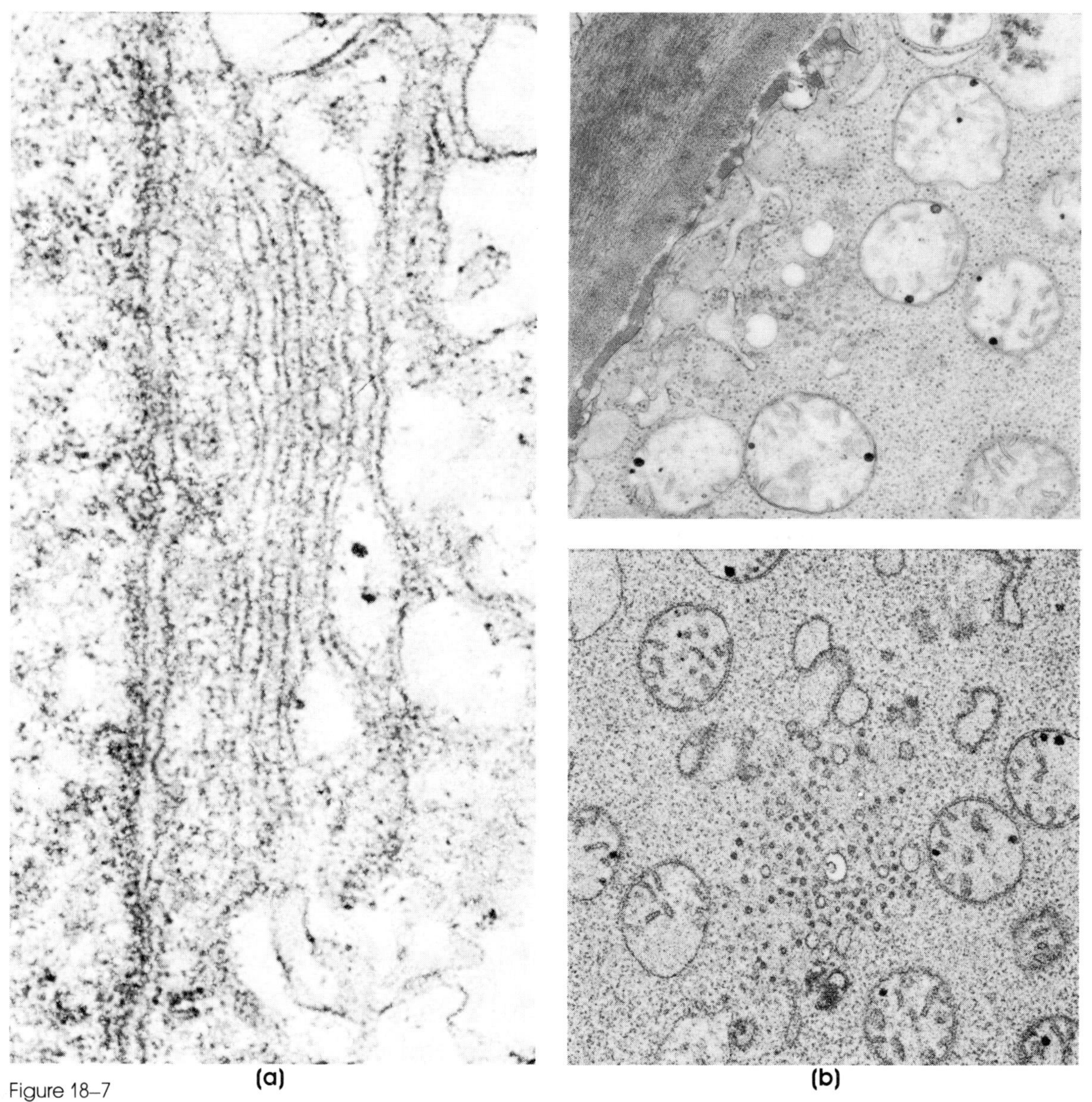

Figure 18–7
(*a*) Golgi body adjacent to the nuclear envelope. (Electron photomicrograph courtesy of Dr. E. G. Pollock.) (*b*) Two photomicrographs showing *zones of exclusion* (clusters of tiny vesicles) which are believed to give rise to dictyosomes (Golgi bodies). (Courtesy of Drs. H. H. Mollenhauer and J. Morré, in *Origin and Continuity of Cell Organelles* (Reinert and Ursprung, eds.); copyright 1971, Springer-Verlag.)

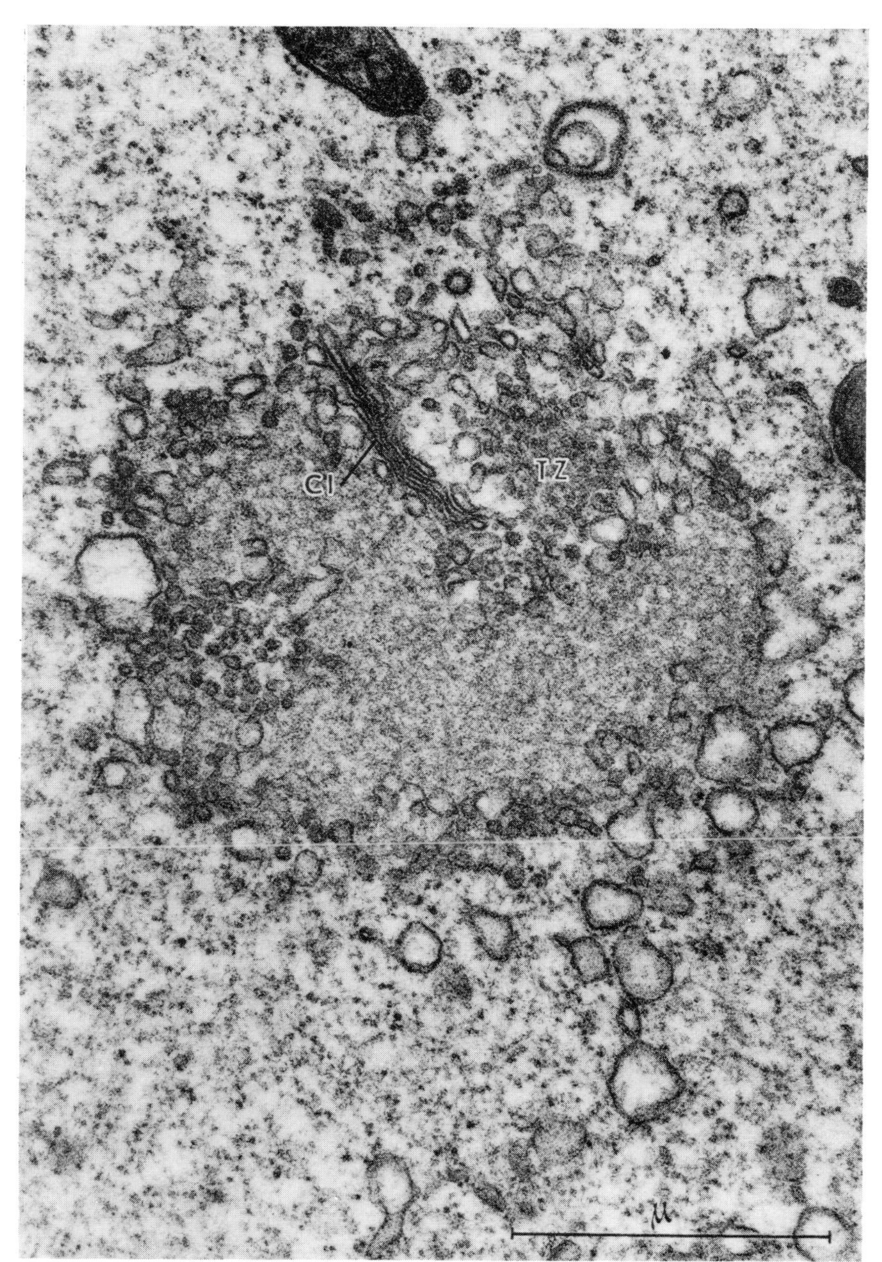

Figure 18–8
Development of the Golgi apparatus in the frog oocyte. In advance of the appearance of a definitive Golgi apparatus, transition zones (TZ) containing numerous small vesicles are seen in the zone of exclusion along with the first cisternae (CI). (Photomicrograph courtesy of Drs. R. Ward and E. Ward; copyright 1968. *J. Microscopie 7,* 1007.)

known. The division of dictyosomes is reasonably well supported by observations of the multinucleate alga *Botrydium granulatum* in mitosis. Just before the spindle forms, a single dictyosome can be observed at each pole of the cell, but by late metaphase, the dictyosomes have doubled. Two dictyosomes with the centriole between them may now be seen at each end of the spindle.

Development of the Golgi Apparatus

Because it is not possible to observe living Golgi bodies clearly, it has not been possible to directly follow the developmental sequence of a single dictyosome. However, developmental sequences have been worked out by observing dictyosomes of cells at different stages of growth and then correlating differences in dictyosome appearance with other developmental changes in the cell. During growth of the ciliate *Tetrahymena pyriformis*, individual smooth-surfaced cisternae are present about the oral region of the cell with small tubules between them. When the cells approach starvation (in the stationary phase of growth), stacks of lamellae form in the same region, apparently from the cisternae. During *conjugation*, these cisternae are modified and assume the appearance of a typical Golgi apparatus. When fission occurs and the cells return to feeding and growth conditions, the Golgi apparatus disappears and isolated cisternae reappear.

The association of dictyosomes to form a complex Golgi apparatus could result either from aggregation of existing dictyosomes or from multiplication of dictyosomes in a confined area. Currently, the balance of evidence supports the first alternative. In spermatids, the dictyosomes fuse rapidly during development to form the Golgi apparatus of the acrosome. During zoospore formation in a fungal mold, the scattered dictyosomes aggregate in the cell. Observations of mammalian cells indicate that the dictyosomes frequently arise near the perinuclear region and migrate to other parts of the cell.

The differentiation of the dictyosome and Golgi apparatus often parallels the differentiation of the cell. In embryonic liver cells, the Golgi apparatus is made up of tubular cisternae. As the cells mature and differentiate, platelike cisternae are formed, and following this stage, secretory vesicles form at the outer edges of the cisternae. This sequence is illustrated in Figure 18–9.

Functions of the Golgi Apparatus

The Golgi apparatus is involved in many different processes in a variety of cells, but in general, the functions are associated with cell **secretion** or with **membrane modifications.** A third general function related to the first two is post-translational protein modification during the final assembly of glycoproteins.

Two sets of experiments bear on the role of the Golgi apparatus in secretion. In 1964 L. Caro and G. Palade showed that the Golgi apparatus in the **acinar** cells of the pancreas is involved in the packaging of enzyme precursors into **zymogen** granules prior to secretion. Caro and Palade injected radioactive amino acids into rats and followed the movements of the "label" using autoradiography. This type of experiment is called a "pulse-chase" because the initial short-term application of labeled amino acids is immediately followed by the more prolonged application of unlabeled forms. Although amino acid metabolism and protein synthesis are not interrupted, the metabolic fate of the labeled amino acids can be traced through the cell with time. As might be expected, after a 3-minute pulse the label appeared almost exclusively in the rough endoplasmic reticulum, since this is the region of protein synthesis. Following the 3-minute pulse, nonlabeled amino acids were added for 17 minutes (e.g., a total of 20 minutes from beginning of the pulse). Although some label was found in the rough endoplasmic reticulum as before, most of the label had shifted to the Golgi apparatus. When the chase was continued for an additional 100 minutes (120 total), almost all the label had left the endoplasmic reticulum and the Golgi apparatus and was now found in the zymogen granules and in the lumen outside of the cells (as a result of the release of the contents of the vesicles through plasma membrane). These experiments showed that the path of the amino acids is first into proteins in the rough ER and that these proteins are then transferred into the cisternae of the Golgi apparatus and then into the zymogen granules.

In 1966 using similar autoradiographic techniques M. Neutra and C. P. Leblond studied the secretion of mucous by the **goblet** cells of the intestinal epithelium. Mucous is a glycoprotein in which glucose and glucose derivatives are linked together forming polysaccharide side chains on the protein molecules (Fig. 18–10). Glucose labeled with tritium was used to follow the assembly and fate of the glycoproteins. Fifteen minutes after injection of the radioactive sugar, the label was most concentrated in the cister-

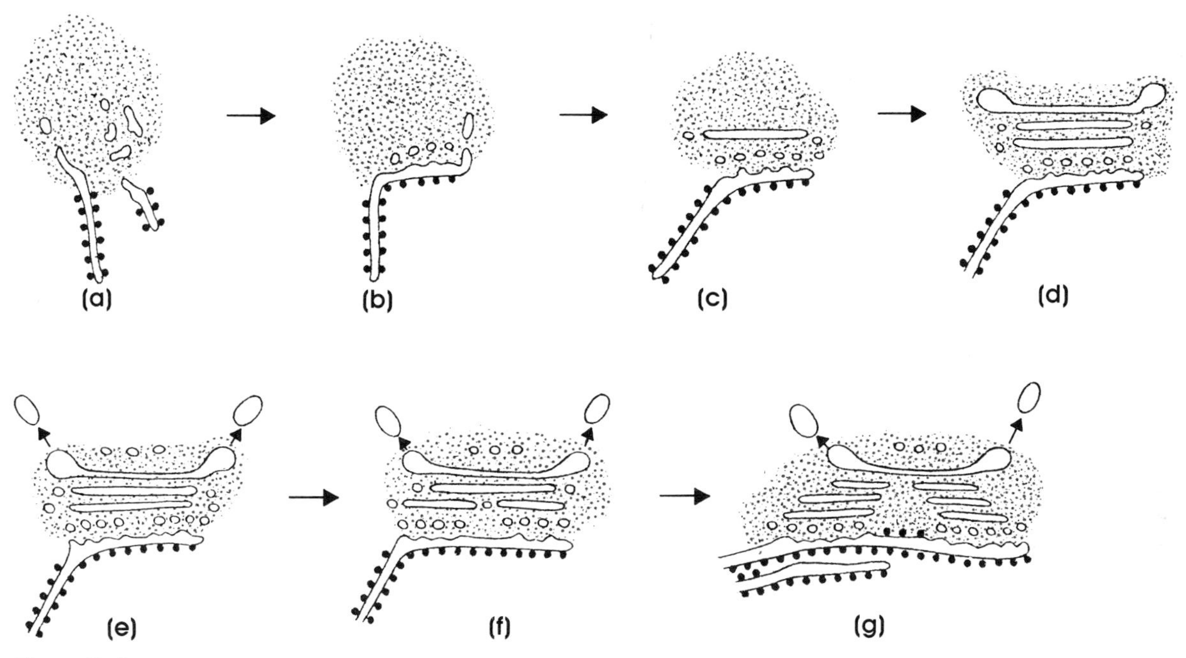

Figure 18–9
Proposed formation of dictyosome from endoplasmic reticulum (*a-c*) and subsequent developmental stages: layered cisternae (*c-d*), formation of secretory vesicles (*e*), division of dictyosome (*f-g*). (Diagram courtesy of Dr. J. Morré et al., in *Origin and Continuity of Cell Organelles* (Reinert and Ursprung, eds.), copyright 1971 Springer-Verlag.)

Figure 18–10
Mucous glycoprotein showing attachment of polysaccharide side chain composed of various derivatives of glucose to the polypeptide chain.

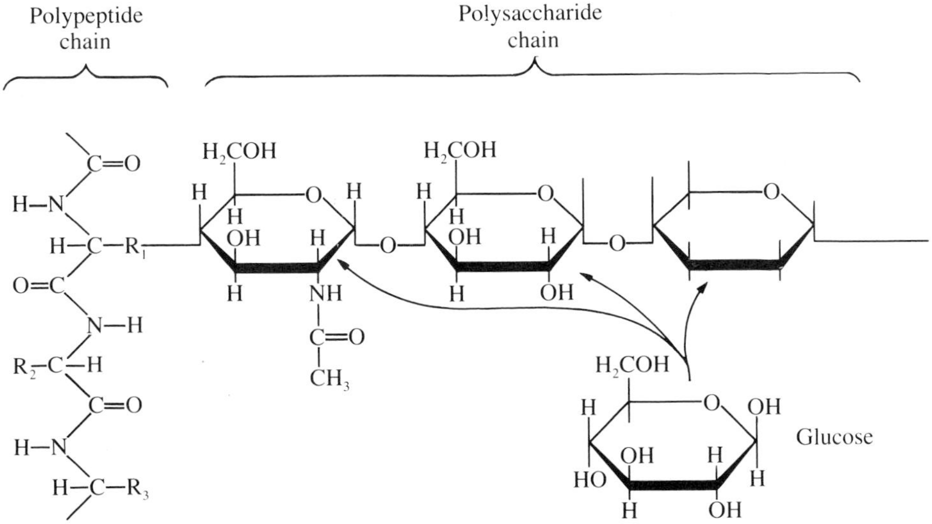

Polypeptide chain

Polysaccharide chain

Glucose

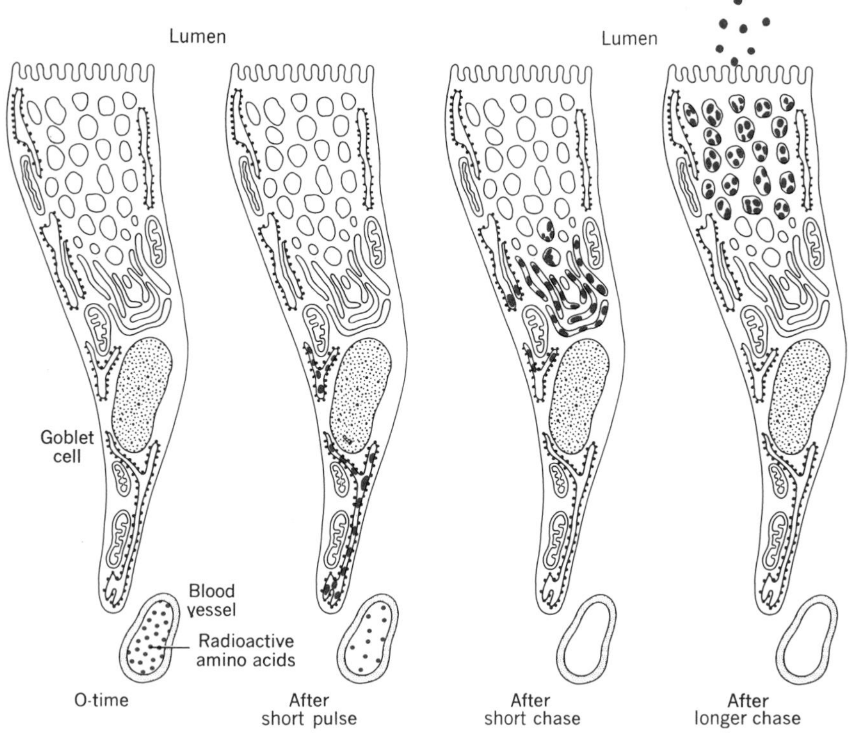

Figure 18–11
Incorporation of amino acids into secretory proteins by the goblet cells of the intestine. Amino acids removed from the bloodstream are used in protein synthesis by rough endoplasmic reticulum. The proteins are conveyed to the Golgi apparatus for incorporation into secretory vesicles. Glycosylation of the proteins (to form glycoproteins) occurs within the Golgi apparatus. Vesicles detach from the maturing face of the Golgi apparatus and migrate to the plasma membrane where they are discharged into the intestinal lumen.

nae of the Golgi apparatus. This label did not enter or associate with the rough endoplasmic reticulum first. After a 20-minute chase the label appeared in the mucous vesicles, and after 4 hours, most had been released through the plasma membrane into the lumen. This experiment not only shows the path of glucose through the cell but also reveals that final stages of assembly of the glycoprotein occur *in the Golgi apparatus*. Using the goblet cell as an example, Figure 18–11 depicts the central role of the Golgi apparatus in the packaging of newly-synthesized proteins into vesicles for secretion. The assembly of large molecules in the Golgi is not unique to goblet cells. Cartilage cells assemble glycoproteins in the cisternae of their Golgi bodies, and sulfate groups have been shown to be added as well. Pectins and cellulose are assembled in the Golgi bodies of plant cells prior to deposition onto the forming cell plate or cell wall.

The participation of the Golgi apparatus in the growth and modification of the cell membrane is well documented, but it is not clear at present whether this is a major or minor means of cell membrane synthesis. The plasma membrane is thicker and has a different composition of phospholipids and sterols than the membranes of the endoplasmic reticulum or the Golgi apparatus. The Golgi membranes seem to be intermediate in thickness between the thin ER membranes and the thick plasma membrane. When cells such as *Trichonympha* (a protozoan) are starved, the dictyosomes and many endoplasmic reticulum membranes disappear. Upon refeeding, the endoplasmic reticulum membranes reappear first, following which vesicles from these membranes appear to branch off and develop into cisternae. With time, the cisternae develop into dictyosomes, and one can observe what appears to be a migration of the outermost cisternae toward the plasma membrane and their fusion with the plasma membrane. It seems reasonable to speculate that the proteins for the membranes are synthesized on the endoplasmic reticulum and the initial membranes formed there as well. Most speculation identifies the smooth endoplasmic reticulum as the site of Golgi membrane assembly. The budding off of vesicles from the smooth ER to form cisternae and dictyosomes is probably accompanied by modification in the lipid components of the membrane. Further modification of both lipids and proteins probably occurs before fusion of dictyosome membranes with the plasma membrane.

Cell-Specific Functions of the Golgi Apparatus

Although a discussion of each of the functions of the Golgi apparatus in different cells is beyond the scope of this book, four important examples are discussed below. A list of other well-studied cases is presented in Table 18–1.

Formation of the Plant Cell Plate and Cell Wall

In plants, the cell plate and cell wall form during anaphase and telophase of mitosis and meiosis II (Chapter 20). During these final stages of nuclear division, the chromosomes have separated into two masses in the cell that will become nuclei. Between these two nuclei, pectin and hemicellulose are deposited bit by bit, forming a plate in the center of the cell, which ultimately expands to the side walls, cutting and separating the protoplasts in two, thereby producing the two daughter cells. Prior to anaphase, the Golgi bodies are found outside the spindle. During anaphase, vesicles appear to be released from the Golgi apparatus, invade the center of the spindle (Fig. 18–12), and aggregate about the spindle fibers. These vesicles provide the carbohydrate that

Table 18–1
Specific Functions of Golgi Structures

Cell	Tissue or Organ	Golgi Function
Exocrine	Pancreas	Secretion of zymogen (proteases, lipases, carbohydrases and nucleases)
Gland cell	Parotid gland	Secretion of zymogen
Goblet cell	Intestinal epithelium	Secretion of mucous and zymogens
Follicle cells	Thyroid gland	Prethyroglobulin
Plasma cells	Blood	Immunoglobulins
Myelocytes, sympathetic ganglia, Schwann cells	Nervous tissue	Sulfation reactions
Endothelial cells	Blood vessels	Sulfation reactions
Liver cells	Liver	Lipid secretion (lipid transformation?)
Alveolar epithelium	Mammary gland	Secretion of milk proteins (and lactose?)
Paneth cells	Intestines	Secretion of proteins (chitinase?)
Brunner's gland cell	Intestines	Synthesis and secretion of mucopolysaccharides, enzymes, hormones
Connective tissue	Amblystoma limb	Synthesis (?) and secretion of collagen
Cornea	Avian eye	Secretion of collagen
Plant cells	Most	Secretion of pectin and cellulose

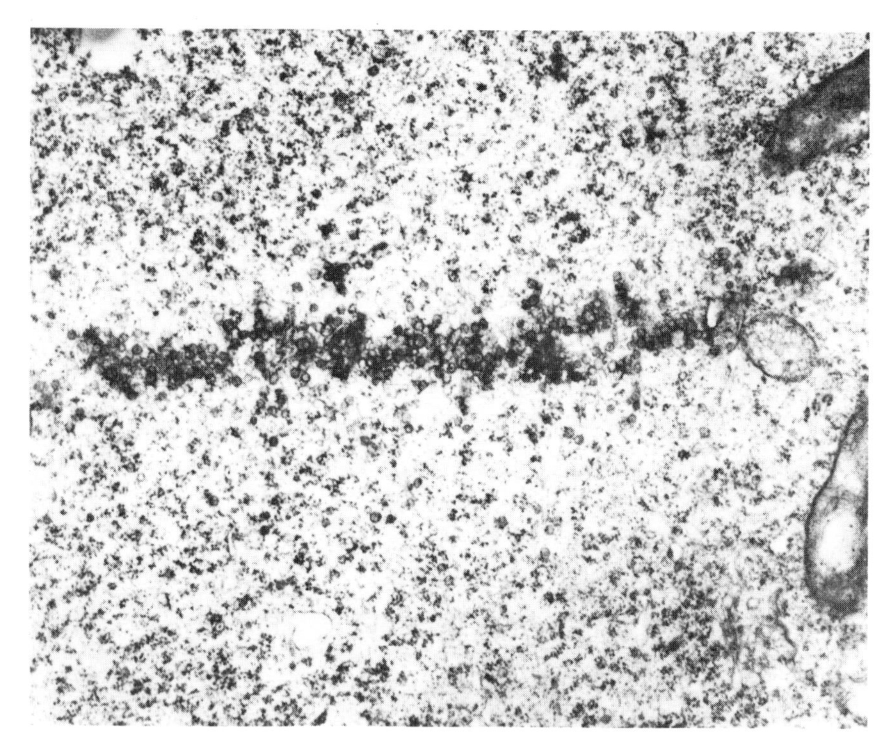

Figure 18–12
Clusters of small vesicles released from the Golgi apparatus during anaphase in *Zea mays* root apex cell. The vesicles migate toward the equatorial region where they contribute polysaccharides to the forming cell plate. (Photomicrograph courtesy of Dr. G. Whaley *et al.*, copyright Academic Press, *J. Utrastruc. Res. 15,* 173, 1966.)

forms the cell plate and eventually the wall. The nature of the carbohydrate secreted by the vesicles is controversial. Some cell biologists believe that cellulose fibers are preformed and secreted, while others believe that the final stages of cellulose synthesis occur after secretion. In either case, the Golgi apparatus is clearly involved in the secretion of the carbohydrate that forms the wall between the two cell halves.

The plasma membrane of plant cells does not pinch inward or grow inward during cell division as it does in animal cells. Instead, the membrane is formed on both sides of the developing cell plate and grows outward with it. Formation of the membrane results from the fusion of the vesicle membranes arising from the Golgi apparatus.

Neurosecretions

Nerve cells were the first cells described by Golgi in 1880 to contain the "internal reticular apparatus" (i.e., Golgi apparatus). Since then, many studies have been conducted on the neurosecretions of this system, but the clarification that has been achieved with gland cells is still lacking. A great variety of substances are secreted by nerve tissue cells, including hormones (noradrenaline, histamine, vasopressin, oxytocin, luteinizing hormone, follicle-stimulating hormone, etc.) and other substances. Perhaps one of the best known neurosecretions is *acetylcholine,* which is frequently described as the synaptic transmitter substance. It is this compound that is released by the **end-bulb** of many neurons (nerve cells) and that crosses the gap, called a **synapse,** between that neuron and the next (Chapter 23). Acetylcholine is known to be present in vesicles (called synaptic vesicles) in the end-bulb. When a nerve impulse reaches the nerve cell endings, the vesicles all discharge through the plasma membrane and can be easily observed. When the acetylcholine reaches the plasma membrane of the neuron at the other side of the synapse (Fig. 18–13), a new impulse is generated. The origin of the synaptic vesicles and the site of synthesis of the acetylcholine are not clear. Many different compounds are present in separate vesicles in the end-bulb; also present are many lysosomes. Whether the Golgi apparatus gives rise to each type of vesicle and perhaps also the lysosomes or whether the lysosomes themselves give rise to some of the vesicles remains unresolved. At the present time, most investigators favor the notion that

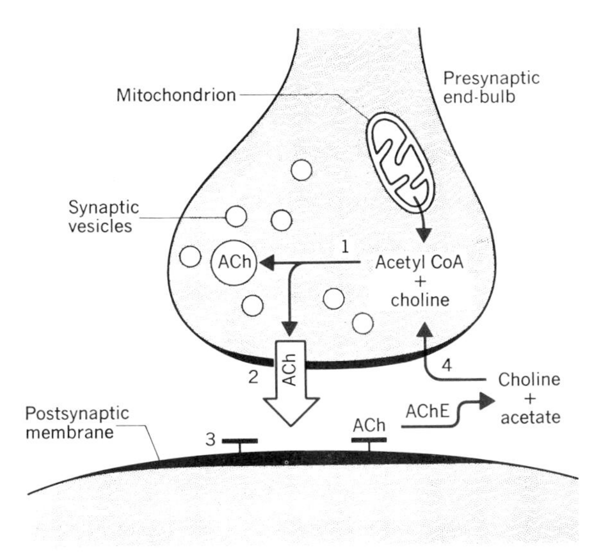

Figure 18–13
Diagrammatic representation of synapse showing chemical transmission of nerve impulse. Formation of acetylcholine in end plate (1), release from synaptic vesicles (2), sites of stimulation on the postsynaptic membrane (3), hydrolysis of acetylcholine by esterase and reabsorption of products (4). ACh, acetylcholine; AChE, acetylcholine esterase.

the Golgi apparatus is the source of the synaptic vesicles. Acetylcholine released by synaptic vesicles into the synapse is broken down into acetate and choline by the enzyme *acetylcholine esterase*. These products are reused for acetylcholine synthesis. Whether the resynthesis occurs in the synapse, in the end-bulb cytoplasm after reabsorption, or in the Golgi apparatus is not known.

Interrelationship Between Golgi, Lysosomes, and Vacuoles

In Chapter 19 the structure of the lysosome is described, as well as the hydrolytic nature of the organellar contents. The membranes of lysosomes, vacuoles, and Golgi bodies are similar (although not identical) to those of smooth endoplasmic reticulum. For this reason, it has been suggested that they may be derived either directly or indirectly from the smooth ER. There is good evidence that dictyosomes accumulate hydrolytic enzymes in their outer, more mature regions, and this has led to the proposal that lysosomes are derived directly from the Golgi apparatus (Chapter 19). Some vacuoles in plants also have been found to contain small amounts of hydrolytic enzymes, and thus their derivation from the Golgi apparatus has been proposed.

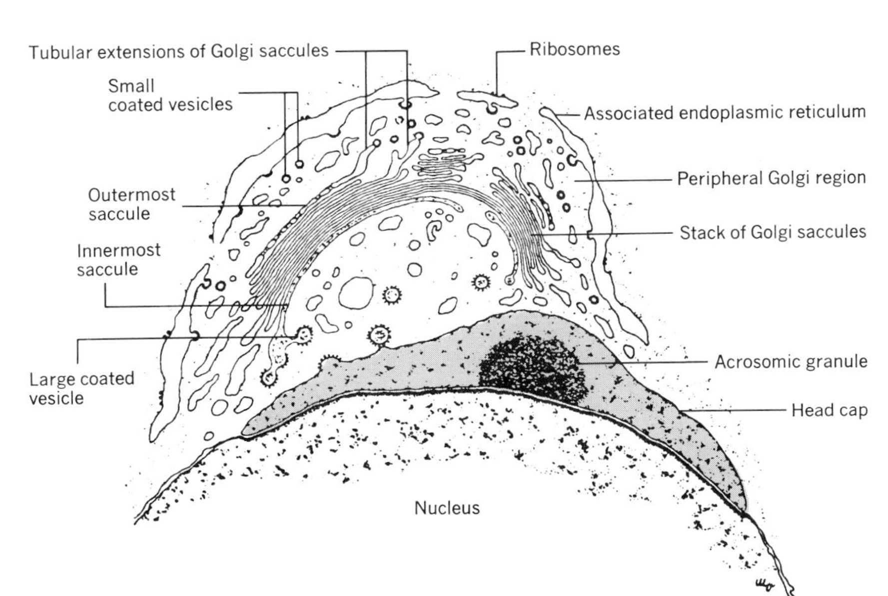

Figure 18–14
Contributions of the Golgi apparatus to the forming acrosome during spermatid development. The uppermost Golgi elements are part of the proximal or forming face; the vesicles contributing to the acrosomal membrane are derived primarily from the distal or maturing face. (Copyright Wistar Institute; F. Susi *et al., Am. J. Anat. 130,* 262, 1971.)

Acrosome Development in Sperm

The development of the acrosome of sperm cells is a good example of the involvement of both the membrane and contents of Golgi cisternae in the formation of another organelle. The acrosome is a membrane-bound structure at the anterior end of sperm cells of most animals. It is a part of the membrane of the acrosome that appears to contribute to the recognition and binding of the sperm to the egg cell in fertilization. The acrosome contains hydrolytic enzymes of which hyaluronidase is the most prevalent and which contributes to the breakdown of the protective surfaces of the egg. As shown in Figure 18–14, the singular large Golgi body buds off large coated vesicles that migrate to the forming acrosome. At the surface of the acrosome, the coated vesicle fuses, contributing its membrane to the growing membrane of the acrosome and the contents to the developing acrosomic granule.

Since the acrosomic granule is made up of hydrolytic enzymes, some have postulated that the acrosome is really a giant lysosome. As the acrosome expands, the Golgi becomes reduced in size, and in many, mature sperm disappears entirely. The outer membrane of the acrosome merges with that of the plasma membrane. In mouse sperm, it has been shown that the area of the plasma membrane that has fused with the acrosomal membrane contains a large number of concanavalin A binding sites. The increased number of carbohydrates (glycoproteins) in the membrane is attributed to the Golgi origin of the membrane.

Summary

The membrane-lined **cisternae** of the Golgi apparatus develop either from the outer portion of the nuclear envelope or from the endoplasmic reticulum. The cisternae may form a stack of vesicles that acts as a unit (called a **dictyosome**), with vesicles joining the stack on one edge, maturing, and leaving the stack at the other edge. A cluster of dictyosomes forms the functioning **Golgi apparatus** of the cell.

The Golgi apparatus functions in cell **secretion** and as a site for assembly of **glycoproteins** and other membranous components. In plant cells, Golgi bodies are associated with the formation of the **cell plate,** and in nerve and gland cells, with various **neurosecretions.** The Golgi apparatus also gives rise to **vacuoles,** including those that become **lysosomes** and certain **microbodies.**

References and Suggested Reading

Articles and Reviews

Beams, H., and Kessel, R., The Golgi apparatus: Structure and function. *Int. Rev. Cytol. 23* 209 (1968).

Neutra, M., and Leblond, C., The Golgi apparatus. *Sci. Am. 220* (2), 100 (Feb. 1969).

Northcote, D., The Golgi apparatus, *Endeavour 30,* 26 (1970).

Northcote, D., Chemistry of the plant cell wall. *Annu. Rev. Plant Physiol. 23,* 113 (1972).

Susi, R., LeBlond, C., and Clermont, Y., Changes in the Golgi apparatus during spermiogenesis in the rat. *Am. J. Anat. 130,* 251 (1971).

Books, Monographs, and Symposia

Morré, D., Mollenhauer, H., and Bracker, C., Origin and continuity of Golgi apparatus, in *Origin and Continuity of Cell Organelles* (J. Reinhert and H. Ursprung, eds.), Springer-Verlag, Berlin, 1971.

Waley, W. G., *The Golgi Apparatus,* Springer-Verlag, New York, 1975.

Chapter 19
LYSOSOMES AND MICROBODIES

Our knowledge of the structure, composition, and function of lysosomes and microbodies is considerably more recent than that of most other cell organelles. Although a variety of small oval bodies seen in plant and animal cells (including what are now termed lysosomes) had been variously called "microbodies" or "cytosomes" for many years, the diversity of their composition and action was not recognized until the 1950s. The "discovery" of lysosomes and microbodies during the 1950s may be attributed to the growing sophistication of electron microscopy, the application of gentler procedures for dispersing tissue and cells, and the development of improved methods for separating, fractionating, and chemically characterizing the subcellular complexes released from disrupted cells. At the present time, lysosomes are recognized as a separate category of organelles, whereas the microbodies include two major types: *peroxisomes* and *glyoxysomes*.

Lysosomes

The existence of lysosomes was subtly suggested for the first time in 1949 in the results of a series of experiments by C. de Duve and his co-workers that were designed to identify the cellular locus of the two enzymes *glucose-6-phosphatase* and *acid phosphatase*. In these experiments, liver tissue homogenates were separated into nuclear, mitochondrial, microsomal, and soluble fractions by differential centrifugation (see Chapter 12) and enzyme assays were performed on each of the collected fractions. Although results with glucose-6-phosphatase clearly indicated that this enzyme was bound to particles sedimenting with the microsome fraction, observations on the distribution of acid phosphatase were at first rather confusing. The confusion centered around three seemingly peculiar but nonetheless reproducible findings: (1) the acid phosphatase activities of tissue homogenates prepared for centrifugation using a glass tube and close-fitting plunger (Dounce homogenizer) were about one-tenth the value observed when tissue was more vigorously dispersed in water using a Waring blender; (2) the total (i.e., combined) enzyme activity of the isolated centrifugal fractions analyzed following differential centrifugation was about twice the activity of the original homogenate; and (3) after storage for several days in a freezer, both the enzyme activity of the homogenate and the collected fractions, especially the mitochondrial fraction, increased dramatically (Table 19–1).

These observations were explained when de Duve showed that acid phosphatase activity was confined to sedimentable particles, the surrounding membranes of which limited the accessibility of the substrate (beta-glycerophosphate) used in the enzyme assay. Only when these membranes were disrupted and the acid phosphatase released from the particles was the enzyme activity demonstrable. This occurred during vigorous dispersion in the Waring blender, which

Table 19–1
Distribution of Acid Phosphatase Enzyme Activity in Liver Tissue Fractions Prepared by Differential Centrifugation

Fraction	Acid Phosphatase Activity (μg phosphate released/20 min)	
	Before Freezing and Storage	After Freezing and Storage for 5 Days
Whole homogenate	10	89
Nuclear fraction	2	10
Mitochondrial fraction	7	46
Microsomal fraction	6	10
Soluble fraction	6	9

Source. Based on the published results of J. Berthet and C. de Duve, *Biochem. J. 50,* 174 (1951), and C. de Duve. The lysosome in retrospect, in *Lysosomes in Biology and Pathology* (J. T. Dingle and H. B. Fell, eds.), North-Holland Publishing Co., Amsterdam, 1969, p. 6.

disrupted virtually all particles present. In contrast, only about 10% of the acid-phosphatase-containing particles were disrupted by homogenization with the Dounce homogenizer, thus accounting for the low activity of the enzyme in the homogenate (Table 19–1). Some added enzyme activity was released during and following centrifugal fractionation, but much larger quantities of enzyme were released by the membrane disruption that occurred during freezing and thawing.

At first, de Duve and his co-workers did not recognize that the acid phosphatase activity was associated with a distinct population of cellular particles. Instead, on the basis of the observed "latent" activity of the mitochondrial fraction obtained by differential centrifugation (compare values before and after freezing in Table 19–1), de Duve believed that acid phosphatase resided within the mitochondria. Continued studies during the early 1950s in which the mitochondrial fraction was further divided centrifugally into a number of subfractions revealed that acid phosphatase was absent from fractions containing rapidly sedimenting mitochondria but was present in high concentrations in fractions containing slowly sedimenting mitochondria. This observation, together with a newly developed appreciation of the potential contamination of sediments occurring during differential centrifugation (see Chapter 12), led de Duve to suspect that the acid phosphatase might, in fact, be associated with a special class of particles distinct from the mitochondria. Added credence was given to this idea by finding that four other acid hydrolases, namely *beta-*

Table 19–2
Some Enzymes Present in Lysosomes

Enzyme	Substrate(S)
Proteases and peptidases	
Cathepsin A, B, C, D and E	Various proteins and peptides
Collagenase	Collagen
Arylamidase	Amino acid arylamides
Peptidase	Peptides
Nucleases	
Acid ribonuclease	RNA
Acid deoxyribonuclease	DNA
Phosphatases	
Acid phosphatase	Phosphate monoesters
Phosphodiesterase	Oligonucleotides, phosphodiesters
Phosphatidic acid phosphatase	Phosphatidic acids
Enzymes acting on carbohydrate chains of glycoproteins and glycolipids	
Beta-galactosidase	Beta-galactosides
Acetylhexosaminidase	Acetylhexosaminides, heparin sulfate
Beta-glucosidase	Beta-glucosides
Alpha-glucosidase	Glycogen
Alpha-mannosidase	Alpha-mannosides
Sialidase	Sialic acid derivatives
Enzymes acting on glycosaminoglycans	
Lysozyme	Mucopolysaccharides, bacterial cell walls
Hyaluronidase	Hyaluronic acid, chondroitin sulfates
Beta-glucuronidase	Polysaccharides, mucopolysaccharides
Arylsulfatase, A, B	Arylsulfates, cerebroside sulfates, chondroitin sulfate
Enzymes acting on lipids	
Phospholipase	Lecithin, phosphatidyl ethanolamine
Esterase	Fatty acid esters
Sphingomyelinase	Sphingomyelin

glucuronidase, cathepsin, acid ribonuclease, and *acid deoxyribonuclease,* were distributed through the centrifugal fractions in an identical manner. Thus, five hydrolytic enzymes, each having an acid pH optimum and acting on completely different substrates, appeared to be present in the same cell particle. On the basis of the *lytic* effects of all of these enzymes, de Duve named the particles "lysosomes." A number of additional enzymes have subsequently been identified in lysosomes (Table 19–2). Most substances present in cells including proteins, polysaccharides, nucleic acids, and lipids are broken down by these enzymes.

It is interesting to note that the initial postulation of the existence of lysosomes was made by de Duve purely on biochemical grounds. However, in 1955, A. Novikoff, working with de Duve, examined centrifugal fractions rich in acid phosphatase activity using the electron microscope and provided the first morphological evidence supporting the existence of these particles. The lysosomes were identified as small, dense membrane-enclosed particles distinct from the mitochondria.

In recent years, sophisticated centrifugal methods have been devised for obtaining preparations rich in lysosomes. Nearly all preparations obtained by differential centrifugation are contaminated with quantities of mitochondria. Although the *average* sedimentation coefficient for mitochondria is greater than that of lysosomes, mitochondria are polydispersed with respect to size, so that the smaller mitochondria invariably sediment with the lysosomes. Moreover, in tissues containing peroxisomes (such as liver and kidney), the range of sedimentation coefficients for these organelles is almost identical to that of the lysosomes. Consequently, it is virtually impossible to obtain lysosome preparations that do not also contain large numbers of peroxisomes. Somewhat greater success is obtained when isopycnic density gradient centrifugation is used in the last stages of the isolation procedure, for the equilibrium densities of lysosomes (1.22 g/cm^3), mitochondria (1.19 g/cm^3) and peroxisomes (1.23–1.25 g/cm^3) in sucrose density gradients are slightly different. Most density gradient procedures used to prepare lysosomes are modifications of the technique developed by W. C. Schneider and depicted in Figure 19–1. Using this technique, most of the mitochondria are banded isopycnically at a density of about 1.19 g/cm^3, while most of the lysosomes form a separate zone at about 1.22 g/cm^3 and can be recovered independently from the density gradient.

By far the greatest purity of lysosomes is obtained from tissues of animals previously treated with Triton WR-1339 (a polyethylene glycol derivative of polymerized *p-tert*-octyl phenol), dextran (a polymer of glucose), and Thorotrast (colloidal thorium hydroxide). These compounds are rapidly incorporated in large quantities by the cell's lysosomes, significantly altering their density. For example, the incorporation of Triton WR-1339 reduces the average density of the lysosome from 1.22 to about 1.10 g/cm^3. It is interesting to note that although the density of lysosomes incorporating Triton WR-1339 is significantly reduced, their size is increased; the result is that Triton WR-1339-loaded lysosomes have the same sedimentation coefficient as do normal lysosomes but have a lower density.

The latent enzymic effect originally noted by de Duve and his co-workers is still employed as a major criterion in evaluating the effectiveness of any lysosome isolation. Accordingly, the lysosome preparation is incubated under the appropriate conditions with the hydrolase substrate before and after treatments known to disrupt the lysosome membranes. If the original preparation contains intact lysosomes, then no substrate is hydrolyzed before treatment (most substrates of the lysosomal hydrolases are unable to permeate the lysosome's membrane); however, disruption of the membrane (by sonification, repeated freezing and thawing, addition of lytic agents such as bile salts, digitonin, Triton X-100, etc.) and release of the lysosomal enzymes are quickly followed by hydrolysis of the added substrates.

Structure and Forms of Lysosomes

Lysosomes are a structurally heterogeneous group of organelles varying dramatically in size and morphology. As a result, it is difficult to identify lysosomes strictly on the basis of morphological criteria. When lysosome-rich fractions were initially isolated centrifugally by de Duve and Novikoff and examined with the electron microscope, it was found that the suspected lysosomes were generally smaller than mitochondria. Typically, they varied in diameter from about 0.1 to 0.8 μ, were bounded by a single membrane, and were usually somewhat electron-dense. Identification of lysosomes in sections of whole cells is considerably more difficult because other small, dense organelles are also bounded by a single membrane. The application of cytochemical procedures at the level of the electron microscope in which the lysosomes are identified

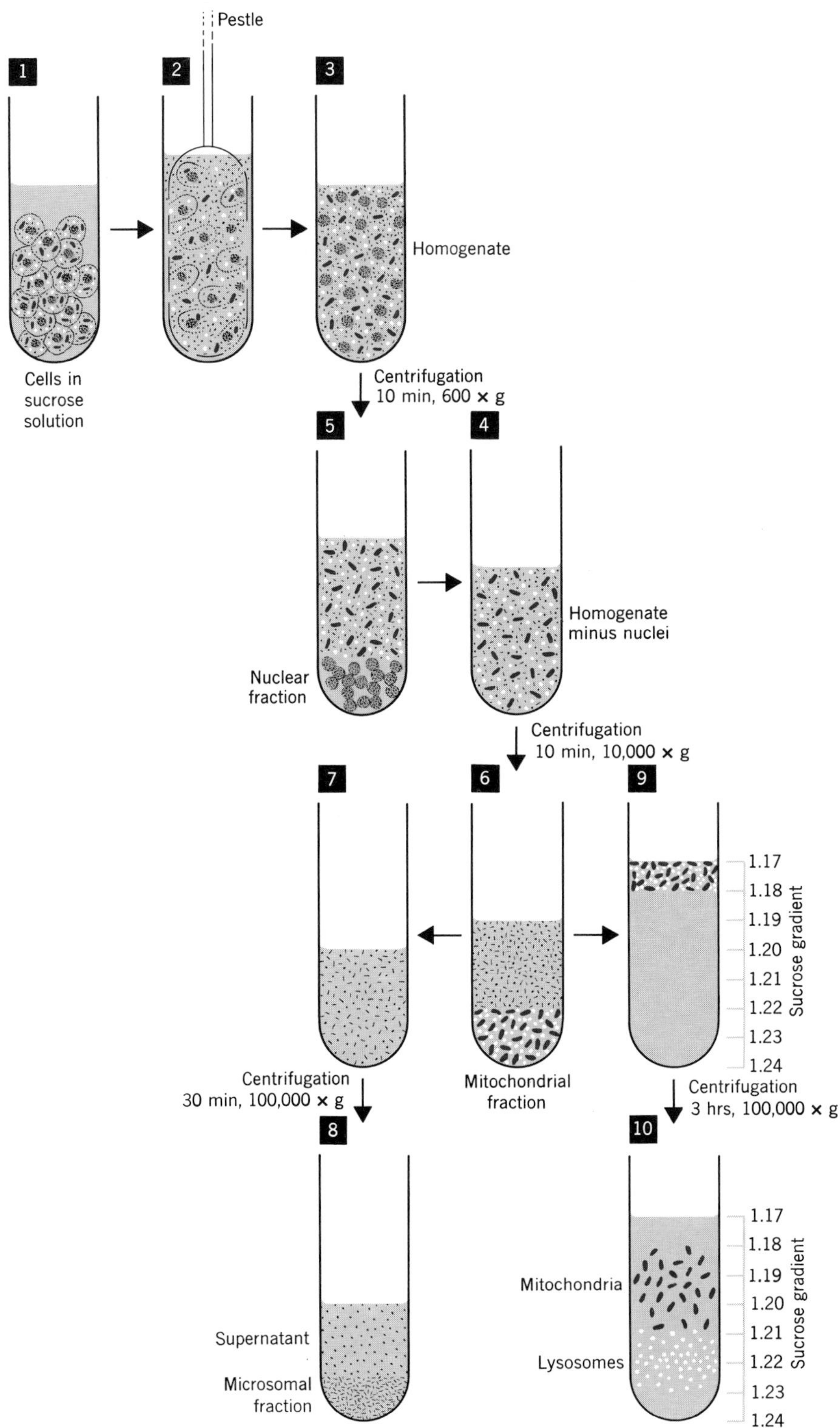

Figure 19–1
Steps in the isolation of lysosomes by sucrose density gradient centrifugation.

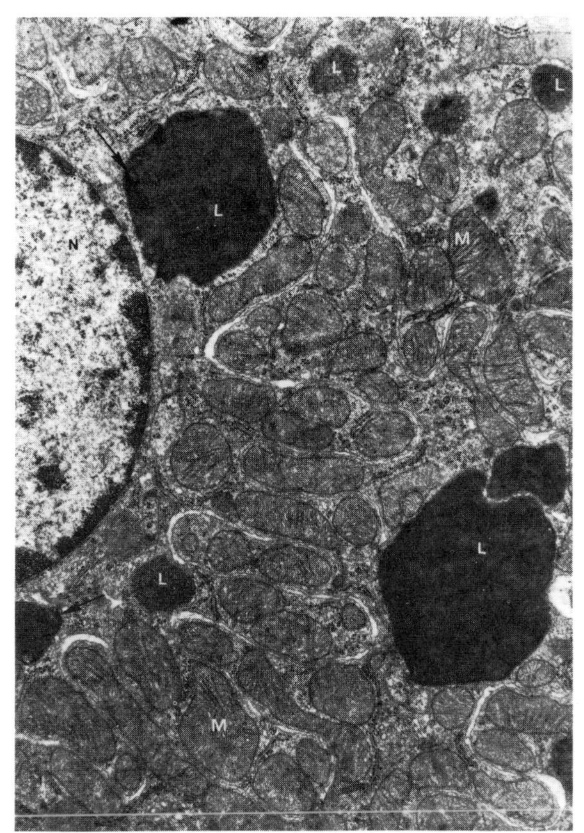

Figure 19–2
Electron photomicrograph of rat kidney treated for phosphatase activity to reveal lysosomes (L). The presence of small bodies (arrows) within lysosomes indicates that these are secondary lysosomes. *N*, nucleus; *M*, mitochondria. (Courtesy of Dr. E. G. Pollock.)

on the basis of their enzyme content is much more reliable. Notable among such procedures is that introduced in 1952 by G. Gomori and which is routinely employed in variously modified forms for the identification of lysosomes on the basis of their high acid phosphatase content. In the Gomori method, the tissue to be examined is incubated at pH 5.0 in a medium containing *beta*-glycerophosphate (a substrate for acid phosphatase) and a lead salt (such as lead nitrate). Phosphate enzymatically cleaved from the substrate during incubation combines with the lead ions to form insoluble lead phosphate which precipitates at the locus of enzyme activity. Since the lead phosphate is electron-dense, lysosomes appear as particularly dark, granular organelles in the electron microscope (Fig. 19–2). For identification with the light microscope, ammonium sulfide may be employed to convert the lead phosphate produced to the black lead sulfide. The Gomori reaction may be carried out with fixed and sectioned material, as well as with fresh tissue, albeit with reduced efficiency as a result of some enzyme inactivation during and following fixation.

Several different lysosomal forms have been identified within individual cells including (1) **primary lysosomes,** (2) **secondary lysosomes,** and (3) **residual bodies.**

Primary Lysosomes. Primary lysosomes, or **protolysosomes,** are newly produced organelles bounded by a single membrane and varying greatly in size. The primary lysosome is a virgin particle in that its digestive enzymes have not yet taken part in hydrolysis.

Secondary Lysosomes. Two different kinds of secondary lysosomes can be identified: **heterophagic vacuoles** (also called **heterolysosomes** or **phagolysosomes**) and **autophagic vacuoles** (also called **autolysosomes**). Heterophagic vacuoles are formed by the *fusion* (see below) of primary lysosomes with cytoplasmic vacuoles containing *extracellular* substances brought into the cell by any of a variety of endocytic processes (see Chapter 15). Following fusion, the hydrolases of the primary lysosome are released into the vacuole (called a **phagosome**). Autophagic vacuoles contain particles isolated from the cell's own cytoplasm, including mitochondria and smooth and rough fragments of the endoplasmic reticulum. The autodigestion of cellular organelles is a normal event during cell growth and repair and is especially prevalent in differentiating and dedifferentiating tissues and tissues under stress. Autophagic vacuoles containing partially degraded mitochondria are shown in Figure 19–3. The formation of heterophagic and autophagic vacuoles is soon followed by enzymatic digestion of the vacuolar contents. As digestion proceeds, it becomes increasingly difficult to identify the nature of the original secondary lysosome and the more general term **digestive vacuole** is used to describe the organelle at this stage.

Residual Bodies. Endocytosed substances and parts of autophagocytosed organelles that are not digested within the secondary lysosomes and transferred to the cytoplasm are retained (usually temporarily) within the vacuoles as residues. Lysosomes containing such residues are called residual bodies (sometimes also called **telolysosomes** or **dense bodies**). The undigested residues often take the

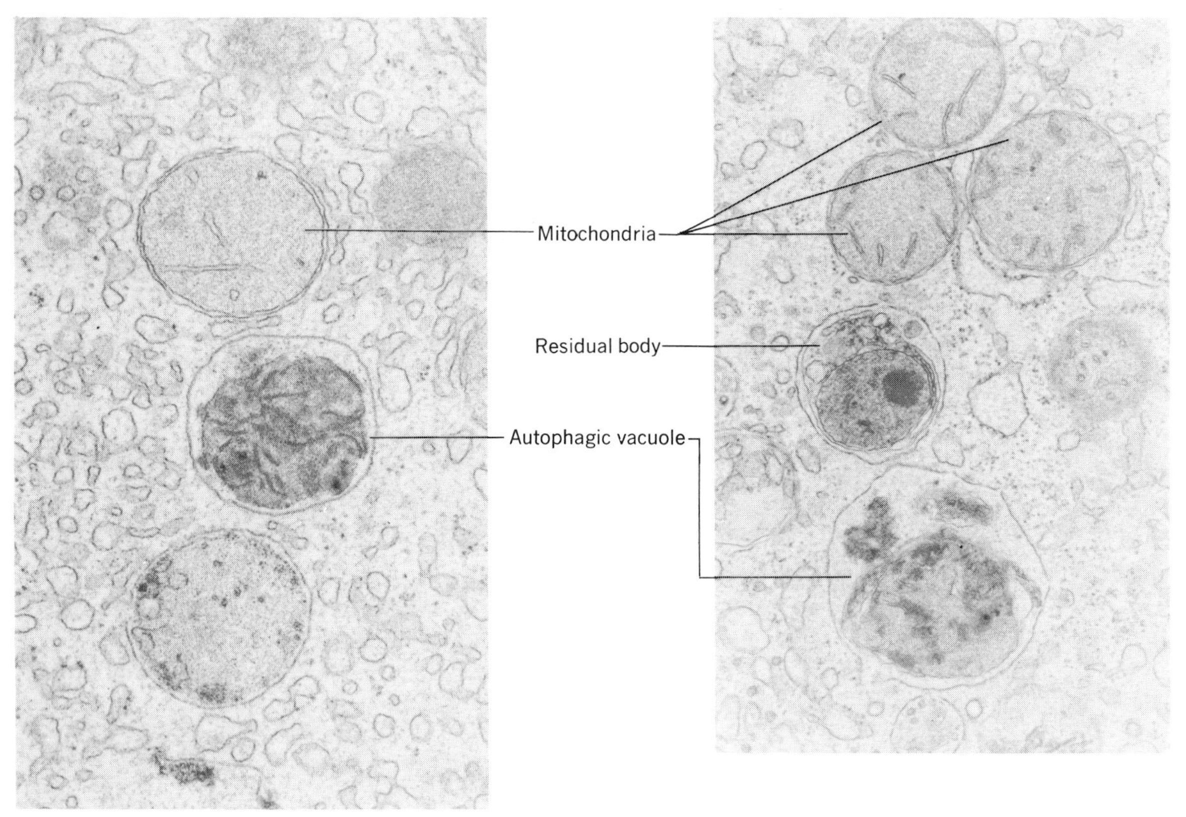

Figure 19–3
Autophagic vacuoles containing a partially degraded mitochondria. (Electron photomicrographs courtesy of Dr. Z. Hruban.)

form of whorls of membranes, grains, amorphous masses, ferritinlike particles, or myelin figures (Fig. 19–4). Residual bodies often fail to display the degree of hydrolytic activity associated with the primary and secondary lysosomes.

Formation and Function of Lysosomes (The "Vacuolar System")

Lysosomal enzymes are concerned with the degradation of metabolites and not with cellular synthetic or transfer reactions. The specific cellular origin of the lysosomal acid hydrolases and the mechanism by which they are incorporated into primary lysosomes are still uncertain. On the basis of numerous cytochemical observations, two theories have gained widespread acceptance. According to Novikoff, the acid hydrolases destined to become part of the lysosome are synthesized on ribosomes in the vicinity of the Golgi bodies (Chapter 18). These hydrolases are then transferred through the cisternae of the smooth ER to the Golgi bodies, where they are enclosed within small vesicles that detach from the peripheral edges of the Golgi complex to form the primary lysosomes. This concept, depicted diagrammatically in Figure 19–5, is supported by a large number of observations made with a variety of tissues. The process is similar to the formation of zymogen granules for secretion by Golgi bodies.

An alternative proposal based on more limited observations suggests that primary lysosomes may be produced directly by dilations of rough endoplasmic reticulum. To reconcile these observations, it has been suggested that the Golgi apparatus may be involved in the formation of primary lysosomes in cells rich in *smooth* endoplasmic reticulum, whereas direct formation by *rough* endoplasmic

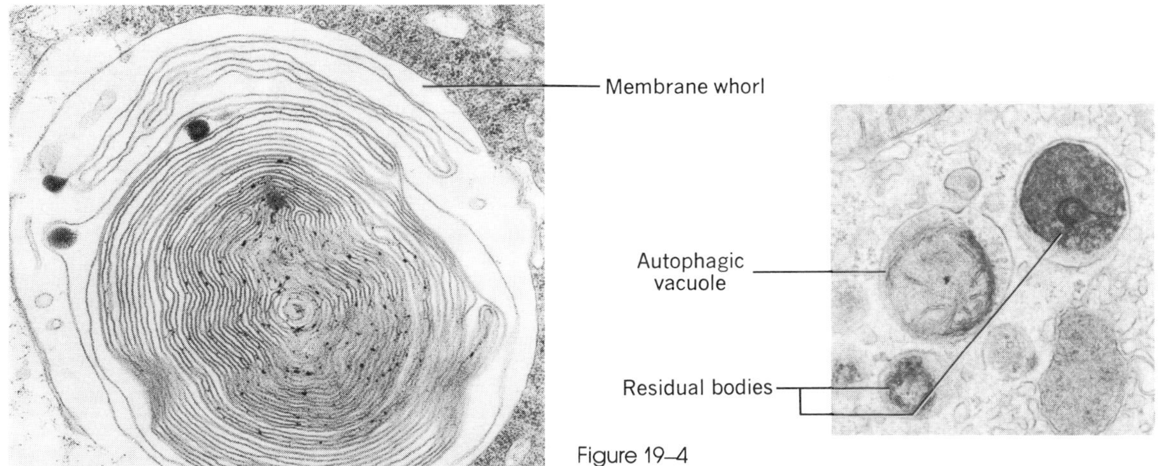

Figure 19–4
***Left:* a membrane whorl. (Electron photomicrograph courtesy of Dr. E. G. Pollock.) *Right:* residual bodies. (Electron photomicrograph courtesy of Dr. Z. Hruban.)**

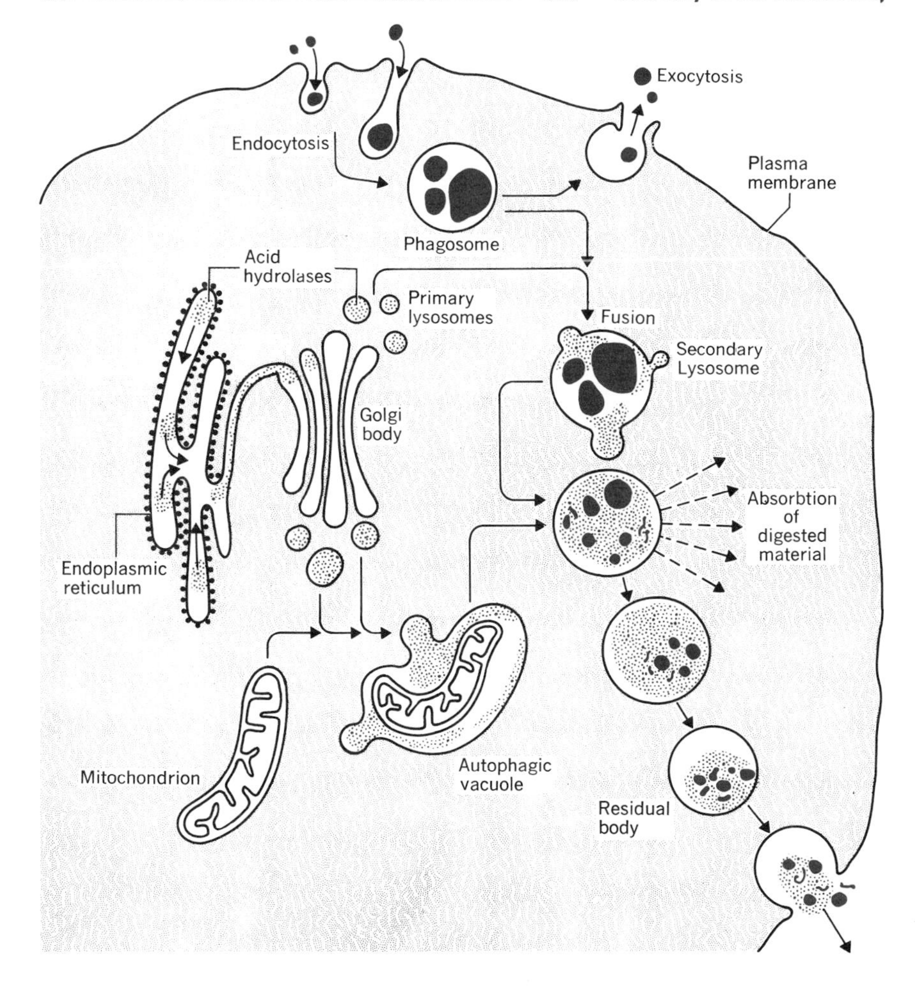

Figure 19–5
Formation and function of lysosomes in cellular heterophagy and autophagy.

reticulum occurs in cells in which that form of cytomembrane predominates.

Heterophagy. Extracellular materials brought into the cell by endocytosis are enclosed within vacuoles called *phagosomes*. These materials may later be rejected unaltered by exocytosis, or the phagosomes may fuse with one or more primary lysosomes that empty their digestive hydrolases into the newly formed particle (now called a *secondary* lysosome, Fig. 19–5). Lysosomal digestion of endocytosed material is termed **heterophagy.**

The fusion of primary lysosomes with phagosomes has been demonstrated in vivo in a number of tissues using various exogenous *markers* introduced into the organism. These markers, which include horseradish peroxidase, ferritin, and hemoglobin, are engulfed by the tissue cells and are later detected within secondary lysosomes along with lysosomal hydrolases.

Z. A. Cohn and B. Benson employed ^{3}H-labeled leucine to trace the fate of newly synthesized hydrolases in peritoneal phagocytes of mice. They found that pinocytic activity was greatly increased when these cells were incubated in blood serum. Autoradiographic analysis revealed that the labeled hydrolases appeared first in the Golgi region of the cells and later within pinocytic vesicles (i.e., phagosomes). Their observations support the proposal that secondary lysosomes are formed by the fusion of phagosomes and primary lysosomes. Moreover, Cohn and Benson also found that the rate at which hydrolases were produced by the cells was related to the level of pinocytic activity, suggesting that the production of primary lysosomes may somehow be regulated by endocytosis (see below).

In some cells, several small primary lysosomes may fuse with a single large phagosome; in other cells, large primary lysosomes sequentially fuse with a number of small phagosomes. The contents of the secondary lysosome change dramatically with time as (1) the contents of the lysosome are enzymatically degraded, (2) new materials are introduced through fusion of additional phagosomes, and (3) additional hydrolases are added by the fusion of new primary lysosomes. The hydrolases in the secondary lysosome break down the endocytosed materials, producing a variety of useful substances (e.g., amino acids, sugars, etc.) as well as some useless waste products. It is generally agreed that usable materials make their way across the membrane of the secondary lysosome and enter the cell cytoplasm, where they participate in cellular metabolism. This transfer probably takes the form of passive diffusion or facilitated or active transport. Eventually, digestion and absorption are terminated, leaving only residues and denatured enzymes within the vacuole, which is now referred to as a **residual body.** In many cells, residual bodies fuse with the plasma membrane, and this is followed by exocytosis (Fig. 19–5). In some cells (especially those of higher organisms), residual bodies accumulate within the cytoplasm or continue to increase in size, eventually interfering with the normal activities of the cell and resulting in cell death. Progressive lysosome engorgement is believed to be involved in the aging process.

Autophagy. The isolation and digestion of portions of a cell's own cytoplasmic constituents by its lysosomes occurs in normal cells and is termed **autophagy** (Fig. 19–5). The phenomenon is most dramatic in the tissues of organs undergoing regression (e.g., changes in the uterus following delivery, during metamorphosis in insects, etc.). Autophagic vacuoles containing partially degraded mitochondria, smooth and rough endoplasmic reticulum, microbodies, glycogen particles, or other cytoplasmic structures are frequently observed in tissue sections examined with the electron microscope. Cellular autophagy results in a continuous turnover of mitochondria in liver tissue. The half-life of the liver mitochondrion is about 10 days and corresponds to the destruction of one mitochondrion per liver cell every 15 minutes.

Distribution of Lysosomes

Since their initial discovery in mammalian liver, lysosomes have been identified in many different cells and tissues; some of these are listed in Table 19–3. The greatest variety of tissues found to contain lysosomes occurs in animals. Although most studies have been carried out using mammalian tissues, lysosomes have been identified in insects, marine invertebrates, fish, amphibians, reptiles, and birds. Lysosomes are particularly numerous in epithelial cells of absorptive, secretory, and excretory organs (liver, kidneys, etc.). They are also present in large numbers in the epithelial cells of the intestines, lungs, and uterus. Phagocytic cells and cells of the reticuloendothelial system (e.g., bone marrow, spleen, and liver) have also been found to contain large numbers of lysosomes. Few lysosomes occur in muscle cells or in acinar cells of the pancreas. Lysosomes are produced by certain cells in tissue culture (HeLa cells,

Table 19–3
Cells and Tissues Containing Lysosomes

Protozoa	Nerve cells
Amoeba	Brain
Campanella	Intestinal epithelium
Tetrahymena	Lung epithelium
Paramecium	Uterine epithelium
Euglena	Macrophages of spleen, bone marrow, liver and connective tissue
Plants	Thyroid gland
Onion seeds	Adrenal gland
Corn seedlings	Bone
Tobacco seedlings	Urinary bladder
Tissue culture cells	Uterus
HeLa cells	Ovaries
Fibroblasts	Blood (leukocytes and platelets)
Monocytes	
Macrophages	
Chick cells	
Lymphocytes	
Animal tissues	
Liver	
Kidney	

monocytes, lymphocytes, etc.). Although it has a number of functions not shared by lysosomes of animal cells, the large vacuole of many plant cells is a modified lysosome. Some of the various roles played by the lysosomes are summarized in Table 19–4.

Leukocytes, especially granulocytes, are a particularly rich source of lysosomes, and this is related to their physiological role as scavengers of microorganisms or other foreign particles in the blood. Following phagocytosis of a bacterium by a leukocyte, numerous lysosomes fuse with the endocytic vacuole containing the microorganism and initiate its digestion. The lysosomes of granular leukocytes are especially large and readily visible by light microscopy. Once the lysosome content of the leukocyte is exhausted, the blood cell dies.

Lysosome Precursors in Bacteria

Although bacterial cells do not possess lysosomes, they do contain a variety of hydrolases that are believed to be localized in the space between the cell wall and the cell membrane. These hydrolases may be synthesized by ribosomes attached to the cell membrane and then dispatched through it. The bacterial hydrolases play a digestive role, breaking down complex substrates in the cell's environment and providing smaller molecules required for cell growth. Bacterial hydrolases also participate in sporulation and autolysis. Although the latter process destroys the individual cells involved, it is highly beneficial to the bacterial population as a whole, for it provides for the survival of a small number of cells under unfavorable environmental conditions. Infolding of the bacterial membrane to form internalized extracellular pockets containing both hydrolases and their substrates would provide the "evolutionary link" with lysosomes of animal and plant cells.

Regulation of Lysosome Production

As noted earlier, the mechanism proposed for primary lysosome formation is strikingly similar to that proposed for zymogen granule formation in pancreatic cells and other instances of secretory protein synthesis. This similarity does not seem so unusual when one considers the following. The enzymatic contents of the primary lysosomes are discharged into vacuoles derived from the plasma membrane and containing extracellular materials. Consequently, the mechanism is similar to secretion except that the extracellular space into which the secretory products pass is internalized as a vacuole (i.e., phagosome). In secretory cells, the pro-

Table 19–4
Some Functions of Lysosomes

1. Nutrition via a digestive role in protozoa and many metazoan cells
2. Nutrition via cellular autophagy during unfavorable environmental conditions
3. Lysis of organelles during cellular differentiation and metamorphosis
4. Scavenging of worn-out cell parts and denatured proteins
5. Destruction of aged red blood cells and dead cells
6. Defense against invading bacteria and viruses by circulating macrophages
7. Dissolution of blood clots and thrombi
8. Keratinization of skin
9. Secretion of hydrolases by sperm for egg penetration during fertilization
10. Yolk digestion during embryonic development
11. Bone resorption
12. Reabsorption in kidney and urinary bladder

duction of new secretory products is regulated by a feedback mechanism in which secretion itself acts as a stimulus for the production of additional secretory materials. The experiments of Cohn and Benson described previously demonstrated the relationship that exists between endocytic activity and lysosomal enzyme synthesis. It has therefore been suggested that the passage of phagosomes into the Golgi regions of the cell is followed by the discharge of some primary lysosomes and that this triggers the synthesis of new acid hydrolases.

Disposition and Action of the Lysosomal Hydrolases

Many of the lysosome's enzymes are released into the surrounding environment when these organelles are physically or chemically disrupted. Those enzymes that are so readily solubilized are believed to be located in the interior of the organelle. Other lysosomal hydrolases cannot be solubilized or are extracted with great difficulty and are thought to be an integral part of the lysosome membrane together with other proteins and lipids. Some of the enzymes known to be present in the lysosomes are listed in Table 19–2; it is to be noted that while this list is extensive, it is by no means complete.

All the substrates of lysosomal enzymes are either polymers or complex compounds and include proteins, DNA, RNA, polysaccharides, carbohydrate side chains of glycoproteins and glycolipids, lipids, and phosphates. The lysosomal breakdown of proteins into amino acids illustrates how these enzymes act in concert. The initial hydrolysis of protein is effected by *cathepsins D* and *E* and also by *collagenase*. These enzymes cleave peptide bonds and produce peptide fragments of varying length. These peptides, together with previously undigested proteins, are further hydrolyzed to individual amino acids by *cathepsins A* and *B*. *Cathepsin C*, *arylamidase*, and the lysosomal *dipeptidases* act on specific peptides, producing additional amino acids.

The breakdown of DNA and RNA is initiated by the enzymes *acid deoxyribonuclease* and *acid ribonuclease*. The resulting oligonucleotides are then degraded first by *phosphodiesterase* and then by *acid phosphatase*, producing nucleosides and inorganic phosphate. Lysosomes also possess all the enzymes necessary for hydrolysis of lipids and polysaccharides.

As noted earlier, some lysosomal enzymes are part of the membrane encasing the organelle. These are believed to form some sort of protective lining, for it is difficult to understand how the membranous portions of autophagocytosed mitochondria and endoplasmic reticulum are so readily hydrolyzed while the lysosome membrane remains impervious. In addition to having acid pH optima, the lysosomal enzymes are particularly resistant to autolysis. Among the enzymes found to be integral parts of the lysosome membrane are *acetylglucosaminidase*, *glucosidase*, and *sialidase*. *Arylsulfatase*, *acid phosphatase*, *ribonuclease*, and *glucuronidase* may also be bound to the membrane under certain conditions.

Enzymes freed from disrupted lysosomes exhibit a wide variation in stability. Some retain their activity for only a few hours following tissue disruption; others are stable for months when appropriately refrigerated. Of the few lysosomal enzymes isolated and characterized to date, several have been shown to be glycoproteins including *cathepsin C*, *acid deoxyribonuclease*, *glucuronidase*, and *acetylglucosaminidase*.

Microbodies

As noted at the beginning of the chapter, the term "microbody" has been used by cell biologists and cytologists for many years to describe a variety of different small cellular oganelles. More recently, it has been restricted to organelles possessing **oxidase, peroxidase,** or **catalase** enzyme activity. Organelles possessing these activities are typically small ovoid structures having a diameter of about 0.5–1.5 μ and containing an amorphous granular matrix and, occasionally, crystalloid inclusions. The organelles vary somewhat in structure, appearance, and function from one tissue to another and from species to species. Two distinct but related forms of microbodies are common in animal and plant cells and in microorganisms; these are *peroxisomes* and *glyoxysomes*.

Peroxisomes

The modern usage of the term "microbody" dates back to 1954 and the work of J. Rhodin, who described the structure and properties of these organelles in the mouse kidney. Since then, organelles of similar organization have been reported in many other tissues of both animals and plants.

In 1969, de Duve showed that microbodies of rat liver contained a number of *oxidases* that transfer hydrogen atoms to molecular oxygen, thereby forming hydrogen peroxide (Fig. 19–6). de Duve coined the term **peroxisomes** for these organelles, although a true peroxidatic activity is generally demonstrable only in vitro. In vivo, conditions favor the removal (or degradation) of hydrogen peroxide by *catalase* rather than by a *peroxidase*. However, since hydrogen peroxide is an intermediate in the reactions, the term "peroxisome" may be appropriate. The chemical and enzymatic relationships between an oxidase, peroxidase, and catalase are shown in Figure 19–6.

A number of enzymes are characteristically present in peroxisomes including *uric acid oxidase, D-amino acid oxidase, α-hydroxyacid oxidase, NADH-glyoxylate reductase, NADP-isocitrate dehydrogenase,* and *catalase*. When *uric acid oxidase* is present in large amounts, it frequently takes the form of a paracrystalline "nucleoid" at the center of the organelle. The functions of peroxisomes in animal cells are not entirely clear. Peroxisomal *catalase* is thought to be involved in the degradation of H_2O_2, which is extremely toxic, the source of the peroxide being other peroxisomal reactions (e.g., those catalyzed by *oxidases*) or cytosol-produced H_2O_2. Uric acid oxidase may be important in degrading purines. The abundance of peroxisomes in cells engaged in lipid metabolism suggests that these organelles may be involved in **gluconeogenesis.**

We saw earlier that one of the characteristic features of lysosomal enzyme activity is its *latency*. No latency is exhibited by the peroxisomal enzymes, as relatively large molecules (including the peroxisomal enzyme substrates) readily permeate the peroxisome membrane.

Figure 19–6
Chemical interrelationship between *oxidase, peroxidase,* and *catalase.*

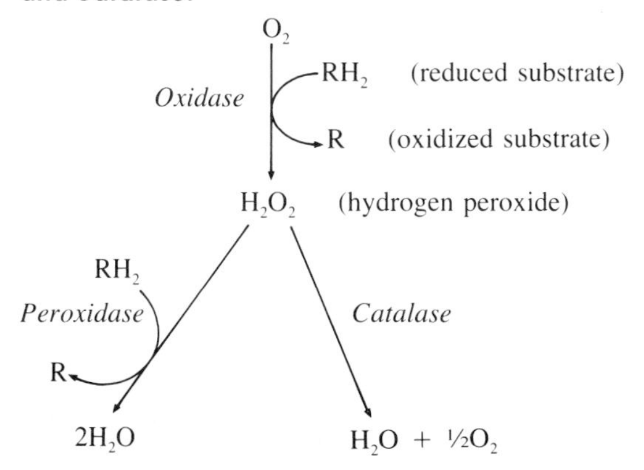

Isolation of Peroxisomes. The sedimentation coefficients and densities of peroxisomes are close to those of lysosomes and are not significantly different from mitochondria. Their similarities to lysosomes account for the fact that for some time peroxisomal enzyme activities were ascribed to lysosomes. Density gradient centrifugation of the "mitochondrial fraction" prepared by preliminary differential centrifugation is the method of choice for isolating peroxisomes. The greatest success in peroxisome purification is obtained if the lysosomes are first allowed to accumulate Triton WR-1339 (see earlier discussion). Triton-loaded lysosomes are considerably less dense than normal lysosomes and so are easily "floated" away from the peroxisomes during density gradient centrifugation.

Formation of Peroxisomes. It is generally thought that peroxisomes arise as outgrowths of the endoplasmic reticulum and that the peroxisomal enzymes dispatched from attached ribosomes into the cisternae make their way to these outgrowths prior to physical separation. Studies on peroxisomal catalase indicate that the subunits of the enzyme and the heme are assembled to form the functional molecule *after* entering the microbody.

Glyoxysomes

In 1967, R. W. Briedenbach and H. Beevers discovered that microbodies of certain plant tissues contained enzymes of the **glyoxylate cycle** (see Chapter 10, Fig. 10–5) in addition to peroxisomal enzymes. They used the term **glyoxysomes** for these particles. Glyoxysomes not only contain the glyoxylate bypass enzymes *isocitrate lyase* and *malate synthetase* but also contain several of the essential enzymes of the Krebs cycle (Chapter 10), which therefore function simultaneously in both groups of organelles.

The relationship between the Krebs cycle and the glyoxylate cycle is shown in Figure 19–7. Both cycles employ the same reactions to produce isocitrate from acetyl CoA and oxalacetate, but beyond this point the pathways differ. In the Krebs cycle, isocitrate is successively decarboxylated to produce succinate and two molecules of CO_2. In the glyoxylate cycle, isocitrate is converted to succinate and

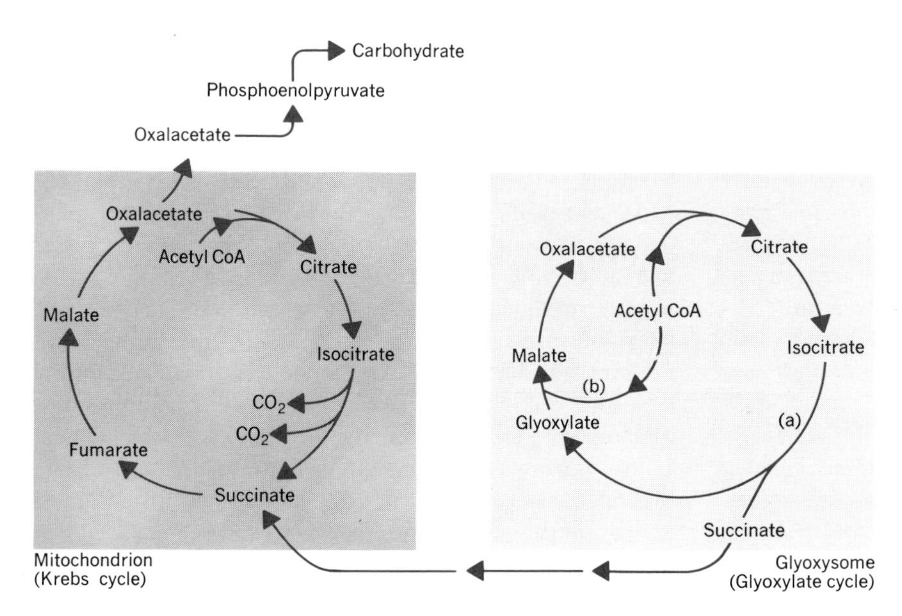

Figure 19–7
Comparison of the Krebs cycle in mitochondria and the glyoxylate cycle in glyoxysomes. Enzymes (*a*) and (*b*) are *isocitrate lyase* and *malate synthetase*, respectively.

glyoxylate. Therefore, instead of being lost as two molecules of CO_2, the 2-carbon glyoxylate condenses with another acetyl CoA to form the 4-carbon dicarboxylic acid, malate. The four carbon atoms of the two acetyl CoA molecules are thus conserved as one 4-carbon compound which, after conversion to succinate and migration to the mitochondrion, may be converted to oxalacetate. The oxalacetate may then be utilized in gluconeogenesis.

Oxalacetate formed in mitochondria from glyoxysomal succinate is presumed to serve as a direct precursor of phosphoenol pyruvate (PEP). The conversion of PEP to carbohydrate occurs by the reversal of the steps of glycolysis (see Chapter 10). Glyoxysome-containing tissues are thus able to convert simple 2-carbon sources such as acetate into carbohydrate. In some tissues, such as the fat-storage cells in seeds, the acetate is obtained through the degradation of fatty acids (Chapter 10). The glyoxylate cycle is especially significant for cells growing exclusively on acetate or fatty acids (e.g., a number of microorganisms), where the cycle acts as a source of 4-carbon dicarboxylic acids.

Distribution and Origin of Glyoxysomes. Glyoxysomes are not as widespread as peroxisomes. They are primarily found in the endosperm, cotyledon, and aleurone tissues of plants. Microbodies possessing glyoxysomelike activities have also been reported to occur in a number of microorganisms including *Euglena, Chlorella, Neurospora,* and *Polytomella*. Reports have appeared in the literature from time to time indicating that DNA is present in glyoxysomes, raising the possibility that these organelles possess some degree of autonomy. However, this notion is not generally accepted. Instead, because of their regular intimate association with the endoplasmic reticulum, glyoxysomes, like peroxysomes, are believed to be produced as outgrowths of the ER.

In concluding this discussion of microbodies, it is important to note that some organelles fitting the general microscopic description of microbodies do not clearly fit into either the peroxisome or glyoxysome category when evaluated in terms of their enzymatic activities. It is entirely possible that microbodies may be associated with varying activities depending upon the specialization of the cell and that microbodies exist whose actions and cellular functions remain to be determined. It is clear that one cannot name particles solely on the basis of microscopic characteristics and expect that all will function identically.

Summary

The realization that certain hydrolase activities are associated with a distinct class of organelles called **lysosomes** is quite recent. Previously, these activities were believed to be localized in mitochondria. Lysosomes function in the intracellular digestion of poorly functioning or superfluous organelles, as well as in endocytosed materials. Several forms of lysosomes may be identified including **primary** lysosomes, **secondary** lysosomes (e.g., heterophagic and autophagic vacuoles), and **residual bodies.** Primary lysosomes are formed at the peripheral edges of the Golgi complex, their hydrolase content derived through the cisternae of the endoplasmic reticulum. Fusion of primary lysosomes with **phagosomes** forms the secondary lysosomes in which digestion occurs. Usable products of this digestive activity are transferred to the cytoplasm. Undigested or unabsorbed materials remain in the residual bodies, which may accumulate in the cell or fuse with the plasma membrane during **exocytosis.**

Two distinct but related classes of microbodies occur in cells; these are **peroxisomes** and **glyoxysomes.** Peroxisomes are probably formed as outgrowths of the endoplasmic reticulum and contain a number of **oxidases** that produce hydrogen peroxide during their degradative activity. The potentially harmful peroxides are further degraded by peroxisomal catalase. Glyoxysomes are found primarily in plant cells and contain enzymes of the Krebs cycle and **glyoxylate bypass,** in addition to peroxisomal enzymes.

References and Suggested Reading

Articles and Reviews

de Duve, C., The lysosome. *Sci. Am. 208*(5), 64 (May 1963).

de Duve, C., The peroxisome: A new cytoplasmic organelle. *Proc. R. Soc. Lond. 173,* 71 (1969).

Leighton, F., Poole, B., Beaufay, H., Baudhuin, P., Coffey, J. W., Fowler, S., and de Duve, C., The large-scale separation of peroxisomes, mitochondria, and lysosomes from the livers of rats injected with Triton WR-1339. *J. Cell. Biol 37,* 482 (1968).

Books, Monographs, and Symposia

Breidenbach, R. W., Microbodies, in *Plant Biochemistry* (J. Bonner and J. E. Varner, eds.), 3rd ed., Academic Press, New York, 1976.

Dean, R. T., *Lysosomes,* Camelot Press, Ltd., Southampton, U.K., 1977.

de Duve, C., and Wattiaux, R., Function of lysosomes, in *Annual Review of Physiology* (V. E. Hall, A. C. Giese, and R. R. Sonnenschein, eds.), Annual Reviews, Inc., Palo Alto, Calif., 1966.

de Reuck, A. V. S., and Cameron, M. P. (eds.), *CIBA Foundation Symposium on Lysosomes,* Little, Brown, and Co., Boston, 1963.

Dingle, J. T., and Fell, H. B. (eds.), *Lysosomes in Biology and Pathology* (Parts I and II), North-Holland Publishing Co., Amsterdam, 1969.

Dingle, J. T., *Lysosomes.* North-Holland Publishing Co., Amsterdam, 1977.

Hruban, Z., and Rechcigl, M., *Microbodies and Related Particles,* Academic Press, New York, 1969.

Strauss, W., Lysosomes, phagosomes and related particles, in *Enzyme Cytology* (D. B. Roodyn, ed.), Academic Press, London, 1967.

Chapter 20
THE CELL NUCLEUS

The structure and function of the nucleus are discussed in this chapter. Our present knowledge of nuclear structures and their functions has been derived principally through the application of modern techniques, especially electron microscopy (Chapter 1) and radioactive tracer methods (Chapter 14). The functions of the nucleus are primarily concerned with the replication and transcription of its nucleic acids (Chapter 7), which ultimately results in regulatory control of the cell by proteins (Chapter 4).

The nucleus of the eucaryotic cell is delimited by a pair of membranes called the **nuclear envelope.** The *outer* and *inner* membranes of this envelope are continuous with each other only around the margins of large pores, which penetrate both membranes. Elsewhere, the two membranes are separated by an intermembrane space. Ribosomes may be attached to the outer surface (or cytosol side) of the outer nuclear membrane (which also may fold out into the cytoplasm), but ribosomes are not considered to be nuclear structures.

Within the nucleus are a number of structures and defined regions (Fig. 20–1). The structures and their functions are described in the sections that follow.

Chromatin

During *mitosis* and *meiosis* (discussed later in the chapter), chromosomes *condense* by compaction of their constituent DNA and protein. As a result, they are easily stained so that they can be studied during nuclear division. During *interphase,* the chromosomes are not condensed, and it is difficult, if not impossible, to distinguish individual chromosomes. However, there are regions of the nucleus called **chromocenters,** that do stain deeply even during interphase. In 1928, E. Heitz identified these chromocenters as portions of chromosomes that remain condensed throughout the cell cycle.

Chromosome material that stains with basic dyes is called **chromatin.** The location and intensity of the staining permits differentiation of two different forms, or states, of chromatin. Portions of the chromosomes that stain lightly with basic dyes and are associated with relatively uncondensed chromosomes are termed **euchromatin.** Darkly staining regions associated with condensed portions of interphase (and early prophase) chromosomes are called **heterochromatin.** In addition, there is usually some condensed chromatin around the nucleolus, called **perinucleolar chromatin,** and some inside the nucleolus, called **intranucleolar chromatin.** The perinucleolar and intranucleolar chromatin appear to be connected and jointly are referred to as **nucleolar chromatin.**

The chromatin-containing regions of interphase nuclei vary among cells, but a few generalizations are possible. Dense clumps of deeply staining chromatin often occur in close contact with the inner nuclear membrane and are

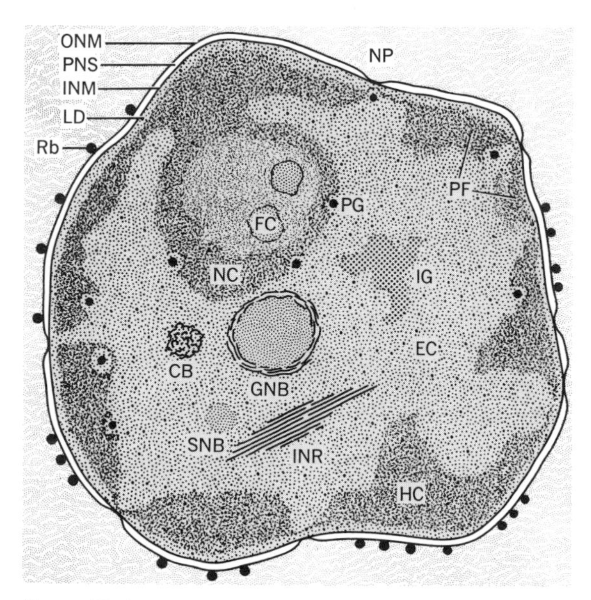

Figure 20–1
Diagram of a section through the nucleus, locating the main structures and defined regions. CB = coiled body, EC = euchromatin, FC = fibrillar center of nucleolus, GNB = granular nuclear body, HC = heterochromatin, IG = interchromatin granules, INM = inner nuclear membrane, INR = intranuclear rodlet, LD = lamina densa, NC = nucleolus-associated chromatin, NP = nuclear pore, ONM = outer nuclear membrane, PF = perichromatin fibrils, PG = perichromatin granules, PNS = perinuclear space, Rb = ribosome, SNB = simple nuclear body.

sometimes called **condensed peripheral chromatin** (Fig. 20–2). Between the peripheral chromatin and the nucleolar chromatin is a region of lightly staining chromatin called **dispersed chromatin.**

There is some reluctance at present to describe peripheral chromatin as heterochromatin and dispersed chromatin as euchromatin because some of the functions now ascribed to euchromatin and heterochromatin have not yet been identified as occurring at the specific sites where peripheral and dispersed chromatin are found. Although heterochromatin has been shown to contain very few, if any, structural genes, it has been identified with gene expression. When "euchromatic" genes of a known function in organisms such as *Drosophila* are relocated into a position adjacent to the heterochromatin, their phenotypic expression is modified. The exact role of the heterochromatin is not known, but because it is found in all nuclei, it is presumed to have a basic function in gene expression. During prophase in nuclei that contain few chromosomes, it is possible to identify the heterochromatic regions of specific chromosomes. The heterochromatin is also characterized by its high content of repetitive DNA sequences, and its replication in the cell cycle occurs at a later time than euchromatin. Euchromatin is believed to be the material containing the structural genes and can be expressed when decomposed during interphase. Euchromatin that temporarily has been physiologically inactivated by condensation so that its gene content cannot be expressed is sometimes called **facultative heterochromatin,** as opposed to perpetually condensed **constitutive heterochromatin.**

Structure and Composition of Chromatin

Electron-microscopic techniques have not yet been sufficiently perfected to allow clear observation of the structure

Figure 20–2
Regions of the interphase nucleus showing the location of the various types of chromatin: (1) condensed peripheral chromatin or heterochromatin; (2) perichromatin or loosened chromatin; (3) interchromatin region—dispersed chromatin or euchromatin; (4) condensed perinucleolar chromatin; nucleolar organizer; (5) intranucleolar chromatin; (6) nucleolar diffuse chromatin with ribonuclear protein.

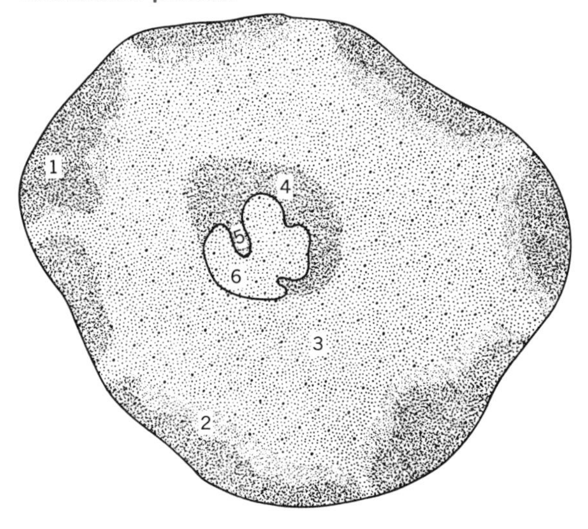

Table 20–1
Chemical Composition of Typical Plant Chromatin

	Content Relative to DNA (%)				Template Activity of Chromatin
Source of Chromatin	DNA	Histone	Nonhistone Protein	RNA	
Pea vegetative bud	1.0	1.30	0.10	0.11	6
Pea embryonic axis	1.0	1.03	0.29	0.26	12
Pea growing cotyledon	1.0	0.76	0.36	0.13	32

Source. From J. Bonner and J. E. Varner, *Plant Biochemistry*, Academic Press, New York, 1976, p. 48.

of chromatin in the interphase nucleus. However, in certain pathological states, the heterochromatin is loosened, and a fibrillar substructure is observed. When a well-developed lamina densa is present, the fibrillar heterochromatin takes on a special granular appearance near the inner membrane. This special association has led investigators to speculate that there is a connection between the chromosomes and the nuclear membrane. The fibrils vary in size; most appear to be about 100 Å in diameter, but a number are thicker (about 150–250 Å). Nucleolar chromatin is also fibrillar in substructure, with fiber diameters of about 100 Å.

As noted above, chromatin is any chromosome material that stains with basic stains. In terms of chemical composition, this includes DNA and the closely associated basic proteins, among which the **histones** are the most common (Table 20–1). Basic dyes probably stain nuclear RNA as well, but most investigators consider chromatin to be composed of DNA and histones.

The weight ratio in chromatin of histone to DNA varies from about 0.8 to 1.3, with an average of about 1.1. The ratio varies not only with the species of organism but also with the tissue. It appears that histones are associated with the chromatin of all eucaryotic organisms except fungi, which resemble procaryotic organisms in this respect.

The ultrastructure of chromatin is thought to involve a repeating pattern of bodies that are associations of DNA and histone. These bodies, called **nucleosomes** (Fig. 20–3), are composed of an octamer of histones around which about 200 base pairs of DNA are coiled. The octamer consists of four pairs of histones: two copies each of H2A and H2B, which are rich in lysine, and two copies each of H3 and H4, which are rich in arginine. H1, another histone, is frequently associated with the nucleosome and may function in the coiling of the nucleosome chain but is not a part of the nucleosome. The number of base pairs of DNA associated with each nucleosome of the chain varies with organism and tissue from which the nucleosomes are extracted. The range is about 140 to 240 base pairs per nucleosome. Both thin and thick chromatin filaments have been described by investigators. The thin chromatin filament (about 100 Å in diameter) is probably a linear array of the nucleosomes. The thick (300 Å diameter) chromatin filament appears to be a spiral of the thin filament with a 100 Å pitch and a 100 Å hole down the central core.

Sites of DNA Replication

The sites of DNA replication in the nucleus have been determined by autoradiography and electron microscopy of cells synchronized in their growth cycle through the action of certain drugs. Although there is some concern that the drugs used to induce synchrony may alter the process, studies made with randomly growing cells tend to support the findings made with synchronously growing cells. Once the cells are synchronized, tritiated thymidine (a precursor in DNA synthesis) is added to the growth medium at selected times, and the uptake and sites of deposition of the labeled thymidine are followed.

By use of these techniques, replication has been shown to begin at random sites in the dispersed chromatin (euchromatin) during the *S* phase of the growth cycle (Chapter 2). By the late *S* phase, replication shifts to the periphery (heterochromatic areas) of the nucleus. The site and time of replication of nucleolar DNA are not yet fully understood. Some evidence indicates that rDNA replication occurs during the late *S* phase. However, tritiated thymidine is incorporated into nucleolar chromatin even after short pulses of the label. The observation that rDNA replication occurs later in the *S* phase could be associated with the more prevalent condensed nucleolar chromatin.

Sites of Transcription

Our understanding of the mechanism of transcription at the molecular level in the nucleolus is fairly detailed (Chapter 21). Site-specific localization of the process is also rapidly becoming clear. Initiation of transcription in the nucleolus is at the border between the intranucleolar chromatin and the remaining (granular) portion of the nucleolus (Fig. 20–4). When tritiated uridine is introduced, the label first appears in this border region. It then accumulates in the fibrillar region between the intranucleolar chromatin and the granular region and then moves into the granular region. The fibers of the fibrillar region are believed to be precursors of the granules of the granular region. Dispersal of the label in the granular region is fairly uniform. The nucleolar fibers may be correlated with the RNA that sediments at 45 *S*, and the granules, with the RNA that sediments at 23 *S* (Chapter 21).

The sites of nuclear transcription outside the nucleolus is not at all certain. Some evidence suggests that all areas of the nucleus may undergo transcription, including both condensed and decondensed (dispersed) chromatin. How-

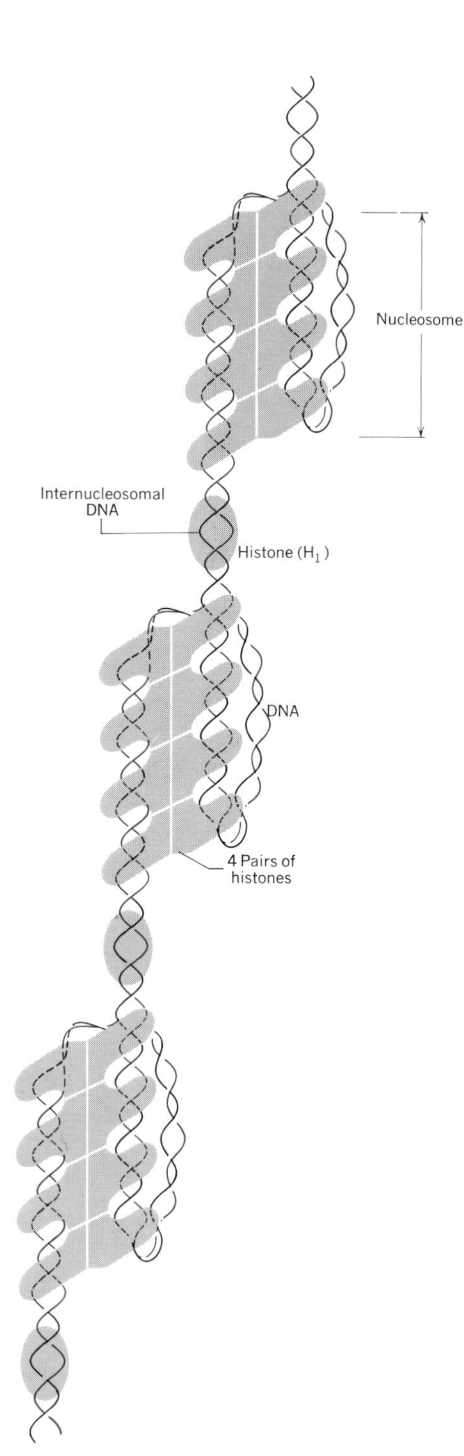

Figure 20–3
(*a*) Diagram of part of a chain of nucleosomes held together by the coiling of DNA. Shape and size of nucleosomes are not to scale and represent only the basic concept of structure. (*b*) Electron photomicrograph of nucleosomes (arrows). (Courtesy of Dr. F. Puvion-Dutilleul.)

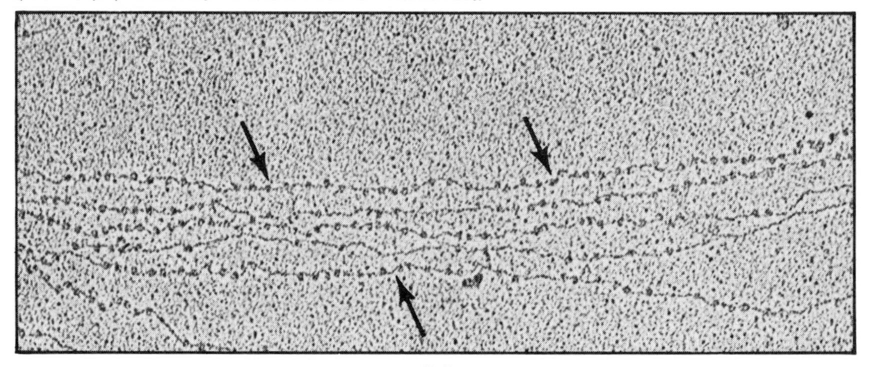

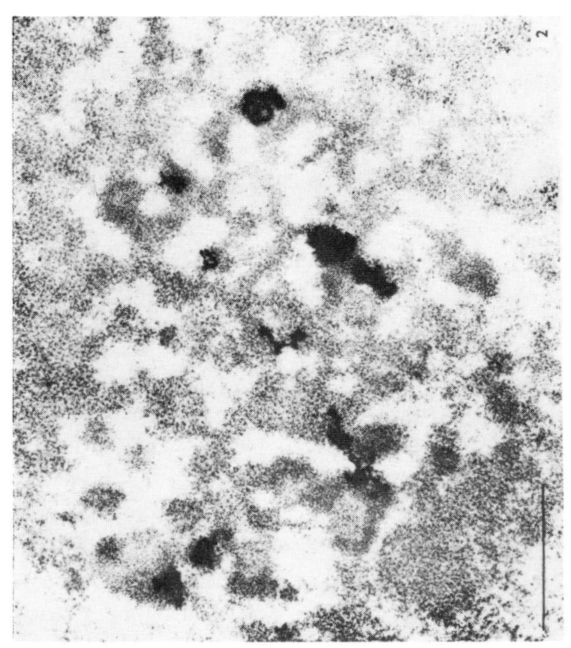

Figure 20–4
Nucleolar transcription. Autoradiogram of a cell labeled with tritiated uridine for 5 minutes. Newly synthesized RNA (black spots) appears to be restricted to the junction between chromatin and ribonucleoprotein. (Courtesy of H. Busch; copyright Academic Press, *Exp. Cell Res. 52,* 431, 1968.)

ever, there is other evidence indicating that transcription begins in the *perichromatin region* (i.e., the region between the peripheral condensed chromatin and the dispersed chromatin, Fig. 20–2).

The Nucleolus

The nucleolus was one of the first subcellular organelles to be identified by microscopy. It was initially described by Fontana in 1774. Light microscopy using basic dyes has shown nucleoli to vary greatly in size, shape, and number, depending on the physiological state of the cell. Lymphocytes, which have little ribosome synthesis, have nucleoli reduced to crescent-shaped structures. Cells actively synthesizing ribosomes are enlarged, revealing "Giant-size" nucleoli. Nucleoli have been described with vacuoles and with dense areas called **nucleolini.** The electron microscope has shown that the nucleolus is made up of granular and fibrillar elements, as well as the intranucleolar and perinucleolar chromatin described above (Fig. 20–5). The nucleolus is not bordered by a membrane.

The function of the nucleolus is described in detail in Chapter 21, which deals with protein synthesis. It is clear that the primary function of the nucleolus is the synthesis of most of the rRNA species found in the large and small subunits of the ribosomes and the packaging of these rRNAs with ribosomal proteins to form preribosomal particles. It has been suggested that they may function in the formation of polysomes, but their interaction with mRNA is not known.

The Nuclear Envelope

The presence of a membrane separating the nuclear material from the cytoplasm is one of the characteristics distinguishing eucaryotic organisms from procaryotic organisms. The existence of a "membrane" delimiting the nucleus was first demonstrated by O. Hertwig in 1893. However, little interest was addressed to this "membrane" until studies with the electron microscope revealed that it was not simply a single membrane, but rather a double membrane in which the outer membrane had features that clearly distinguished it from the inner membrane. The two membranes lie close together, one surrounding the other. The two membranes fuse together at the **nuclear pores** but elsewhere are separated by the **perinuclear space.** The folded appearance at the nuclear pores and the perinuclear space between the membranes aptly justifies the term **nuclear envelope** to describe the entire structure.

The functions of the nuclear envelope are diverse. Clearly, it acts to compartmentalize the cytoplasm and nucleoplasm (Fig. 20–6). Mitochondria, Golgi bodies, vacuoles, lysosomes, chloroplasts, and other cytoplasmic organelles usually do not enter the nucleus, and the chromosomes and nucleoli rarely exit into the cytoplasm. However, ribosome precursors from the nucleolus and mRNA and tRNA molecules do leave the nucleus through the nuclear pores. Also molecules enter the nucleus from the cytoplasm, and a number of nucleocytoplasmic effects result; some of these are described later in the chapter.

The nuclear evelope is not just a physical barrier, but also functions in connection with cellular activities on each side. Ribosomes may be attached to the outer membrane, which thereby takes on a structural appearance like rough

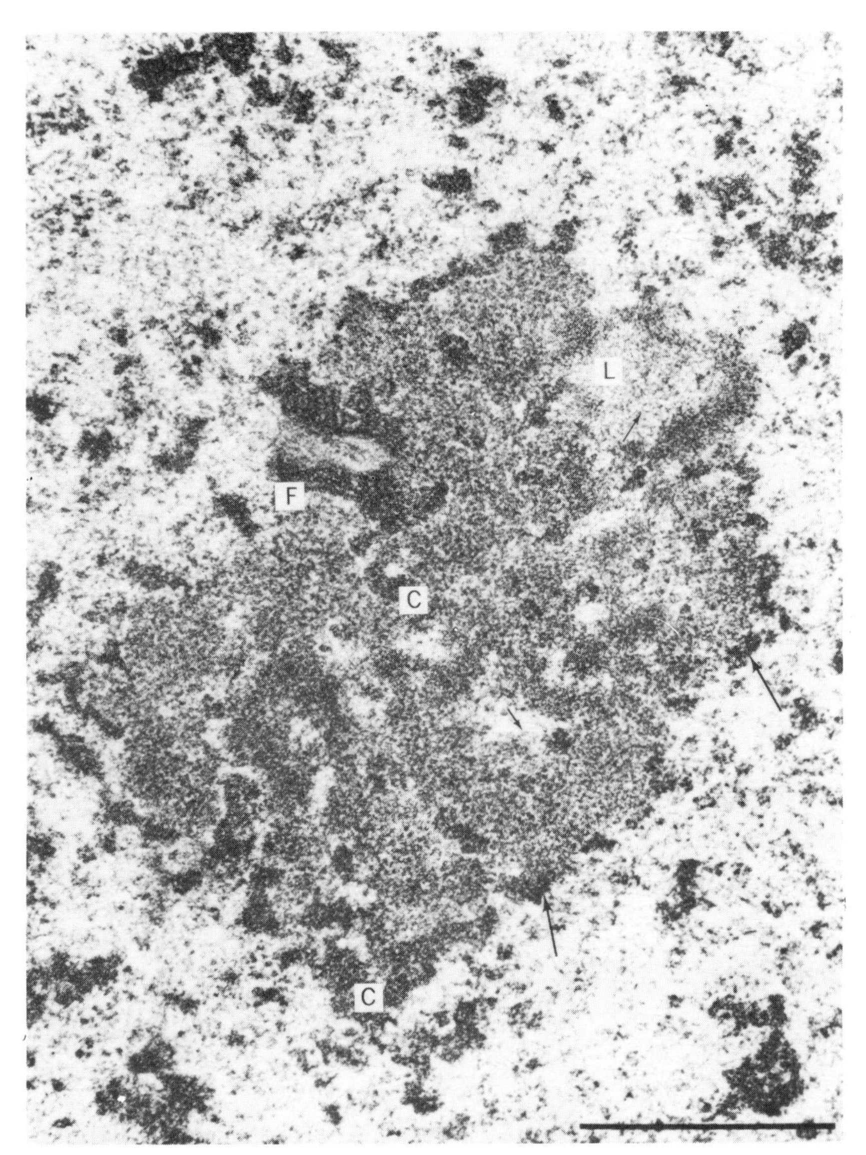

Figure 20–5
A nucleolus showing light areas (L) filled with chromatin clumps (short arrows), fibrillar centers (F) condensed chromatin (C), and perinucleolar chromatin (long arrows). (Courtesy H. Busch; copyright 1970 Academic Press, *The Nucleolus*, p. 69.)

endoplasmic reticulum (Chapter 21). Electron micrographs reveal that the outer membrane is continuous with the endoplasmic reticulum (Fig. 20–7), and there is evidence that the endoplasmic reticulum forms as an outgrowth of the outer nuclear membrane. In some plants, the network of cytoplasmic membranes extending from the nuclear membrane is continuous with membranes of chloroplasts. In effect, the chloroplast is anchored to the nuclear envelope, and channels (cisternae) between the two structures provide for the exchange of genetic information.

The outer membrane of the nuclear envelope also appears to give rise to Golgi bodies (Chapter 18) and, in so doing, functions much like smooth endoplasmic reticulum. Figure 20–8 shows the Golgi apparatus of a cell in close proximity to the nuclear envelope. The forming surface of the Golgi body faces the nuclear envelope and appears to develop by fusion of vesicles formed and pinched off from the outer membrane.

The inner membrane of the nuclear envelope has a special function associated with the chromosomes. In the interphase

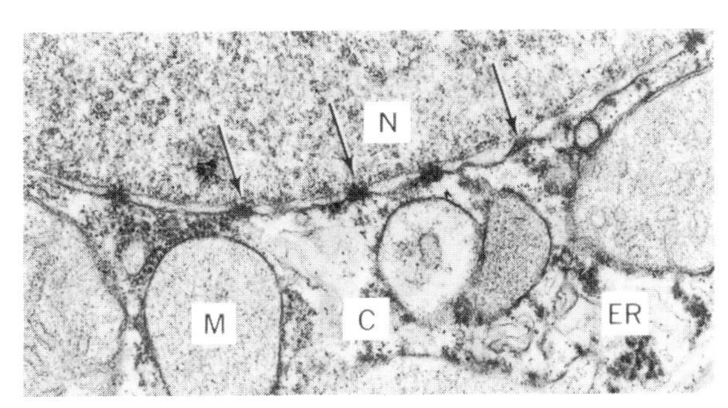

Figure 20–6
Nuclear envelope. C, cytoplasm; N, nucleus (arrows point to nuclear pores with plugs); M, mitochondrion; ER, endoplasmic reticulum. (Courtesy of Dr. E. G. Pollock.)

Figure 20–7
Arrows show connections between the outer membrane of the nuclear envelope (NE) and the endoplasmic reticulum (ER). N, nucleus; C, cytoplasm. (Courtesy of R. Chao.)

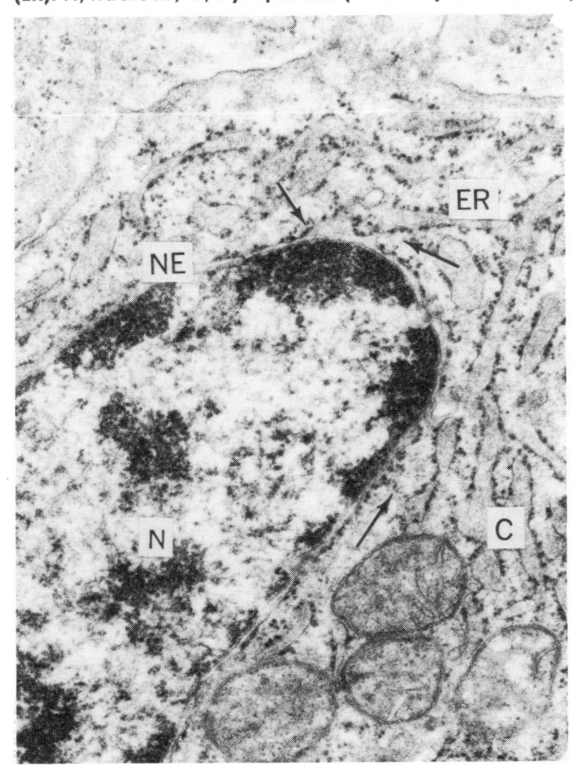

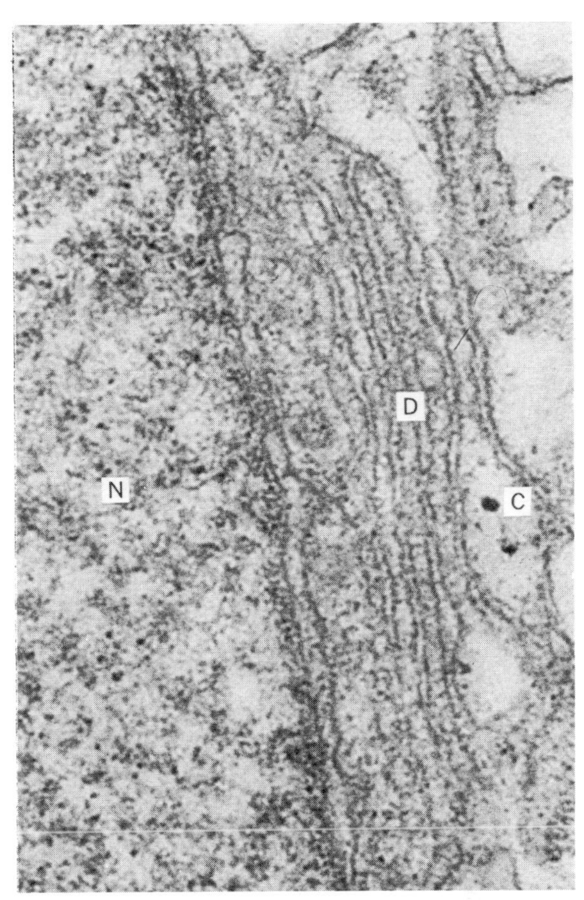

Figure 20–8
Relationship between the nuclear envelope and developing Golgi apparatus (D). N, nucleus; C, cisterna. (Photomicrograph courtesy of Dr. E. G. Pollock.)

nuclei of many cells, the heterochromatin is closely appressed to the inner membrane. This association does not occur at or near the nuclear pores; instead, the chromosomes seem to be firmly attached to the interpore sections. The purpose of the connections is not known, but suggestions being studied currently include (1) a role in chromosome replication, (2) a role in orienting the chromosomes during interphase and early prophase of mitosis and meiosis, and (3) a site of formation of the nuclear envelope.

It seems reasonable to expect that some materials are

exchanged between the nucleoplasm and the cytoplasm through open nuclear pores. Such exchanges presumably involve large molecular complexes such as ribonucleoprotein particles. In addition to this route of exchange, there are others. For example, small ions and molecules readily exchange between the two sols by permeating the nuclear membranes, and except for a few special cases, electrical potential differences across the nuclear envelope are negligible. Larger molecules and particles may pass through the membrane by formation of small pockets and vesicles that transverse the envelope and empty on the other side. It is also possible that small sections of the entire envelope evaginate or break away and undergo dissolution (as is known to occur during mitosis and meiosis). These exchange mechanisms are summarized in Figure 20–9.

Pores are not unique to nuclear membranes, occurring also in the plasma membrane and the membranes of the endoplasmic reticulum. However, the pores of the nuclear membrane possess a special character. The number of nuclear pores varies widely among cell types but, in general, ranges from 1 to about 60 pores per square micrometer. Unlike pores in the plasma membrane, the nuclear pore is a complex of structures. The *orifice* is essentially circular,

Figure 20–9
Possible pathways for the transport of materials from the nucleus to the cytoplasm. (1) Active and/or passive transport; (2) transport by vesicle formation; (3) transport through the inner membrane followed by invagination of the outer membrane; (4) transport through a pore; and (5) envelope evagination followed by dissolution.

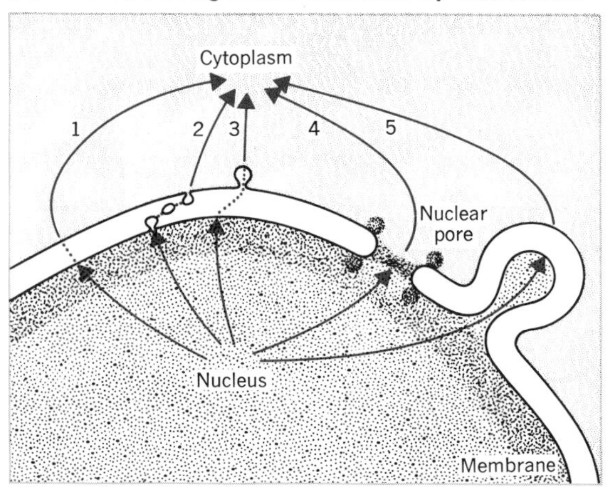

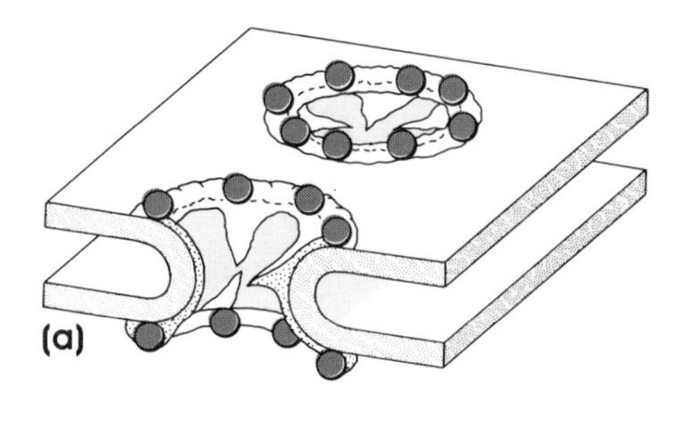

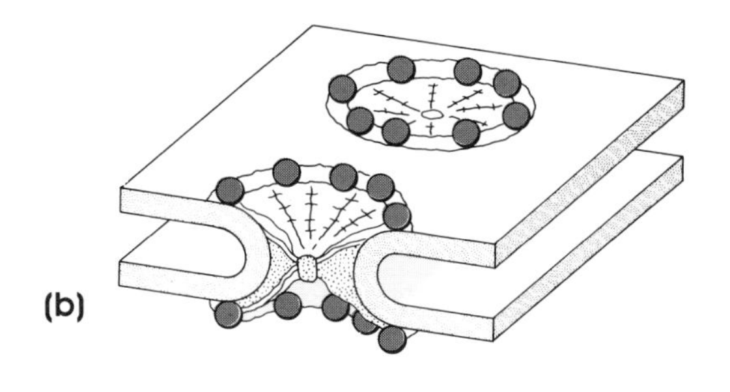

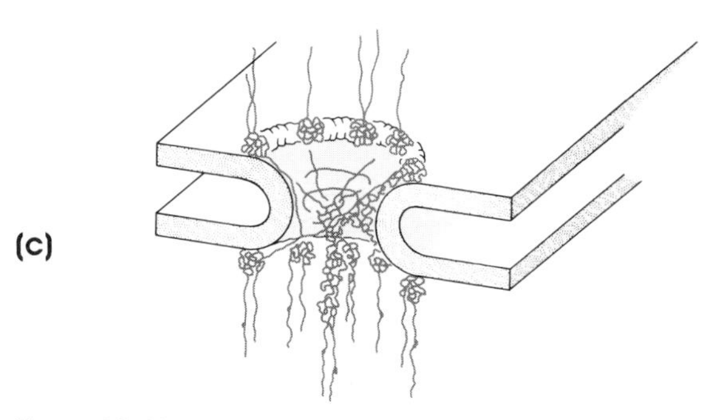

Figure 20–10
Models of nuclear pores illustrating the general shape of the annulus and the eight symmetrically-positioned subunits. Projects into the orifice of the pore may be fingerlike (*a*), conical (*b*) or fibrous (*c*).

with occasional evidence of a polygonal shape. The inner pore diameter is markedly constant for a given cell type but may yield variable values depending on the technique used to fix and examine the cell. The usual range is 600–1000 Å, with most nuclear envelopes having pore orifice diameters of about 725 Å.

The inner and outer rims of the pore are made up of nonmembranous material, which gives the pore the appearance of lips (Figs. 20–10 and 20–11). Seen in freeze-etched preparations, these appear as rings and are therefore called **annuli** (sing. *annulus*). The annulus is not uniform but has eight granular *subunits* symmetrically placed about it. The lumen of the pore frequently contains distinct substructures. Some are conical projections from the sides into the center, while others are fingerlike or fibrous.

Other Nuclear Structures

A number of small granules, fibrils, and tubules have been described in the nucleus. Many of these are structural variations of the chromatin or nucleolus already described. They are usually identified by their location or structure; for example, "interchromatin granule," "perichromatin granule," "fibrillar centers," "coiled body," and so forth. Most of these structures, while well described and chemically analyzed have unknown functions.

The **lamina densa,** located immediately within the inner nuclear membrane, is a region that appears to lack specific structure and does not contain DNA or RNA. Its proposed functions are to support the nuclear membrane and facilitate transport across the membrane.

Figure 20–11
Freeze fracture photomicrograph of the surfaces of the nuclear envelope and the nuclear pores; see book cover. (Courtesy of Dr. E. G. Pollock.)

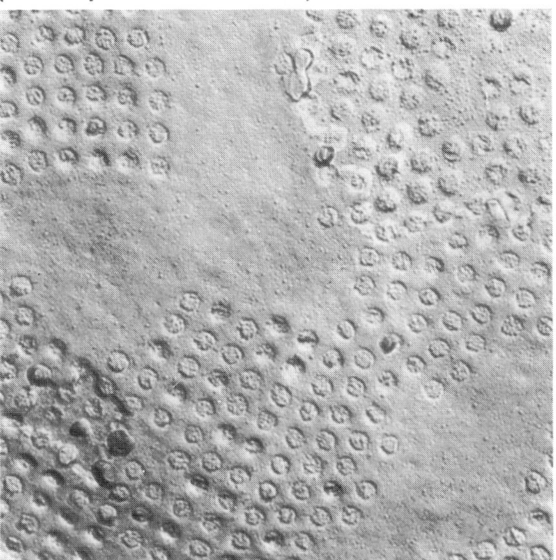

Chromosomes

Chromosomes were first described by numerous investigators in the period between 1875 and 1880, and the name "chromosome" was introduced by Waldeyer around 1880. These structures readily take up basic dyes and are easily seen between the poles of a cell during division. The material composing the chromosome, **chromatin,** is highly condensed during the division stages, so that the chromosomes can readily be counted and individually described when spread on a microscope slide. The number of chromosomes per nucleus varies greatly among the various animal and plant species, but for each specific species a constant number of chromosomes, each exhibiting specific shape and size at metaphase, (Fig. 20–12) can be identified.

Chromosome shape and size change during the stages of nuclear division (mitosis and meiosis) and during interphase. Each chromosome has two arms, one on each side of the primary constriction or **centromere** (also called **kinetochore**). The metaphase chromosome has doubled so that it appears to have two sets of arms (Fig. 20–13). *Secondary constrictions* associated with nucleoli are observed in some chromosomes and are called **nucleolar organizing regions** (NOR). *Teritary constrictions* are seen in nearly all chromosomes. Although their significance is not understood, they help to distinguish one chromosome from another. During interphase the chromatin is spread out (decondensed), so that the structure of the chromosome is virtually impossible to observe.

Mitosis

During nuclear division, there is a progressive change in the structure and appearance of the chromosomes. This process, although continuous, is divided into stages as shown in Figure 20–14. **Prophase,** the first stage of mitosis, includes (1) the sequence of events involving shortening (i.e. condensation) of the chromosomes, (2) the dispersion and disappearance of the nucleolus and the nuclear envelope, and (3) the formation and appearance of microtubules that form the *spindle*. If, prior to prophase, the cell contained a centriole, then the centriole replicates during this phase, with the two centrioles moving to opposite ends of

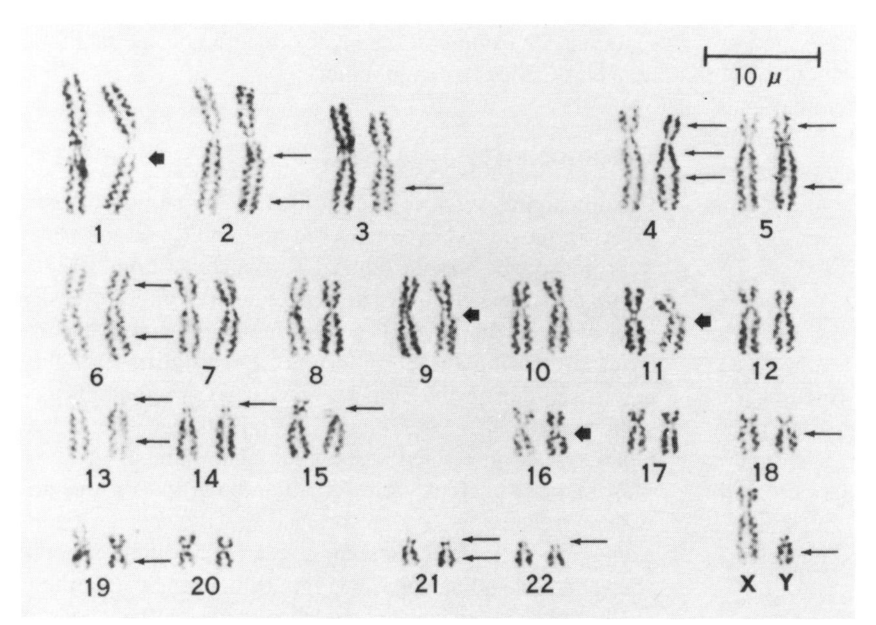

Figure 20–12
Metaphase chromosomes of a male human cell as seen with the light microscope. The colchicine treatment used in the preparation of the chromosomes and the centromere (or primary constriction) create the characteristic "X" appearance of the arms. (Courtesy Y. Ohnuki, copyright 1968 Springer-Verlag, *Chromosoma 25,* 416.)

the spindle. The chromosomes become distinguishable by light microscopy as a result of their progressive shortening and thickening during prophase and eventually can be seen to be composed of two *sister chromatids* held together by the centromere. The sister chromatids result from the replication of the chromosomal DNA that occurred during the interphase or *S* phase of the cell cycle (Chapter 2). The metaphase chromosome, then, is really two fully duplicated, identical chromosomes, temporarily joined at the centromere.

Toward the end of prophase (sometimes called *prometaphase,* since the second phase is *metaphase*), the chromosomes migrate toward the center of the spindle. The cause of the movement is not known but is somewhat erratic. At **metaphase,** the centromeres of each chromosome are aligned midway across the spindle on a plane called the **equatorial plate.** At this point, the centromeres of the paired chromatids attach to separate spindle fibers (attachment fibers). Other spindle fibers run from pole to pole but do not attach to the chromosomes. Simultaneously, the centromeres of each chromosome finish their duplication and begin migrating toward opposite poles of the spindle, marking the onset of anaphase. The cause of the movement is not yet resolved. J. McIntosh and co-workers have proposed that the movement is due to the interaction and sliding of the spindle microtubules past one another, the sister chromatids being pulled to opposite poles. R. Dietz proposed a model based on the progressive removal of elements from the spindle fibers, thus causing them to shorten and draw the chromosome toward the pole. Working with the blood lily, A. Bajer obtained evidence that the spindle microtubules radiating from each centriole interact with a second set of microtubules arranged in a transverse plane near the equatorial plate. The alternate attraction of these fibers to one pole and then the other leads to the gradual separation of the chromosomes; this mechanism has aptly been called the ''zipper hypothesis.''

The phase characterized by the separation of centromeres and movement of chromosomes to the poles is called **anaphase.** Once the sister chromatids are separated from each other, they are called chromosomes and behave and function as such. In effect, the number of chromosomes in the cell has really been doubled ever since the *S* phase. However, during anaphase, a process begins that divides the cell in half, separating the two complements of chromosomes. The process of cell division or *cytokinesis* is distinct from nu-

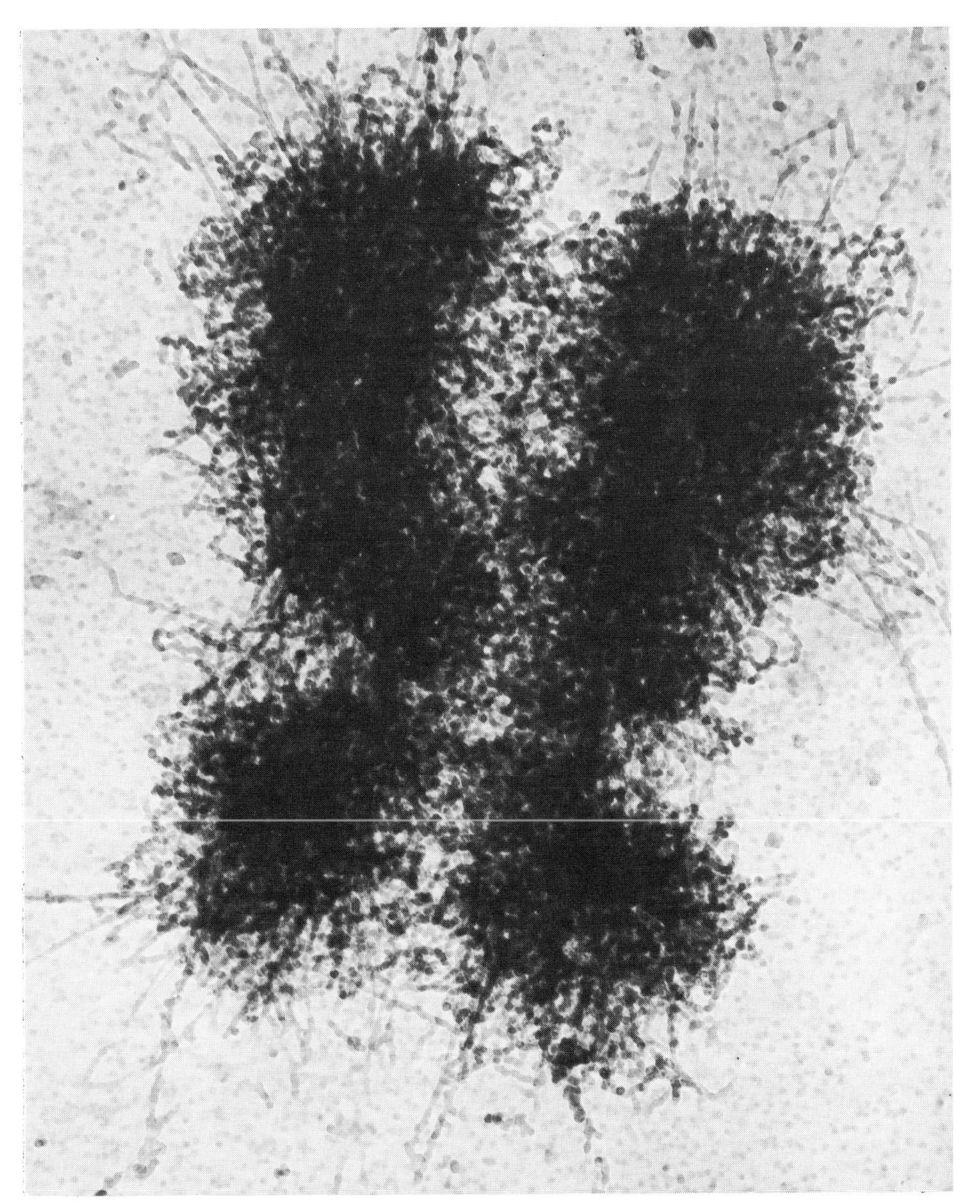

Figure 20–13
Electron photomicrograph of a human metaphase chromosome. (Courtesy Gunther F. Bahr, Armed Forces Institute of Pathology.)

clear division but is normally synchronized to occur during the later stages of mitosis.

The final phase of mitosis, **telophase,** involves the decondensation of the chromosomes once they have reached the poles of the spindle. During this phase, the nucleoli reform, together with a nuclear envelope about each chromosome group.

The gross morphology of the chromosomes during each

(a) Interphase

(b) Early prophase

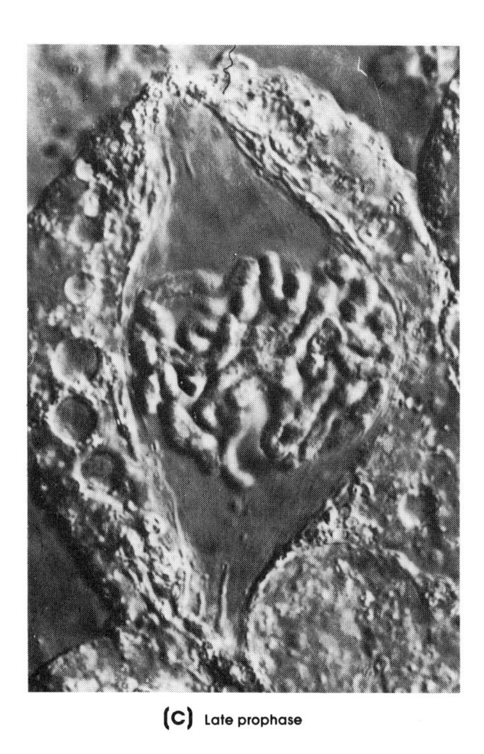

(c) Late prophase

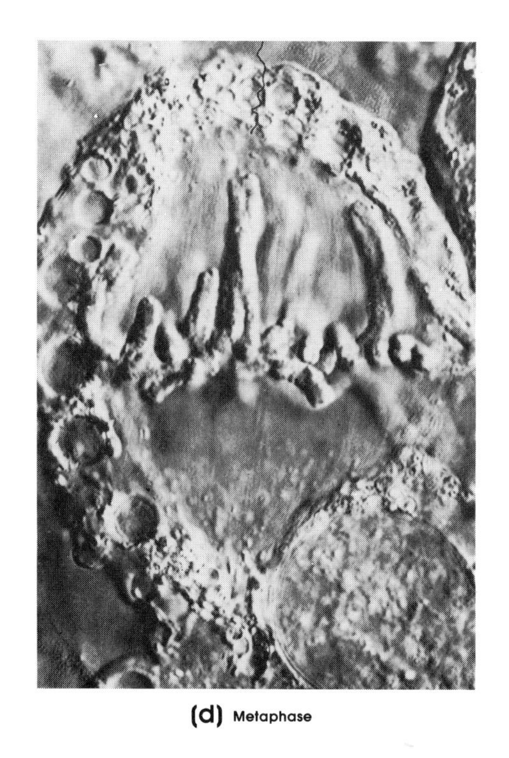

(d) Metaphase

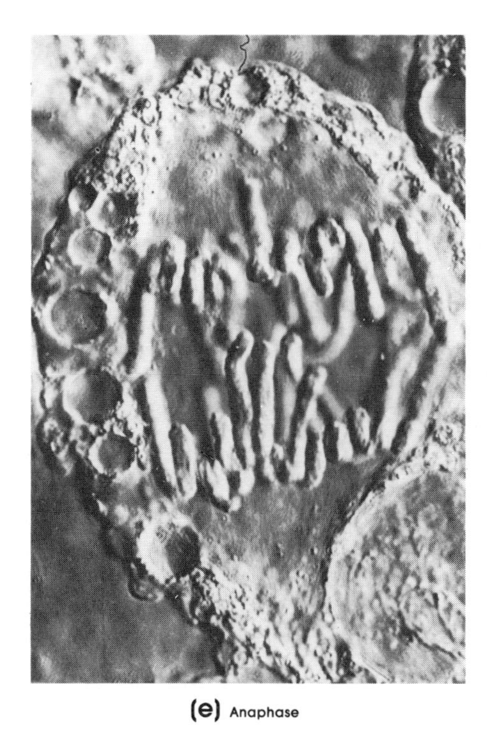

(e) Anaphase

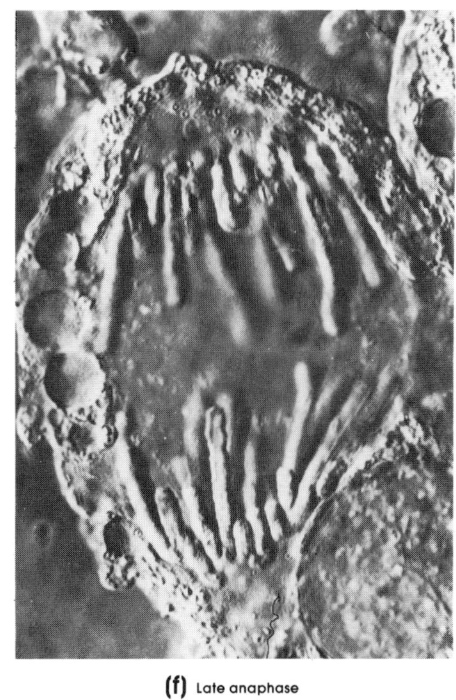

(f) Late anaphase

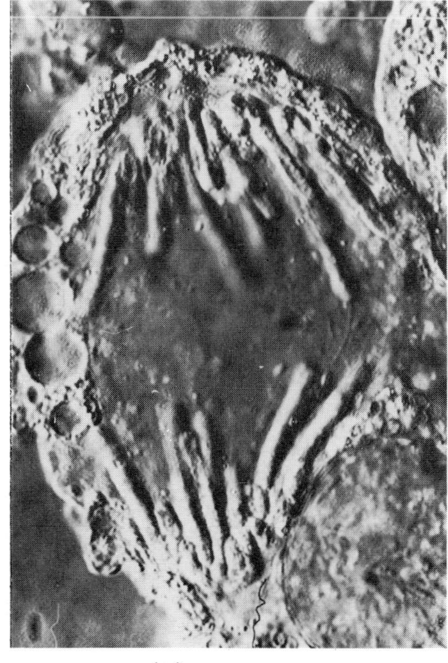

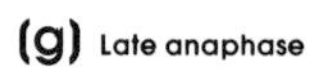

(g) Late anaphase

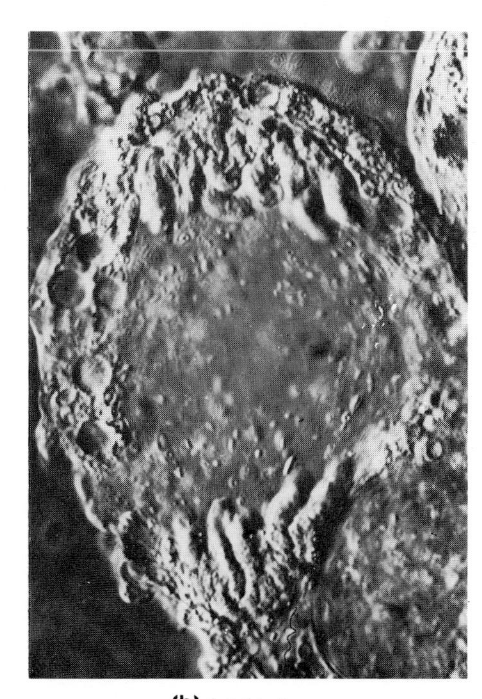

(h) Early telophase

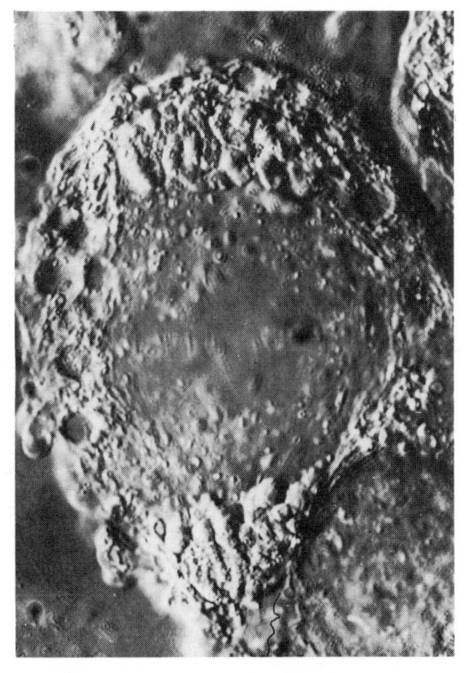

(i) Telophase and beginning of cell plate formation

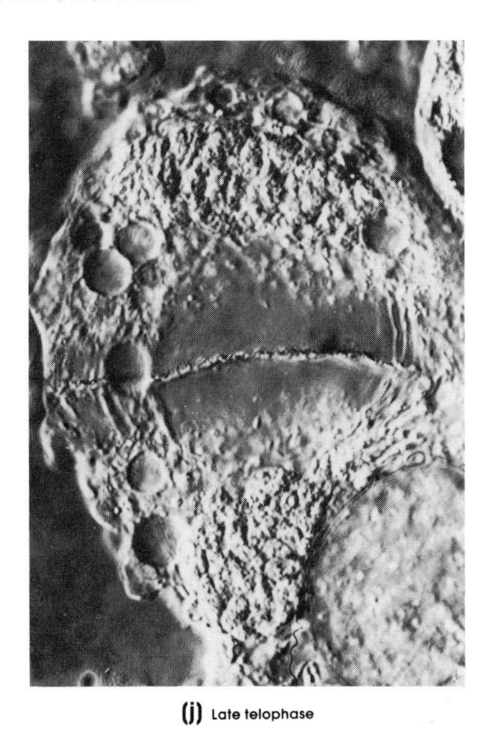

(j) Late telophase

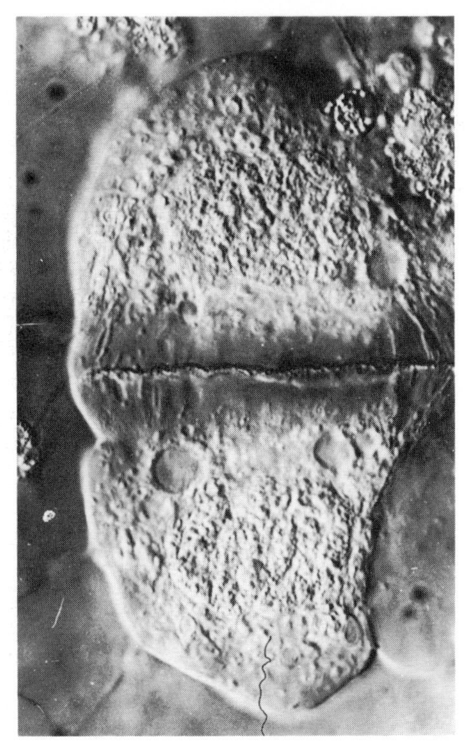

(k) Interphase

Figure 20–14
Mitosis in a plant cell (*Haemanthus*, the blood lily). Courtesy of Dr. A. S. Bajer.)

of the stages of mitosis is shown in Figure 20–14. During prophase and metaphase, each mitotic chromosome consists of two sister chromatids held together at the centromere. The chromosomes become shorter and thicker as prophase proceeds. By late anaphase, the chromosome is a single, not a double, structure, with one centromere as a result of the separation that occurred at the beginning of anaphase. In telophase, the chromosomes begin to decondense so that by the beginning of interphase, they exist as single, thread-like fibers. This single fiber structure persists through the G_1 period of the cell cycle's interphase. About midway through interphase, the DNA composing the backbone of the fiber undergoes replication (the *S* period of the cell cycle); as a result, two identical strands of DNA are present in each chromosome through the later stages of interphase (G_2 period of the cell cycle, Fig 2–10). The two strands of fibers are held together at the centromeres.

The structure of the chromosomes is ultimately due to

the filamentous nature of the double-helix DNA molecules (Chapter 7). In each chromosome, the DNA is complexed with histones and other proteins, but the overall shape and characteristics of the chromosome reflect the nature and properties of DNA rather than protein.

Ultrastructure of the Chromosome

The condensed chromosomes visible as distinct bodies during mitosis are composed of an organized array of **chromatin fibers** (Fig. 20–13), but in their decondensed state give rise to a highly filamentous network. Each chromatin fiber is believed to contain one double-helix molecule of DNA. The average diameter of the chromatin fiber is about 100 Å, whereas the diameter of the DNA double helix is only 20 Å. The difference is due to supercoiling of the DNA molecule and the presence of large quantities of proteins (primarily histones), some enzymes, and RNA in the fiber.

Chromatin fibers fall into two classes, based on their diameter. **Type A** chromatid fibers are narrow strands of high electron density, whereas **type B** fibers are thicker and show light diagonal cross-striations, indicating that this chromatin may be supercoiled. Chromatin with varying thicknesses, presumably alternating areas of type A and B fibers, have also been described.

The **packing ratio** of interphase chromatin fibers averages about 56:1. (The packing ratio is the length of DNA divided by the length of the chromatin fiber.) The packing ratios of metaphase chromatids are usually greater than 100:1. Thus, it is estimated that during prophase the chromatin fiber undergoes a twofold condensation, since the total amount of DNA remains unchanged. Since the chromatin fiber also twists and folds, the chromosome appears to shorten much more than twofold.

In addition to its condensation, the total mass of the chromosome doubles during prophase. The increase is caused by the accumulation of more non-DNA constituents, most specifically nonhistone proteins. The increase is reflected by measurements of type B fibers of interphase and metaphase chromosomes. Interphase type B fibers have diameters of 230–250 Å, whereas metaphase type B fibers have diameters that usually exceed 300 Å.

The folding of fibers that accompanies chromosome condensation during prophase does not appear to be regular or consistent. The fibers are twisted in an apparently random manner, producing configurations that are probably not duplicated in other chromosomes (not even homologous chromsomes) or the same chromosome during successive mitotic divisions.

Metaphase Chromosome-Chromosome Associations

Although the chromosomes arranged on the equatorial plate at metaphase are usually described as random and independent, it has been known for some time that certain specific associations occur. There may be size assortment on the spindle with the long chromosomes on the outside and the short on the inside. Certain chromosomes are known to group together or have arms directed toward one another. G. Hoskins has shown that human chromosomes are bound together by chromosome-to-chromosome "connectives." The connectives are composed of DNA and protein and are strong enough to hold the chromosomes together even when teased out of a cell. Fibers linking separate chromosomes have been visualized using electron microscopy.

Polytene Chromosomes

In the salivary glands of dipteran flies, of which the genus *Drosophila* has been most extensively studied, and in certain other tissues as well, the interphase nucleus is characterized by extremely large chromosomes. These so-called giant chromosomes are actually several hundred parallel and tightly packed copies of the *same* chromosome (Fig. 20–15). Because of their multiple structure, they are called *polytene* chromosomes. Along the length of the polytene chromosome, disks or bands of variably staining intensities can be distinguished. The identification of the genetic content of these bands (and interband regions) has played a major part in the genetic analysis of dipteran organisms.

Some of the bands of giant chromosomes appear to be swollen or "puffed," the specific bands exhibiting puffing varying from one tissue to another even within the same organism. It is now well established that the puffed bands of these interphase chromosomes correspond to regions in which the DNA is being actively transcribed into mRNA (Chapter 21).

Bacterial and Viral Chromosomes

The "chromosomes" of bacteria and viruses are single nucleic acid molecules. In most cases, they consist of a double

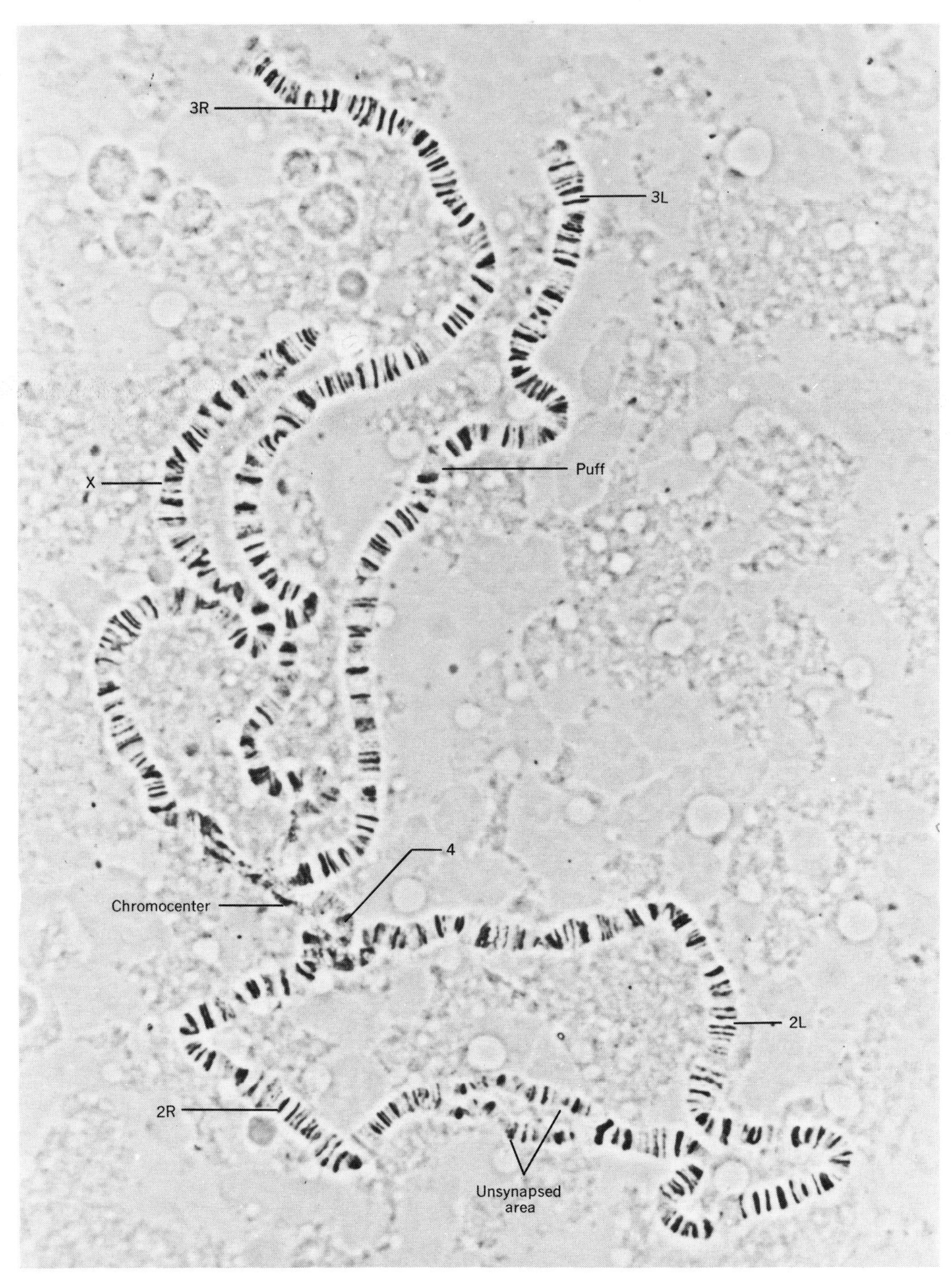

Figure 20–15
Giant polytene chromosomes from a diploid salivary gland cell of *Drosophila melanogaster*. The chromosomes are labeled (X, 2R, 2L, 3R, 3L, and 4). A puff is shown on 3L and an unsynapsed region of 2R reveals the two chromosomes. (Courtesy of Dr. G. Lefevre.)

helix of DNA, but in certain viruses they consist of only a single strand of DNA or RNA. The DNA may be linear or circular, depending on the organism.

Bacteria. The chromosomes of *E. coli* and *B. sublilis* have been studied more extensively than those of any other bacteria. In both species the chromosome is a single, circular DNA molecule more than a millimeter long. The circular molecule is packed into a dense mass less than 1 μ in diameter. The circular DNA can be released intact by proper osmotic treatment of the cell (Fig. 20–16). The bacterial chromosome is attached either to the plasma membrane or to an infolding of the membrane called a *mesosome* (Chapter 1). Attachment to a membrane is necessary for separation of daughter chromosomes, since these cells do not form a spindle. Within the nucleoid, the DNA appears to be folded

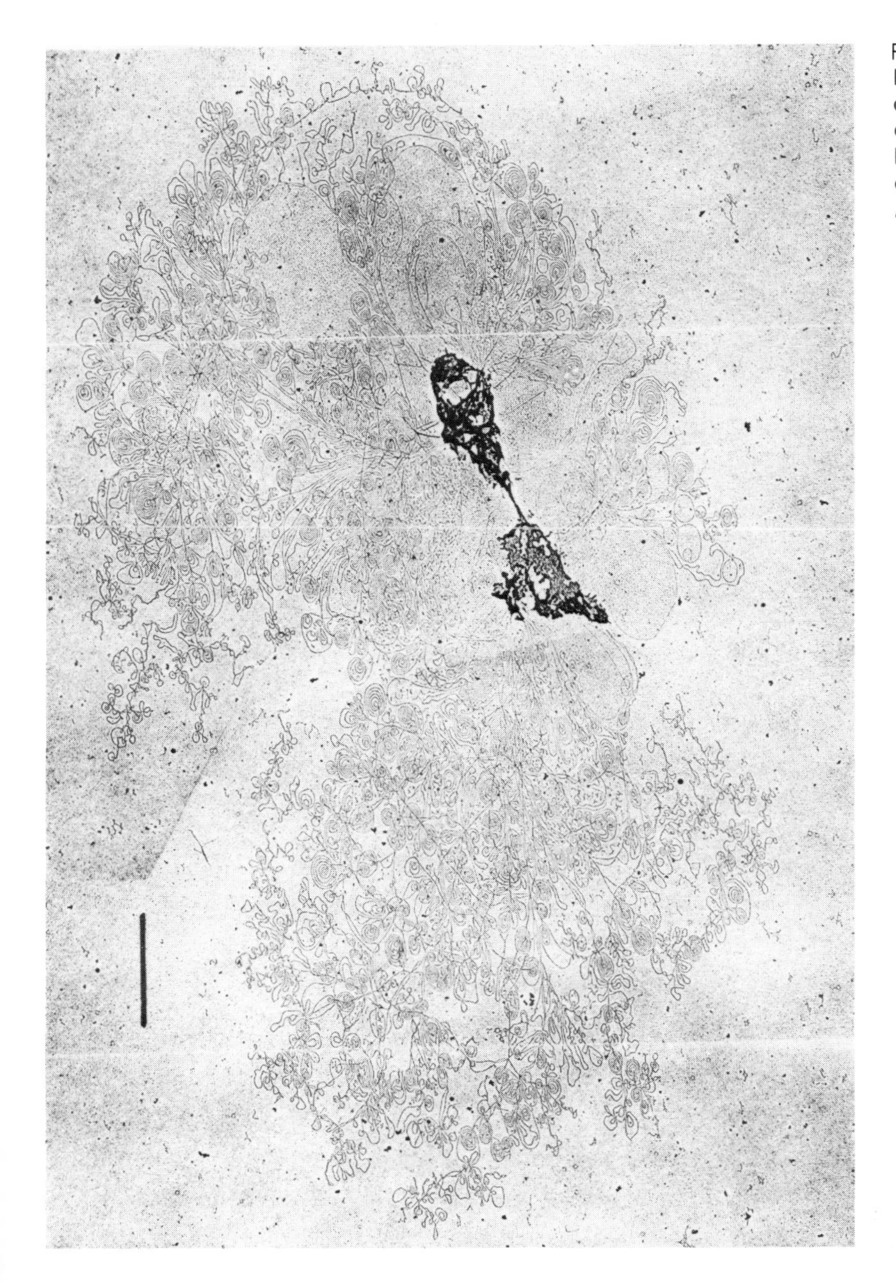

Figure 20–16
Electron photomicrograph of the single chromosome released from *E. coli* by osmotic lysis. The reference bar is 1μ long. (Courtesy of Dr. A Worcel, copyright 1974 Academic Press, *J. Mol. Biol. 82,* 108.)

back and forth in a semiparallel manner, rather than being coiled and supercoiled, although DNA from osmotically exploded cells has a tendency to coil upon release.

Viruses. The *tobacco mosaic virus* (TMV) was the first to be crystallized and its structure determined in detail. TMV is a cylindrical tube 300 Å long and 180 Å in diameter (Fig. 20–17). Its "chromosome" consists of a single molecule of RNA of some 6400 nucleotides and has a length of about 33,000 Å. The RNA is wound in a helix about the central aqueous core of the particle with 49 nucleotides per turn (inner diameter, 34 Å; outer diameter, 80 Å; pitch, 23 Å).The RNA is held in this arrangement by an outer sheath of 2,130 identical protein units.

The *T-even* and *lambda* phages of bacteria are DNA viruses in which the DNA is a single, linear, double-stranded molecule (Fig. 20–18). The lambda phage DNA is about 17 μ long, and the T2 and T4 phage DNAs are about 100 μ long. The lambda phage DNA assumes a circular shape after injection into the host bacterium. Numerous other viruses have been studied (Table 20–2).

Figure 20–17
Tobacco mosaic virus (Photomicrograph courtesy of Dr. F. A. Eiserling.)

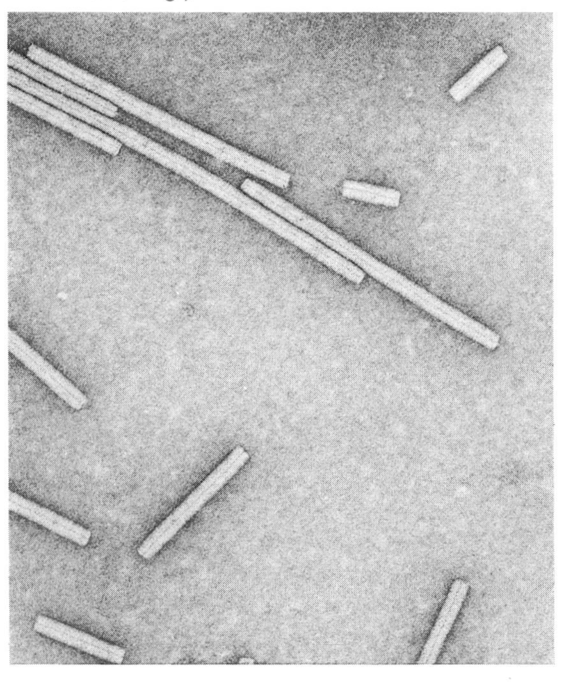

Replication

The **replication** of DNA has been studied for 25 years. The superb work of J. D. Watson, F. H. C. Crick, and M. H. F. Wilkins in 1953 established the double-stranded nature of DNA (Chapter 7) and the basic manner in which the molecule might be duplicated. A basic understanding now exists about the mechanism of DNA replication, but much of the detailed information is still in the process of refinement and amplification. Let us first consider the basic information.

Replication as a "semiconservative" process

The double-stranded DNA helix acts as a "template" or framework for the synthesis of a second identical DNA molecule. If both strands of the original or "parent" molecule recombined after replication and the new molecule were composed of the two newly synthesized strands, the process would be termed *conservative*. However, this is *not* the case. Instead, the two strands of the parent DNA molecule are separated, and each acts as a framework for building a new and complementary other half. Therefore, in the two resulting DNA molecules, one strand of each is from the parent DNA molecule and one strand of each is a newly synthesized polynucleotide chain.

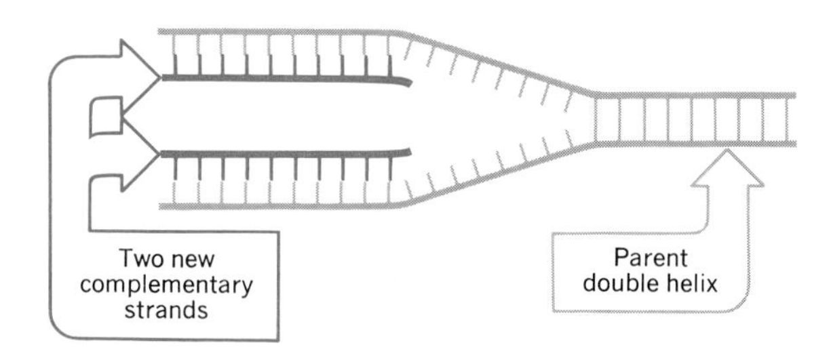

This method of replication is termed **semiconservative** and is the only known method occurring in plants, animals, microorganisms, and DNA viruses.

Replication by the Addition of Nucleotides in the 5′ ⟶ 3′ direction

The DNA molecule is made up of two polynucleotide chains interwoven to form a double helix. Each nucleotide within a chain is joined to the next by a phosphodiester bond

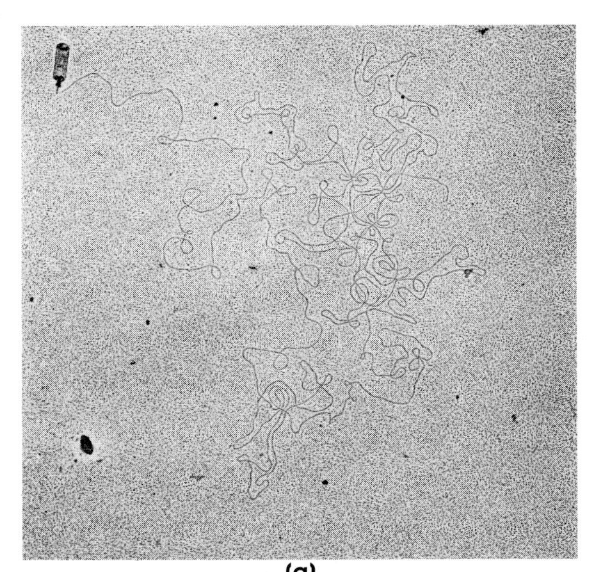

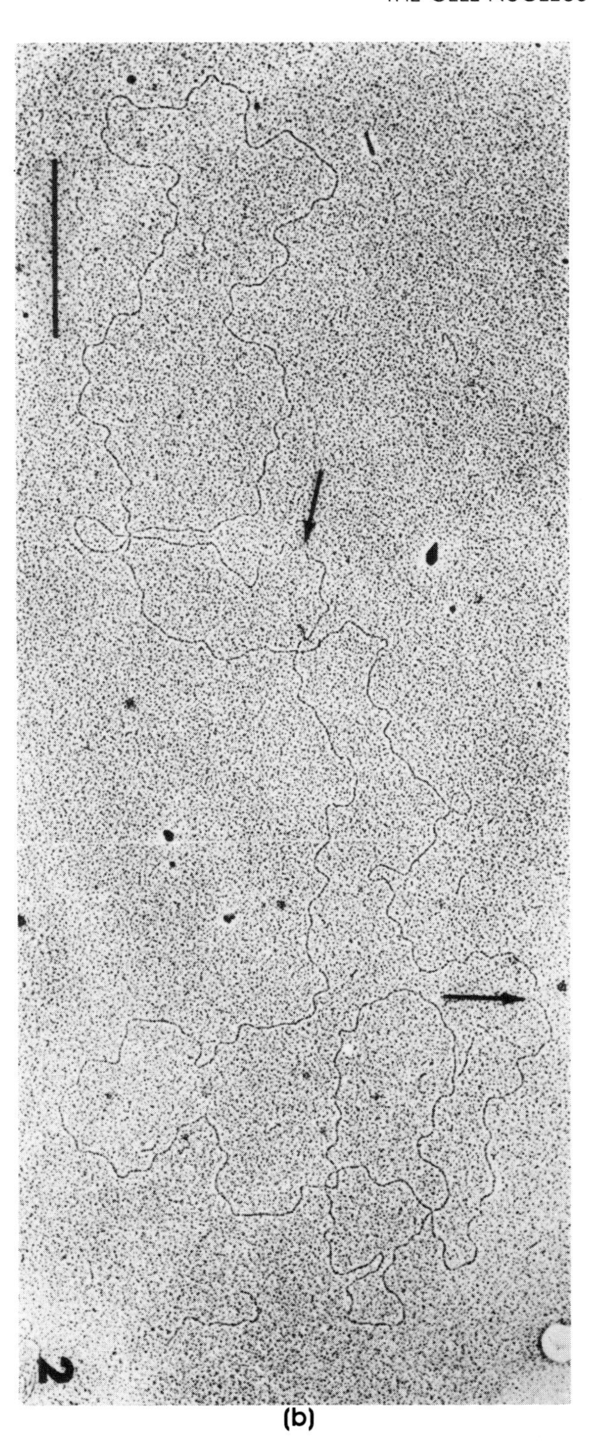

Figure 20–18
(a) DNA molecule released from a T_4 bacteriophage (Courtesy of Dr. F. A. Eiserling.)
(b) Lambda phage DNA. The reference bar is 1μ long; arrows show small discontinuities in the molecule. (Courtesy of Dr. H. Ris, copyright 1963 by Cold Spring Harbor Laboratory: *C.S.H. Sym. Quant. Biol. 28*, 2.)

Table 20–2
Sizes and other characteristics of certain viruses

Virus	Related Viruses	Particle Shape	DNA Mol. Wt. × 10^{-6}	DNA Number of Base Pairs, in kb	DNA Shape	Comments
SV 40	Polyoma	Polyhedron	3.4	5.1	Duplex, circular	Animal cell host
φX174	S13	Polyhedron	1.8	5.4	Single strand, circular	Duplex replicative form
M13	fd, f1	Filament	1.9	5.7	Single strand, circular	Duplex replicative form
T7	T3	Head, short tail	23	35.4	Duplex, linear	Terminal redundancy
λ	φ80, 434, P2, 186	Head, tail	32	49	Duplex, linear	Cohesive ends form replicative circles, lysogenic
T5		Head, tail	76	115	Duplex, linear	Nicked
T4	T2, T6	Head, tail	120	180	Duplex, linear	Terminal redundancy; permuted
R17	MS2, f2 Qβ	Polyhedron	1.0	3.0	Single strand, linear	RNA, not DNA

Source. From A. Kornberg, *DNA Synthesis*, W. H. Freeman and Co., San Francisco. Copyright © 1974.

linking the 3′ carbon of its deoxyribose to the 5′ carbon of the deoxyribose of the next nucleotide (Chapter 7). At one end of the chain, there is a phosphate group attached to the 3′ carbon of the last nucleotide, and at the other end, there is phosphate group on the 5′ carbon of the last nucleotide. The two chains that form the double helix run in opposite directions from the 3′ to the 5′ end and are said to be *antiparallel*. When replication occurs, the two antiparellel chains separate (Fig. 20–19), and nucleotides are added one at a time to the free 3′ position of the growing polynucleotide, and thus the forming polynucleotide grows from its 5′ end toward its 3′ end.

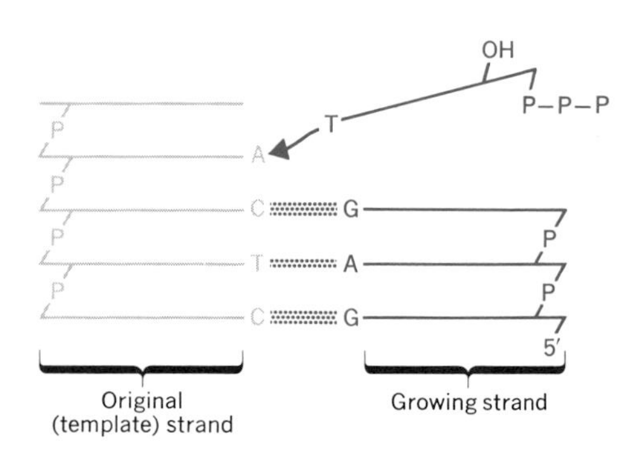

Replication in One or Two Directions

Replication begins at one point on the chromosome following the separation of the two DNA strands at that point. The addition of nucleotides forming the two new strands occurs successively along both parental strands starting from that point. According to the *unidirectional* model, this growth proceeds in one direction; along one chain nucleotides are added to the 3′ end, but on the antiparallel chain, nucleotides are added in the reverse direction (i.e., synthesis still occuring in the 5′ ⟶ 3′ direction). The new short segments of oligonucleotides are subsequently joined together.

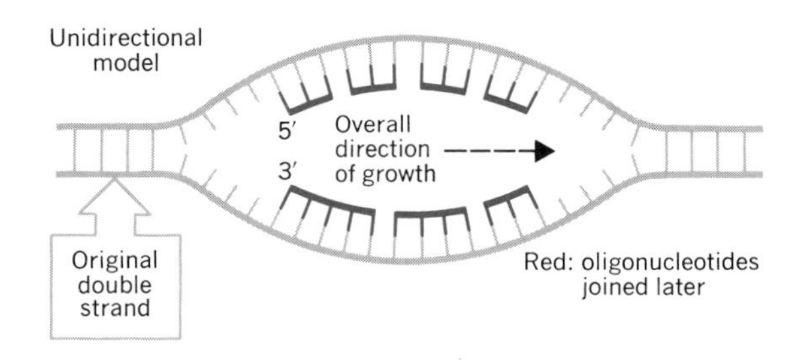

Replication that proceeds in two directions from a point, *bidirectional replication*, shows a similar pattern; nucleotides are added to the 3′ ends only, short segments of

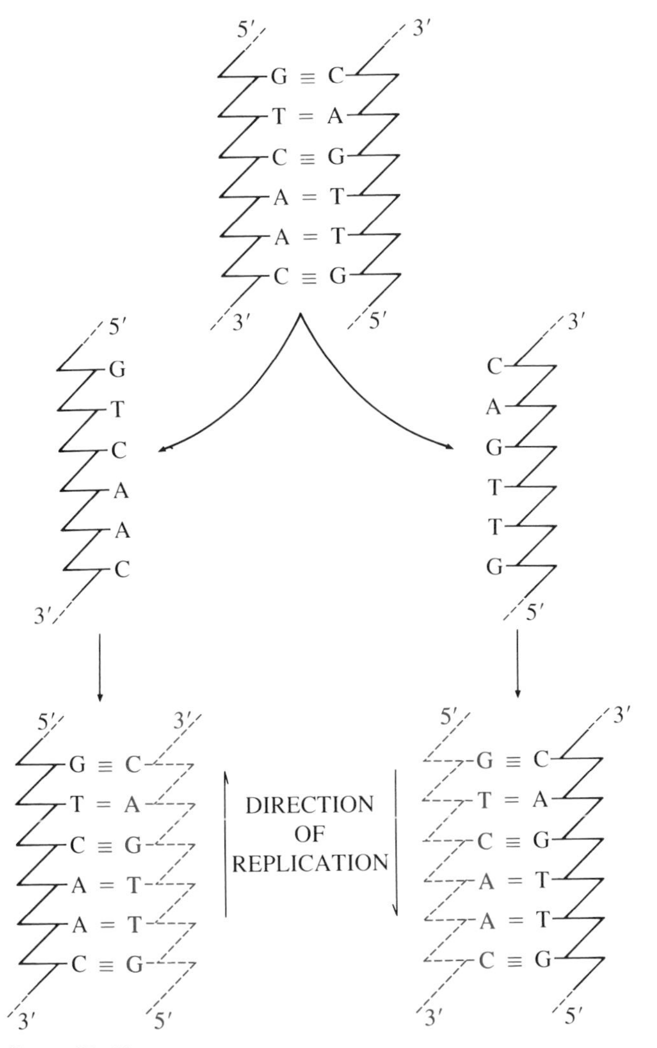

Figure 20–19
Diagram of DNA replication.

oligonucleotide being formed, and are then joined along each growning strand. The points of growth are called *forks*.

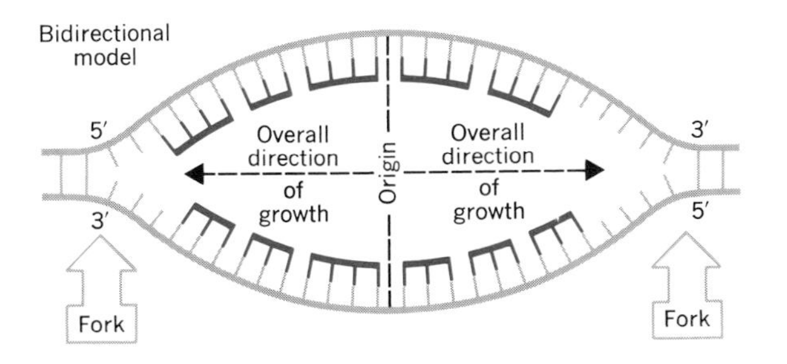

The oligonucleotide fragments formed (sometimes called Okazaki fragments after their discoverer) develop on the one chain and are fused later. These fragments can be about 100 nucleotides long in animal cells and about 1000 to 2000 nucleotides long in procaryotes.

"Swivel" Mechanism

The two interhelical polynucleotides of the chromosome must be untwisted to allow replication and formation of two new DNA molecules. In the *E. coli* chromosome, which is about 1mm long and replicates in about 40 minutes, the DNA double helix would have to unwind at 7500 revolutions per minute. How the ''swivel'' actually works is not known with certainty, but enzymes have been proposed that first break the chain ahead of the replication fork (''nickase''), allow rotation about a phosphodiester bond, and then reseal the chain (''ligase'').

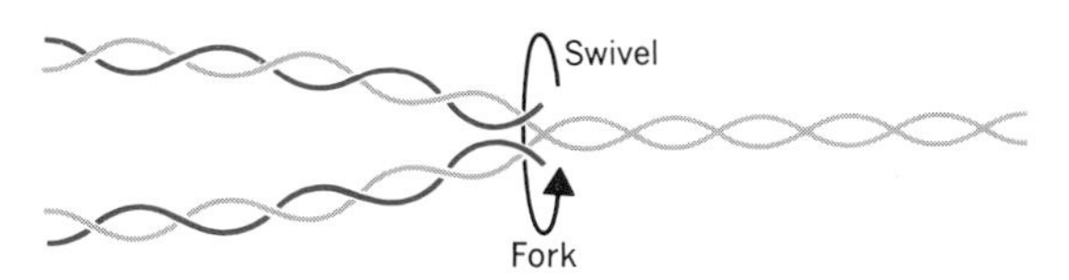

Enzymatic Replication

Some of the enzymes of replication are known, while others are still only postulated. Known to exist in *E. coli* cells are three DNA polymerases called polymerase I, II, and III. **Polymerase I** was the first discovered in *E. coli* by A. Kornberg and called **DNA-dependent DNA polymerase.** Since then, membrane-bound **polymerase II** has also been found in *E. coli*. Both enzymes elongate a polynucleotide chain by adding a nucleotide to the 3′ position of the last deoxyribose in the chain. The nucleotide, which is in the form of a triphosphate, loses its terminal two phosphates (pyrophosphate, PP_i) on addition to the chain. The pyrophosphate is then hydrolyzed by *pyrophosphatase*. The coupled reactions have a $\Delta G°$ of -7.5 kcal/mole.

$$(\text{Polynucleotide})_n + \text{nucleotide triphosphate} \xrightarrow{\text{polymerase}} (\text{Polynucleotide})_{n+1} + PP_i \qquad \Delta G° = 0.5 \text{ kcal/mole}$$

$$PP_i \xrightarrow{\text{pyrophosphatase}} 2\,P_i \qquad \Delta G° = -8.0 \text{ kcal/mole}$$

There is much recent evidence that these two polymerases are, in fact, DNA *repair enzymes* rather than major replicases and that the more recently discovered **polymerase III** is actually the direct catalyst of DNA replication.

Endonucleases, nickases, ligases, enzymes for initiating replication, and enzymes for terminating replication are all known to exist and are under considerable study. The functions of some are discussed below.

RNA Primers

How the initial nucleotides become associated with the DNA template is not known, but it appears that a first step is the formation of a 50 to 100 unit RNA polynucleotide, to the 3′ end of which the replicases add deoxyribonucleotides. Presumably, the RNA segment is removed by cleavage before linking the 3′ end of the DNA to the 5′ end upon completion of replication.

Replication in Procaryotes

Much of the pioneering work in DNA replication in procaryotes has been done with the circular DNA of *E. coli*. Replication begins at one point on the molecule and proceeds in both directions. The point of initiation is believed to be the point at which the chromosome is attached to the inner surface of the plasma membrane. Although such a membrane site has not yet been isolated, it seems reasonable to expect that such an attachment site could also facilitate distribution of the replicated DNA strands into the two daughter cells as a partitioning membrane forms during cytokinesis.

If autoradiographic techniques are used to study the replication of the chromosome, two loops or rings are seen. In such studies, *E. coil* cells are initially incubated for about 30 minutes in ^{3}H-thymidine (a DNA precursor), which uniformly labels their DNA. The cells are then incubated in the radioactive label for a second but shorter interval, during which a second round of DNA replication begins but is not completed. The strands of the chromosome are partially separated to accommodate the synthesis of new DNA (Fig. 20–20), and additional labeled thymidine is incorporated. If the incubation is stopped at a point where replication has proceeded about halfway around the circular DNA molecule and the DNA is examined by autoradiography, structures having the shape of the Greek letter "theta" (θ) are seen. The darker, more heavily labeled loop reflects the ^{3}H-thymidine incorporated in the partially completed second round of replication.

R. Okazaki showed that in *E. coli,* replication proceeds in short segments of 1000 to 2000 nucleotides. Subsequently, these pieces are joined together by the action of DNA ligase, which catalyzes the formation of the phosphodiester bond. The existence of the DNA ligase was first demonstrated in 1967, and since then the enzyme has been purified and is now being used in in vitro studies of DNA replication.

The initiation of the synthesis of these short DNA segments is not catalyzed by any known DNA polymerase. Instead, the first event is **transcription** of a short length of DNA catalyzed by *DNA-directed RNA polymerase*. This produces an RNA unit that acts as a *primer* for addition by DNA polymerase of deoxyribonucleotides to the 3′ end of the RNA. After synthesis of the short segment of DNA, the RNA primer is excised and replaced by DNA polymerase action.

Many questions about replication remain unanswered; for example, what initiates RNA primer transcription? What terminates this transcription after about 10% of the replication fragment has been formed? What mechanism excises the RNA primer?

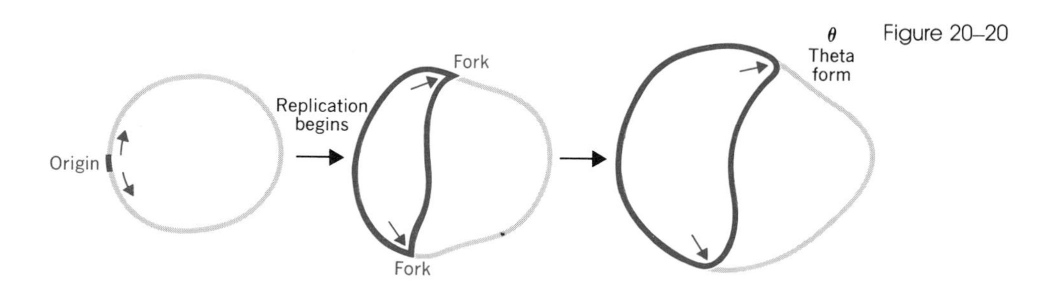

Figure 20–20

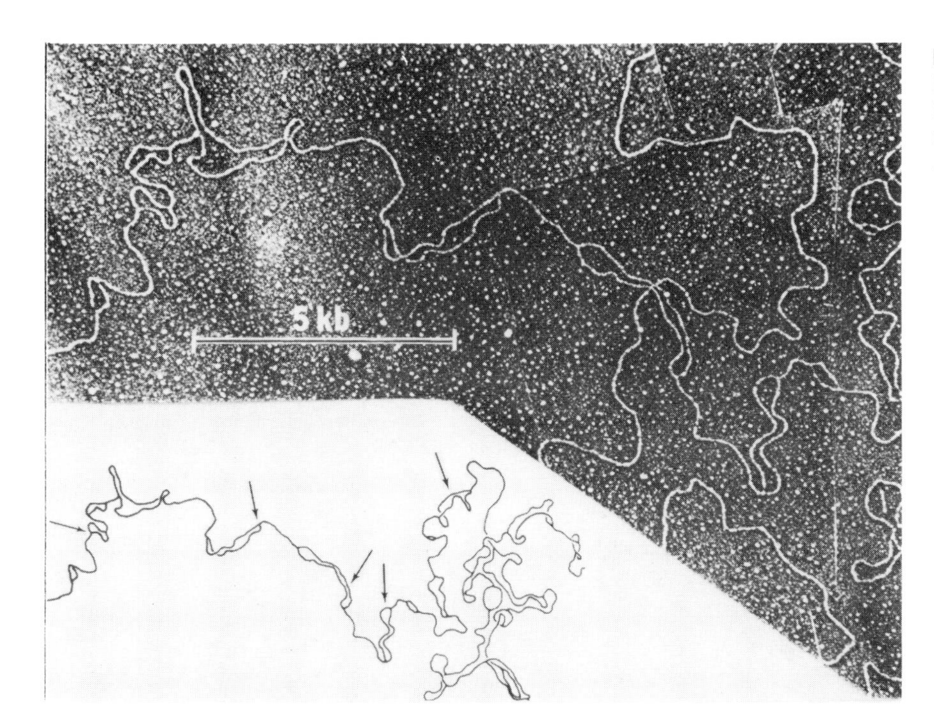

Figure 20–21
Electron micrograph of *Drosophila* DNA showing multiple sites of replication (arrows). (Courtesy of *Proc. Natl. Acad. Sci. 71,* 135. 1974.)

Replication in Eucaryotes

Most of the mechanisms described for procaryotes also exist in eucaryotes. Of interest is the fact that DNA polymerases extracted from bacteria can promote replication in eucaryotes. Also, ligases extracted from one group of organisms can be effective in others.

There is good evidence that eucaryotic chromosomes are attached to the inner membrane of the nuclear envelope, an aspect that is similar to the attachment of the procaryotic chromosome to the cell membrane. However, the function of the attachment to the nuclear envelope is not clear. Replication in a eucaryotic chromosome begins simultaneously at many sites along the chromosome (Fig. 20–21). Replication begins with transcription of RNA primers and continues with synthesis of the polydeoxynucleotides added in the $5' \longrightarrow 3'$ direction and by formation of short segments. The RNA primers are excised and the segments of polydeoxynucleotides joined together.

Replication of Viral DNA

Viruses alone are unable to replicate their own DNA but utilize the metabolic machinery of the host cell, directing its activities through material introduced into the host on infection. The degree to which the virus relies on the host cell is related to the size of the virus (Table 20–2). Small viruses like ϕX174, which has only 9 known genes and 5375 nucleotides, contain information for the formation of proteins for their coat and rely entirely on the host cell to provide the machinery for DNA replication. Larger phages, like T4, which attacks *E. coli,* contain around 20 genes, coding for the enzymes of their own replicative process as well as for coat proteins.

When lysogenic phages (like λ phage) infect a host cell (i.e., *E. coli*) that is metabolically active, a virulent infection is produced and gives a lytic response. Under these conditions, the DNA of the phage penetrates the host cell and replication occurs by an independent process like that of the T4 phage. However, if the λ phage infects a host cell that is metabolically *inactive* (i.e., glucose is exhausted and cyclic AMP is elevated), a *temperate* infection occurs causing a *lysogenic response*. The inserted viral DNA is integrated into the host bacterial DNA and is replicated along with the host DNA, using the native mechanism of the host cell. A temperate infection does not cause a detectable change in the host, and the inserted phage DNA (now called *prophage*) may be replicated and carried through many generations of the host.

The Virus Life Cycle

There are 6 basic steps in the life cycle of a virus:

1. *Virus adsorption*. The virus is adsorbed to the surface of the host cell. The specificity between the virus and the host is achieved through the recognition by specific viral adsorption proteins called *pilot proteins,* for specific host cell-surface receptors. The pilot protein not only attaches the virus to the host but may also orient the viral DNA (or RNA in the case of the RNA viruses) for insertion into the most appropriate part of the cell.

2. *DNA (or RNA) insertion*. Penetration of the viral DNA through the host cell envelope and cell membrane and into the interior of the cell may be aided by the pilot proteins, which may be carried into the cell with the DNA.

3. *Expression of viral nucleic acid*. Once in the host cell, the viral DNA is expressed in some manner. In the case of the small single-stranded viruses like ϕX174 and M13, a second strand complementary to the single strand is synthesized, yielding a duplex form. For the double-stranded DNA viruses, certain segments necessary for virus replication are transcribed and translated, while for the RNA viruses, certain portions are direclty translated.

4. *Replication of virus nucleic acid*. The mechanism of DNA replication varies with the type of virus, but ultimately many complete double (or single) strands are produced for final ''packaging.''

5. *Virus assembly*. The mature DNA strands are assembled within a shell or coat of proteins.

6. *Release*. Finally, the assembled viruses are released from the host cell. The stages are summarized in Figure 20–22. In some cases, viruses emerge from the host cell in *buds* and do not cause serious disruption of the cell surface; in other cases, the host cell ruptures (lyses), spilling the viruses into the surrounding medium.

LAMBDA (λ) Phage

For a number of lysogenic phages, *E. coli* is the host cell. One of these—the **lambda (λ) phage**—is interesting because its double-stranded DNA has single-stranded tails at each end of the molecule that are complementary to each other (Fig. 20–23). When the λ phage DNA penetrates *E. coli,* the complementary bases at the ends pair (i.e., ''anneal'') and form a circular DNA molecule. If the λ DNA is inserted into a metabolically active cell, replication of the DNA occurs immediately and is followed by lysis and

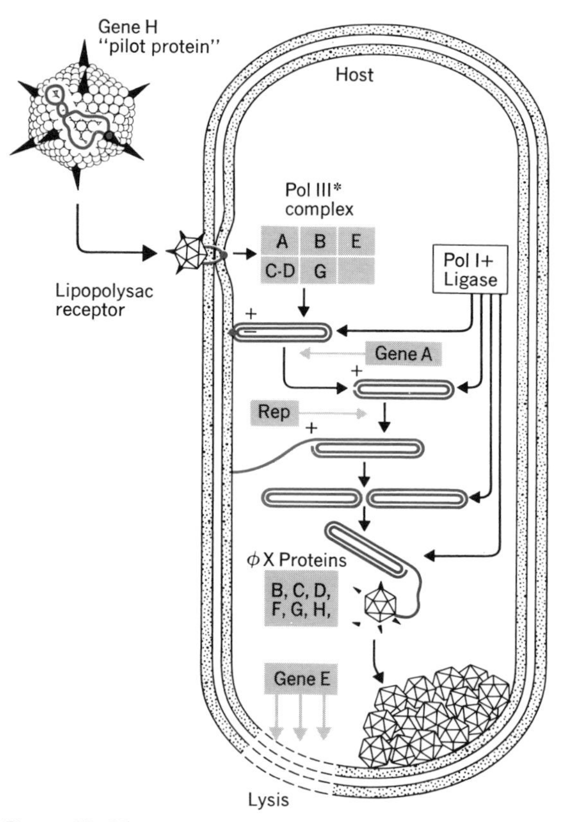

Figure 20–22
øX174 life cycle. Virus attaches to host and inserts its single-stranded circular DNA (i.e., "+" strand) together with pilot protein. Viral DNA contains 9 genes, A-J. Host enzymes (e.g., Pol I, Pol III*, ligase, and "rep") together with viral proteins produced by transcription and translation of certain viral genes) initially produce complementary (i.e., "−") strand and then many more duplex structures. Virus coat proteins produced by transcription and translation of complementary strands assemble to form polyhedral capsid which encapsulates only the + strands of each duplex. Rupture of host cell wall by E gene proteins releases viruses. Role of gene J is unknown. (Redrawn from A. Kornberg, *DNA Synthesis,* W. H. Freeman and Co., copyright 1974.)

release. On the other hand, in metabolically inactive cells, integration of the λ DNA into the *E. coli* chromosome occurs. This integration is very specific; it requires the product (called **integrase**) of one of the λ genes *("int")* and is inserted at a specific site in the *E. coli* chromosome called *"att."* The *E. Coli* chromosome is broken, the λ DNA inserted, and the λ prophage is formed and may be replicated along with the *E. coli* DNA over a thousand generations (Fig. 20–24).

Excision of the λ DNA from the *E. coli* chromosome is also an unusual phenomenon and requires the *int* gene and *xis* gene. On rare but detectable occasions during excision, a part of the host chromosome may be removed along with the λ DNA. The genes removed with the λ DNA are those for galactose utilization *("gal")* and biotin synthesis *("bio")*. These genes occur on each side of the *att* site. The transfer of host genes such as *gal* and *bio* by viruses to new host cells is called **transduction.** When the new cell receives these genes, it is said to be *transduced,* and **gene recombination** (see below) has occurred.

Plasmids and Episomes

In many bacteria, there are relatively small DNA molecules called **plasmids** that are not part of the bacterial chromosome. Plasmids are double-stranded, circular, and supercoiled. By size, they can be divided into two groups; (1) those of about 5×10^6 daltons and (2) those of about 10^8 daltons. There are about 20 copies of the smaller-sized plasmids per cell and 1 to 2 copies of the larger-sized plasmids. Some of the plasmids can integrate into the chromosome and are then called **episomes.** The number of plasmids per cell appears to remain constant from one generation to the next. Some plasmids can be transferred from one cell to another by viral transduction.

Figure 20–23
Structure of the linear form of λ phage DNA showing the overlapping ends and the conversion of the linear form into the circular form by annealing of the ends. (Redrawn from Arthur Kornberg, *DNA Synthesis,* W. H. Freeman and Company, copyright 1974.)

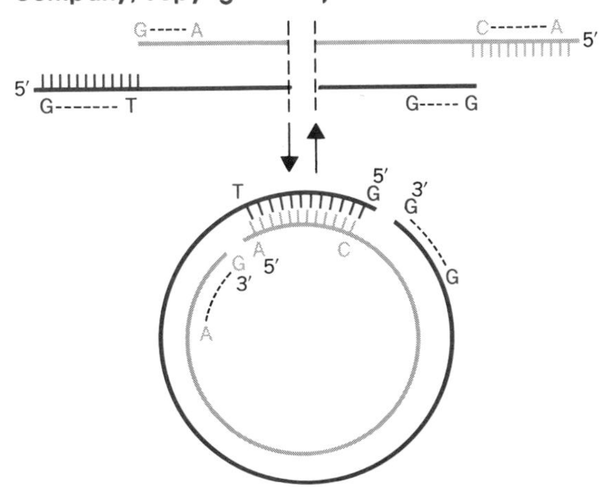

Recombination

Recombination refers to the exchange of genetic information between chromosomes. As a result, genes occur in new combinations in successive cell generations. In eucaryotic cells, recombination occurs during *meiosis* (described below) in which crossing over is the principal event that produces recombination of genes. Haploid organisms have no ready mechanism for recombination, but gene rearrangement may occur by transduction, transformation, or conjugation.

Bacterial **transformation** (see Chapter 7) is a type of recombination. O. Avery, C. MacLeod, and M. McCarty showed that when DNA was extracted from virulent *Diplococcus pneumoniae* (pneumococcus) and mixed with living cells of a nonvirulent strain of the same species, the cells absorbed the DNA and became virulent. It was later shown that the absorbed DNA became incorporated into the genome and was replicated and passed along to subsequent generations.

Recombinant DNA

Within the last few years, technology has provided the means to achieve controlled recombination in certain cells. It is now possible to extract DNA (even selected genes) from one cell species and insert it into the genome of different species. (DNA selected from one cell species for insertion into the genome of another is called **recombinant DNA.**)

The technical advances that have resulted in this ability are founded on the discovery of a variety of **restriction enzymes** that can be contracted from various organisms but that are effective generally on DNA. These enzymes cleave DNA at selected sites (Table 20–3). In addition, the means

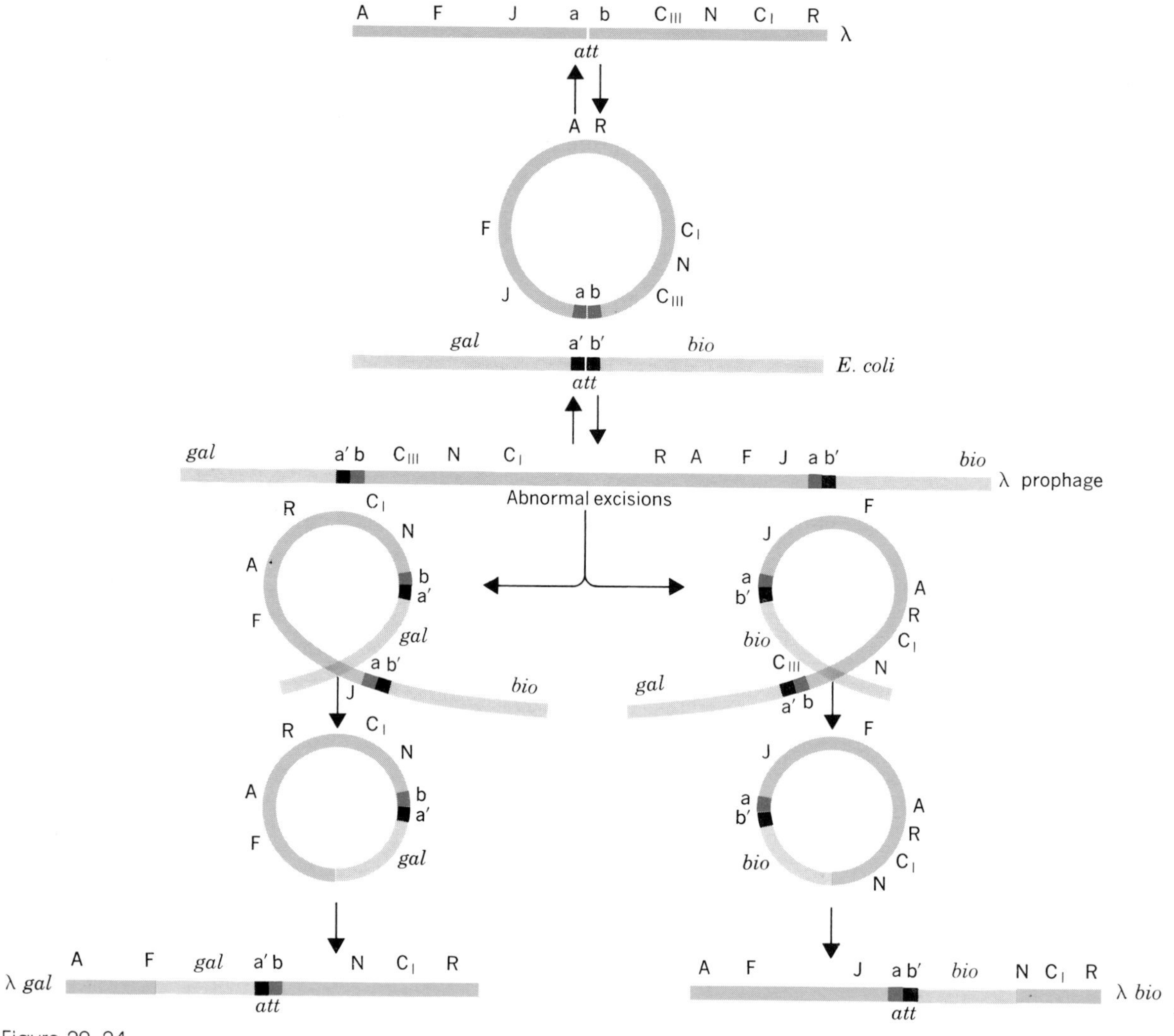

Figure 20–24
Diagram of mechanics of intergration and excision of λ prophage DNA into chromosome of *E. coli*. (Redrawn from Arthur Kornberg, *DNA Synthesis*, W. H. Freeman and Company, copyright © 1974.)

are now at hand for splicing extracted pieces together in various combinations. Finally, it is possible to attach the restructured DNA (recombinant DNA), even in relatively large pieces, to the DNA of living cells so that the recombinant DNA is replicated along with the cells normal genome. As a result, a whole cell line or culture with the recombinant DNA (called a **clone**) is produced (Fig. 20–25).

The potential advantages for mankind made possible through the discrete utilization of recombinant DNA technology (such as providing more abundant crops, more effective pharmaceutical products, even solutions to energy

Table 20–3
Sequence Specificities of Some Commonly Used Restriction Enzymes

Designation	Source	Sequence
*Eco*RI	*E. coli* RY 13	5′···T/AG ↓ pAATTCA/T···3′
*Hind*III	*Haemophilus influenzae* Rd	5′···A ↓ pAGCTT···3′
*Hae*II	*Haemophilus aegyptius*	5′···PuGCGC ↓ pPy···3′
*Hpa*II	*Haemophilus parainfluenzae*	5′···C ↓ pCGG···3′
*Hha*I	*Haemophilus haemolyticus*	5′···GCG ↓ pC···3′
*Bam*HI	*Bacillus amylolique faciens* H	5′···G ↓ pGATCC···3′
*Sal*I	*Streptomyces albus* G	?
*Pst*I	*Providencia stuartii*	?

Source. Reproduced with permission from the *Annual Review of Biochemistry*, Vol. 46. Copyright © 1977 by Annual Reviews Inc.

problems and the eradication of cancer) are presently being argued against the potential of human disaster through the chance production of ecological imbalances with "new" organisms, new pathogens, or even the creation of new biological weapons for militarists and terrorists. The potential for disaster has frequently been exaggerated, especially by the scientifically uninformed community, but the need for caution and effective control of recombinant DNA experiments is recognized by everyone. At the time of this writing, "adequate" control *versus* "excessive" control of research in this area has become a political and ethical issue throughout the world.

Presently, cloning has been carried out with *E. coli* and yeast, and the vehicle has been either a plasmid or a λ phage. Recombinant DNA experiments in eucaryotes have employed kidney tissue culture cells of African "green monkeys." The claims that have been made so far relative to transformation of cells of higher organisms by microbial DNA or a somewhat sustained expression of microbial DNA by plant and animal cells are still controversial. Most current research in the area of recombinant DNA is concerned with methodology and techniques, although a number of applications of great biological importance should be forthcoming soon.

Figure 20–25
Diagram of mechanics for splicing foreign DNA into plasmid DNA and subsequent propagation to form clones of recombinant DNA. (see text.)

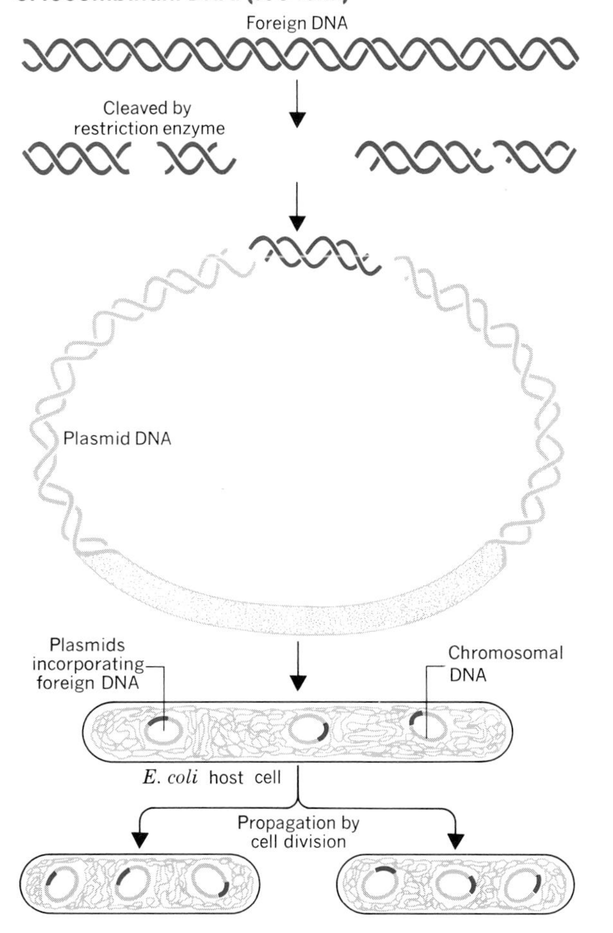

Meiosis

Meiosis is a form of nuclear division of fundamental importance among sexually reproducing organisms. An in-depth discussion of meiosis on a cellular as well as a genetic basis is beyond the scope of this book and is normally treated at length in textbooks on genetics (see References). However, for the sake of completeness in this discussion of nuclear phenomena, we will consider the major meiotic events and their implications.

Meiosis occurs in eucaryotic organisms whose cells contain the *diploid* number of chromosomes. Diploid infers "double" in the sense that the genetic information in any one chromosome can be found in an identical (or modified) form in a second chromosome in the nucleus. The two chromosomes forming such pairs are said to be homologs.

As noted earlier in this chapter, human diploid cells contain 46 chromosomes, or 23 homologous pairs. The 46 chromosomes of the zygote formed at fertilization are derived equally from the sperm and egg cells, each of these gametes contributing one member of each homologous pair. Mitosis then produces the billions of cells that ultimately make up the whole organism.

Only certain cells of diploid organisms undergo meiosis, the mechanism being restricted to the reproductive tissues (i.e., ovaries and testes), which produce the eggs and sperm.

In meiosis, a diploid nucleus divides twice, producing four daughter nuclei each with half the number of chromosomes of the parental diploid cell. Although only half the diploid number are present (i.e., the "haploid" number), the set is complete, for each nucleus contains one member of each pair of homologous chromosomes. The homologous chromosomes generally assort themselves randomly at anaphase between the developing daughter nuclei, and this accounts for part of the genetic variation that characterizes sexually reproducing organisms. Additional genetic variation occurs during prophase by a process called *crossing over*. The genetic implications of random assortment and crossing over are principal subjects of genetics courses.

For convenience, the events of meiosis, like mitosis, are divided into a number of phases (prophase, metaphase, etc.). However, since meiosis is characterized by *two* successive rounds of nuclear division we distinguish the phases as prophase I, prophase II, metaphase I, metaphase II, and so on.

The initial stages of the first round of meiosis are similar in many respects to prophase of mitosis (i.e., the chromosomes condense, the nuclear membrane and nucleoli dissolve and disappear, etc.). Prophase I of meiosis is distinguished by the following five stages (see also Fig. 20–26):

Prophase I

1. *Leptotene stage* (leptonema). The threadlike structure of the chromosomes becomes apparent.

2. *Zygotene stage* (zygonema). Homologous chromosomes align themselves side by side by a process known as **synapsis.** The *allelic* genes (i.e., those encoding products of similar or identical function) become situated adjacent to one another. The unit consisting of two synapsed, duplicated homologous chromosomes is called a *bivalent* chromosome, or *tetrad*. Since DNA replication occurred prior to the onset of prophase I, each homologous chromosome actually consists of two identical sister chromatids. Therefore, altogether, a bivalent chromosome contains four chromatids.

As synapsis between the homologous chromosomes progresses, a protein framework develops between adjacent homologous chromatids, which join together at one or more points forming *chiasmata* (sing., chiasma). Chiasmata result from the cleavage by endonucleases of the DNA at corresponding positions in two nonsister chromatids, followed by a crossing over and reunion of the DNA chains under the influence of DNA polymerases (Fig. 20–27).

3. *Pachytene stage* (pachynema). Chromatids become increasingly distinct.

4. *Diplotene stage* (diplonema). Homologous chromatids separate from each other, except at points where crossing over has taken place (i.e., at chiasmata).

5. *Diakinesis*. Chromosome condensation is completed, yielding short, thick structures that are readily distinguishable.

Metaphase I

The tetrads align in the center of the spindle with the centromeres on the equatorial plate.

Anaphase I

Homologous chromosomes (but *not* sister chromatids) separate and move to opposite poles of the spindle.

Telophase I

Chromosomes aggregate at the poles so that two nuclear areas are distinguished.

Interkinesis (or **interphase**)

This is the period between the end of the first telophase and the onset of prophase II. The two nuclei produced by the

Prophase I

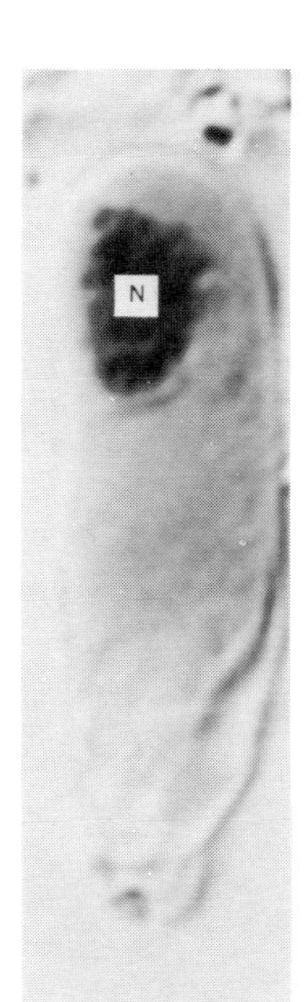

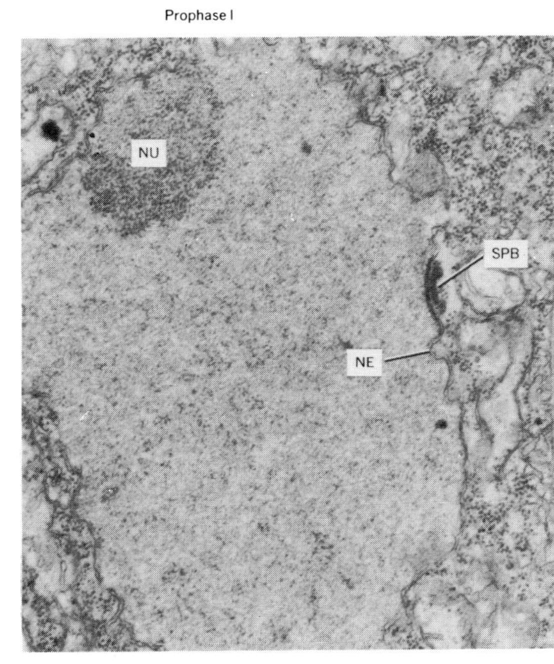

Metaphase I

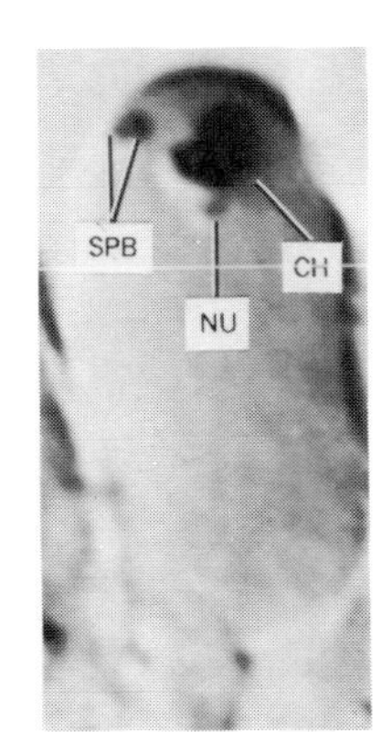

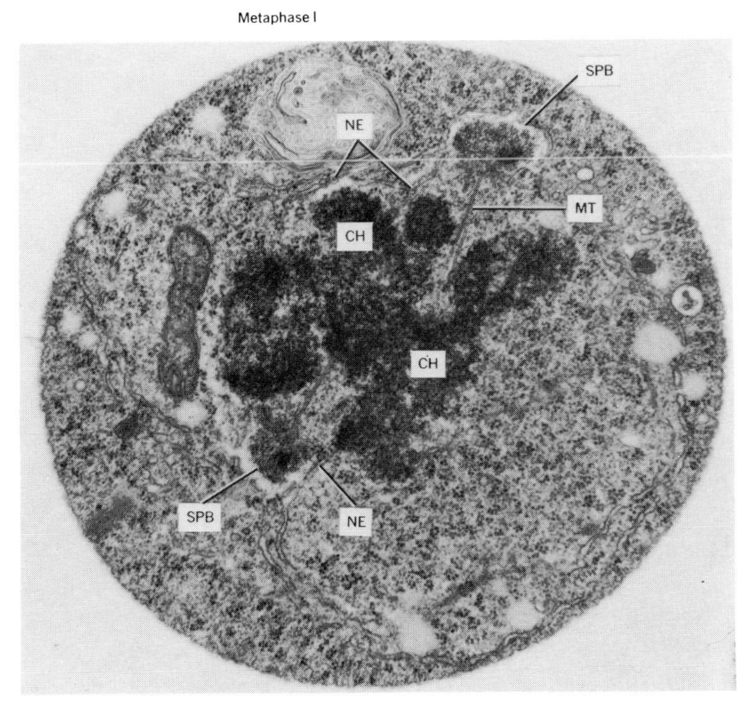

Figure 20–26
Phases of meiosis as seen in the basidium of *Pholiota* using light microscopy and transmission electron microscopy. The phases are described in the text. CH, chromatin; CM, continuous microtubules; ER, endoplasmic reticulum; KC, kinetochore; M, membranes; MT, microtubules; N, nucleus; NE, nuclear envelope; NU, nucleolus; PER, perinuclear endoplasmic reticulum; SPB, spindle pole body. (Photomicrographs courtesy of Dr. K. Wells, copyright 1978 Springer-Verlag *Protoplasma 94,* 85.)

Anaphase I

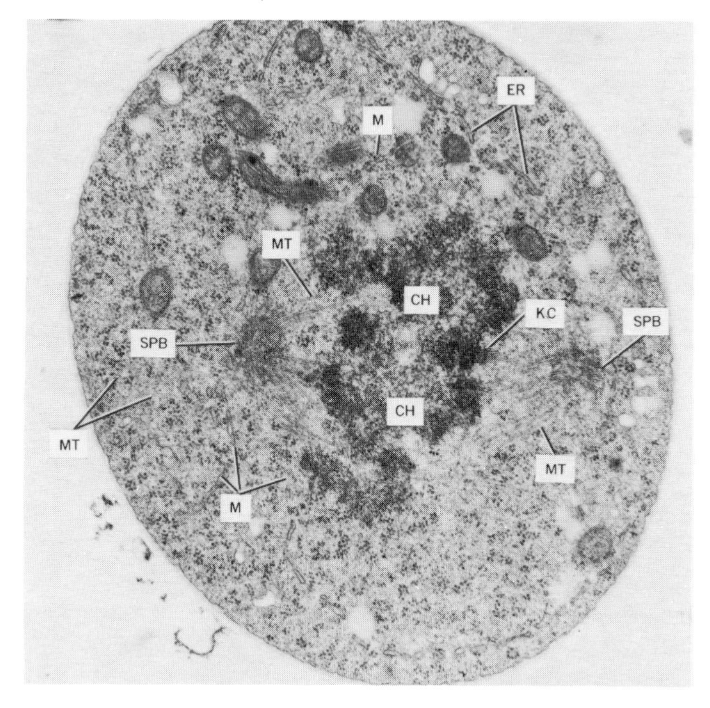

Telophase I

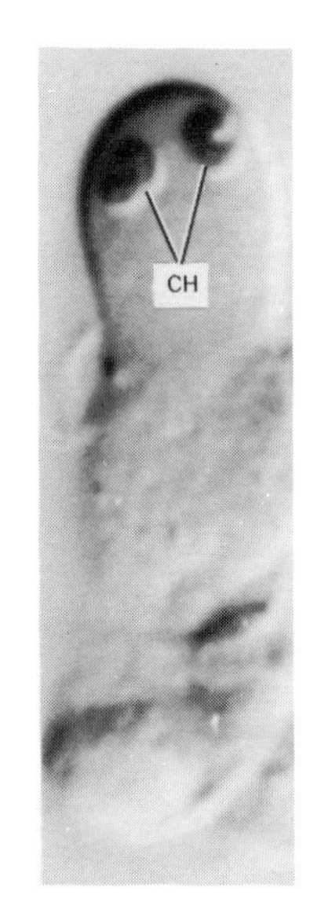

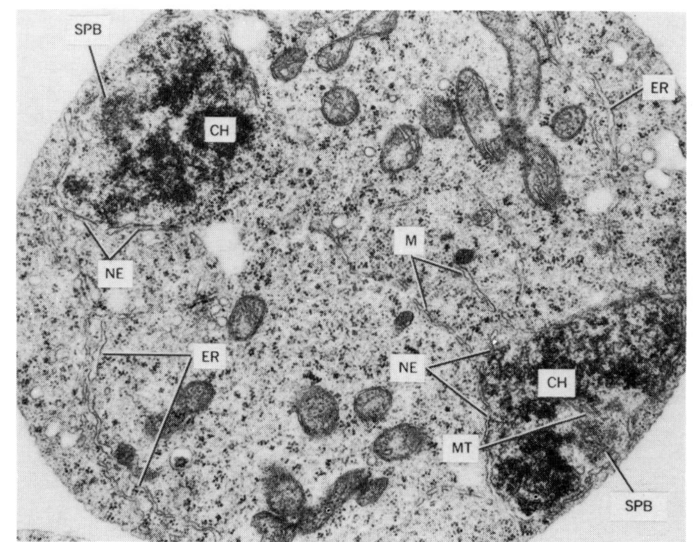

Interphase I

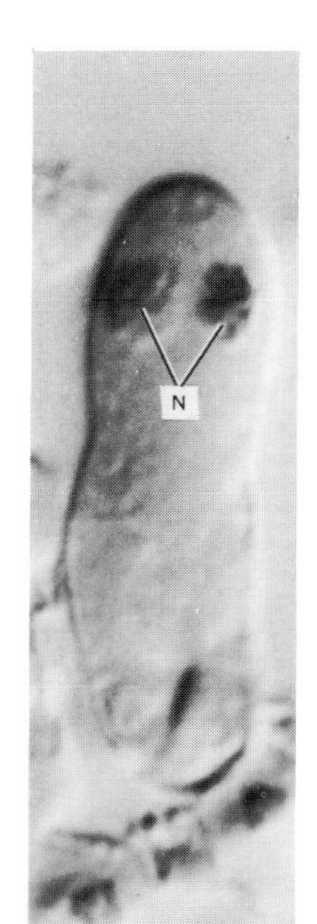

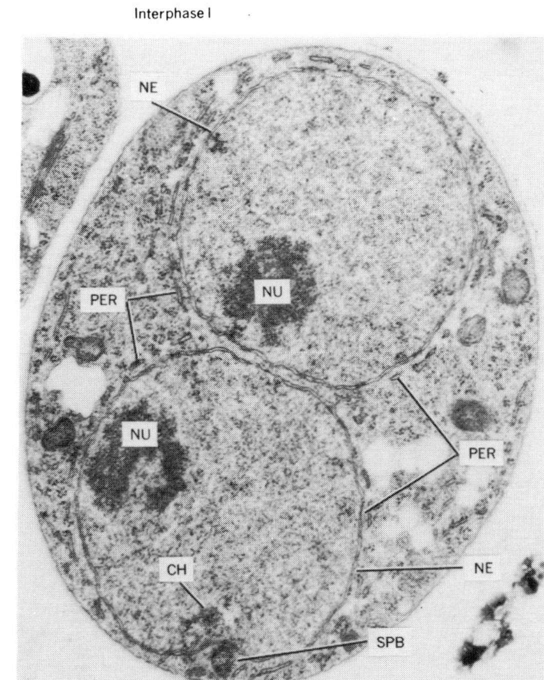

Metaphase — anaphase II

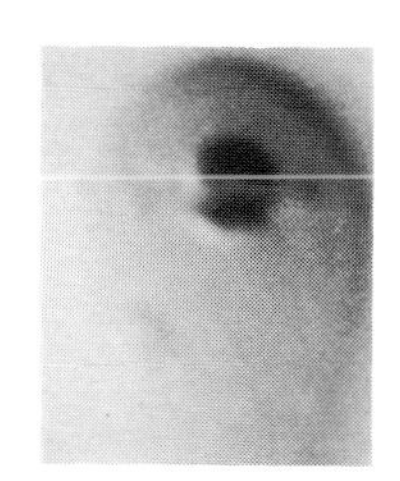

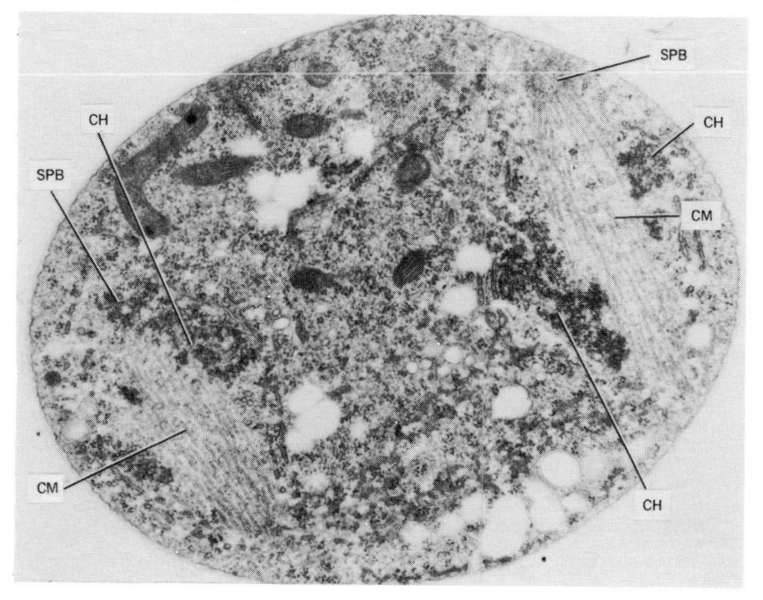

Anaphase II

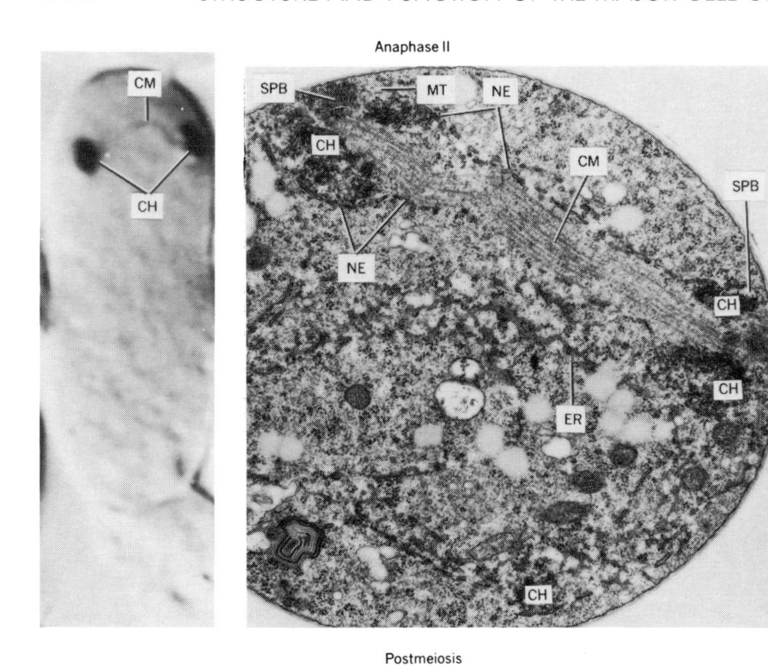

Postmeiosis

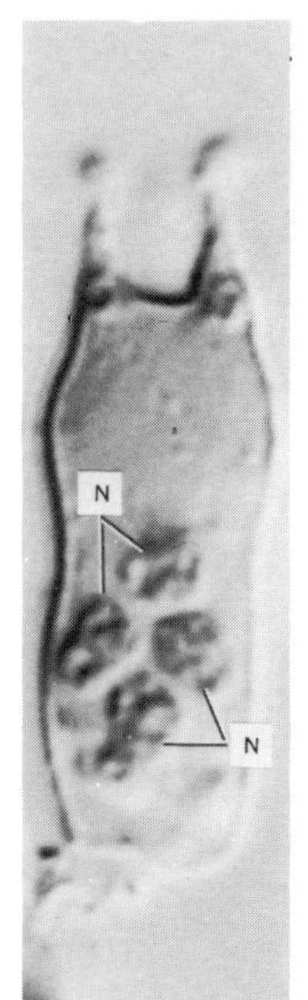

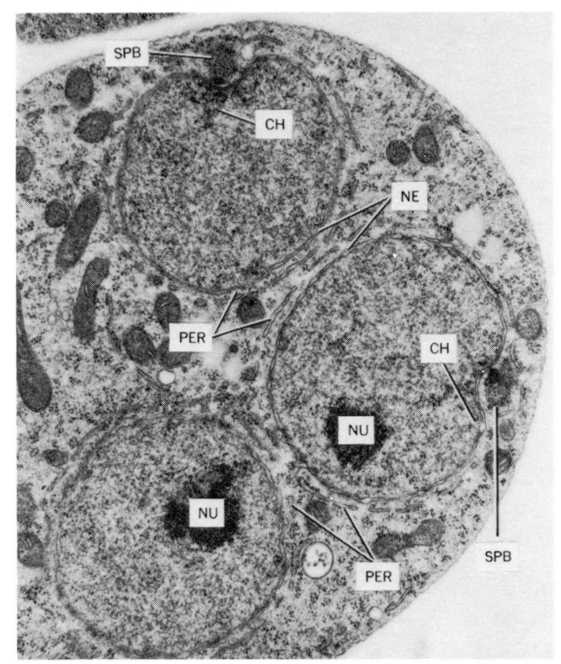

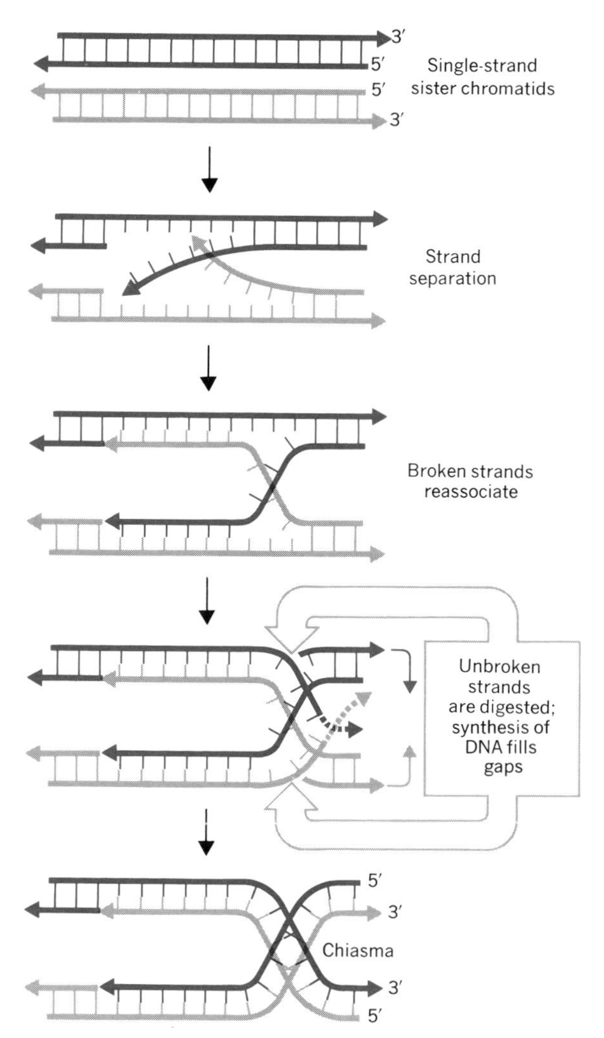

Figure 20–27
Model of crossing-over and chiasmata formation as it would appear at the DNA level in a bivalent eucaryotic chromosome.

first meiotic division do *not* engage in DNA replication during this interphase, which is usually quite short.

Prophase II
The events characterizing this phase are similar to mitotic prophase, although each cell nucleus has only half the number of chromosomes as does a cell in prophase I; that is, it is already *haploid*. Each chromosome remains composed of two chromatids.

Metaphase II
The events occurring in this phase are similar to those in mitotic metaphase.

Anaphase II
The events occurring in this phase are similar to those in mitotic anaphase, as chromatids now separate and pass to opposite poles.

Telophase II
The events occurring in this phase are similar to those in mitotic telophase.

Meiosis produces four cells, each with the haploid number of chromosomes. In many higher animals and some plants, meiosis in the female reproductive tissues is accompanied by an uneven division of the cytoplasm, in which case one of the two nuclei formed during telophase I forms a nonfunctional *polar body* and may not enter prophase II (Fig. 20–28). In some organisms (such as humans) the polar body completes meiosis, but the two smaller polar bodies produced during telophase II are likewise nonfunctional. The second division of the larger cell produced during telophase I is also unequal and produces an additional polar body. During the production of spermatozoa in the male reproductive tissues, division of the cytoplasm is equal, but remarkable cytoplasmic differentiation of the four spherical haploid spermatids produced after meiosis is required (Fig. 20–28) before functional, flagellated spermatozoa are produced.

Cytokinesis

In most instances, there is one nucleus per cell, with the ratio being perpetuated by the synchronization of division of the cytoplasm with nuclear division. Mitotic divisions without cytoplasmic divisions occur in many species to produce multinucleate or *coenocytic* cells. Striated muscle cells, *mycelia* of molds, and some phloem cells are examples of multinucleate cells. Few cells exist without a nucleus. The sieve cells of plants lack a nucleus but are always in close proximity to a nucleated companion cell. Mammalian erythrocytes lose their nucleus during the final stages of development in the bone marrow.

In general, animal cells and many lower plant cells divide by a pinching or furrowing action that squeezes the cell into two parts about the dividing nucleus. In plant cells, which possess a cell wall, the division of the cell occurs by the

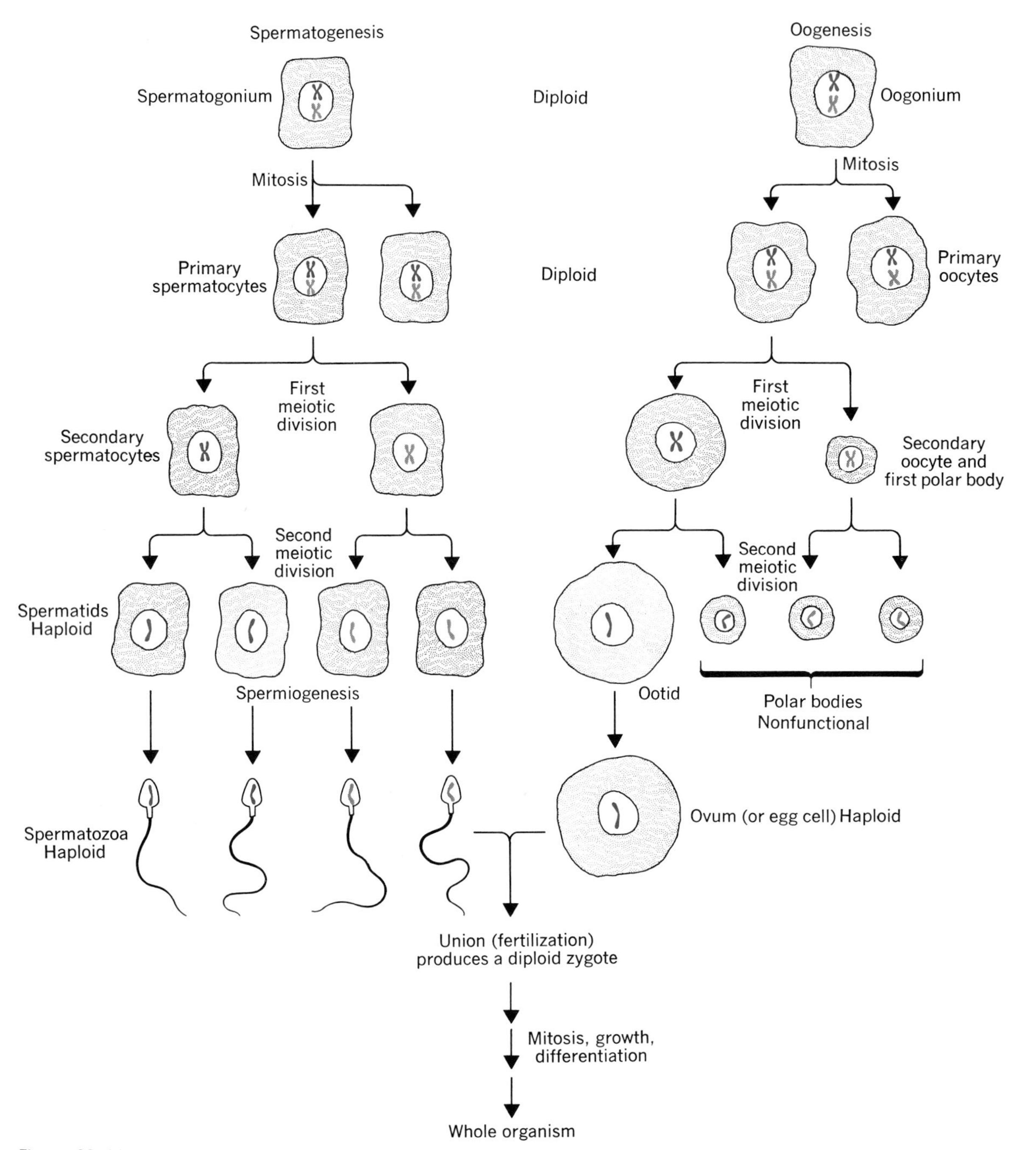

Figure 20–28
Gametogenesis (production of egg and sperm cells) in higher animals.

formation of a **cell plate.** This plate begins forming at the center of the spindle toward the end of anaphase. It expands outward until the cell is cleaved in half. **Furrowing** appears to be aided by the formation of *microfilaments* (discussed in Chapter 22). The cell plate is formed by the progressive accumulation of carbohydrate and lipid materials derived from the cisternae of the Golgi apparatus (Chapter 18).

Summary

Eucaryotic nuclei are delimited by a pair of **enveloping membranes** that are continuous with each other around the margins of **nuclear pores.** Within the nucleus are the **chromosomes, nucleoli, nucleoplasm, ribosomelike particles,** and a variety of irregularly shaped particles. The chromosomes during **mitosis** and **meiosis** assume a distinctive shape as a result of the condensation of their constituent DNA and proteins. Metaphase chromosomes are characterized by a **primary constriction** or **centromere** and a varying number of **secondary constrictions.** The arms of the chromosome contain a framework of a **double helix of DNA** about which are packed a number of proteins. Each strand of DNA, called a **chromatin fiber,** is **supercoiled.** The amount of coiling is responsible for the observation of two types of chromatin fibers, the narrower **type A** and the broader **type B.** The **packing ratio,** or length of DNA divided by the chromatin fiber length, is about 56:1 for interphase chromosomes and 100:1 for metaphase chromosomes.

During interphase the chromatin fibers show varying degrees of **condensation.** The more condensed portions of interphase chromatin are called **heterochromatin,** and the less condensed regions, **euchromatin.** Euchromatin appears to contain most of the **structural genes,** whereas heterochromatin may be more concerned with the control of **gene expression.** The chromatin fibers have a repeating pattern of subunits called **nucleosomes,** which are specific associations of histones with DNA. DNA replication begins during the *S* phase of the cell cycle at random sites in the euchromatin and later shifts to the heterochromatin.

The chromosomes of bacteria and viruses are single nucleic acid molecules. They may be single-stranded DNA or RNA or may consist of a double helix of DNA. These chromosomes are not contained within a nuclear membrane but in bacteria may be attached to the cell membrane or an infolding of the cell membrane called a **mesosome.**

Replication is a **semiconservative process,** each strand of the double helix of DNA serving as a **template** for formation of new strands. Replication begins at one point on the chromosome and proceeds along both strands, adding nucleotides to both in the $5' \longrightarrow 3'$ direction. A **swivel mechanism,** consisting of enzymes for breaking the polynucleotide strands **(nickases)** and resealing the strands **(ligases),** prevents twisting of the DNA during replication.

Much of the pioneering work on replication was done with procaryotes, especially the circular DNA molecule of *E. coli*. Technology has reached the stage where **recombination** (the exchange of genetic information between chromosomes, resulting in new combinations of genes in successive generations) can be accomplished experimentally using chromosomes of different organisms. The ethical and moral implications for scientists are acute and the focus of considerable controversy.

References and Suggested Reading

Articles and Reviews

Dubochet, J., and Noll, M. Nucleosome arcs and helices. *Science 202,* 280 (1978).

Grobstein, C., The recombinant-DNA debate. *Sci. Am. 237*(1), 22 (July 1977).

Khorana, H. G., Total synthesis of a gene. *Science 203,* 614 (1979).

Kornberg, R. D., Structure of chromatin. *Annu. Rev. Biochem. 46,* 931 (1977).

Landy, A., and Ross, W., Viral integration and excision: Structure of the lambda "att" sites. *Science 197,* 1147 (1977).

Sinsheimer, R. L., Recombinant DNA. *Annu. Rev. Biochem. 46,* 415 (1977).

Weissback, A., Eucaryotic DNA polymerases. *Annu. Rev. Biochem. 46,* 25 (1977).

Books, Monographs, and Symposia

Adams, R. L. P., Burdon, R. H., Campbell, A. M., and Smellie, R. M. S., *Davidson's Biochemistry of the Nucleic Acids,* 8th ed., Academic Press, New York, 1976.

Bonner, J., and Varner, J. E., *Plant Biochemistry,* Academic Press, New York, 1976.

Busch, H. (editor), *The Cell Nucleus,* Vols. I, II, III, Academic Press, New York, 1974.

DuPraw, E. J., *DNA and Chromosomes,* Holt, Rinehart and Winston, Inc., New York, 1970.

Gardner, E. J., *Principles of Genetics,* John Wiley & Sons, Inc., New York, 1972.

Kornberg, A., *DNA Synthesis,* W. H. Freeman and Co., San Francisco, 1974.

Lehninger, A. L., *Biochemistry,* 2nd ed., Worth Publishers, Inc., New York, 1975.

Chapter 21 RIBOSOMES AND THE SYNTHESIS OF PROTEINS

Until the 1930s, it was the prevailing view that DNA was found only in animal cells and RNA only in plant cells. This view was dispelled by a number of findings in the 1930s that definitively established that both DNA and RNA are present in animal and plant cells. Moreover, J. Brachet and T. Caspersson showed that the bulk of the RNA was present in the cytoplasm and that cells actively engaged in protein synthesis (such as pancreas cells and the silk-gland cells of silk worms) contain greater amounts of RNA than cells that do not actively produce protein. Albert Claude showed in the 1940s that the cytoplasmic RNA was included in tiny particles of ribonucleoprotein later to be called "ribosomes."

Protein Turnover In Cells

The rate of breakdown and replacement of protein in cells was badly misunderstood prior to 1939. In growing animals generally and in secretory tissues in particular (e.g., the liver, pancreas, and endocrine glands), active synthesis of protein was known. However, the amount of protein synthesis taking place in other tissues of the adult was believed to be very low and confined to that necessary to replace protein lost through damaged or dying cells. These small protein losses, together with the catabolism of dietary amino acids, were believed to be responsible for the urea and ammonia measurable in urine. Proteins were thus regarded as highly stable constituents lasting virtually the entire lifetime of the cell.

The first serious challenge to the "wear and tear" view of protein turnover came as a result of the work of R. Schoenheimer in 1938. Schoenheimer synthesized a number of amino acids in which the ^{15}N content of the alpha-amino nitrogen was considerably increased over the natural amount of this isotope. Schoenheimer then injected ^{15}N-containing glycine and leucine into rats and noted that these labeled amino acids were incorporated into the proteins of many tissues very rapidly. Although ^{15}N is not a radioactive isotope of nitrogen, it may nonetheless be distinguished chemically from the more common ^{14}N form and is called a "heavy" isotope of nitrogen. The results clearly indicated that protein synthesis in adult animals is not restricted to growing and secretory tissues but occurs in nearly all cells and that tissue proteins are in a continuous state of metabolic flux, being broken down and replaced by newly synthesized molecules.

Although the radioactive isotope of carbon, ^{14}C, was produced in the Berkeley cyclotron in 1940, it was not until 1947 that ^{14}C-labeled amino acids became available. The availability of radioactive amino acids was followed by a series of classical tracer experiments by H. Borsook, T. Hultin, P. Zamecnik, and P. Siekevitz which verified the findings of Schoenheimer that most tissues readily incorporate amino acids into protein and also added crucial details to the newly emerging view of protein synthesis and metabolic turnover.

The first attempts to determine the subcellular site of amino acid incorporation into protein were carried out in 1950 by Borsook. Minutes after injecting ^{14}C-labeled amino acids into the bloodstreams of guinea pigs, Borsook removed the animals' livers and, using the technique of differential centrifugation (see Chapter 12), prepared subcellular fractions of the tissue. Borsook showed that it was the **microsomal fraction** that contained the highest degree of radioactivity and suggested that the microsomes were the repository of the cell's protein-synthesizing apparatus. In the same year, Hultin demonstrated that it was the microsomal fraction of chick liver tissue homogenates that incorporated intravenously injected ^{15}N-glycine into protein.

By 1952, Siekevitz and Zamecnik had been able to demonstrate the in vitro incorporation of ^{14}C-labeled amino acids into liver cell proteins by both tissue slices and tissue homogenates. By measuring and comparing protein synthetic activity in cell-free whole homogenates, individual cell subfractions, and various combinations of subfractions, Siekevitz showed that the incorporation of amino acids into proteins by microsomes was dependent on an energy source provided by the mitochondrial subfraction and required enzymes and other factors present in the cytosol. The demonstration that amino acid incorporation into protein required metabolic energy laid to rest a view popular in the 1940s that polypeptide synthesis might be brought about by the reversal of protein hydrolysis. It is especially interesting to note that Siekevitz demonstrated the existence in the cytosol of a $MgCl_2$-precipitable factor required for protein synthesis. Since $MgCl_2$ was known to precipitate RNA, Siekevitz suggested that RNA might somehow be involved in protein synthesis, a fact not fully recognized until many years later.

The studies described above established the general cytological and chemical basis of protein biosynthesis. Exhaustive research since the 1950s by dozens of groups of investigators has revealed the step-by-step, reaction-by-reaction details of the process and has given us an astounding insight into the molecular organization of the cell's protein-synthesizing apparatus. Although much of this chapter is devoted to the examination of ribosome structure and to the chemical events that accompany protein synthesis, much of what is to be presented is better understood if *first* placed in perspective with a brief, preliminary overview of the subject; this will set the foundation for the more comprehensive study that follows. For simplicity, this synopsis will be concerned only with *cytoplasmic* protein synthesis in eucaryotic cells.

A Preliminary Overview of Protein Biosynthesis

The variety and specific amino acid compositions of the proteins synthesized by a cell are ultimately governed by the cellular DNA. This DNA is enzymatically *transcribed* in the cell nucleus to produce a host of RNAs, including ribosomal RNA (rRNA), messenger RNA (mRNA), and transfer RNA (tRNA). The base sequences of these RNAs are *complementary* to the base sequences of the DNA molecules transcribed. rRNA is ultimately incorporated into the cytoplasmic ribosomes, which may be *free* in the cytosol or *attached* to the surface of the intracellular membrane network that faces the cytosol. Each ribosome consists of two parts or **subunits**—a *small* subunit and a *large* subunit. The small subunit binds mRNA entering the cytosol from the nucleus, and the functional complex is completed with the subsequent addition of the large subunit. Attached ribosomes are linked to the endoplasmic reticulum via the large subunit.

The nucleotides of mRNA are arranged as a linear sequence of **codons,** each codon consisting of three successive nitrogeneous bases (also known as *triplet*). The codon sequence of each mRNA molecule contains all the information necessary to (1) properly *initiate* polypeptide synthesis on the ribosome, (2) designate the specific *sequence* of amino acids to be incorporated (i.e., the primary structure of the polypeptide), and (3) *terminate* polypeptide synthesis and *release* the completed polypeptide. Table 21–1 shows the various codons of mRNA and their meanings in protein synthesis. This is called the "genetic code." The code is said to be *degenerate* because in certain instances, a single amino acid may be coded for by more than one codon.

Molecules of tRNA entering the cytosol from the nucleus combine with amino acids; this is a molecule-specific association in that each amino acid species is enzymatically combined with a particular type (or species) of tRNA. The products, called **aminoacyl-tRNA,** represent the form in which amino acids are incorporated into newly synthesized protein. Each species of tRNA contains, among other functional groups, an **anticodon** (a sequence of three bases) that

Table 21–1
The Genetic Code

First Base	Second Base U	C	A	G	Third Base
U	phe	ser	tyr	cys	U
	phe	ser	tyr	cys	C
	leu	ser	"stop" (ochre)	"stop" (opal)	A
	leu	ser	"stop" (amber)	try	G
C	leu	pro	his	arg	U
	leu	pro	his	arg	C
	leu	pro	gln	arg	A
	leu	pro	gln	arg	G
A	ile	thr	asn	ser	U
	ile	thr	asn	ser	C
	ile	thr	lys	arg	A
	met ("start")	thr	lys	arg	G
G	val	ala	asp	gly	U
	val	ala	asp	gly	C
	val	ala	glu	gly	A
	val ("start")	ala	glu	gly	G

is recognized by a corresponding (probably complementary) codon of mRNA and ensures that the correct amino acid will be incorporated into its proper position in the primary structure of the polypeptide being synthesized.

Once the mRNA-ribosome complex has been formed, amino acids bound to their specific tRNA molecules are sequentially brought to the ribosome and incorporated into the growing polypeptide chain. This process, called *translation,* is believed to take place by an orderly and linear movement of the mRNA along the ribosome (or vice versa) so that each codon is translated in sequence. The elongation of the polypeptide chain takes place by a series of enzyme-catalyzed reactions occurring on two adjacent sites of the ribosome; these are the **amino acid** (or **acceptor**) site and the **peptide** (or **donor**) site. To understand the process of elongation, consider an intermediate stage in the synthesis of a polypeptide. At this time, the growing polypeptide chain is attached to the peptide side of the ribosome by a molecule of tRNA and is termed peptidyl-tRNA. The mRNA codon located in the vacant amino acid site specifies the form of aminoacyl-tRNA that can be bound there. With a new aminoacyl-tRNA in position in the amino acid site, the bond linking the growing polypeptide to its tRNA is broken and replaced by a peptide bond with the amino acid of aminoacyl-tRNA. This leaves the peptidyl-tRNA (which is now one amino acid longer) temporarily in the amino acid site. The tRNA molecule released in the process reenters the cytosol where it may combine with another amino acid to be used in protein synthesis. Formation of the peptide bond is followed by a shift of the peptidyl-tRNA to the peptide site, once again leaving the amino acid site vacant. This shift is accompanied by the movement of the ribosome and/or mRNA so that the next codon is in position in the amino acid site and may now be translated. Thus, tRNA molecules employed in bringing amino acids to the ribosome are transiently bound first to the amino acid site and then to the peptide site before returning to the cytosol.

Amino acids are sequentially added to the growing polypeptide until its primary structure is complete. Once the end of the message coded in the strand of mRNA is reached, the completed protein is released from the ribosome. The ribosome separates from the mRNA and dissociates into its two subunits; these may be used again in another round of protein synthesis.

Many mRNA molecules are large enough to be simultaneously translated by a number of ribosomes. These ribosomes move in a series along the mRNA, translating its coded message into a number of *identical* proteins. The release of one ribosome at the end of the message is accompanied by the attachment of a new ribosome at the beginning of the message. Such strings of ribosomes are called **polysomes,** and most protein synthesis that takes place in cells occurs on these structures. Although each mRNA molecule may be attached to several ribosomes, each ribosome synthesizes but a single protein chain before dissociating into its subunits. The mechanics of protein synthesis described briefly here for perspective only is treated in detail in later sections of this chapter.

Structure, Composition and Assembly of Ribosomes

In this section, we are concerned with the organization, composition, and assembly of the cytoplasmic ribosomes of procaryotic and eucaryotic cells. Organellar ribosomes (e.g., chloroplast and mitochondrial ribosomes) will be considered separately later in the chapter. Although functionally analogous, many differences exist between the ribosomes of procaryotic and eucaryotic cells (Table 21–2). Considerably more is known about the structure and composition of bacterial ribosomes than ribosomes of eucaryotic cells, as will become evident during the discussion that follows. Most of the work on procaryotic ribosomes has been carried out using *E. coli*. Although some variations are observed among the procaryotes, findings using *E. coli* are generally representative.

Ribosomes in the cytoplasm of eucaryotic cells have a sedimentation coefficient of about 80 *S* (M. W., about 4.5 $\times$ 10^6) and are composed of 40 *S* and 60 *S* subunits. In procaryotic cells, ribosomes are typically about 70 *S* (M. W., about 2.7 $\times$ 10^6) and are formed from 30 *S* and 50 *S* subunits. The complete ribosome formed by combination of the subunits is also referred to as a **monomer.** Although ribosomes from both procaryotic and eucaryotic sources are about 30 to 45% protein (by weight), with the remainder being ribonucleic acid, the specific protein and RNA components of these two major classes of ribosomes differ (Table 21–2 and Fig. 21–1); carbohydrate and lipid are virtually absent. Magnesium ions (and perhaps other cations) play an important role in maintaining the structure of the ribosome. Dissociation into subunits occurs when

Table 21–2
Properties and Composition of Eucaryotic and Procaryotic Ribosomes

	Eucaryotes	Procaryotes
Monomers		
Sedimentation coefficient	80 *S*	70 *S*
Molecular weight	4.5×10^6	2.7×10^6
Number of RNAs	4	3
Number of Proteins	70	55
Small subunit		
Sedimentation coefficient	40 *S*	30 *S*
Molecular weight	1.5×10^6	0.9×10^6
RNAs present	18 *S* (M.W., 0.7×10^6) (2110 nucleotides)	16 *S* (M.W., 0.6×10^6) (1600 nucleotides)
Number of proteins	30 (total M.W., 0.78×10^6)	21 (total M.W. 0.3×10^6)
Large subunit		
Sedimentation coefficient	60 *S*	50 *S*
Molecular weight	3×10^6	1.7×10^6
RNAs present	5 *S* (M.W., 3.2×10^4) (120 nucleotides) 5.8 *S* (M.W., 5×10^4) (150 nucleotides) 28 *S* (M.W., 1.7×10^6) (5000 nucleotides)	5 *S* (M.W. 3.2×10^4) (120 nucleotides) 23 *S* (M.W. 1.1×10^6) (3200 nucleotides)
Number of proteins	40 (total M.W., 1.37×10^6)	34 (total M.W. 0.5×10^6)

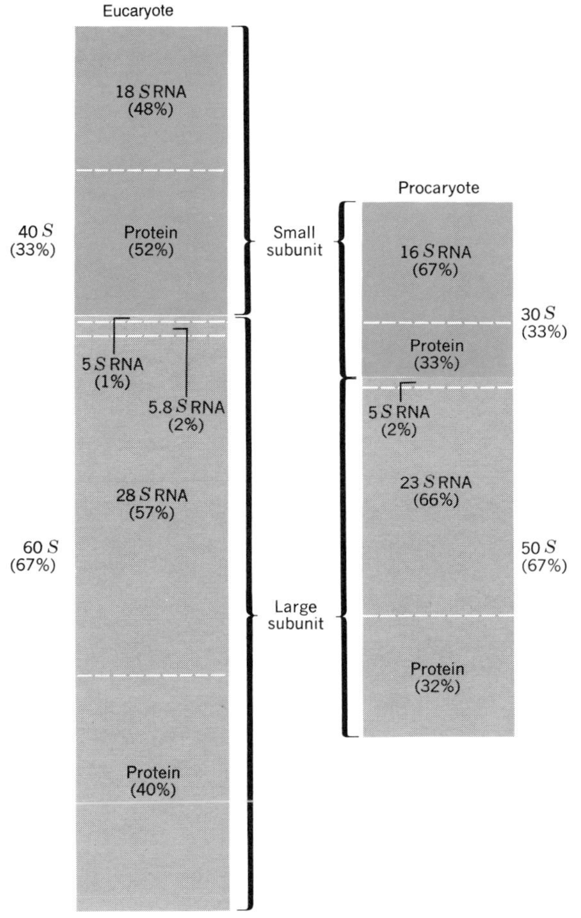

Figure 21–1
Geometric representation of the relative sizes, RNA, and protein complements of eucaryotic and procaryotic ribosomes. Each area corresponds to the molecular weight of that component. (Numbers in parentheses are the weight percentages of the components.)

Mg^{++} is removed. The precise role (or roles) of Mg^{++} remains uncertain, although interaction with ionized phosphate of subunit RNA is presumed.

Procaryotic Ribosomes

RNA Content. The small subunit of procaryote ribosomes contains one molecule of an RNA called 16 *S* RNA (M. W., 0.6×10^6) while the large subunit contains two RNA molecules, a 23 *S* RNA (M. W., 1.1×10^6) and a 5 *S* RNA (M. W., 3.2×10^4) (see Table 21–2). All three rRNAs are products of closely linked genes transcribed in the sequence 16 *S* → 23 *S* → 5 *S*. This assures an equal proportion of each unit. A polynucleotide containing the 16 *S transcript* is enzymatically cleaved (by an *endoribonuclease*) from the growing RNA strand once transcription has entered the 23 *S* region of the DNA (referred to as *r*DNA). A second polynucleotide containing the 23 *S* transcripts is similarly released once the 5 *S* region is reached. A final product contains the 5 *S* transcript. The initial transcription products are successively "trimmed" to form the 16 *S*, 23 *S*, and 5 *S* RNAs finally incorporated into the ribosomal subunits. Figure 21–2 presents the scheme of maturation of the procaryotic rRNAs. For clarity, the incorporation of the ribosomal proteins is not shown. Ribosomal proteins combine with the tRNAs at various stages of subunit assembly: some are incorporated during "pre-rRNA" transcription, others following pre-rRNA cleavage from the growing polynucleotide and during trimming, and still others once the mature rRNA products are formed. Certain proteins bind to the rRNAs only transiently and are not found in the fully assembled subunits.

Multiple copies of the rRNA genes occur in the genomes of procaryotic (and eucaryotic) cells (Table 21–3); this is known as **reiteration.** In *E. coli* the number of rRNA genes is estimated to be between 5 and 10 and accounts for about 0.4% of the cell's total DNA. The primary structures of the three procaryotic rRNAs have been extensively studied. 5 *S* RNA was identified in 1963 and, being the smallest of the three rRNAs (about 120 nucleotides), was sequenced first (in 1967). Figure 21–3 compares the primary structures of several 5 *S* RNAs. All the sequences are compatible with the existence of a "stem" formed by base-pairing of the 3′ and 5′ ends of the molecules.

The sequencing of 16 *S* RNA (1600 nucleotides) and 23 *S* RNA (3200 nucleotides) is rapidly approaching completion by studies being carried out in a number of research laboratories. The known sequence for 16 *S* RNA is shown in Figure 21–4. Methylation of certain bases in the sequence of 16 *S* RNA (and also in the sequence of 23 *S* RNA) occurs while transcription is taking place. No methylation of 5 *S* RNA nucleotides occurs. Unlike 5 *S* RNA in which duplication of certain sequences occurs, no repeated sequences are found in 16 *S* and 23 *S* RNA. Although the rRNAs are single linear polynucleotides, they contain a number of dou-

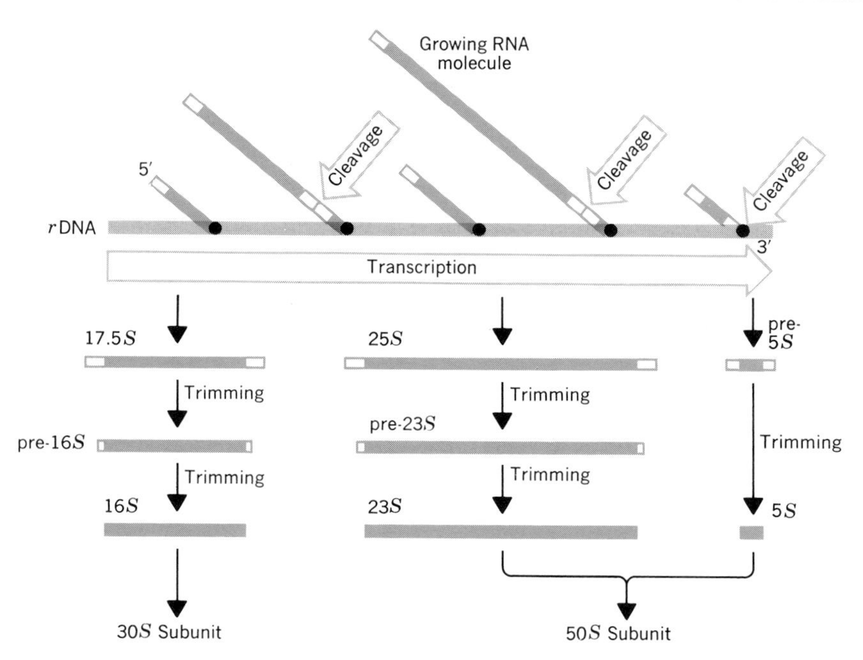

Figure 21–2
"Processing" of procaryotic rRNAs. (See text for explanation.) Each rectangular block represents an rRNA molecule. Colored portions correspond to regions of the rRNA that become final products and are incorporated into ribosomal subunits. Open portions are "spacer" segments transcribed from rDNA but eliminated during processing; these spacers represent about 20% of the original RNA transcript.

ble-helical regions that form ''hairpins'' stabilized by conventional, complementary base-pairing (Chapter 7). Several **palindromes** (base sequences reading the same from either the 5′ or 3′ ends) exist in 16 *S* RNA, and these may play a role in restricting the formation of the double-helical regions. In 16 *S* RNA, a 7-nucleotide segment of the chain at the 3′ end is believed to interact with mRNA, leading to its binding during the initiation of translation. The 5 *S* and 23 *S* RNAs interact with one another in the large subunit, and both appear to be involved in amino acyl-tRNA and peptidyl-tRNA binding during polypeptide chain elongation. Since several proteins of the small subunit interact with 23 *S* RNA, the latter may also have a role in subunit association.

It is generally believed that ribosomal RNA transcripts are not translated into protein (i.e., rRNAs cannot serve as messengers); however, ribosomal proteins are the products of a typical transcription-translation process.

Table 21–3
Reiteration of rRNA Genes in Various Cells

Cell or Tissue	Number of Genes Per Genome
Liver	750
HeLa cells	1100
Xenopus (toad)	900
Drosophila melanogaster	260
Tobacco leaves	1500
Saccharomyces cerevisiae	140
E. coli	5–10
Bacillus subtilis	9–10
B. megaterium	35–45

Protein Content. Nomura, Kurland, and others have established that the small procaryotic ribosomal subunit contains 21 proteins molecules (identified as S1, S2, S3 . . . S21) and the large subunit 34 proteins (L1, L2, L3 . . . L34). All the ribosomal proteins have been isolated and characterized (Table 21–4). The small subunit proteins range in molecular weight from 10,900 (S17) to 65,000 (S1); the large subunit proteins vary in molecular weight from 9600 (L34) to 31,500 (L2). Most of the ribosomal proteins are basic in nature, being rich in basic amino acids and having isoelectric points around pH 10 or higher. About 33 of the 55 ribosomal proteins have been fully sequenced (13 from the small and 20 from the large subunit). This, together with the RNA observations described earlier and the discussion of protein synthesis later in the chapter, suggests that the procaryotic ribosome may well be the first

1 10 20 30 40 50 60
E.coli UGCCUGGCGGCC-GUAGCGCGGUGGUCCCACCUGACCCCAUGCCGAACUCAGAAGUGAAACGCCG
1 10 20 30 40 50 60
P. fluorescens UGUUCUUUGACGAGUAGUCCCAUUGGAACACCUGAUCCCAUCCCGAACUCAGAGGUGAAACGAUG

1 10 20 30 40 50 60
Yeast G-GUUGCGGCCAUACCAUCU-AG-AAAGCACCGUUCUCCGUCCGAUAACCUGUAGUUAAGC-UGGUA
1 10 20 30 40 50 60
Chicken GC-CUACGGCCAUCCCACCCCUGUAACG--CCCGAUCUGGUCUGAUCUCGGA-AGCUAAGCAGGGUC
1 10 20 30 40 50 60
Human GU-CUACGGCCAUACCACCC-UG-AACGCGCCCGAUCUCGUCUGAUCUCGGA-AGCUAAGCAGGGUC

70 80 90 100 110 120
E.coli UAGC-GCCGAUGGUAGUGUGGGGU---CUCCCCAUGCG-AGAGUAGGGAACUGCCAGGCAUOH
70 80 90 100 110 120
P. fluorescens CAUC-GCCGAUGGUAGUGUGGGGU---UUCCCCAUGUCAAGAUCUCG--ACCAUAGAGCAUOH

70 80 90 100 110 120
Yeast AGAGCCUGACCGAGUACUCUACUGGGUG-ACCAUACGC----GAA-ACCUAGGUGCUGCAAUCUOH
70 80 90 100 110 120
Chicken -GGGCCUGGUU-AGUACUUGGAUGGGAG-ACCGCCUGG----GAAUACC-GGGUGCUGUAGGCUUOH
70 80 90 100 110 120
Human -GGGCCUGGUU-AGUACUUGGAUGGGAG-ACCGCCUGG----GAAUACC-GGGUGCUGUAGGCUU(U)OH

Figure 21–3
Comparison of the primary structures of *E. coli*, *Pseudomonas fluorescens*, yeast, chicken, and human ribosomal 5 *S* RNA. Homologous sequences within the procaryote and within the eucaryote RNAs are underlined.

organelle completely understood in terms of molecular structure and function. An exhaustive analysis of the primary structures of procaryotic ribosomal proteins in order to evaluate their degree of *homology* (see Chapter 4) indicates that these proteins did *not* have a common evolutionary ancestor. Homologies among them do not occur more often than would be expected on a random basis.

Whittman, Traut, Stöffler, Kurland, Nomura, and others have studied the relationships between the three rRNAs and the ribosomal proteins and have shown that some 28 proteins bind specifically and *directly* to the rRNAs (i.e., the **primary** binding proteins); 14 bind to 16 *S* RNA, 3 bind to 5 *S* RNA, and 11 bind to 23 *S* RNA (Table 21–5). Those proteins that do not bind directly to rRNA (i.e., the **secondary** binding proteins) presumably interact with the primary binding proteins in the assembled ribosome.

Best understood is the RNA-protein interaction in the 30 *S* subunit. Figure 21–4*a* shows the approximate regions within the primary structure with which the primary binding proteins associate; the relative positions of the secondary binding proteins are also indicated. In addition to RNA-protein interaction, there is considerable protein-protein in-

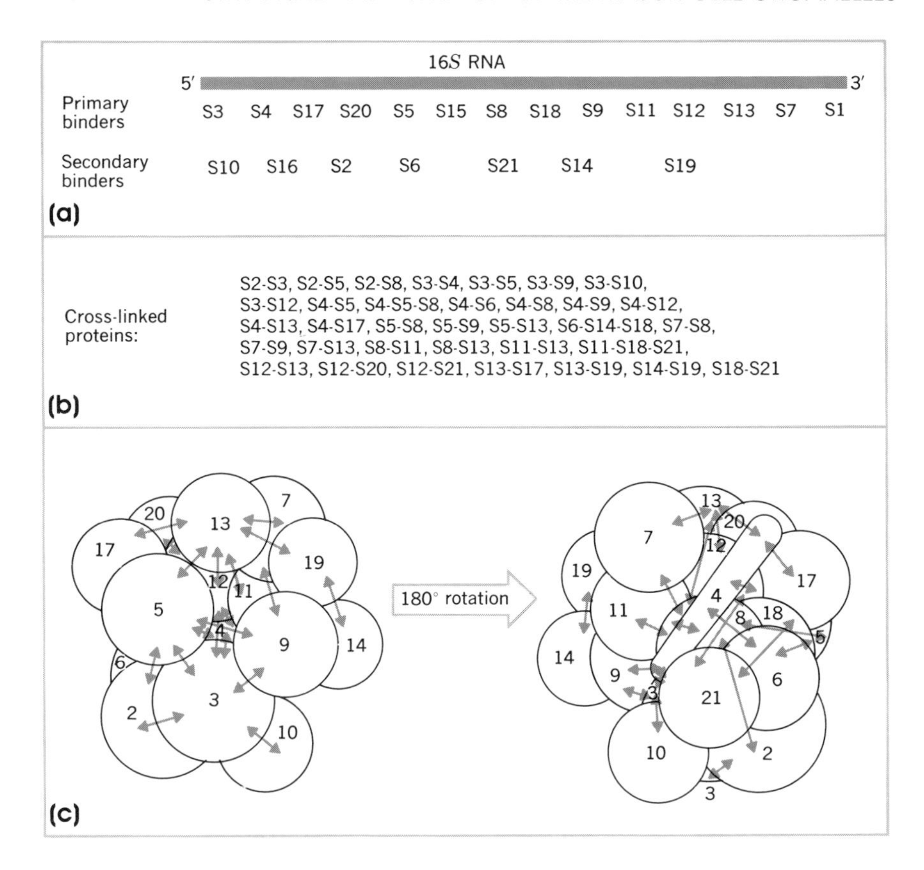

Figure 21–4
16 *S* RNA–protein and protein-protein interactions in the 30 *S* ribosomal subunit. (*a*) Relative positions of the primary and secondary binding proteins. (*b*) Cross-linked proteins. (*c*) Model of spatial relationships among the proteins, seen from opposite sides.

teraction, including interactions among the primary binding proteins and with the secondary binders. Figure 21–4*b* lists the known protein-protein interactions. This information, together with the known sizes of the proteins and other parameters, allows the construction of a scheme depicting the spatial relationships among the proteins that make up the small subunit (Fig. 21–4*c*). For simplicity, the 16 *S* RNA molecule is omitted from the scheme, but it is believed that the backbone of the polynucleotide winds its way among the proteins, with interactions occurring between hairpin turns of the RNA and surface residues of the protein molecules.

The association of the RNA and protein complements of the 50 *S* ribosomal subunits is not so completely understood as the 30 *S* subunits; however, the situation appears to be analogous in that certain proteins bind to specific regions of the 23 *S* and 5 *S* RNAs and protein-protein interactions are numerous.

Each ribosomal subunit contains *no more than one copy* of each of the S and L proteins, and not all ribosomes in a population contain all the proteins. The interface between the large and small subunits of the ribosome is an important functional area. Proteins S9, S11, S12, S15, S20, L26, and L27 are located at the interface; moreover, it appears that S20 and L26 are identical proteins and can associate with either subunit. Table 21–6 lists the supposed functions of many of the ribosomal proteins in the process of translation; this table should be examined again in connection with the description of the translation mechanism given later in the chapter.

Assembly of Procaryotic Ribosomes

Since all of the proteins and RNAs of the procaryotic ribosome subunits may be isolated, it is possible through recombination studies to examine the assembly process.

Nomura and others have shown that the assembly of individual subunits and their association to form functional ribosomes (i.e., ribosomes capable of translating mRNA into protein) occurs *spontaneously* in vitro when all the individual rRNAs and protein components are available. Thus the ribosome is capable of *self-assembly,* and this is believed to be the mechanism in situ. The assembly is promoted by the unique and complementary structures of the ribosomal protein and RNA molecules and proceeds through the formation of hydrogen bonds and hydrophobic interactions. There is *order* to the assembly in that certain proteins combine with the rRNAs prior to the addition of others (see Figure 21–4*a*). Cooperativity also exists, since addition of

Table 21–4
Proteins of the 30 *S* and 50 *S* Ribosome Subunits

Small (30 *S*) Subunit		Large (50 *S*) Subunit	
Protein	Molecular Weight	Protein	Molecular Weight
S1	65,000	L1	26,700
S2	28,300	L2	31,500
S3	28,200	L3	27,000
S4	26,700	L4	25,800
S5	19,600	L5	22,000
S6	15,600	L6	22,200
S7	22,700	L7	13,400
S8	15,500	L8	17,300
S9	16,200	L9	17,300
S10	12,400	L10	19,000
S11	15,500	L11	19,600
S12	17,200	L12	13,200
S13	14,900	L13	17,800
S14	14,000	L14	16,200
S15	12,500	L15	17,500
S16	11,700	L16	17,900
S17	10,900	L17	16,700
S18	12,200	L18	14,300
S19	13,100	L19	14,900
S20	12,000	L20	17,200
S21	12,200	L21	13,900
		L22	14,800
		L23	12,700
		L24	14,300
		L25	12,000
		L26	12,000
		L27	12,700
		L28	12,300
		L29	12,000
		L30	11,200
		L31	10,000
		L32	10,500
		L33	10,500
		L34	9,600

Table 21–5
Primary Binding Proteins of the 30 *S* Ribosomal Subunit

16 *S* RNA	5 *S* RNA	23 *S* RNA
S1	L5	L1
S3	L18	L2
S4	L25	L3
S5		L4
S7		L6
S8		L13
S9		L16
S11		L17
S12		L20
S13		L23
S15		L24
S17		
S18		
S20		

Table 21–6
Supposed Functions of the Ribosomal Proteins During Protein Synthesis

Function	Proteins Involved
mRNA binding	S1, S4, S18, S21
Initiation	S1, S6, S9, S11, S12, S13, S14, S18, S19
f-met-tRNA binding	S1, S2, S3, S5, S6, S10, S12, S13, S14, S19, S20, S21
Codon recognition	S3, S4, S5, S11, S12
Function of *A* and *P* sites	S1, S2, S3, S10, S14, S19, S20, S21, L1, L11, L13, L14, L15, L16, L27, L32
Aminoacyl-tRNA binding	S3, S6, S9, S11, S14, S18, S19, S21 L6, L16
GTPase center	S2, S5, S9, S11 L7, L11, L12
Termination	L11, L16

certain proteins to the growing subunit facilitates addition and binding of others.

No self-assembly takes place when *L* proteins are added to 16 *S* RNA or when *S* proteins are added to 5 *S* and 23 *S* RNA. However, it is interesting to note that RNA from the 30 *S* subunit of one procaryotic species will combine with the *S* proteins of another procaryote to form functional subunits. The same is true for 50 *S* subunit proteins and RNAs from different procaryotes. Assembly of *hybrid subunits* and formation of functional monomers from these occur in spite of the fact that ribosomal proteins and RNAs from different procaryotes have different primary structures. It is clear that their secondary and tertiary structures, which are very similar, are more important in guiding rRNA-protein interactions. Although some proteins from yeast, reticulocyte, and rat liver cell ribosomes can be replaced by *E. coli* ribosomal proteins, *hybrid monomers* formed from these procaryotic/eucaryotic subunits will not function in protein synthesis.

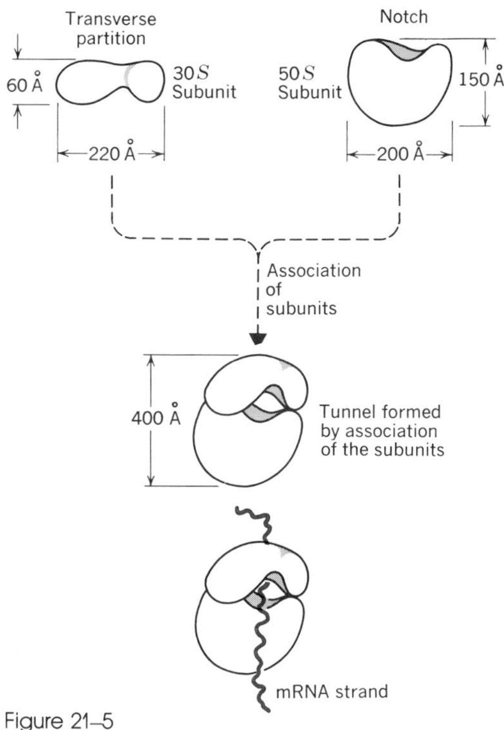

Figure 21–5
Model of the procaryote ribosome.

Model of the Procaryotic Ribosome

In spite of all that is known about the composition of ribosomes and the interaction of its molecular components, it is still difficult to propose a viable model of ribosome structure. Although electron microscopy has been immensely helpful in working out the gross structure and organization of other organelles, ribosomes are small enough to elude detailed analysis. Moreover, most techniques used to isolate and then prepare ribosomes for electron microscopy unavoidably alter the ribosome's native shape and organization.

Notwithstanding these limitations, several reasonable proposals can be made about the structure of the ribosome monomer and its subunits based upon the available electron-microscopic data, results of small-angle x-ray analysis, and of course, chemical studies. The 30 *S* subunit approximates an oblate ellipsoid of revolution having dimensions of 60 Å × 200 Å × 220 Å. A transverse partition or groove encircles the long axis of the subunit, dividing it into segments of one-third and two-thirds (Fig. 21–5*a*). The 50 *S* subunit is somewhat more spherical, having dimensions of 150 Å × 200 Å × 200 Å, and possesses a flattened or notched region on one surface (Fig. 21–5*a*). Association of the subunits to form the 70 *S* monomer (Fig. 21–5*b*) is accompanied by a deformation of the 30 *S* subunit at its transverse partition. The subunits are thereby joined in two regions on either side of a tunnel formed by the 50 *S* subunit notch and the 30 *S* subunit groove. The 70 *S* monomer has a maximum diameter of about 400 Å. There is considerable morphological and biochemical evidence supporting the idea that the tunnel in the monomer accommodates messenger RNA and the aminoacyl-tRNAs during protein synthesis (Fig. 21–5*c*). For example, (1) in many electron photomicrographs of polyribosomes, the thin mRNA strand seems to "disappear" into the ribosomes; (2) in vitro experiments have shown that when the synthetic messenger polyU is associated with the 70 *S* monomer, the polynucleotide is protected from ribonuclease attack over a length of about 70 to 120 nucleotides; and (3) transfer RNA is protected from cleavage by nucleases when associated with the ribosome. The observation that nascent (i.e., growing) polypeptides are also protected from proteolysis suggests that they, too, are located within the ribosome—in either the same or perhaps a separate channel (see later).

Genes for Ribosomal RNA and Protein

The genome of *E. coli* and other procaryotes consists of a single, long, circular DNA molecule tightly packed into the nuclear region of the cell. The *E. coli* chromosome is about 1300 nm long and appears to contain at least three separate regions coding for rRNA. Each region contains closely linked 5 *S*, 23 *S*, and 16 *S* rDNA genes. Since some 5 to 10 copies of each gene occur in the genome, more than one copy of each gene is likely present in each rDNA region.

Genes coding for ribosomal proteins are present in at least two separate regions of the *E. coli* chromosome. The same regions appear also to contain genes for *RNA polymerases,* some *transfer RNAs,* and the *elongation factors* required for protein biosynthesis (see later). The genes are distributed among at least four operons (Chapter 11), each operon containing genes for a dozen or more proteins (Fig. 21–6).

Eucaryotic Ribosomes

The cytoplasmic ribosomes of eucaryotic cells differ from those of procaryotes in both size and chemical composition (Table 21–2, Fig. 21–1). The monomer has a sedimentation coefficient of 80 *S* and is formed from 40 *S* and 60 *S* subunits. In addition, ribosomes occur in two states in the cytoplasm. They may be associated with cellular membranes such as those of the endoplasmic reticulum (i.e., "attached" ribosomes) and engaged in the synthesis of secretory or vesicle proteins, or they may be freely distributed in the cytosol and synthesize proteins retained within the cell. The functional differences between attached and free ribosomes will be pursued later, but let us turn first to a consideration of the chemical and morphological characteristics of eucaryotic ribosomes.

Figure 21–6
One of the ribosomal protein operons of the *E. coli* genome. Note that genes for both *S* and *L* proteins occur in the same operon. P, promotor gene.

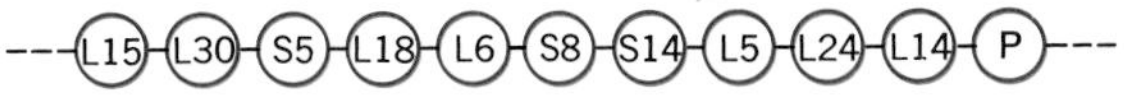

---(L15)-(L30)-(S5)-(L18)-(L6)-(S8)-(S14)-(L5)-(L24)-(L14)-(P)---

RNA Content. The small subunit of the eucaryotic ribosome contains one molecule of 18 *S* RNA (M.W. 0.7×10^6), while the large subunit contains 28 *S* (M.W. 1.7×10^6), 5 *S* (M.W. 3.2×10^4) and 5.8 *S* (M.W. 5×10^4) RNAs. Hence, in addition to molecular weight or size differences, a major distinction between the RNA complements of procaryotic and eucaryotic ribosomes is the presence of an additional molecule of RNA in the large subunit of eucaryotes. Of the four rRNAs, the 5.8 *S* molecule has only recently been discovered and characterized (5.8 *S* RNA has variously been referred to previously as lRNA, 7 *S* RNA and 5.5 *S* RNA). The 5.8 *S* RNA eluded earlier identification because of its intimate association with 28 *S* RNA in the ribosome.

18 *S*, 5.8 *S*, and 28 *S* rRNAs are the transcription products of closely linked genes in the chromosomes of the *nucleolar organizing region* (NOR) of the cell nucleus. Considerable redundancy exists since hundreds, perhaps even thousands, of copies of these rRNA genes are believed to be present (see Table 21–3). The genes for 5 *S* RNA are *not* present in the NOR but occur elsewhere in the nucleus. Consequently, unlike procaryotes in which the 5 *S* RNA genes are linked to the genes for other rRNAs, the 5 *S* RNA genes of eucaryotes occur separately in the nucleus. This difference, together with other observations to be noted later, support a contention that the 5 *S* rRNAs of procaryotic and eucaryotic ribosomes are not analogous; instead, it is the eucaryotic 5.8 *S* RNA that is the "counterpart" of procaryotic 5 *S* RNA.

Figure 21–7 depicts the transcription and post-transcriptional modification of eucaryotic rRNAs. It should be noted that 5 *S* RNA is a *primary* transcription product and is not the product of post-transcriptional trimming (another distinction from procaryotic 5 *S* RNA; see Fig. 21–2). Whereas the precursors of the procaryotic rRNAs are sequentially cleaved from the growing transcript, a *single,* high-molecular-weight transcript, 45 *S* RNA, containing the precursors of 18 *S*, 5.8 *S*, and 28 *S* rRNAs is produced in eucaryotes. About half of the 45 *S* RNA molecule is represented by *spacer* sequences that are trimmed during final processing. The first processing step (Fig. 21–7) divides the 45 *S* RNA into two parts; the larger of these (41 *S* RNA) eventually gives rise to 5.8 *S* and 28 *S* RNA, while 18 *S* RNA is derived from the smaller product. Not shown in Figure 21–7 but discussed later is the incorporation of the ribosomal proteins.

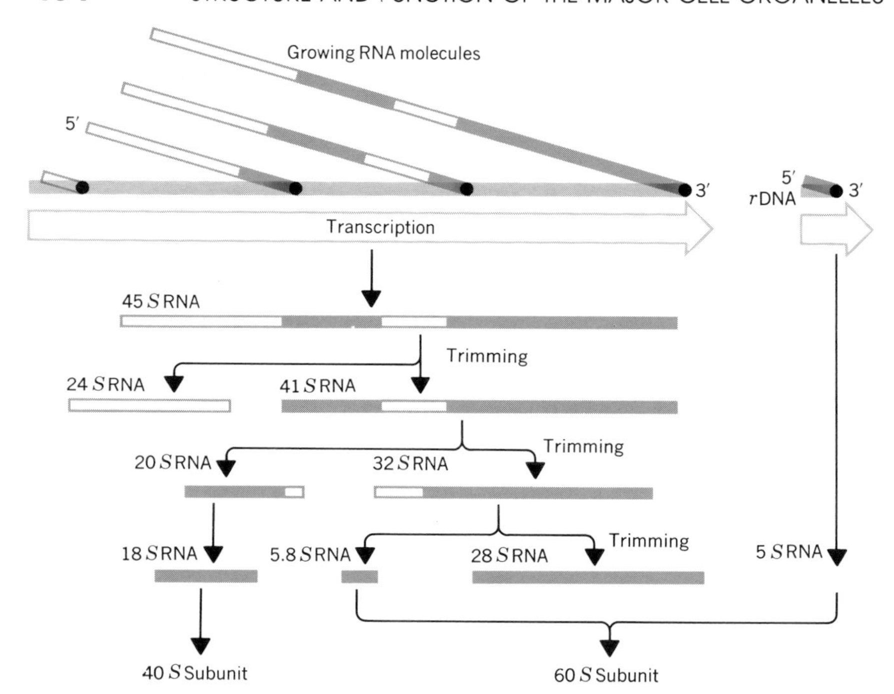

Figure 21–7
Processing of eucaryotic rRNAs. (See text for explanation.) Each rectangular block represents an rRNA molecule. Colored portions correspond to regions of the rRNA that become final products and are incorporated into the ribosomal subunits. Open portions are "spacer" segments transcribed from rDNA but eliminated during processing; these spacers represent about 50% of the original RNA transcript.

It is natural when comparing procaryotic and eucaryotic cells to look for structures or molecules of similar or even identical function. With regard to ribosome structure and composition, the analogy of 16 *S* RNA (of procaryotes) and 18 *S* RNA (of eucaryotes) is obvious, since both are parts of the small subunits of ribosomes and also have other features in common. Similarly, an analogy exists between 23 *S* RNA (of procaryotes) and 28 *S* RNA (of eucaryotes). But, what about procaryotic 5 *S* RNA and eucaryotic 5 *S* and 5.8 *S* RNA? Eucaryotic 5 *S* RNA is similar in size (about 120 nucleotides) to procaryotic 5 *S* RNA and also lacks modified nucleotides. In contrast, eucaryotic 5.8 *S* RNA is larger (about 150 nucleotides) and contains small numbers of modified nucleotides. Notwithstanding these differences, there is significant albeit not yet conclusive evidence for the contention that eucaryotic 5.8 *S* (not 5 *S*!) RNA is analogous to procaryotic 5 *S* RNA. For example, (1) the additional nucleotides of 5.8 *S* RNA occur for the most part in two sections at the 5′ and 3′ ends of the polynucleotide chain; the central portion reveals primary nucleotide sequences more closely related to procaryotic 5 *S* RNA than to eucaryotic 5 *S* RNA; (2) as noted earlier; procaryotic 5 *S* and eucaryotic 5.8 *S* RNAs are transcription products of closely linked rRNA genes and undergo post-transcriptional processing; and (3) there is evidence to support the proposal that 5.8 *S* RNA, like procaryotic 5 *S* RNA, interacts at the A site of the ribosome, whereas eucaryotic 5 *S* RNA interacts with tRNA during the initiation phase of protein synthesis.

Protein Content. Various studies have established that the small subunits of eucaryotic ribosomes contain 30 proteins (S1, S2, S3 etc.), and the large subunits, 40 proteins (L1, L2, L3, etc.) (Table 21–7). The proteins of eucaryotic ribosomes are not only more numerous but also have greater average molecular weights (Table 21–8). From a chemical standpoint, eucaryotic ribosomal proteins have similar general properties as those in procaryotes (e.g., rich in basic amino acids, high isoelectric point, etc.). Certain eucaryotic and procaryotic ribosomal proteins reveal homologous regions, and these homologous proteins appear also to be functionally similar.

Nucleolar Organizing Region. Eucaryotic cells contain several hundred copies of the genes encoding for rRNA. These genes are arranged in a tandem fashion on one or

Table 21–7
Ribosomal Proteins of Liver Cells

Small Subunit		Large Subunit	
Symbol	Molecular Weight	Symbol	Molecular Weight
S1	44,000	L1	38,600
S2	41,000	L2	32,400
S3	38,100	L3	53,000
S4	35,300	L4	53,700
S5	29,800	L5	45,800
S6	38,500	L6	48,500
S7	31,300	L7	38,300
S8[a]	32,500	L8	35,800
S9	27,200	L9	31,800
S10	27,300	L10	34,800
S11	26,300	L11	26,500
S12	38,500	L12	23,000
S13	21,200	L13	33,900
S14	24,900	L14	32,300
S15	25,300	L15	30,000
S16	20,500	L16	24,800
S17	22,500	L17	29,500
S18	21,500	L18	29,000
S19	20,500	L19	32,000
S20	10,100	L20[b]	27,000
S21	18,800	L21	28,000
S22	—	L22	—
S23	23,900	L23	23,300
S24	22,600	L24	24,000
S25	22,100	L25	23,700
S26	19,100	L26	25,600
S27	16,900	L27	21,900
S28	11,500	L28	22,700
S29	10,000	L29	24,000
S30	12,400	L30	21,700
		L31	20,800
		L32	20,700
		L33	22,000
		L34	14,500
		L35	19,700
		L36	18,100
		L37	16,800
		L38	10,500
		L39	10,000
		L40	25,500

[a]S8 is likely the same protein as L13.
[b]L20 may be a small subunit protein, designated S31.

Table 21–8
Average Molecular Weights of Procaryotic and Eucaryotic Ribosome Proteins

	Procaryote	Eucaryote
Small subunit	18,900	25,300
Large subunit	16,400	28,100

more chromosomes of the nucleus. The DNA sequences between successive rRNA genes cannot be transcribed and represent *spacer* DNA. The rRNA genes and the spacer segments are usually looped off the main axis of the chromosome and are referred to as the **nucleolar organizing region** (NOR). It is here that most of the rRNA is synthesized. The NOR coalesces with nuclear proteins and forms the visible bodies known as **nucleoli.** Most eucaryotic cells contain one or a few nucleoli, but certain egg cells are a striking exception. The oocytes of amphibians (e.g., the clawed toad, *Xenopus laevis*) are extremely large cells and are engaged in the synthesis of especially large quantities of cellular protein. These cells produce large numbers of ribosomes in order to provide the means to sustain such quantitative protein synthesis. Accordingly, it is not unusual to find hundreds or thousands of nucleoli (and NORs) in the nuclei of these cells. Such large numbers of nucleoli are the result of gene amplification—the differential replication of the rRNA genes of the genome. The ribosomes produced in the oocyte serve its needs for protein synthesis from the period prior to fertilization through the first few weeks of embryonic development.

By gently dispersing nuclear fractions isolated from oocytes of the amphibian *Triturus viridescens* and "spreading" the material on grids, O. L. Miller and B. R. Beatty in 1969 were able to obtain photomicrographs of transcription *in progress*. Since then, the same approach has been extended by a number of other investigators to mammalian oocytes and to spermatocytes and embryo cells from various organisms. The visualization of transcriptional activity is achieved most easily with spread *nucleoli* because of the high degree of rDNA gene amplification (Fig. 21–8). The tandem rDNA genes are serially transcribed by RNA polymerases to produce 45 *S* rRNA. The rRNA (apparently complexed with protein) appears as a series of fibrils of varying length extending radially from an axial, linear DNA fiber (Fig. 21–9). These feather-shaped or "Christmas tree"

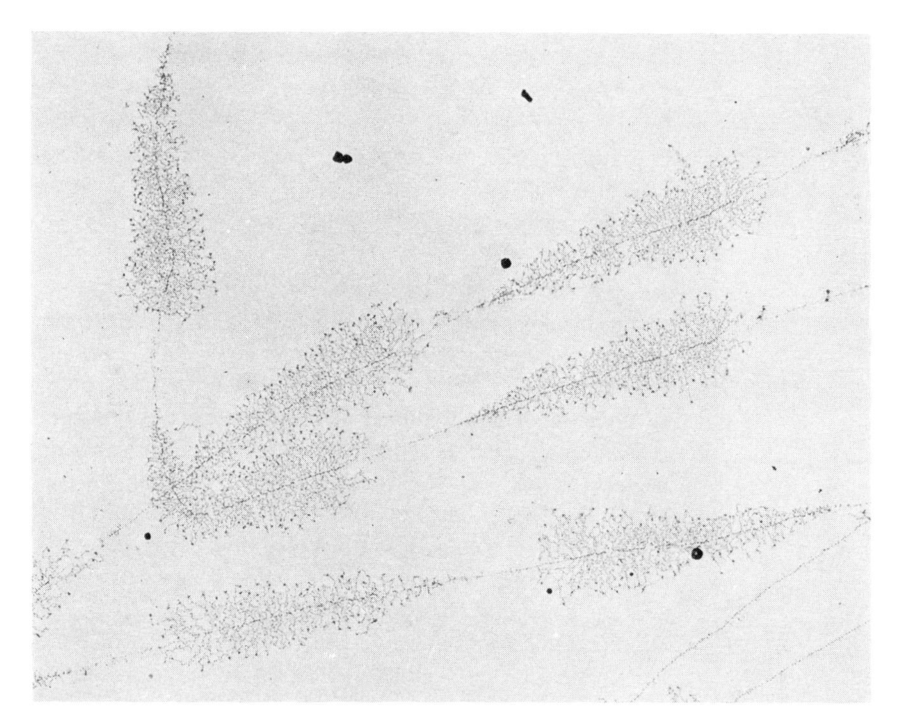

Figure 21–8
Visualization of transcription of rDNA genes. The thin axial DNA fiber is being transcribed simultaneously by a number of RNA polymerase enzymes (small black dots; see also Fig. 21–9). The transcripts (ribonucleoprotein complexes) appear as fine fibrils extending radially away from the DNA axis. Magnification, 18,000 X. (Electron photomicrograph courtesy of Dr. O. L. Miller, from O. L. Miller, and B. R. Beatty, *Science 164,* 956, 1969. Copyright 1969 by the American Association for the Advancement of Science.)

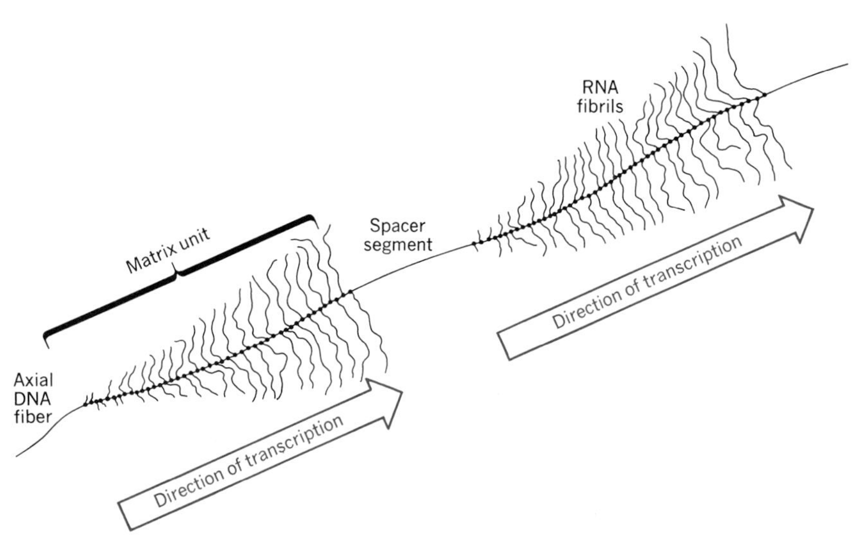

regions are called **matrix units.** The spaces between successive matrix units are nontranscribed *spacer* segments. The ribonucleoprotein (RNP) fibrils are seen to be in various stages of completion. The short fibrils near the tip of each feather are RNA molecules whose synthesis has only just begun, while the longest fibrils represent RNA molecules whose synthesis is almost complete. Hence, *the direction of rDNA transcription is apparent in the photomicrograph.*

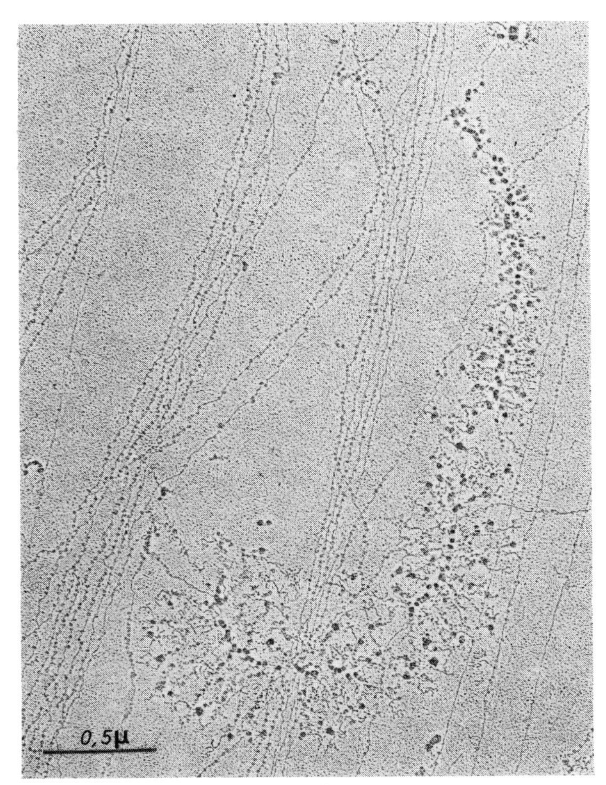

Transcriptionally inactive DNA

Transcriptionally active DNA

Nucleosomes

RNA transcripts (fibrils)

RNA polymerase

"Terminal knobs"

Direction of transcription

Bare DNA fiber

Figure 21–9
High magnification electron photomicrograph of matrix unit and neighboring non-transcribed DNA fibers. See legend of Figure 21–8 and text for explanation. (Photo courtesy of Dr. F. Puvion-Dutilleul.)

In high magnification views (Fig. 21–9) even the *RNA polymerase* enzyme molecules carrying out the transcription of the DNA are visible along the axial DNA fiber.

Success in visualizing transcription has not been restricted to nucleolar genes. Almost identical results have been obtained with nonnucleolar chromatin. Here, however, the RNA transcripts represent messenger RNA.

Dispersed and spread nuclear fractions contain nontranscribing DNA as well as matrix units (Fig. 21–9). The succession of *nucleosomes* (Chapter 20) reveals itself as a series of beadlike structures along the DNA fiber. Regions in which DNA is undergoing replication (called *replicons*) can also be seen (Fig. 21–10). S. L. McKnight and O. L. Miller have shown that DNA of homologous "daughter" fibers of the replicon also occurs as chains of nucleosomes, suggesting that replication may not require dissociation of nucleosomes or that nucleosomes are almost immediately reformed. Transcriptional activity can be identified within a replicon (Fig. 21–11), indicating that the newly synthesized DNA is almost immediately available for transcription. The growing RNA fibrils are seen in homologous regions of *both* chromatid arms of the replicon.

Assembly of Eucaryotic Ribosomes. The assembly of eucaryotic ribosomes is more complex than that of procaryotes; the principal stages of the process are outlined in Figure 21–12. Transcribed 45 *S* RNA combines with proteins in the nucleolus to form ribonucleoprotein complexes (RNP). However, not all the protein molecules of the complex become a part of the completed ribosomal subunit.

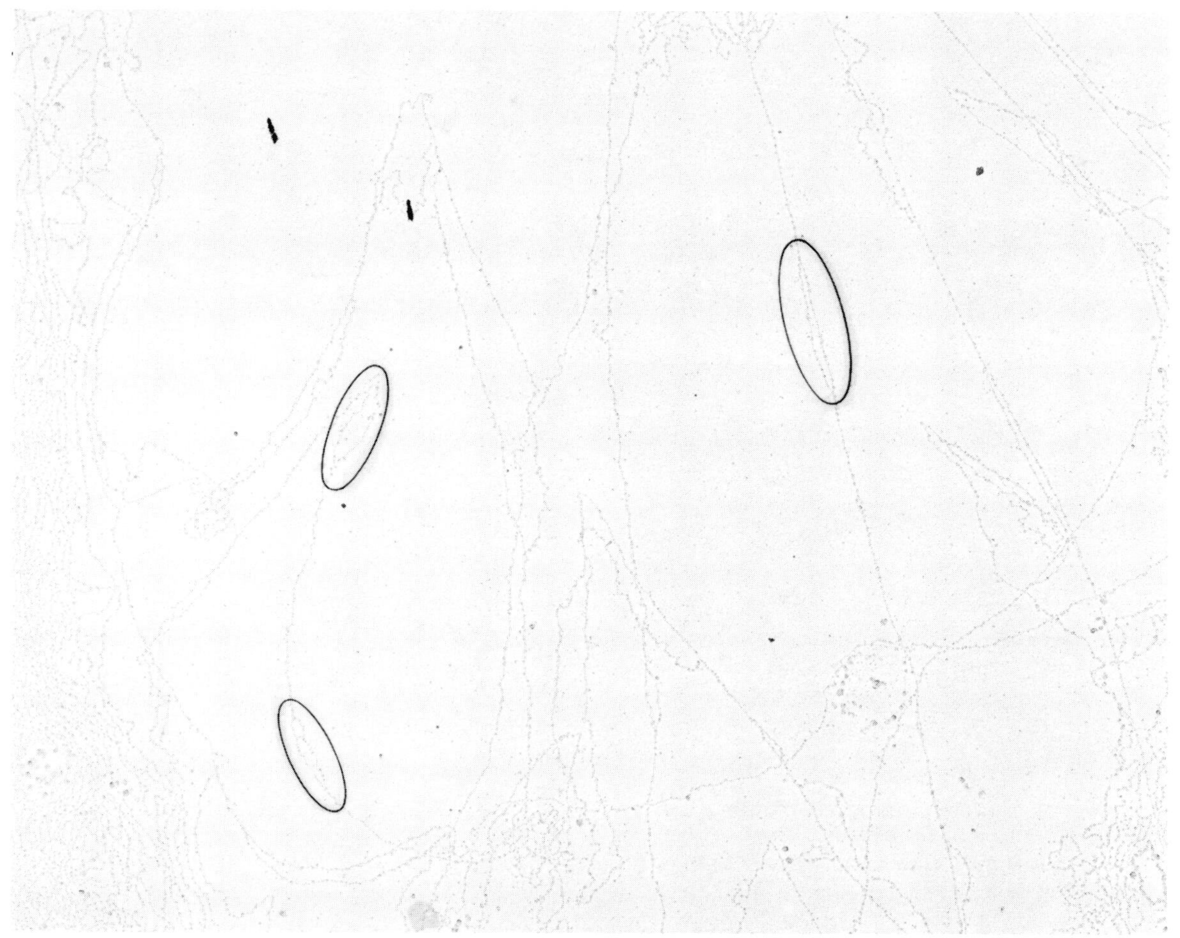

Figure 21–10
DNA fibers from dispersed nuclei exhibiting regions of replication (i.e., *replicons*). Three replicons are specifically identified in the figure. Magnification, 26,000 X. (Electron photomicrograph courtesy of Dr. S. L. McKnight.)

Instead, certain proteins are released as RNA processing ensues; these "nucleolar proteins" return to a nucleolar pool and are reutilized. Those proteins that are retained during processing and become part of the completed subunits are, of course, legitimately called "ribosomal proteins." Enzymatic cleavage of the RNP complex during processing produces three classes of fragments. One fragment contains spacer RNA and nucleolar proteins. (It should be noted that the spacer RNA is produced by transcription of rDNA and *not* the spacer DNA between genes.) The spacer RNA is hydrolyzed, and the free nucleolar proteins return to the pool. A second RNP fragment contains a complex of 18 *S* RNA and certain ribosomal proteins that give rise to 40 *S* ribosome subunits in the cytoplasm. The third RNP fragment, which contains 28 *S* and 5.8 *S* RNA and ribosomal proteins, combines with 5 *S* RNA transcribed from extranucleolar rRNA genes, and the complex exits the nucleus to give rise to 60 *S* subunits in the cytoplasm. Like the genes for 45 *S* RNA, the extranucleolar 5 *S* RNA genes occur in multiple tandem copies. Among the various pro-

Figure 21–11
Visualization of homologous transcriptional activity in the two daughter chromatids of a replicon. Magnifications: upper photomicrograph, 15,000 X; lower photomicrograph, 27,000 X. (Courtesy of Dr. S. L. McKnight.)

teins synthesized in the cytoplasm using ribosome subunits derived from the nucleus are the ribosomal proteins themselves. These apparently reenter the nucleus for incorporation into new RNP complexes.

Model of Eucaryotic Ribosomes

In spite of the differences in overall sizes (as manifested in the greater molecular weights, sedimentation constants, sizes, and numbers of rRNAs and proteins), the ribosomes of eucaryotes are remarkably similar in morphology to those of procaryotes.

The 40 *S* subunit approximates a slightly flattened ellipsoid of revolution, having dimensions of 115 Å × 140 Å × 230 Å. As in 30 *S* subunits of procaryote ribosomes, the 40 *S* eucaryote subunit is divided into segments of one-third and two-thirds by a transverse groove (Fig. 21–13). The 60 *S* subunits is generally rounder in shape, having a diameter of about 200 Å. One side of the large subunit is somewhat flattened, with a notch that becomes confluent with the transverse groove of the small subunit during the formation of the monomer (Fig. 21–13). The resulting channel through the ribosome is believed to accommodate the mRNA strand during translation.

Free and Attached Ribosomes

The cytoplasmic ribosomes of eucaryotic cells can be divided into two classes: (1) **attached** ribosomes and (2) **free** ribosomes (Fig. 21–14). Attached ribosomes are ribosomes

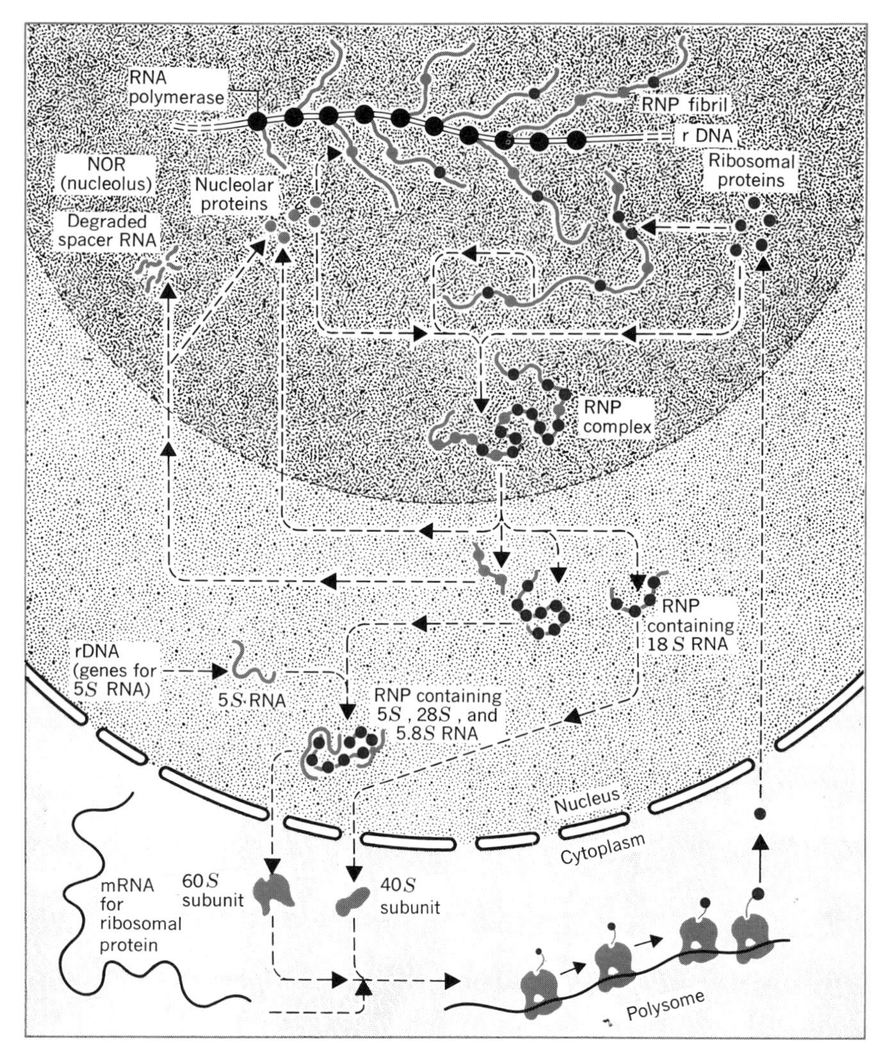

Figure 21–12
Synthesis and assembly of the components of eucaryotic ribosomes.

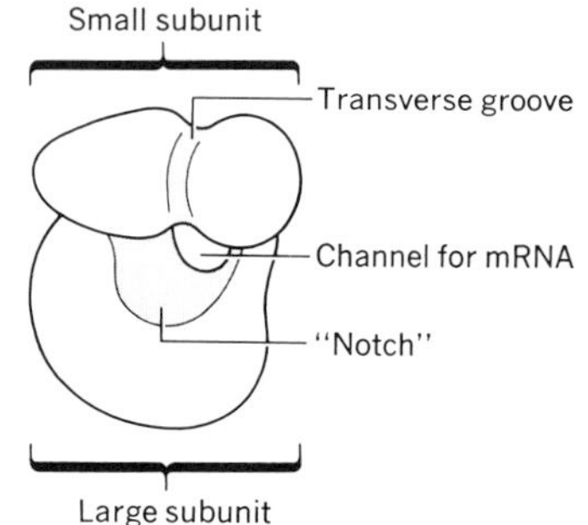

Figure 21–13
Model of the eucaryotic cytoplasmic ribosome.

associated with intracellular membranes, primarily the endoplasmic reticulum, whereas free ribosomes are distributed through the hyaloplasm or cytosol. Although all animal and plant cells contain both attached and free ribosomes, the proportion of each varies from one tissue to another and can be caused to shift within a single tissue in response to the administration of certain substances, notably hormones and growth factors.

Membranes of the endoplasmic reticulum (ER) that contain attached ribosomes constitute what is called "rough" ER (or RER), while intracellular membranes that are devoid of ribosomes are called "smooth" ER (SER) (see Chapter 1). The ribosomes of RER are attached to the hyaloplasmic surface of membranes (as opposed to the lumenal or cisternal surface). Attachment to the membrane occurs through the large (60 *S*) subunit.

For many years, there has been considerable controversy about the functions of attached and free ribosomes. The currently accepted view suggests that proteins destined to be secreted from the cell or to be incorporated into such intracellular bodies as lysosomes and peroxisomes (which may or may not release their contents to the cell exterior)

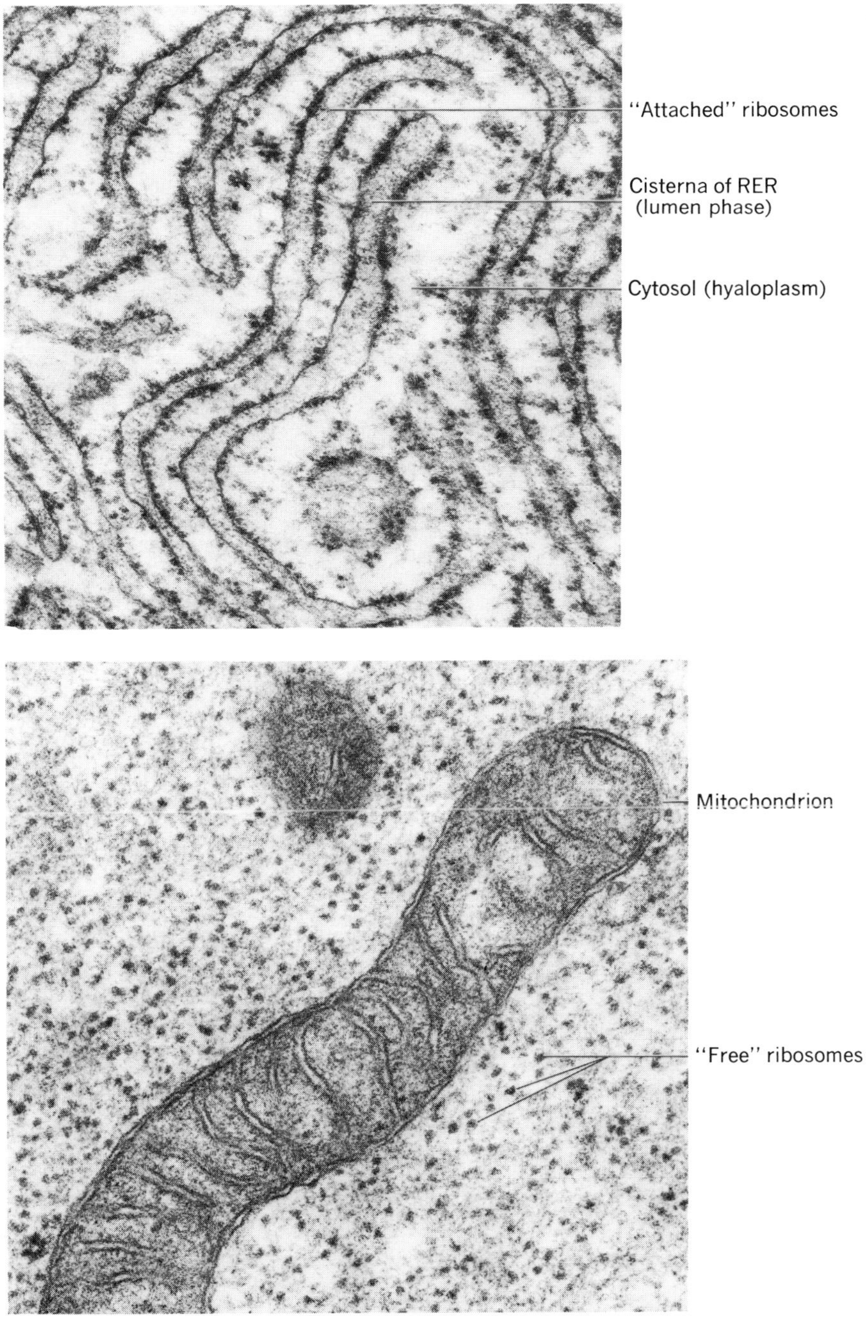

Figure 21–14
"Attached" (above) and "free" (below) ribosomes. (Electron photomicrographs courtesy of R. Chao.)

are synthesized on attached ribosomes, whereas most (but not all) proteins destined for the cytosol are synthesized on free ribosomes. For example, many of the proteins circulating in the blood plasma are derived via secretion by the liver, and these plasma proteins are known to be synthesized exclusively by the attached ribosomes of the liver cells.

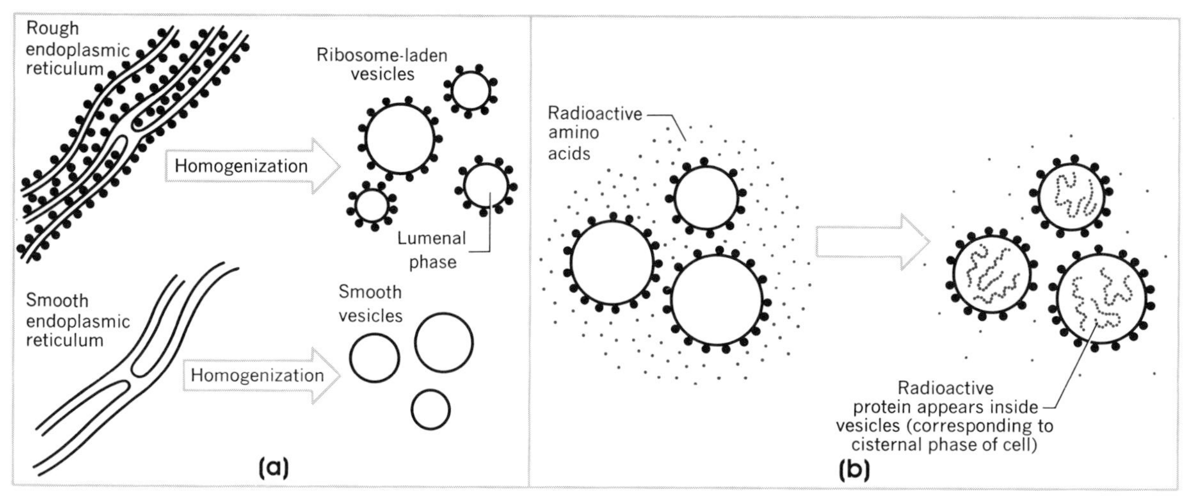

Figure 21–15
(*a*) Production of microsomal vesicles during homogenization of endoplasmic reticulum. (*b*) Vectorial synthesis of protein by ribosome-laden vesicles.

Thyroglobulin, which is secreted by the thyroid gland, is also synthesized by attached ribosomes. So too are the milk proteins produced by mammary gland cells. There is also a good deal of evidence indicating that some of the proteins that make up the membranes of intracellular organelles may be synthesized on ribosomes attached to the endoplasmic reticulum. Included in this category would be integral and extrinsic proteins asymmetrically distributed in the exterior half of these membranes (see Chapter 15).

When cells are disrupted, the sheets of endoplasmic reticulum are broken into small vesicular fragments (Fig. 21–15*a*), which may be isolated by centrifugation with the *microsomal phase*. As noted in Chapter 12, this phase is quite heterogeneous and contains a variety of small particles in addition to fragmented endoplasmic reticulum. Fragmentation of the endoplasmic reticulum produces two kinds of vesicles; fragments of RER form vesicles whose outer surface is studded with ribosomes, while SER vesicles are free of ribosomes. The volume within the vesicle corresponds to the lumenal or cisternal phase of the cell.

In a series of elegant experiments, D. Sabatini and C. M. Redman examined protein synthesis in vitro by ribosome-laden vesicles. They were able to demonstrate that radioactive amino acids incorporated into protein by the ribosomes did not appear in the suspending medium but were recovered instead within the vesicles (Fig. 21–15*b*). These results argue strongly in favor of the proposal that attached ribosomes synthesize proteins for secretion, since the interior of the vesicles corresponds to the cisternal phase of the cell. Sabatini and Redman called this directional synthesis of protein **vectorial synthesis.** Once the protein is released into the cisternae, it is transported to the Golgi apparatus for packaging (Chapters 18 and 19).

Although it is widely accepted that proteins destined to be secreted from the cell are synthesized on membrane-bound ribosomes, there are several pieces of evidence that indicate that membrane-bound ribosomes may have other functions as well. For example, J. R. Tata and others, working with muscle and nerve tissue, have observed the synthesis by RER ribosomes of small quantities of proteins for intracellular utilization. The rapid proliferation of RER during periods of active cell growth also suggest a nonsecretory function for attached ribosomes. The enzyme *serine dehydrogenase* has been shown to be specifically synthesized by attached ribosomes in liver cells, and yet this is an intracellular enzyme. Preliminary evidence suggests that some mitochondrial enzymes may be differentially synthesized by RER ribosomes.

Several independent lines of investigation support the idea that attached and free ribosomes may be structurally

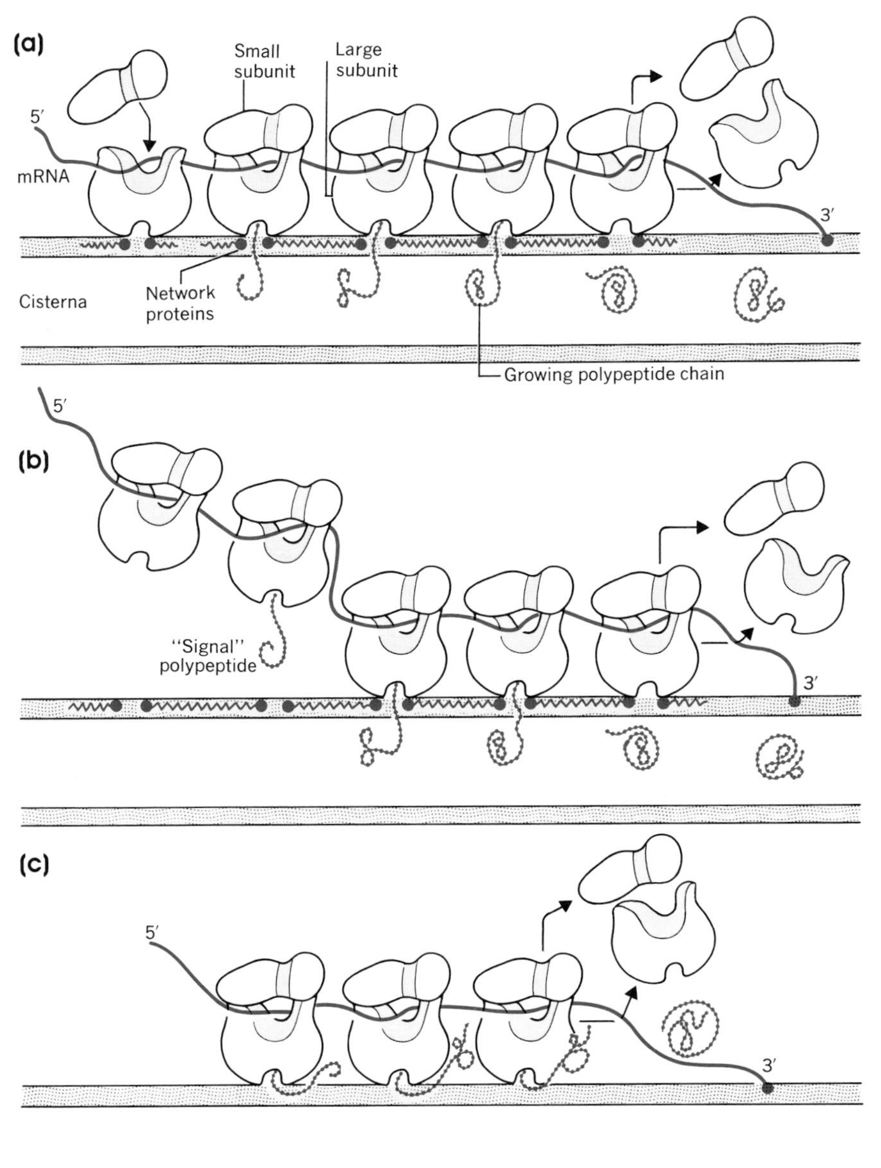

Figure 21–16
Models for the synthesis of protein by attached ribosomes. (See text for details.)

different. The **reticulocyte** (a developmental form produced during the differentiation of the mature red blood cell; see later) is primarily engaged in the synthesis of hemoglobin molecules, but small amounts of other cellular proteins are also produced. The globin chains for hemoglobin are synthesized on free ribosomes, (i.e., there is no endoplasmic reticulum in reticulocytes), while other cell proteins appear to be synthesized on ribosomes attached to the internal surface of the plasma membrane. These two ribosome populations have been isolated and separately studied; discontinuous electrophoretic analysis of dissociated ribosomes reveal several important differences in their constituent proteins.

Working with liver tissue, Sabatini and his colleagues

have shown that ER membranes to which ribosomes are attached contain two proteins absent in ribosome-free ER membranes. Biochemical studies coupled with freeze-fracture electron microscopy suggest that these membrane proteins form an interconnecting network of binding sites for ribosomes. The network is characteristic of RER but absent in SER. Figure 21–16 depicts three contemporary models for the translation of mRNA by attached ribosomes. According to one of these models (Fig. 21–16*a*), the attachment of the ribosome to the membrane *precedes* the initiation of polypeptide chain synthesis. An alternative mechanism, for which there is growing experimental evidence involves synthesis of a "signal" region of the polypeptide before the ribosome attaches to the membrane. The signal amino acid sequence interacts with membrane proteins, directing the association of the large subunit with receptor sites in the membrane. In both Figures 21–16*a* and *b* the protein is discharged into the cisterna of the ER. Figure 21–16*c* depicts the model that accounts for the synthesis of cytoplasmic proteins by attached ribosomes. In the latter case, the protein is not discharged through the membrane. Protein is presumed to exit the ribosome through a channel in the large subunit. In all the models of Figure 21–16, mRNA is shown attached to the membrane at its 3′ end.

Ribosomes of Organelles

The mitochondria and chloroplasts of eucaryotic cells contain their own DNA and protein-synthesizing apparatus. Although reports of the presence of DNA in mitochondria and chloroplasts appeared periodically in the scientific literature since the 1920s, such an astounding proposition was not generally accepted until the 1960s, when H. Ris and M. Nass independently verified the existence of DNA fibrils in animal and plant cells using electron-microscopic methods originally developed to visualize DNA in procaryotic cells (Fig. 21–17).

Little attention was given to the possibility that mitochondria and chloroplasts might contain ribosomes and other elements involved in protein synthesis until the presence of DNA in these organelles was established. With such an impetus provided, it did not take long before ribosomes were indeed identified in and isolated from chloroplasts and mitochondria.

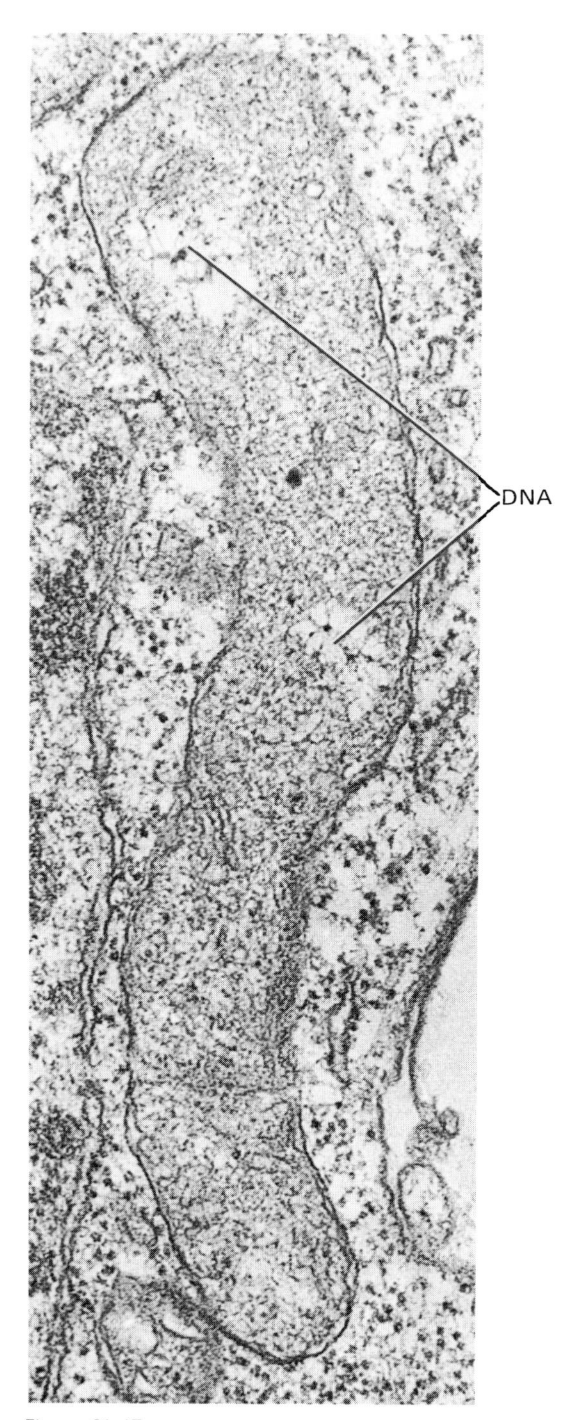

Figure 21–17
DNA fibrils in mitochondria (Electron photomicrograph courtesy of Dr. E. G. Pollock.)

Table 21–9
Properties of Chloroplast and Mitochondrial Ribosomes

	Sedimentation Coefficient	
	Particle	RNAs
Chloroplast ribosomes		
Monomer	70 *S*	
Large subunit	50 *S*	5 *S* and 23 *S*
Small subunit	33 *S*	16 *S*
Mitochondrial ribosomes		
Animal cells		
Monomer	50–60 *S*	
Large Subunit	40–45 *S*	16–18 *S*
Small Subunit	30–35 *S*	12–13 *S*
Yeast, fungi, and protists		
Monomer	70–80 *S*	
Large subunit	50–55 *S*	21–24 *S*
Small subunit	32–38 *S*	14–16 *S*
Higher plant cells		
Monomer	70–80 *S*	
Large subunit	50–60 *S*	>23 *S*
Small subunit	40–44 *S*	>16 *S*

Chloroplast Ribosomes

Chloroplast ribosomes have a sedimentation coefficient of 70 *S* and consist of 50 *S* and 33 *S* subunits. In this respect, chloroplast ribosomes are similar to those of procaryotic cells but distinct from eucaryotic cytosol ribosomes. The large subunit of the chloroplast ribosome contains 5 *S* and 23 *S* RNAs and the small subunit 16 *S* RNA (Table 21–9).

Mitochondrial Ribosomes

Unlike chloroplast ribosomes, which are similar in all groups of organisms studied, mitochondrial ribosomes are quite heterogeneous. With respect to their disposition within the mitochondrion, the ribosomes occur either free in the *matrix* of the organelle or are associated with the cristael membranes. (Ribosomes attached to the cytosol side of the outer mitochondrial membrane are cytoplasmic ribosomes engaged in vectorial synthesis of certain intramitochondrial proteins; see below).

Although mitochondrial ribosomes from all sources studied consist of two subunits, there is considerable variation in the sizes of the subunits and the monomers formed from them (Table 21–9). Mitochondrial ribosomes of yeast, fungi, protists, and higher plants are characterized by a sedimentation coefficient of 70 to 80 *S*, whereas those of animal cells have a sedimentation coefficient of 50 to 60 *S* and are therefore unusually small. Unlike the ribosomes of chloroplasts, procaryotic cells, and eucaryotic hyaloplasm, mitochondrial ribosomes contain only *two* species of rRNA.

Protein Synthesis in Chloroplasts and Mitochondria

The amounts of DNA present in chloroplasts and mitochondria is only about 10 to 15% of that necessary to encode for the hundreds of different proteins present in these organelles. Therefore, most of the proteins of chloroplasts and mitochondria are the products of genetic information in the cell nucleus. Experimentally, ribosomal RNAs of chloroplasts and mitochondria can be shown to hybridize with their organellar DNA, and it is generally agreed that these RNAs are synthesized in the organelles. The origins of only a small number of chloroplast and mitochondrial proteins have been established to date, but the results are interesting and perhaps also surprising. Most, if not all, of the proteins that make up the organelles' ribosomes are synthesized in the cytoplasm of the cell and apparently make their way to the organelle for assembly (along with rRNA) into ribosomes. Entry into the organelle may be via vectorial discharge. The synthesis of several organelle enzymes, including *ribulosediphosphate carboxylase* (of chloroplasts) and *cytochrome oxidase,* and an *ATPase* (of mitochondria) appears to be a ''joint operation'' of the organelle and the cytoplasm. For example, of the seven polypeptides that make up cytochrome oxidase, four are synthesized on cytosol ribosomes and three on mitochondrial ribosomes. There is some evidence that a few of the chloroplast membrane proteins are synthesized on chloroplast ribosomes.

Much remains to be learned about the total function of organelle ribosomes; however, it is clear that while most chloroplast and mitochondrial proteins are the products of the nucleocytoplasmic protein-synthesizing machinery of the cell, a small number of proteins and portions of certain multisubunit enzymes are the products of genetic material intrinsic to the organelle and the result of translation on organelle ribosomes.

Mechanism of Protein Synthesis

Thousands of experiments carried out during the past 20 years involving hundreds of scientists have slowly and painstakingly revealed the intricate details of the mechanism by which proteins are synthesized in procaryotic and eucaryotic cells. The overwhelming majority of these studies were conducted using two particular kinds of cells, namely the bacterium *E. coli* and the mammalian *reticulocyte*. Bacterial cells represent highly desirable sources for studying protein synthesis, since the cells themselves are readily obtained and conveniently cultured in the laboratory. Moreover, since the ribosomes of bacteria are not attached to intracellular membranes, they are readily isolated from disrupted cells.

The mammalian immature red blood cell or **reticulocyte** has been the overwhelming favorite among scientists studying protein synthesis in eucaryotic cells for a number of very important reasons; these are best appreciated by briefly considering the origin and features of this unusual cell.

In mammals, red blood cells are produced in the bone marrow and pass through a number of characteristic developmental stages before the mature red cell or **erythrocyte** enters the circulating blood (see Chapter 23). In its early stages of development, the red blood cell possesses most of the structural elements that characterize typical animal cells (nucleus, mitochondria, lysosomes, endoplasmic reticulum, etc.), but nearly all of these structures are degraded and lost by the reticulocyte stage. The reticulocyte has only a small number of ribosomes, soluble enzymes, and other soluble constituents with which it completes the synthesis of hemoglobin begun at earlier stages. Indeed, hemoglobin accounts for nearly all the protein being synthesized by the cell, and in this respect the reticulocyte is a more desirable source for studying protein synthesis than bacteria which synthesize many different proteins. Because the reticulocyte contains no organelles other than ribosomes, the latter are readily isolated following lysis of the cell. The cell is called a reticulocyte because its cytoplasm displays a fine reticulum when stained with certain basic dyes (such as methylene blue), the reticulum being formed in part by precipitation of the residual cytoplasmic RNA and ribosomes.

In the bone marrow, the reticulocyte is transformed to the mature erythrocyte by the loss of its remaining ribosomes and the termination or completion of hemoglobin synthesis. The erythrocyte is then released from the bone marrow and enters the circulating blood.

The separation of reticulocytes from other bone marrow cells in order to follow protein (i.e., hemoglobin) synthesis is no simple matter, and for this reason, marrow tissue is rarely used as the source of reticulocytes. Instead, reticulocytes are obtained using the following procedure. The experimental animal (usually a rabbit or rat) is rendered severly anemic either by removing a large portion of its blood or by introducing a hemolytic agent (typically, phenylhydrazine) into its bloodstream. The hemolytic agent quickly produces an extensive intravascular hemolysis. In either instance, the resulting anemia is followed within several days by a marked increase in red blood cell production in the bone marrow and the premature release of reticulocytes into the circulating blood. The bloodstream becomes literally flooded with reticulocytes, which may account for nearly all circulating red blood cells in severely anemic animals. Thus, large numbers of reticulocytes can easily be obtained by removing a blood sample from these anemic animals, and it is therefore unnecessary to employ bone marrow tissue itself.

Protein synthesis can be studied using either intact cells or a ''cell-free'' system. That is, under appropriate experimental conditions, not only do *whole* cells incorporate amino acids into new protein but disrupted cells or simply isolated ribosomes supplemented with all the requisite soluble components (e.g., amino acids, tRNA, mRNA, enzymes, cofactors, etc.) also carry out protein synthesis. In recent years, other cells including HeLa cells, liver cells, and yeast cells have been employed to study protein synthesis, and these studies have confirmed most of the observations originally made using *E. coli* and reticulocytes. Some important differences do exist in the mechanism of protein synthesis in procaryotic and eucaryotic cells, but the overall process is fundamentally the same. Those differences that do exist will be noted below as we consider the details of this process.

Protein synthesis involves a number of distinct and sequential steps including (1) **activation** of amino acids, (2) formation of an **initiation complex** between messenger RNA and the ribosome subunits, (3) polypeptide chain **initiation,** (4) chain **elongation,** (5) chain termination and **release** of the completed polypeptide, and (6) **dissociation** of the messenger RNA-ribosome complex. Before we consider each of these stages individually, it is worthwhile for perspective to discuss first the pioneering studies of H. M. Dintzis, who in the early 1960s established the *linearity* of chain elongation and determined the *direction* in which polypeptide chain assembly takes place. His work not only provided an insight into the complexity of the process yet to be revealed, but his brilliant selection of methodology served as a model and guide for many of the studies subsequently carried out by other scientists investigating the mechanics of protein synthesis.

Linearity and Direction of Polypeptide Chain Assembly—the Experiments of H. M. Dintzis

Dintzis' experiments were carried out to determine whether assembly of a polypeptide chain (1) began at one terminus and proceeded in order toward the other (and if so, which end was synthesized first), (2) began near the middle and then grew toward both termini simultaneously, or (3) occurred simultaneously at several (or many) points along the polypeptide chain with the eventual linking up of all segments. For his studies, Dintzis used reticulocytes isolated from the blood of rabbits made anemic by the injection of phenylhydrazine.

To follow hemoglobin synthesis in these cells, Dintzis used radioisotopically labeled leucine. Leucine was selected for the following reason: this particular amino acid is more-or-less uniformly distributed through the primary structure of human alpha and beta globin chains (see Chapter 4), and although the primary structure of rabbit hemoglobin had not been worked out at the time, there was good reason to suspect that it would be quite similar to that of human hemoglobin. This supposition was subsequently verified.

As a starting point, Dintzis proposed a model for protein synthesis according to which the polypeptide chains are assembled in sequence beginning at one end. Therefore, at any chosen instant in time, say t_0, we would expect to find polypeptide chains of various lengths (i.e., varying degrees of completion) attached to the mRNA-ribosome complexes in the cell. Such partially completed polypeptide chains are called **nascent** chains. If at t_0 cells are briefly incubated in a medium that permits continued nascent chain growth, and if radioactive amino acids are included in that medium, then we would expect that at time t_1, each nascent chain would have increased in length by adding a section containing the radioactive label. (This type of experiment is called a "pulse label" experiment.) If the time period in which the cells are incubated in labeled medium is sufficiently short in comparison to the time required for synthesis of a whole polypeptide, then only a few chains will be completed in this interval and released from the ribosome (i.e., those nascent polypeptide chains that were already near completion at the time the incubation was begun). The remaining chains will still be attached to their respective mRNA-ribosome complexes. For short incubations, the radioactivity present in completed, and released chains should therefore be confined to those regions of the chain synthesized *last*.

It should be clear from the preceding discussion that the longer the time interval of incubation in labeled medium (i.e., from t_0 to t_2 or t_3, etc.), the more chains that will be completed and released and therefore the further "back" along the polypeptide's primary structure will radioactive label be found. That is, for any period of incubation equal to or shorter than the time required for complete assembly of whole polypeptide chains, one should find a "gradient of radioactivity" among those chains that were completed in that time interval, such that the highest radioactivity is toward the end of the chain synthesized last and the lowest radioactivity is toward the end synthesized first. Figure 21–18 depicts this concept in diagrammatic terms.

If incubation in the presence of radioactive amino acids is carried out for a period of time greater than that required for the assembly of a whole chain, then not only will all of those chains already begun at t_0 be completed using the label but *new* chains containing the label throughout will also be synthesized and released. Therefore, after prolonged incubation in the presence of labeled amino acids, we would expect to find *no* "gradient of radioactivity" in completed chains. Instead, the radioactive amino acids would be more-or-less distributed uniformly through the entire length of the polypeptide.

To test this model, it is necessary to isolate the polypeptide chains synthesized during the periods of incubation of varying length and to determine and compare the amounts of radioactive amino acids in various segments along their total lengths.

Dintzis incubated the rabbit reticulocytes at 15°C in a medium that included ^3H-labeled leucine, as well as all the other materials necessary for continued hemoglobin synthesis. He selected 15°C rather than 37°C (the normal environmental temperature of these mammalian cells) because at this lower temperature, protein synthesis is sufficiently

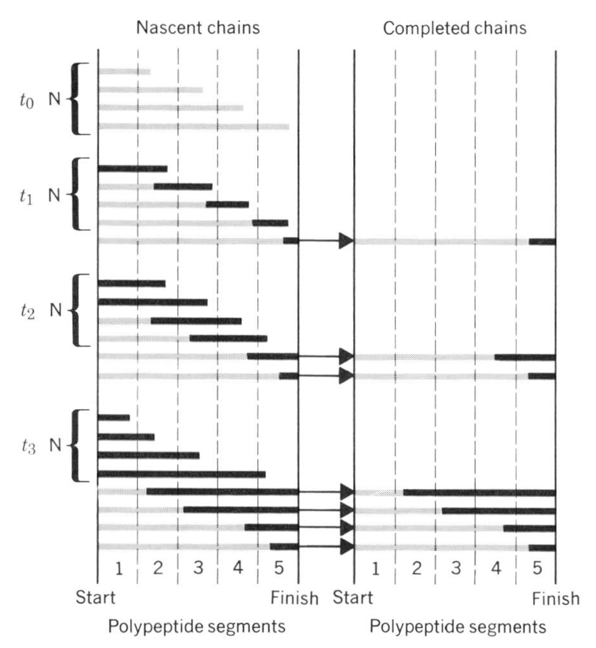

Figure 21–18

Model according to Dintzis for the linear growth of nascent polypeptide chains. Nascent (N), unlabeled chains at time t_0 are shown in black. Radioactive segments added in *separate* experiments by times t_1, t_2, etc. are shown in color. Polypeptide chains completed and released from ribosomes are shown to the right, while those remaining nascent are indicated by N. In this hypothetical polypeptide, trypsin digestion produces five polypeptide segments of various lengths. (See text discussion for further explanation.)

slowed down to permit these experiments to be carried out more easily. After incubation for varying periods of short duration, the reticulocytes were removed, washed, and lysed and the lysate separated into three fractions by centrifugation. A low-speed centrifugation removed plasma membranes and other large particulate debris from the lysate, while a high-speed centrifugation of this supernatant provided a pellet containing ribosomes with nascent polypeptides and a second supernatant containing hemoglobin—including those molecules nascent at t_0 but completed and released during the period of incubation.

The hemoglobin from these experiments was dissociated into its constituent heme and globin parts and the globin separated into alpha and beta chains by ion-exchange column chromatography. The alpha and beta chains were then treated with the proteolytic enzyme *trypsin* which cleaves peptide bonds on the alpha-carboxyl side of lysine and arginine residues. Therefore, each chain is split into a specific number of peptide segments. For clarity, only five such segments are depicted in Figure 21–18, although rabbit alpha and beta globin chains regularly provided 35 segments. These peptides were then separated from one another by a technique called "fingerprinting" which combines paper electrophoresis with paper chromatography to produce a two-dimensional distribution of separate peptides across the sheet of filter paper. Having separated the peptide fragments from one another, the next task was to determine their specific ^{3}H-labeled leucine contents and to compare the results after varying periods of incubation in labeled medium. Not all peptide fragments produced by trypsin digestion of globin chains would be expected to contain leucine residues, and in fact, only nine fragments in each chain did. Obtaining quantitative data on the specific radioactivity in each peptide fragment, comparing the results for several experiments, and determining whether these results support or contradict the proposed model for chain assembly required that two major problems be solved. One problem is that the total yield of peptide fragments unavoidably varied from one experiment to the other as a result of differential losses at each stage in the isolation procedures. The other problem is that if the model is correct, the radioactivity of an ^{3}H-leucine-containing peptide fragment would vary not only as a function of its position along the primary structure of the globin chain but would depend also on the number of leucine residues in that fragment. That is, it is necessary to compensate in some manner for the differential numbers of leucine residues in each peptide.

Dintzis solved these problems by using what he called an "internal standard." At the end of each pulse incubation experiment, he added hemoglobin that was uniformly labeled with ^{14}C-leucine to the ^{3}H-leucine-containing samples, and the *mixture* was then carried through the stages of digestion and fingerprinting. The uniformly labeled hemoglobin was prepared by long-term incubation of reticulocytes in medium containing ^{14}C-leucine; this yielded a preparation in which, for each hemoglobin molecule, either *all* the leucine positions were occupied by the radioactive form of the amino acid or *none* of them was labeled. (This would depend upon whether the hemoglobin molecule was synthesized *before* or *after* the recticulocytes were placed in the labeled medium.) By expressing the radioactivity of each resulting peptide fragment as the ratio

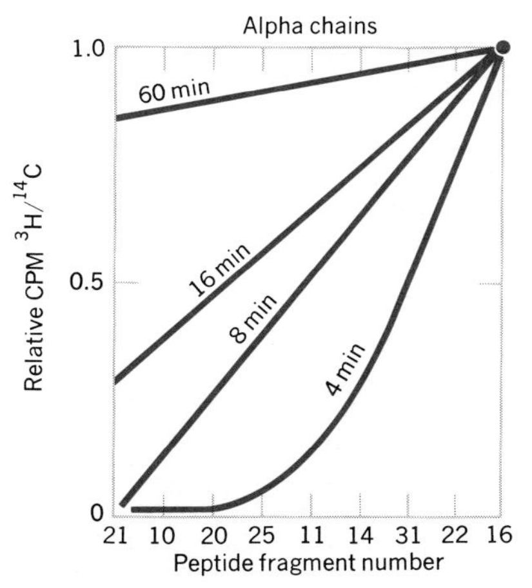

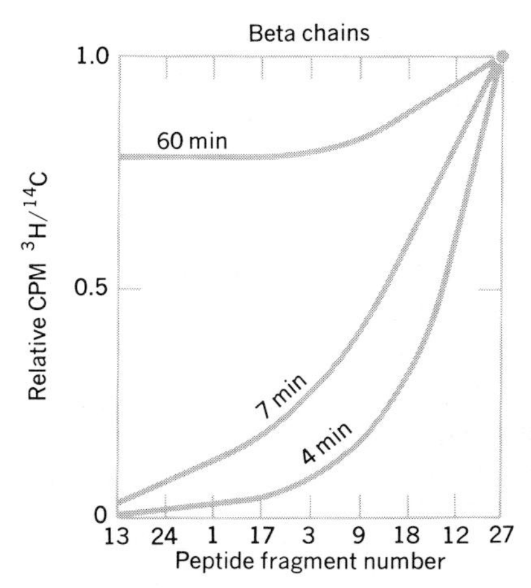

Figure 21–19
Distributions of radioactivity among the peptides produced by tryptic digestion of alpha and beta chains of rabbit hemoglobin released from reticulocyte ribosomes after various periods of incubation at 15°C. (See text for explanation.)

^{3}H-leucine:^{14}C-leucine, Dintzis simultaneously circumvented both the problems of *differential losses* of peptides during the course of the experiment and the *differential numbers* of leucine residues in the peptide fragments. Internal standards of this kind are now almost routinely used in experiments of this nature.

The kinds of results obtained by Dintzis are shown in Figure 21–19. The numbers assigned to each peptide in order to identify and compare them in different "fingerprints" are more or less arbitrary. Dintzis arranged the data for each of the labeled peptides from very short (e.g., 4-minute) incubations to yield a curve showing increasing radioactivity as a function of a selected peptide sequence. This in itself is not significant, since any collection of numerical data can be arranged in increasing order. What was important was that if the selected peptide sequence was then held constant when plotting the data from a series of incubations of longer duration, the resulting curves gave rise to a family of decreasing slopes (Fig. 21–19). Examining the graph for alpha chains in Figure 21–19, it can be seen that after a 4-minute incubation, appreciable radioactivity was recovered in only four peptide fragments (i.e., numbers 14, 31, 22, and 16). As a result of longer incubations, proportionately more radioactivity was found in these peptides and in others that were not labeled at 4 minutes. Even after 60 minutes of incubation, a gradient of radioactivity still persisted for the peptide sequence, although by this time the slope was approaching zero. These data are consistent with the model of linear growth proposed by Dintzis and are not consistent with the other alternatives. Accordingly, peptide fragment 16 would be near the end of the polypeptide chain synthesized last and peptide fragment 21 would be near the end of the chain synthesized first. It can also be seen from the data in Figure 21–19 that after a 7-minute incubation in radioactive medium, some labeled leucine was found in all the peptides, indicating that only 7 minutes are required for the synthesis of a complete alpha or beta globin chain at 15°C. An extrapolation of this to the normal environmental temperature for these cells would suggest that about 1.5 minutes are required for complete chain assembly at 37°C. Since the alpha and beta globin chains of rabbit hemoglobin each contain about 150 amino acids, this corresponds to the elongation of the chain at an average rate of about two amino acids per second.

It was subsequently found that peptide fragment 16 from the alpha globin chain digest contained the *C-terminal* amino acid. This indicated that it is the C-terminus that is synthesized last and that synthesis must therefore begin with the *N-terminal* amino acid of the polypeptide.

At about the same time that Dintzis was carrying out his experiments, J. Bishop, J. Leahy, and R. Schweet, also working with rabbit reticulocytes, reached similar conclusions about the direction of polypeptide chain assembly. The N-terminal amino acid of the globin chains is valine. Bishop, Leahy, and Schweet incubated reticulocytes in ^{14}C-labeled valine and then isolated the reticulocyte ribosomes together with their nascent globin chains. The ribosomes were then incubated in an in vitro system providing for continued growth of the nascent chains but containing ^{12}C-valine (i.e., ordinary valine). After incubation, the amount of N-terminal ^{14}C-valine was compared with that in other regions of the globin chains completed and released from the ribosomes and was found to be significantly higher. These results supported the concept that chain growth began at the N-terminus. A. Yoshida and T. Tobita, studying the synthesis of an amylase from bacteria, and R. E. Canfield and C. B. Anfinsen, studying egg white lysozyme synthesis, also reached similar conclusions about the direction of protein synthesis.

In addition to showing the regular progression of polypeptide addition to growing globin chains, the data of Dintzis (and later Naughton and Dintzis, and others) also reveal differences in the *instantaneous* rates of chain elongation along the polypeptide. This was first suggested by S. W. Englander and L. A. Page, who proposed that curves such as those obtained in the pulse label experiments of Dintzis were also *profiles* of nascent chain lengths at t_0. They noted that the increment of radioactivity between one leucine-containing peptide and another reflected the number of nascent chains having their *growing ends between these two leucines at* t_0 and that the slopes at various points along the curve are *inversely proportional* to the rates of chain growth through these points. Consider as an example the 7-minute curve for beta chains shown in Figure 21–19. According to Englander and Page, the rate of chain elongation through the region of the chain containing peptide 1 would be considerably faster than the rate of chain growth through the region containing peptide 12. (Compare the slopes of the curve in these two regions.) If a uniform (i.e., constant) rate of growth occurred over the entire length of the polypeptide, the data resulting from pulse label experiments would yield a family of straight lines. Using bone marrow cells and carrying out pulse label experiments similar to those of Dintzis, R. M. Winslow and V. M. Ingram achieved similar findings for the synthesis of alpha and beta chains of human hemoglobin A, namely that the rate of chain growth is greater during the synthesis of the first half of the polypeptide than during the synthesis of the second half. Some hypothetical curves based on pulse label experiments of the type conducted by Dintzis, Winslow and Ingram, and others are shown in Figure 21–20, along with an explanation of what these curves imply about the rates

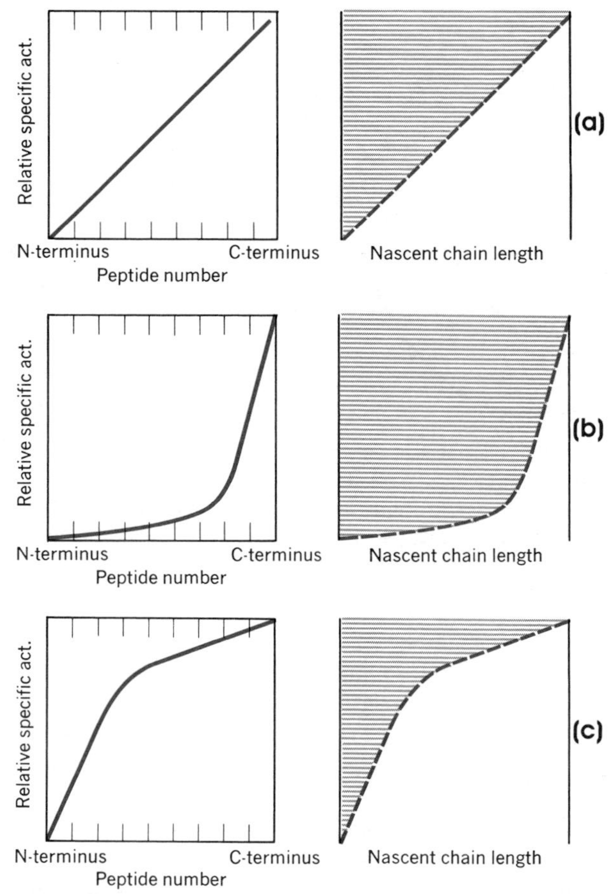

Figure 21–20
Hypothetical curves from pulse-label experiments similar to those of Dintzis showing implications of these curves regarding the rates of chain growth and nascent peptide chain lengths at t_0. Curve *a* would result from a situation in which there were equal numbers of nascent chains of all lengths at t_0, the growth rate being constant. Curve *b* (which is similar to Dintzis' results with hemoglobin) would result from a case in which there were more long nascent chains than short nascent chains, the growth rate being rapid at first but slowing down toward the end of the chain. Curve *c* would result from a case in which there were more short chains than long chains, the growth rate being slow at first but increasing toward the end of the chain.

of chain growth and the instantaneous distributions of nascent chain lengths on the cell's ribosomes. Recently, A. J. Morris showed that the decreases in the rate of assembly of globin chains occur specifically in the vicinity of amino acids 40, 57, 89, and 120–145 of the primary structure. Since it is clear from the hemoglobin data that polypeptide chain growth does not necessarily proceed at a constant rate, we would expect that in polysomes, the distances between successive ribosomes along the strand of messenger RNA would not be equal but would reflect the differential rates of translation of the mRNA code; electron-microscopic studies indicate that this is, in fact, the case.

There are several possible explanations for these findings. For example, it could be that not all mRNA codons are translated at the same rate and that the rate of chain elongation over a given region of the polypeptide's primary structure depends on the types and amounts of amino acids and tRNAs present in the cell. It is also believed that at least for some proteins the assumption of tertiary structure begins during the course of chain growth, and this, too, might influence the rate of chain elongation.

Table 21–10
Nucleosides of tRNA

Nucleoside	Symbol
Unmodified	
Adenosine	A
Uridine	U
Guanosine	G
Cytidine	C
Modified	
Inosine	I
Methylinosine	I^{Me}
Methylguanosine	G^{Me}
Dimethylguanosine	G^{DiMe}
2′-O-methylguanosine	$G^{2'-OMe}$
Thymine riboside	T
Pseudouridine	ψ
Dihydrouridine	D
Acetylcytidine	C^{Ac}
Methylcytidine	C^{Me}
Methyladenosine	A^{Me}

Processing and Structure of Transfer RNA

The first stage in the incorporation of an amino acid into a growing polypeptide chain involves the "activation" of the amino acid, that is, the enzymatic attachment of the amino acid to a specific transfer RNA molecule capable of inserting that amino acid into its appropriate position in the polypeptide chain being assembled on the mRNA-ribosome complex. Each tRNA molecule is specific for a particular amino acid. In a given tissue or cell, amino acid specific tRNA can exist in multiple forms called *isoaccepting* species. For example, *E. coli* contains five different (i.e., "isoaccepting") tRNAs capable of combining with leucine. Altogether, there are about 60 to 80 different tRNAs in a cell or tissue.

tRNA Processing. Transfer RNA is produced in precursor form by *RNA polymerase* transcription of DNA. The precursor, which may contain the base sequence of more than one tRNA, has extra nucleotides at the 3′ and 5′ ends and also internally; these are subsequently cleaved by **endonucleases** and **exonucleases.** All mature tRNAs contain the sequence C-C-A at the 3′ end of the molecule, and this segment may be added following transcription by an appropriate *nucleotidyl transferase*. Nucleosides at various positions in the primary structure may be modified enzymatically (see Table 21–10) to produce the final tRNA product capable of aminoacylation. In some species of tRNA, as many as 16% of the bases may be modified.

Structure of tRNA. The first successful complete purification of tRNA was achieved by R. W. Holley using countercurrent distribution, and in 1965, Holley reported the primary structure of yeast alanine tRNA. The primary structure was determined using small polynucleotide fragments produced by enzymatic digestion of the isolated tRNA by pancreatic *ribonuclease* and *phosphodiesterase*. Since Holley's pioneering work, more than 75 tRNAs have been fully sequenced, and all exhibit similar primary, secondary, and tertiary structures.

The tRNAs contain a linear sequence of 75 to 85 nucleotides that can be arranged to form the classical "cloverleaf" pattern shown in Figure 21–21*a* and originally proposed by Holley. There are five folded regions: the **amino acid arm,** the **dihydrouridine (DHU) arm,** the **anticodon arm,** the **TψC arm,** and the **extra** (or **variable**) **arm.** Each arm consists of a double-helical *stem* stablized by base-pairing.

All except the amino acid arm possess a *loop* region containing unpaired bases. With only a few exceptions, the base-pairing that creates the secondary structure of the helical regions is of the conventional Watson-Crick type involving hydrogen bonds between A and U and between G and C. To distinguish these bonds from other hydrogen bonds important in maintaining the tertiary structure of tRNA, the hydrogen bonds of tRNA are denoted as *secondary* hydrogen bonds and as *tertiary* hydrogen bonds accordingly.

As seen in Figure 21–21*a*, certain positions are invariant or semi-invariant among all tRNAs so far sequenced. (The term "semi-invariant" is used to denote a position invariably occupied by the same *type* of base: purine or pyrimidine.) For example, the four unpaired bases that terminate the sequence at the 3′ end of the molecule are always purine-C-C-A; the last of these bases (i.e., adenine) forms the bond with the amino acid (see later). Most of the invariant and semi-invariant positions are found in the DHU loop and in the TψC loop. The invariant residues form hydrogen bonds with one another that are crucial to the maintenance of the characteristic tertiary structure of the tRNAs and also provide recognition sites for interactions with enzymes and with the ribosome. The loop of the anticodon arm contains seven bases, three of which form the *anticodon*.

One of the characteristic features of tRNA is that a large proportion of the nucleosides are modified. Table 21–10 lists some of the more than 40 modified nucleosides regularly occurring in the tRNA. Most modifications involve methylation of the regular bases (i.e., A, U, G, and C) or methylation of the 2′ hydroxyl oxygen of the riboses. The role (or roles) of the modified bases are not known with certainty, but suggested roles include the *prevention* of base pairing (1) within the tRNA molecule in order to provide a characteristic tertiary structure and (2) between tRNA and mRNA during translation. It is interesting that the base at the 3′ side of the anticodon is nearly always a modified purine when the first base of the corresponding codon is either A or U.

During translation, the three bases of the anticodon form hydrogen bonds with the corresponding codon bases of mRNA. An examination of Table 21–1 reveals that a single amino acid may be coded for by two or more codon sequences. For example, the codon for alanine may be GCU, GCC, GCA, or GCG—the *third* base being seemingly unimportant. Keeping in mind that during translation the tRNA and mRNA molecules are antiparallel, it would appear that the base occupying the *first* position of the anticodon may recognize one or more different bases occupying the *third* position in the codon (Table 21–11). (Remember that the numbering of the polynucleotide chain beings at the 5′ end and finishes at the 3′ end.) Francis Crick refers to this base as the "wobble" base, implying that it may orient in different ways in order to accommodate the appropriate base-pairing.

Table 21–11
Degeneracy in Codon-Anticodon Recognition

Base Occupying First Position of tRNA Anticodon	Base Occupying Third Position of mRNA Codon
I	U, C, or A
A	U only
G	U or C
C	G only
U	A or G

The most widely accepted model for the tertiary structure of tRNA is that proposed by S. H. Kim and is based principally on x-ray crystallographic studies of phenylalanine tRNA from yeast cells (Fig. 21–21*b*). The molecule has an L shape, with all double-helical regions being right-handed and antiparallel. The amino acid and TψC stems form one continuous double helix, and the DHU and anticodon stems form another. The two helices are perpendicular to each other, thereby forming the L, with the anticodon and C-C-A termini at opposite ends. The molecule is about 20 Å thick, which corresponds to the diameter of the RNA double helix. Figure 21–21*c* relates to the tertiary structure of tRNA to

Figure 21–21
Structure of tRNA. (*a*) Cloverleaf pattern showing secondary structure resulting from hydrogen bonding (i.e., •—•) in the helical stems of each arm. Invariant and semi-invariant positions are indicated with the nucleoside symbol (see Table 21–10). Pu, purine, Py, pyrimidine; G*, guanosine or 2′-O-methylguanosine; A*, adenosine or 1-methyladenosine; α and β are variable regions containing up to four nucleosides. (*b*) Tertiary structure of tRNA proposed by Kim; (*c*) Rearrangement of the cloverleaf secondary structure to more clearly show the L-shape tertiary structure; ribbonlike regions form helical segments through hydrogen bonding. (*d*) Stereo pair of tRNA (by permission of Dr. S. H. Kim).

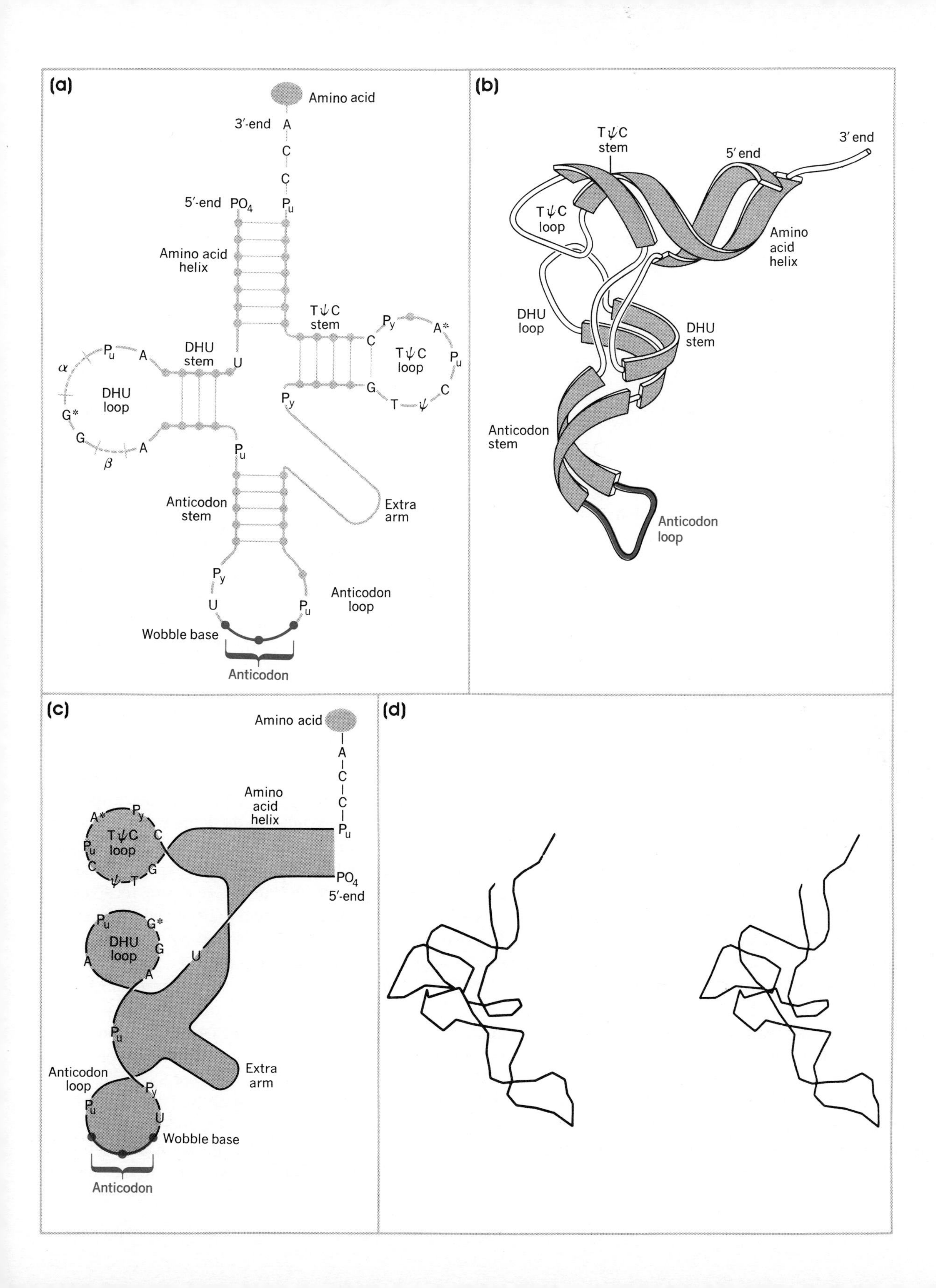

(a)
Amino acid
3′-end
A
C
C
Pu
5′-end
PO4
Amino acid helix
TψC stem
Py
A*
TψC loop
Pu
C
G
T
ψ
DHU stem
U
Pu
A
α
DHU loop
G*
G
β
A
Py
Pu
Anticodon stem
Extra arm
Py
U
Pu
Anticodon loop
Wobble base
Anticodon
(b)
TψC stem
3′ end
5′ end
TψC loop
Amino acid helix
DHU loop
DHU stem
Anticodon stem
Anticodon loop
(c)
Amino acid
A
C
C
Pu
Amino acid helix
A*
Py
TψC loop
C
Pu
C
G
ψ
T
PO4
5′-end
Pu
G*
DHU loop
G
A
A
U
Pu
Extra arm
Anticodon loop
Py
Pu
U
Wobble base
Anticodon
(d)

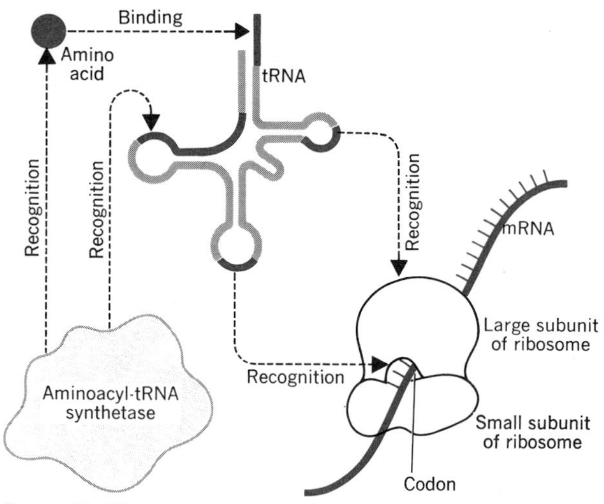

Figure 21–22
Relationship between various regions of tRNA and the enzymes, ribosomal proteins, and RNAs with which tRNA interacts during protein synthesis.

the cloverleaf secondary structure. The three-dimensional appearance of the molecule is presented in Figure 21–21*d*.

Different regions of a tRNA molecule appear to serve as recognition and binding sites for various enzymes, ribosomal proteins, and other RNAs that interact with tRNA during the various stages of protein synthesis (Fig. 21–22). Although the amino acid is bound to the adenosine nucleoside at the 3′ end of the tRNA molecule, this region appears to have little if anything to do with codon-anticodon recognition or binding. This was elegantly demonstrated in a classic experiment by F. Lipmann and F. Chapeville, who prepared a tRNA specific for cysteine (abbreviated $tRNA^{Cys}$) which they then enzymatically combined with ^{14}C-cysteine to form cysteinyl-$tRNA^{Cys}$ (or cys-$tRNA^{Cys}$). They treated the cys-$tRNA^{Cys}$ with Raney nickel which removes the sulfhydryl group from the cysteine residue, leaving ala-$tRNA^{Cys}$ (i.e., a tRNA molecule specific for cysteine but containing bound ^{14}C-labeled alanine instead). When the radioactive ala-$tRNA^{Cys}$ was employed in an in vitro protein-synthesizing system using a synthetic messenger RNA that coded for the production of a polypeptide rich in cysteine, the radioactive alanine residues were incorporated in place of cysteine. This showed that the specificity for codon recognition resided with the tRNA molecule and not with the amino acid.

Activation of Amino Acids

Amino acid activation involves two major steps (Fig. 21–23). In the first, the alpha-carboxyl group of the amino acid reacts with ATP to form an *aminoacyl-adenylate* and pyrophosphate. For each species of amino acid, there is at least one specific enzyme, called an *aminoacyl-tRNA synthetase* (or *ligase*) that catalyzed the reaction, and a number of these amino-acid-activating enzymes have been isolated and studied. The aminoacyl-adenylate formed is not released from the enzyme but remains complexed to it, presumably by a linkage between the enzyme and the R group of the amino acid. In the second step, the aminoacyl-AMP complex recognizes and reacts with a molecule of tRNA specific for that amino acid to form *aminoacyl-tRNA,* and the enzyme and AMP are released. The reaction between tRNA and the amino acid involves esterification to the 2′ or 3′ hydroxyl group of ribose in the terminal adenosine unit of tRNA by the alpha-carboxyl carbon atom of the amino acid. These reactions occur in the cytosol. The aminoacyl-tRNA thus formed can now participate in protein synthesis on the mRNA-ribosome complexes. A number of investigators contributed to the elucidation of the above reaction sequences, but most notable among them are P. Zamecnik, M. B. Hoagland, P. Berg, and E. J. Ofengand.

Formation of the Initiation Complex

Amino Acid and Peptide Sites of the Ribosome. In both procaryotes and eucaryotes, two sites on the intact ribosome are involved in protein synthesis; these are called the *amino acid site* and the *peptide site*. The peptide site is the region of the ribosome to which the growing polypeptide chain is bound by tRNA, while the amino acid site receives the tRNA bearing the next amino acid to be added to the chain. The two sites are believed to be shared by each of the two subunits making up the intact ribosome.

At first, it was believed that no special mechanism was required in order to initiate chain growth. It was proposed that a ribosome would simply attach to the 5′ end of the mRNA and proceed to translate successive codons. It is clear now that this is not the case and that a specific mechanism exists for initiating translation and which prevents *out-of-phase* translation of the mRNA code.

The problem of out-of-phase translation warrants further consideration. Suppose that a section of mRNA contains

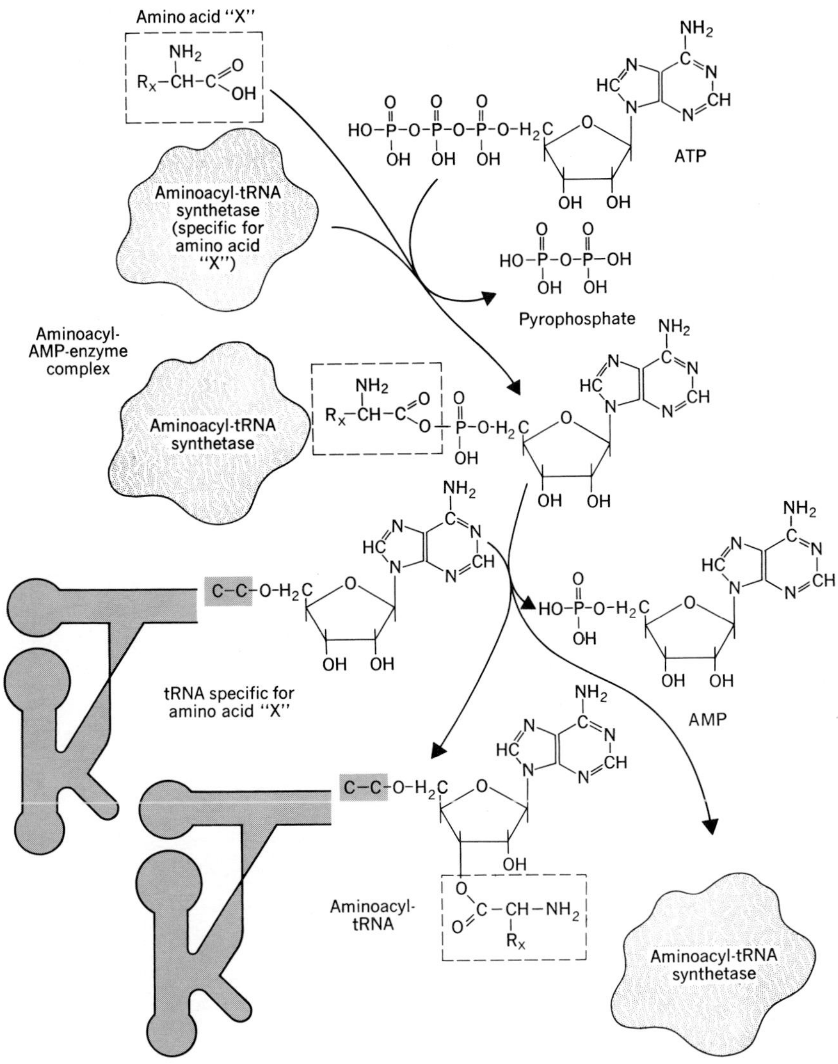

Figure 21–23
The reaction sequence of amino acid activation. Although in the final product shown the amino acid is esterified to the 3′ hydroxyl position of the terminal ribose of tRNA, an equilibrium mixture of 2′ and 3′ esters may be formed.

the following codon sequence:

. . . U G U A A G G C U A G A . . .

This section of mRNA would therefore code for the addition of the amino acid sequence cys—lys—ala—arg to the growing polypeptide chain (see Table 21–1 for the genetic code). However, suppose that this section of mRNA was translated out-of-phase as follows:

. . . U G U A A G G C U A G A . . .

In this case, the sequence val—arg—leu would be incorrectly incorporated into the primary structure of the protein. The need for a mechanism that ensures that translation is begun in-phase and carried out in-phase is apparent.

Characteristics of mRNA. In the eucaryotic cells so far studied, mRNA is typically about 1500 nucleotides long and consists of *both* translated and nontranslated regions. Since mRNA of this size could maximally code for a polypeptide 500 amino acids long, it is most likely that eucaryotic mRNA is **monocistronic** (or **monogenic**); that is, the mRNA contains the codon sequence for no more than *one* polypeptide. Messenger RNAs of procaryotic cells are quite variable in length, since some mRNAs are transcribed from more than one gene; these are termed **polycistronic** (or **polygenic**) mRNAs. For example, five of the enzymes involved in tryptophan synthesis in *E. coli* are encoded in a single polycistronic mRNA produced by transcription of five closely linked genes (the mRNA contains more than 7000 nucleotides). Other procaryotic mRNAs code for as many as 10 enzymes; interestingly, the various enzymes encoded in a polycistronic message are part of the *same* metabolic pathway.

The 5′-phosphate and 3′-OH ends of all eucaryotic mRNAs are similar. The 5′ end contains the sequence $m^7GpppN'_{m}pN''_{m}p$. . . , referred to as the "cap" (Fig. 21–24), and the 3′ end of mRNA contains a poly A sequence 20 to 250 nucleotides long known as the "tail." Not all the remainder of the mRNA molecule may be translated (Fig. 21–24). For example, in addition to the cap and poly A segments, hemoglobin mRNA contains a 150-nucleotide segment near poly A that is not translated into globin. Hemoglobin alpha and beta chain mRNAs have an estimated molecular weight of 200,000 to 220,000 and contain 650 to 670 nucleotides. The alpha and beta globin chains are coded for by 423 and 444 nucleotides, respectively; the remainder of the mRNA contains a poly A tail 50 to 75 nucleotides long, a nontranslated sequence of 150 to 175 nucleotides and, of course, a cap segment. The special chemical nature of the cap region of mRNA is apparently involved in ensuring the proper initiation of translation.

In eucaryotic cells, mRNA is produced by the transcription of nuclear DNA. Although little is known about mRNA processing, it is clear that a pre-mRNA is modified prior to its exit from the nucleus. The structural genes for the globin chains of hemoglobin contain many more base pairs than required to produce the globin-mRNA described above. The coding sequences are interrupted one or more times by intervening sequences whose transcripts do not appear in the final message. In the beta globin chain gene, one of these intervening sequences contains more than 500

Figure 21–24
***Top:* Cap region of mRNA. *Bottom:* Various sections of typical mRNA.**

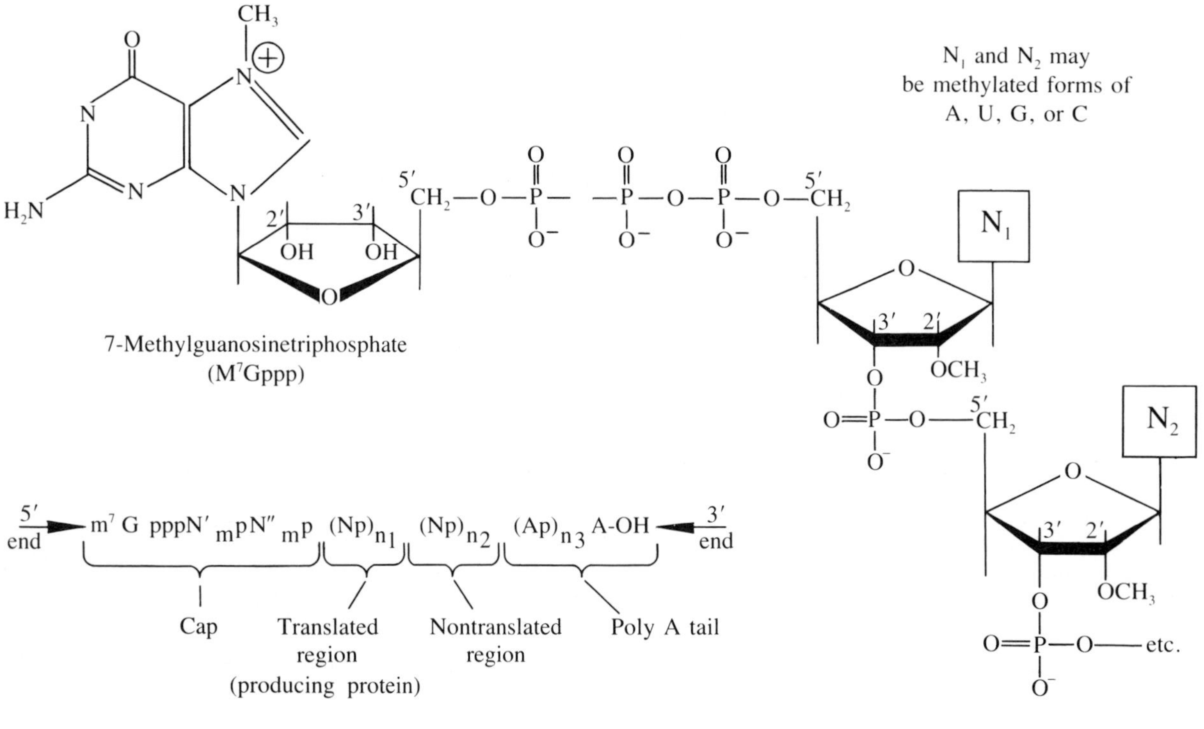

7-Methylguanosinetriphosphate (M^7Gppp)

N_1 and N_2 may be methylated forms of A, U, G, or C

5′ end → m^7 G pppN′ $_m$pN″ $_m$p $(Np)_{n1}$ $(Np)_{n2}$ $(Ap)_{n3}$ A-OH ← 3′ end

Cap — Translated region (producing protein) — Nontranslated region — Poly A tail

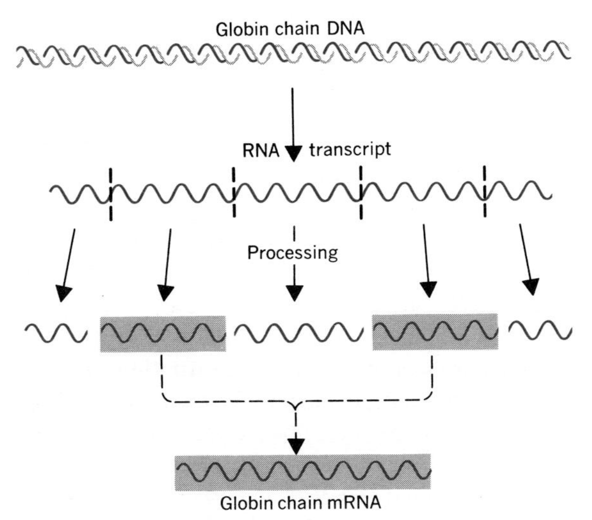

Figure 21–25
Processing of mRNA.

base pairs. The entire gene is transcribed into a 15 *S* message, but during processing, certain internal RNA sequences are eliminated and the coding segments precisely rejoined (Fig. 21–25).

Role of Formylmethionine. In 1963, J. P. Waller reported the amazing finding that nearly one-half of all proteins in *E. coli* cells have the amino acid methionine in the N-terminal position. (Remember that protein synthesis begins at the N-terminus!) Then, in 1964, K. A. Marcker and F. Sanger discovered an unusual species of aminoacyl-tRNA in *E. coli*—***N*-formylmethionyl-tRNA**—and suggested that this molecule may play a role in the special mechanism of chain elongation because the presence of the *N*-formyl group in the amino acid (leaving only the alpha-carboxyl group available for peptide bond formation) would restrict this residue to the N-terminus. The structural formulae of *N*-formylmethionyl-tRNA and methionyl-tRNA are shown in Figure 21–26.

Two transfer RNA molecules specific for methionine are present in *E. coli*, but only one of these can participate in the subsequent enzymatic formylation of the methionine residue. These tRNAs may be denoted as $tRNA^{Met}$ and $tRNA^{Met}_{f}$. The formation of *N*-formylmethionyl-tRNA occurs as follows:

$$\text{(1) met} + \text{tRNA}^{\text{Met}}_{\text{f}} \xrightarrow[\text{tRNA synthetase}]{\text{specific aminoacyl-}} \text{met-tRNA}^{\text{Met}}_{\text{f}}$$

$$\text{(2) met-tRNA}^{\text{Met}}_{\text{f}} + \text{formate} \xrightarrow[\text{enzyme}]{\text{formylating}} \text{N-formylmet-tRNA}^{\text{Met}}_{\text{f}}$$

The codon for methionine is A U G (Table 21–1), and when this codon occurs anywhere except at the beginning of mRNA, it codes for met-$tRNA^{Met}$. However, when A U G is the first codon of the mRNA, it codes for *N*-formylmet-$tRNA^{Met}_{f}$ and chain initiation. For this reason, the A U G codon is also called the **initiator** or **start** codon. The picture is somewhat complicated by the fact that the codon G U G, which is a codon for valine, also codes for *N*-formylmet-$tRNA^{Met}$ when G U G occurs at the *beginning* of the message; anywhere else in the mRNA molecule, G U G codes for valine. The interaction that takes place between the aminoacyl-tRNA anticodon, and the mRNA codon thus depends on both base-pairing and the location of the codon in mRNA. It is now clear that *N*-formylmet's role as the initiating amino acid in protein synthesis is not restricted to *E. coli* but is a characteristic of procaryotes in general.

The process of initiation in eucaryotes is fundamentally similar to that in procaryotes. As in procaryotes, there are at least two methionine tRNAs that recognize the A U G codon of mRNA; however, only one these tRNAs can participate in chain initiation. The initiator methionyl-tRNA (met-$tRNA^{Met}_{i}$) can be enzymatically formylated in vitro,

Figure 21–26
Methionyl-tRNA and formylmethionyl-tRNA

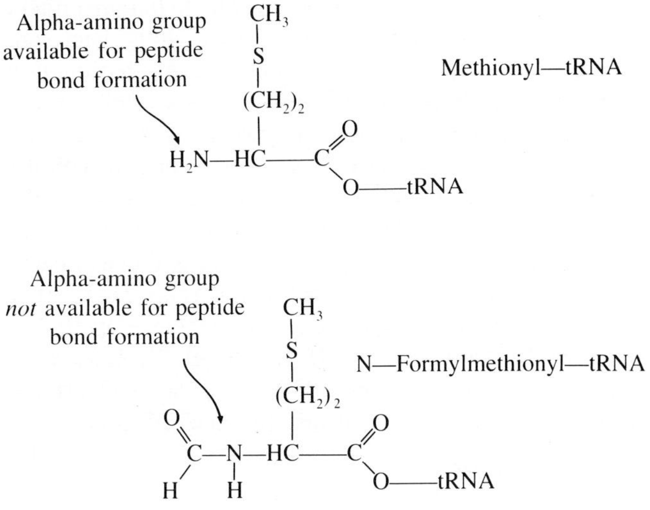

Methionyl—tRNA

Alpha-amino group available for peptide bond formation

$$H_2N{-}HC(CH_2)_2{-}S{-}CH_3 \quad {-}C(=O){-}O{-}tRNA$$

N—Formylmethionyl—tRNA

Alpha-amino group *not* available for peptide bond formation

$$H{-}C(=O){-}N(H){-}HC(CH_2)_2{-}S{-}CH_3 \quad {-}C(=O){-}O{-}tRNA$$

Table 21–12
Soluble Factors Required for Protein Synthesis

	Procaryotes	Eucaryotes
Initiation factors	IF-1	
	IF-2	eIF-2
		eIF-2′
		$eIF\text{-}2a_1$[a]
		$eIF\text{-}2a_2$
		$eIF\text{-}2a_3$
	IF-3	eIF-3
Elongation factors	EF-Tu	EF-1
	EF-Ts	
	EF-G	EF-2
Termination (release) factors	RF-1	RF
	RF-2	
	RF-3	

[a]$eIF\text{-}2a_1$, $eIF\text{-}2a_2$, and $eIF\text{-}2a_3$ are called *accessory factors*.

although this does *not* appear to take place under native circumstances, and there are no formylating enzymes present in the eucaryotic cytosol. The other methionyl-tRNA ($met\text{-}tRNA^{Met}$) recognizes A U G codons located internally in mRNA. Of special interest is the observation that the initiation of protein synthesis in mitochondria and chloroplasts takes place in much the same manner as in procaryotes. The initiating aminoacyl-tRNA is formylated using *formylase*, which is present in the organelles but absent from the cytosol.

Initiation Factors. A number of factors present in the *soluble* phase of the cell are required in order to *initiate* protein synthesis; others, to be discussed more fully later, are needed for polypeptide *elongation* and for *termination* (Table 21–12). Procaryote initiation factors were discovered by the observation that washed *E. coli* ribosomes could not translate natural messengers unless supplemented with the wash. Eucaryote initiation factors were discovered by the similar observation that washed reticulocyte ribosomes would not initiate globin synthesis unless the wash was added back. Among the factors listed in Table 21–12, IF-1 of procaryotes has no equivalent in eucaryotes, but IF-2 is functionally equivalent to eIF-2 and eIF-2′. Eucaryotic accessory initiation factors $eIF\text{-}2a_1$, $eIF\text{-}2a_2$, and $eIF\text{-}2a_3$ have no procaryotic equivalents. IF-3 and eIF-3 are functionally identical.

In eucaryotes, polypeptide synthesis is initiated when a complex formed between $met\text{-}tRNA^{Met}_i$, eIF-2, and GTP associates with the peptide site (*P* site) of the small ribosomal subunit. This is followed by eIF-3-stimulated attachment of mRNA to the small subunit, with the initiation A U G codon near the 5′ end of the mRNA molecule aligned at the peptide site (Fig. 21–27). The next mRNA codon aligns at the amino acid site (*A* site). Addition of the large subunit follows release of eIF-2 and may require accessory factors $eIF\text{-}2a_1$, $eIF\text{-}2a_2$ and $eIF\text{-}2a_3$. This is followed by the hydrolysis of GTP and the release of GDP and P_i. The final product, called the 80 *S* **initiation complex** may now proceed to translate the remainder of the message. Figure 21–27 serves also to describe initiation in procaryotic cells if (1) $eIF\text{-}2a_1$, $eIF\text{-}2a_2$, and $eIF\text{-}2a_3$ are deleted, (2) IF-2 and IF-3 are substituted for eIF-2 and eIF-3, and (3) $fmet\text{-}tRNA^{Met}_f$ is substituted for $met\text{-}tRNA^{Met}_i$. IF-1 of procaryotes may assist in mRNA binding to the 30 *S* subunit and $fmet\text{-}tRNA^{Met}_f$ insertion.

In certain cells, additional factors may be required for the formation of an initiation complex. For example, the initiation of globin synthesis in reticulocytes depends on the availability of heme; consequently, the synthesis of globin chains and heme are tightly coordinated.

As noted earlier, nearly half of all proteins in *E. coli* have a methionine residue in the N-terminal position. Since it is not a formylated residue, the formyl group must be removed from the methionine either during or immediately following polypeptide synthesis. Indeed, enzymes that are able to remove formate from formylmethionine residues of polypeptides are present in *E. coli* and other procaryotic cells. For those proteins that have amino acids other than methionine in the N-terminus (i.e., the majority of cell proteins), a mechanism must exist for removing methionine from the end of the polypeptide, and accordingly, aminopeptidases, which specifically cleave the peptide bond between methionine and the second amino acid of the polypeptide chain, have been identified. Consequently, for most *E. coli* proteins, formylmethionine is removed from the end of each polypeptide chain by enzymatically cleaving first the formyl group and subsequently the methionine residue itself.

Only a small percentage of eucaryotic proteins have methionine as the N-terminal amino acid. Like procaryotes, eucaryotes possess an amino peptidase that removes methionine from the N-terminus of growing polypeptides.

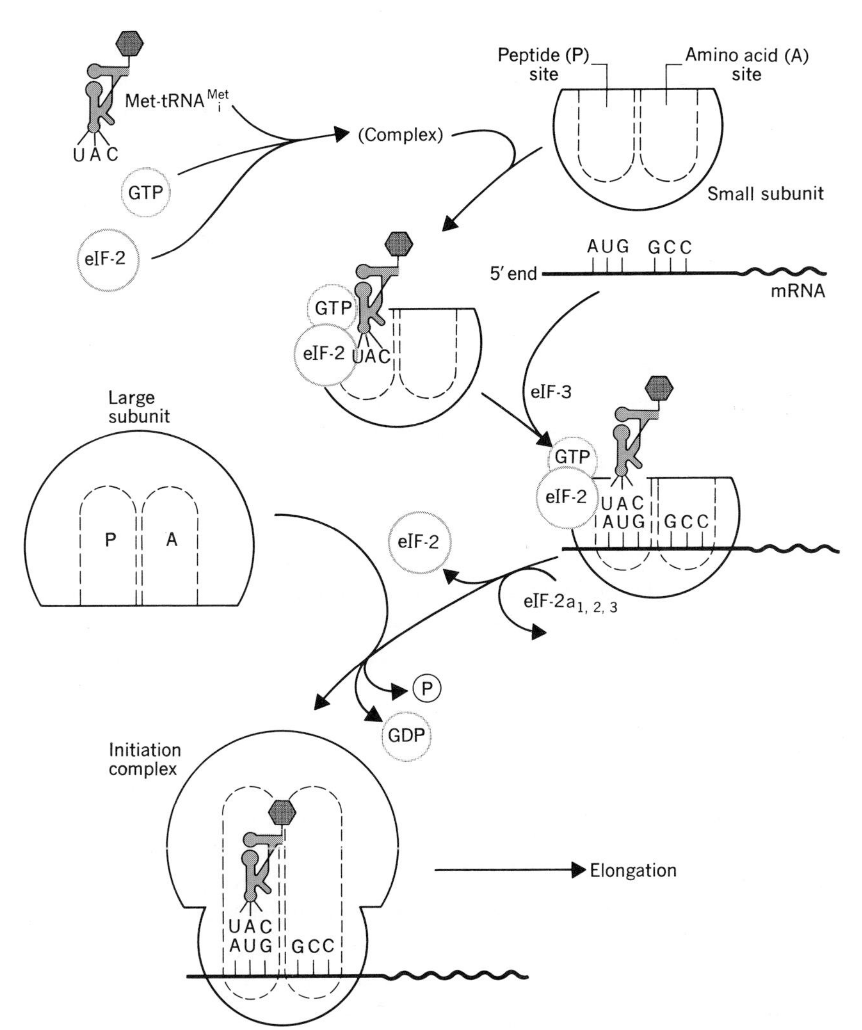

Figure 21–27
Stages in the formation of the initiation complex in eucaryotic cells (see text for explanation). For diagrammatic simplicity, the small and large subunits of the ribosome are shown as disk-like surfaces. The reader should take note that most of these events (as well as those depicted in Figs. 21–28, 21–29, and 21–30) occur *within* the channel(s) formed by the association of the subunits. The second codon, G C C (which codes for ala), is shown for illustrative purposes only.

Studies with hemoglobin in which valine is the N-terminal residue of both the alpha and beta chains have revealed that the methionine which initially occupies the N-terminal position is removed only after the polypeptide is about 30 residues long. Until that point, the peptide bond linking methionine to valine is apparently protected in some fashion from cleavage by the enzyme. Indeed, other proteolytic enzymes, including papain, trypsin, chymotrypsin, and pronase, are unable to hydrolyze the peptide bonds of short nascent chains but can act on the outer segments of longer chains. This observation (and similar observations made for other proteins) is in accord with the current model of ribosomal structure in which the growing polypeptide chain is protected over that portion of its length residing in the interior of the organelle.

Influence of Magnesium Ions. Before we proceed to a discussion of chain elongation, some brief comments are warranted regarding the roles that magnesium ions (Mg^{++}) play in protein synthesis, for it has long been known that a critical, low level of Mg^{++} is necessary in order for protein synthesis to proceed normally. If the level of Mg^{++} falls below this, ribosomes dissociate into their subunits and protein synthesis ceases. On the other hand, an experimental increase in the Mg^{++} level above that which normally exists in cells is accompanied by all kinds of aberrations of the normal requirements for chain initiation and elongation. For example, protein synthesis can be carried out in vitro using synthetic messengers such as poly U (the resulting polypeptide being polyphenylalanine). Indeed, most of the early studies of protein synthesis were carried out using synthetic messengers. Such syntheses occurred in spite of the absence of caps, initiator codons, tails, etc., but *only* if the in vitro system was supplemented with a magnesium ion concentration far in excess of that which occurs normally in the cell.

Magnesium ions are thought to play at least two specific roles: (1) Mg^{++} seems to be a *cofactor* for several of the enzymes that mediate initiation and chain growth, and (2) Mg^{++} probably forms *salt bridges* (ionic bonds) with RNA and, in this manner, links the two ribosomal subunits togther through their respective rRNAs; in some way Mg^{++} may also assist in binding mRNA to the ribosome.

```
               O                    O
               ||                   ||
RNA· · ·—O—P—O—Ribose—O—P—O—Ribose—· · ·
               |                    |
               O−                   O−
               ⋮                    ⋮
Salt bridges   Mg++                 Mg++
               ⋮                    ⋮
               O−                   O−
               |                    |
RNA· · ·—O—P—O—Ribose—O—P—O—Ribose—· · ·
               ||                   ||
               O                    O
```

Chain Elongation

Once the initiator aminoacyl-tRNA is located in the peptide site of the ribosome, chain **elongation** ensues. Addition of the second and subsequent aminoacyl-tRNAs follow a similar pattern. GTP reacts with a soluble phase elongation factor, EF-1, to form a complex which then combines with aminoacyl-tRNA. The resulting EF-1-GTP aminoacyl-tRNA combination interacts with the ribosome so that the aminoacyl-tRNA becomes bound to the vacant amino acid site. This step is accompanied by hydrolysis of GTP and the release of inorganic phosphate and an EF-1-GDP complex (the latter can be recycled to EF-1-GTP using additional GTP) (Fig. 21–28). Occupation of both the *P* and *A* sites of the ribosome is followed by the formation of a peptide bond between the amino acid bound to tRNA in the *P* site and the amino acid that just entered the *A* site. In forming this bond, the alpha-carboxyl group of the amino acid attached to the terminal adenosine unit of tRNA in the *P* site is transferred to the free alpha-amino group of the amino acid held by its tRNA in the *A* site (Fig. 21–29). The formation of the peptide bond is catalyzed by the enzyme *peptide synthetase,* one of the proteins of the large subunit, and temporarily leaves polypeptidyl-tRNA in the *A* site. A second elongation factor, EF-2 (also known as *translocase*), catalyzes a complex rearrangement of the ribosome in which the free tRNA at the *P* site is released and the peptidyl-tRNA and mRNA codon are shifted to the vacated *P* site. The ribosome is thus moved one codon further along the message, with peptidyl-tRNA of the *P* site elongated by one amino acid. The translocation step is accompanied by the hydrolysis of another molecule of GTP. With translocation completed, the ribosome is once again ready to accept aminoacyl-tRNA in the free *A* site. These steps of the elongation reactions are repeated for each new codon of the message entering the *A* site.

Chain elongation in procaryotic cells involves *three* soluble elongation factors: EF-Ts, EF-Tu, and EF-G. Binding of aminoacyl-tRNA to the *A* site requires an EF-Tu·GTP complex, and binding is followed by release of EF-Tu·GDP and inorganic phosphate. EF-Tu·GTP is replenished by EF-Ts-catalyzed transphosphorylation using GTP as substrate:

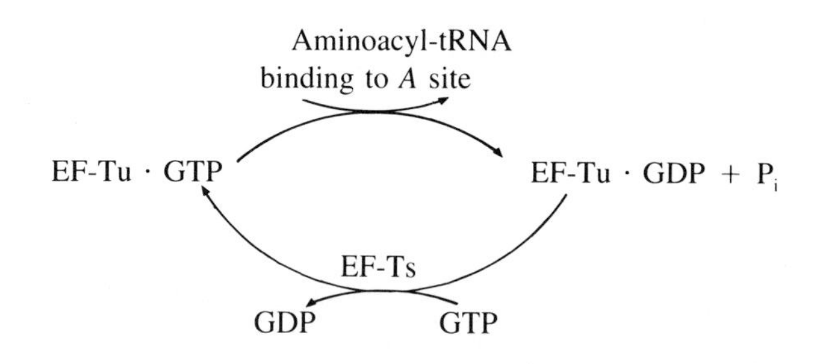

EF-G of procaryotes functions in the same manner as EF-2 of eucaryotes.

The enzymatic cleavage of met and/or formate from the N-terminus of procaryote proteins and met from the N-terminus of eucaryotic proteins takes place after a number of rounds of elongation have already been completed (Fig. 21–30). This leaves the amino acid coded for by the *second* mRNA codon in the "new" N-terminus.

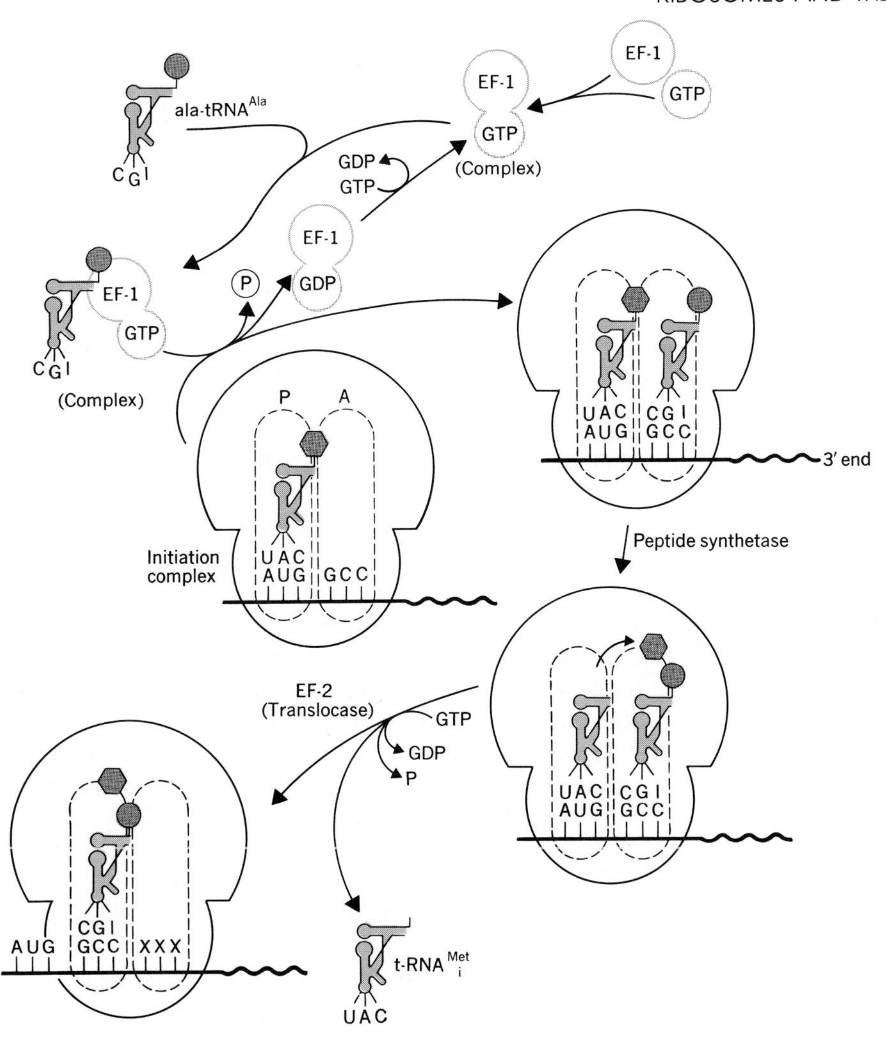

Figure 21–28
Chain elongation in eucaryotic cells. Elongation begins with addition to the initiation complex, but the reactions are the same whenever peptidyl-tRNA occupies the *P* site. For illustrative purposes only, G C C is shown as the second codon of the message. (See text and legend of Fig. 21–27 for other pertinent information.)

Chain Termination

Chain **termination,** like initiation, involves a specific mechanism and does not occur automatically once the ribosome reaches the end of the message. An examination of Table 21–1 reveals that there are three triplets (sometimes referred to as the "nonsense" triplets) that do not code for any amino acid; these are the "amber" codon, U A G; the "ochre" codon, U A A; and the "opal" codon, U G A. Studies in both procaryotic and eucaryotic cells have implicated these codons in the process of chain termination.

The RNA from certain viruses that infect *E. coli* undergoes a mutation, with the result that virus coat protein is synthesized as incomplete polypeptide chains in an *E. coli*

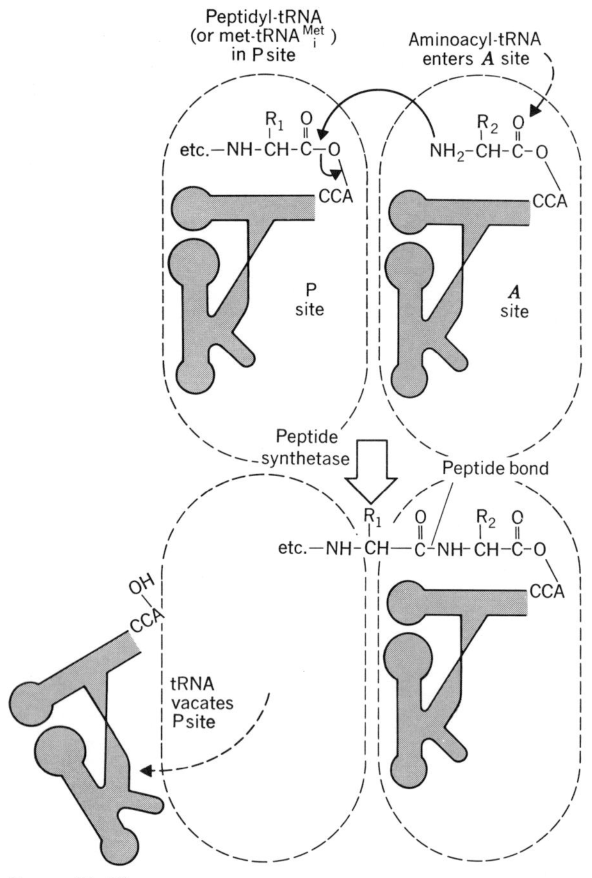

Figure 21–29
Action of peptide synthetase.

in vitro cell-free system. The incomplete polypeptides contain the N-terminus but not the C-terminus, suggesting that normal synthesis was interrupted and the partially completed chains released from the ribosomes. These mutations are suppressible; that is, *E. coli* mutants can be found that support the continued elongation of these polypeptides. However, in the resulting polypeptides, serine (code word U C G) or tryptophan (code word U G G) replaces glutamine (code word C A G). This has been interpreted to mean that the phage mutation involved the change of codon C A G to U A G (i.e., C was mutated to U) which in normal *E. coli* cells resulted in incomplete chains, while in the mutant *E. coli* strain the U A G was being read as the codon for serine or tryptophan. Observations of this sort indicated that the normal role for the U A G codon is chain termination. Other studies with *E. coli* suggest similar roles for the U A A and U G A codons.

In humans, a point mutation in which the first position of the UAA codon at the end of the translated region of alpha globin chain mRNA is altered results in the translation of a major segment of the normally untranslated region (i.e., the ribosome continues past the terminator into the region near the 3′ end of the message that normally remains untranslated). The result is the production of alpha chains containing 31 extra amino acids at the C-terminal end of the polypeptide. Hemoglobins formed using these mutant alpha chains function abnormally (the abnormal hemoglobin is known as Hb "constant spring"). Similar point mutations have recently been identified for beta globin chains.

Once the C-terminal amino acid has been added to the end of the polypeptide, the polypeptidyl-tRNA is translocated from the *A* site to the *P* site of the ribosome, as described earlier. This moves one of the nonsense or terminator codons into position in the *A* site (Fig. 21–31). The terminator is not recognized by a particular tRNA or other RNA species. Instead, release of the completed polypeptide from its tRNA requires participation of a soluble protein called **release factor** (RF). One release factor has been identified in eucaryotic cells and three in procaryotes.

Binding of release factor and GTP to the free *A* site is followed by the activation of the peptidyl synthetase and translocase systems. The bond linking the completed polypeptide to tRNA is hydrolyzed, the polypeptide and tRNA released from the ribosome, and RF and GTP moved into the *P* site. Hydrolysis of GTP is followed by release of RF, GDP, and inorganic phosphate. At this time, the ribosome dissociates into its subunits, freeing mRNA.

Many procaryotic and viral mRNAs are polycistronic. They contain several initiator and terminator signals. The ribosome may dissociate from the mRNA upon reaching any of the terminator codons, and different subunits may attach at the adjacent initiator codon. Initiation may occur with different frequencies so that production of the different polypeptides encoded in the polycistronic message can occur at different rates.

For many proteins, the release of the completed polypeptide chain is followed by the spontaneous assumption of its functional secondary and tertiary structure. For others, all or part of the final secondary and tertiary structure is assumed *as the primary structure is being laid down*. In

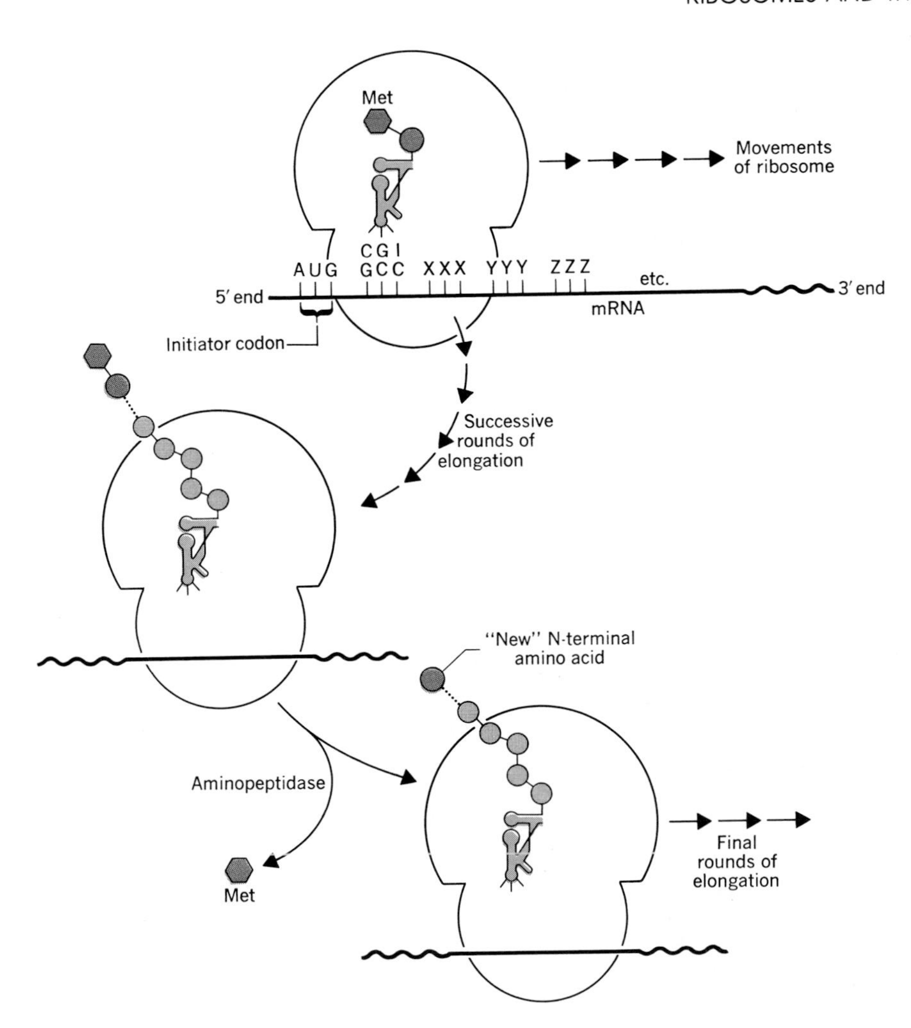

Figure 21–30
Removal of met from the N-terminal position of growing polypeptide chain during elongation rounds in eucaryotic cells.

some cases, it is conceivable that the tertiary structure that is most favored thermodynamically is *not* the functional structure. Therefore, the *progressive assumption* of tertiary structure *during* elongation effectively reduces the number of possible alternative shapes that could be assumed by the polypeptide following release.

The completion and release of the alpha and beta globin chains of hemoglobin have been intensively studied. Upon completion of translation of alpha chain mRNA, the alpha chains are released from the ribosomes. However, beta chain synthesis is not completed nor is the completed chain released until an available alpha chain combines with the partially completed nascent beta chain. When this occurs, the remaining portion of the beta chain mRNA is translated, and the completed beta chain is then released as an alpha-beta dimer (i.e., $\alpha\beta$). Alpha-beta dimers combine in the cytoplasm to form the tetrameric molecule. Consequently, a very small number of free alpha chains may be detected in maturing red blood cells, and these represent recently completed chains about to combine with nascent beta chains on the beta chain mRNA-ribosome complexes of the cell. This phenomenon occurs not only during the synthesis of the globin chains of hemoglobin but also in the synthesis of the light and heavy chains of the immuno-

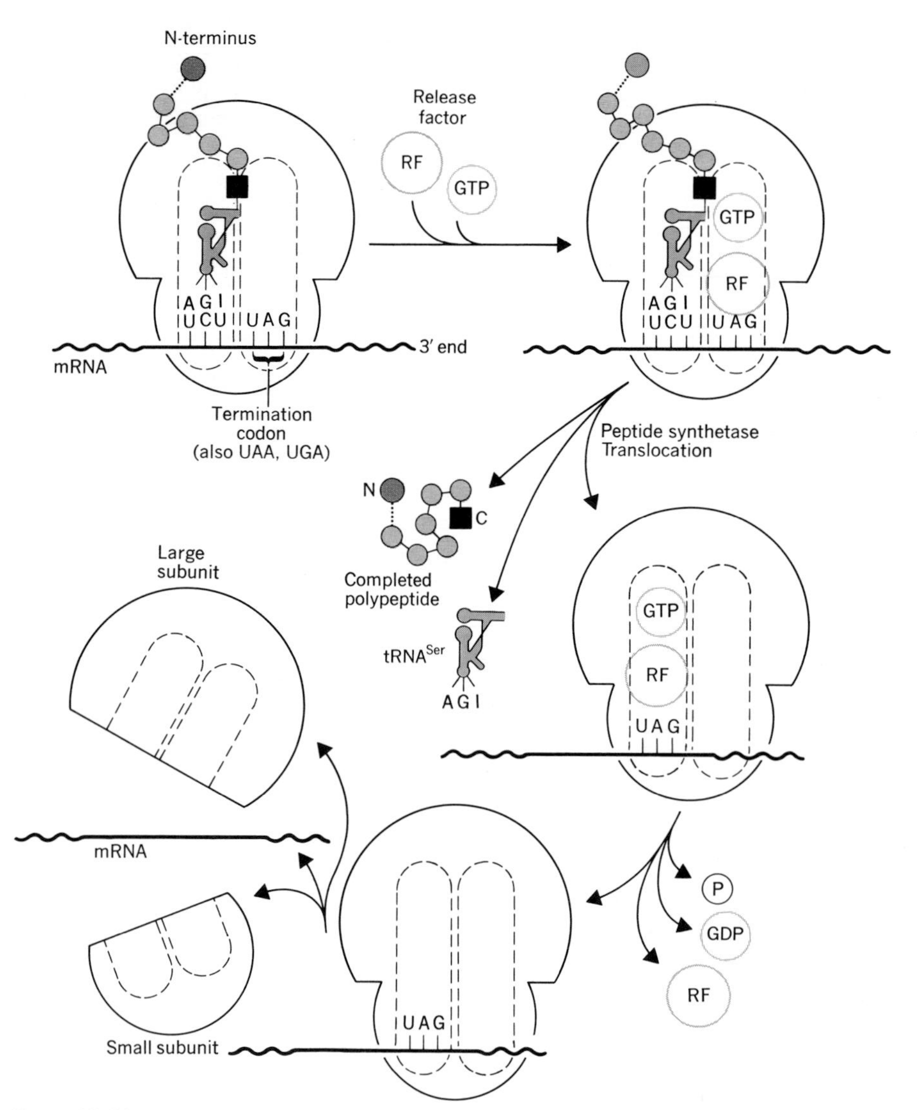

Figure 21–31
Stages of chain termination. (See text and legend of Fig. 21–27 for additional pertinent information.) Serine is shown as the C-terminal amino acid for illustrative purposes only.

globulins. Indeed, the phenomenon may be a general one for proteins composed of two or more polypeptide chains.

Polyribosomes (Polysomes)

Since the globin chains of hemoglobin each contain about 150 amino acids, their messenger RNAs must contain at least 450 nucleotides. Each nucleotide yields a linear translation of 3.4 Å (Chapter 7), so that the mRNA would be at least 1,500 Å long. In contrast the diameter of a ribosome is only about 240–400 Å. These observations lead Rich, Warner, Knopf, and Hall in 1962 to propose that the translation of a single mRNA might be carried out simultaneously by several ribosomes (i.e., polyribosomes) attached to and moving in succession along the message. For example, in the case of globin chain synthesis, four or more ribosomes could be attached to the mRNA.

Rich and his co-workers incubated rabbit reticulocytes for short periods in a medium containing ^{14}C-labeled amino acids. During this brief incubation, radioactive segments were added to each nascent chain. Following this, the cells were lysed and the lysate fractionated by centrifugation through a sucrose density gradient. Fractions collected from the gradient at the conclusion of centrifugation were examined in two ways: (1) the distribution of ribosomes through the gradient was determined by measuring the ultraviolet light absorption of the ribosomal RNA, and (2) the distribution of nascent polypeptides was determined from the radioactivity of the collected fractions. Typical results are shown in Figure 21–32. Two UVL-absorbing regions were identified in the density gradient. The first (i.e., least rapidly sedimenting) peak (fractions 24 to 29), which corresponded to particles of about 80 *S* and represented *single* ribosomes, had no radioactivity associated with it. Instead, the radioactivity was distributed over a region of the gradient containing more rapidly sedimenting (i.e., larger) particles (i.e., fractions 10 to 20). This indicated that protein synthesis in reticulocytes took place on structures that were larger than individual ribosomes, and Rich suggested that these were groups of ribosomes held together by mRNA.

When the enzyme ribonuclease was added to the lysate prior to centrifugation, the rapidly sedimenting peak disappeared while the first peak increased in size and was now associated with the radioactivity. This result, of course, supported Rich's proposal. Further confirmation came from

Figure 21–32
Distribution of ^{14}C-labeled amino acids and UVL-absorbing material in reticulocyte lysates subjected to density gradient centrifugation. (See text for explanation.)

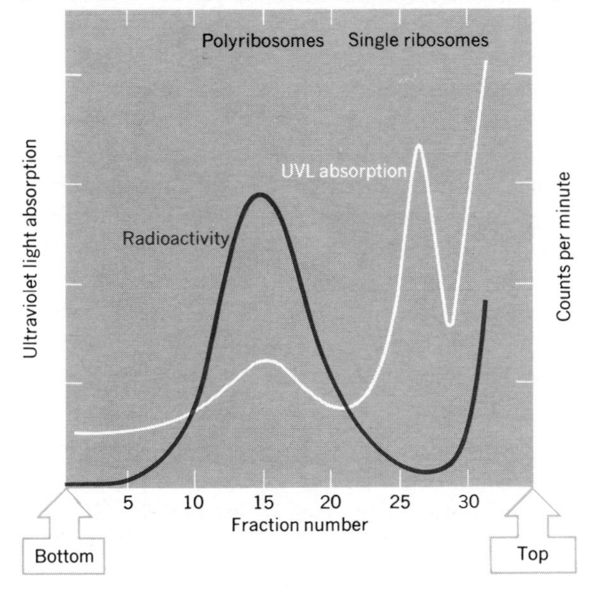

Figure 21–33
Polysomes. The messenger RNA strands linking neighboring ribosomes of each polysome are indicated by arrows. Magnification, 200,000 X. (Courtesy of Dr. O. L. Miller, from O. L. Miller, B. A. Hamkalo, and C. A. Thomas, *Science 169*, 394, 1970. Copyright © 1970 by the American Association for the Advancement of Science.)

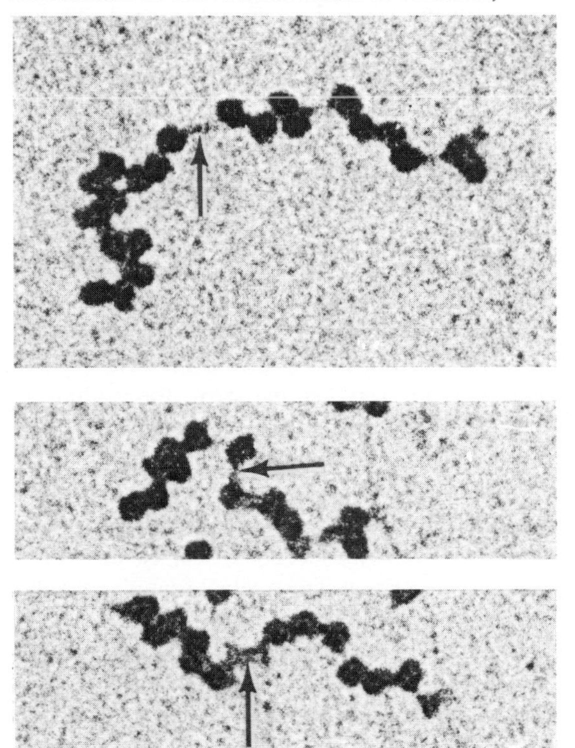

electron-microscopic examination of the fractions, which revealed that the first peak contained single ribosomes, while the rapidly sedimenting peak contained clusters of ribosomes—the further down the gradient the sample was withdrawn for microscopic examination, the larger was the observed cluster size. The predominant size cluster contained five ribosomes (called a *pentamer*), with smaller numbers of clusters containing six ribosomes *(hexamers)* and four ribosomes (tetramers).

Subsequently, electron-microscopic studies using negative staining techniques showed that the ribosomes were connected by a thin thread about 10–15 Å thick and about 1500 Å long and were separated by gaps varying from 50 to 150 Å. This corresponded to the diameter of an RNA molecule and the approximate length predicted for the globin messenger. Some polysomes are shown in Figure 21–33.

In the model for polysome function originally proposed by Rich, the several ribosomes move along the mRNA strand, each synthesizing a polypeptide chain. When a ribosome reaches the end of the message, it detaches, while at the other end, another ribosome attaches to the mRNA.

Although it is now clear that polysome function in globin chain synthesis is not precisely as Rich predicted, the fundamentals of his model remain valid. The size of a polysome depends on both the length of the mRNA being translated and the amount of time required for initiation, elongation, and termination. For example, recent studies by S. H. Boyer have established that polysomes engaged in the synthesis of alpha globin contain an average of four ribosomes, whereas beta globin chain polysomes contain an average of six ribosomes. This occurs despite the fact that both globin chains are about the same size. The difference is due to the lower frequency with which initiation occurs on alpha globin chain mRNA (αmRNA) (which in the case of globin chain synthesis is the *rate-limiting* step). Elongation and termination occur at about the same rates for both globin chains. It was noted earlier that equal amounts of alpha and beta chains are produced in normal erythrocytes. If this is so, one might ask how such a balance is maintained in view of the polysomal differences just noted. The balance results in part from the presence in these cells of larger quantities of αmRNA than βmRNA, which thus compensates for differences in the frequency of initiation. The larger quantities of αmRNA result, in turn, from the presence of *two pairs* of alpha chain structural genes and only one pair of beta chain structural genes (see Chapter 4).

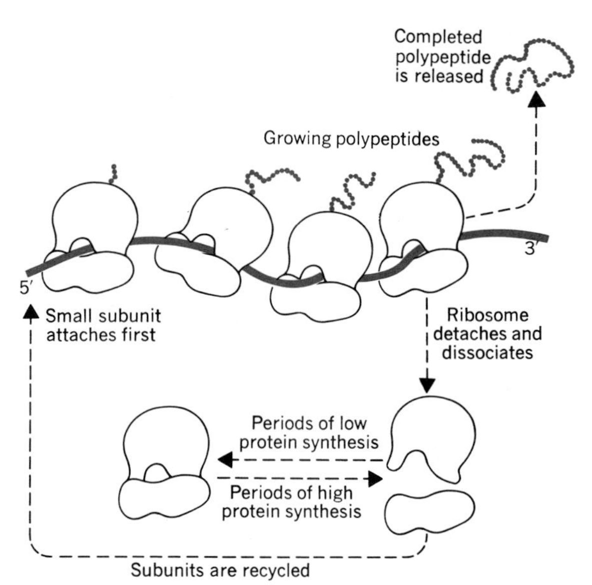

Figure 21–34
The polyribosome cycle.

The current model for polysome function in eucaryotic cells is shown in Figure 21–34 and does not differ dramatically from that originally proposed by Rich. The detachment of the ribosome from mRNA is accompanied by the release of the completed (and probably folded) polypeptide chain. The ribosome immediately dissociates into its small and large subunits, which then enter a common cell subunit pool. During periods of reduced protein synthesis, subunits combine to form a pool of inactive but intact ribosomes (monosomes), but during periods of active cellular protein synthesis, these ribosomes again dissociate into subunits. As noted earlier, the small subunit attaches to the 5′ end of the mRNA before the large subunit. There is some evidence that during periods of active protein synthesis, subunits recently released from the messenger are *preferentially* reused for translation because of their closer proximity to the 5′ end of mRNA than other subunits randomly spread through the cell.

Kinetics of Transcription and Translation in Maturing Erythrocytes. Because so much is known about hemoglobin and the development of erythrocytes, it is possible to carry out a number of calculations that give us some insight into the frequency and rate of transcription and translation. Since there are about 5.4×10^9 erythrocytes and 0.16 g of hemoglobin in each 1 cm^3 of human blood (these values are readily determined in the laboratory), this implies that a single red blood cell contains

$$0.16 \text{ g of hemoglobin per } 5.4 \times 10^9 \text{ cells} = 3 \times 10^{-11} \text{ g of hemoglobin} \qquad (21\text{–}1)$$

The molecular weight of human hemoglobin A is about 64,500, and one gram-molecular weight (gMW) would contain 6.02×10^{23} molecules of hemoglobin (i.e., Avogadro's number); hence a single red blood cell contains

$$\frac{(3 \times 10^{-11} \text{ g Hb / cell})(6.02 \times 10^{23} \text{ Hb molecules / gMW})}{(6.45 \times 10^4 \text{ g Hb / g MW})}$$

$$= 2.8 \times 10^8 \text{ Molecules of hemoglobin} \qquad (21\text{–}2)$$

It has already been noted that the synthesis of hemoglobin in the maturing red blood cell occurs over a period of 3 days. Knowing this, and using the result of equation 21–2 above, it is possible to calculate the *average* number of hemoglobin molecules whose synthesis is completed *each second* during this period of differentiation. It would be

$$\frac{2.8 \times 10^8 \text{ molecules of Hb}}{(3 \text{ days})(24 \text{ hr / day})(60 \text{ min / hr})(60 \text{ sec / min})} = 1.1 \times 10^3 \text{ molecules of hemoglobin completed per second} \qquad (21\text{–}3)$$

Therefore, an average of 1100 molecules of hemoglobin are completed in each second of the 3-day maturation period. This, of course, assumes that synthesis is continuous and also uniform over the entire 3 days, which is not actually the case. Therefore, the real value would be greater than this during the period of peak synthesis and lower at other times. However, for purposes of this discussion we can assume continuous and uniform synthesis.

Since there are four globin chains in each hemoglobin molecule, there would be $4(1.1. \times 10^3)$ or 4400 globin chains completed per second in a single red blood cell. The experiments of Dintzis and others indicate that between 60 and 90 seconds are required for the synthesis of a single globin chain. If for convenience we use the larger value, then during any given second there would be

$$(90 \text{ sec})(4400 \text{ chains/sec}) = 396{,}000 \text{ chains} \qquad (21\text{–}4)$$

of globin production in the cell. Since the most common form of alpha chain polysome in the cell is the tetramer, this means that there must be 198,000/4 or 49,500 molecules of αmRNA in the cell. The most common beta chain polysome is the hexamer, implying that there are 198,000/6 or 33,000 molecules of βmRNA. This assumes that each molecule of mRNA is stable for the whole 3-day maturation period and is available at the outset of hemoglobin synthesis. Since neither assumption is entirely valid, the actual amounts of the globin mRNAs in the cell are most likely considerably higher during peak hemoglobin synthesis. Each erythroblast contains four structural genes for alpha chains and two structural genes for beta chains; therefore, each alpha chain structural gene would have to be transcribed into αmRNA 12,375 times (i.e., 49,500/4), and each beta chain structural gene would have to be transcribed into βmRNA 16,500 times (i.e., 33,000/2).

If the developing red blood cells contains an average of 49,500 molecules of αmRNA, and these are used to produce sufficient alpha globin chains for 2.8×10^8 molecules of hemoglobin, then the average αmRNA is translated 11,313 times. That is,

$$\frac{(2.8 \times 10^8 \text{ molecules Hb})(2\alpha \text{ globin chains / Hb})}{(49{,}500 \text{ molecules } \alpha\text{mRNA})} \qquad (21\text{–}5)$$

= 11,313 alpha globin chains per αmRNA. The corresponding translation frequency for βmRNA would be

$$\frac{(2.8 \times 10^8 \text{ molecules Hb})(2\beta \text{ globin chains / Hb})}{(33{,}000 \text{ molecules } \beta\text{mRNA})} \qquad (21\text{–}6)$$

= 16,969 times.

These figures indicate that the globin mRNAs are extremely stable (recent experimental evidence points to a half-life of at least several hours), especially in comparison with procaryotic mRNAs, which may be translated only once.

In all of the above calculations, we have been considering *averages* only. The synthesis of hemoglobin in the maturing erythrocyte is *not* uniform or continuous throughout development. Instead, hemoglobin synthesis reaches a maximum in early development. Therefore, during the period of maximum synthesis more molecules of mRNA would be required, and the number of globin chains produced from a single mRNA molecule would be lower. Despite this, the high frequency of transcription and translation in this cell is clearly apparent.

In eucaryotic cells, transcription occurs in the cell nucleus and translation occurs later in the extranuclear hyaloplasm. In procaryotes, which have no nucleus, not only do transcription and translation occur in the same region of the cell but they also occur at the same time.

In bacteria such as *E. coli,* transcription of DNA is accompanied by the translation of the *nascent messenger RNA*. The mRNA is synthesized beginning with its 5′ end, and as soon as the mRNA strand is long enough, a ribosome attaches to the messenger and begins translation. As mRNA synthesis proceeds, more ribosomes attach to the elongating strand to form a polysome. O. L. Miller and others have provided elegant proof of this in the form of electron micrographs such as the one shown in Figure 21–35. In this figure, a portion of the *E. coli* chromosomal DNA appears as a thin filament being actively transcribed by a number of RNA polymerase molecules into mRNA strands. (RNA polymerase is the enzyme that transcribes DNA into RNA.) To each of these mRNA strands, ribosomes have attached to form polysomes. In Figure 21–35, it appears that transcription proceeds from left to right along the DNA, since the mRNA length (and therefore polysome size) exhibits a general increase in that direction. Unfortunately, electron micrographs do not reveal the growing nascent polypeptide chains on the ribosomes, since the amino acids that make up the polypeptides are too small to be resolved.

CoTranslational and Post-Translational Protein Modification

CoTranslational Modification

In many instances, a number of changes are made in the structure and organization of polypeptide chains *during* their synthesis; these are called **cotranslational modifications** and include (1) **deformylation,** (2) **amino acid cleavage,** (3) **side-chain alteration,** (4) **disulfide bridge formation,** (5) **sugar addition,** and (6) **tertiary folding.**

Deformylation. In procaryotes and in eucaryotic mitochondria and chloroplasts, the formyl group of the N-terminal methionine of the growing polypeptide chain is enzymatically cleaved (see earlier).

Amino Acid Cleavage. In both procaryotes and eucaryotes, N-terminal methionine and occasionally other amino acids as well are enzymatically cleaved from the free N-terminus by an *aminopeptidase*.

Side-Chain Alteration. The R groups of certain amino acids are often altered following inclusion of the amino acid into the growing polypeptide chain. For example, during the synthesis of *collagen,* certain proline and lysine residues are hydroxylated (to form hydroxyproline and hydroxylysine, respectively). Other amino acid side chains may be phosphorylated (e.g., serine).

Disulfide Bridge Formation. Juxtaposed sulfhydryl groups of cysteine residues may be oxidized to form disulfide bridges. This normally occurs after tertiary folding (see

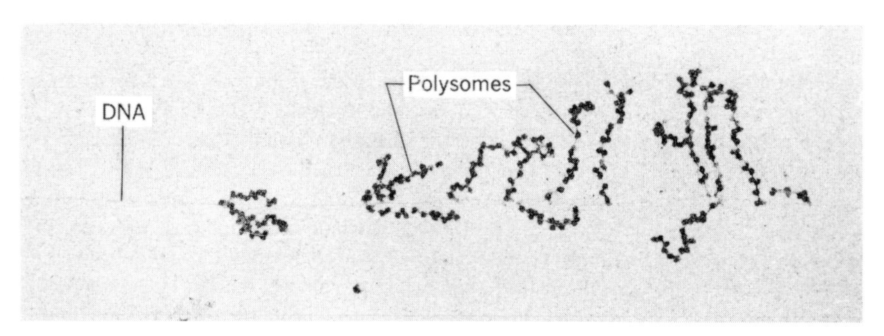

Figure 21–35
Visualization of the simultaneous transcription and translation of of *E. coli* chromosomal DNA. Magnification, 43,000 X. (Courtesy of Drs. O. L. Miller and B. A. Hamkalo, from O. L. Miller, B. A. Hamkalo, and C. A. Thomas, *Science 169,* 394, 1970. Copyright © 1970 by the American Association for the Advancement of Science.)

below) orients these R groups into the necessary steric positions.

Sugar Addition. Sugars may be enzymatically attached to certain amino acids during the synthesis and completion of various *glycoproteins*.

Tertiary Folding. Although some proteins may spontaneously fold to form their biologically active tertiary structure following completion and release of the polypeptide from the ribosome (e.g., *ribonuclease,* see Chapter 4), others undergo tertiary folding during translation. In *E. coli,* tertiary folding of enzymatic polypeptides during their synthesis endows these nascent proteins with catalytic properties prior to termination and release.

Post-Translational Modifications **Post-translational modifications** are changes that occur in protein structure *after* completion and release of the polypeptide have taken place. Some of the modifications already described as cotranslational may also occur following translation. For example, enzymatic hydroxylation and phosphorylation of amino acid side chains, the formation of disulfide bridges, and the addition of sugars to certain residues may occur following release of the completed polypeptide. Moreover, tertiary folding, although begun during translation, is completed following polypeptide release. Some modifications, however, are characteristically post-translational; included in this category are (1) **peptide cleavage,** (2) **quaternary association,** and (3) **addition of prosthetic groups.**

Figure 21–36
Post-translational cleavage of *proinsulin* to form *insulin*.

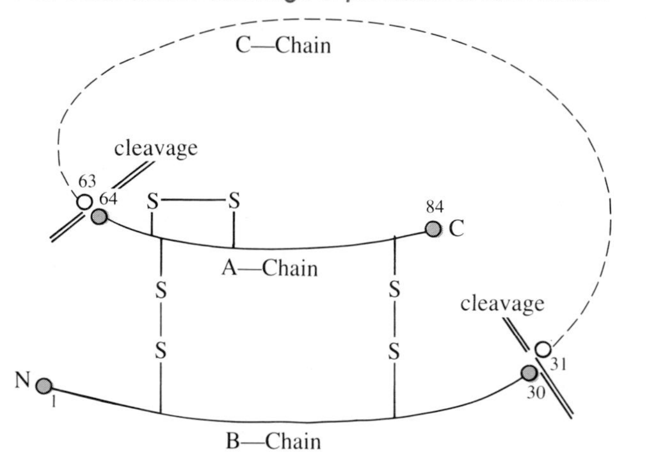

Peptide Cleavage. For some proteins, major changes in structure in the form of cleavage of specific bonds and removal of sections of the polypeptide occur following translation. For example, the *A* and *B* polypeptide chains that comprise the insulin molecule (Chapter 4) are produced by post-translational cleavage of a single translation product (Fig. 21–36) called *proinsulin,* which has no hormonal activity.

The activation of the zymogen *chymotrypsinogen* to form the digestive enzyme *chymotrypsin* (see Chapter 8) serves to illustrate the level of complexity that post-translational peptide cleavage can assume. Chymotrypsinogen is broken into two polypeptides by the enzyme *trypsin,* the product (which is still linked by disulfide bridges) being π-chymotrypsin (Fig. 21–37). π-chymotrypsin acts to catalyze its own conversion to the active digestive enzyme α-chymotrypsin. This activation involves the removal of two dipeptides of π-chymotrypsin, producing a product consisting of three interconnected polypeptide chains. Activation of chymotrypsinogen occurs in the small intestine, and this is where protein digestion takes place, whereas the pancreas, which is the site of production of the enzyme, secretes it in the zymogen form.

The post-translational modifications that produce insulin and chymotrypsin also demonstrate that *a protein consisting of more than one polypeptide chain may not be encoded by a corresponding number of mRNAs (or genes!)* but may be encoded by a single mRNA (or gene). Post-translational peptide cleavage may produce a series of separate polypeptide chains that ultimately make up the final protein product. Indeed, evidence is at hand that indicates that in some procaryotes and in the case of certain viruses, a single polypeptide chain may be cleaved to produce several individual proteins.

Quaternary Association. As noted in Chapter 4 and also earlier in this chapter, some proteins that possess quaternary structure are assembled by the spontaneous interaction of individual polypeptide chains. In the case of hemoglobin,

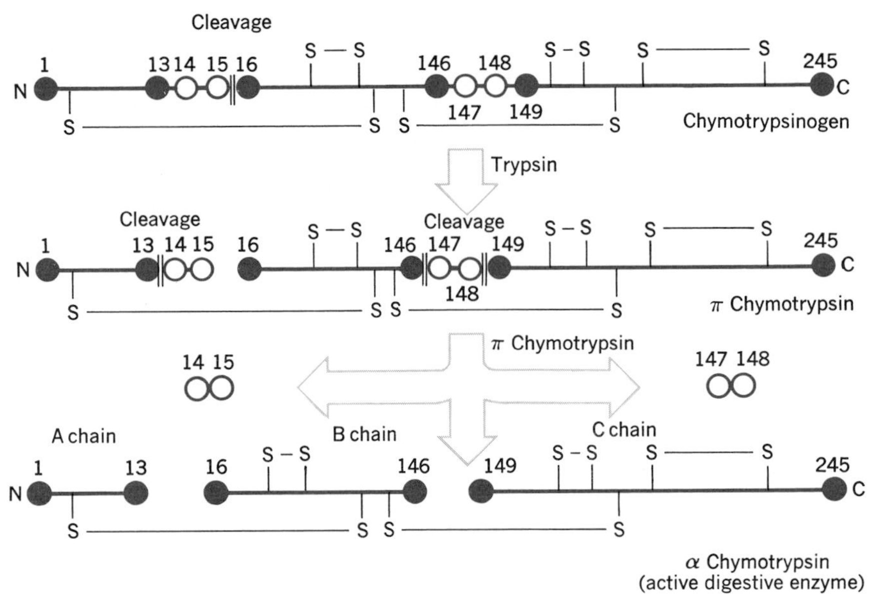

Figure 21–37
Post-translational cleavage of *chymotrypsinogen* to form *chymotrypsin*.

for example, separate alpha and beta chains spontaneously combine to form asymmetric dimers, and these combine to form the funtional tetramer. Formation of the dimer may be post-translational for the alpha chain and cotranslational for the beta chain (i.e., the completed and released alpha chains combine with growing nascent beta chains). Assumption of quaternary structure is accompanied by the formation of stabilizing bonds between neighboring protein subunits and modification of the individual tertiary structures they previously possessed.

Addition of Prosthetic Groups. The prosthetic groups of enzymes and other proteins are attached following release of the completed polypeptide chains, and attachment may be spontaneous or catalyzed enzymatically. In the case of hemoglobin, the insertion of the heme groups occurs *after quaternary association is complete* and begins with the alpha globin chains.

The completion of the hemoglobin molecule by the successive attachment of its four heme groups brings to a conclusion a synthetic process about which more is known than for any other so complex a protein. In this chapter, we have seen that (1) heme regulates the initiation of globin chain synthesis; (2) globin chain elongation is not uniform but that alpha and beta chains are produced in equal amounts; (3) completed alpha chains combine with nascent beta chains; (4) asymmetric dimers spontaneously associate to form tetramers; and (5) heme insertion is sequential. To this must be added the long-known fact that heme acts as a negative effector of its own synthesis through feedback inhibition of an enzyme catalyzing an initial step in the heme biosynthetic pathway. In this fashion, the production of more heme than can be utilized for assembly of hemoglobin is avoided. Acting in concert, all of these mechanisms provide for the 1:1:1 ratio of alpha chain, beta chain, and heme group synthesis in the maturing red blood cell. Figure 21–38 summarizes our existing knowledge of the regulation of the synthesis of the hemoglobin protein.

Transfer RNA Specialization

When W. F. Anderson and J. M. Gilbert reported in 1969 that addition of isolated tRNA fractions to a reticulocyte cell-free system synthesizing globin chains altered the bal-

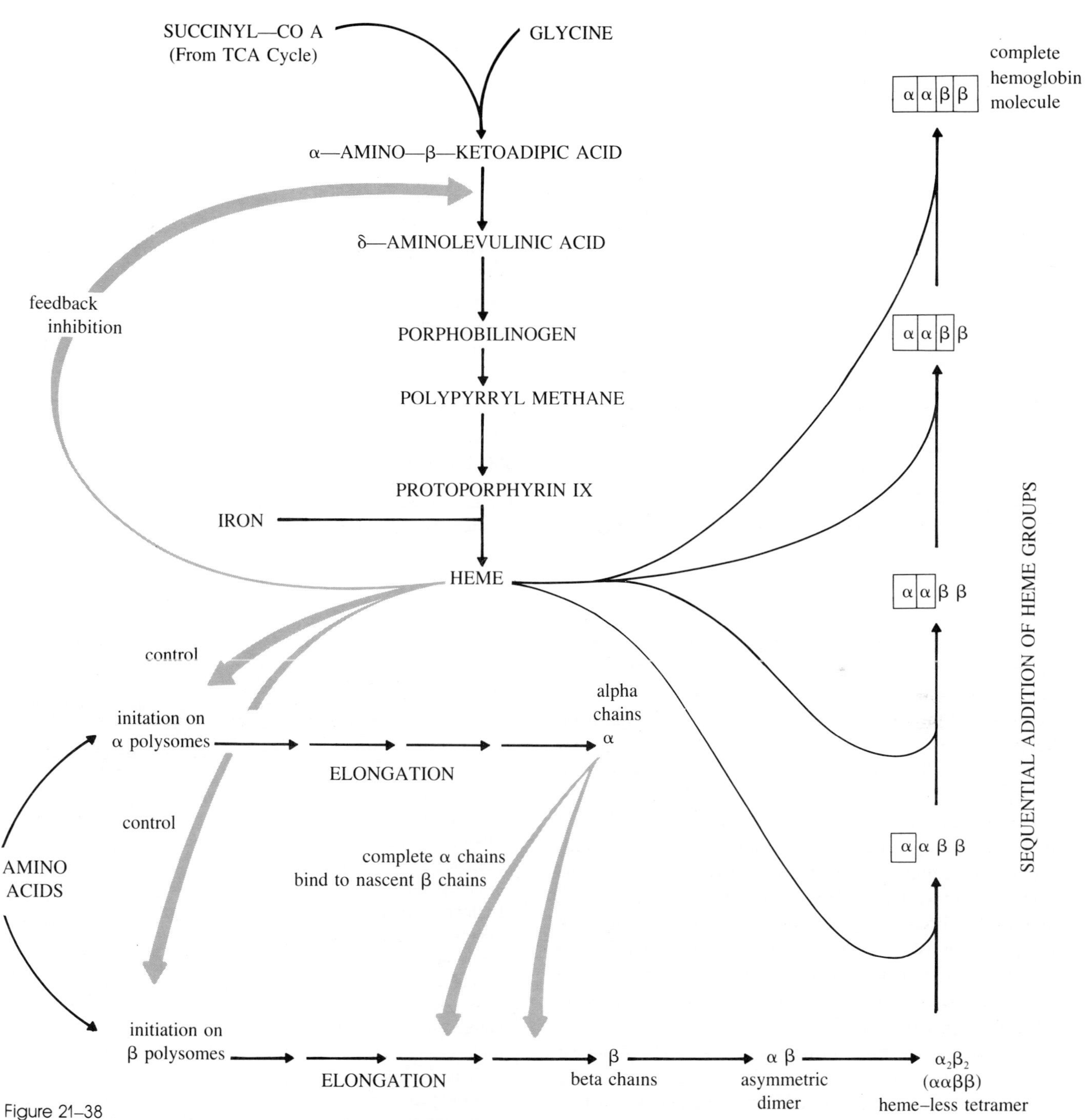

Figure 21–38

Synthesis of the hemoglobin molecule and its regulation. Arrows in black are synthetic pathways; bold arrows in color show mechanisms of control.

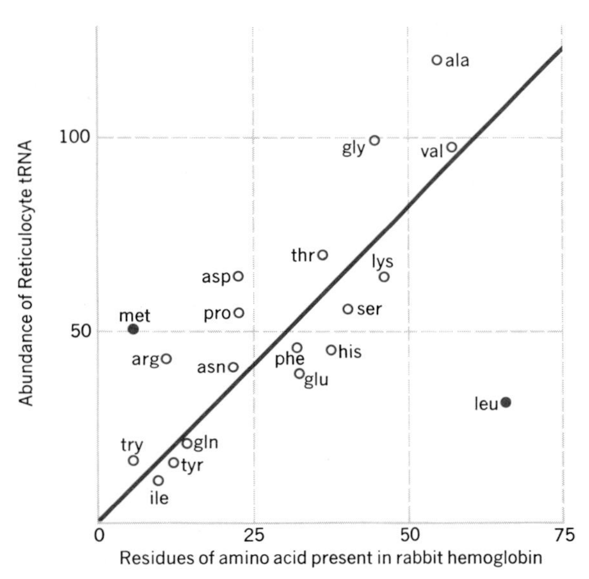

Figure 21–39
Transfer RNA specialization in the reticulocyte.

ance of alpha and beta globin chain production, the question was raised as to whether cells might have a rather specialized complement of tRNA. That this is indeed the case was borne out by the extensive studies of D. W. E. Smith on the amounts and types of tRNAs present in the reticulocyte. As Figure 21–39 shows, a direct relationship exists between the frequency with which each type of amino acid occurs in the hemoglobin molecule and the abundance in the cell of the tRNAs specific for that amino acid species. This implies that cellular mechanisms exist that coordinate the production of various tRNAs according to the types and amounts of different amino acids present in the proteins of that cell—a most striking implication! This notion is supported by additional evidence of tRNA specialization in other cells, including silk-gland cells, lymphocytes, and cells of the pancreas and liver.

In Figure 21–39, the coordinates for met and leu appear to be exceptions to the linear distribution. For met, there appears to be an excess of $tRNA^{Met}$ (and $tRNA^{Met}_{i}$) (the additional met residues involved in chain initiation are already taken into account in the data), while for leu, there is a shortage of $tRNA^{Leu}$. The shortage of leucine tRNA in reticulocytes is believed to be yet another rate-limiting control factor for the production of globin chains in the maturing cell.

Inhibitors of Protein Synthesis

Many substances are known to act as inhibitors of various stages of protein synthesis. Included among these are a number of *antibiotics* produced by one strain of microorganism and lethal to other strains of the same or a different species. Some of the best understood inhibitors of protein synthesis are listed in Table 21–13. Because the actions of many of these inhibitors are quite specific, they have proved extremely useful tools in the step-by-step elucidation of the mechanism of protein synthesis.

Inhibitors of *Both* Procaryotic and Eucaryotic Protein Synthesis

Aurintricarboxylic acid inhibits formation of the initiation complex by preventing the association of mRNA with the

Table 21–13
Inhibitors of Protein Synthesis in Procaryotic and Eucaryotic Cells

Inhibitor	Effective in: Procaryotes	Eucaryotes
Anisomycin	−	+
Aurintricarboxylic acid	+	+
Chloramphenicol	+	−
Colicin E3	+	−
Cycloheximide	−	+
Diphtheria toxin	−	+
Edeine	+	+
Erythromycin	+	−
Fusidic acid	+	+
Pactamycin	−	+
Puromycin	+	+
Ricin	−	+
Sodium fluoride	−	+
Sparsomycin	−	+
Streptomycin	+	−
Tetracycline	+	+
Trichodermin	−	+

small ribosomal subunit. Inhibitors of initiation are readily distinguished from inhibitors blocking other stages of protein synthesis because of the *delay effect* that follows their administration. That is, protein synthesis continues for a short time after administration of the inhibitor, since peptide chains whose growth had already begun are unaffected and grow to completion. *Edeine,* a polypeptide isolated from *Bacillus brevis,* inhibits the binding of aminoacyl-tRNA and *N*-formylmet-$tRNA^{Met}_{f}$ (in procaryotes) to the small subunit. *Fusidic acid* is a steroidal antibiotic; in procaryotes, it inhibits the binding of aminoacyl-tRNA to the ribosome, whereas in eucaryotes, it inhibits translocation by reacting with elongation factor.

Puromycin, an inhibitor of protein synthesis, was one of the first of such inhibitors to have its specific effect determined. This antibiotic *mimics* aminoacyl-tRNA and binds to the free *A* site of ribosomes engaged in protein synthesis. Catalytic formation of a bond between the nascent polypeptide and puromycin is followed by the release of the *peptidyl-puromycin* from the ribosome, since no further elongation is possible. The specific effects of puromycin have been used to advantage for studies of nascent chain length, the kinetics of chain elongation, and the identification of the effects of other antibiotics. *Tetracycline* inhibits protein synthesis by blocking aminoacyl-tRNA binding to the small subunit.

Inhibitors Specific for Procaryotes

Chloramphenicol (chloromycetin) binds to the large subunit of procaryotic ribosomes and interferes with the functioning of *peptide synthetase,* thereby inhibiting chain elongation. *Colicin E3* inhibits protein synthesis in procaryotes by interfering in some manner with the functioning of the small subunit. *Erythromycin* binds to ribosomes that are *not* engaged in protein synthesis, preventing their potential participation, but does not bind to ribosomes containing nascent chains (i.e., ribosomes that are part of a functioning polysome). *Streptomycin* was one of the earliest discovered antibiotics and was employed as an agent against bacterial infection for many years before its specific chemical actions were known. Streptomycin binds to protein S12 of the small ribosome subunit, causing release of N-formylmet-$tRNA^{Met}_{f}$ from initiation complexes (thereby preventing initiation of chain growth) and also causing misreading of the codons of mRNA by ribosomes already involved in chain elongation.

Inhibitors Specific for Eucaryotes

Anisomycin is an antibiotic produced by *Streptomyces* that inhibits peptide bond formation when bound to the small ribosomal subunit. *Cycloheximide* binds to the large subunit, preventing the translocation of tRNA in the *A* site to the *P* site. *Diphtheria toxin* (produced by a strain of *Corynebacterium diphtheriae*) inhibits protein synthesis through its action on EF-2 *(translocase)*. EF-2 exists in cells in two forms—ribosome-bound and free. Diphtheria toxin acts enzymatically to alter free EF-2, rendering the factor inactive. Ribosome-bound EF-2 is not susceptible to inactivation by the toxin.

Pactamycin (produced by a strain of *Streptomyces*) binds to free small subunits (not to small subunits already part of polysomes) where it prevents initiation by inhibiting binding of met-$tRNA^{Met}_{i}$ and formation of the initiation complex. The toxic effects of *ricin* (a protein present in the castor bean) have been known for nearly a century. Ricin consist of two polypeptide chains (linked by disulfide bridges), one of which acts as the inhibitor once incorporated into the cell. Ricin acts on the large subunit, preventing formation of the 80 *S* initiation complex. Like ricin, *sodium fluoride* acts as an inhibitor of initiation; NaF blocks addition of the large subunit to mRNA.

Sparsomycin, another antibiotic produced by *Streptomyces,* inhibits the association of the amino acid moiety of aminoacyl-tRNA from binding to the large subunit and, in so doing, blocks peptide synthetase. *Trichodermin* is the only chemical compound so far identified as a specific inhibitor of the termination stage of polypeptide synthesis.

Inhibitors of Organellar Protein Synthesis

Protein synthesis by mitochondrial and chloroplast ribosomes is also subject to inhibition by certain antibiotics and other chemicals. Shortly after the initial demonstration of organellar protein synthesis, it was found that chloramphenicol, a strong inhibitor of procaryote protein synthesis, blocks synthesis in mitochondria and chloroplasts; while cycloheximide, which blocks eucaryote cytoplasmic ribosomal protein synthesis, is without effect on mitochondrial and chloroplast synthesis. These observations provided added credence for the notion that procaryotic cells, mitochondria, and chloroplasts have a common evolutionary origin. It is now clear, however, that the picture is considerably more complex. For example, streptomycin, which inhibits procaryotic protein synthesis, fails to inhibit mitochondrial protein synthesis in yeast cells. Other antibiotics inhibit mitochondrial protein synthesis but have no effect on procaryotes. Erythromycin inhibits the synthesis of proteins in procaryotes, yeast mitochondria, and chloroplasts but fails to block protein synthesis in mammalian mitochondria. The latter observation suggests that the nature of mitochondrial protein synthesis varies among different groups of eucaryotes. In general, mitochondria from higher eucaryotes are more resistant to inhibitors of procaryotic protein synthesis than are mitochondria from lower eucaryotes. In chloroplasts, protein synthesis is inhibited by the same agents that block this process in procaryotic cells.

The differential sensitivity of eucaryotic cytoplasmic and mitochondrial ribosomes to specific inhibitors provides a means for examining the sources of certain mitochondrial proteins. The synthesis of a mitochondrial protein in the presence of cycloheximide (a cytoplasmic inhibitor) indicates that the mitochondria are the source of the protein, whereas synthesis of the protein in the presence of chloramphenicol indicates that the mitochondrial protein is produced in the cytoplasm and then moves to the mitochondria. Experimentally, determinations of this sort are carried out by incubating cells in media containing both radioactively labeled amino acids and inhibitor. The synthesis of the mitochondrial protein is manifested by the appearance of radioactivity in the proteins later isolated from the mitochondria.

Using this approach, it has been possible to show that perhaps 85 to 90% of all mitochondrial ribosomal proteins are synthesized in the cytoplasm and then make their way into the mitochondria where, together with mitochondrial rRNA, they are assembled into ribosomes. Of the seven polypeptides that make up the enzyme *cytochrome oxidase*, four are synthesized in the cytoplasm and three are synthesized in the mitochondria.

Summary

Using isotopes of nitogen and carbon as tracers of amino acid metabolism, a number of investigators working in the 1940s and early 1950s showed that protein turnover in tissues was not simply a function of "wear and tear." Instead, cell protein undergoes continuous breakdown and synthesis, the latter being a property of the **microsomal fraction** of tissue homogenates. It is the **ribosomes** in the microsomal phase that carry out the assembly of the polypeptide chains that make up proteins, and these tiny organelles are currently the object of intensive biochemical and electron-microscopic analyses. The ribosomes of procaryotic and eucaryotic cells are made up of two **subunits,** each subunit containing a specific combination of **proteins** and **ribosomal** ribonucleic acids (rRNA). Association of these subunits with **messenger** RNA (mRNA) is followed by **translation** of the mRNA base sequence (i.e., its **codons**) into the primary structure of polypeptides. **Transfer** RNAs (tRNA) combine with specific amino acids and enter the ribosome-mRNA complex. The succession of **aminoacyl-tRNA** species entering the complex and donating amino acid residues to the growing polypeptide is prescribed by the mRNA codon sequence. All the RNAs are **transcribed** from DNA and **processed** (**cleaved, trimmed,** or otherwise modified) prior to becoming functionally active.

Ribosomes may be **free** in the cytoplasm or **attached** to intracellular membranes, each variety functioning in the assembly of different proteins. Individual polypeptides are assembled beginning at the N-terminus and proceeding toward the C-terminus. The rate of chain elongation may vary in different regions of the polypeptide, and for proteins composed of two or more chains, release of one completed chain from the ribosome may be dependent on association of a second, completed chain. In most instances, a single message is translated by several ribosomes traveling in close succession along the mRNA molecule; such complexes are called **polysomes.** Initiation and termination of chain growth, as well as the elongation cycle, require specific factors and are catalyzed by specific enzymes that are either constituents of the ribosomes or dissolved in the cytoplasm. Modifications of a protein may occur during or following translation. **Cotranslational** modifications include **deformylation** at the N-terminus, amino acid **cleavage, side-chain alteration,** formation of **disulfide bridges, addition of sugars,** and **tertiary folding.** Modifications that are characteristically **post-translational** include **peptide cleavage, quaternary association,** and **addition of prosthetic groups.**

Many substances, including a number of **antibiotics,** act as inhibitors of protein synthesis. Some inhibitors are effective only in procaryotes and some only in eucaryotes. Still others inhibit protein synthesis in both procaryotic and eucaryotic cells.

References and Suggested Reading

Articles and Reviews

Boyer, S. H., Smith, K. D., and Noyes, A. N., Immunological purification and characterization of hemoglobin chain-synthesizing polysomes. *Hemoglobin: Comparative Molecular Biology Models for the Study of Disease, Annals, N.Y. Acad. Sci. 241*, 204 (1974).

Brawerman, G., Characteristics and significance of the polyadenylate sequence in mammalian messenger RNA, in *Progress in Nucleic Acid Research and Molecular Biology*, Vol. 17 (W. E. Cohn, ed.), Academic Press, New York, 1976, p. 118.

Brimacombe, R., Nierhaus, K. H., Garrett, R. A., and Wittman, H. G., The ribosome of *Escherichia coli*, in *Progress in Nucleic Acid Research and Molecular Biology*, Vol. 18 (W. E. Cohn, ed.), Academic Press, New York, 1976, p.1.

Brimacombe, R., Stoffler, G., and Wittman, H. G., Ribosome structure, in *Annual Review of Biochemistry*, Vol. 47 (E. E. Snell *et al.*, eds.), Annual Reviews, Inc., Palo Alto, 1978, p. 217.

Burka, E. R., Protein synthesis by membrane-bound reticulocyte ribosomes. *Hemoglobin: Comparative Molecular Biology Models for the Study of Disease, Annals N.Y. Acad. Sci. 241*, 191 (1974).

Caskey, C. T., The universal RNA genetic code. *Q. Rev. Biophys. 3*, 295 (1970).

Clark, B. F. C., and Marcker, K. A., How proteins start. *Sci. Am. 218* (1), 36 (Jan., 1968).

Darnell, J. E., mRNA structure and function, in *Progress in Nucleic Acid Research and Molecular Biology*, Vol. 19 (W. E. Cohn and E. Volkin, eds.), Academic Press, New York, 1976, p. 493.

Dintzis, H. M., Assembly of the peptide chains of hemoglobin. *Proc. Natl. Acad. Sci. 47*, 247 (1961).

Englander, S. W., and Page, L. A. Interpretation of data on sequential labeling of growing polypeptides. *Biochem. Biophys. Res. Commun. 19*, 565 (1965).

Erdmann, V. A., Structure and function of 5 S and 5.8 S RNA, in *Progress in Nucleic Acid Research and Molecular Biology*, Vol. 18 (W. E. Cohn, ed.), Academic Press, New York, 1976, p. 45.

Furuichi, Y., Muthukrishnan, J. T., and Shatkin, A. J., Caps in eukaryotic mRNAs, in *Progress in Nucleic Acid Research and Molecular Biology*, Vol. 19 (W. E. Cohn and E. Volkin, eds.), Academic Press, New York, 1976, p. 3.

Goodenough, U. W., and Levine, R. P., The genetic activity of mitochondria and chloroplasts. *Sci. Am. 223*, 22 (May, 1970).

Holley, R. W., The nucleotide sequence of a nucleic acid. *Sci. Am. 214* (2), 30 (Feb., 1966).

Hunt, T., The control of globin synthesis in rabbit reticulocytes. *Hemoglobin: Comparative Molecular Biology Models for the Study of Disease, Annals, N.Y. Acad. Sci. 241*, 223 (1974).

Hunt, T., Control of globin synthesis. *Haemoglobin: Structure, Function and Synthesis, Br. Med. Bull. 32*, 257 (1976).

Itano, H. A., Genetic regulation of peptide synthesis in hemoglobin. *J. Cell. Physiol. 67* (suppl. 1), 65 (1966).

Kearns, D. R., High-resolution nuclear magnetic resonance investigations of the structure of tRNA in solution, in *Progress in Nucleic Acid Research and Molecular Biology*, Vol. 18 (W. E. Cohn, ed.), Academic Press, New York, 1976, p. 91.

Kim, S. H., Three-dimensional structure of transfer RNA, in *Progress in Nucleic Acid Research and Molecular Biology*, Vol. 17 (W. E. Cohn, ed.), Academic Press, New York, 1976, p. 182.

Lodish, H. F., Translation control of protein synthesis, in *Annual Review of Biochemistry*, Vol. 45 (E. E. Snell et al., eds.), Annual Reviews Inc., Palo Alto, 1976, p. 39.

Marcker, K., and Sanger, F., N-formyl-methionyl-s-RNA. *J. Mol. Biol. 8*, 835 (1964).

McKnight, S. L., and Miller, O. L., Electron microscopic analysis of chromatin regulation in the cellular blastoderm *Drosophila melanogaster* embryo. *Cell 12*, 795 (1977).

Miller, O. L., The visualization of genes in action. *Sci. Am. 228* (3), 34 (Mar., 1973).

Miller, O. L., and Beatty, B. R., Visualization of nucleolar genes. *Science 164*, 955 (1969).

Miller, O. L., Hamkalo, B. A., and Thomas, C. A., Visualization of bacterial genes in action. *Science 169*, 392 (1970).

Morris, A. J., Slabaugh, R. C., and Protzel, A., Size characteristics of nascent globin chains (peptidyl tRNA) in the reticulocyte. *Hemoglobin: Comparative Molecular Biology Models for the Study of Disease, Annals N.Y. Acad. Sci. 241*, 310 (1974).

Nikolaev, N., and Hadjiolov, A. A., Maturation of ribosomal ribonucleic acids and the biogenesis of ribosomes, *Prog. Biophy. Mol. Biol. 31*, 95 (1976).

Nomura, M., Ribosomes. *Sci. Am. 221* (4), 28 (Oct., 1969).

Perry, R. P., Processing of RNA, in *Annual Review of Biochemistry*, Vol. 45 (E. E. Snell et al., eds.), Annual Reviews Inc., Palo Alto, 1976, p. 630.

Pestka, S., Insights into protein biosynthesis and ribosome function through inhibitors, in *Progress in Nucleic Acid Research and Molecular Biology*, Vol. 17 (W. E. Cohn, ed.), Academic Press, New York, 1976, p. 217.

Proudfoot, N. J., and Brownlee, G. G., Nucleotide sequences of globin messenger RNA. *Haemoglobin: Structure, Function and Synthesis, Br. Med. Bull. 32*, 251 (1976).

Puvion-Dutilleul, F., Bachellerie, J-P., Zalta, J-P., and Bernhard W., Morphology of ribosomal transcription units in isolated subnuclear fractions of mammalian cells. *Rev. Biol. Cell. 30*, 183 (1977).

Rich, A., Polyribosomes. *Sci. Am. 209* (6), 44 (Dec. 1963).

Rich, A., and Kim, S. H., The three dimensional structure of tRNA. *Sci. Am.* 238(1), 52 (Jan. 1978).

Rich, A., and RajBhandary, U. L., Transfer RNA: Molecular structure, sequence and properties, in *Annual Review of Biochemistry,* Vol. 45 (E. E. Snell, et al., eds.), Annual Reviews, Inc., Palo Alto, 1976, p. 805.

Rich, A., Warner, J. R., and Goodman, H. M., The structure and function of polyribosomes. *Cold Spring Harbor Symp. Quant. Biol. 28,* 269 (1963).

Satir, B., The final steps in secretion. *Sci. Am. 233*(3), 28 (Oct., 1975).

Shore, G. C., and Tata, J. R., Functions for polyribosome-membrane interactions in protein synthesis. *Biochim. Biophys. Acta 472,* 197 (1977).

Smith, D. W. E., Reticulocyte transfer RNA and hemoglobin synthesis. *Science 190,* 529 (1975).

Sommer, A., and Traut, R. R., Identification of neighboring protein pairs in the *Escherichia coli* 30 S ribosomal subunit by cross-linking with methyl-4-mercaptobutyrimidate. *J. Mol. Biol. 106,* 995 (1976).

Spitnik-Elson, P., and Elson, D., Studies on the ribosome and its components, in *Progress in Nucleic Acid Research and Molecular Biology,* Vol. 17 (W. E. Cohn, ed.), Academic Press, New York, 1976, p. 77.

Weissbach, H., and Ochoa, S., Soluble factors required for eukaryotic protein synthesis, in *Annual Review of Biochemistry,* Vol. 45 (E. E. Snell *et al.,* eds.), Annual Reviews Inc., Palo Alto, 1976, p. 191.

Winslow, R. M., and Ingram, V. M., Peptide chain synthesis of human hemoglobins A and A_2. *J. Biol. Chem. 241,* 1144 (1966).

Books, Monographs and Symposia

Lewin, B. M., *The Molecular Basis of Gene Expression,* Wiley Interscience, London, 1970.

Nomura, M., Tissieres, A., and Lengyel, P. (editors), *Ribosomes.* Cold Spring Harbor Monograph Series, Cold Spring Harbor Laboratory, N.Y., 1974.

Masters, C. J., and Holmes, R. S., *Haemoglobin, Isoenzymes and Tissue Differentiation. Frontiers of Biology,* Vol. 42, North-Holland Publishing Co., Amsterdam, 1975.

McConkey, E. H. (editor), *Protein Synthesis,* Vol. 1, Marcel Dekker, Inc., New York, 1971.

McConkey, E. H. (editor), Protein Synthesis, Vol. 2, Marcel-Dekker, Inc., New York, 1976.

Stewart, P. R., and Letham, D. S. (editors), *The Ribonucleic Acids,* Springer-Verlag, New York, 1973.

Tedeschi, H., *Mitochondria: Structure, Biogenesis and Trasducing Function. Cell Biology Monographs,* Vol. 4, Springer-Verlag, Vienna, 1976.

Chapter 22
CILIA, FLAGELLA, MICROFILAMENTS, AND MICROTUBULES

Microfilaments and microtubules are two distinctly different kinds of structures found in most cells. Although different, both are associated with cell movement and support and in some cases are so closely associated with the same functions in the cell that they are collectively referred to as the **microtubule-microfilament system** (Fig. 22–1). The presence of these two types of unbranched, elongated structures in cells has been known since before the turn of the century, but only within the past decade have the techniques for proper fixation and staining been worked out so that good descriptions could be obtained from electron photomicrographs. Microtubules in particular eluded description because their proteins depolymerize at low temperatures, and for many years, standard fixation of cells for electron microscopy was carried out at 0°C.

Microfilaments are elongated, unbranched, proteinaceous strands. The microfilaments usually consist of bundles or groups of proteins sometimes wound in a helix. In most cells the microfilaments are 30–60 Å in diamenter but in other cells the microfilaments may be much thicker. Two types, thick and thin microfilaments, are recognized. The **myosin** filaments of striated muscle cells are about 100 Å in diameter, while in amoebae and slime molds the thick microfilaments are 150–250 Å wide and are tapered at the ends. Striated muscle cells contain two types of microfilaments (usually called myofilaments); myosin, the so-called thick filaments, and actin, the thin filaments. Actin, or proteins closely related to actin, are found in most eucaryotic cells and are not restricted to muscle. In striated muscle cells, the thin F-actin filaments are composed of two strands of protein or protofilaments coiled around each other in a double helix (Fig. 22–2). Each protofilament is made up of monomers of G-actin which have been polymerized.

Microfilaments are disassociated by low concentrations of *cytochalasin B* (Fig. 22–3), a substance derived from the mold *Helminthosporium dematoiderum*. Cells treated with this metabolite lose certain functions attributed to the actions of microfilaments. Some of the more common processes sensitive to cytochalasin B and associated with microfilaments are *phagocytosis, pinocytosis, exocytosis* (Chapter 15), *cytokinesis* (later in this chapter), and *cytoplasmic streaming* (in plant cells). If the concentration of added cytochalasin B is sufficiently high, the actin microfilaments of muscle cells are dissociated and muscle contraction is prevented.

Microtubules differ from microfilaments in several fundamental respects (Table 22–1): (1) they are composed of different kinds of proteins, (2) they possess a different structure, and (3) they are dissociated by a different set of compounds. In microtubules, the protein monomer is *tubulin*, a heterodimer containing two subunits designated α and β. Each subunit consists of a single polypeptide, each the same size (54,000 daltons) but different in composition. In electron photomicrographs, the chains of globular subunits ap-

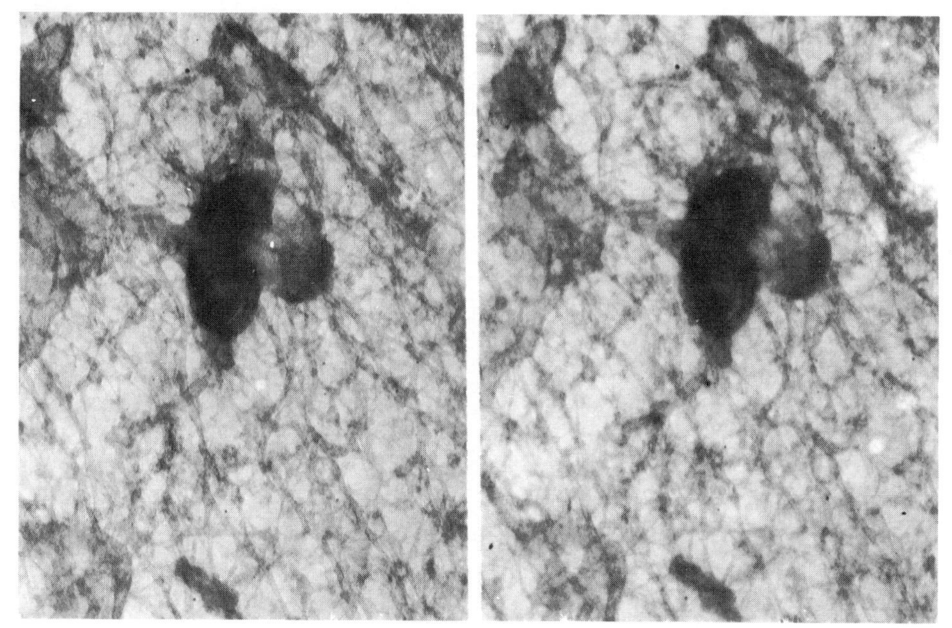

Figure 22–1
Stereo view of a human cell showing numerous microtubules and their interconnections with other elements in the cytoplasm. (Courtesy of K. Porter and J. J. Wolosewick. Copyright *Am. J. Anat. 147,* 303, 1976.)

Figure 22–3
Cytochalasin B structure.

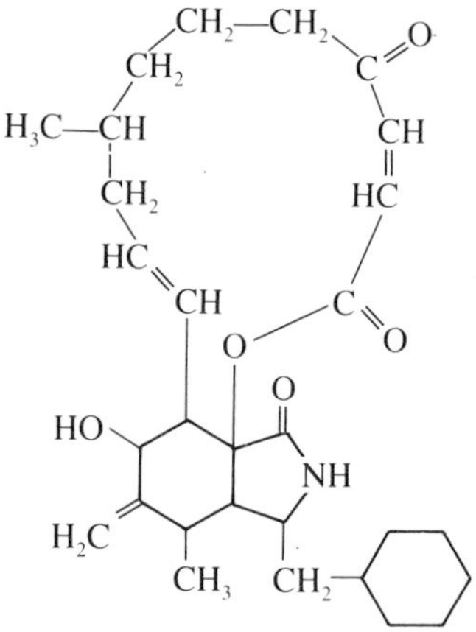

Figure 22–2
The thin filament of striated muscle consisting of two chains of polymerized G-actin monomers.

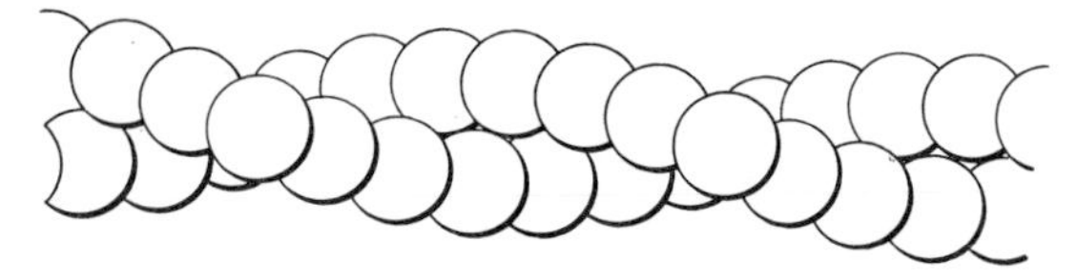

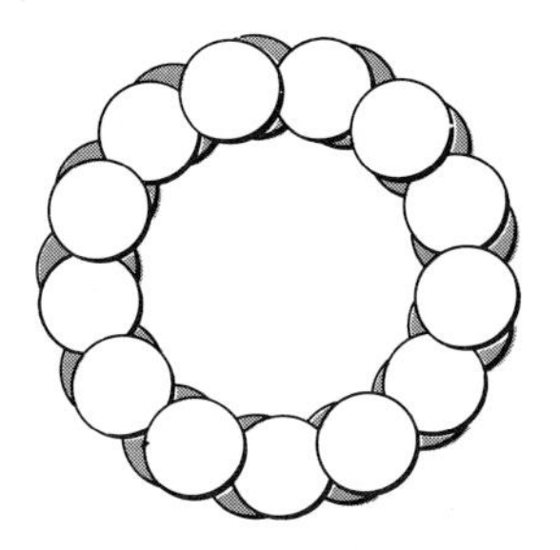

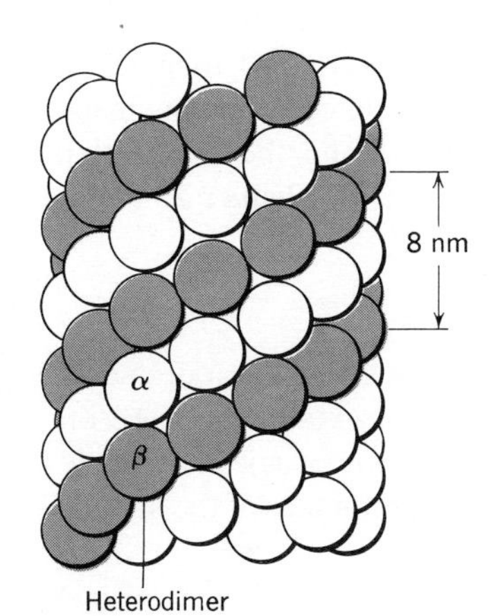

Figure 22–4
Microtubule model seen in cross section and laterally. The circumference of the tubule usually has 13 subunits (see text for variations). Although subunits appear to be linear chains that are parallel to the axis of the tubule, the paired subunits (α + β heterodimers) join to form a helix about the central, hollow core.

Table 22–1
Comparison of the properties of most microfilaments and microtubules

	Microfilament	Microtubule
Diameter	30–60 Å	100–250 Å
Structure	Double-helical protofilament	Hollow tube of 13 protofilaments
Protein	Actin or actinlike protein	Tubulin
Disassociating or inhibiting agent	Cytochalasin B	Colchicine, vinblastine, or vincristine
Subunit binding agent	ATP	GTP

Figure 22–5
Microtubule inhibitors. (*a*) colchicine, (*b*) vincristine. (Vinblastine has a methyl group in place of the formyl group.)

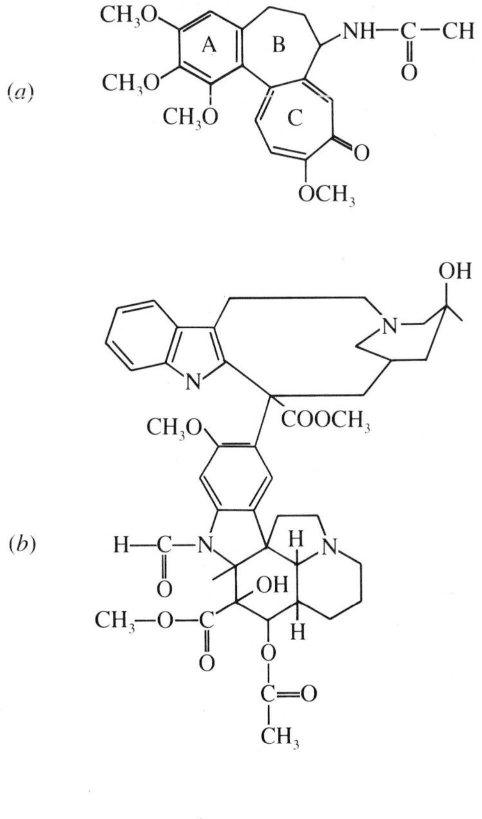

(*a*)

(*b*)

pear to run parallel to each other as well as to the axis of the microtubule (Fig. 22–4). However, the α and β subunits form a heterodimer and the arrangement of these paired subunits suggests a helical assembly of the heterodimers about the hollow central core.

In the assembly of the subunits, the α and β polypeptides are each first activated by binding to GTP. The activated subunits assemble onto other subunits. As each subunit becomes bound to the growing tubule, GTP undergoes hydrolysis, with the resutling GDP and phosphate remaining tightly bound. *Vincristine* and *vinblastine* (Fig. 22–5) interfere with this assembly process, causing precipitation of the tubulin in a polymeric, three-dimensional disarray. *Colchicine* (Fig. 22–5), another inhibitor of microtubule formation, acts by binding to the tubulin and preventing polymerization. A similar assembly process of protein units and ATP is seen in the formation of actin microfilaments (Chapter 23).

Microtubules are associated with a wide variety of functions in cells. In addition conferring shape to some cells, microtubules function to mechanically extrude material (collagen from connective tissue) from cells, separate chromosomes during mitosis and meiosis, and act as part of the motion mechanism of flagella and cilia.

Distribution and Functions of Microfilaments

Muscle Cells

Striated, smooth, and **cardiac** muscle cells contain vast numbers of microfilaments that function during the contraction of these cells. There are two basic types of microfilaments in these muscle cells; **thin filaments** composed mainly of *F-actin* and **thick filaments** made up of *myosin*. In muscle cells, the two types of microfilaments interact with each other through **cross-bridges** that enable them to slide past one another and effect a shortening or contraction of the cell. In striated muscle cells (primarily those cells that make up the muscles that move the skeleton), the number and geometric arrangement of the two microfilaments are greatest. Equally spaced about each thick filament are six thin filaments, and equally spaced about each thin filament are three thick filaments (Fig. 22–6). Units of several hundred thick and thin filaments are grouped together to form a **myofibril.** Each striated muscle cell contains many myofibrils. The structure of muscle and the contraction process are considered in more detail in Chapter 23.

Cytokinesis

Cell division in animal cells involves two separate mechanisms: **mitosis** and **cell cleavage,** or **cytokinesis.** The division and separation of daughter chromosomes is brought about in part by the action of the microtubules of the spindle

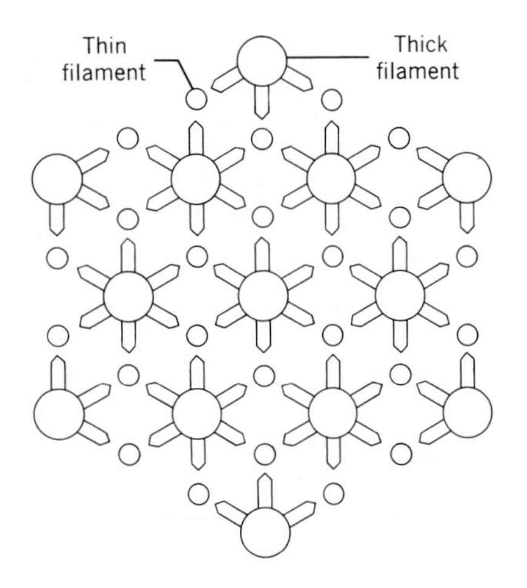

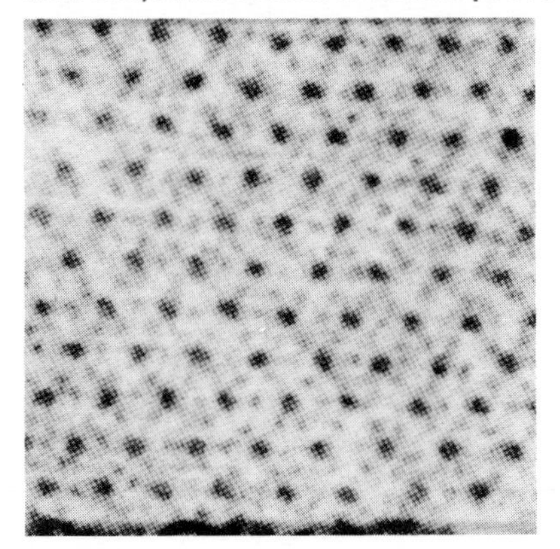

Figure 22–6
Cross section of a striated muscle myofibril showing geometric arrangement of thick and thin filaments. (Courtesy of D. Fawcett. Copyright 1966 W. B. Saunders and Co., *Atlas of Fine Structure* p. 57.

(discussed later in this chapter). The separation of the chromosomes may also be aided by the constriction of the cell or infolding of the plasma membrane forming a girdle about the spindle. Cytokinesis usually begins toward the end of anaphase. In animal cells, two events characterize the initiation of cytokinesis: (1) the cell begins to constrict about the midline of the spindle, and (2) dense material begins to collect about the peripheral spindle fibers also along the midline. Both processes continue as the plasma membrane moves inward, causing the cell to assume a "dumbbell" shape. The material collecting at the midline of the spindle becomes quite dense, forming a structure known as the **midbody** (Fig. 22–7). Just before the infolding edges of the plasma membrane meet and fuse, the midbody fades and disappears.

The furrowing or pinching-in of the plasma membrane is reminiscent of the action of a purse string or of a rubber band tightening about a soft object. The origin of the force that causes the constriction is still being debated, but the involvement of microfilaments in the process is supported by their regular presence in the area of constriction and by the observation that cytochalasin B inhibits the process. The presence of specialized organelles just inside the plasma membrane in the area of constriction can be seen in Figure 22–8. These electron photomicrographs by T. E. Schroeder show an *Arbacia* sea urchin egg at various stages of cleavage. In Fig. 22–8*a*, cleavage has not begun, but the area in which it will occur is framed. The details of the frame in part *(a)* are shown in part *(d)*, and it can be seen that no specialized organelles have yet formed. Figure 22–8*b* shows the same area 6 minutes after the beginning of cleavage. Just inside the plasma membrane at the bottom of the furrow, a dense collection of organelles has formed a structure called the contractile ring (cr). An enlargement of the framed area is shown in Figure 22–8*e*, but it is not possible to identify the organelles as microfilaments in the photograph because the long axis of the microfilaments runs perpendicular to the plane of the section (actually, the microfilaments are arranged in a circle about the furrow). In part *(c)* the cleavage is essentially complete; a few spindle fibers joining the two cells remain, and there is some indication of the fading midbody, but the contractile ring has disappeared.

The molecular mechanism involved in the constriction has not yet been determined with certainty, but it is speculated that the mechanism involves sliding actin and myosin filaments (much like that known to occur during muscle cell contraction).

Plasma Membrane Movement

Intestinal epithelial cells have many small projections called **microvilli** that extend into the digestive cavity. The microvilli increase the surface area of the intestine, thus enhancing absorption of the digested food. Microvilli also cyclically shorten and extend into the lumen of the intestine, a phenomenon that probably facilitates food absorption. Numerous actin microfilaments can be seen attached to the plasma membrane at the tip and side of each microvillus (Fig. 22–9). M. S. Mooseker has shown that the actin microfilaments of intestinal epithelium are attached at one end to an α-actininlike protein of the plasma membrane, just as the muscle thin filaments are attached to α-actinin of the Z-membrane. Mooseker also found that there is an association of the actin microfilaments with myosin filaments in the *terminal* web region of the cell. The movement of the microvilli may involve a mechanism much like that employed during myofibril contraction. It naturally follows that the movements of the microvilli may involve a mechanism similar to that which results in the contraction and relaxation of myofibrils.

The Formation of Pseudopodia and Amoeboid Movement.

Amoebae, slime molds, white blood cells, and a number of other cells achieve movement by the formation of pseudopodia. A **pseudopod** is a portion of a cell that flows forward forming an extension into which the remaining cytoplasm of the cell subsequently follows. More than one pseudopod at a time can form, but continued cell movement in one direction requires the reversal of the process in the nondominant pseudopodia. In pseudopod-forming cells, the outer portion of the cytoplasm is thick or gel-like and is called **ectoplasm;** the more internal sol-like cytoplasm is called **endoplasm.** During cell movement, the endoplasm flows forward into the pseudopod, but as it reaches the anterior end of the pseudopod, it is transformed into gel, thereby forming part of the new ectoplasm. At the rear of the moving cell the ectoplasm *solates,* moves into the cell interior, and becomes endoplasm (Fig. 22–10). There are two schools of thought concerning the origin of the force that causes the movement. One hypothesis (put forth by R. D. Allen and others) proposes that the endo-

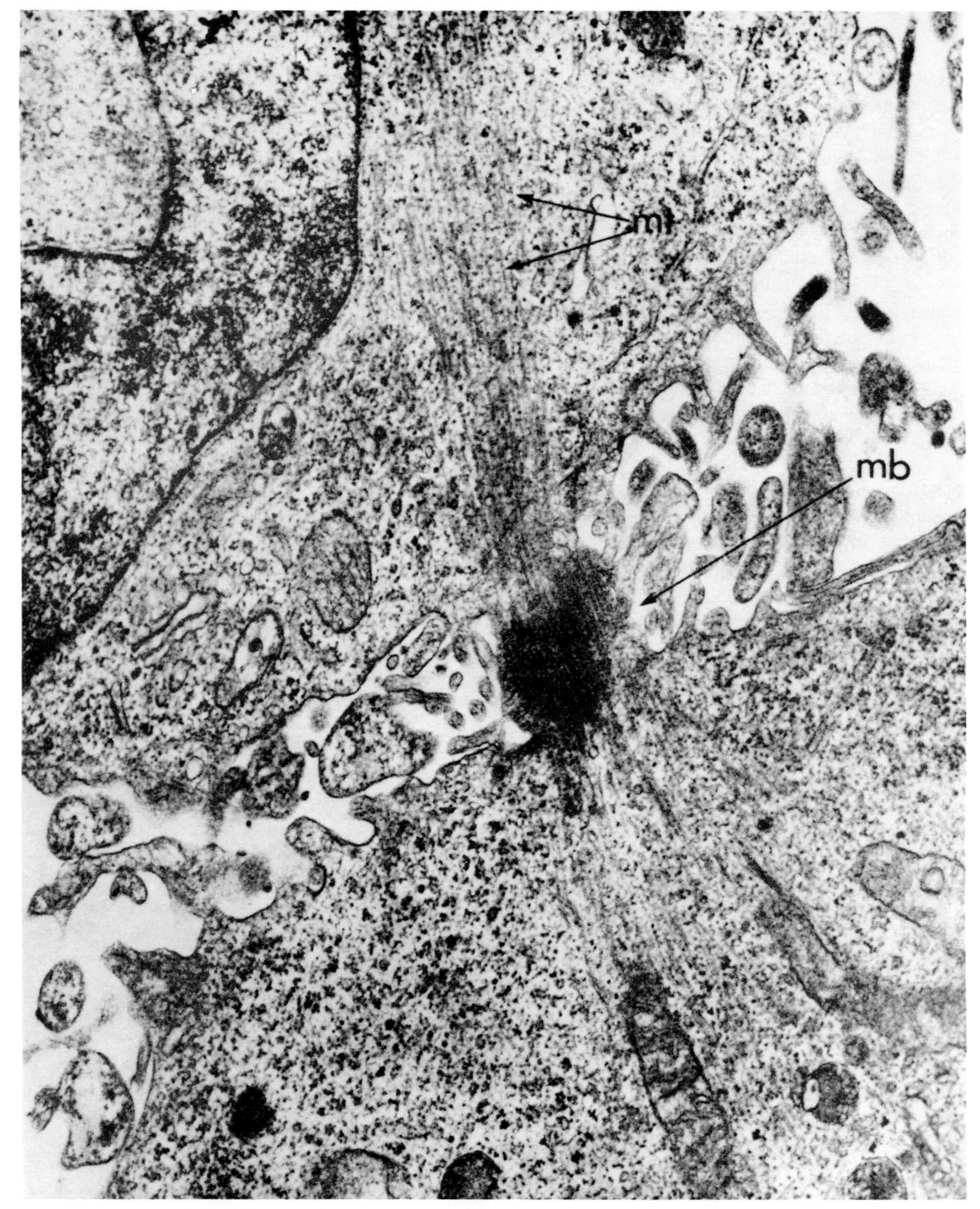

Figure 22–7

Final phase of cytokinesis in an animal cell. Only a few microtubules (*m*) of the spindle remain. Note the midbody (*mb*). (Courtesy of B. Byers. Copyright 1968 Springer-Verlag, *Protoplasma 66*, 423.)

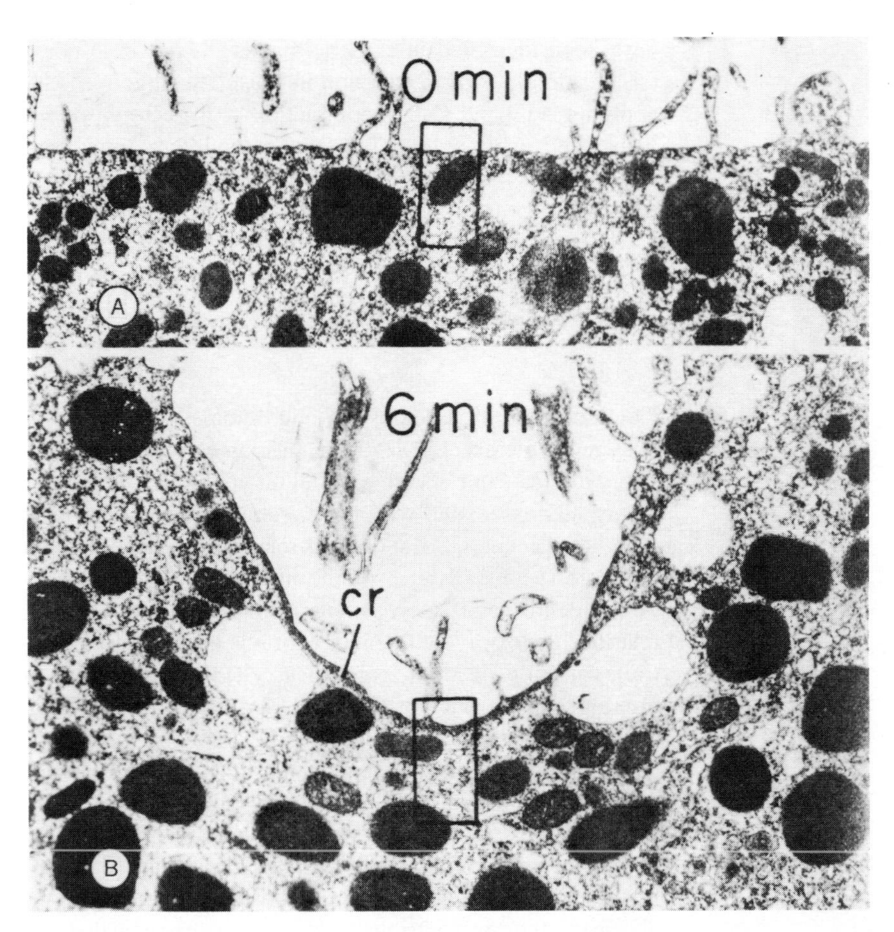

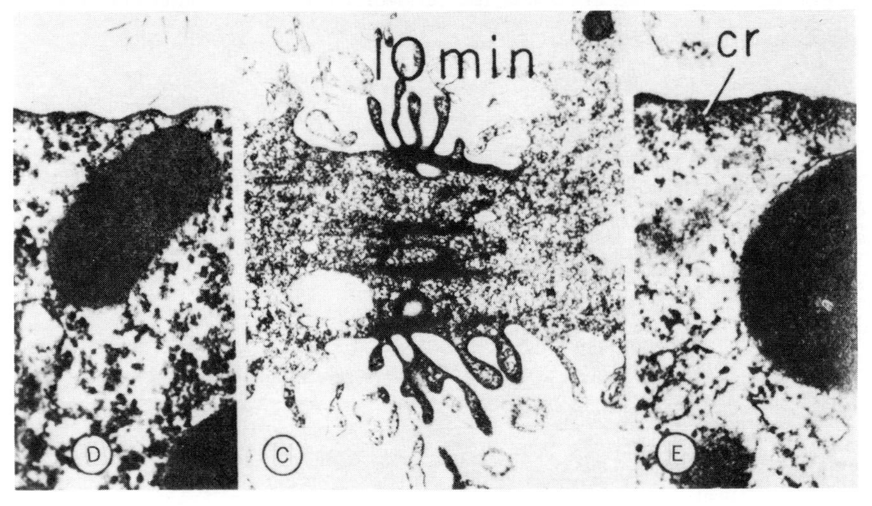

Figure 22–8
Cytokinesis in *Arbacia* sea urchin egg. (See text for details.) (Courtesy of T. E. Schroeder. Copyright 1976 Cold Spring Harbor Laboratory, *Cell Motility*, Book A, p. 268.)

Figure 22–9
Microvilli of chicken intestinal epithelium (Photomicrograph courtesy of Dr. E. G. Pollock.)

plasm in the anterior region of a pseudopod undergoes contraction as it becomes ectoplasm. These contractions progressively *pull* the ''viscoelastic'' endoplasm forward. The second (and older) hypothesis is that the ectoplasm toward the rear of the cell contracts, *pushing* the endoplasm forward. In the latter case, the streaming of the cytoplasm would be passive.

Regardless of the site of application of the force, a variety of evidence indicates that the process involves contraction and microfilaments. For example, cytochalasin B inhibits amoeboid movement. Actin and myosin microfilaments have been identified in amoeboid cells, as has an ATPase (an enzyme known to function in muscle contraction). By applying ATP and Ca^{++}, contraction of the ectoplasm can be induced. Although not fully understood, the contractions that underlie amoeboid movement appear to be similar to those that occur in muscle cells.

Distribution and Functions of Microtubules

Centrioles

Centrioles are an assembly of microtubules. These organelles are typically 150–250 nm in diameter and are composed of nine triple sets of parallel microtubules. Centrioles are usually associated with the nucleus and, although their function is generally described as related to nuclear division or, more specifically, to spindle fiber formation, the function of centrioles is more extensive than this. Indeed, their relationship to spindle fiber formation is questionable. Centrioles usually occur in pairs in a cell and are located in proximity to the nucleus (Fig. 22–11). As nuclear division begins, the centriole microtubules separate and migrate to opposite sides of the nucleus. Subsequently, as the chromosomes condense and the nuclear envelope disappears (Chapter 20), spindle fibers appear, extending from the area of one centriole through the cell to the other centriole. Because of the close physical relationship between the spindle fibers and the centriole, it has been suggested that the

Figure 22–10
Amoeboid movement.

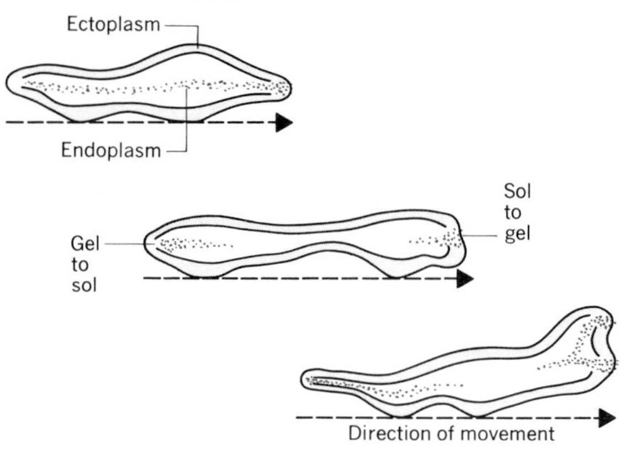

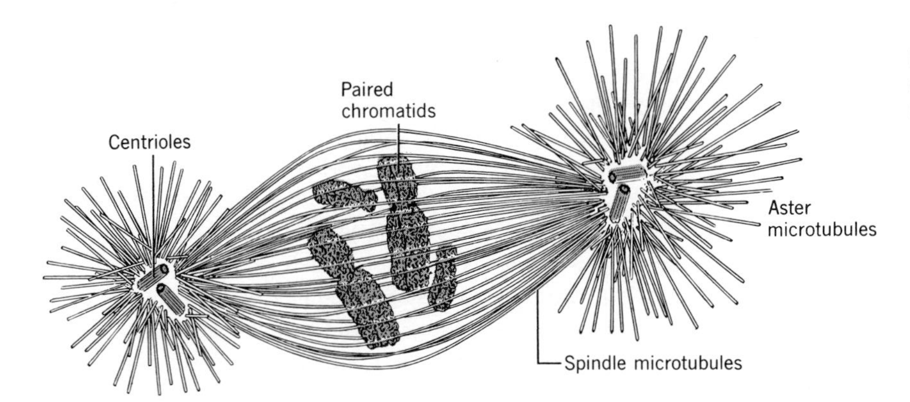

Figure 22–11
Diagram of centriole, aster mictotubules, spindle microtubules, and chromosomes during nuclear division.

centrioles may be associated with the spindle fiber microtubules. However, a number of observations argue against this idea. For example, the cells of cone-bearing and flowering plants do not have centrioles but do form spindles. Ferns have centrioles only in those cells developing flagella, and yet all fern cells undergoing mitosis have spindle fibers. The spindle fibers do not contact the centrioles, and no spindle fiber or aster microtubules are seen arising from a centriole.

Centrioles do play some role in the formation of microtubules found in flagella and cilia. In these locomotor organelles, the centrioles become the **basal bodies** or **kinetosomes** (also *blepharoplasts, basal granules,* or *basal corpuscles*), which are structures located at the base of the flagellum or cilium. As described later in the chapter, basal bodies appear to act as organizing centers for tubulin and the assembly of flagellar microtubules.

Structure of the Centriole. Centriole structure is basically the same in the cells of all species so far studied. The generalization used to be made that centrioles were found in animal cells but not in plant cells. It is clear from the discussion above that this is not true. Most algal cells (but not red algae), moss, some fern cells, and most animal cells have centrioles, whereas cone-bearing plants, flowering plants, red algae, and some nonflagellated or nonciliated protozoans (like amoebae) do not have centrioles. Some species of amoebae have a flagellated stage as well as an amoeboid stage; a centriole develops during the flagellated stage but disappears during the amoeboid stage.

The most notable characteristic of centriole structure is the nine sets of microtubules; each set contains three microtubules. The microtubules are arranged like vanes on a pinwheel when seen in cross section (Fig. 22–12) and form the basic framework of the cylindrical structure of the centriole. Although there is no surrounding membrane, the nine triplets appear to be imbedded in an electron-dense material that does not extend into the center of the cylinder. While the diameter of the centriole varies only slightly (150–250 nm), length is much more variable (160–800 nm). Centrioles usually occur in pairs, with their long axes oriented perpendicular to each other, but they do not touch each other, and there is no connecting material.

All nine triplets are identical. The innermost (or *A*) microtubule of each set is a complete, round tubule, but the middle *(B)* and outer *(C)* microtubules share the wall of the preceding tubule and are incomplete. Also, the *C* tubules may not run the full length of the centriole as do the *A* and *B* tubules. The triplets, although generally parallel to each other, may be closer together at the proximal end of the centriole (that end when observed ''end-on'' which has the triplets tilted inward in a clockwise direction, as shown in Fig. 22–12). The triplets may also spiral somewhat about the axis of the centriole or individually twist. Less twisting of the microtubules is observed in the basal body form of the centriole. Dense material connecting *A* and *C* tubules has frequently been observed. Strands of material extend inward from each *A* tubule and join together at the central *hub*. These strands, when seen in cross section, give the centriole the appearance of a cartwheel (Fig. 22–12*b*). More dense material occurs in the center of the centrioles when it takes the form of a basal body.

The idea prevalent several years ago that new centrioles were formed from preexisting centrioles (i.e., one acted as

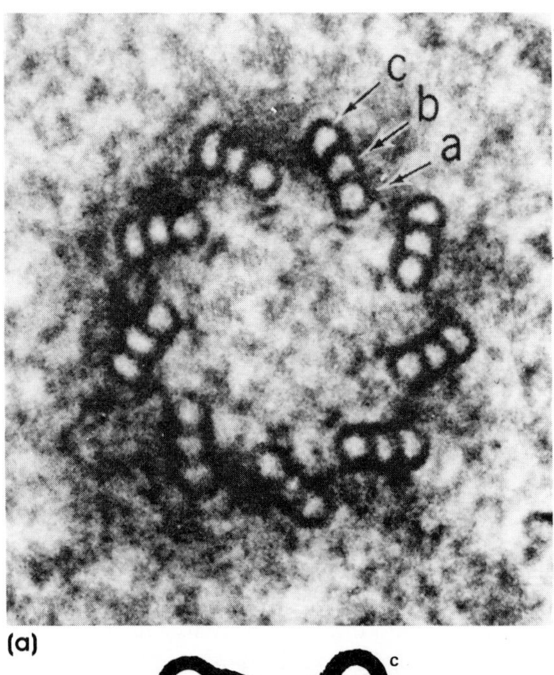

(a)

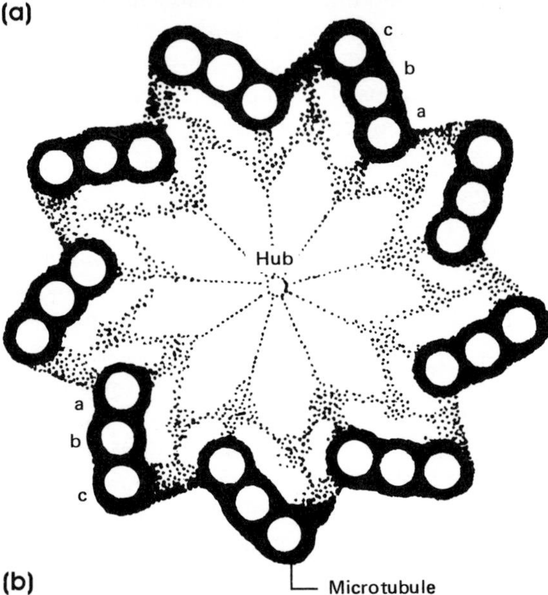

(b)

Figure 22–12
(*a*) Cross section of a centriole as revealed by electron microscopy. The spokes of the "pinwheel" turn inward in a clockwise manner, indicating that this view is from the proximal end of the centriole. (Courtesy D. Fawcett. Copyright 1966 W. B. Saunders and Co., *Atlas of Fine Structure*, p. 185.) (*b*) Diagram of a centriole showing the nine triplets that form the "pinwheel" and the electron-dense granular connectives.

the template for the assembly of another) is no longer accepted. Also incorrect is the suggestion that centrioles divide. The actual development of centrioles can be inferred from electron photomicrographs, and it is clear that centrioles arise de novo, without association with preexisting centrioles, basal bodies, or any other particle. However, **procentriole granules** have been described. New centrioles develop near preexisting ones, usually at right angles to one another, and separated by 50–100 nm. Development begins near the proximal end of the older centriole.

In ciliates, development of new centrioles (or ultimately, basal bodies) is usually detected first by the presence of a small, amorphous, electron-dense body. About this body develop a number of microtubule-like *procentrioles*. In cases where a single centriole develops, the procentriole forms within the amorphous body. During development, the electron-dense body disappears, as if it were being used to produce the developing procentriole.

Basal body development has been studied in the ciliates *Paramecium* and *Tetrahymena* and in tracheal epithelia of *Xenopus* and chicks. The order of development is virtually identical in all of these. In *Paramecium,* for example, development of the basal body begins with the formation of a single microtubule in the amorphous mass. Microtubules are added one at time until there is an equally spaced ring of nine (Fig. 22–13). There is some evidence that a ''connector'' exists between the microtubules, which could act to set the distance between them. Each of the nine microtubules in the ring are *A* tubules. The *B* tubules develop next and, finally, the *C* tubules. Before the tubules reach the doublet stage, the resulting cylinder is rarely longer than 70 nm, but after this stage, the tubules elongate. At the same time, the hub and ''cartwheel'' is added in the center (Fig. 22–12*b*). The *A*-*C* links are not formed until the end of development.

The basal bodies act as crystallization centers for the development of the tubules of cilia and flagella. Basal bodies form as described above usually well within the cell and often adjacent to preexisting centrioles. When not associated with a flagellum and cilium, the structures are called centrioles, but when they migrate to just within the plasma membrane and act as a center for flagellum or cilium development, they are called basal bodies (or kinetosomes, etc.).

The synthetic functions of centrioles and basal bodies are not clear, since it does not seem reasonable that they act

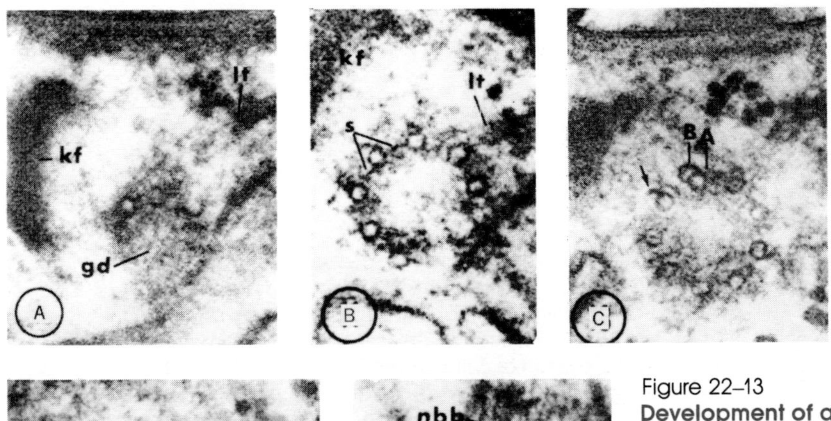

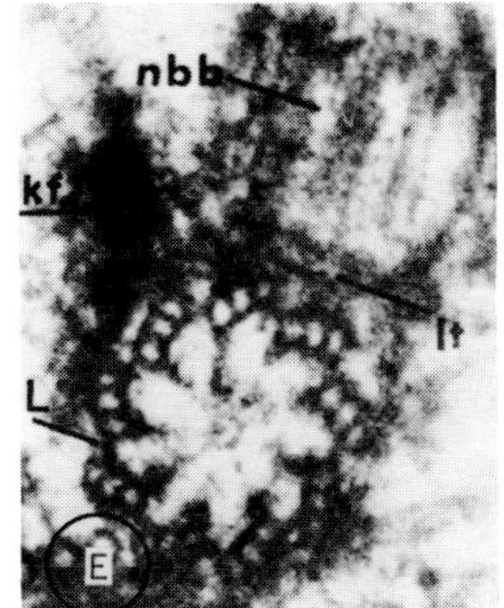

Figure 22–13
Development of a basal body in *Paramecium*. (*a*) The first microtubules form; (*b*) a developing ring of seven microtubules; (*c*) ring containing nine *A*-microtubules, with some *B*-tubules forming; (*d*) complete set of nine triplets; (*e*) mature basal body with new basal body (nbb) seen in longitudinal section above. (Courtesy of R. Dippell. *Proc. Nat. Acad. Sci. 61,* 461, 1968.)

as sites for protein synthesis. It has been suggested that these bodies may contain DNA and may carry out transcription or RNA synthesis. However, the few reports of the occurrence of DNA in the centrioles have been severely criticized. Skepticism also exists about the presence of RNA in centrioles, despite a number of studies showing that specific structural changes occur in the centrioles of cells treated with RNAase. (In *Paramecium,* for example, the dense inner core disappears.)

Cilia and Flagella

Cilia and flagella are specialized extensions of the cell (Fig. 22–14), projecting from the cell surface into the surrounding medium, where they whip back and forth or create a corkscrew action. In some instances, ciliary or flagellar movements propel the cells through their environment. The propulsion of the animal sperm cell by the whiplike movements of its "tail" is a classic example, although technically the sperm tail is not a flagellum but does have a similar basic structure. In other cases, the cell remains stationary (as in tissue), and the surrounding medium is moved past the cell by the beating of its cilia (as in the epithelial lining of the trachea or the collar cells lining the internal chambers of sponges).

Cilia are generally shorter than flagella (i.e., 5–10 nm vs. 150 nm or longer), but cilia are found in larger numbers per cell. Flagella may occur alone or in small groups, and infrequently they are present in large numbers, such as in a few protozoa and sperm of more advanced plants. The distinction is somewhat arbitrary, because other than differences in their lengths, the structure and action of cilia and flagella of eucaryotic cells are identical. (Bacterial flagella differ in structure; see later.)

A cilium or flagellum is composed of three major parts: a central **axoneme** (a semirigid structure extending from the cell body through the long axis of the cilium), the surrounding plasma membrane, and some cytoplasm (Fig. 22–15). The axonemal elements of nearly all cilia and flagella (as well as the tails of sperm cells) contain the same arrangement of microtubules. The microtubules are arranged in the now well-known "9 + 2" pattern. In the center of the axoneme are two microtubules that run the length of the cilium and are joined together by a bridge (Fig. 22–16).

(a)

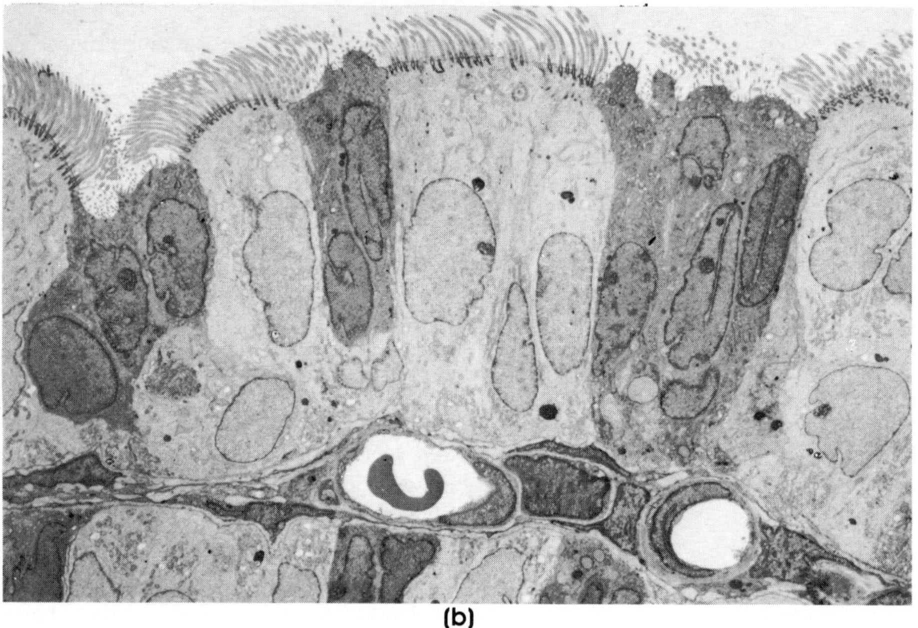

(b)

Figure 22–14
(*a*) Scanning electron photomicrograph of cilia on the dorsal surface of *Paramecium.* The cilia beat rhythmically creating a wavelike appearance (a *metachronal* wave). (Courtesy of E. Vivier. Copyright © 1974 Elsevier Publ. Co. *Paramecium*, p. 14.) (*b*) Cilia at the lumenal surface of epithelial cells lining the oviduct. (Courtesy of J. Rhodin, *Histology—A Text and Atlas* (Oxford University Press, Copyright © 1974, New York.) In Paramecium, the beating cilia propel the cell, whereas the cilia of the oviduct epithelium serve to push the egg released from the ovary toward the uterus.

Projections form the central microtubules occurring periodically along their length form what appears to be an enclosing *sheath*. Each of the central microtubules is composed of 13 protein filaments.

Nine doublet microtubules surround the central sheath. One microtubule of each doublet (i.e., the **A subfiber**) is composed of 13 protein filaments. The adjoining **B subfiber** is "incomplete," consisting of 11 protein filaments (Fig. 22–16). **Radial spokes** occuring at regular intervals along the axoneme extend from each A subfiber inward to the

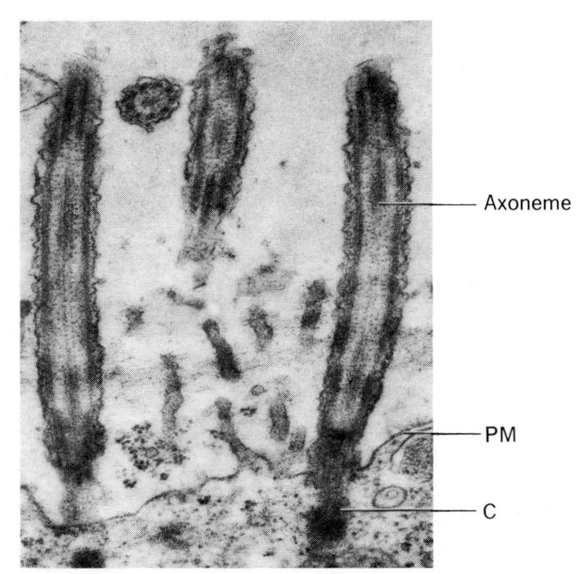

Figure 22–15
Basal ends of cilia showing the axoneme, the cell's plasma membrane (PM) which runs over the surface of the cilia, and the centriole (C). (Photomicrograph courtesy of R. Chao.)

central sheath. Adjacent A and B subfibers are joined by **interdoublet links;** these links occur irregularly along the length of the axoneme. Each A subfiber has sets of two "arms" composed of an enzyme called *dynein*. The *outer* dynein arm points away from the center of the axoneme, while the *inner* is directed somewhat into the axoneme.

Each beat of a cilium or flagellum involves the same pattern of microtubule movement. The beat may be divided into two phases, the **power** or **effective** stroke and the **recovery** stroke (Fig. 22–17). The power stroke occurs in a single plane, and by convention, the doublets are numbered clockwise in relation to this plane (Fig. 22–18).

The sliding microtubule model of ciliary movement is accepted by most investigators. In this model, the doublet microtubules slide past one another in such a manner as to produce localized bending of the cilium. This activity is accompanied by ATP hydrolysis. The localized bending takes the form of a wave that begins at the base of the organelle and proceeds toward the tip. The localized bending is produced through the cyclic formation and breakage of links between the dynein arms of one doublet and the neighboring doublet. The protein filaments that make up

Figure 22–16
Diagram of a cilium or flagellum seen in cross-section. (See text for explanation.) (Courtesy of F. D. Warner and P. Satir. Modified from *J. Cell Biol. 63*, 40. Copyright 1974 by The Rockefeller Press.)

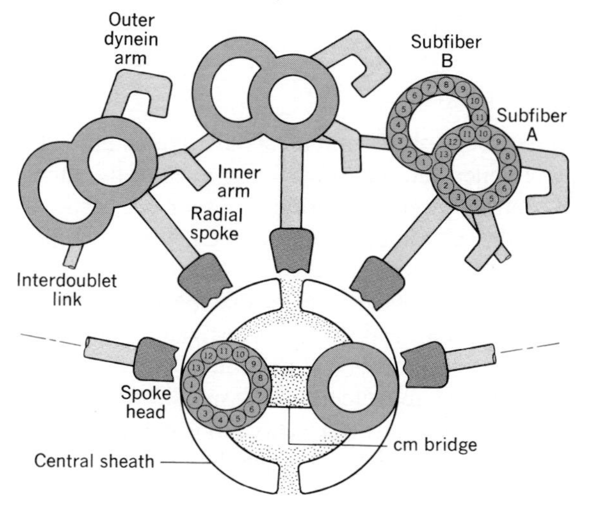

Figure 22–17
Two typical patterns of flagellar and ciliary motion. (*a*) successive waves move toward the tip of the flagellum propelling the cell (e.g., sperm) in the opposite direction. (*b*) the power stroke of a cilium is similar to the action of an oar in a rowboat; the recovery stroke (color) may not take place in the same plane as the power stroke. The large arrows show the direction of movement of the surrounding liquid.

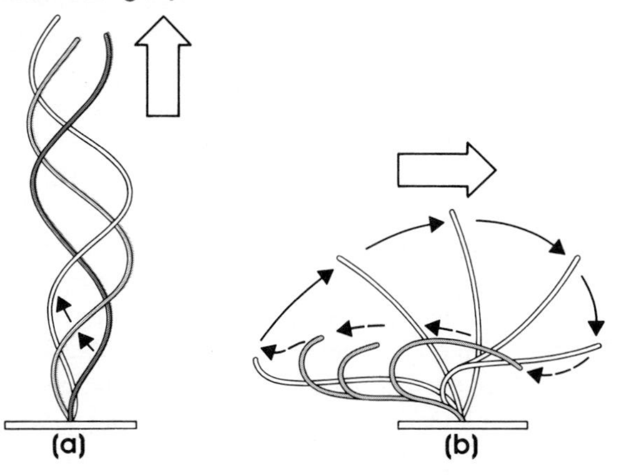

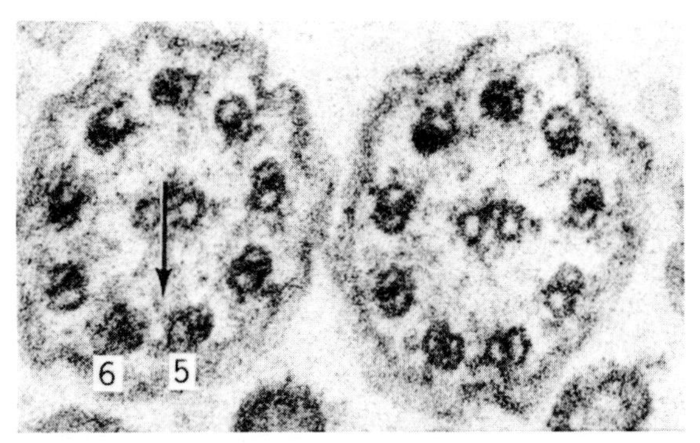

Figure 22–18
Cross-section of cilia viewed from the basal end. The direction of the power stroke (arrow) determines the numbering of the doublets. (Photomicrograph courtesy of R. Chao.)

each doublet are rows of *tubulin* molecules that apparently contain the sites that interact with the dynein. The fact that the sliding of microtubules past one another results in bending and *not* contraction of the cilium may be explained by the behavior of the radial spokes that connect the outer nine doublets to the central sheath. In straight regions of the axoneme, the radial spokes are aligned perpendicular to the doublets from which they arise, whereas in the bent regions they are tilted and stretched (Fig. 22–19). Firm attachment of the radial spokes at both ends provides the *resistance* necessary to produce bending as the doublets slide past one another. Indeed, if the radial spokes of sperm tails are destroyed by exposure to trypsin, addition of ATP results in the axonemes becoming *longer* and *thinner,* since microtubule sliding is no longer resisted. In effect, sliding is *uncoupled* from bending by elimination of the connections between the doublets and the central sheath.

The biochemical reactions taking place in conjunction with ciliary movement are generally considered to be similar to the reactions occurring during muscle contraction (Chapter 23). Analogies are evident between the dynein-tubulin system and the actin-myosin system. However, whereas Ca^{++} appears to activate to the actin-myosin system, these ions have the opposite effect on the dynein-tubulin system. The regulation of the Ca^{++} level in a cilium or flagellum probably involves the plasma membrane surrounding the axoneme. Under normal circumstances (i.e., during periods of continuous beating), the internal Ca^{++} level is low (about 0.1 μM) while Mg^{++} (necessary to stimulate the ATPase of the membrane) remains in the millimole range. When the membrane is depolarized, the Ca^{++} level inside the cilium increases and beating ceases. ATP is clearly the source of energy for movement and is produced by cellular respiration. In many cells, mitochondria are located adjacent to the basal body of the cilium or flagellum, and ATP diffuses toward the tip of the organelle. In sperm, a large mitochondria is an integral part of the tail (Fig. 22–20) and is wrapped in a spiral about the middle piece of the axoneme.

Bacterial Flagella. Flagella of bacteria are very different from cilia and flagella of eucaryotic cells. There is no surrounding plasma membrane, no axoneme of microtubules, and no sliding filaments. The bacterial flagellum consists of a **spiral filament** about 13.5 nm in diameter and 10–15 μ long, composed of the protein *flagellin*. The filament is attached to a ''hook,'' which in turn is connected to a *rod* penetrating the bacterium wall and membrane (Fig. 22–21). A number of rings connect the rod with the membrane and wall layers.

Bacterial flagella work by rotation of the rod and hook, which causes the filament to spin. When the filament spins in a clockwise direction, the cell is propelled smoothly, but when the spin is counterclockwise, an irregular tumbling motion of the cell is observed. The source of energy for the rotation (believed to be generated at the cell membrane) is an intermediate in oxidative phosphorylation and is not ATP. It is presumed to be the proton motive force discussed in Chapter 10.

The Mitotic Spindle

The **spindle fibers** of the mitotic spindle are composed of both microtubules and microfilaments. Most of the studies of chromosome movement during mitosis have been concerned with the microtubules and, at the present time, it is believed that these units alone can account for the movement. However, most workers agree that the microfilaments may also be involved, and several investigators support the idea that both microtubules and microfilaments must act to effect chromosome movement. The spindle fibers cause three distinct chromosome movements during mitosis (Chapter 20): (1) orientation of sister chromatids, (2) alignment of the kinetochores on the metaphase plate, and (3) division

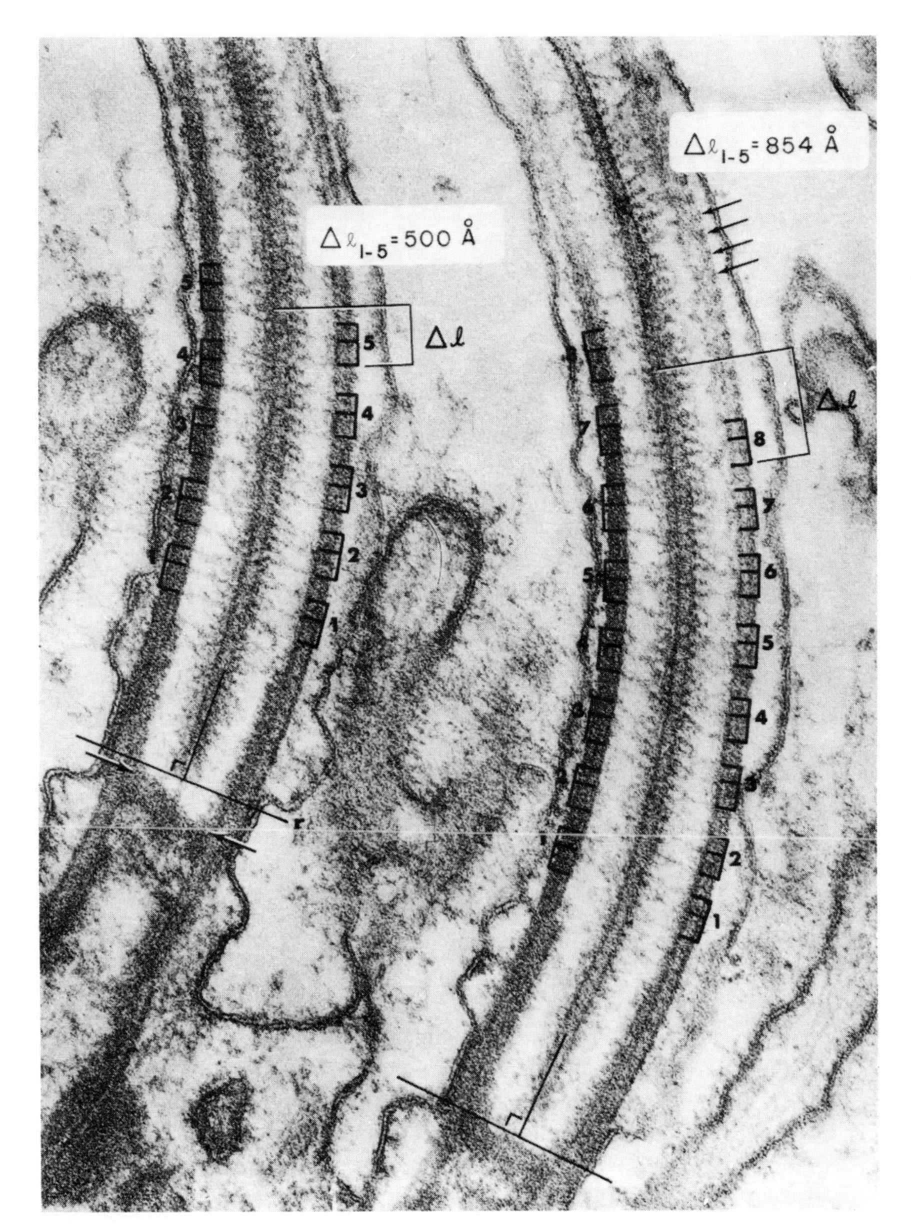

Figure 22–19
Median longitudinal section through the basal end of beating cilia. Successive radial spokes occur in groups of three (numbered 1, 2, 3, 4, etc.). The spokes in each group are directly opposite one another in a straight cilium and are perpendicular to the doublet microtubules. In a bending cilium, the radial spokes are tilted and stretched. In the cilium to the right, displacement of the spokes results in group 4 being opposite group 5. Δl is a measure of the amount of displacement taking place as diametrically opposite doublets slide tipward on the inner concave side and toward the base on the outer convex side. (Courtesy of R. Warner and P. Satir, *J. Cell Biol. 63,* 25. Copyright © 1974 The Rockefeller Press.)

and separation of kinetochores and movement of sister chromatids (segregation) to opposite poles of the spindle.

Three kinds of microtubules occur in the spindle: (1) the **kinetochore microtubules,** which terminate in a kinetochore; (2) the **polar microtubules,** which terminate at the poles; and (3) the **free microtubules,** which do not terminate in either a pole or a kinetochore. All three types can be dissociated into tubulin subunits by colchicine or cold temperatures. Also present in the spindle are microfilaments that are believed to be composed, at least in part, of actin. Immunofluorescence techniques indicate that the actin is present between the chromosomes and the poles of the spindle but is not present in the interzone between separating anaphase chromosomes.

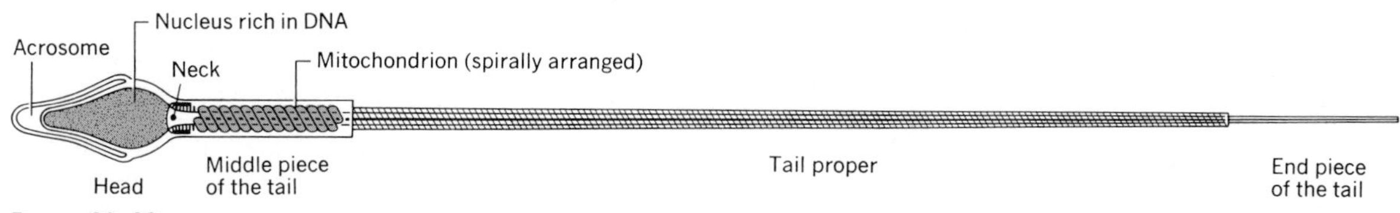

Figure 22–20
A mature human sperm cell.

Over the years, several models have been proposed to account for the movements of the chromosomes during anaphase. For example, it has been suggested that the chromosomes are *pushed* apart by spindle fibers developing between kinetochores, that they are *pulled* apart by spindle fibers extending between the kinetochores and the poles of the spindle, and that chromosomes *migrate* along spindle fibers. There is no strong evidence to support the movement of the chromosomes along the spindle fibers. Although individual spindle fibers do not stretch or contract per se, they do change in length through polymerization and depolymerization. S. Inoué has shown that spindle fibers are in a state of continuous flux, the free microtubules alternately growing and decreasing in length. His in vitro studies indicate that the microtubules assemble preferentially at one end and disperse preferentially at the other, so that the microtubule grows at one end and is dissolved at the other.

The kinetochore microtubules are required for movement, for if they are dissociated by clochicine or cold temperatures, chromosome movement stops until the dissolving agent is removed or the temperature raised and the kinetochore fiber allowed to reform. During movement, the kinetochore fiber becomes shorter, while the fibers between the chromosomes become longer.

Actin filaments alone cannot cause the movement of the chromosomes. This is demonstrated by the fact that agents like colchicine that dissociate microtubules but not actin stop chromosome movement. An interaction between actin and tubulin of the microtubules has not been demonstrated, and myosin is not present in the spindle; consequently, the role of actin remains unclear.

Figure 22–21
Diagram of the basal region of a bacterial flagellum. (Courtesy of J. Adler. Copyright © 1971 American Society of Microbiology, *J. Bact. 105,* 395.)

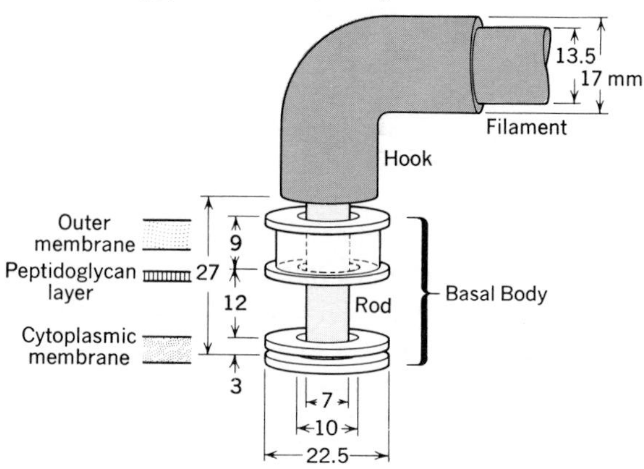

Other Cell Movements

The references at the end of this chapter may be consulted for additional examples of the diverse forms of cell and cytoplasmic movement. Regardless of whether the movements are "internal" (such as cyclosis in plant cells) or result in a major change in shape or position of the cell, the present evidence indicates that microfilaments and/or microtubules are fundamental to these activities. In most cases, an interaction between proteins—such as in the actin-myosin or dynein-tubulin systems—with the simultaneous involvement of an ATPase is the underlying biochemical phenomenon.

Summary

Microfilaments and **microtubules** frequently function together in cells to bring about movement or contribute to the cell's structural framework. The microfilaments are unbranched bundles of **actin** or actinlike proteins wound in a double helix. These proteins are disassociated by low concentrations of **cytochalasin.** Microfilaments have been shown to play a role in **phagocytosis, pinocytosis, exocytosis, cytokinesis,** and **cytoplasmic streaming.**

Microtubules are cylindrical structures whose walls are composed of heterodimers of α and β **tubulin.** These globular proteins appear to be arranged in 13 chains that run parallel to each other and to the hollow axis of the tube. Microtubules are dissociated by **colchicine, vinblastine,** or **vincristine.** Microtubules are involved in the function of **spindle fibers, flagella,** and **cilia** and in **mechanical extrusion** of materials such as collagen from cells. **Centrioles** consist of nine triple sets of parallel microtubules.

References and Suggested Reading

Articles and Reviews

Goldman, R., Berg, G., Bushnell, A., Chang, C-M., Dickerman, L., Hopkins, N., Miller, M., Pollack, R., and Wang, E., Fibrillar systems in cell motility, in *Locomotion of Tissue Cells, Ciba Foundation Symposium,* Vol. 14, Associated Scientific Publishers, New York, 1973, p. 83.

Inoué, S., Fussler, E., Salmon, E., and Ellis, G., Functional organization of mitotic microtubules. *Biophys. J. 15,* 725 (1975).

Nagai, R., and Rebhun, L., Cytoplasmic microfilaments in streaming Nitella cells. *J. Ultrastruc. Res. 14,* 571 (1966).

Roberts, K., Cytoplasmic microtubules and their functions. *Prog. Biophys. Mol. Biol. 28,* 273 (1974).

Satir, P., How cilia move. *Sci. Am. 231* (4), 44 (April 1974).

Books, Monographs and Symposia

Borgers, M., and DeBrabander, M., *Microtubules and Microtubule Inhibitors,* North-Holland Publishing Co., New York, 1975.

Goldman, R., Pollard, T. and Rosenbaum, J., *Cell Motility.* Book A: *Motility, Muscle and Non-muscle Cells.* Book B: *Actin, Myosin and Associated Proteins.* Book C: *Microtubules and Related Proteins,* Cold Spring Harbor Laboratory, Cold Spring, Harbor, N.Y., 1976.

Reinert, J., and Ursprung, H., *Origin and Continuity of Cell Organelles.* Springer-Verlag, Berlin, 1971.

Part 6
SPECIAL CELL FUNCTIONS

Chapter 23
CELL DIFFERENTIATION AND SPECIALIZATION

Differentiation is but one aspect of the more general field of developmental biology. Developmental biologists are concerned with the changes that organisms and their cells and molecules undergo in making the transition from cells unspecialized in structure and/or function to forms having a permanent and specific structure and function. Four component processes characterize these changes: *determination, growth, differentiation,* and *morphogenesis*.

Determination and growth are the initial processes. New cells are usually not "committed" to a specific function. For example, a fertilized egg cell divides many times to produce a ball of cells called a **morula.** Normally, the morula develops into a single embryo, each cell or group of cells giving rise to specific organs or tissues. However, one cell separated from the others at the morula stage is capable of growing into a complete embryo. Even at later stages of embryo development, cells that normally become epidermal tissue (i.e., ectoderm) can be surgically transplanted to another part of the embryo and there develop into mesodermal tissue. At some early stage of organismal development, all cells of a given species have the potential to develop into any of a variety of different tissue and cell types of that species. This potential is based on the genetic composition of the chromosomes and utlimately on the specific sequence of DNA nucleotides in each gene. The generalization can also be made that at some point in development, cells become committed to a specific course of differentiation. The process that establishes the fate of a cell is called **determination.** During determination, some alternative modes of gene expression become permanently "turned off" while others are sequentially expressed, further and further restricting the course of differentiation of the cell. Prior to and during determination the cell may increase its biomass through **growth.**

During **differentiation** the cell acquires new properties. These can be *structural* (such as the formation of actin and myosin microfilaments) or *biochemical* (as in the appearance of enzymes of a new metabolic pathway). Differentiation may also take the form of *loss* of preexisting structures or biochemical processes. For example, in the differentiation of mammalian red blood cells, the nucleus and other organelles are lost, together with the biochemical processes that these structures provided. The gross result of the internal structural differentiation of the cell, its growth, and the effects of environmental factors is **morphogenesis**—the generation of new form or shape.

Above all, development and differentiation rests with DNA molecules in the nucleus. Before a cell can develop into a hair cell of a mammal, a feather cell of a bird, or a scale cell of a reptile, there must be genes in the nucleus of that cell whose transcription and translation into the appropriate enzymes and other proteins allow the cell to differentiate in that direction. Moreover, given the proper genetic complement, conditions must allow these genes to be expressed. Gene expression during development is regulated at three levels. First, there are the basic biochemical (i.e., molecular) regulatory mechanisms, such as mass action, feedback control, and allosteric enzyme function (see Chapter 11). The second level of control is effected through the interrelationship between the nucleus, the cytoplasm, and the cytoplasmic organelles. The third level of control involves the interactions between the cell and its environment. A variety of experiments have demonstrated the effects of the nucleus and cytoplasm on one another.

Nucleocytoplasmic Relationships

Serial Transplantation of Frog Embryo Nuclei

A series of distinctive nuclear transplantation studies were made by R. Briggs and T. J. King in the early 1950s. Frog egg cells (which are haploid) can be enucleated and induced to develop *parthenogenetically*. When the nucleus is withdrawn from a cell in a normally developing embryo at the blastula stage (i.e., a diploid nucleus) and placed in the enucleated egg cell, a normal embryo develops. This demonstrates that determination has not yet taken place in the nucleus from the blastula stage cell. Instead, the cell still contains the potential to express all the genes necessary to produce the different types of cells of the complete embryo. The totipotency of the blastula cell nucleus can be perpetuated even over a series of experimental transfers; that is, if a nucleus is withdrawn from a blastula stage cell following the first transfer (Fig. 23-1, blastula donor A) and used to induce a second anucleate egg, the embryonic development of that egg will proceed normally; if a nucleus is removed from this blastula stage and transferred to a third anucleate egg, normally embyronic development again occurs, and so on.

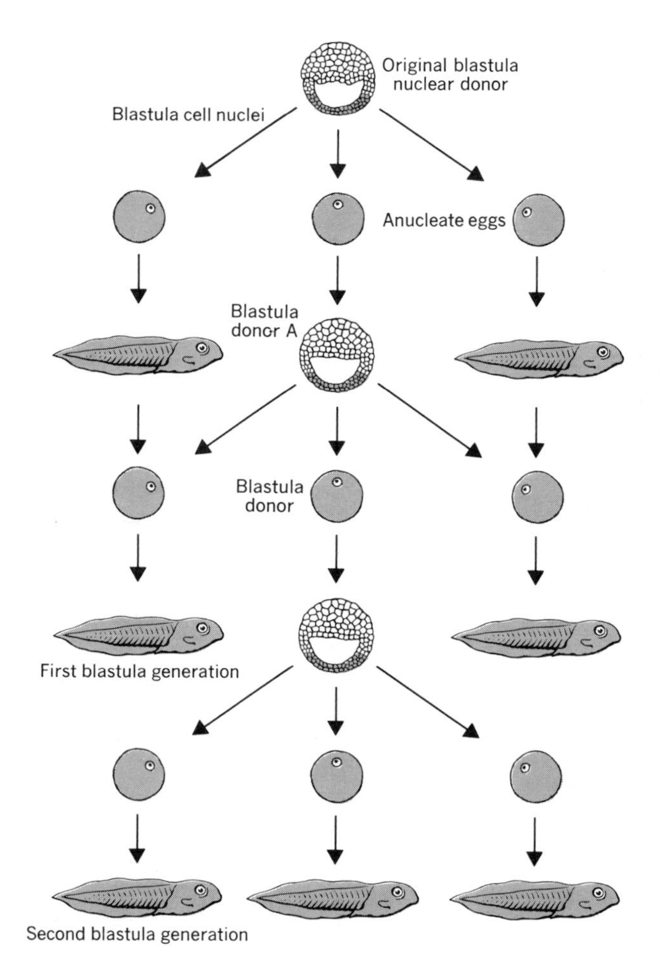

Figure 23–1
Effects of serial transplantation of nuclei from blastula stage cells to anucleate egg cells.

However, if a nucleus is taken from a cell at a later embryonic development stage (such as from the prospective midgut tissue of a late gastrula stage) and transferred to an anucleate egg, there is a different result. An anucleate egg is induced to develop when it receives the gastrula stage nucleus, but the development is abnormal. In the example just given, the ectodermal tissues of the developing embryo are usually deficient, while the endodermal and, to some extent, the mesodermal tissues develop normally. Obviously, some significant difference exists between the nuclei of blastula stage and gastrula stage cells. The nuclear change is permanent, as shown by the results of serial transfer experiments (Fig. 23–2). The embryos produced by transplantation from each generation of gastrulas are abnormal. The expression of the genes of the gastrula cell's nucleus has been restricted, and these restrictions are not altered by placing the nucleus in undifferentiated cytoplasm (i.e., the anucleate egg). The daughter nuclei produced by mitosis following transplantation retain this restricted gene expression. The causes of restricted gene expression in the later stages of development of the frog embryo are unknown, but it is likely that the *position* of a cell in the gastrula and the nature and concentrations of metabolites accumulating in its cytoplasm are at least partially responsible. The cytoplasmic metabolites are the result of both the cell's own metabolism and absorption from the external surroundings, and these substances result in the turning off of specific genes.

Experiments with *Acetabularia*

Classic experiments with the unicellular alga *Acetabularia* demonstrated that the cytoplasm has a direct effect on gene expression. This unicellular plant contains a single nucleus and has the shape of a long tube with rootlike branches at the base. The nucleus is located in one of these branches. As the cell matures, an umbrella-shaped cap develops at the top of the cell. The nucleus ultimately divides (by meiosis and then mitosis) to produce many daughter nuclei that migrate into the cap and become encysted. Eventually, the cysts are released, break open, and the gametes swim out, conjugate, and start a new cycle (Fig. 23–3).

Acetabularia has good powers of regeneration, for if the cap is cut off, the cell will grow a new cap. This can be repeated many times with the same organism. If the cap is cut off and the nucleus then removed, a new cap will be regenerated once, but the regeneration will not occur if the experiment is repeated with the same anucleate organism. The regeneration of the cap in the absence of a nucleus indicates that nuclear information had previously been transferred to the cytoplasm and has persisted there, at least through the time required to form the new cap. The information in the cytoplasm is in the form of messenger RNA.

Different shape caps are formed by different species of *Acetabularia*. *A. mediterranea* has a complete cap, and *A. crenulata* has a fingerlike cap. The interactions of the cytoplasm and nucleus are strikingly illustrated by grafting experiments using these two species (Fig. 23–4). If the foot of *A. crenulata* (which contains the nucleus) is joined to an anucleate stalk of *A. mediterranea,* the two pieces fuse and a cap develops at the end of the stalk. The cap assumes characteristics of both species, presumably resulting from the mRNA cytoplasmic information originally

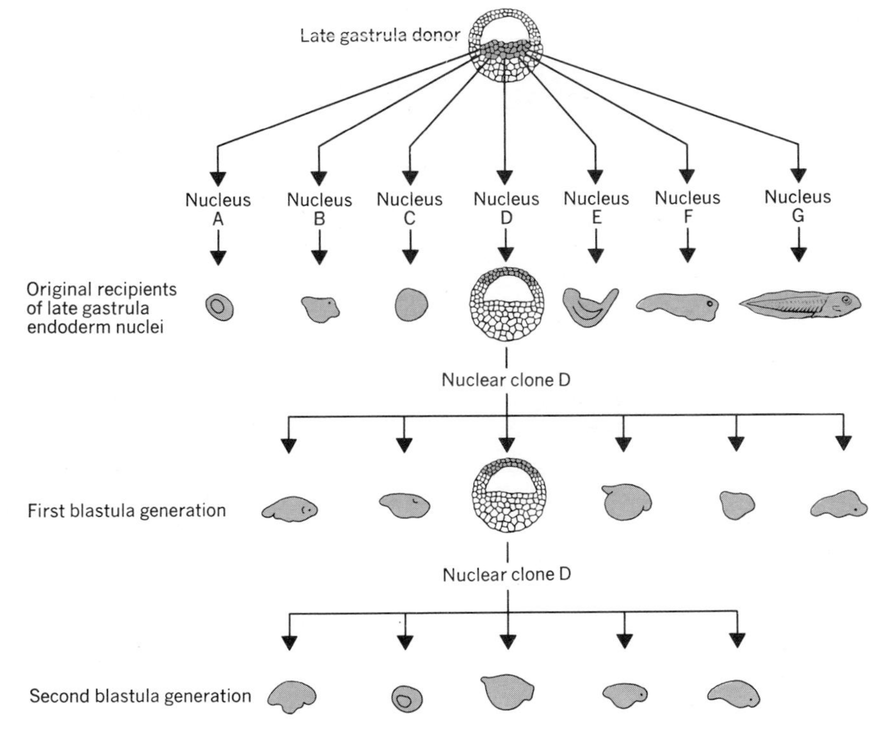

Figure 23–2
Effects of serial transplantation of nuclei from late gastrula-stage cells to anucleate egg cells.

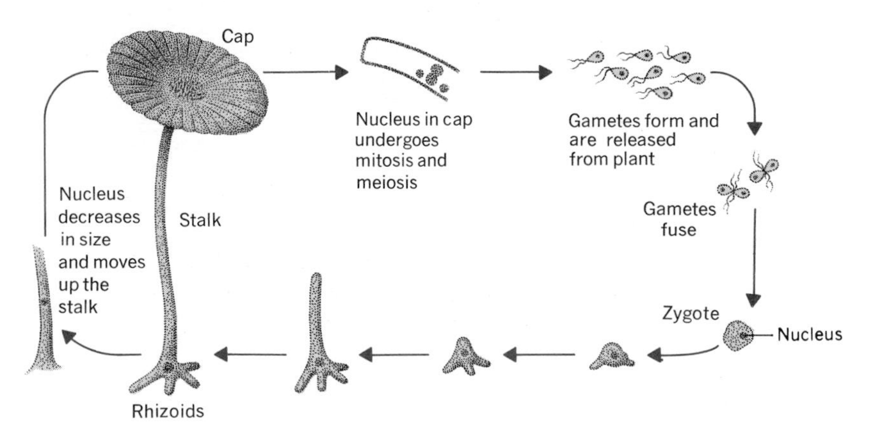

Figure 23–3
Life cycle of *Acetabularia.*

present in the *A. mediterranea* stalk and the influence of perhaps both the nucleus and some cytoplasmic mRNA of the *A. crenulata* foot. If the regenerated cap from the two grafted cell parts is removed and a second cap allowed to form, the new cap has the *A. crenulata* shape. Thus, any cap information (mRNA) in the original cytoplasm of *A.*

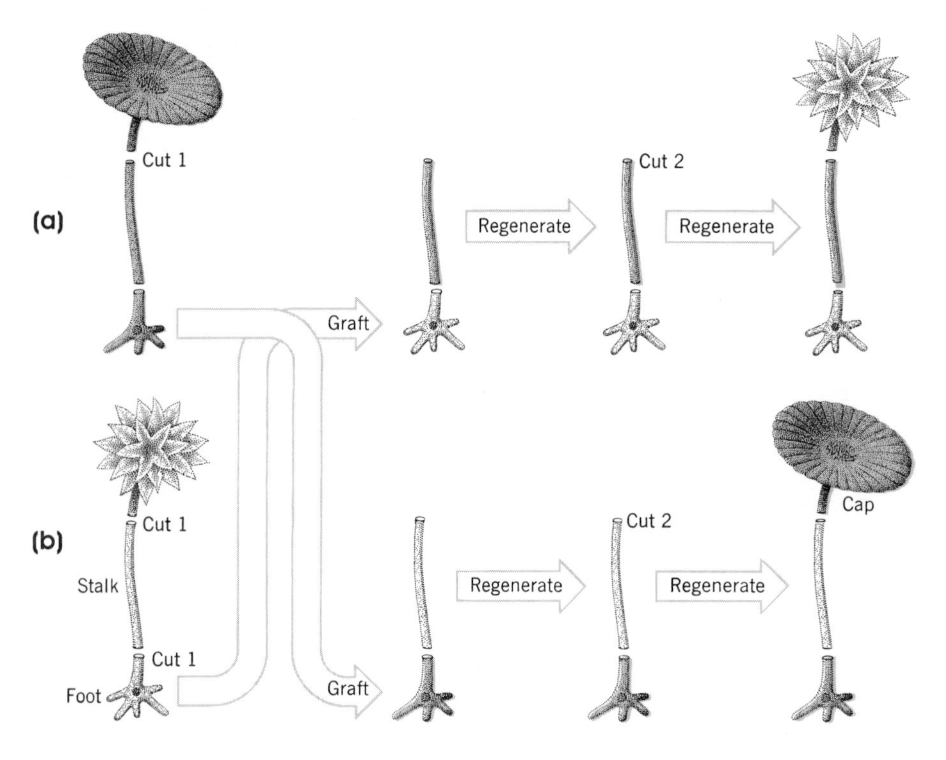

Figure 23–4
Effects of grafting experiments with nucleate and anucleate cell segments of *Acetabularia crenulata (b)* and *A. mediterranea (a)*. (See text for description and results of grafting.)

mediterranea is ''lost'' after one regeneration, whereas the cap information (mRNA) of the *A. crenulata* nucleus continues to be formed and expressed.

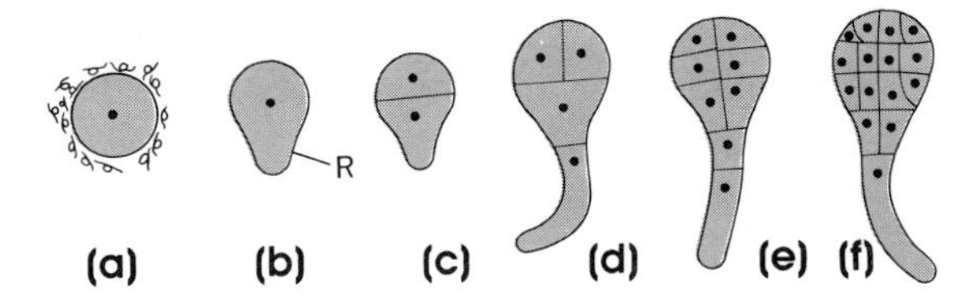

Figure 23–6
Development of early stages of *Fucus* embryos. (*a*) Fertilization; (*b*) development of rhizoid (R); (*c*) first division; (*d-f*) subsequent divisions.

Environmental Effects on Differentiation

A cell's neighbors and the surrounding environment have a direct effect upon differentiation. Since the fates of the late blastula–early gastrula stage cells of frog embryos (as well as embryos of a number of other species) have been determined, a map can be made of the types of tissues to be formed by each group of cells (Fig. 23–5). If at the late blastula stage, cells that normally form the eye lens are surgically exchanged with cells that form the gut, embryonic development is still normal. The transplanted lens cells were influenced by their new position in the embryo and develop into gut tissue. Likewise, the transplanted presumptive gut cells develop into lens tissue.

The cells that make up plant embryos develop from undetermined cells and soon differentiate into roots, stems, leaves, and flowers. The control of differentiation in higher plants is still unclear, but the direct effects of environmental factors on simple plants like *Fucus,* a brown marine alga, can be vividly demonstrated. The fertilized *Fucus* egg cell has no apparent polarity. Fifteen hours after fertilization, a *rhizoid* (rootlike growth) develops on one side, and this is followed by nuclear division and formation of a wall that separates the rhizoid from the *apical* cell. This first division establishes a polarity in the embryo (Fig. 23–6). It appears that environmental factors influence the side of the fertilized egg on which the rhizoid will develop. In a cluster of fertilized *Fucus* eggs, the rhizoid develops on that side of each egg that is closest to the center of the group. In a temperature gradient, the rhizoids develop on the warm side. In a pH gradient, they develop on the more acid side, and in white light, they form on the shaded side.

As we have seen, the expression of a cell's genes is influenced by its cytoplasm, the adjacent cells, and the environment. The cellular differentiation that results from these influences produces a number of uniquely specialized cells. *Specialized cell structure reflects specialized cell functions, and specialized function is founded on the differentiation of specialized structure*—a relationship that is universally recognized.

Among specialized cell types, muscle cells, nerve cells, and red blood cells have been more extensively studied than any other form. Some of their specializations are considered in the following sections.

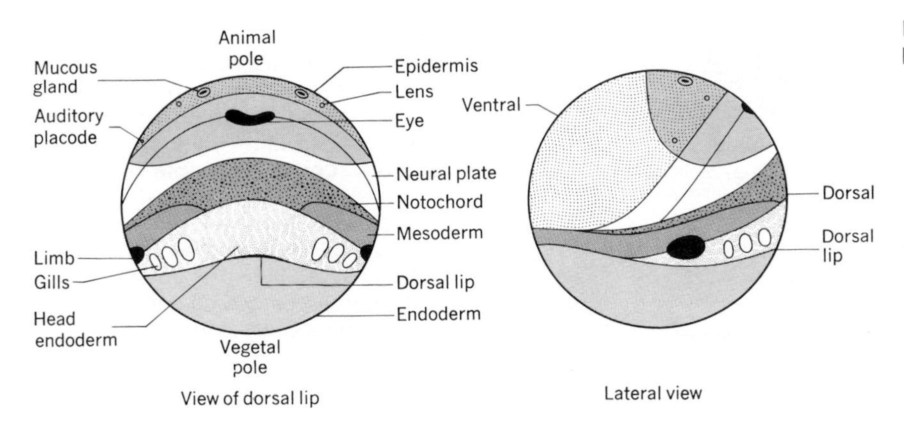

Figure 23–5
Map of the fate of blastula-stage cells.

Red Blood Cells

The differentiation of mammalian red blood cells in bone marrow (a process called **erythropoiesis**) is one of the more striking examples of specialization occurring in nature. In adults, erythropoiesis begins with a **pluripotent stem cell** that gives rise in 7 to 10 days to mature, hemoglobin-filled erythrocytes. As noted earlier in the chapter, differentiation and specialization of cells may be accompanied not only by the acquisition and development of specialized structures but also by the *loss* of internal structures of physiological properties. The latter is the case during erythrocyte differentiation, for mature red blood cells lack nuclei, mitochondria, endoplasmic reticulum, Golgi bodies, ribosomes, and most other typical cell organelles.

The mature erythrocyte is a simple cell, delimited at its periphery by a plasma membrane and containing internally a highly concentrated, para-crystalline array of hemoglobin molecules for oxygen transport. The apparent simplicity of the maturing erythrocyte is the principal reason for its selection over most other kinds of cells as the preferred object for studying plasma membrane structure and protein (i.e., hemoglobin) structure and biosynthesis. These subjects, together with the contributions made by intensive study of the erythrocyte, are treated in some depth in Chapters 4, 15, and 21 and will not be dealt with here.

Erythropoiesis

The development and differentiation of the mammalian red blood cell is depicted in Figure 23–7. Development takes place in the extrasinusoidal stroma of the bone marrow and begins with pluripotent stem cells capable of proliferating granulocytic leukocytes (white blood cells), as well as erythrocytes. When primitive stem cells undergo division, one of the daughter cells remains undifferentiated and pluripotent, so that depletion of marrow stem cells does not normally take place. The erythropoietic activity of the bone marrow is under hormonal influence, increasing or decreasing according to the level of circulating **erythropoietin** (produced in the cortical region of the kidneys and secreted into the bloodsteam). The erythrocyte progenitors show an increasing sensitivity to erythropoietin through the pro-erythroblast stage and a parallel, ever-decreasing proliferative potential (Fig. 23–8). As a result, by the pro-erythroblast stage, the cells are irreversibly committed to the maturation sequence leading to erythrocytes.

The mRNAs for the various globin chains of hemoglobin appear and increase in quantity in the proerythroblast and erythroblast stages, and this is followed by the synthesis and accumulation of hemoglobin. By the late erythroblast and normoblast stages, the synthesis of hemoglobin accounts for more than 95% of all protein synthesis occurring in the cells. Hemoglobin synthesis is concluded in the reticulocyte stage and is accompanied by the progressive dissolution of internal cell structure (including the loss of the nucleus) and residual nucleic acids. Late reticulocytes leave the bone marrow and enter the circulating blood, where in the ensuing hours, they lose their granulation and are transformed into the biconcave discs typical of mature eryth-

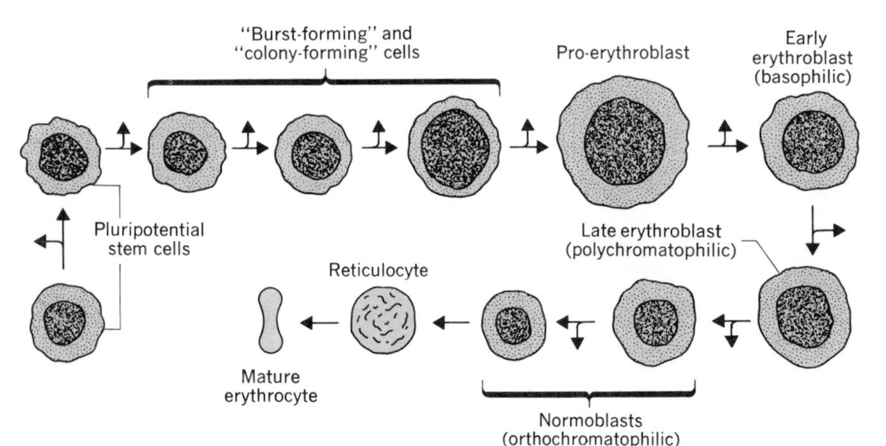

Figure 23–7
Differentiation and maturation scheme of the mammalian erythrocyte. Differentiation begins with pluripotent stem cells in the bone marrow and ends 7 to 10 days later with the release of reticulocytes or mature erythrocytes into the bloodstream. Each double-headed arrow represents a mitotic division into two daughter cells only one of which is shown. Normoblasts become reticulocytes and then erythrocytes without division (the cell nucleus is lost during the normoblast stage).

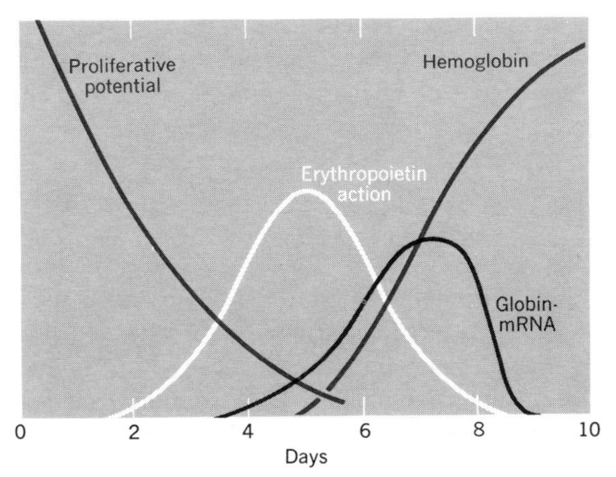

Figure 23–8
Relationship between the time course of differentiation and maturation of erythrocytes in the bone marrow and the proliferative potential of progenitor cells, sensitivity to erythropoietin influence, accumulation of mRNA for globin chains and the buildup of cellular hemoglobin.

rocytes. Complete differentiation and maturation from stem cell to the erythrocyte takes from 7 to 10 days.

Because of its highly differentiated state, the mature erythrocyte is incapable of further proliferation. In humans, the average life span is 120 days. Since there are 5 billion red blood cells in each milliliter of blood, a few simple calculations* quickly reveal that in a typical adult the differentiation and maturation of 3 million erythrocytes is completed *each second*! Obviously, an appreciable proportion of the body's energy and resources is continuously consumed to support erythropoietic activity. This is in stark contrast with other highly differentiated cells such as those of muscle and nerve, whose proliferation ceases shortly after birth.

*Since about 8% of the body weight is blood and blood contains 5 billion red cells per milliliter, a person weighing 78 kg (172 lb) contains 6.24 kg or about 6.3 liters of blood. Therefore, altogether there would be 3.15×10^{13} circulating erythrocytes. Since all of these are turned over in an interval of 120 days or 1.04×10^{7} seconds, this means that 3.03×10^{6} red blood cells reach full maturity and are released into the bloodsteam each second.

Genetic and Molecular Basis of Erythrocyte Differentiation

Because of the intensity with which the red blood cell has been studied by biochemists, cell biologists, molecular geneticists, physiologists, and others, the differentiation and specialization of this cell is better understood in molecular terms than any other; accordingly, erythrocyte differentiation may serve as a model for the genetic and molecular events that underlie cell differentiation generally.

The structural genes coding for the various globin chains of hemoglobin represent only a miniscule fraction of the total genetic complement of the pluripotent progenitor cells. (In humans, there are two pairs of alleles (i.e., four genes) for alpha globin chains, one pair for beta chains, and two pairs for gamma chains. These genes account for less than 0.0001% of the total DNA content of the nucleus, and yet globin chain synthesis accounts for more than 95% of all protein synthesis taking place in the cell by the later stages of maturation. It is clear that during differentiation, the *stem cells and their progeny are committed to the highly selective expression of only a small number of genes.*

Organization of the Globin Chain Genes in Chromatin. Cellular DNA includes not only structural genes but also sequences vital to DNA organization within the chromosomes and the coordination of gene expression during differentiation. About 60% of all the nucleotide sequences in the DNA of stem cells belongs to the "unique sequence" class in which only a few copies of each sequence are present. The structural genes for the globin chains belong to this class. Highly repetitive DNA accounts for about 10% and consists of short sequences repeated in tandem many thousands of times. These regions are not transcribed and are believed to be located in condensed chromatin at the chromomeres and other chromosomal constrictions (see Chapter 20). The balance of the DNA—about 30%—is moderately repetitive (repeated up to several hundred times) and includes the genes for histones and the rRNAs (see Chapter 21).

Some moderately repetitive DNA sequences are located adjacent to the structural genes for globin and other proteins and may be involved in the coordination of gene activity. For example, by binding to the moderately repetitive sequences, regulatory substances could "turn on" the transcription of *physiologically related* genes. In the case of the red blood cell, such a gene set would be represented by the globin chain genes, the genes for the heme synthetic enzymes, and the genes for the cell surface antigens (which serve as the basis for blood typing). Models accounting for gene regulation at this level were considered in Chapter 11.

As noted in Chapter 20, the genes of eucaryotic cells are associated with histones and other basic proteins to form complexes called chromatin composed of repeating *nucleosome* units. Euchromatin has the more open structure and is transcriptionally active, while condensed heterochromatin is not transcribed. Experiments with developing erythrocytes indicate that the globin chain genes are included in nucleosomes of condensed chromatin during the early stages of differentiation and that these are converted to the open form just before globin mRNA synthesis begins. The transcription of globin DNA sequences by RNA polymerase that occurs when these genes are "turned on" can be traced to a modification of the associated histones and an interaction with nonhistone proteins. A selective expression of globin mRNA genes occurs in erythroblasts, with the result that globin mRNA is produced in amounts 100 times greater than expected on the basis of the proportion of total template DNA represented by the globin genes. Globin chain synthesis parallels the appearance of globin mRNAs in erythroblasts.

Although globin chain synthesis accounts for the overwhelming majority of all protein synthetic activity in the maturing erythrocyte, a number of other structural genes are expressed (e.g., those for the enzymes of glycolysis, enzymes for initiation, elongation and termination of globin chain assembly, enzymes of the metabolic pathway for heme synthesis, enzymes and structural proteins of the plasma membrane, and so on). Although all protein synthesis comes to a halt during reticulocyte maturation, a red cell retains a limited metabolic capacity after release from the bone marrow. For example, glycolytic activity provides the ATP needed to maintain the sodium and potassium pumps of the erythrocyte membrane (see Chapter 15) and for other energy-requiring processes. What limited metabolism is retained by the mature cell serves to sustain it during its 120-day and 700-mile journey through the circulatory system.

The progressively selective expression of genes in the maturing red blood cell is accompanied by the loss of internal structure so that in the mature state the only remaining organelle is the plasma membrane itself. Internally, the cell consists primarily of a highly concentrated (30% by weight) crystal-like arrangement of hemoglobin molecules suited to the cell's principal function—oxygen transport.

Morphological and Physiological Specialization of Red Blood Cells

Little cell growth occurs during the periods between successive mitotic divisions of erythropoiesis. As a result, the mature erythrocyte is among the smallest cells of the body. In humans, the average red blood cell has a volume of about 100 μ^3 and is disk-shaped (Fig. 23–9). This rather unusual cell shape has fascinated scientists for more than two centuries. How so specific a shape is maintained in the absence of cytoskeletal elements, microtubules, or microfilaments is still being intensively studied today. Two points of view dominate this controversial area. One hypothesis is that the biconcave shape results from the chemical nature, arrangement and *interaction* of the proteins and lipids in the plasma membrane. In spite of positive hydrodynamic pressure inside the cell—a pressure that would be expected to impose spherical shape to a body encapsulated by a flexible membrane—tensional and compressional forces in the membrane are translated into biconcavities on opposing membrane surfaces. Such an explanation has been substantiated using artificial membranes as models.

Figure 23–9
Scanning electron photomicrograph revealing the disc shape of erythrocytes. (Photomicrograph courtesy of Dr. M. A. Lichtman.)

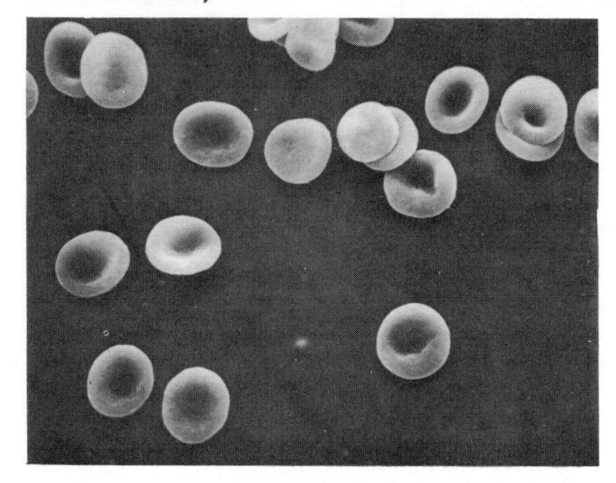

A second and broadly supported view is that the internal para-crystalline organization of hemoglobin molecules imposes an overall shape on the cell. Such a view is supported by the finding that erythrocytes of individuals with *abnormal* hemoglobins (hemoglobins containing one or more amino acid substitutions) usually lack the normal biconcavity. The most notorious example is the "sickle" or crescent shape of erythrocytes in individuals with *sickle-cell anemia*. In this genetically determined disease, a single substitution occurs in the beta globin chains of the hemoglobin, and under conditions of oxygen depravation or shortage, the relative positions of neighboring hemoglobin molecules within the cell change. These changes alter the normal shape change to the cell.

Oxygen enters and leaves the erythrocyte by diffusion and the biconcave shape facilitates oxygen flux by increasing the surface area-to-volume ratio of the cell. Oxygen inside the cell forms a reversible combination with hemoglobin (see Chapter 4), binding to hemoglobin as the blood circulates through the capillary networks of the lungs (where the net oxygen flux is directed into the erythrocyte) and being released as the blood circulates through oxygen-deficient tissues. The disk shape of the erythrocyte induces the formation of *rouleaux* or long *stacks* of cells, with the result that far greater numbers of cells can pass through the narrow capillaries when arranged in such a regimented manner than when the cells are freely and independently suspended. Experimental elimination of the biconcave shape using hypotonic solutions eliminates the ability of erythrocytes to form rouleaux.

Muscle Cells

Vertebrates possess three types of muscle tissue: (1) *smooth* muscle, the contractile elements of most of the digestive system and most visceral organs, (2) *cardiac* muscle, found only in the heart, and (3) *striated* (or *skeletal*) muscle, responsible for most of the gross movements of the body. Each of these tissues is composed of cells called **muscle fibers,** which contain microfilaments of actin, myosin, and other proteins responsible for the contractile nature of muscle. The microfilaments are arranged differently in each of the tissue types. The arrangement is most highly organized in striated muscle, and it is this type of muscle tissue that has been most extensively studied.

The organization of striated muscle is shown diagram-

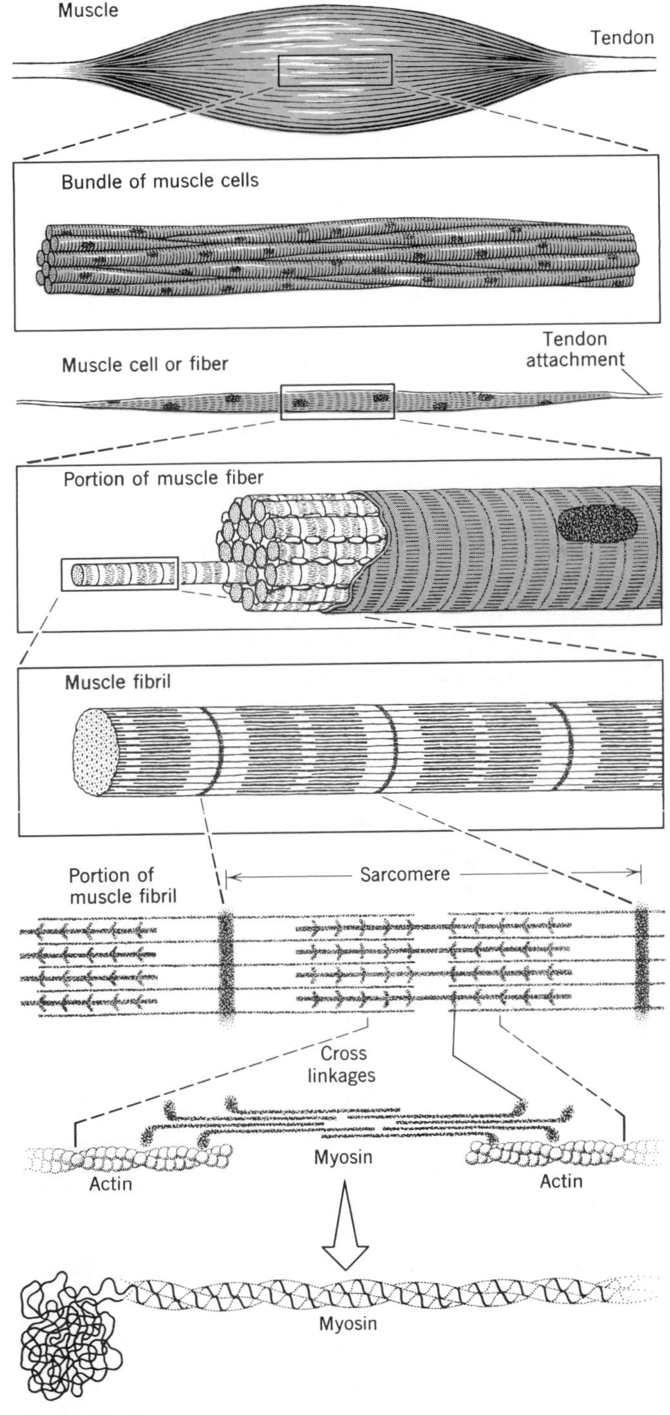

Figure 23–10
Organization of striated muscle.

matically in Figure 23–10. During development, the muscle fiber is formed by the end-to-end fusion of many cells into a continuous tubelike structure. This explains the fact that striated muscle fibers are multinucleate and contain many more mitochondria than most other cells. The plasma membrane of the muscle fiber is called the *sarcolemma;* in addition to its exceptional length, the sarcolemma is characterized by the numerous porelike invaginations that extend into the sarcoplasm (cytoplasm) at right angles to the long axis of the cell. These transverse or **T-system** extensions of the sacrolemma (Fig. 23–11) make contact with most of the internal **myofibrils**—the contractile elements of the cell. Each myofibril contains a large number of *thick* (150 Å diameter) and *thin* (60 Å diameter) microfilaments called **myofilaments** interconnected by protein bands. The myofilaments are parallel to each other and to the long axis of the cell. Seen in cross-section (Fig. 23–12), the myofilaments are arranged in a repeating geometric pattern. The thick filaments are equidistant from each other with each surrounded by 6 thin filaments in an hexagonal array (Fig. 23–12b). The thick and thin filaments slide past one another during contraction. When examined microscopically in longitudinal section, each myofibril reveals areas of different density (Fig. 23–13). The alternating

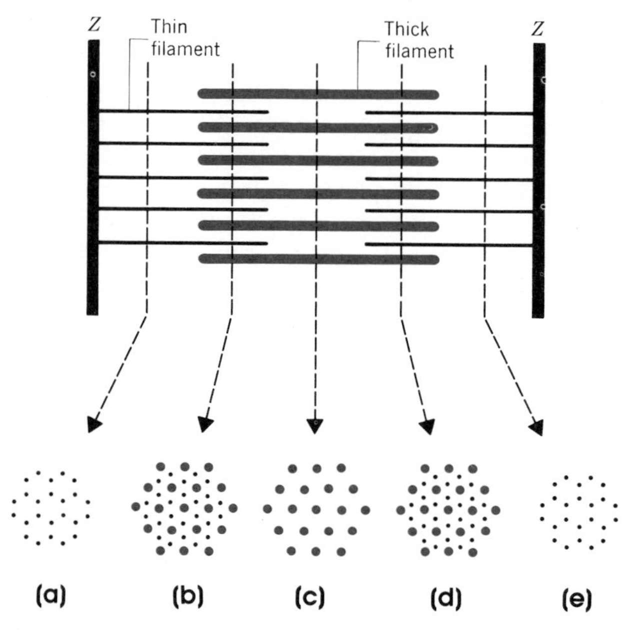

Figure 23–12
Cross sections at various positions in the sarcomere. Thin filaments are arranged parallel to thick filaments and form an hexagonal array in regions of overlap (i.e., regions *b* and *d*).

Figure 23–11
The sarcoplasmic reticulum and T-system of striated muscle.

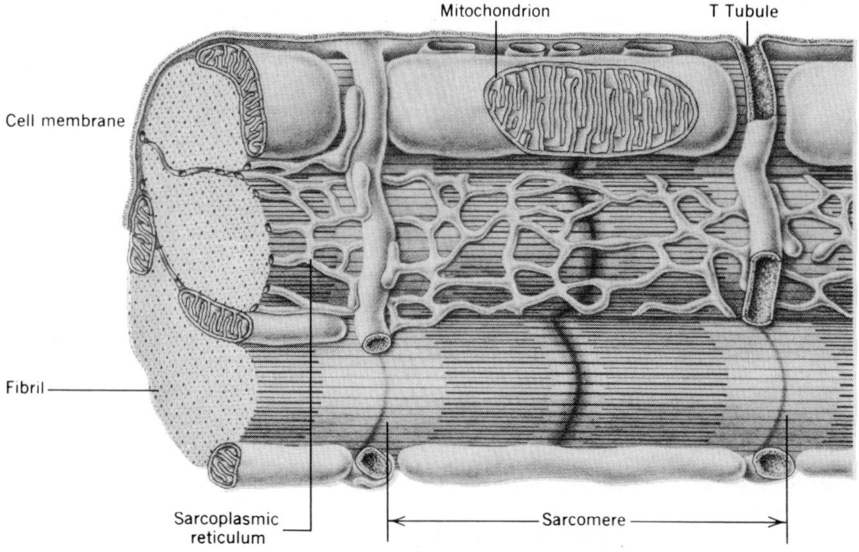

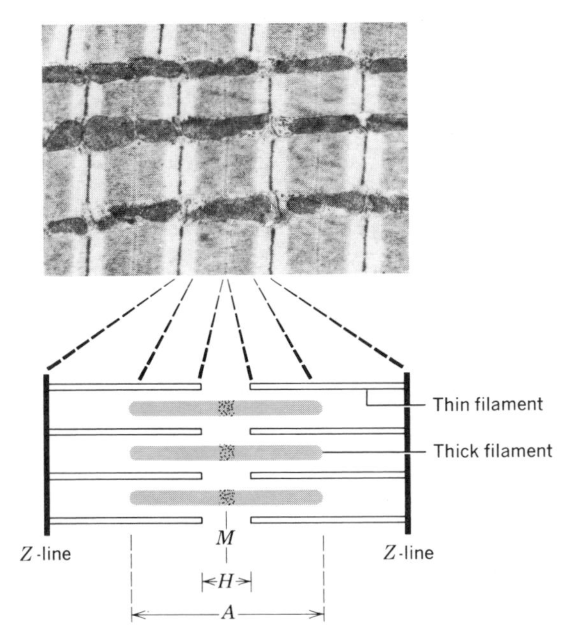

Figure 23–13
Relationship between the light and dark zones seen in electron photomicrographs of striated muscle and the arrangement of thick and thin filaments. (Photomicrograph courtesy of R. Chao.)

light and dark areas are called the *A* (i.e., **anisotropic**) and *I* (i.e., **isotropic**) bands. The *Z* and *M* lines are septa that extend across the myofibril. Thick myofilaments are attached to the *M* lines and extend in both directions overlapping the ends of the thin filaments. The overlapping fibers cause an increase in density in this region. The *H* zone is that part of an *A* band that has only thick filaments, while the *I* band has only thin filaments. As contraction occurs, the thick and thin filaments slide past each other, so that the *I* band and *H* zone disappear (Fig. 23–14). The *Z* lines are brought closer together, and the myofibril as a whole becomes thicker. The portion of a fibril between one *Z* line and the next is a **sarcomere.**

The thick filaments are composed of a fibrous protein called **myosin.** The myosin molecule contains a ''head'' and ''tail'' portion (Fig. 23–15). The tail is formed by two alpha helical polypeptide chains twisted around each other to form a right-handed superhelix (Chapter 4). The chains extend into and form a portion of the head along with two other short polypeptide chains. Myosin can be broken into three pieces with proteolytic enzymes. Treatment with trypsin releases part of the tail [i.e., **light meromyosin** (LMM)] from the molecule. Light meromyosin has no ATPase activity and cannot combine with actin. The remaining portion of the molecule, called **heavy meromyosin** (HMM), contains ATPase activity and binds actin. Treatment of HMM with the enzyme *papain* severs the head (the portion containing the ATPase activity) from the tail.

In thick filaments, the myosin molecules are arranged with their tails parallel to each other and their heads projecting away from the long axis of the filament at intervals (Fig. 23–16). No heads are present in the center of the filament, this region coinciding with the *H*-zone. The heads of the myosin molecules form cross-bridges with adjacent actin filaments.

Thin filaments are composed primarily of actin but also present are small amounts of *tropomyosin, troponin, α-actinin,* and *β-actinin* (see Fig. 23–17). In the ionic environment of the cell, the actin exists in the fibrous (i.e., F-actin) form consisting of two chains of actin monomers

Figure 23–14
Changes in the distribution of thick and thin filaments in neighboring sarcomeres during contraction and relaxation.

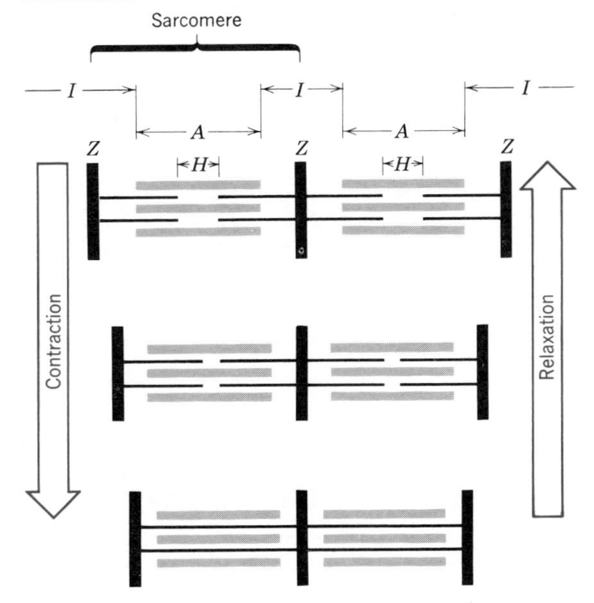

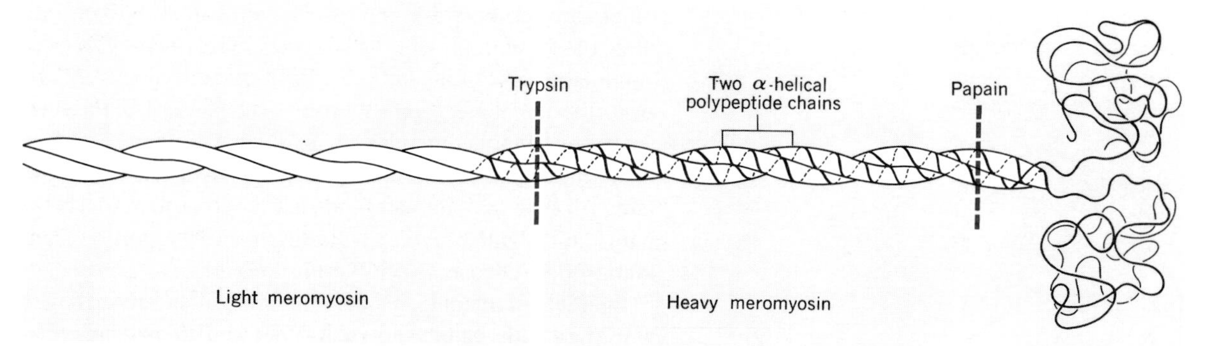

Figure 23–15
The myosin molecule. Trypsin splits the molecule into two parts: *light* meromyosin and *heavy* meromyosin. Papain severs the head (the region containing ATPase activity) from the tail.

Figure 23–16
(*a*) Diagram of platelet myosin. (Courtesy T. D. Pollard. Copyright 1975, Society of General Physiologists, *Molecules and Cell Movement,* S. Inoue and R. Stephens, eds.) (*b*) electron photomicrograph of platelet myosin. Arrowheads indicate myosin heads, and the symbol I delimits the length of the bare shaft. D indicates the diameter of the shaft and the brackets the length of the filament. (*c*) model of the filament with dimensions given in nanometers. (*b* and *c,* courtesy T, D. Pollard. Copyright 1975, The Rockefeller Press, *J. Cell Biol. 67,* 72.)

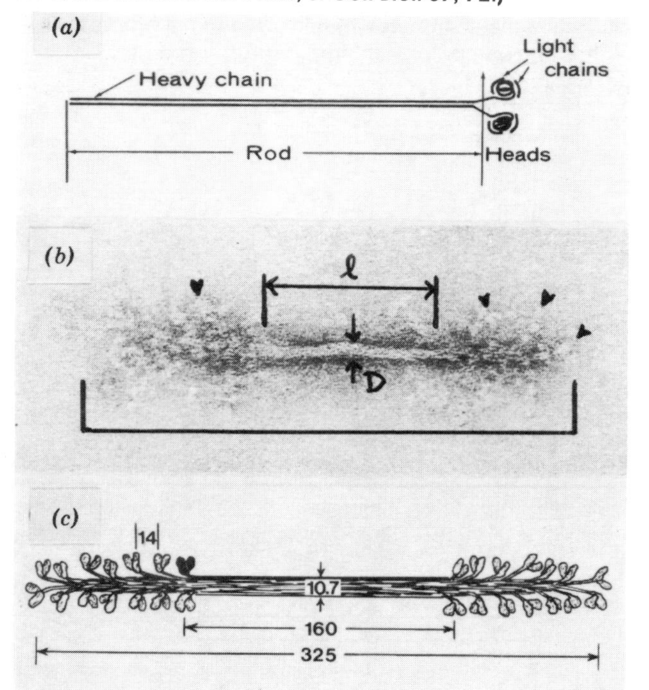

Figure 23–17
Association of the thick and thin filaments in striated muscle. The myosin head is oriented at an angle to the long axis of the thick filament and forms a cross-bridge with the thin filament. The thin filaments are composed of two helical rows of actin monomers. The regulatory proteins, tropomyosin and troponin, occur in regular association with the actin.

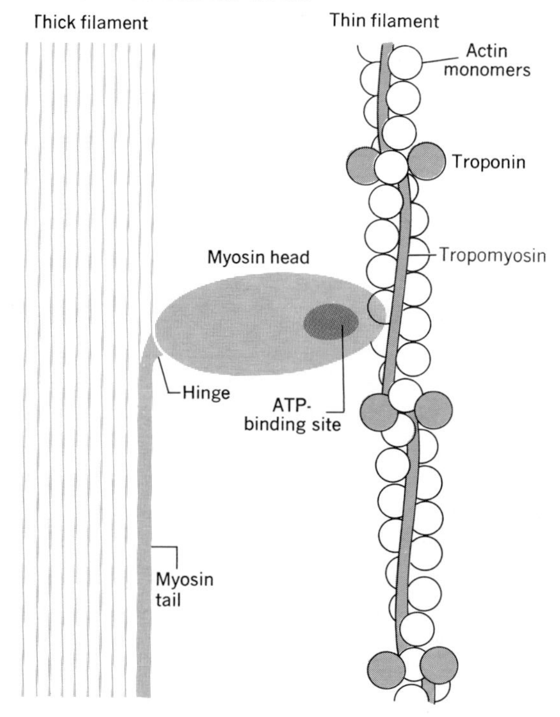

coiled about each other. When extracted from muscle tissue and dialyzed, actin becomes globular (i.e., G-actin) as the monomers separate. The G-actin monomer binds one molecule of ATP and one Mg^{++}. The interaction of the ATP bound to actin and the ATPase of myosin is only one step in the overall process of muscle contraction discussed below. The proteins tropomyosin and troponin are arranged at intervals along the actin filament and serve as regulatory agents in contraction. Actinin may function in polymerization of the actin monomers.

Muscle Contraction

When mixed together, purified myosin and actin will bind to each other by forming cross-bridges between the two molecules. Experimentally, this is noted by an increase in viscosity of the mixture. Although the nature of the cross-bridges is not known with certainty, it is believed to be between the actin monomer and a binding site in the head of the myosin molecule. When ATP and Mg^{++} are added, the actin and myosin disengage (the viscosity is lowered), and the ATP is then hydrolyzed by the ATPase present in the head of the myosin. When ATP hydrolysis is complete, the actin and myosin recombine (the viscosity again increases). In situ, the sequence of events is believed to occur over and over again, causing the actin to slide past the myosin. Thus, by sequentially forming and breaking the cross-bridges, the cell's fibrils become shorter and thicker. The full cycle from relaxation to contraction and back to relaxation involves four important processes: (1) triggering (or initiating) contraction, (2) the "power stroke" (i.e., contraction per se), (3) relaxation, and (4) the generation of energy for contraction. Each of these will now be considered.

Triggering or Initiating Contraction.

Prior to contraction there is no evidence of cross-bridges between the thick and thin myofilaments. The ATP and Mg^{++} concentration of the sarcoplasm is high, while the concentration of Ca^{++} is low. The head of each myosin molecule is oriented at 90° to the tail of the molecule (Fig. 23–18, step 1) and contains two tightly bound ADP and phosphate molecules. The actin binding site on the myosin head cannot bind actin because the actin is blocked by tropomyosin.

Normally, contraction of a muscle cell is initiated when an impulse from a nerve fiber causes depolarization of the sarcolemma at the **myoneural junction** (i.e., the junctions between muscle and nerve cells). The depolarization quickly spreads across the sarcolemma and down through the complex system of invaginations forming the sarcoplasmic reticulum or T-system. Depolarization is accompanied by a change in the permeability of the sacroplasmic reticulum membranes to ions. Of special importance are the resulting movements of calcium ions, which are rapidly released into the sarcoplasm from storage vesicles elaborated by the sarcoplasmic reticulum. Calcium ions promote the formation of cross-bridges between the myosin and actin filaments, thereby initiating contraction.

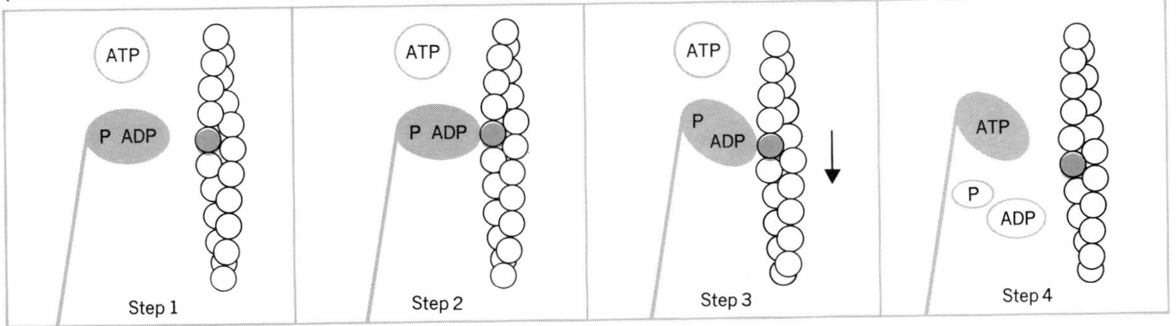

Figure 23–18
Steps in the contraction of muscle. For reference, the particular actin monomer involved in the round of contraction reactions is shown in color. (See text for additional details.)

The Power Stroke. The free Ca^{++} entering the sarcoplasm binds to troponin, causing a conformational change in this protein. The binding site on a G-actin monomer is consequently exposed and combines with a site on the myosin head (Fig. 23–18, step 2). Upon binding to actin, the myosin head undergoes a conformational change, altering to 45° its angle to the tail of the rest of the myosin molecule. As a result, the actin filament is caused to slide about 120 Å. This movement is the power stroke (Fig. 23-18, steps 3 and 4). This small movement occurs at almost the same moment for thousands of pairs of actin and myosin molecules and positions a new set of myosin heads and actin binding sites near one another. The process is repeated over and over again, causing each actin filament to move a total distance of about 0.25 μ.

Before each successive power stroke (altering the head angle of myosin), the bridge formed between actin and the myosin head during the previous cycle must be severed. Coincident with the power stroke is the release of the bound ADP and phosphate from the myosin head and the binding and hydrolysis of the adjacent ATP. This frees the myosin from the actin following the power stroke. Hydrolysis of ATP has been proposed as the cause of bending of the myosin head. Hydrolysis releases H^+, and the localized acidification of the protein's environment in the area of the hinge could produce the conformational change that underlies the power stroke. The hydrolysis of the newly bound ATP reestablishes the energized state of the myosin in preparation for a new reaction round with actin.

Relaxation. The cessation of nervous stimulation allows the sarcolemma to reestablish its polarity and regain its normal permeability. As the membrane repolarizes, Ca^{++}, which entered the sarcoplasm, is actively transported back into the vesicles of the sarcoplasmic reticulum. This active transport relies on an ATP-dependent Ca^{++} "pump." The energy for the transport is derived from the hydrolysis of ATP. The pump is Ca^{++}–stimulated ATPase, and extraction of this enzyme from the sarcoplasmic reticulum reveales that it constitutes a large proportion of the sarcoplasmic membrane protein.

With the removal of Ca^{++} from the sarcoplasm, the configuration of troponin again changes, and the binding site on the actin is again shielded from the head of the myosin. The actin and myosin no longer form cross-bridges, and contraction stops. The thick and thin fibers slide back to their positions as a result of the contractions of *antagonistic* muscles (muscles pulling in the opposite direction), the weight of the muscle itself, and the elasticity of the sarcolemma.

Generation of Energy for Contraction. ATP energy is required during contraction (i.e., during the movements of the head of the myosin molecules) and during relaxation (when pumping Ca^{++} back into the sarcoplasmic vesicles). At least two molecules of ATP are required for each cross-bridge formed, and an additional ATP is hydrolyzed in order to pump away the Ca^{++} associated with the troponin. In effect, two-thirds of the ATP consumed is used for contraction and one-third for relaxation.

Both glycolysis and oxidative respiration can provide ATP to muscle cells, but the *immediate* source of ATP during contraction is phosphate from the high-energy compound phosphocreatine (Fig. 23–19). This compound, which is in high concentration in muscle cells, rapidly transfers high-energy phosphate to ADP through the action of *creatine kinase*.

$$\text{Phosphocreatine} + \text{ADP} \longrightarrow \text{creatine} + \text{ATP}$$

The equilibrium lies far to the right maintaining an almost constant level of ATP in the muscle fiber during normal levels of muscular contraction. In experimental situations, it is possible to deplete the phosphocreatine supply by causing repeated contractions while blocking glycolysis and respiration.

The regeneration of the depleted phosphocreatine is brought about first by the synthesis of ATP via glycolysis or oxidative respiration and then through the action of creatine kinase. There are two basic types of skeletal muscle tissue—*red* and *white*—and the relative proportions of ATP contributed by glycolysis and respiration vary with each type. The red (or *slow*) *fibers* contain large amounts of

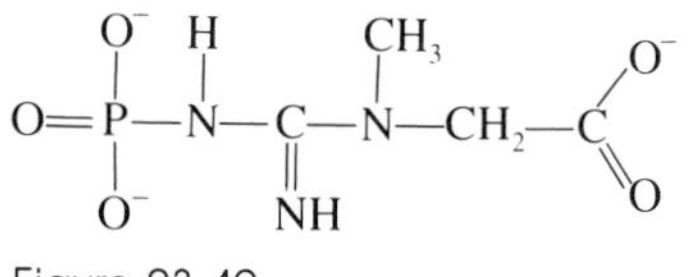

$$O{=}\underset{\underset{O^-}{|}}{\overset{\overset{O^-}{|}}{P}}{-}\overset{\overset{H}{|}}{N}{-}\underset{\underset{NH}{\|}}{C}{-}\overset{\overset{CH_3}{|}}{N}{-}CH_2{-}C\begin{matrix}\nearrow O^- \\ \searrow\!\!\searrow O\end{matrix}$$

Figure 23–19
Structure of phosphocreatine.

myoglobin and cytochromes and utilize oxidative phosphorylation to phosphorylate ADP. Fatty acids are the major substrate and are oxidized to acetyl-CoA. Acetyl CoA enters the Krebs cycle reactions in mitochondria, producing CO_2 and the reducing power for ATP generation. Red muscle tissue contains numerous mitochondria, whereas white (or *fast) fibers* contain few mitochondria and little myoglobin. In these fibers, glycolysis is the primary means of ADP rephosphorylation.

Red fibers are found in greater number in muscles with slow, rhythmic contraction patterns, such as heart muscle and the flight muscles of birds. White fibers predominate in quickly contracting muscles, such as the jumping muscles of frogs. The proportion of red to white fibers in the leg muscles of humans appears to be determined genetically and is believed to account for individual differences between long distance and sprint runners.

During vigorous activity, blood supplying oxygen to and removing wastes from the muscles may be inadequate. Oxygen depletion of muscle promotes glycolysis and results in the accumulation of lactic acid. Upon relaxation, oxygen is resupplied to the muscle and lactic acid removed. A prolonged period of muscle exertion may consume oxygen more rapidly than it can be absorbed by blood passing through the lungs. As a result, breathing at a rate above normal continues even after exercise is stopped. The extra oxygen consumed during this period balances the lactic acid produced by glycolysis and is called the **oxygen debt.** Excess lactic acid is transported to the liver, where it is "recycled" first into glucose and then into glycogen.

Nerve Cells or Neurons

The tissues of the nervous system contain a variety of cells, but one of the most highly differentiated and specialized is the neuron or nerve fiber itself. The primary and specialized function of these cells is the conduction and transmission of impulses from one part of an organism to another. In some instances, the impulse may travel several feet, and a single neuron may bridge the entire distance. In certain respects, neurons are similar to most other cells (Fig. 23–20), and the **cell body** contains the typical spectrum of organelles. It is the **processes** or extensions that make neurons easily distinguishable from other cells. Processes that receive transmitted impulses and conduct them toward the cell body are called **dendrites,** and processes that conduct impulses away from the cell body are called **axons.** The axons terminate in a special structure called the **end-plate,** which is responsible for transmission of the impulse across the gap (synapse) to the dendritic endings of the next neuron or across the myoneural junction to a muscle cell. A **nerve** is formed by a bundle of many neurons.

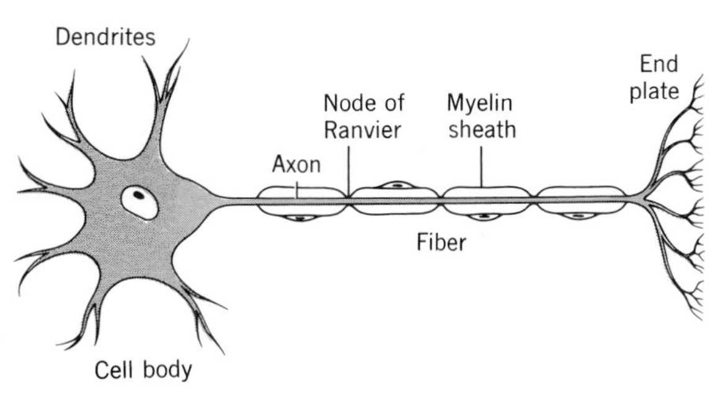

Figure 23–20
Diagram of a neuron, nerve cell, or nerve fiber.

Neurons begin developing from the neural tube during early embryonic development and continue development in many animals until well after birth. The development of motor neurons that have elongated axons is best understood. These cells begin as ventral cells in the neural tube. They have the capacity to migrate by amoeboid movement and to spin out long processes. Initially, the young cells gather in compact groups along the neural tube. Then the cytoplasm of the cells flows outward in a strandlike manner with amoeboidlike activity at the end of the strand. The long, thin strand becomes the axon of the neuron. Branching may occur; the direction of development is influenced by the adjacent tissues.

The "impulse," which normally begins at the dendrites and spreads wavelike across the entire cell to the end of the axon, is a transitory physicochemical change in the state of the cell membrane. Experimentally, an impulse can usually be started anywhere on the cell surface by applying a variety of stimuli, including electric shock, pressure (pinching), heat, cold, pH change, light, and various chemicals. Once initiated, the impulse is propagated along the membrane without dependence on a continuing stimulus. The speed with which the impulse travels along the fiber is not dependent on the strength of the stimulus; that is, it does not travel faster if initiated by a stronger stimulus. The rate of movement of the impulse varies with the type of neuron but is in the range 2–100 m/sec, a speed far too slow to compare the impulse to the movement of an electron through a wire during electrical flow.

As the impulse passes along the cell membrane, two major changes occur. One is a change in electrial potential across the membrane, called the **action potential,** and the second is a change in membrane permeability.

Action Potentials

There is an electrical potential across the plasma membrane of all cells but that of the neuron is highly developed. The axoplasm or cytoplasm of the axon and the extracellular fluid surrounding the cell contain a number of ions and act as good electrical conductors. The plasma membrane is a weak insulator but does offer a much higher resistance than the two fluids. When one member of a pair of microelectrodes attached to a voltage recording device is inserted into the axoplasm, and the other electrode is placed in the extracellular fluid, an electrical potential across the membrane is measured (Fig. 23–21). When a cell is not conducting an impulse, the potential is called the **resting potential** and is constant for a cell. For most nerve cells, the potential is 50–90 mv, with the inside surface of the membrane negative with respect to the outside. When a cell is stimulated and the resulting impulse passes the region of the electrodes, a brief change in potential is recorded. By placing electrodes at several points along the axon, it is possible to follow the transitory change in potential as it spreads along the process. As the **action potential** passes each recording electrode (Fig. 23–22), the internal voltage rapidly changes from − 90 mv to 0 mv, and then to +20 mv or +30 mv. Within 0.5 to 1.0 msec, the membrane recovers, and the −90 mv resting potential is restored. The impulse sweeping along the axon represents a transitory change in potential of 110–120 mv.

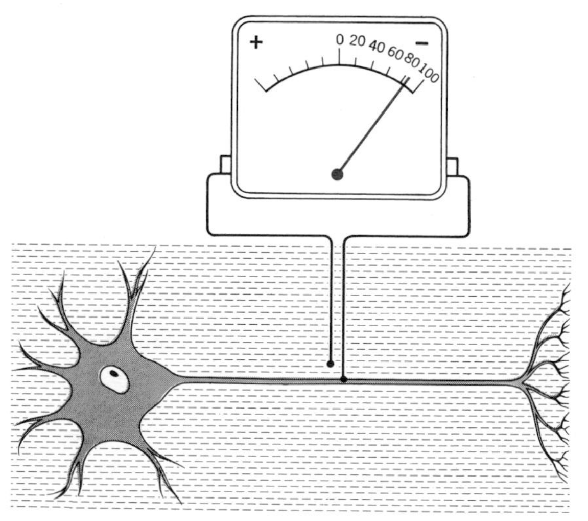

Figure 23–21
Measuring the *resting potential* of a nerve cell. One electrode is placed in the layer of fluid bathing the surface of the axon and the other inserted through the plasma membrane into the axoplasm.

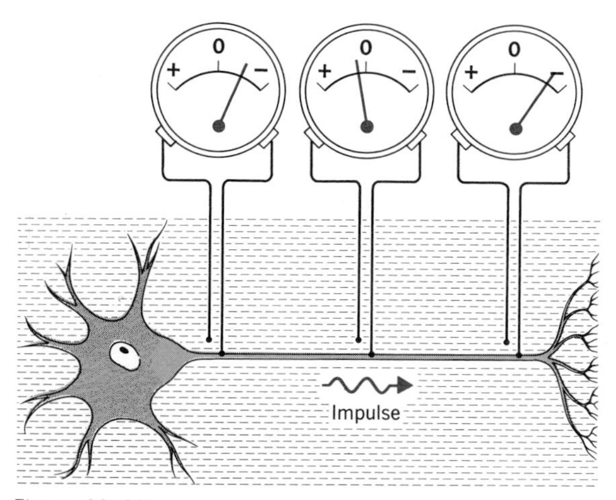

Figure 23–22
The transitory change in potential occurring across the nerve cell membrane as an impulse is conducted along the fiber can be measured using a series of recording devices placed at several points along the fiber. At each point, one electrode is placed in the fluid at the axon surface, while the other is inserted through the plasma membrane into the axoplasm.

Ion Gradients Across the Membrane

The electrical potential across the membrane is not caused by free electrons—biological materials rarely have a significant number of free electrons. The potential is caused by an unequal distribution of certain ions. The imbalance between anions and cations that produces the −90 mv potential is too small to accurately measure chemically but an analysis of the types of ions present in the axoplasm and extracellular fluid shows some significant differences in their proportions (Table 23–1). The extracellular fluid contains Na^+, Cl^-, and HCO^-_3 in higher concentration than the axoplasm, while the axoplasm contains higher concentrations of K^+ and organic anions than the extracellular fluid. The total quantities of cations and anions on each side of the membrane are in balance. The fact that the totals in Table 23–1 balance across the membrane is simply the result of the inadequate sensitivity of the chemical tests.

In order to maintain the ion concentration differences across the membrane, metabolic energy must be expended by the cell, and the membrane structure must not be disrupted. Any event that reduces cell respiration and ATP production will allow Na^+ to enter the cell and K^+ to leave in accordance with their respective concentration gradients. Injury to membrane will cause a similar flux of ions. Metabolic energy is used to pump Na^+ out of the cell and K^+ into the cell in order to maintain the ion concentration differences. Although the *Na^+-pump* mechanism is not fully understood, a number of models have been proposed to explain the experimental findings. One model (Chapter 15; Fig. 15–34) is based on changes in the tertiary structure of an enzyme carrier. Sodium ions are bound to the surface of the enzyme inside the cell along with ATP, and K^+ is bound to the outer surface. The binding causes a conformational protein change that displaces the ions. The ATP is hydrolyzed, and the displaced ions are released.

Initiation of the Action Potential

Application of a stimulus to the neuron causes a temporary inhibition of the Na^+/K^+ pump in the membrane at the point of stimulation and is followed by the diffusion of Na^+ from the exterior into the axoplasm. K^+ diffuses from the cell, but the net rate of outward K^+ movement is less than that of Na^+. As a result, there is a net inward passage of cations, which causes the potential across the membrane to change and ultimately produce the +20 mv potential difference between the inside and the outside (see above). The inhibition of the Na^+-pump and the dramatic flux of ions is temporary. Normally within 0.5 msec, the original integrity and polarity of the membrane is reestablished as Na^+ which entered the cell is pumped out and K^+ is pumped in.

If the stimulus is sufficiently intense (i.e., reaches a *threshold* level), an action potential is initiated. Any stimulus above the threshold level also initiates an action potential, but if the stimulus intensity is below threshold (i.e., *subthreshold*) the membrane recovers without initiating the action potential. The "all-or-none" nature of the response is directly related to the ability to inhibit the Na^+/K^+ pump, causing the membrane to depolarize.

Table 23–1
Ion Composition of Axoplasm and Extracellular Fluid of Mammalian Nerve-Muscle Preparations (Ion_o/Ion_i = ratio of external to internal ion concentration)

	Extra-cellular ion (mM/liter)	Intra-cellular ion (mM/liter)	Ion_o/Ion_i
Na^+	145	12	12.1
K^+	4	155	1/39
Other cations	5	—	—
Cl^-	120	4	30
HCO_3^-	27	8	3.4
Other anions	7	155[a]	—
Potential	0 mv	−90 mv	—

[a]Value for organic anions.

Conduction of the Action Potential

Once depolarization has occurred, current flows between the depolarized and adjacent region of the axon membrane (which is normally polarized) (Fig. 23–23). The flow is outward and through the adjacent resting portion of the membrane, then inward and through the depolarized region. The flow of current inhibits the Na^+/K^+ pump in the resting region and thereby propagates the action potential further along the axon. Recovery of the Na^+ pump in the previously depolarized region reestablishes the normal resting potential across the membrane. The cycle repeats itself over the entire length of the nerve fiber.

The speed at which the impulse is propagated along the nerve fiber is directly dependent upon the diameter of the axon, because the axon behaves like a resistive-capacitative (RC) circuit. In RC circuits, the time course for voltage change varies with the product of the resistance and capacitance. An axon with large diameter offers less resistance than a smaller one and therefore depolarizes more readily and conducts faster. The rapid conduction speed observed in squid axons (25 m/sec) is due to the large axon diameter (1 mm or more!). The rate of impulse conduction in mammalian nerve fibers exceeds that in the squid, but the increased speed is gained by a change in capacitance and not by an increase in axon diameter. The capacitance is changed

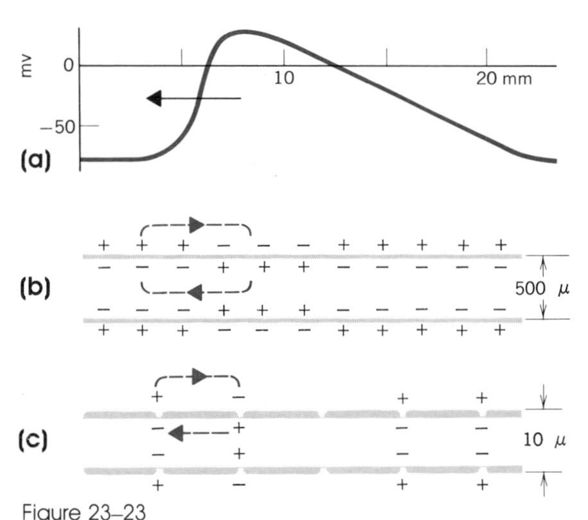

Figure 23–23

The speed of the impulse is related to the diameter of the axon; however, large-diameter unmyelinated fibers (*a*) and narrow myelinated fibers (*b*) conduct equally rapidly because in myelinated fibers the impulse skips from one node of Ranvier to the next. (Courtesy J. W. Woodbury, copyright 1976 by W. B. Saunders Co., *Introduction to Basic Neurology.*)

by the presence of myelin (a good insulator) in special cells that are wrapped around the axon forming a sheath (Fig. 23–24). The outer surface of the axon membrane is exposed to the extracellular fluid only at the *nodes of Ranvier* (Fig. 23–20), which are about 2 mm apart. The internode areas retain few charges at resting potential—probably fewer than at the nodes. Thus, there are far fewer charges to be neutralized, and conduction is considerably faster. In effect, the impulse skips from depolarized node to adjacent resting node, a process called **saltatory conduction.**

Synaptic Transmission

When the impulse or action potential reaches the end-plate of the axon, **neurotransmitters** are secreted. The secretions (described in Chapter 19) diffuse across the synaptic space and initiate an action potential in the next neuron. The neurotransmitter released by the axon end-plate at the myoneural junction leads to the formation of an action potential in the muscle cell membrane, which similarly causes a wave of depolarization to spread through the membrane. In the case of muscle cells, Ca^{++} enter the sarcoplasm on depolarization of the sarcoplasmic reticulum and this is followed by contraction.

Summary

Differentiation is the mechanism resulting in structural and biochemical change occurring in cells and is one stage in the development of organisms. The potential to differentiate depends upon the genetic composition of the DNA in the cell and the proper conditions for the expression of the genetic information. Not only must the genetic information be **transcribed** and **translated** into appropriate proteins and enzymes, but the conditions must also be proper for the **metabolic regulatory mechanisms** to function. A third level of control involves the interactions between the cytoplasm and the nucleus, the **nucleocytoplasmic relationship.**

The cytoplasm of the cell accumulates a variety of products of metabolism, materials that have entered from the environment or adjacent cells, and products of nuclear activity such as mRNA. All of these materials can enter the nucleus and affect gene expression, initiating the transcription of some DNA segments and halting the transcription of other areas. Thus the cell changes or differentiates with time.

The development of red blood cells (erythrocytes) illustrates the sequential cause and effect mechanisms of differentiation. Stem cells in the bone marrow may give rise to red or white blood cells. Increasing levels of the hormone **erythropoietin** traveling in the bloodstream to the marrow cause the erythrocyte progenator cells to form mRNAs for the globin chains of hemoglobin. The cells become committed to be red blood cells, as 95% of all protein synthesis is directed to hemoglobin. The buildup of hemoglobin is then followed by the dissolution of cell structures such as the nucleus and residual nucleic acids. The cells are irreversibly differentiated into red blood cells.

Muscle cells are uniquely differentiated into a mechanism for contraction. Myofibrils of actin and myosin are geometrically differentiated parallel to the long axis of the cell and spacially arranged about one another so that high-energy bonds can form between the filaments. When stimulation is applied or ATP and Mg^{++} are added, the bonds between actin and myosin strands progressively break. Enzymes and proteins precisely located during differentiation react

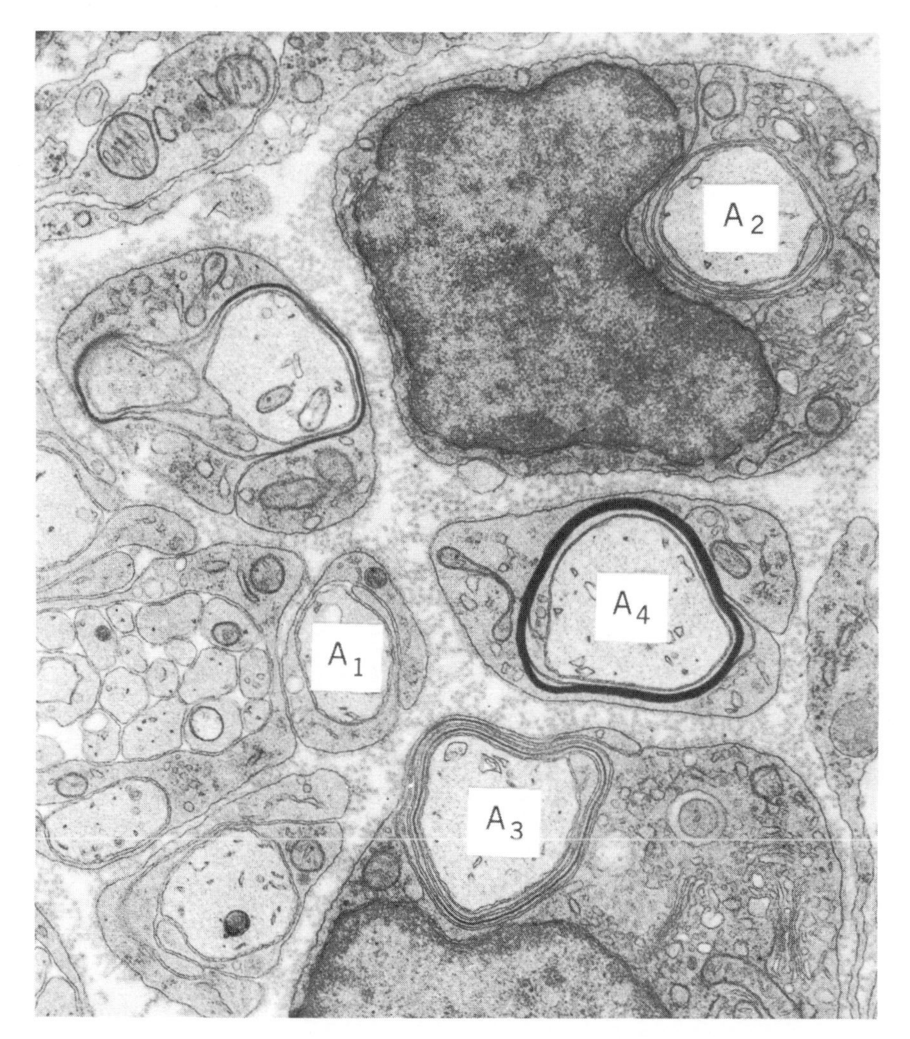

Figure 23–24
Myelin-filled Schwann cells surrounding axons (A). As the cells develop (A_1, A_2, A_3, A_4) the amount of myelin increases, and the Schwann cells twist around the axons forming many layers of membranes and myelin. (Courtesy H. Webster. Copyright 1976 W. B. Saunders Co., *The Fine Structure of the Nervous System.*)

and contract, causing the filaments to slide past one another and bring about overall shortening of the cell.

Nerve cells are also examples of highly specialized differentiation. The cells are elongated and have developed membrane structures that can achieve a selected differential permeability. This enables the cell to maintain an electrical potential as well as to propagate an action potential along the length of the cell.

References and Suggested Reading

Articles and Reviews

Baker, P. F., The nerve axon. *Sci. Am. 214*(3), 74 (March 1966).

Huxley, H. E., Muscle contraction and cell motility. *Nature 243,* 445 (1973).

King, T. J., and Briggs, R., Changes in the nuclei of differentiating gastrula cells as demonstrated by nuclear transplantation. *Proc. Natl. Acad. Sci. 41,* 321 (1955).

Nienhuis, A. W., and Benz, E. J., Regulation of hemoglobin synthesis during the development of the red cell. *N. Engl. J. Med. 297,* 1318–1328 (1977).

Patterson, P. H., Potter, D. D., and Furshpan, E. J., The chemical differentiation of nerve cells. *Sci. Am. 239* (1) 50 (July 1978).

Weber, A., and Murray, J. M., Molecular control mechanism in muscle contraction. *Physiol. Rev. 53,* 612 (1973).

Werz, G., Determination and realization of morphogenesis in *Acetabularia. Brookhaven Symp. Biol.,* Upton, N.Y. (1965).

Books, Monographs and Symposia

Aidley, D. J., *The Physiology of Excitable Cells,* Cambridge University Press, Cambridge, 1971.

Bessis, M., Weed, R. I., and Leblond, P. F. (editors), *Red Cell Shape,* Springer-Verlag, New York, 1973.

Goldman, R., Pollard, T., and Rosenbaum, J., *Cell Motility.* Book A and B. Cold Spring Harbor Conferences on Cell Proliferation. Vol. 3, Cold Spring Harbor Laboratory, Cold Spring Harbor, N.Y., 1976.

Hodgkin, A. L., *The Conduction of the Nervous Impulse,* Charles C. Thomas, Publishers, Springfield, Ill., 1964.

Ruch, T. C., and Patton, J. D., (editors), *Physiology and Biophysics,* W. B. Saunders Co., Philadelphia, 1965.

GLOSSARY

Acetyl coenzyme A An intermediate in energy-transferring reactions in metabolism, part of initial step of TCA cycle reactions.

Actin A protein found in thin filaments of striated muscle, microfilaments, and many nonmuscle cells.

α-Actinin A protein (M.W. ~ 95,000) found in the Z-line of muscle fibers.

Actinomycin D Antibiotic that inhibits the elongation of RNA chains.

Activation energy Energy required by a system to allow a chemical reaction to proceed.

Active site Region of the enzyme that binds and alters the substrate molecule.

Active transport An energy-requiring movement of molecules across a membrane.

Actomyosin A complex of two proteins, actin and myosin; the basic contractile element in muscle.

Adenosine triphosphatase ATPase; the enzyme that hydrolyzes ATP to form ADP and inorganic phosphate.

Adenosine triphosphate ATP; a nucleoside triphosphate; a high-energy intermediate in energy-transferring metabolism.

Aerobes Cells that live in and utilize oxygen.

Affinity chromatography A technique for separation of molecules. Molecules are attached to an insoluble (e.g., sepharose) matrix. Only those molecules that show affinity to the bound molecule (e.g., an antibody for its antigen) are retained. These trapped molecules can be subsequently eluted.

Allele One of the alternative forms of a gene.

Allosteric effectors Small molecules, usually metabolites, that bind to allosteric proteins at a site other than the active site so as to cause a change in protein shape.

Allosteric enzymes Enzymes whose activity is modulated by the binding of allosteric effectors at sites other than the active site.

Aminoacyl adenylate In protein synthesis, an activated compound that is an intermediate in the formation of a covalent bond between an amino acid and its tRNA adaptor.

Aminoacyl synthetase Any one of at least 20 different enzymes that catalyze (1) the reaction of a specific amino acid with ATP to form aminoacyl-AMP (activated amino acids) and pyrophosphate and (2) the transfer of the activated amino acid to tRNA forming aminoacyl-tRNA and free AMP.

Anaerobes Cells that can live without oxygen.

Anaphase Stage of mitosis or meiosis in which the chromosomes move toward opposite ends of the spindle.

Anaplerotic Reactions that replenish intermediates depleted by other metabolic pathways.

Aneuploidy Chromosome number that is not an exact multiple of the haploid number.

Angstrom (Å) A unit of length usually for describing molecular dimensions; equal to 10^{-8} cm.

Anticodon The three-base group on a tRNA molecule that recognizes and pairs with a three-base codon of mRNA.

Antigen A substance, usually a protein, that upon injection into a vertebrate is capable of stimulating the production of neutralizing antibodies.

Apoenzyme The protein component of an enzyme; apoenzyme + coenzyme = holoenzyme.

Aster The region at the poles of a dividing cell, composed of microtubules, a clear zone, and a pair of centrioles.

ATP See adenosine triphosphate.

ATPase See adenosine triphosphatase.

Attenuator region A region of DNA within an operon at which most RNA polymerase molecules stop transcription. Receipt of a specific antitermination factor will cause transcription to proceed.

Autophagic vacuoles Membrane-lined vacuoles containing morphologically recognizable cytoplasmic components. They include autolysosomes (which are secondary lysosomes) and autophagosomes (which are vesicles sequestering cytoplasmic organelles).

Autophagy The process of sequestration of intracellular components in vacuoles.

Autoradiography Determination of the location and geometry of radioactive components introduced into cells by means of exposure of photographic emulsions placed in contact with the cells.

Autotrophic cells Cell that can synthesize macromolecules from simple nutrient molecules, such as carbon dioxide, ammonia, and water.

Bacteriophage A virus requiring a bacterial host for its replication.

Basal body An organelle located at the base of cilia and believed to be involved in the organization of ciliary microtubules.

β-Galactosidase An enzyme catalyzing the hydrolysis of lactose into glucose and galactose.

Bivalent A synapsed pair of homologous chromosomes.

Calorie A unit of energy; the amount of heat required to raise the temperature of 1.0 g of water from 14.5° to 15.5°C.

Carcinogen An agent that induces cancer.

Carrier A transport protein within the membrane that binds temporarily with another molecule being transported across the membrane.

Catabolite repression Decreased synthesis of specific enzymes in bacteria grown on glucose or other good catabolite source. Caused by low levels of cyclic AMP in such cells.

Catalyst An agent that increases the rate of chemical reaction without altering the equilibrium point of that reaction.

C_4 cycle A CO_2-reducing pathway in photosynthesis; also known as the Hatch-Slack pathway.

Cell culture A population of cells grown in vitro.

Cell cycle The sequence of events in dividing cells including the G_1, S, G_2, and M periods.

Cell division Formation of two daughter cells from a parent cell by enclosure of the two nuclei in separate cell compartments.

Cell-free extract A fluid containing most of the suspended organelles and soluble molecules of a cell, made by breaking cells and removing the remaining whole cells.

Cellular affinity Tendency of cells to adhere specifically to cells of the same type. The property is lost in cancer cells.

Cell wall Rigid or semirigid structure enclosing the protoplast of most plant and procaryotic cells.

Centriole Microtubule-containing organelle located at the spindle poles in dividing cells or forming the basal portion of a cilium or flagellum.

Centromere The primary constriction of the chromosome to which the spindle fibers attach and which is required for chromosome movement to the poles at anaphase.

Chiasma Site of DNA exchange between two chromatids of a bivalent.

Chlorophyll Light-capturing pigment, located in chloroplast thylakoids or in procaryotic cells.

Chloroplasts Membranous structures containing chlorophyll which are present in the cytoplasm of eucaryotic photosynthetic cells; site of photosynthesis.

Chromatid One-half of a replicated chromosome, joined to the other chromatid at the centromere region.

Chromatin The nuclear material easily stained for light microscopy. Seen as dense masses in transmission electron photomicrographs.

Chromatin fiber The elongated deoxyribonucleoprotein molecule of the chromosome.

Chromomere A beadlike or knobby region of chromosomes often seen in early stages of meiosis.

Chromosome The gene-containing structure in the nucleus or nucleoid.

Cilium (pl. **cilia**) A whiplike organelle produced by a centriole; causes motion.

Cisterna (pl. **cisternae**) A flattened, membrane-bordered channel.

Clone A group of cells that have descended from a single cell by mitosis.

Codon A sequence of three nucleotides that code for an amino acid or chain termination.

Coenzyme A small organic molecule associated with the protein portion of a holoenzyme, weakly bound at the active site of the enzyme, and required for enzyme activity.

Coenzyme A A small organic molecule that participates in energy-transfer reactions, usually as a carrier of activated metabolites (e.g., acetate).

Colchicine An alkaloid that binds tubulin on a molar basis and thereby causes breakdown of microtubules.

Colinearity The spatial correlation between codons in DNA and amino acids in the polypeptide translated from the DNA.

Colony A group of continguous cells, usually derived from a single cell, growing on a solid surface.

Complement A series of blood serum proteins which, when activated, lyse foreign cells.

Complementary base-pairing Specific hydrogen bond interactions between a particular purine and a particular pyrimidine in nucleic acids; for example, guanine and cytosine, adenine and thymine, or adenine and uracil.

Concanavalin A A lectin.

Constitutive enzyme An enzyme synthesized at a constant rate.

Copolymer A polymeric molecule containing more than one kind of monomer unit.

Corepressor A metabolite that combines with repressor protein and blocks transcription of messenger RNA.

Coupled reactions Two chemical reactions that have a common intermediate through which energy can be transferred from one reaction to the other.

Coupling factor F_1 factor; the headpiece of the mitochondrial inner membrane subunit that has ATPase activity.

Covalent bond Interaction between atoms by sharing electrons.

Cristae Foldings of the mitochondrial inner membrane and the site of enzymes of oxidative phosphorylation and electron transport.

Crossing over Exchange of homologous chromosome segments leading to recombination of linked genes.

Cyclic adenosine monophosphate Cyclic AMP; adenosine monophosphate with phosphate group bonded between 3′ and 5′ carbon atoms to form cyclic molecule; this nucleotide is active in regulating numerous reactions in cells.

Cycloheximide An inhibitor of protein biosynthesis.

Cytochrome oxidase Cytochrome a-a_3; the terminal enzyme of aerobic respiration that transfers electrons to oxygen.

Cytochromes Electron-transport intermediates containing heme or related prosthetic groups that undergo valency changes of the iron atom.

Cytogenetics The study of biological systems using the combined methods of cytology and genetics.

Cytoplasm The protoplasmic contents of the cell, exclusive of the nucleus.

Cytosol The unstructured portion of the cytoplasm in which the organelles are suspended; the cytoplasmic fluid.

Dalton Unit of molecular weight approximately equal to the weight of a hydrogen atom.

Deletion Loss of part of a chromosome or DNA molecule from the genome.

Denaturation Change in the native configuration of a macromolecule resulting from heat treatment, extreme pH changes, chemical treatment, or other denaturing agents. It is usually accompanied by loss of biological activity.

Deoxyribonucleic acid (DNA) The genetic material.

Diakinesis Last of the stages of phophase in meiosis I.

Dictyosome A stack of cisternae that forms part of the Golgi apparatus.

Diffusion The net overall movement of molecules in the direction of a lesser concentration.

Dimer Structure resulting from association of two identical subunits.

Diploid A cell or individual or species having two sets of homologous chromosomes in the nucleus of somatic cells.

Diplotene A stage of prophase in meiosis I.

Disulfide bond Covalent bond between two sulfur atoms in separate amino acids of a protein.

DNA polymerase I Enzyme found to catalyze the formation of the 3′—5′ phosphodiester bonds of DNA. It possesses 3′ to 5′ single-strand proofreading and 5′ to 3′ double-strand exonuclease activities, for use in DNA repair, its chief biological function.

Duplication An extra copy of one or more genes in the chromosome complement.

Dynein Protein component, in microtubule doublets of the cilium or flagellum.

Effector A regulatory metabolite that activates or inhibits an enzyme by binding to an allosteric site on the enzyme.

Electron carriers Intermediates such as flavoproteins and cytochromes that reversibly gain or lose electrons.

Electron transport The movement of electrons from substrates to oxygen catalyzed by the oxidative respiratory chain intermediates.

Electrophoresis A method of separating macromolecules or particles according to their charge, size, and shape as they migrate through a gel or other medium in an electrical field.

Endergonic reaction A chemical reaction with a positive standard free energy change; an energy-consuming reaction.

Endocytosis Intake of solutes or particles by enclosure in a portion of plasma membrane bringing these materials into the cell.

Endoplasmic reticulum ER; folded membrane system distributed within the cytoplasm of eucaryotic cells; frequently has attached ribosomes (rough ER).

Endothermic process A process that absorbs heat.

End-product repression A control mechanism in which the synthesis of an enzyme required for a metabolic pathway is inhibited by the final product of that metabolic pathway, thereby stopping further pathway reactions.

Entropy The randomness or disorder of a system.

Enzymes The protein catalysts of biological systems.

Eucaryotic cells Cells having nuclear membranes and membrane-surrounded organelles.

Euchromatin Noncondensed, active chromosomes or chromosome regions of the interphase nucleus.

Exergonic reaction A reaction with a negative standard free energy change; an energy releasing reaction.

Excited state The energy-enhanced state of an atom or molecule existing after an electron has been moved from its normal stable orbital to an outer orbital having a higher energy level.

Exocytosis A mode of transport of substances out of the cell by enclosure in a vesicle, fusion with the plasma membrane, and subsequent expulsion to the outside.

Exothermic process A process in which heat is evolved.

Facilitated diffusion Assisted transport of molecules across the membrane along a concentration gradient.

Fatty acid Long hydrocarbon chain components of many lipids.

Feedback (end-product) inhibition Inhibition of the first enzyme in a metabolic pathway by the end product of that pathway.

Fermentation Oxidation of carbohydrate in non-oxygen-requiring pathways such as glycolysis.

First law of thermodynamics Energy can be neither created nor destroyed; statement of the principle of the conservation of energy.

Flagellum (pl. **flagella**) Elongated organelle produced by a centriole; ultrastructurally similar to a cilium but usually longer than a cilium.

Flavin adenine dinucleotide FAD; an electron carrier molecule that acts in energy-transfer reactions as a coenzyme; the reduced form of the redox couple is $FADH_2$.

Fluid mosaic membrane Model of cell membranes that postulates the distribution of proteins in a phospholipid bilayer and permits movements of particles within the membrane.

Fluorescent antibody technique Detection of selected antigens in cells by staining with a specific antibody conjugated with a fluorescent dye.

Formamide A small organic molecule used in double-helical DNA denaturation. Formamide combines with the free NH_2 groups of adenine and prevents the formation of A-T base pairs.

Formylmethionyl-tRNA fmet-tRNA; the initial aminoacyl-transfer RNA complex that reacts with the small ribosome subunit at the beginning of polypeptide chain synthesis.

Free energy A component of the total energy of a system that can do work under conditions of constant temperature and pressure.

Freeze-fracture Procedure for preparing materials for electron microscopy by rapid freezing and fracturing of the tissue; the exposed fracture faces are used to create a replica that is observed and photographed in the electron microscope; the fracture faces may or may not be further sublimed before the replica is made.

Furrowing A cell division mechanism that involves a pinching-in, or cleavage, to form two daughter cells from the parent cell.

Galactosidase, β (beta) See β (beta) galactosidase.

Gap junction Nexus; portions of the plasma membranes of adjacent cells that contain a space between the two membranes that permits cell-to-cell communication.

Gene A portion of a chromosome that codes for RNA.

Generation time The time necessary for growing cells to double their numbers or mass.

Genetic information The information contained in a sequence of nucleotide bases in a DNA or RNA molecule.

Genetic map The arrangement of mutable sites on a chromosome as deduced from genetic recombination experiments.

Genome The genes associated with a haploid set of chromosomes.

Genotype The genetic constitution of an organism (as contrasted with its physical appearance or phenotype).

Gluconeogenesis Synthesis of carbohydrates from noncarbohydrate precursors such as fats or proteins.

Glycolipids Lipids that contain polar, hydrophilic carbohydrate groups.

Glycolysis The process of glucose catabolism.

Glycoprotein A conjugated protein containing one or more sugar residues.

Glyoxylate cycle An anaplerotic pathway replenishing intermediary metabolites.

Golgi apparatus A region of the cytoplasm that functions in processing and packaging components for secretion from the cell.

Granulocytes Leukocytes with distinct cytoplasmic granules. Includes eosinophils, basophils, and neutrophils.

Group-transfer reactions Reactions (excluding oxidations or reductions) in which molecules exchange function groups.

Growth curve The change in the number of cells or protoplasmic mass in a growing culture as a function of time.

Growth factor A specific substance that must be present in the growth medium to permit a cell to multiply.

Hairpin loops Regions of double helix formed by the pairing of two contiguous complementary stretches of bases on the same single DNA or RNA strand.

Haploid Cell or individual having one copy of each chromosome.

Haptens Small nonantigenic molecules that are capable of stimulating specific antibody synthesis when chemically coupled to a larger molecule.

Heavy isotope Form of atoms containing greater than the common number of neutrons and thus more dense than the commonly observed isotope (e.g., ^{15}N, ^{13}C).

HeLa cells An established line of human cervical carcinoma (cancer) cells derived from *He*len *La*ne.

Helix A spiral structure with a repeating pattern described by two simultaneous operations—rotation and translation. It is the natural conformation of many regular biological polymers.

Heme An iron-containing porphyrin that serves as a prosthetic group in hemoglobins and in enzymes such as catalase and cytochromes.

Hemoglobin Protein carrier of oxygen found in red blood cells; composed of two pairs of identical polypeptide chains and an iron-containing heme group.

Heterochromatin Highly compacted chromatin regions of chromosomes during interphase.

Heterotrophic cells Cells that require complex nutrient molecules such as glucose, amino acids, etc. from which to obtain energy and to build their own macromolecules.

High-energy bond A bond that yields a large (at least 5 kcal/mole) amount of free energy upon hydrolysis.

High-energy phosphate compound A phosphorylated compound having a highly negative standard free energy of hydrolysis.

Histone A protein component of the chromosome having a high content of the basic amino acids arginine and lysine.

Holoenzyme The complete form of an enzyme.

Homologous Having the same or similar gene content.

Homologous chromosomes Chromosomes that pair during meiosis, have the same morphology, and contain genes governing the same characteristics.

Hormone A chemical substance synthesized in one organ that, in small amounts, modulates biochemical functions in the cells of another tissue or organ.

Hydrogen bond An electrostatic force between one electronegative atom and a hydrogen atom covalently linked to a second electronegative atom.

Hydrolysis The cleavage of a molecule into two or more molecules by the addition of a water molecule.

Hydrophilic Molecules or parts of molecules that readily associate with water; usually containing polar groups that form hydrogen bonds in water.

Hydrophobic bond The association of nonpolar groups with each other in aqueous solution.

Idiotype The binding specificity of an immunoglobin for a specific antigen.

Immunoglobins Y-shaped protein molecules that bind to and neutralize antigens.

Inducible enzymes Enzymes whose rate of production can be increased by the presence of inducers in the cell.

Intermediary metabolism The chemical reactions in a cell that transform food molecules into molecules needed as a source of energy and as precursors for cell growth.

Interphase The state of the eucaryotic nucleus when it is not engaged in mitosis or meiosis; consists of G_1, S, and G_2 periods in cycling cells.

Inversion Structural rearrangement of part of a chromosome so that genes within that part end up in inverse order.

In vitro (Latin: "in glass") Experiments done on isolated cells, tissues, or cell-free extracts rather than in situ, in place within the organism.

In vivo (Latin: "in life") Experiments done on or with intact living organisms.

Ion pumps Systems that actively transport molecules across a membrane by expelling one substance out of the cell and thereby helping to drive many kinds of molecules into the cell along an energy gradient.

Isomers Alternative molecular forms of a chemical compound.

Isopycnic density gradient centrifugation A method used to separate macromolecules and cell components on the basis of differences that cause them to come to rest at equilibrium in a region of the gradient that has a density of solute corresponding to their own buoyant density in the solute.

Isotopes Alternative nuclear forms of an atom, all having the same atomic number (proton number) but different atomic weights (neutron number varies).

Isozymes Alternative molecular forms of an enzyme.

Karyotype A photograph or diagram of a complete complement of chromosomes from a cell or individual.

Kinetochore Body that attaches laterally to the chromosomal centromere and is the site of chromosomal tubule attachment.

Krebs cycle Most common pathway for oxidative metabolism of pyruvic acid, which is an end-product of glucose fermentation; also known as the citric acid cycle or the tricarboxylic acid cycle.

Label (radioactive) A radioactive atom, introduced into a molecule to facilitate observation of its metabolic transformations.

Lampbrush chromosome Giant diplotene chromosome found in an oocyte nucleus, with loops projecting in pairs from most chromomeres. Loops are sites of active gene expression.

Lectins Cell-agglutinating proteins. Most lectins are isolated from plant seeds.

Leptotene The first of the prophase I stages in meiosis, before chromosome synapsis begins.

Ligase Enzyme that joins together the parts of single strands of DNA between the 5′ end of one strand and the 3′ end of another.

Lipid Class of organic compounds that are poorly soluble or insoluble in water but soluble in nonaqueous (organic) solvents such as ether.

Lipid bilayer An early model for the structure of cell membranes based upon the hydrophobic interactions between phospholipids. The polar head groups face outwardly, while the hydrophobic tails are clustered in the interior.

Lysis The bursting of a cell by the destruction of its cell membrane.

Lysogenic bacterium A bacterium that contains a prophage.

Lysogenic viruses Viruses that can become prophages.

Lysosomes Intracellular granules that contain a large variety of hydrolytic enzymes; these fuse with ingested food vacuoles and break down their contents.

Lysozymes Enzymes that degrade the polysaccharides found in the cell walls of certain bacteria.

Lytic infection Viral infection leading to lysis of cell.

Lytic viruses Viruses whose proliferation within the host cell leads to the cell's lysis.

Macromolecules Molecules having molecular weights in the range of a few thousand to hundreds of millions of molecular weight units (Daltons).

Macrophage Large, phagocytic white blood cell.

Matrix The essentially unstructured substance of a cell or organelle consisting of a suspension of molecules and particles in a watery medium.

Meiosis The reduction division of the nucleus in sexual organisms that produces daughter nuclei having half the number of chromosomes as the original nucleus.

Melting The separation of the two strands of duplex DNA to form single strands by disruption of hydrogen bonds between the duplex strands.

Meromyosin, heavy Portion of the myosin molecule with ATPase activity and Ca^{2+}-binding properties produced by trypsin digestion of myosin.

Mesosome An extensively infolded portion of the procaryotic plasma membrane that functions in respiration and cell division.

Messenger RNA (mRNA) The complementary copy of DNA that is made during transcription and that codes for protein during translation.

Metabolic pathway A set of consecutive cellular enzymatic reactions that converts one molecule to another.

Metaphase The stage of mitosis or meiosis when chromosomes are aligned along the equatorial plane of the spindle.

Microbody A membrane-bounded cytoplasmic organelle with varied enzyme content and functions; may contain catalase and the enzymes of the glyoxylate cycle.

Microfilaments Long, intracellular fibers that contain polymerized actin and that are thought to function in maintenance of cell structure and movement.

Micron (μ) A unit of length convenient for describing cellular dimensions; it is equal to 10^{-3} cm or 10^{4} Å.

Microsome A membrane-rich fraction of a tissue homogenate produced during centrifugation.

Microtubule An unbranched cylindrical assembly of protofilaments involved in cell movement phenomena; spindle fibers, ciliary subfibers, and centriole subfibers are microtubules.

Microvilli Fingerlike projections of plasma membranes of cells.

Mitochondria Membrane-surrounded organelles of aerobic cells that contain respiratory enzyme systems.

Mitosis The division of the nucleus that produces two daughter nuclei exactly like the original parental nucleus; somatic nuclear division.

Monolayer A single layer of cells, molecules, or other particles.

Monomer The basic subunit from which, by repetition of a single reaction, polymers are made. For example, amino acids (monomers) yield polypeptides (polymers).

Multienzyme system A group of enzymes active in the sequential steps of a metabolic pathway and in physical proximity to one another.

Mutagens Physical or chemical agents, such as radiation, heat, or alkylating or deaminating agents, that raise the natural frequency of mutation.

Mutation A change in the gene structure of a chromosome.

Myoblasts Precursor cells that aggregate to form the multinucleated striated muscle cell.

Myofibril Parallel units of a muscle fiber composed of bundles of myofilaments.

Myofilament Individual thick (myosin) and thin (actin) filaments of the myofibril.

Myosin Protein molecules, each composed of two coiled subunits (M.W. $\sim$ 220,000), that can aggregate to form a thick filament, globular at each end.

NAD, NADP Nicotinamide adenine dinucleotide and nicotinamide adenine dinucleotide phosphate, carriers of electrons in many enzymatic oxidation-reduction reactions.

Negative control Prevention of biological activity through a specific molecule; an example is inhibition of mRNA initiation by binding of specific repressor to specific sites along a DNA molecule.

Neutral fats Glycerides; fatty acid esters of glycerol; a major storage form of fats.

Nuclear envelope The double membrane surrounding the eukaryotic nucleus.

Nucleic acid Polymer of nucleotides in an unbranched chain; DNA and RNA.

Nucleoid A region of segregated DNA in prokaryotic cells not separated from the cytoplasm by a membrane.

Nucleolar organizing region (NOR) The specific part of the nucleolar organizing chromosome containing rRNA genes.

Nucleolus Spherical structure found in nucleus of eucaryotic cells. Involved in rRNA synthesis and ribosome formation.

Nucleoplasm The unstructured matrix portion of the nucleus in which the chromosomes and nucleoli are suspended.

Nucleoside Molecule containing a nitrogenous base linked to a pentose sugar.

Nucleosome (Nu particles) Spherical (100 Å diameter) masses seen along partially dissociated chromatin.

Nucleotide A nucleoside phosphate, a nitrogenous base linked to a pentose sugar linked to phosphate.

Nucleus The major membrane-bordered compartment of the eucaryotic cell containing the chromosomes and nucleoli.

Open system A system that exchanges matter as well as energy with its surroundings.

Operator A specific nucleotide sequence in the operon that binds repressor and exerts control over transcription of adjacent structural gene(s).

Operon A cluster of associated genes and recognition sites that participate in regulating and specifying amino acid polymerization into polypeptides; includes regulatory gene, promotor site, operator site, and structural gene(s).

Organelle A discrete structural differentiation of the cell containing particular enzymes and performing particular functions for the whole cell, e.g., mitochondria, ribosomes, etc.

Oxidant An oxidizing agent that loses electrons, or hydrogens, to a reducing agent or reductant.

Oxidation The loss of electrons from an atom, ion, or a compound.

Oxidative phosphorylation The enzymatic phosphorylation of ADP to ATP that is coupled to electron transport along the respiratory chain to oxygen.

Pachytene A stage of prophase I of meiosis characterized by synapsis of homologous chromosomes.

Peptide bond A covalent bond between two amino acids in which the alpha-amino group of one amino acid is bonded to the alpha-carboxyl group of the other.

Permease A type of carrier protein situated in the plasma membrane and involved in transport of specific substrate molecules across that membrane.

Peroxisomes Intracellular organelles that contain a fine granular matrix and often crystal-like cores. They contain enzymes involved in hydrogen peroxide metabolism, including catalase. They may be important in purine degradation, photorespiration, and the glyoxylate cycle.

pH Measure of hydrogen ion concentration in aqueous solutions.

Phagocytosis A form of endocytosis in which large amounts of particulate material, even whole cells, are taken up into large vesicles.

Phase-contrast microscope An instrument that translates differences in the phase of transmitted or reflected light into gradations of contrast.

Phenotype The observable properties of an organism; produced by the interaction of genotype and the environment.

Phosphodiester linkage A covalent linkage involving esterfication to phosphoric acid.

Photophosphorylation Process of formation of ATP from ADP and inorganic phosphate in the light reactions of photosynthesis; occurs by a cyclic or noncyclic pathway involving photosystems I and II.

Photorespiration Uptake of oxygen and release of carbon dioxide by photosynthetic cells or whole plants in the light.

Photosynthesis The enzymatic conversion of light energy into chemical energy by forming carbohydrates and oxygen from CO_2 and H_2O in green plant cells.

Photosynthetic phosphorylation The enzymatic formation of ATP from ADP in green plants coupled to light-dependent transport of electrons from excited chlorophyll.

Photosystem I (PS I) A photochemical reaction system in photosynthesis; coupled with photosystem II.

Photosystem II (PS II) A photochemical reaction system in photosynthesis; coupled to photosystem I.

Phycobilin An accessory photosynthetic pigment present in red and blue-green algae.

Pinocytosis Endocytosis of soluble materials into small vesicles.

Plaque Round, clear areas in a confluent sheet of cells; results from the killing or lysis of clusters of cells by several cycles of virus growth.

Plasmalemma Plasma membrane of the cell.

Plasmids Cytoplasmic, autonomously replicating chromosomal elements found in bacteria.

Plasmodesmata Cytoplasmic channels through the cell walls connecting the protoplasts of adjacent plant cells.

Plastid Eucaryotic organelle that stores pigments or carbohydrates.

Polyacrylamide gel electrophoresis A method of molecular separation that relies on the differential migration of molecules, usually proteins or polynucleotides, through a polyacrylamide matrix upon application of an electrical potential.

Polymer An association of monomer units into a large molecule.

Polymerase Enzyme catalyzing the synthesis of DNA or RNA from nucleoside triphosphate precursors.

Polynucleotide A linear sequence of nucleotides in which the sugar of one nucleotide is linked through a phosphate group to the sugar on the adjacent nucleotide.

Polynucleotide ligase Enzyme that covalently links DNA backbone chains.

Polynucleotide phosphorylase A bacterial enzyme that catalyzes the polymerization of ribonucleoside diphosphates to yield phosphate and RNA.

Polypeptide A long, unbranched polymer of amino acids.

Polyploid Cell or individual having an excess of one or more whole complements of chromosomes.

Polysome Polyribosome; an aggregation of ribosomes, connected by a strand of messenger RNA.

Polytene chromosome Giant chromosome composed of many fibrils (up to 2000) arising from successive rounds of chromatid duplication. Pairing of many identical chromomeres gives rise to characteristic banding pattern.

Pore An opening in a membrane or other structure; often referring to the nuclear pore complex of the nuclear envelope.

Positive control Control by a regulatory protein required for gene expression.

Primary constriction Location of centromere on chromosome.

Primary protein structure The number of polypeptide chains in a protein, the sequence of amino acids within them, and the location of interchain and intrachain disulfide bridges.

Primer A structure that serves as a growing point for polymerization.

Procaryote Simple unicellar organism, such as bacterium or blue-green alga, with no nuclear membrane.

Procentriole An immature centriole.

Prometaphase The stage of nuclear division when condensing of the chromosomes occurs just before their alignment on the equatorial plane of the spindle.

Promotor A specific nucleotide sequence in the operon to which RNA polymerase binds.

Prophase The first stage of mitosis or meiosis, after DNA replication and before chromosomes align on the equatorial plane of the spindle.

Proplastid An immature plastid.

Prosthetic groups Coenzymes that are bound to their enzymes.

Protamines A class of proteins rich in the basic amino acid arginine. They are found complexed to the DNA of sperm in many invertebrates and fish.

Protist Unicellular eucaryotic organisms such as protozoa, euglenoids, and algae.

Protoplasm The living material of the cell.

Protoplast The living structure of the cell, made of protoplasm, contained within but including the plasma membrane.

Provirus The state of a virus in which it is integrated into the genome of a host cell and is transmitted from one cell generation to another.

Puff A region of expanded chromosome undergoing active transcription, usually observed in giant polytene chromosomes.

Pulse chase experiment A radioactively labeled compound is added to living cells or a cell extract (pulse), and a short time later, an excess of unlabeled compound is added. Samples are then taken at periods after the pulse to follow the course of the label as a compound is metabolized (chase).

Purine Parent compound of the nitrogen-containing bases adenine and guanine.

Puromycin Antibiotic that inhibits polypeptide synthesis by competing with aminoacyl tRNAs for the *A* binding site.

Pyrimidine Parent compound of the nitrogen-containing bases cytosine, thymine, and uracil.

Quantum The energy of a photon.

Quaternary structure The manner in which the separate polypeptide chains of a protein are held together and oriented with respect to one another in space.

Radioactive isotope Isotope with an unstable nucleus that stabilizes itself by emitting ionizing radiation; important as tracers in biology.

Reannealing Renaturation; specifically, the restoration of duplex DNA regions through complementary base pairing of single stranded DNA molecules.

Redox couple Compounds that occur in both the oxidized and reduced forms and that are participants in oxidation-reduction reactions, such as NAD^+-NADH.

Reductant A reducing agent that accepts electrons or hydrogens in oxidation-reduction reactions.

Reduction Reactions involving gain of electrons or hydrogens.

Regulation The modulation of metabolism or gene action through control mechanisms.

Regulatory genes Genes whose primary function is to control the rate of synthesis of the products of other genes.

Release factor Specific macromolecule involved in the reading of the "stop" signal during protein synthesis.

Renaturation The return of a protein or nucleic acid from a denatured and nonfunctioning state to its "native" functioning configuration.

Repetitive DNA Repeated sequences of nucleotides that may occur in great numbers of reiterated copies in a chromosome complement.

Replicating fork Y-shaped region of chromosome that acts as growing point in DNA replication.

Replicating forms (RF) The structure of a nucleic acid during its replication; most frequently used to refer to double-helical intermediates in the replication of single-stranded DNA and RNA viruses.

Repressible enzyme Enzyme synthesized in the absence of its substrate which then represses further synthesis of the enzyme.

Repressor A protein product of the regulator gene of the operon that binds to the operator site and prevents transcription of structural genes.

Residual bodies Secondary lysosomes containing undigested residues, membrane fragments, and whorls.

Respiration The oxidative breakdown and release of energy from molecules by reaction with oxygen in aerobic cells.

Restriction enzymes Components of the restriction-modification cellular defense system against foreign nucleic acids. These enzymes cleave unmodified, double-stranded DNA at specific sequences that exhibit twofold symmetry about a point.

Reticulocyte Immature red blood cell still capable of limited hemoglobin synthesis.

Reverse transcriptase An enzyme coded by certain RNA viruses that is able to make complementary single-stranded DNA chains from RNA templates and then to convert these DNA chains to double-helical form.

Ribonucleic acid (RNA) Nucleic acids that function in transcription and translation of DNA.

Ribosomal DNA (rDNA) The genes at the nucleolar organizing region that code for ribosomal RNA.

Ribosomal RNA (rRNA) Ribonucleic acids that are part of the ribosome structure and that function in protein synthesis.

Ribosomes Small cellular particles made up of rRNA and protein. Ribosomes are the site of protein synthesis; in eucaryotic cells, they are often attached to the endoplasmic reticulum.

RNA (ribonucleic acid) A polymer of ribonucleotides.

RNA polymerase Enzyme that catalyzes the formation of RNA from ribonucleoside triphosphates, using DNA as a template.

Rough ER (RER) Portion of the endoplasmic reticulum bearing ribosomes.

Sarcolemma The plasma membrane of a muscle cell or fiber.

Sarcomere The contractile unit of muscle fiber, extending from one Z-line to an adjacent Z-line.

Sarcoplasm The cytoplasm of a muscle cell or fiber.

Sarcoplasmic reticulum Endoplasmic reticulum of a muscle cell or fiber.

Scanning electron microscope (SEM Electron-microscopic technique that permits observation of the surface structure (with a three-dimensional effect) rather than just thin sections.

Secondary constriction Any pinched-in site along a chromosome other than the primary constriction at the centromere.

Secondary structure Structure of a polypeptide chain describing the location, extent, and types of helices (as well as non-helical regions).

Second law of thermodynamics The principle that all physical and chemical change proceeds in a direction such that the entropy of the universe increases.

Secretion Release of cellular products into the extracellular space.

Sedimentation coefficient A quantitative measure of the rate of sedimentation of a given substance through water at 20°C in a unit centrifugal field; expressed in *Svedberg* units, *S*.

Semiconservative replication The usual mode of duplex DNA synthesis resulting in daughter duplex molecules that contain one parental strand and one newly formed strand.

Serum protein Protein found in the serum (cell-free) component of blood, includes globulins, immunoglobulins, albumin, clotting factors, and enzymes.

Sex chromosome Any chromosome involved in sex determination; such as the X and Y chromosomes.

Sliding filament mechanism A model used to explain the structural basis of cell movements, such as muscle contraction and bending of cilia and flagella.

Smooth ER (SER) Portion of the endoplasmic reticulum devoid of ribosomes.

S period Interval during the cell cycle in which DNA replication occurs.

Spindle Aggregation of microtubules during nuclear division that functions in the alignment and movement of chromosomes at anaphase.

Spindle fiber A microtubule in mitotically or meiotically dividing cells that extends from one pole to an attachment in the centromere region of a chromosome or that extends from pole to pole.

Spontaneous process A process accompanied by a decrease in free energy.

Standard electrode potential E_0; the oxidation-reduction potential of a substance relative to a hydrogen electrode; expressed in volts.

Standard free-energy change ΔG^o; a thermodynamic constant representing the difference between the standard free energy of the reactants and the standard free energy of the products of a reaction; energy-requiring reactions have a positive ΔG^o, while energy-releasing reactions have a negative ΔG^o.

Standard state Most stable form of a pure substance at 1.0 atmosphere pressure and 25°C (298 K). For reactions occurring in solution, the standard state of a solute is a 1.0 M solution.

Steady state A nonequilibrium state of an open system through which matter is flowing and in which all components remain in constant concentration.

Stereoisomers Molecules that have the same structural formula but different spatial arrangement of dissimilar groups bonded to a common atom. Stereoisomers have differences in their crystal structures and differ in the direction in which they rotate polarized light; they also differ in their ability to be used in an enzyme-catalyzed reaction.

Steric (Stereochemical) Relating the arrangement in space of the atoms in molecules.

Steroids Compounds that are derivatives of a tetracyclic structure composed of a cyclopentane ring fused to a substituted phenanthrene nucleus.

Streptomycin An antibiotic isolated from *Streptomyces griseus* (a soil bacterium) that binds specifically to bacterial 30 *S* ribosomal subunits, thereby blocking protein biosynthesis.

Stroma Unstructured matrix of the chloroplast that bathes the grana and stroma thylakoids.

Substrate Specific compound acted upon in the active site of an enzyme.

Supercoils Twisted forms taken by covalently closed, circular, double-stranded DNA molecules when purification has removed the protein components of the chromosome, thereby slightly changing the pitch of the double helix.

Suppressor gene A gene that can reverse the phenotypic effect of a variety of other genes.

Svedberg unit The unit of sedimentation equal to 10^{-13} seconds. The number of *S* units of a molecule or particle in a given centrifugal field is related to the weight, shape, and density of the molecule or particle.

Synapsis Specific pairing of homologous chromosomes, typically during zygotene of prophase I in meiosis.

Synaptinemal complex A complex structural component situated between a pair of synapsed homologous chromosomes during pachytene of meiosis I.

System An isolated collection of matter and energy. All other matter and energy in the universe apart from the system is said to be outside the system or its surroundings.

Telophase Stage of nuclear division when the nucleus reestablishes its interphase structure.

Temperature-sensitive mutation Mutation yielding a protein that is functional at low (high) temperature but that is inactivated by temperature elevation (lowering).

Template A macromolecular pattern that can be used for the synthesis of another, complementary macromolecule.

Tertiary structure The three-dimensional folding of a polypeptide chain into a complex structural form, brought about by interactions among side chains of amino acids.

Thermodynamics The branch of physical science that deals with exchanges of the energy in collections inherent in matter.

Thylakoid A closed membrane sac that may be disk-shaped in grana or may be greatly elongated in a chloroplast; light-requiring reactions of photosynthesis take place here.

T_m Midpoint melting temperature; temperature at the midpoint of transition of a preparation of duplex DNA molecules to single strands during melting.

Transcription Process by which the base sequence of DNA is copied into a complementary RNA molecule.

Transfer RNA (tRNA) The RNA molecule that carries an amino acid to a specific codon in messenger RNA during translation.

Transformation The genetic modification induced by the incorporation into a cell of DNA from another source.

Translation Process by which amino acids are assembled into a polypeptide on the ribosome, under the direction of the base sequence transcribed from DNA into messenger RNA.

Translocation A structural rearrangement involving parts of or entire non-homologous chromosomes.

Tritium ^3H; a radioactive isotope of hydrogen; extremely important in tracer studies.

Tropomyosin A muscle protein that associates with actin to form long, thin fibers; plays a role in the regulation of muscle contraction.

T-system Invaginations of the sarcolemma in muscle fibers of striated muscle, producing a system of transverse tubular infoldings.

Tubulin Globular protein subunits (M.W. 55,00 and 57,000) whose regular helical packing forms the hollow, cylindrical microtubules.

Ultracentrifuge Centrifuge capable of rotor speeds up to 75,000 rpm and able to rapidly sediment tiny particles and macromolecules.

Ultraviolet light Electromagnetic radiation having a wavelength shorter than that of visible light (3900–2000 Å). Causes DNA base-pair mutations and chromosome breaks.

Uncoupling agent A substance (example, 2,4-dinitrophenol) that can uncouple phosphorylation of ADP form electron transport; the energy is therefore released as heat.

Unit membrane Membrane showing a railroad-track or dark-light-dark pattern of electron density in the electron microscope; a model of membrane structure proposing that a phospholipid bilayer is coated on its outer and inner surfaces by proteins.

Vacuole A membrane-enclosed sac in the cell cytoplasm filled with molecules and particles in a watery medium, frequent in plant cells.

Van der Waals force A weak, attractive force between atoms; particularly important in hydrophobic bonding of amino acids in proteins.

Vesicle A small, spherical, membrane-bordered element.

Viruses Infectious, disease-causing particles that require a host cell for replication and that contain either DNA or RNA as their genetic material.

Weak bonds Forces between atoms that are weaker than the forces involved in a covalent bond such as ionic bonds, hydrogen bonds, and Van der Waals forces.

Wobble Ability of third base in tRNA anticodon (5′ end) to hydrogen bond with any two or three bases at 3′ end of codon. Thus, a single tRNA species can recognize several different codons.

X-ray crystallography The use of x-ray scattering by crystals to determine the three-dimensional structure of molecules, especially proteins and nucleic acids.

Zygote The product of fusion of two gametes; the cell from which a new individual develops in each sexual generation.

Zygotene Stage during prophase of meiosis I in which homologous chromosomes undergo synapsis.

Zymogen A digestive enzyme precursor lacking catalytic activity in this form, for example, pepsinogen (converted to active pepsin).

INDEX

Accessory pigments, 370, 371, 378
Acetabularia, 391, 529-531
Acetabularia experiments, 529
Acetolactate synthetase, 166
Acetylcholine, 400, 401
Acetylcholine esterase, 146, 401
Acetyl coenzyme A, 345, 547
Acetylgalactosamine, 104, 118
Acetylglucosamine, 104, 114, 118, 156
Acetylmuramic acid, 156
Acetylneuraminic acid, 104
Acid phosphatase, 403, 404
Acid ribonuclease, 405
Acids, 67
Aconitase, 346
Acrosome, 396, 401, 402, 522
ACTH, 78, 88
Actin, 79, 311, 547
  F-actin, 507
  G-actin, 507, 508
α-Actinin, 511, 537, 547
β-Actinin, 547
Actinomycin D, 391, 547
Action potential, 542-544
Activation energy, 547
Active site, 150-165
Active transport, 71, 322-329, 547
Actomyosin, 547
Acylation, 123
Acyl carrier protein, 199
Acyl enzyme, 160
Adenine, 131-133, 136
Adenosine diphosphate (ADP), 71
Adenosine monophosphate (AMP), 71, 72
  cyclic, 216
Adenosine nucleotide, 71
Adenosine triphosphatase, 547
Adenosine triphosphate (ATP), 162, 331, 547
  balance sheet, 361
  breakdown, 332
  glycolysis, 361
  hydrolysis, 179
  Krebs cycle, 361
  production, 380
  structure, 71
  synthesis, 179
Adipose tissue, 124
ADP, *see* Adenosine diphosphate
ADP-ATP exchange, 354
Adrenaline, 169
Adrenocorticotrophic hormone, *see* ACTH
Adrenocorticotrophin, see ACTH
Adsorbents, 265
Adsorption, 263
Aerobacter aerogenes, 52
Aerobes, 547
Aerobic glycolysis, 203
Affinity chromatography, 250, 269, 547
Agarose, 269
Agglutination, 91, 92, 306
Agglutinins, 307
Alanine, 78, 96

Albumin, 254
Alcohol fermentation, 188
Alcohol fractionation, 128
Aldohexose, 112
Aldolase, 209
Aldose, 109, 110
Algae, 114
Alimentary digestion, 143
Alkaline sink, 356
Allantoin, 162
Alleles, 162, 547
Allelic genes, 444
Alleloenzymes, 162
Allelozymes, 162
Allen, R., 511
Allolactose, 213
Allophycoxanthin, 371
Allosteric effectors, 209, 210, 220, 547
Allosteric enzymes, 165-170, 547
  feedback control, 216
  monovalent, 210
  polyvalent, 210
Allosterism, 151, 155
Alloway, J., 128
Alpha (α) amino group, 86
Alpha carbon, 78
Alpha carboxyl carbon, 161
Alpha carboxyl group, 86, 161
Alpha helix, 84, 85, 91, 97
Alpha rays (particles), 166, 279, 288
Altmann, R., 127, 331
Amberlite, 266
Amide, 159
Amidic forms, 78
Amination, 174
Amino acids, 78, 105, 139
  abbreviations, 87
  acidic, 78
  activation of, 484, 485
  amidic, 79
  aromatic, 79, 80
  basic, 79, 80
  cleavage, 498
  C-terminal, 479
  D-form, 78, 80
  general formula, 78
  hydrophobic, 80, 90
  hydrophilic, 80
  invariant, 97, 100
  L-form, 78, 80
  metabolism, 72
  neutral, 79, 80, 90
  nonpolar, 96
  N-terminal, 479
  secondary, 79
  semi-invariant, 97, 100
  side chain alteration, 498
  stereo diagram, 80
  sulfur-containing, 79
  synthesis, 211
Aminoacyl adenylate, 547
Aminoacyl synthetase, 547
Amino alcohol, 122
Amino group, 78
Ammonia, 146
Amoeboid movement, 511, 514
AMP, *see* Adenosine monophosphate
Amphibolic pathways, 203
Amphibolic reactions, 358
Amylase, 146
Amylopectin, 117-119
Amyloplasts, 386
Amylose, 117-119
Amytal, 353
Anabaena, 40
Anabolic pathways, 358
Anabolism, 173, 174, 187, 205
Anaerobic respiration, 193
Analytical ultracentrifuge, 228-230
Anaphase, 60, 426, 547
Anaplerotic reactions, 203, 358, 359, 547
Anderson, N., 236
Anderson, T., 129
Anderson, W., 500
Aneuploidy, 547
Anfinsen, C., 94, 480
Angstrom (Å), 547
Anion exchangers, 266
Anisomycin, 503
Anisotropic bands, 537
Annelids, 114
Annulus, 424, 425
Anomers, 110
Anoxia, 209
Antibiotic, 78, 503, 504
Antibodies, 91, 260, 306, 307, 502, 504
Anticodon, 454, 482, 547
Antigen, 91, 92, 307, 547
Antigen-antibody reaction, 91, 259, 306, 308
Anti-immunoglobulin antibodies (AIA), 308
Antimycin A, 353
Antiparallel polynucleotides, 135, 139
Antiserum, 260, 269
Apical cell, 531
Apoenzyme, 161, 547
Apoferritin, 91
Apparent dissociation constant, 68
Arabinose, 104
Arginase, 152
Arginine, 152
Arthropod shells, 113, 118
Artichokes, 118

Ascending chromatography, 264
Ascorbate, 162
Ascorbic acid oxidase, 162
Asparagine, 78, 104
Aspartate transcarbamylase, 164, 166, 168
Aspartic acid, 78, 89, 93, 152, 164
Aspartic semialdehyde, 166
Aspartokinase, 210
Aspartyl transcarbamylase, 165
Astasia longa, 56
Aster, 515
Asymmetric carbon, 109
Asymmetric dimer, 96
ATP, *see* Adenosine triphosphate
ATPase, 354
Attenuator region, 548
Aurintricarboxylic acid, 502
Autolysosomes, 407
Autophagic vacuoles, 407, 408, 548
Autophagy, 409, 410, 548
Autoradiography, 58, 277, 285, 548
Autotroph, 173, 174, 184
Autotrophic bacteria, 37
Autotrophic cells, 548
Avery, O., 128, 441
Avian leukemia virus, 47
Axon, 541
Axoneme, 517, 518
Axoplasm, 543

Bacillus subtilis, 37, 38
Bacillus thuringiensis, 42
Background count, 283
Bacteria, 24, 37, 58, 91, 115
  capsule, 39
  cell wall, 37, 39, 156, 158
  chromosome, 433
  flagella, 38, 520
  photosynthesis, 385
  viruses of, *see* Virus
Bacteriophage, *see* Virus
Bajer, A., 426
Barajas, L., 339
Barlund, H., 318
Basal body, 34, 515, 517, 548
Base, 67
Batch culture, 55, 56
Beams, J., 228
Beatty, B., 465
Beevers, H., 413
Belt desmosomes, 310, 311
Benda, 331
Bensley, R., 333
Benson, B., 410, 412
Benzene, 121
Berg, P., 484
Bessis, M., 326
Beta-alanine, 78
Beta (β) galactosidase, 213, 548, 550
Beta (β) galactoside permease, 213
Beta lactoglobulin, 254
Beta (β) oxidation, 187, 199, 205
Beta-pleated sheet, 86, 156
Beta rays (particles), 182, 279, 280, 288
Beta thiogalactoside acetyltransferase, 213
Bimolecular lipid leaflet model, 293, 327
Bimolecular reactions, 143, 144
Biocytin, 72
Bioluminescence, 180
Bishop, J., 479, 480
Bivalent, 548
Blake, C., 156
Blastocladiella emersonia, 333, 334
Blastula, 528, 531
Blepharoplast, 515
Blood, 122, 123
  coagulation, 105, 143, 163
  pigments, 77
  plasma, 163
Blood group glycoprotein, 105
Blue-green algae, 24, 39, 40
Boat form (glucose), 112
Boiling point, 63
Bonds, 70
  coordination, 161
  covalent, 145, 154
  disulfide, 86, 89, 159, 163
  electrostatic, 84, 89, 90, 96, 151, 165
  ester, 131, 158
  glycosidic, 112, 146, 156
  hydrogen, 83, 89, 90, 96, 133, 156
  hydrophobic, 89, 90, 96, 135
  ionic, 158
  modification of, 220
  peptide, 159, 161-163
  phosphodiester, 135
  salt, 89
  van der Waals, 90, 156
Bone, 113, 118, 148
Bonner, W., 245
Borsook, H., 453
Botrydium granulatum, 396
Boundary lipid, 304
Boyer, S., 496
Brailsford, J., 310
Brain, 78, 124
Brakke, M., 233
Branched pathway, 208
Branton, D., 298
Bretscher, M., 301
Briedenbach, R., 413
Briggs, R., 528

Britten, R., 219
Britten-Davidson Model, 220
Brownian motion, 4
Brown, R., 4
Budding, 334
Buffers, 67-71
Bulk transport, 324
Bull, B., 310
Bundle sheath cells, 384
Buoyant force, 227
Burke, D., 149

C_3 pathway, 383, 384
C_4 pathway, 383, 384, 548
C_4 plants, 384
Calorie, 70, 548
Calvin cycle, 383-385
Calvin, M., 380
Cancer, 138, 203, 443
Canfield, R., 480
Canham, P., 310
Capillaries, 27, 100
Capsid, 42, 46
Capsule, 37, 128
Carbamyl aspartate, 164, 165
Carbamyl phosphate, 164, 166
Carbohydrate, 75, 103
  metabolism, 143, 190-199
  synthesis, 71
Carbonic acid, 67
Carbonic anhydrase, 146, 148, 162
Carbonium ion, 158
Carbon-platinum replica, 18
Carboxylation, 72
Carboxyl group, 78, 83, 121, 158
Carboxymethylcellulose, 266
Carboxypeptidase, 156, 161, 162
Carcinogen, 548
Caro, L., 396
Carotenoid, 183, 370-372
Carrier-solute complex, 321
Cartilage, 113, 118
Cashel, 217
Catabolic repression, 214, 548
Catabolism, 173, 187, 205
Catabolite gene–activator protein (CAP), 216
Catalase, 148, 161, 357, 413
Catalysis, 139, 143, 154, 161
Catalyst, 77, 548
Catalytic residue, 154, 159
Cathepsin, 405, 412
Cation exchangers, 266
Cavitation, 226
Cell(s)
  age of, 57
  body, 541
  cleavage, 510
  continuous culture of, 54
  culture, 548
  cycle, 55-59, 391, 548
  densities, 236
  differentiation, 527-531
  disruption, 225
  division, 49, 548
  doctrine, 3, 4
  fractionation, 225-247
  fractions, 124
  growth, 49, 58
  harvesting, 245
  history of, 4
  membranes, 296
  oxidations, 360
  quantitation, 53
  sedimentation coefficient of, 236
  separation of, 241
  size, 54, 57, 58
  specialization, 531-545
  synchronous culture, 55
Cell-free extract, 548
Cell plate, 399, 402, 451
Cellophane membrane, 251, 252
Cellular activity, control, 127
Cellular affinity, 548
Cellular membranes, 296
Cellulose, 113, 118, 119
Cellulose acetate, 255
Cellulose nitrate, 251
Cell wall, 35, 36, 113-119, 156, 158, 548
  bacteria, 37, 39, 156, 158
  formation, 399
  fungal, 113, 114
Central Dogma, 138
Central lamella, 311
Centrifugal elutriation, 242, 243
Centrifugation
  analytical, 228, 252
  continuous-flow, 245, 246
  density gradient, 233
  differential, 231, 232, 246
  equilibrium isopycnic, 235, 236
  preparative, 231
  rate, 234, 235
  theory, 226
  two-dimensional, 235
  zonal, 235, 242
Centriole, 34, 514-516, 548
Centromere, 425, 548
Cephalin, 122, 123
Cerebrocide, 124
Cerenkov radiation, 285
Ceruloplasmin, 105
Chair form (glucose), 112

Chance, B., 351
Chaos chaos, 333
Chapeville, F., 484
Chargaff, E., 133, 135
Chase, M., 128, 129
Chelate, 71
Chelating agent, 249
Chemical-coupling hypothesis, 355
Chemiosmotic-coupling hypothesis, 356
Chemo-organotroph, 173
Chemostat, 54
Chemotroph, 173
Chiasma (Chiasmata, pl.), 444, 449, 548
Chitin, 113, 118
Chloramphenicol, 503, 504
Chlorella, 58, 363, 380
Chlorocruorin, 91
Chloroform extraction, 128
Chloromycetin, 503
Chlorophyll, 35, 363, 369, 548
  a, 369, 371, 376
  absorption of light, 369, 370, 371, 375
  b, 369, 371
  bacteriochlorophyll, 369, 370, 375
  c, 369, 371
  d, 369
  electronic states, 375
  p_{700}, 386
  structure, 370
  trap, 379
Chloroplast, 35, 194, 219, 331, 363-387, 548
  chemical composition, 368, 370
  development, 372, 373, 374
  fine structure, 365
  isolation, 364
  protein synthesis in, 475
  ribosomes, 475
  shape and size, 364
Cholesterol, 124, 292
Choline, 123
Chondrioids, 37
Chondroitin, 113, 118
Chromatid, 426, 444, 548
Chromatin, 25, 218, 417-419, 451, 548
  dispersed, 418
  euchromatin, 417
  facultative, 418
  fibers, 431, 548
  heterochromatin, 417
  location, 418
  structure, 418, 419
Chromatography, 272
  affinity, 250, 269
  gas, 250
  ion-exchange, 250, 266-269
  paper, 250, 263-265
  thin layer, 250, 265
Chromatophore, 37, 374, 385
Chromocenter, 417
Chromomere, 548
Chromoplasts, 386
Chromoprotein, 103, 106
Chromosomes, 60, 130, 425, 548
  bacterial, 431, 433
  bivalent, 444
  centromere, 425
  metaphase, 426, 427
  polytene, 431, 432
  secondary constriction, 425
  tertiary constriction, 425
  ultrastructure, 431
  viral, 431
Chymotrypsin, 156, 159, 160-163, 500
Cilia, 34, 517-523, 548
Cisterna(ae), 390, 391, 402, 548
Citrulline, 78
Claude, A., 231, 333
Clone, 442, 548
Code, 139, 455
  degeneracy of, 482
Codon, 454, 548
  initiator, 487
  start, 487
Coenzymes, 72, 161, 162, 548
Cofactors, 161, 162
CO_2 fixation, 381, 386
  C_3 pathway, 383
  C_4 pathway, 383
  CAM, 383
  Hatch-Slack, 383
  pathways, 383
Cohn, Z., 410, 412
Coiled body, 425
Colchicine, 509, 510, 548
Cole, 292
Colinearity, 548
Collagen, 77, 86
Collagenase, 161, 412
Collander, R., 318
Collision probability, 144
Collodion, 251
Colony, 548
Compartmentalization, 219, 220
Competitive inhibition, 150, 151, 155, 321
Complement, 549
Complementary base pairing, 134, 135, 549
Complementary base sequences, 454
Complementary nucleotides, 135-137
Concanavalin A, 307, 549
Condensed peripheral chromatin, 418
Conformational change, 166
Conformational-coupling hypothesis, 355

Conformational states, 341, 342
Conjugate acid-base pair, 68
Conjugation, 396
Conjugated proteins, 103, 106
Consden, R., 263
Constitutive enzyme, 213, 549
Constitutive heterochromatin, 418
Constitutive mutant, 214
Contact inhibition, 53, 306
Contact residues, 153, 154
Contractile proteins, 77
Contractile ring, 511
Contraction (muscle)
  energy, 540
  power stroke, 540
  steps, 539
  trigger, 539
Cooperativity, 100, 155, 165-168
Coordinate induction, 214
Copolymer, 549
Copper, 162
Copper transport, 105
Corepressor, 216, 549
Corey, R., 81, 86
Cortical granules, 16
Corticosterone, 124
Cotranslational modifications, 504
Cotransport, 324
Countercurrent distribution, 233, 250, 261-263
Coupled reactions, 177, 188, 549
Coupling factor, 355, 549
Covalent bond, 64, 70, 549
Craig apparatus, 261
Craig, L., 261
Crassulacean acid metabolism (CAM), 383, 385
Creatine, 540
Creatine kinase, 179, 540
Crenation, 316
Crick, F., 86, 133-135, 482
Crista(ae), 25, 27, 337, 338, 360, 549
Crossing over, 449, 549
Cross-linked agarose, 271
Cross-linked dextrans, 270, 271
C-terminus (polypeptide), 87
Culture fractionation, 57
Curie, 283
Cuticle, 114
Cyclic adenosine monophosphate (cyclic AMP), 72, 166, 216, 549
Cyclic photophosphorylation, 378, 379
Cyclohexane, 124
Cycloheximide, 503, 504, 549
Cysteine, 86, 94, 128
Cystine, 86
Cytidine triphosphate (CTP), 164-168
Cytochalasin B, 507, 508, 511
Cytochrome, 103, 549
  b_{559}, 378
  c, 78, 254
Cytochrome oxidase, 353
Cytogenetics, 549
Cytokinesis, 60, 438, 449, 510-513
Cytoplasm, 549
Cytoplasmic streaming, 507
Cytosine, 131, 133, 136
Cytoskeleton, 309, 328
Cytosol, 125, 540
Cytosomes, 403

Dalton, 549
Dandelions, 118
Danielli-Davson Membrane Model, 294, 295
Dark reactions, 375, 380-385
Davidson, E., 219
Davis, B., 256
Davson, H., 292, 294
Death phase, 52
Decarboxylase, 144, 162
Declining phase, 52
deDuve, C., 403-405, 413
Defense mechanism, 91
Dehydroascorbate, 162
Dehydrogenase, 144, 162
Dehydrogenation, 72
Dehydration synthesis, 80
Deletion, 549
Denaturation, 95, 153, 549
Dendrite, 541
Dense bodies, 407
Density-dependent inhibition, 53
Density gradients, 56, 246
2-Deoxyadenosine-5′-phosphate, 132
Deoxyribonuclease, 128
Deoxyribonucleic acid (DNA), 93, 105, 128-139, 549
  amount per cell, 130
  antiparallel, 436
  complementary base pairing, 134
  discovery, 127
  double stranded, circular, 46
  double stranded, linear, 46
  fibrils, 474
  model, 133, 134
  negative strand, 136
  plus strand, 136
  recombinant, 441
  replication, 135, 136
  single-stranded, 46, 136
  structure, 133, 134
  viral, 129, 136
Deoxyribose, 131, 132, 136
dePetris, S., 308
deSaussure, N., 374

Descending chromatography, 263
Desmosomes, 26, 27, 311, 328
Determination, 527
Developmental biology, 527
Diakinesis, 444, 549
Dialysis, 250, 251
Dictyosome, 28, 36, 390-396, 549
Dicumarol, 354
Diethylaminoethylcellulose, 266
Dietz, R., 426
Differential cell agglutinability, 242
Differentiation, 527, 531
Diffusion, 22, 252, 314, 329, 549
  coefficient, 317
  constant, 252
  facilitated, 321, 322
  kinetics, 321
  mediated, 321
Digestive vacuole, 407
Dihydrouridine, 139
Dihydroxyacetone, 109, 110
Dimorphic chloroplasts, 384
2,4-dinitrophenol, 354
Dintzis, H., 477-480
Diplococcus pneumoniae, 128
Diploid, 549
Diplonema, 444
Diplotene, 444, 549
Dipole, 64, 70
Disaccharide, 109, 112
Dispersed chromatin, 48
Dispersion, 63
Dissociating agent, 249
Dissociation constant (K), 68, 147
Distribution coefficient, 318
Disulfide bond, 549
Disulfide bridge, 86, 89-91
DNA, *see* Deoxyribonucleic acid
DNA-dependent DNA polymerase, 437
DNA-directed RNA polymerase, 438
DNA polymerase I, 549
Donnan, F., 320
Double helix, 134, 135, 139
Double-label experiments, 285
Double reciprocal plot, 150
Doyle, W., 325
DPN (see NAD)
Drosophila, 431, 432
Duodenum, 163
Duplication, 549
Dutrochet, R., 4
Dynamic proteins, 77
Dynein, 549

E_0, 179
Ectoplasm, 511
Edeine, 503
Edidin, M., 302
EDTA, 302
Effectors, 165-166, 210, 549
Einstein-Stark law, 182
Elaioplasts, 386
Elastase, 161
Electrical potential, 253, 258
Electrogenic pump, 324
Electromagnetic radiation, 180-182, 279
Electron carriers, 549
Electronegative atom, 70
Electron microscopy, 93, 115, 287
Electronic cell counter, 53, 54
Electron spin resonance (ESR), 95
Electron transport, 549
  energetics, 352
  inhibitors, 352
  pathway, 350
  system (e-t-s), 331, 347-350, 362
Electrophoresis, 242-244, 253, 272, 549
  continuous-flow, 261
  density gradient, 261
  discontinuous (disc), 250, 256, 258, 357
  immuno-, 250, 259
  moving-boundary, 250, 255
  zone, 250, 255, 265
Electrostatic bonds, 84, 89
Elutriation, 246
Embedding, 8
End bulb, 400
Endergonic reactions, 71, 177, 184, 204, 331, 549
Endocytosis, 325, 329, 410, 549
Endomembrane system, 389
Endonuclease, 481
Endoplasm, 511
Endoplasmic reticulum, 16, 27, 549
  and Golgi, 390
  and ribosomes, 471
Endothermic process, 549
End plate, 541
End-product inhibition, 210
End-product repression, 549
Energy, 173
  calculations, 204
  changes, 174, 176
  cycles, 175
  electrical, 174
  free, 176, 177
  kinetics, 174
  potential, 174
Englander, S., 480
Enhancement, 376
Enterokinase, 162
Enthalpy, 176
Entropy, 94, 176, 549

Enzymes, 77, 103, 139, 549
  action of, 161
  activity, 150-154, 208, 209, 220
  atomic structure, 155
  binding sites, 165
  constitutive, 213
  crystallization, 143
  induced, 213
  induction of, 218, 220
  inhibition of, 190
  inhibitors, 150, 151
  isolation of, 143
  kinetics, 146, 209
  production, 190
  regulation, 163
  repression, 214
  structure, 151, 161
  synthesis, 131, 213
  tertiary structure, 151
Enzyme-substrate affinity, 209
Enzyme-substrate complex, 146, 148, 153
Epichlorohydrin, 231
Epinephrine, 72, 169
Episome, 441
Epithelium, 304
Epon, 8
Equatorial plate, 426
Equilibrium constant, 145, 146, 207
Erythroblast islands, 326
Erythrocytes, 19, 22, 99, 291, 304, 532
  cell surface properties, 309
  differentiation, 532
  morphology, 534
  proliferation rate, 533
  scanning EM, 534
  structure, 310
Erythromycin, 503, 504
Erythropoiesis, 532
Erythropoietin, 77, 78, 105, 532
Erythrose, 110
Erythrulose, 110
Escherichia coli, 37, 38, 46, 65, 128
Essential amino acids, 202
Ester, 131, 159
Esterification, 178
Estrogen, 124
Ethanolamine, 123, 260
Ether, 121, 122
Etioplasts, 386
Eucaryotic cells, 24, 105, 549
Euchromatin, 417, 549
Euglena, 118
Evans, W., 297
Excited state, 550
Excision, 441
Exclusion limit, 270, 271
Exergonic reactions, 71, 178, 184, 204, 550
Exocytosis, 325-329, 415, 507, 550
Exonuclease, 481
Exothermic process, 550
Exponential growth, 50, 51
Extrinsic proteins, 298, 300
Eye, 184

Facilitated diffusion, 321, 322, 550
Facultative heterochromatin, 418
FAD, *see* Flavin adenine dinucleotide
Faraday, 180
Fat-soluble vitamin, 72
Fat storage, 124
Fatty acids, 105, 121, 125, 199, 550
  biosynthesis, 201
  chain elongation, 357
  oxidation, 202, 357
  saturated, 121
  unsaturated, 121
Feedback inhibition, 164, 210, 550
Feedforward stimulation, 210
Fenestrae, 390
Fermentation, 188, 193, 205, 560
Fernandez-Morán, 339
Ferredoxin, 378
Ferritin, 308, 326
Fertilization, 118, 130
Feulgen reaction, 130
Fibrillar centers, 425
Fibrin, 77, 163
Fibrinogen, 105, 163, 254
Fibrinolysin, 161
Fibrous proteins, 77
Fick's equation, 317
Ficoll, 234
Filtration, 56, 242
Fingerprinting, 478
First law of thermodynamics, 550
First-order kinetics, 147
Fischer, E., 78, 153
Fixation, 8
Flagella(um), 34, 517-520, 550
  bacterial, 38, 520
Flagellin, 38, 520
Flame ionization, 271
Flavin adenine dinucleotide (FAD), 72, 74, 161, 349, 550
Flavin coenzymes, 72
Flavin-linked dehydrogenases, 349, 362
Fletcher, 334
Fluid mosiac membrane model, 298, 327, 550
Fluorescein, 7, 244
Fluorescence-activated cell sorting, 242-245
Fluorescent antibody, 302, 550
Fluorescent dyes, 244
Fluorochrome, 7

Flavin mononucleotide (FMN), 72, 162, 349
Fol, H., 130
Follicle-stimulating hormone (FSH), 105, 400
Fontana, 421
Food cups, 327
Food vacuoles, 327
Formaldehyde, 8
Formamide, 550
Formylmethionine, 487
Formylmethionyl-tRNA, 487, 550
Fox, C., 304
Fraenkel-Conrat, H., 129, 130
Franklin, R., 133
Free energy, 176, 177, 550
Freeze fracturing, 14, 33, 34, 299, 300, 340, 550
  membranes, 298
Freezing point, 64
Fructokinase, 178
Fructose, 110, 112, 115, 118, 146
Fructose-6-phosphate, 143, 166
Frye, D., 302
FSH, *see* Follicle-stimulating hormone
Fucose, 104
Fucoxanthin, 370
Fucus, 336, 531
Fulwyler, L., 245
Fumerase, 91, 165
Fumaric acid, 151, 162
Fungal cell wall, 113, 114
Fungi, 114
Furanose, 109, 118
Furrowing, 451, 511, 550
Fusidic acid, 503

ΔG, ΔG°, $\Delta G^{\circ\prime}$, 177
Galactosamine, 118
Galactose, 104, 109, 110, 113
Galactosidase, β(beta), 213, 548, 550
Galileo, 3
Gallant, 217
Gametogenesis, 450
Gamma-aminobutyric acid, 78
Gamma globulin, 92
Gamma rays, 181, 182, 279
Gap junctions, 310-313, 328, 550
Gas chromatography, 250, 271-272
Gastrula, 528
Geiger-Müller counters, 282
Gel filtration, 270, 272
Gene, 138, 550
  expression, 528
  recombination, 441
  regulation, 218
Generation time, 51, 58, 550
Genetic block, 190
Genetic code, 455
Genome, 138, 218, 550
Genotype, 550
Gesner, C., 3
Giant chromosomes, 431
Gibbs-Donnan Effect, 320, 321
Gibbs, J., 320
Gilbert, J., 500
Glass electrode, 68
Globin chain genes, 533
Globular protein, 77, 86, 89
Glucocorticoids, 218
Glucokinase, 178
Gluconeogenesis, 196-198, 205, 413, 550
Glucosamine, 118
Glucose, 109-113, 166, 169
Glucose catabolism, 360
Glucose oxidase, 105
Glucose-6-phosphate, 143, 403
Glucosyl donor, 199
Glucuronic acid, 114, 118
Glucuronidase, 405
Glutamic acid, 78, 93, 152, 161
Glutamic acid decarboxylase, 152
Glutamine, 78, 93, 167
Glutamine synthetase, 210
Glyceraldehyde, 109-111
Glycerides, 105
Glycerol, 121, 122
Glycerophosphate, 148
Glycerophosphatide, 121, 122, 124
Glycine, 78, 86, 96
Glycogen, 113, 115, 117
  metabolism, 169, 170, 190-192
  synthesis, 199, 200
Glycolipid, 121, 123, 550
Glycolysis, 191, 193, 351, 352, 550
Glycolytic enzymes, 210
Glycophorin-A, 309
Glycoproteins, 103-106, 550
Glycosidic bond, 112-118
Glyoxylate metabolism, 194, 197, 357, 413
Glyoxysomes, 31, 351, 412-415
Goblet cell, 393, 396, 398
Golgi apparatus, 28, 30, 390-397
Golgi, C., 389
Gomori, G., 407
Gordon, A., 263
Gorter, E., 292, 293, 299, 309
G_1 phase (cell cycle), 59
G_2 phase (cell cycle), 59
Grabar, P., 259
Grana(um), 37, 365-368, 378
Granulocytes, 411, 550
Green, D., 333, 350
Grendel, F., 292, 293, 297, 309
Grew, N., 4

Griffith, F., 128
Group-transfer reactions, 550
Grow-off method, 56
Growth cycle of cells, 50-52
Growth factor, 71, 550
Guanidine, 95
Guanine, 131-133, 136
Guanosine nucleotide, 71

Hackenbrock, R., 341
Hairpin loops, 550
Half-life, 280, 281
Haptens, 550
Harvey, E., 4, 292, 294
Hatch-Slack pathway, 383, 384
Haworth formula, 111
Heat of fusion, 63
Heat of vaporization, 63
Heitz, E., 417
HeLa cells, 551
Helixes (in proteins), 83-86
Heme, 93, 97, 98, 161, 162, 551
Hemerythrin, 91, 103
Hemocyanin, 77, 78
Hemoglobin, 91-96
  A_2, 100
  abnormal, 535
  cooperativity in, 99
  human forms, 100
  invariably ionized positions, 99
  invariably nonpolar positions, 99
  ontogeny of, 100
  oxygen dissociation curve, 100
  stereo view, 98
  synthesis, 501, 502, 532
Henderson-Hasselbach relationship, 69
Herriott, R., 129
Hershey, A., 128, 129
Hertwig, O., 130, 421
Herzenberg, L., 245
Heterochromatin, 417, 423, 551
Heterogeneous nuclear RNA, 219
Heterokaryon, 302
Heterolysosomes, 407
Heterophage, 407-410
Heteropolysaccharide, 113, 114
Heterotroph, 173, 184
Heterotrophic cells, 551
Hexa-N-acetylglucosamine, 148
Hexokinase, 148, 162
Hexose monophosphate shunt, 194, 195
High-energy bond, 551
High-energy phosphate, 72, 551
Hill, R., 377
Histamine, 400
Histidine, 96, 159
Histone, 105, 128, 318, 419, 551
Hoagland, M., 484
Hogeboom, G., 333
Holley, R., 263, 481
Holoenzyme, 551
Homoserine, 78, 166
Homoserine dehydrogenase, 166
Hooke, R., 4
Hormones, 78, 124, 163, 218, 220, 551
Hoskins, G., 431
Host cell, 138
H subunit, 162
Hulett, H., 245
Hultin, T., 453
Hyaloplasm, 27
Hyaluronic acid, 114, 118
Hyaluronidase, 402
Hydration spheres, 64, 66, 71
Hydrogen bond, 64, 89, 551
Hydrogen ion concentration, 68
Hydrolase, 144
Hydrolysis, 156-159, 178, 551
Hydronium ion, 67
Hydrophobic bonds, 70, 71, 89, 90, 551
Hydroxylapatite, 266
Hydroxylysine, 104
Hydroxyproline, 86, 104
Hyperbolic kinetics, 99, 100
Hypertonic solutions, 316, 317
Hypotonic solutions, 315

Idiotype, 551
Immune response, 91
Immune system, 305, 307
Immunoglobulins, 91, 92, 269, 307, 308, 551
Impulse, 541-544. *See also* Action potential
Induced enzymes, 213
Induced-fit hypothesis, 155
Inducible enzymes, 551
Ingram, V., 480
Inoculum, 52
Inositol, 123
Inosonation, 225, 249
Insulin, 91, 270, 499
Integrase, 441
Integrator gene, 219
Intercalated disc, 311
Interchromatin granule, 425
Interkinesis, 444
Intermediate junctions, 310
Internal reticular apparatus, 389, 400
Interphase, 58, 444, 551
Intranucleolar chromatin, 417
Intrinsic proteins, 298, 300
Inulin, 115
Inversion, 551

Invertase, 146, 148
In vitro, 551
In vivo, 551
Ionic bond, 70
Ionic compound, 63
Ionic-dipole interaction, 70, 71
Ionic strength, 250
Ionization, 64, 182
Ionophores, 354
Ions, 67, 75
  functions, 67
  pairs, 280, 282
  pumps, 551
Iron transport, 105
Islet cells, 91
Isoaccepting tRNA, 481
Isocitric acid, 166
Isocitric dehydrogenase, 166, 346
Isocitric lyase, 194, 551
Isoelectric focusing, 260, 261
Isoelectric pH, 251, 260
Isoenzyme, *see* Isozyme
Isoleucine, 166
Isomerase, 143
Isomers, 551
Isotonic solutions, 315
Isotopes, 275, 287, 551
  chart, 189
  decay, 279
  half-life, 189
  tracers, 190
Isotropic bands, 537
Isozymes, 162, 210, 213, 551

Jacob, F., 214, 216
Janssens, F. and Z., 3

K, *see* Dissociation constant
K^1, *see* Apparent dissociation constant
K_{eq}, *see* Equilibrium constant
K_M, *see* Michaelis-Menten constant
Kamen, M., 377
Karyotype, 551
Keilin, D., 350
Kennedy, E., 333, 343, 353
Keratin, 77, 86
∝-Ketoglutarate, 166, 346
Ketohexoase, 112
Ketose, 109
Kim, S., 482
Kinetic energy, 64, 144, 153
Kinetochore, 425, 521, 551
Kinetosome, 515
King, T., 350, 528
Kohlrausch, F., 256
Kollicker, 331
Koshland, D., 155
Krebs cycle, 188, 343-347, 351, 414

Lac operon, 216, 217
Lactic dehydrogenase, 162, 212, 213
Lactose, 113
Lamella(ae), 35, 365
Lamina densa, 419, 425
Lampbrush chromosome, 551
Langmuir, I., 293
Langmuir trough, 292
Latent state, 138
Leahy, J., 479, 480
Leblond, C., 396
Lecithin, 122, 292
Lectins, 306, 307, 551
Lehninger, A., 333, 343, 353
Lenard, J., 308
Leptonema, 444
Leptotene, 444, 551
Leucine, 93, 96
Leucocytes, 411, 532
Leucoplasts, 387
Lewis, W., 325
Ligands, 71
Ligase, 551
Light cycles, 55
Light reactions, 375-380
Lindahl, P., 242
Lineweaver-Burke plot, 150
Lineweaver, H., 149
Linoleic acid, 121, 122
Lipid, 121-125, 552
  metabolism, 143
Lipoic acid, 162
Lipoproteins, 104, 105
Lithotroph, 173
Lock and key hypothesis, 153, 154
Locomotor organelles, 34
Lumenal phase, 27
Lutein, 370
Luteinizing hormone, 105, 400
Luteol, 371
Lymph, 91
Lymphocyte, 19, 21, 307
Lysine, 89, 152
Lysis, 138, 552
Lipmann, F., 484
Lysogenic bacterium/virus, 552
Lysogenic infection, 439
Lysosomes, 28, 32, 401, 403-412
Lysozyme, 153-159, 552
Lytic infection, 552
Lytic viruses, 552

McCarty, M., 128, 441

McConnell, H., 304
McEwen, C., 242
McIntosh, J., 426
McKnight, S., 467
MacLeod, C., 128, 441
Macrophage, 552
Macula adherens, 310
Magnesium ion, 162
Malate-aspartate shuttle, 360
Malate synthetase, 413
Malonic acid, 151
Malonyl reaction, 199
Malpighi, M., 3
Maltose, 112, 113
Mannan, 113
Mannose, 104, 110
Marker, K., 487
Martin, A., 263
Mass action, 207, 219
Mast S., 325
Matrix of mitochondria, 336, 360, 552
Matrix unit, 466, 467
Maxwell-Boltzmann distribution, 144, 145
Mediated diffusion, 321
Meiosis, 443, 444, 449, 552
Melting point, 63
Membrane asymmetry, 302
Membrane lipids, 303
  asymmetry, 305
  erythrocyte, 305
  mobility, 303
Membrane proteins, 104
  asymmetric distribution, 300
  enzymatic properties, 302
  extrinsic, 300
  integral, 300, 301
  intrinsic, 300
  isolation, 302
  mobility, 302
  peripheral, 300
Membranes
  cellophane, 251, 252
  faces of, 18, 299
  functions, 122
  modifications, 396
  see also Plasma membrane
Membrane spheres, 339, 340
Membrane whorls, 409
Menten, M., 146, 147
Meromysin, 537, 552
Mesophyll tissue, 385
Mesosome, 37, 433, 552
Messenger RNA (mRNA), 138, 216, 485-487
Metabolic pathways, 187, 203, 552
  branch points, 164
  control, 169
  efficiency, 204
Metabolic turnover, 124
Metabolism, 4
  carbohydrate, 143, 190
  cancer cell, 203
  control mechanisms, 188
  lipid, 143, 199
  major pathways of, 188
  nitrogen, 202
  regulation of, 151, 209
Metal ions, 161, 162
Metamorphosis, 100
Metaphase, 60, 426, 552
Metchnikoff, E., 327
Methionine, 128
Methylguanosine, 139
Metric measurements, 46
Micelle, 63
Michaelis, L., 146, 147, 333
Michaelis-Menten constant, 147-149, 209
Michaelis-Menten kinetics, 165-168
Microbodies, 31, 403, 412-415, 552
Microfilaments, 507-514, 552
Microscopy, 5-9, 18-20
Microsomal fraction, 124, 472
Microsomal vesicles, 472
Microtome, 9
Microtrabeculae, 23
Microtubules, 34, 35, 507-510, 514-517, 552
  distribution, 514
  function, 514
  model, 509
Microvilli, 19, 23, 24, 311-313
Midbody, 511
Middle lamella, 35
Miescher, F., 127
Miller, O., 465, 467, 498
Mitochondria, 28, 29, 194
  discovery, 331
  distribution, 333
  division stages, 336
  enzymes of, 338
  evolution, 336
  functions, 357
  lipid content, 125
  matrix of, 194, 336, 337
  membranes of, 336
  origin, 333
  permeability, 358
  protein synthesis, 475
  ribosomes, 336, 475
  structure, 333-337
  transport systems in, 359
Mitosis, 56, 425-431
Mitotic index, 55
Modulator, 210

Molar ratios, 136
Molecularity, 143
Molecular sieving, 270, 272
Monod, F., 214, 215, 216, 219
Monogenic, 486
Monomer, 552
Monomolecular reactions, 143, 144
Monosaccharide, 118
Mooseker, M., 511
Morris, A., 481
Morton, R., 481
Morula, 527
Movement
  amoeboid, 511, 514
  cell, 522
  ciliary, 519
  flagellar, 519
  plasma membrane, 511
M-phase, 58
Mucigen granules, 390
Mucin, 104, 105
Mucopeptide, 37
Mulder, G., 77
Multienzyme system, 552
Muscle, 100, 335, 510, 535-539
Mutagens, 552
Mutation, 552
Mycoplasmas, 24, 41
Myelin, 123, 304
Myoblasts, 552
Myofibril, 552
Myofilaments, 507, 535
  thick, 510, 516
  thin, 510, 536
Myoglobin, 78, 99, 100
Myoneural junction, 539
Myosin, 507, 538, 552

NAD *see* Nicotinamide adenine dinucleotide
NADP, *see* Nicotinamide adenine dinucleotide phosphate
Nass, M., 474
Negative control, 552
Negative staining, 13
Neoxanthin, 370
Nernst equation, 348
Nerve cell, 400, 541
Neuraminic acid, 124
Neuron, 541
Neurosecretions, 400, 402
Neurotransmitter, 544
Neutral fat, 122, 125, 552
Neutra, M., 396
Nexus, 310, 313
Niacin, 72, 74, 161
Nickases, 451
Nicolson, G., 298, 302
Nicotinamide, *see* Niacin
Nicotinamide adenine dinucleotide (NAD), 71-73, 161, 162, 349, 552
Nicotinamide adenine dinucleotide phosphate (NADP), 71-73, 161, 349, 552
Nitrogen bases, 133, 136, 139
Nitrogen metabolism, 202
Node of Ranvier, 544
Nomura, M., 458
Noncompetitive inhibition, 151, 155
Noncyclic photophosphorylation, 378, 379
Nonhistone, 105, 218
Nonsulfur purple bacteria, 386
Noradrenaline, 400
North, A., 156
Novick, A., 54
Novikoff, A., 405, 408
Nuclear envelope, 25, 31-36, 417, 421-425, 552
Nuclear pore, 25, 31-36, 421, 424, 451
Nuclear transmutation, 288
Nucleic acid, 75, 77, 131-139
  biosynthesis, 143
  cellular role, 127
  composition and types, 128, 131
  structure, 131
  viral, 43
Nuclein, 127, 128
Nucleocytoplasmic relations, 528, 544
Nucleoids, 37, 433, 553
Nucleolar chromatin, 417
Nucleolar organizing region (NOR), 425, 463-465, 553
Nucleolini, 421
Nucleolus, 25, 34, 36, 420-422, 451, 465
Nucleoplasm, 421, 553
Nucleoproteins, 105
Nucleoside, 131, 553
Nucleoside triphosphate sugars, 199
Nucleosome (Nu particles), 419, 420, 467, 534, 553
Nucleotides, 71, 131-139, 553
Nucleotidyl transferase, 481
Nucleus, 31, 36, 417-451, 553
  transplantation of, 528
Numerical aperture, 5

Objective lens, 5
Ocular lens, 5
Offengand, E., 484
Okazaki, R., 438
Oleic acid, 121, 122
Oligomycin, 354, 355
Oligosaccharide, 104, 190
Omega (ω) oxidation, 199
Ommatidia, 21
Oncogenic substance, 138
Open system, 553
Operator, 215, 553
Operon, 215-219, 553

common abbreviations, 216
Oppenheimer, S., 306
Opsin, 183, 184
Optical isomer, 109
Orci, L., 311
Organelle, 553
Organotroph, 173
Ornstein, L., 256
Osmium tetroxide, 115, 294
Osmometer, 315
Osmosis, 314, 329
Osmotic barrier, 37
Osmotic lysis, 242, 315
Osmotic pressure, 226, 314
Ovalbumin, 104, 105
Overloading, 190
Overton, E., 292
Oxidant, 553
Oxidase, 72
Oxidation, 178, 553
Oxidation-reduction, 72, 348
Oxidative phosphorylation, 342, 343, 353-357, 553
Oxygen debt, 541
Oxytocin, 400

P_{700}, 376
P_{trap}, 379
Pachynema, 444
Pachytene, 444, 553
Packing ratio, 431
Pactamycin, 503
Page, L., 480
Palade, G., 396
Palasade parenchyma, 363, 364
Palindromes, 458
Pancreatic lipase, 152
Pantothenic acid, 72, 74, 78, 161
Papain, 156
Paramecium caudatum, 52
Paramylum, 113, 118
Parasite, 91
Pardee, A., 214
Parthenogenesis, 528
Parvovirus, 46
Passive carriers, 358
Passive transport, 313
Pauling, L., 81, 86
Pentose, 109, 131, 136
Pentose phosphate pathway, 194, 195
Pepsin (pepsinogen), 127, 152, 162, 163
Peptide bond, 80, 106, 553
Peptides, 148, 161
Peptide synthetase, 492
Perhydrocyclopentanophenanthrene, 124
Perichromatin granule, 425
Perichromatin region, 421
Perinuclear space, 31, 421
Perinucleolar chromatin, 417
Periodic table, *inside front cover*
Peripheral proteins, 298, 300
Peritrichous bacteria, 38
Permeability, 314, 339
constant, 317, 318
factors influencing, 318
tables, 318
Permease, 322, 553
Peroxidase, 105, 161, 162, 413
Peroxisomes, 31, 412, 413, 553
Perutz, M., 96
Peterson, E., 266
pH, 68, 89, 553
Phagocytosis, 325, 327-329, 507, 533
Phagosomes (phagolysosomes), 327, 407, 410
Phase-contrast microscope, 553
Phenotype, 553
Phenylalanine, 79
Phillips, D., 156
Phosphatase, 148, 407
Phosphate ester linkage, 72
Phosphate turnover, 179
Phosphate-water exchange, 354
Phosphatidic acid, 121, 122
Phosphatidyl choline, 122, 123
Phosphatidyl ethanolamine, 122, 123
Phosphatidyl inositol, 122, 123
Phosphatidyl serine, 122, 123
Phosphoarginine, 179
Phosphocreatine, 179, 540
Phosphodiester linkage, 553
Phosphofructokinase, 166, 169
Phosphoglucoisomerase, 209
Phosphogluconate pathway, 194, 195, 205
balance sheet, 196
Phosphoglyceric acid, 383
Phosphohexose isomerase, 143
Phospholipid, 105, 124, 292, 304
Phosphoric acid, 131
Phosphorolysis, 190
Phosphorylase, 165, 166
Phosphoryl choline, 123
Phosphotungstic acid, 115
Photochemical reaction, 375
Photolithotroph, 173
Photolysis, 386
Photon, 180
Photophosphorylation, 553
Photorespiration, 553
Photosynthesis, 35, 174, 183, 553
action spectrum, 376
bacterial, 385
C_3, 383, 384
C_4, 383, 384, 548

electron transport in, 379
path of carbon in, 383
primary photochemical events in, 376
Photosynthetic phosphorylation, 553
Photosystems, 376
I, 376-379, 553
II, 376-379, 553
light absorption, 377
Phototroph, 173
Phycobilin, 553
Phycobilosome, 40
Phycocyanin, 371
Phycoerythrin, 371
Phytohemagglutinins, 306
Pickels, E., 228
Pilot protein, 440
Pinocytic vesicles, 25, 326, 410
Pinocytosis, 325-329, 507, 553
Planar group, 81, 82
Planchet, 282, 283
Planck's law, 182
Plant cells, 34
Plaque, 553
Plasma, 92, 99
Plasmalemma, 553
Plasmalogen, 121-123
Plasma membrane, 25, 26, 36, 291-330
antibodies, 306
early studies of, 291
enzymes of (table), 303
fluid-mosaic model of, 298
intercellular junctions, 310
lectins, 306
lipid asymmetry of, 308
lipids in, 292, 304
movement, 511
protein asymmetry, 308
surface carbohydrate, 307
trilaminar appearance, 295, 297
unit membrane model, 295, 297
Plasmids, 441, 553
Plasmodesmata, 35, 36, 310, 313, 553
Plasmolysis, 316, 317
Plastids, 361, 386, 553
Plastoquinone, 377
Pleuropneumonia-like organisms (PPLO), 24, 41
Ploidy, 130
Pluripotent stem cell, 532
Polar body, 449
Poliomyelitis, 138
Polycistronic mRNA, 216
Polymer, 553
Polymerase I, 437, 438
Polymetaphosphate, 179
Polynucleotide ligase, 554
Polynucleotide phosphorylase, 554
Polynucleotides, 134-139, 553
Polyoma, 46
Polypeptides
structure, 80-89, 154-162
synthesis, 476-499
see also Proteins
Polyribonucleotide, 158
Polyribosomes, *see* Polysomes
Polysomes, 455, 494, 495, 504, 554
cycle, 496
Polysaccharides, 77, 109-119, 128, 190
Polytene chromosomes, 431, 554
Polytomella agilis, 56, 58, 118, 333, 334
POPOP, 284
Power stroke, 519
ppGpp, 217, 219
PPLO, *see* Pleuropneumonia-like organism
PPO, 284
pppGpp, 217
Precursor-product relationships, 277
Pressure cell, 225, 246
Priestly, J., 374
Primary constriction, 554
Primary light reaction, 182
Primer molecule, 199
Priming reaction, 199
Procaryotic cells, 37
Procentriole, 516, 554
Proenzyme, 162, 163
Progesterone, 124
Proinsulin, 163
Proline, 79
Prometaphase, 554
Promotor, 554
Promotor locus, 216, 217
Prophage, 440
Prophase, 60, 425, 554
Proplastid, 361, 374, 386, 554
Prosthetic group, 103, 161, 162, 554
Protamine, 128, 554
Protease, 161
Proteins, 75-89
amino acid content, 78, 79
biosynthesis, 454, 476-479, 490-495
biosynthesis inhibitors, 502
conjugated, 103
cotranslational modification of, 498
isoelectric points, 254
molecular weights of, 78
post-translational modification of, 499
primary structure of, 86
quaternary structure of, 90
secondary structure of, 88
tertiary structure of, 89, 90
turnover, 453
see also Polypeptides

Proteolysis, 160, 163
Proteolytic enzyme, 163
Proteoplasts, 386
Proteus mirabilus, 38
Protist, 554
Protofilaments, 507
Protolysosomes, 407
Protomer, 166, 167
Proton donor, 67
Proton transport, 352
Protoplasm, 4, 554
Protoplast, 554
Protozoa, 37, 113, 115, 117
Provirus, 43, 138, 554
Pseudopodia, 327, 511
Pseudouridine, 139
Ptericidin, 353
Pulse-chase experiment, 396
Pulse label, 480
Purine, 131, 132, 134, 158, 554
Purkinje, J., 4
Puromycin, 503, 554
Purple sulfur bacteria, 386
Pyranose, 109, 118
Pyridine-linked dehydrogenase, 349, 362
Pyridine nucleotide, 71
Pyridoxamine phosphate, 72
Pyridoxyl phosphate, 72
Pyrimidines, 131-134, 158, 159, 164
Pyruvate, 166, 193, 194, 205
Pyruvate dehydrogenase, 345

Quantasomes, 367
Quantum (quanta), 182, 554
Quinone, 377

Racker, E., 350
Radiation, 275-288
Radioactive isotopes, 275-278, 554
Radioactive ligand, 301
Raff, M., 308
Ramachandran, G., 86
Ravazzola, M., 311
Reaction kinetics, 143, 144
Reaction rates, 145, 146
Re-annealing, 554
Receptor protein, 302
Recombinant DNA, 441, 442
Recombination, 441
Red blood cell, *see* Erythrocyte
Red drop, 376
Redman, C., 472
Redox couple, 179, 348, 554
Redox potential, 348, 349, 378
Reductant, 554
Reduction, 554
Reference electrode, 68
Regeneration, 529
Regulation, 554
Regulatory gene, 214, 215, 554
Regulatory metabolites, 212, 214
Regulatory site, 165-167
Reiteration, 457, 458
Relaxed mutants, 217
Release factor, 554
Renaturation, 554
Reovirus, 46
Repetitive DNA, 554
Replica, 299
Replicating fork, 554
Replicating forms (RF), 554
Replication of DNA, 58, 137, 138
  addition of nucleotides, 434
  conservative, 434
  enzymes, 437
  eucaryotic, 439
  procaryotic, 438
  semiconservative, 434
  sites of, 419
  in two directions, 436, 438
  in viruses, 439
Replicons, 467, 468
Repressible enzyme, 554
Repression, 210, 218
Repressor-inducer complex, 215
Repressors, 214, 555
Residual bodies, 407-410, 555
Respiration, 193, 342, 555
Respiratory particles, 337
Resting potential, 542
Restriction enzymes, 441, 443, 555
Reticulocyte, 472, 476, 555
Reticuloendothelial system, 91, 410
Retinal, 183
Reverse transcriptase, 555
Reverse transcription, 138
Rhodin, J., 412
Rhodopsin, 183, 184
Rhopheocytosis, 327, 329
Riboflavin, 72, 74, 161
Ribonuclease, 94, 158, 159
Ribonucleic acid (RNA), 93, 128-139, 555
  mRNA, 138, 216
  negative strand, 138
  plus strand, 138
  polymerase, 216, 467, 481
  replication of, 137
  RNA, 138, 555
  soluble, 139
  structure, 136
  synthesis, 136
  tRNA, 138

virus, 136, 137
Ribose, 112, 131, 132
Ribosomal DNA (rDNA), 555
Ribosomes, 28, 455-475, 504, 555
acceptor site, 455
amino acid site, 455, 484
assembly, 456, 460, 461, 467, 470
attached, 370, 471
composition of, 455, 456
donor site, 455
eucaryotic, 463
free, 469, 471
initiation complex, 484
model, 462, 469, 470
monomers, 456
peptide site, 455, 484
procaryotic, 457
protein content, 458, 464, 465
RNA content, 457
sedimentation coefficient, 456
size, 457
structure, 455
subunits of, 454
Ribothymidine, 139
Ribulose, 110
Ribulose diphosphate, 382
Rich, A., 86, 496
Ricin, 503
Ris, H., 474
RNA, *see* Ribonucleic acid
RNA polymerase, 555
RNA primer, 438
Robertson, J., 294
Robertson's unit membrane model, 294-296
Rod cells, 183
Roodyn, D., 334
Roseman, S., 306
Rotenone, 353
Rothman, J., 308
Rotor (centrifuge), 230-237
Rough colonies, 128
Rough ER, 27, 555
Rouleaux, 309, 310, 535
Rous sarcoma virus, 46
rRNA
genes for, 463
protein interactions, 459
synthesis, 217
Ruben, S., 377, 380

Sabatini, D., 472, 473
Sachs, J., 374
Salicylanilides, 354
Salivary amylase, 152
Saltatory conduction, 544
Salt bonds, 89
Salting in, 250, 272
Salting out, 250, 251, 272
Sanger, F., 91, 487
Sarcolemma, 536, 555
T-system, 536
Sarcomere, 537, 555
Sarcoplasm, 555
Sarcoplasmic reticulum, 555
Saturation kinetics, 321
Schizosaccharomyces pombe, 58
Schleiden, M., 4
Schmitt, F., 294
Schneider, W., 405
Schroeder, T., 511
Schwann cell, 545
Schwann, T., 4
Schweet, R., 479-480
Scintillation counters, 282-288
Secondary amino groups, 152
Secondary constriction, 425, 555
Secondary light reactions, 182
Second law of thermodynamics, 555
Secretion, 390, 396, 402, 555
Secretory vesicles, 390
Sectioning, 9
Selective permeability, 294, 314
Selective solubility, 292
Self-absorption, 283
Semiconservative replication, 555
Semipermeable, 314
Semipermeable membrane, 251
Sendai virus, 302
Sensor sites, 219
Sephadex, 266, 270
Sepharose, 270
Sequestered metal, 71
Serine, 79
Serine dehydrogenase, 472
Serine protease, 161
Serum protein, 555
Sex chromosome, 555
Sex hormones, 182
Shadow casting, 9, 12
Sheeler, P., 238
Shoenheimer, R., 453
Sickle-cell anemia, 535
Siekevitz, P., 453, 454
Silk, 86
Simonsiella, 20, 37, 38, 39
Singer, S., 298, 350
Single state, 375
Sjostrand, F., 339
Sliding filament mechanism, 555
Smith, D., 502
Smooth colonies, 128
Smooth ER, 28, 555

Soap, 63
Sober, H., 266
Sodium dodecyl sulfate (SDS), 302
Sodium fluoride, 503
Sodium-potassium pump, 322, 323
Sodium pump, 297, 543
Solomon, A., 319
Soluble phase, 231
Soluble RNA (sRNA), *see* Transfer RNA
Somatic cell, 130
Sonifier, 225
Sørensen, 68
Soybean, 148
Sparsomycin, 503
Specific activity, 277
Specific heat, 63
Specific ionization, 280
Spectrin, 309
S-period, 555
Sperm, 127, 128, 130, 402, 522
S-phase, 58
Spheres of hydration, 64, 319
Sphingolipid, 121, 122, 123
Sphingomyelin, 123
Sphingosine, 122, 123
Spindle, 425, 511, 515, 520
Spindle fiber, 555
Spiral filament, 520
Spongy parenchyma, 363, 364
Spot desmosomes, 310, 311
Standard electrode potential, 179, 555
Standard free energy, 177, 178
Standard free energy change, 555
Standard redox potentials. 180
Standard state, 555
Sta-put devices, 243, 244
Starch, 113, 116, 118, 255
  metabolism, 190, 199, 200, 205
State-3 respiration, 354
State-4 respiration, 354
Stationary phase, 52
Steady state, 148, 555
Steady-state concentrations, 205
Steck, T., 301
Stedman, P., 218
Stereoisomers, 555
Sterinochloroplasts, 386
Steroid, 121, 124, 125
  physiological properties, 124
Sterols, 105
  synthesis of, 187
Stillmark, H., 306
Stimulus to nerve, 543
Stoffler, H., 459
Stokes' law, 227
Strasburger, E., 130
Streptococci, 38
Streptomyces griseus, 353
Streptomycin, 503, 556
Stringent control, 217
Stroma, 35, 365-370, 380, 556
Stromacenters, 367
Stroma lamellae, 37
Structural gene, 214, 216
Structural proteins, 77, 302
Substrate, 145, 556
  activation, 145
  effector, 208
Substrate level phosphorylation, 331, 361
Substrate oxidation, 342
Succinic acid, 151, 162
Succinic acid dehydrogenase, 151, 162
Succinyl CoA, 346
Sucrase, 148
Sucrose, 148
  hydrolysis, 177
  structure, 112
Sugar beet chloroplast, 365
Sulfhydryl group, 151
Sulfolipid, 370
Supercoils, 556
Super helix, 86
Superoxide dismutase, 357, 358
Suppressor gene, 556
Surface tension, 63
Suspension, 63
Svedberg, T., 228, 229
Svedberg unit, 229, 556
Sweet, R., 245
Swivel mechanism, 437
Synapse, 400, 401, 544
Synapsis, 444, 556
Synaptic transmission, 544
Synaptinemal complex, 556
Synchronous culture, 54, 56
Synchronous growth, 50, 55
Synchrony by induction, 55
Synchrony by selection, 55, 56
Synge, R., 263
Synovial fluid, 113, 118
System, 556
Szent-Gyorgyi, A., 350
Szilard, L., 54

T-2 bacteriophage, 12
Tata, J., 472
Telolysosomes, 407
Telophase, 60, 427, 556
Temperate infection, 439
Temperature-sensitive mutation, 556
Teratoma, 19

Terminal bars, 310
Tertiary structure, 556
Tetrad, 444
Tetrahymena pyriformis, 56, 58, 118
Thermodynamics, 174, 176, 184, 556
Thiamine, 161
Thiamine pyrophosphate, 161, 162, 345
Thin-layer chromatography, 250, 265
Threonine, 79, 166
Threonine deaminase, 166
Thrombin, 161
Thunberg, T., 350
Thylakoids, 35, 365-368, 379, 380, 386
Thymine, 131, 133, 136
Thyroxin, 77, 218
Tight junctions, 27, 310, 311, 312, 328
Tissue (fractionation of), 225-247
Tisulius, A., 255
Titration, 68
T_m, 556
TMV, *see* Tobacco mosaic virus
Tobacco, 129
Tobacco chloroplast, 365
Tobacco mosaic virus (TMV), 42, 129, 130, 434
Tobita, T., 480
Tonofilaments, 26, 311
Totipotency, 528
TPN, *see* NADP
Transaldolase, 194
Transcription, 137-139, 214-216, 220, 438, 454
  kinetics, 496-498
  nucleolar, 421
  sites, 420
Transcriptional control mechanisms, 213
Transductions, 184, 441
  chemical, 180
  ionic, 184
  light, 180
Transferase, 144
Transfer RNA (tRNA), 139, 481-485, 556
  anticodon, 482
  arms, 481, 482
  invariant positions, 482
  processing, 481
  semi-invariant positions, 482
  structure, 481, 482, 483
Transformation, 128, 441, 556
Transforming principle, 128
Transhydrogenases, 353
Transketolase, 194
Translation, 137-139, 214-220, 476-479, 490-495
  kinetics, 496-498
  visualization of, 498
Translational control mechanisms, 213
Translocase, 322, 358
Translocation, 556
Transmutations, 275
Transport (across membranes)
  active, 322
  bulk, 324
  NADH/NADPH, 360
  osmosis, 314
  passive, 313
Trap chlorophyll, 386
Tricarboxylic acid cycle (TCA-cycle), *see* Krebs cycle
Trichodermin, 503
Trichonympha, 398
Triglycerides, 121, 199
Trimolecular reactions, 143
Triose, 109
Triplet basis, 454
Triplet state, 375
Trisaccharide, 109
Tritium, 556
Triton, 302
Tropomycin, 537, 556
Troponin, 537
Trypsin, 78, 152, 156, 161, 162, 163
Trypsinogen, 163, 210
Tryptophan, 79, 159
T-system, 536, 539, 556
Tubulin, 507-510, 520
Turgor, 316
Turnover number, 146
Tyrosine, 79, 152

Ubiquinone, 350, 362
Ultracentrifugation, 252, 272
Ultrafiltration, 251, 252, 272
Ultramicrotome, 9, 10
Ultraviolet light, 181, 556
Uncoupling agent, 556
Universal solvent, 63
Unit gravity separation, 242-246
Unit membrane, 338, 556
Uracil, 131, 132, 136
Uranyl acetate, 9
Uranyl nitrate, 9
Urea, 95, 146, 148
Urease, 78, 146, 254
Uric acid, 162
Uric acid oxidase, 162
Uridine diphosphoglucose, 166
Uridine nucleotide, 71
Uridine-5′-phosphate, 132

Vacuolar system, 408
Vacuoles, 35, 37, 401
Valine, 79
van der Waals forces, 70, 71, 556
van der Waals interactions, 90
van Helmont, J., 374

van Leeuwenhoek, A., 4
Vasopressin, 400
Vectorial synthesis, 308, 472
Vinblastine, 509
Vincristine, 509
Violaxanthin, 370, 371
Viral nucleic acid, 138
Viral protein, 138
Virchow, R., 4
Virions, *see* Viruses
Virology, 46
Viruses, 41, 43, 91, 548
  adsorption, 440
  animal, 43
  assembly of, 440
  bacterial, 43
  characteristics of, 436
  chromosome, 434, 435
  coat, 130
  composition, 128
  head, 128, 129
  lambda, 434, 440
  lysogenic, 43
  Newcastle, 138
  ϕX174, 42, 46, 136, 440
  plant, 43
  proliferative cycle, 43, 440
  reovirus, 46, 138
  reproduction, 128
  RNA in, 46
  RNA tumor, 138
  Rous sarcoma, 138
  Sendai, 138, 302
  sizes, 436
  SV 40, 46
  temperate, 43
  T-even and T-odd, 42, 43, 46, 128, 129
  Tobacco Mosaic (TMV), 129, 130, 434
  virulent, 43
Visible light, 181
Vision, 183
Vitamins, 71, 72, 75, 161
  A, 183
  B_2, 161
  D, 124
  table of, 72

Waldeyer, 425
Waller, J., 487
Warburg, O., 333, 350
Ward, E., 391
Ward, R., 391
Watson, J., 133, 134, 135
Wavelength, 180
Weak acid, 68
Weismann, A., 130
Wertz, G., 391
Wilkie, D., 334
Wilkins, M., 133
Williams, C., 259-260
Wilson, E., 130, 334
Wilstätter, 374
Winslow, R., 480
Wobble base, 482, 557
Wood-Werkman reaction, 203
Wool, 86

X-rays, 181, 182
X-ray crystallography, 95, 133, 557
Xylose, 104
Xylulose, 110

Yeast cell wall, 113, 118
Yoshida, A., 480

Zamecnik, P., 453, 454, 484
Zero-order kinetics, 147
Zinc, 161, 162
Zonal rotors, 57, 238-240
Zones of exclusion, 390
Zonula adherens, 310
Zonula occludens, 310
Zwitterion, 89
Zygote, 557
Zygotene, 557
Zymogens, 162, 163, 396, 557

COMMON ABBREVIATIONS

A	Adenine or adenosine
Å	Angstrom
ACTH	Adrenocorticotrophic hormone
AMP, ADP, ATP	Adenosine 5′-mono-, di-, and triphosphate
cAMP	Cyclic AMP
ala	Alanine
arg	Arginine
asn	Asparagine
asp	Aspartic acid
ATPase	Adenosine triphosphatase
C	Cytosine
c	Centi, 10^{-2}
c	Velocity of light (vacuum), 2.997 $\times 10^{10}$ cm sec^{-1}
^{14}C	Carbon 14
cal	Calorie
cm	Centimeter
CMP, CDP, CTP	Cytidine mono-, di-, and triphosphate
CoA, CoA-SH, acyl-CoA, acyl-S-CoA	Coenzyme A and its acyl derivatives
CoQ	Coenzyme Q; ubiquinone
CPM	Counts per minute
CsCl	Cesium Chloride
cys	Cysteine
cyt.	Cytochrome
d	deci, 10^{-1}
DNA	Deoxyribonucleic acid
DNAase	Deoxyribonuclease
DPN^+, DPNH	Same as NAD^+, NADH
E_o	Standard electrode potential
e^-	Electron
EDTA	Ethylenediaminetetraacetic acid
ER	Endoplasmic reticulum
ESR	Electron spin resonance
ETP	Electron transfer particle (from mitochondrial membrane)
$\mathcal{F}$	Faraday
f	femto, 10^{-15}
FAD, $FADH_2$	Flavin adenine dinucleotide, oxidized and reduced forms
FDP	Fructose-1, 6-diphosphate
fmet	*N*-formylmethionine
FMN, $FMNH_2$	Flavin mononucleotide and its reduced form
FP	Flavoprotein
g	Gram, gravity
G	Gauss, guanine, or guanosine
G	Giga, 10^9
$\Delta G^{o\prime}$	Standard free-energy change at pH 7
Gal	D-Galactose
GDH	Glutamate dehydrogenase
GLC	Gas-liquid chromatography
glc	Glucosamine
GlcNAc	*N*-Acetyl-D-glucosamine
gln	Glutamine
glu	Glutamic acid
gly	Glycine
GMP, GDP, GTP	Guanosine mono-,di-, and triphosphate
G3P	Glyceraldehyde-3-phosphate (3-phosphoglyceraldehyde)
G6P	Glucose-6-phosphate
hr	Hour
h	Planck's constant, 6.626 $\times 10^{-34}$ joule-sec 1.584 $\times 10^{-34}$ cal-sec
h	Hecto, 10^2
H^+	Hydrogen ion
^3H	Tritium
Hb	Hemoglobin
his	Histidine
ile	Isoleucine
IMP, IDP, ITP	Inosine mono-, di-, and triphosphate
J	Joule (1.0 J = 0.239 cal = 10^7 erg)
K	Kelvin (degrees)
k	Kilo, 10^3
kcal	Kilocalorie
l	Liter
λ	Phage lambda or wavelength of radiation
LDH	Lactate dehydrogenase
leu	Leucine
LH	Luteinizing hormone